Nickel Sulfide Ores and Impact Melts

T0318768

Nickel Sulfide Ores and Impact Melts
Origin of the Sudbury Igneous Complex

Peter C. Lightfoot

Vale Base Metals
Ontario, Canada

AMSTERDAM • BOSTON • HEIDELBERG • LONDON
NEW YORK • OXFORD • PARIS • SAN DIEGO
SAN FRANCISCO • SINGAPORE • SYDNEY • TOKYO

Elsevier
Radarweg 29, PO Box 211, 1000 AE Amsterdam, Netherlands
The Boulevard, Langford Lane, Kidlington, Oxford OX5 1GB, United Kingdom
50 Hampshire Street, 5th Floor, Cambridge, MA 02139, United States

Notices

Knowledge and best practice in this field are constantly changing. As new research and experience broaden our understanding, changes in research methods, professional practices, or medical treatment may become necessary.

Practitioners and researchers must always rely on their own experience and knowledge in evaluating and using any information, methods, compounds, or experiments described herein. In using such information or methods they should be mindful of their own safety and the safety of others, including parties for whom they have a professional responsibility.

To the fullest extent of the law, neither the Publisher nor the authors, contributors, or editors, assume any liability for any injury and/or damage to persons or property as a matter of products liability, negligence or otherwise, or from any use or operation of any methods, products, instructions, or ideas contained in the material herein.

Library of Congress Cataloging-in-Publication Data
A catalog record for this book is available from the Library of Congress

British Library Cataloguing-in-Publication Data
A catalogue record for this book is available from the British Library

ISBN: 978-0-12-804050-8

For information on all Elsevier publications
visit our website at https://www.elsevier.com/

Working together
to grow libraries in
developing countries

www.elsevier.com • www.bookaid.org

Publisher: Candice Janco
Acquisition Editor: Amy Shapiro
Editorial Project Manager: Tasha Frank
Production Project Manager: Paul Prasad Chandramohan
Designer: Maria Ines Cruz

Typeset by Thomson Digital

*This book is dedicated to my wife, Nancy,
who patiently tolerated my work on Sudbury.
Nancy provided the foundation of happiness
that made this book possible.*

Contents

Preface

Unusually large ore deposits are created by events that focus magma, fluids, energy, and metals into crustal containers and/or structures. These broadly magmatic–hydrothermal deposit types include porphyry copper, iron oxide copper gold, kimberlite-hosted diamond, and magmatic sulfide deposits. All of these deposits are marked by the generation of magmas and/or fluids in the mantle in areas of anomalous heat related to plume events and/or rifting or subduction of oceanic crust beneath continental margins. The energy driving the process comes from the mantle along with much of the mineral wealth in the ores. The world class orthomagmatic Ni-Cu-Co-PGE-Au deposits associated with the Sudbury Igneous Complex (SIC) are quite different in the sense that the energy source and genesis was triggered by a large meteorite impact event, and although there has been extensive work to try to identify mantle contributions, there is no compelling evidence that magma, fluid, energy, or metals come from deep inside the Earth.

This book explores the linkages between sulfide and silicate magmas generated by the 1.85 Ga Sudbury impact event which produced one of the largest Ni-Cu-PGE sulfide-ore-deposit camps which has now been mined for over 100 years. I examine the relationship between crustal melts on the one hand and magmatic sulfide ore deposits on the other. Normally magmatic sulfide ore deposits rich in Ni, Cu, and PGE require a mafic or ultramafic contribution of magma; at first sight the relationships at Sudbury constitute an oxymoron. Melts of average upper crust would be felsic in composition and bereft of metals normally found in ultramafic and mafic magmas that are normally required to form magmatic sulfide ore deposits. This book provides geological and geochemical evidence that the rocks of the SIC were produced by crustal melting, differentiation, and crystallization. The rocks record evidence for the formation of immiscible magmatic sulfide from the melt sheet, concentration of base and precious metals by equilibration of the sulfides with the silicate magma of the melt sheet and the gravitational concentration of these sulfides toward the base of the complex where they form the main ore bodies. The timing of processes can be established from the geological relationships between the rocks, and these observations inform models of ore genesis, and help explain the diversity in composition and scale of mineral zones which range from several million to just a few thousand tons of contained metal.

The impact process provided sufficient energy to melt the upper crust as well as produce the pseudotachylite breccias, syn-impact radial and concentric dykes, and an uneven crater floor covered by an impact melt that crystallized to form the Main Mass of the SIC. The scale on which ore deposits were formed is a direct function of the thickness of the primary melt sheet and the topography of the crater floor. Dense immiscible magmatic sulfides were localized in physical depressions and cracks in the crater floor where it was possible for the sulfides to concentrate. These relationships are examined in ore deposits which formed at the lower contact, in the immediate footwall of contact deposits, and in radial and concentric dykes generated in response to the migration of impact melt into the fractured crust.

The ore deposits of the SIC provide a wealth of variation in geometry, mineralogy, and chemistry that record evidence of the scale of primary segregation and concentration, and they record a protracted history of crystallization that sometimes permitted the segregation and localization of differentiated sulfide melts enriched in Cu, Ni, and PGE (the Footwall Deposits), leaving behind sulfides that are typically rich in Ni but have lesser Cu and PGE contents (the Contact Deposits). Geological relationships in ore bodies developed in radial and concentric structures show evidence for the emplacement of multiple different sulfide melts formed at different stages from the melt sheet; these deposits are often hosted in dykes (the Offset Deposits) and zones of extensive partial melting and brecciation associated with the embryonic development of dykes (in the South Range Breccia Belt). The details of these relationships offer a superb example of magmatic sulfide ore formation which provides new insight into the controls on the genesis and localization of magmatic sulfide ore deposits that can be applied elsewhere.

The preservation of the SIC is very good, and there is an excellent three-dimensional understanding of the complex above a depth approaching 3 km due to mining and exploration activity over 100+ years that provides a basis for the insights in this book. At least three major orogenic events have modified the rocks through deformation and metamorphism. These effects create an additional complexity in understanding the ore bodies. Tectonic displacement and solid-state migration of soft sulfides (kinesis) into spaces has created a superimposed range of relationships in the ores. The effects of these processes are often difficult to distinguish from syn-magmatic emplacement of sulfide melts into structures that were active during crater re-adjustment.

This book aims to unite an understanding of the process of melt sheet evolution with the formation of the magmatic sulfide mineralization. A sequence of events in the evolution of the SIC is constrained by geological and geochemical evidence, and the mechanisms of ore formation are placed into this context. Sudbury provides a wealth of lessons which helps inform studies of other magmatic sulfide deposits, and provides models which will continue to support future discovery of ore deposits at Sudbury. The elegance and simplicity of the basic scientific relationships at Sudbury provides an opportunity to present students with a unified theory of ore genesis in a crustal melt sheet. The theory remains part of focused research activity of Sudbury specialists, but it also provides a more general framework in which geologists can understand entire mineral systems where ore deposits are not treated in isolation from their host rocks.

In Chapter 1 the themes of the geology, impact origin, and the ore deposits of the Sudbury Structure are woven together in this book in a traditional approach, beginning with an account of the importance of Sudbury mines for nickel, copper, and precious metal production in a global context. The history of the Sudbury Mining Camp is discussed in terms of both the discovery of the ore deposits and the vigorous debate surrounding the origin of the host rocks. There has been a major shift in understanding from endogenic models requiring explosive magmatic events to the more recently accepted models of impact tectonics, melt generation, and ore formation. The theories of formation of immiscible magmatic sulfide ore deposits are introduced, with the supporting evidence that shows Sudbury to be part of a group of ore deposits normally associated with mafic and ultramafic rocks.

In Chapter 2 the geology of the Sudbury Structure and the associated igneous rocks are presented in a traditional way, and the information is designed to provide a foundation on which the ore deposits can be understood and their formation placed in the context of a sequence of events following the impact process.

Chapter 3 presents a detailed review of the petrology and geochemistry of the magmatic rocks that comprise the SIC that permits the origin and evolution of the silicate magmas to be more clearly understood in the context of crustal melting and differentiation. Selected topics introduced in this section include the source of the metals, the timing of emplacement of the Offset Dykes, and the differentiation of the rocks comprising the main body (Main Mass) of the SIC.

The ore deposits at Sudbury are described in detail in Chapter 4 as the products of gravitational settling and differentiation of the sulfide liquid. The details of ore body morphology and relationship to the magmatic rocks, breccias, and the footwall of the SIC are provided. Case studies of the discovery, geology, mineralization, and geochemistry are presented for important examples of the main styles of mineralization.

Chapter 5 provides a unified hypothesis for the formation of the SIC and the associated ores; it returns to look at the sequence of events and provides process models that examine the differentiation and evolution of the Sudbury melt, it's sulfide saturation history, and the source of the metals.

Chapter 6 shows how the information is used to support exploration. Sudbury's position amongst large, magmatic sulfide systems such as Noril'sk, Thompson, Jinchuan, and Voisey's Bay is examined, and the similarities and differences in the process of formation of the ore deposits is highlighted. This chapter highlights the potential for future discovery of ore bodies at Sudbury, and it shows where Sudbury is placed in a global context of nickel sulfide and laterite deposits.

The aim of this text is to produce an overall understanding of the geology of the Sudbury impact structure and its ore bodies without writing an encyclopedic text that is inaccessible to students of geology. The elegance and simplicity of the basic ideas in impact geology and ore formation at Sudbury provides students with a grand theory of ore deposit formation with a minimal amount of the jargon and terminology that has evolved during 100 years of investigation.

This book is designed for an audience that includes not just those dedicated to finding the next generation of magmatic sulfide ore deposits, and understanding them, but also students of Earth Sciences who wish to gain a basic understanding of one of the most interesting and enigmatic ore-deposit systems on Earth—Sudbury.

Peter C. Lightfoot
Vale Base Metals and
Laurentian University

Acknowledgments

Many individuals contributed in ways that brought this book to publication, and I name them because they deserve credit for influencing the development of the ideas presented in this book or ensuring that I completed a careful navigation through the approval process.

Vale is thanked for their permission to write the book. Scott Mooney provided the essential support required in Vale to bring this book to completion; Scott reviewed the book proposal, endorsed the concept to Vale, and then reviewed each chapter. I thank Scott for his support and his continued interest in understanding the rocks that comprise the Sudbury enigma. Scott Mooney and Kevin Graham are thanked for working with the author and Elsevier to secure the book contract as well as developing a path to publication. Vale endorsed and encouraged this contribution, and I thank Jennifer Maki, Conor Spollen, and Cory McPhee for their support and endurance through the technical details. I thank Heather Brown for helping to guide me through the development of the arcane three-way publishing contract required to realize this process and publication of the book.

The book proposal was reviewed by Tony Green, Reid Keays, Chris Hawkesworth, Franco Pirajno, Ed Ripley, and Scott Mooney; their input and advice helped me to focus on the applied value of Sudbury geology, and it ensured that I tried harder to extract the most important salient points from the wealth of data and historic work on the impact melts and associated ore deposits.

The readers who have helped me to improve the text through careful reviews comprise scientists and technical experts from academia, government, and industry. My thanks are extended to Sam Davies, Ian Fieldhouse, Anatoliy Franchuk, Lisa Gibson, Sandy Gibson, Reid Keays, Jason Letto, Glenn McDowell, Enrick Tremblay, Ben Vandenburg, and Xu-Ming Yang.

The graphic design, layout, and drafting of all images and diagrams was undertaken by Alex Gagnon on the basis of draft diagrams provided by the author; Alex's skills in graphic design and conveying the geological concepts at the heart of this book make the final product so much better for all of his hard work and attention to detail.

Ben Vandenburg carefully selected and photographed all of the polished samples to illustrate the wealth of detail that supports an understanding of paragenesis; Ben also prepared the photomicrographs of the thin and polished sections for Chapters 3 and 4, and provided expert preparation so that rock samples could be photographed.

Lisa Gibson provided excellent 2D renditions of geological features evident in the 3D renditions of geological models that are difficult to convey without careful thought.

I will forever appreciate the mentoring and encouragement from Terry Morgan of Bishop Gore School. During my undergraduate work at Oxford, I was inspired to work on the geology of ore deposits by Professor Dick Stanton. Professor Tony Naldrett was my MSc supervisor and later a coinvestigator of the Siberian Trap and the associated Noril'sk mineral system. Professor Chris Hawkesworth was my PhD supervisor, and brought flood basalt geology and geochemistry into focus in the Deccan Trap. Both influenced my approach to Sudbury geology and they are thanked for sharing ideas, opinions, and data on Sudbury. I am deeply indebted to Professor Reid Keays who has been a colleague for over two decades and continues to be a source of ideas and inspiration that have contributed to my understanding of Sudbury geology.

The ideas expressed in this book have been developed over a period approaching 25 years of work on the Sudbury Igneous Complex and the associated ore deposits. I have benefited enormously from joint research with Tony Naldrett, Chris Hawkesworth, Reid Keays, Ed Ripley, and Will Doherty. Scientific interactions with many individuals have helped me to formulate these ideas; I am especially grateful to my colleague and friend, Professor Igor Zotov of the Russian Academy of Sciences, Moscow, who I miss greatly. I also extend my thanks to Drs. Doreen Ames, Sarah-Jane Barnes, Steve Barnes, Anthony Cohen, Fernando Corfu, Sarah Dare, Burkhard Dressler, Nick Gorbachev, Mike Lesher, Chusi Li, Des Moser, Gordon Osinski, Ulrich Riller, and Ron Sage.

I have been fortunate to have jointly supervised a number of graduate students and worked with postdoctoral fellows who have contributed to the understanding of magmatic sulfide ore deposits. My thanks goes to Emilie Boutroy, Mark Cooper, Lisa Cupelli, James Darling, Keith Farrell, Anatoliy Franchuk, Jian-Feng Gao, Kathy Hattie, Glen House, Grant Mourre, Mars Napoli, Jon O'Callaghan, Kostas Papapavlou, Steve Prevec, Aaron Venables, Yu-Jian Wang, Sheng-Hong Yang, and Mei-Fu Zhou.

I am privileged to have worked over the past 20 years with many world-class exploration geoscientists in the minerals industry whose dedication and enthusiasm for Sudbury geology has been an outstanding help in bringing applied focus to this book. These experts neither occupy the hallowed halls of academia nor are they experts from the government sector; they are applied exploration geoscientists. In particular I want to thank the following who have given freely of their ideas and helped me to gather or access the best samples and better understand Sudbury and its place in the spectrum of global magmatic sulfide ore deposits: Dick Alcock, Gordon Bailey, Mike Baril, Andy Bite, Cam Bowie, Charlene Brisbois, Sean Dickie, Sherri Digout, Dawn Evans-Lamswood, Catharine Farrow, Vivian Feng, Kevin Fenlon, Ian Fieldhouse, Carrie Forget, Brain Gauvreau, Lisa Gibson, Sandy Gibson, Paul Golightly, Graeme Gribbin, Roger Jackson, Alan King, Sasa Krstic, Andy Lee, Jason Letto, Herb Mackowiak, Hadi Mahoney, Michael McBurnie, Glenn McDowell, Bert McNabb, Chris Meandro, Rogerio Monteiro, Gord Morrison, Kyle Napoli, Mars Napoli, Krystal Oneill, Rob Palkovits, Ryan Paquette, Ed Pattison, Rob Pelkey, Clarence Pickett, Ben Polzer, Ashok Rao, Noelle Shriver, James Siddorn, Gary Sorensen, Peter Stewart, Rob Stewart, Matthew Stewart, Enrick Tremblay, Tony Vanwiechen, Shawna Waberi, Robert Wheeler, Christina Wood, Manqiu Xu, and Xue-Ming Yang.

From Elsevier, I wish to thank Tim Horscroft, Amy Shapiro, Tasha Frank, Marisa LaFleur, and Paul Prasad Chandramohan for all of their advice, input, and hard work on this book.

I thank my parents Irene and Bryan, my wife Nancy, my daughter Megan, and my family Betty, Fred, Isobel, Heather, Rob, Andrew, and Josh. They have all been a constant source of encouragement and understanding.

The author wishes to thank the following organizations and publishers for permission to use materials as follows:

Fig. 3.32a–d are from Ripley et al. (2015) is used with the permission of The Society of Economic Geologists.

Fig. 4.31b is used with the permission of the Archives of Ontario.

Figs. 4.77e and 5.1f are modified from Kenkmann and Dalwigk (2000), and is used with the permission of John Wiley and Sons.

Figs. 5.11a–e, 5.12a,b, 5.13a–e, and 6.22b,c are modified from Lightfoot and Evans-Lamswood (2015), and arc uscd with permission of Elsevier.

Figs. 4.23a–c, 4.49a,b, 4.50a–c, 4.51a–e, 4.52a–c, 4.58a–d, 4.59, and 5.3a–c are modified from publications of the Ontario Geological Survey, and used under license to Elsevier.

Sudbury is a center for production of base and precious metals from sulfide ores; Vale operates an integrated mine-mill-smelter-refinery complex; the photograph shows the smelter complex.

SUDBURY – AN INTRODUCTION TO THE ORE DEPOSITS AND THE IMPACT STRUCTURE

INTRODUCTION TO SUDBURY

Three facts about the Sudbury Structure place it among the crown jewels of planetary geoscience: (1) It is one of the largest, oldest, and best-preserved impact structures on planet Earth; (2) it was and remains the birthplace of important geoscience controversies in igneous petrogenesis, ore deposit geology, and impact cratering; and (3) it is home to Canada's largest mining camp with more than 100 years of mining activity on deposits of high-grade nickel sulfide ores. Quite how these three themes and the many strands of science and discovery that underpin them have evolved into a holistic understanding of the geology of Sudbury make it a classic study in Earth Sciences.

Wagner's views on continental drift, and the theory of plate tectonics were at the root of a revolution in Earth Sciences in the 20th century (Wegener, 1929), but the sudden and catastrophic events that change the planet in seconds to years rather than millions of years are at the crux of a shift in geoscience emphasis away from progressive (albeit rapid) change to sudden and profound shifts in the configuration of planet Earth. The Sudbury Structure is a case study in rapid change. Its formation was triggered by an impact event that lasted a fraction of a second followed by a crustal readjustment period likely lasting much less than 250,000 years. Exploration and mining activities over a period of more than 100 years in the Anthropocene underpin our current three-dimensional understanding of the geology of the shallow part of the Sudbury Structure above ~3 km depth. The path to our current understanding has triggered many epiphanies and quite a few global revolutions in the Earth Sciences. Living in the footprint of an astrobleme, one has a privileged opportunity to understand the complex geology as an outcome of a catastrophic event that happened on a short time scale relative to the slow motion of plate tectonics and mantle plumes. This book is written to explain the importance of geoscience, exploration, and discovery as it relates to not just ore deposits, but the further understanding of the Sudbury impact structure.

The City of Sudbury, located in Northeastern Ontario, Canada, is one of the world's principal sites of global nickel production; it is a city with a mineral industry that has evolved through almost 130 years. Since the discovery of the ore deposits, over 11.1 million metric tons of nickel and 10.8 million metric tons of copper together with by-products of cobalt, silver, gold, and platinum group elements (PGE) have been mined from the ore deposits (modified after Mudd, 2010). This wealth has been and continues to be generated from seven major mine complexes and 21 smaller ore deposits around the outer margin of the SIC which comprises part of the Sudbury Structure; the principal mines are owned and operated by international mining companies (Vale, Glencore, KGHM) and smaller mining companies (eg, Wallbridge). The high-grade ore deposits of the SIC are among the largest known historically mined and future resources of Ni–Cu–PGE sulfides, and comprise the foundation of the economic wealth of one of the largest mining camps in the world. The economic wealth generated at Sudbury in terms of just nickel and copper value at current metal

Nickel Sulfide Ores and Impact Melts. http://dx.doi.org/10.1016/B978-0-12-804050-8.00001-8

prices (± March 2015 nickel price) is close to 215 billion US dollars. The Sudbury Nickel Camp has underpinned the growth of the economy of Canada and Ontario, inspired contributions to the science of magmatic ore deposit geology (Naldrett, 2004), triggered the development of exploration technologies such as airborne geophysics and down-hole electromagnetic geophysical tools (Polzer, 2000; King, 2007), and provided a foundation for the development of mining technologies to handle the challenges of extraction of mineral from deep mines, the process technology for sulfide ores, and the foundation for the future growth of a global service center in the City of Sudbury in Northeastern Ontario.

The Sudbury Nickel Camp is second in the world in terms of contained Ni in sulfide deposit (contained metal in historic production and unmined reserves and resources), behind that of the Noril'sk Camp, located in the low Arctic of Siberia. The latter deposits are controlled by the Russian mining company, Noril'sk Nickel. Fig. 1.1 shows the contained metal in the principal global nickel sulfide deposits and the position of the two main companies operating in the Sudbury Mining Camp. The figure shows only deposits with more than 500 kt contained nickel from past production plus unmined ores; there are remarkably few high-grade nickel sulfide ore deposits, but there are quite a large number of lower-grade deposits (<1% Ni) that are largely undeveloped as mines. Fig. 1.1 shows the

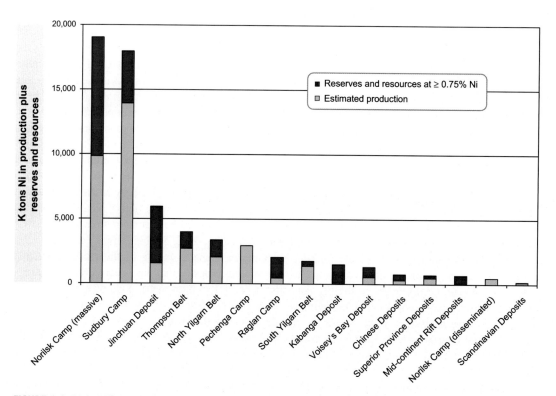

FIGURE 1.1 Global Nickel Production (nickel contained in ore) and Reserves Plus Resources of Major Nickel Sulfide Deposits.

Based on a compilation of global nickel reserves and resources in 2013 and production to the end of 2014.

current amount of contained nickel metal in reserves and resources plus the historic production of nickel for Sudbury (Vale and Glencore data are broken out and shown separately, but collectively they account for 99% of the production from the Sudbury Deposits). Total production plus unmined reserves and resources for the Noril'sk low-grade nickel deposits (Noril'sk Nickel), the Noril'sk high-grade nickel deposits, and other large developed and undeveloped nickel deposits are also shown. Production and untapped high-grade nickel resources from Sudbury (typically with nickel grades between 1.5 and 2.5%) exceed those of the Noril'sk Deposits (which typically have nickel grades of 1–3%), but in terms of overall contained metal in high-grade and low-grade deposits (ie, ≤1% Ni), Noril'sk exceeds Sudbury in contained nickel. Both Sudbury and Noril'sk have more than twice the contained nickel of the next largest deposits which include Pechenga (Kola Peninsula, Russia), Jinchuan (Gansu Province, China), Mt Keith (Western Australia), and Thompson (Manitoba, Canada) (Fig. 1.1; Naldrett, 2004).

The discovery of significant nickel sulfide deposits is quite a rare event as shown in a plot of discovery year versus total metal produced together with reserves and resources (Fig. 1.2); Fig. 1.2 illustrates the fact that large deposits such as Noril'sk and Sudbury are quite unique, whereas more recent discoveries tend to be mid-size or smaller. Mining companies continue to try to expand the reserve and resource base of all of these deposits, but the probability that recent discoveries will grow to the size of

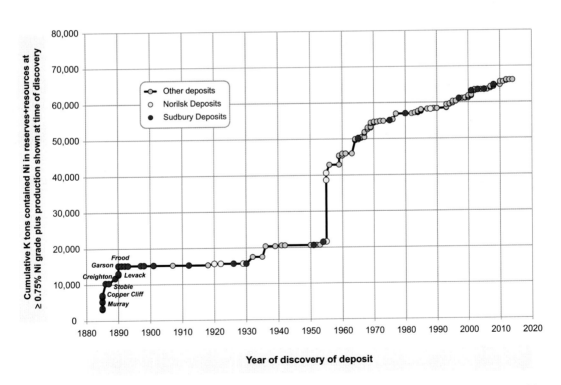

FIGURE 1.2 Past Production (nickel in ore), Unmined Reserves, and Resources of Deposits Plotted Against Initial Year of Discovery. The deposits are broken out into high grade (>1% Ni) and low grade (≤1% Ni).

Based on a compilation of global nickel reserves and resources in 2013 and production to the end of 2014.

Sudbury or Noril'sk is very low, and the global trend has been toward the discovery of smaller deposits through time.

The Sudbury Mining Camp has had an enormous economic and social impact on not only the Sudbury Region but also on the economic development of Canada. The City of Sudbury grew up around the natural resources and mining industries, and is now a center for industry, government, academia, and health. The present-day skyline of Sudbury is still dominated by the Superstack which is 380 m tall (Conroy and Kramer, 1995), and was the tallest freestanding structure at the time. It was commissioned in 1972 to control the release of sulfur dioxide gas that contributed to the original discovery of acidification of lake waters in Kilarney Provincial Park (Beamish and Harvey, 1972; Wren, 2012). The multi-mine head-frames attest to the new depths to which ore is mined and extracted as well as the infrastructure of mills, smelters, and refineries which take raw ore and convert it to metal that is sold to market. Sudbury has inspired some landmark contributions to music like "Sudbury Saturday Night" (Connors, 1967). The growth of the community has largely been based on mining but it has now grown into a new regional center. The Sudbury Neutrino Observatory (SNO) is a recent facility designed to detect solar fission reactions; it utilizes the underground facilities in Vale's Creighton Mine to detect and understand the flux of solar neutrinos at a depth of 2100 m (http://www.sno.phy.queensu.ca). The Nobel Prize in Physics (2015) was awarded to Dr. Takaaki Kajita (Japan) and Dr. Arthur B. McDonald (Canada), for their work at the SNO Laboratory on neutrinos.

In addition to its economic importance, Sudbury offers a unique opportunity to study the unique geological features and events that gave rise to the best-known examples of magmatic sulfide ore deposits. The metal endowment has a direct spatial relationship to a geological structure that was produced 1850 million years ago by an asteroid impact. This impact produced the conditions necessary for the ore deposits to form, and created an assemblage of rocks in the Sudbury Region that have puzzled geologists for over 100 years. The effects of the Sudbury impact event included a tsunami event and fall out of debris recorded in rocks of similar age over 1200 km away from Sudbury (Addison et al., 2010). This book examines the sequence of events that gave rise to the Sudbury Structure and its associated ore deposits. It is not designed to be a textbook in the conventional descriptive sense, but aims to tell the story of how detailed geology, petrology, and geochemistry of the Sudbury Structure help establish a sequence of events in the catastrophic impact event and the processes which resulted in the formation of the Sudbury ore deposits. The book is written from the perspective of a geologist who works with three-dimensional datasets in the exploration of the Sudbury Structure, and who works on the footprint of the Sudbury Mines.

HISTORY OF DISCOVERY AND PRODUCTION AT SUDBURY

The recognition of the geophysical manifestation of mineralization at Sudbury was first made by A.P. Salter in 1857 (Giblin, 1984a,b). Salter observed a deflection in his compass readings in an area close to the current Creighton Mine. The actual discovery of nickel in Sudbury occurred after construction of the Canadian Pacific Railway began near to the future location of Murray Mine (Fig. 1.3). In 1883, Thomas Flanagan noticed copper (nickel) sulfides in the area, which resulted in a detailed geological survey in an area which eventually became Murray Mine.

Early prospecting activities resulted in a number of discoveries (see Table 1.1 for a summary of the main events in the evolution of the Sudbury Mining Camp). The early development of these deposits tended to be ad hoc, with small open pits, drifts, inclined shafts, and surface block caving (Whiteway, 1990; Boldt, 1967). Early in the life of the camp, these ores were roasted and then shipped to foreign destinations for processing in well-established plants in locations such as Swansea in South Wales (Francis, 1881). These shipments underpinned the later parts of the industrial revolution in the United Kingdom and the position of Britain at the center of a global empire; it also triggered enormous environmental damage in industrial areas such as the Lower Swansea Valley. Only later on in the development of the camp were technologies developed to concentrate, smelt, and refine the metal in Sudbury from large open pit mine operations such as Frood-Stobie and Clarabelle and underground mine complexes serviced by shafts (Boldt, 1967; Table 1.1).

The first mine to enter production was Copper Cliff in 1886, with the first local smelting in 1888 (Table 1.1). Geological studies undertaken in 1886 led to the start of production in 1901 at Creighton mine which was owned and operated by the Canadian Copper Company. The discovery of nickel in the Sudbury ores spurred the creation of the International Nickel Company (Thompson and Beasley, 1960). These two divisions became known under the trade name Inco in 1919 (Table 1.1).

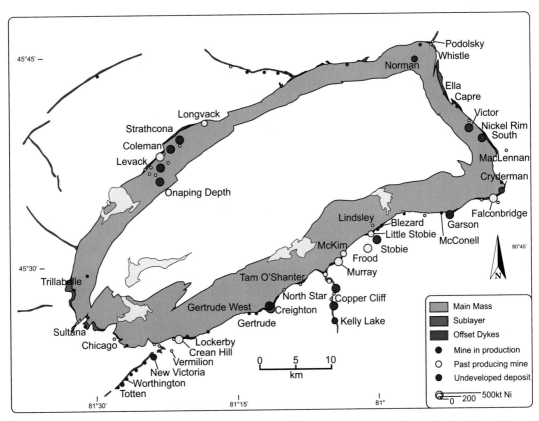

FIGURE 1.3 Distribution of Mines and Mine Complexes of the Sudbury Region (Historic Producing Mines, Present mines, and Undeveloped Deposits Scaled to Contained Nickel)

Table 1.1 Summary of Some of the Important Historic Events in the Discovery, Exploration, and Development of the Sudbury Ore Deposits

Year(s)	Important Events	Description of Landmark Events	Sources of Additional Information (Publications and Web Sources)
1856–57	Sudbury discovery	Government Surveyor, Albert Salter, identified magnetic rocks and presence of Ni and Cu in assays. Technical significance of the observations was not recognized, but the scientific reports stand in recognition of the discovery	Salter (1856), Murray (1857)
1856	Creighton Deposit discovered	Original discovery by Albert P. Salter. Rediscovered in 1886 by Henry Ranger. First ore produced in 1901 from open pit; acquired by Inco in 1925; shafts sunk in 1928–35 and 1969 (a series of progressively deeper shafts to 7137 ft.); mined at depth by ramp; now one of the the deepest mines in North America, and discovery of new ore continues in 2016	Salter (1856)
1883	Murray Deposit discovered	Construction of Canadian Pacific Railway; gossan identified during forest clearing by Thomas Flanagan; this event triggered the acquisition of the future Murray Mine property by Thomas and William Murray. The property was sold to A.H. Vivian and Company of Swanses (Wales), and the property was mined from 1889–94. The British American Nickel Company purchased the deposit in 1912, and Inco acquired the mine in 1925. The property then saw discontinuous production up to 1971 when it was closed down	Giblin (1984a,b)
1884	Frood-Stobie Deposit discovered	Discovered by Thomas Frood; open pit mining commenced in 1889. Shaft developed in 1911, and a large open pit was opened in 1938; in 1965 a shaft was sunk to service the Frood and Stobie Deposits and Frood was merged with Stobie in 2000; Frood Mine remains a part of the Stobie Complex in 2015	Zurbrigg (1957); Hattie (2010); Cutifani (2005); Inco and Falconbridge (2006)
1884	Worthington Deposit discovered	Found along the Canadian Pacific Railroad by F.C. Crean. The mine was in operation between 1885 and 1927 when the workings collapsed. This deposit is the surface expression of Totten Mine	Sudbury Star, October 4, 1927
1884	Copper Cliff Deposit (North Mine) discovered	Discovered by Thomas Frood; the mine was opened by the Canadian Copper Co. in 1886; in 1963 a shaft was sunk on the 138 Ore Body; the mine temporarily closed in 1978 and re-opened in 1984. Copper Cliff was the first property with serious mining in Sudbury; originally considered a copper deposit, but the ores were shown to contain nickel in 1887	Giblin (1984a,b)
1885	Copper Cliff Deposit (South Mine) discovered	The Evans Ore Body was discovered by F.J. Eyre, and worked as a shallow pit from 1889; production of deeper ore bodies commenced in 1970 with a connection to North Mine in 2004	Cochrane (1984)
1885	Crean Hill Deposit discovered	Discovered by Francis Crean in 1885. Production commenced in 1906–14, 1915–19, 1964–78, and 1986–2002	Royal Ontario Nickel Commission (1917)

Year(s)	Important Events	Description of Landmark Events	Sources of Additional Information (Publications and Web Sources)
1885	Little Stobie Deposit discovered	Discovered by James Stobie; mining started by Mond Nickel in 1902. First shaft sunk in 1966 with production commencing in 1971	Davis (1984)
1885	Stobie Deposit discovered	Discovered by James Stobie in 1885. Opened in 1886, and intermittent production until underground mine developed in 1948. This deposit is still mined in 2016	Zurbrigg (1957); Hattie (2010)
1885	Totten Deposit discovered	Francis Crean discovered the mineralization and the patent was granted to Henry Totten who had first named the Totten mineralization along a segment of the Worthington Offset. Production commenced between 1964 and 1974, and then in 2014. This is the most recent mine to be built in the Sudbury Camp	Lightfoot and Farrow (2002); Inco and Falconbridge (2006)
1885	Blezard Deposit discovered	Land was patented in 1888 and production took place between 1889 and 1893; the deposit has not yet been mined at depth, but the plunge extent comprises the Thayer Lindsley Mine which entered production in 1991 by Falconbridge Limited	Royal Ontario Nickel Commission (1917); Inco and Falconbridge (2006)
1886	Copper Cliff Mine development	First organized development and mining and smelting activity by Canadian Copper Company	Giblin (1984a,b)
1886	Victoria Deposit discovered	The deposit was discovered by Henry Ranger. The mine roast yard and smelter were operated by Mond Nickel Company from 1899 to 2923	Royal Ontario Nickel Commission (1917)
1887	New Jersey refinery, USA	Discovery that consignment of material from Copper Cliff Mine to Orford refinery contained nickel which was referred to as the "demon metal" as it was an impurity in the copper	Royal Ontario Nickel Commission (1917)
1888	Sudbury roast-yard construction	Eleven ore-roasting yards were built and operated in Sudbury between 1888 and 1929	Wren (2012)
1889	Levack Deposit discovery	James Stobie and Rinaldo McConnell discovered the mineralization. Mining of the Levack Deposit commenced in 1914. Production from McCreedy West Ore Body commenced in 1974	Royal Ontario Nickel Commission (1917)
1890	Victor Deposit discovery	The property was first patented in 1890, and a small near-surface mine was developed and exhausted in 1960. The depth extent of Victor remains undeveloped in 2015	Royal Ontario Nickel Commission (1917)
1891	Garson Deposit discovered	Discovered by John Thomas Cryderman; mining commenced in 1908	Royal Ontario Nickel Commission (1917)
1892	Coleman Deposit discovery	Discovered by Thomas Baycroft (Coleman) and James Stobie (Coleman East); purchased by Inco in 1942; production commenced in 1970; McCreedy East Deposit accessed via underground ramp from Coleman Mine in 1996	Royal Ontario Nickel Commission (1917)

(Continued)

Table 1.1 **Summary of Some of the Important Historic Events in the Discovery, Exploration, and Development of the Sudbury Ore Deposits** *(cont.)*

Year(s)	Important Events	Description of Landmark Events	Sources of Additional Information (Publications and Web Sources)
1892	Cryderman Deposit discovery	Discovered by Albert Harvey; within the broad footprint of the Falconbridge Deposit. Cryderman is not yet mined in 2016	Royal Ontario Nickel Commission (1917)
1897	Pyrometalurgy begins in Sudbury	Processing of ores by blast furnace in Copper Cliff commenced	Wren (2012)
1897	Whistle Deposit discovery	Discovered by Isaac Whistle; serious production did not commence until 1988, and the mine was closed and decommissioned in 1998	Lightfoot et al. (1997a,c)
1899	Mond Process technology developed	A process was developed to extract and purify nickel by converting nickel oxide into nickel metal using carbon monoxide to form nickel carbonyl which breaks down to nickel metal and carbon monoxide at high temperature	Mond et al. (1890)
1900	Victoria Mine production	Mond's first Canadian mining properties located in Denison Township, was purchased from Ricardo McConnell and associates in 1899. this site renamed the Victoria Mine began development in 1900	Royal Ontario Nickel Commission (1917)
1901	Creighton Mine production	First ore was produced from Creighton Mine open pit; Inco acquired the property in 1925	Cutifani (2005; Inco and Falconbridge (2006); Royal Ontario Nickel Commission (1917); http://www.vale.com/canada/EN/aboutvale/history
1902	Sudbury Region mining companies incorporated	Incorporation of International Nickel Company by amalgamation of smaller producers including Canadian Copper Company and Orford Copper Company	Giblin (1984a,b)
1902	Nickel refinery in Clydach, Wales	A nickel refinery at Clydach, Wales was built by Mond Nickel Co. International Nickel Co. formed on April 1, 1902.	Royal Ontario Nickel Commission (1917)
1908	Garson Mine enters production	First production from Number 1 shaft by Mond Nickel Corporation; acquired by Inco in 1929	Cutifani (2005); Inco and Falconbridge (2006)
1912	Capre Deposit discovered	Discovered by Tom Bennett and purchased by Inco in 1926. Exploration by Inco in 2008 revealed three footwall mineral zones	Stewart and Lightfoot (2010)
1911	Coniston Smelter construction	In 1911 the Mond company began construction of a smelter in Coniston; it operated from 1913 to 1972	Boldt (1967)
1915	O'Donnell Roasting Yard built	The largest and last in a succession of roasting yards built to provide primitive processing of Sudbury ores; built in Graham Township west of Creighton	Royal Ontario Nickel Commission (1917)

Year(s)	Important Events	Description of Landmark Events	Sources of Additional Information (Publications and Web Sources)
1916	New York listing of International Nickel Company	First listing of International Nickel Company on New York Stock Exchange in 1915; the International Nickel Company of Canada was incorporated as a subsidiary company in 1916	http://www.vale.com/canada/EN/aboutvale/history
1918	Port Colborne nickel refinery, Ontario, Canada	Start up of the Port Colborne nickel refinery	http://www.vale.com/canada/EN/aboutvale/history
1919	Inco trademark originates	Trademark name "Inco" first used by the International Nickel Company	http://www.vale.com/canada/EN/aboutvale/history
1924	Acton refinery built in the United Kingdom	Acton Precious Metals Refinery was built	Boldt (1967)
1928	Frood Mine developed	By 1928, the International Nickel Company was developing Frood Mine; it was determined that it and Mond's Frood Extension were part of the same ore body. Mond negotiated an agreement where the interests of the Mond Nickel Company were merged into the International Nickel Company	Boldt (1967)
1928	Falconbridge Smelter built	Falconbridge Smelter constructed adjacent to the Falconbridge Deposit; expanded and upgraded in many stages, and producing 62,093 tons of nickel in matte in 2005	Falconbridge staff (1959); Cutifani (2005); Inco and Falconbridge (2006)
1929	Corporate acquisition	International Nickel Company acquired the Mond Nickel Company, Limited	Giblin (1984a,b)
1928	Falconbridge Nickel Mines incorporated	Incorporation of Falconbridge Nickel Mines Limited and commencement of mining and smelting in Falconbridge; the Falconbridge Mine was closed in 1984	Giblin (1984a,b)
1930	Copper Cliff Smelter built	Inco have operated a smelter in Copper Cliff since 1930; the facility has evolved with many upgrades and by 2005 was able to produce 97,4500 tons of nickel in product	Boldt (1967)
1939	Frood Mine	HRH King George VI and the future Queen Elizabeth visited Sudbury, and the future Queen went underground at Frood Mine	http://www.vale.com/canada/EN/aboutvale/history
1939–1945	World War 2	~700 kt Ni supplied to Allied war efforts	http://www.vale.com/canada/EN/aboutvale/history
1944	Kelly Lake Deposit discovered	Initial discovery of near-surface mineral zone (725); 720 and 740 mineral zones discovered in 1955; exploration between 1995 and 2000 identified depth extent. The Kelly Lake deposit remains undeveloped 2016	Cochrane (1984)
1946	Airborne geophysical methods	First use of airborne electromagnetic survey methods developed by Inco	Boldt (1967)

(Continued)

Table 1.1 Summary of Some of the Important Historic Events in the Discovery, Exploration, and Development of the Sudbury Ore Deposits *(cont.)*

Year(s)	Important Events	Description of Landmark Events	Sources of Additional Information (Publications and Web Sources)
1946	Growth in nickel demand	After the Second World War, the US government began to stockpile nickel, to diversity its supply of noncommunist nickel. Falconbridge was the chief beneficiary of this policy and grew significantly. Between 1946 and 1961, the region's population grew from 70,000 to 137,000	Boldt (1967)
1954	Copper Cliff smelter	Erection of world's tallest smelter chimney (194 m)	http://www.vale.com/canada/EN/aboutvale/history; Cutifani (2005); Inco and Falconbridge (2006)
1956	Discovery of Thompson, Manitoba	Discovery of nickel-rich ores at Thompson by airborne electromagnetic method; subsequent diversification of nickel production into a second camp with production starting in 1961	Thompson and Beasley (1960); http://www.vale.com/canada/EN/aboutvale/history
1968	Indonesia nickel laterite smelting commences	Initial agreement to develop nickel laterite deposits in Sulawsei was reached. The plant was built in 1977 by PT Inco, and the first matte product was shipped for refining in 1979. Additional nickel laterite projects held by Inco and other companies entered production in subsequent years	Mudd (2010)
1959	Frood Mine	HM Queen Elizabeth II and HRH Prince Phillip visit Frood Mine	http://www.vale.com/canada/EN/aboutvale/history
1969	Creighton Mine shaft	Inco completed the 2175 m deep Creighton 9 shaft - the deepest mine shaft in the western hemisphere	http://www.vale.com/canada/EN/aboutvale/history
1970	Coleman mine enters production	First production from contact ore deposits; in 1996 McCreedy East 153 Ore Body developed	Cutifani (2005); Inco and Falconbridge (2006)
1971	Clarabelle Mill	Clarebelle Mill commissioned to treat 35,000 tons ore per day from 6 ore deposits; produced concentrate for Copper Cliff Smelter	Cutifani (2005); Inco and Falconbridge (2006)
1971	Copper Cliff superstack; NASA astronauts visit Sudbury	Inco completes construction of the 381 m superstack chimney. NASA sent Apollo 16 astronauts to study Sudbury geology, and the following year Apollo 17 astronauts visited Sudbury for the same reason	http://www.vale.com/canada/EN/aboutvale/history
1972	Copper Cliff refinery	Operation began at the refinery to process nickel oxides and sulfide from the Copper Cliff Smelter	http://www.vale.com/canada/EN/aboutvale/history
1974	Depth extent of Victor Deposit discovered	Discovery of depth extent of contact ore deposits at Victor	Morrison et al. (1994)
1975	Inco Limited established	International Nickel Company of Canada changed name to Inco Limited	Giblin (1984a,b)

Year(s)	Important Events	Description of Landmark Events	Sources of Additional Information (Publications and Web Sources)
1979	Strike action	9 month strike by Inco employees over production and employment cutbacks impacted production	http://www.financialpost.com/story.html?id=2c7b0cfe-14be-40dc-8d92-8a74685e97782&p=4
1982	Falconbridge Limited established	Falconbridge Nickel Mines Limited changed name to Falconbridge Limited	Giblin (1984a,b)
1989	Victor Footwall Deposit discovered	Discovery of footwall ore deposit at Victor in the East Range of the Sudbury Structure	Morrison et al. (1994)
1991	Copper Cliff SO$_2$ abatement technology implementation	Inco introduced new SO$_2$ abatement technology to smelter complex	http://www.vale.com/canada/EN/aboutvale/history
1995	Discovery of Voisey's Bay, Labrador	Discovery of the Ovoid Deposit; Inco purchased a 25% interest in the project, and in 1996 Inco acquired 100% of Diamond Fields Resources Inc. for $4.3 billion Canadian dollars. It took many years of negotiation with the Government of Newfoundland and Labrador and the Innu and Inuit First Nations before production commenced in 2005	Naldrett et al. (1996a,b); McNish (1999); Lightfoot et al. (2011a,b, 2012)
2000	Discovery of depth extent of Kelly Lake Ore Body	Inco announces details of the discovery of the Kelly Lake Ore Body in the southern segment of the Copper Cliff Offset	Polzer (2000)
2001	Discovery of Nickel Rim South Deposit	Discovery of Nickel Rim South Deposit; mine construction began in 2003 by Falconbridge Limited	McLean et al. (2005)
2001	Acquisition of Inco properties	Junior mining company, FNX, entered into an option agreement with Inco to acquire five properties in the Sudbury Basin	http://www.infomine.com/index/pr/Pa740692.PDF
2006	Xstrata acquires Falconbridge	Xstrata acquired Falconbridge for $22.5 billion CAD	http://www.sudburyminingsolutions.com/project-management-process-nickel-rim-south-ramps-up.html
2006	Vale acquires Inco	Inco Limited acquired by Brazilian miner, Vale for $18.9 billion US; Vale Inco was then established as a Canadian business unit of Vale	http://www.vale.com/canada/EN/aboutvale/history
2009	Nickel Rim South Mine enters production	Falconbridge commences ramp-up of production at Nickel Rim South Mine	http://news.bbc.co.uk/2/hi/business/4794173.stm
2010	New Victoria Deposit discovered	Discovery of a new deposit ~1 km southwest of Victoria Mine; now called the Victoria Project, it was initially referred to as Victoria's secret by the company who explored the property (Quadra-FNX)	Farrow et al. (2011)

(Continued)

Table 1.1 Summary of Some of the Important Historic Events in the Discovery, Exploration, and Development of the Sudbury Ore Deposits *(cont.)*

Year(s)	Important Events	Description of Landmark Events	Sources of Additional Information (Publications and Web Sources)
2010	FNX and Quadra merge	FNX merges with Quadra; the combined company was acquired by KGHM in 2011 for 2.84$ Billion	http://www.bloomberg.com/news/articles/2011-12-06/poland-s-kghm-to-buy-quadra-fnx-for-2-83-billion-to-add-u-s-chile-mines
2013	Glencore merges with Xstrata	Glencore merges with Xstrata	http://www.ft.com/intl/cms/s/0/9d355d82-b31a-11e2-95b3-00144feabdc0.html#axzz3kd5BQRYd
2014	Totten Mine commences production	Production commences in Sudbury's newest mine, owned and operated by Vale	http://www.vale.com/canada/EN/aboutvale/history

Explanation of color code

Original greenfield prospective/exploration discovery
Brownfield prospecting/exploration discovery
Mine development/production started
Fundamental technology development
Major corporate event

Early methods of processing of massive sulfide ore involved the use of open roasting yards where the mined ore was piled on wood and set on fire to burn off the sulfur in the ore and produce a very primitive product that could be shipped and refined. The roast yards were the cause of environmental destruction from the perspective of both the deforestation activity required to support the process, and the sulfur dioxide which killed local vegetation (Gunn, 2011). The early locally roasted product was shipped to the United States for copper refining and at the Orford refinery in New Jersey. The refined copper from the Orford refinery contained a contribution of nickel that had long been recognized as a contaminant in copper production. Nickel-bearing copper has the historic name *Kupernickel*, where the word *Nickel* had the meaning of a "demon metal" as it interfered with the production of a quality copper product.

The recognition of the importance of nickel in steel alloys resulted in a steady growth in the market for nickel, and by 1920 Sudbury produced 80% of the global nickel supply. A steady stream of new discoveries and new mines entered production over the 20th century, and these include the world-class ore deposits at Frood Stobie which began production in the 1920s under the ownership of Inco (Frood Mine) and Mond Nickel (Stobie Mine), eventually to be merged under Inco into the Frood-Stobie Mine Complex. During the 1930s, the copper cliff smelter was completed and this eliminated the need for heap roasting.

Falconbridge was incorporated in 1928 by Thayer Lindsley. Falconbridge acquired mining claims in the town of Falconbridge, and by 1930 the Falconbridge Mine began operation (Falconbridge, 1959). The Falconbridge Mine closed in 1984 once its ore reserves were exhausted.

In the 1940s, nickel production was increased by Inco as nickel was needed in the production of materials to support the war effort. In the 1950s, the construction of the world's tallest smelter chimney was completed, measuring 195 m tall. This would later be overshadowed by the currently used "superstack" which was built in the 1970s and is 380 m tall (shown in the photograph at the front of the chapter). In the 1960s, Inco completed the sinking of what was then the world's deepest shaft, Shaft No. 9 at Creighton Mine which reached a depth of 2175 m.

The Clarabelle Mill was constructed in 1971 and was capable of handling 35,000 tons of ore per day. This allowed multiple mines in Sudbury to produce ore and truck it to one central location. The year 1989 was recorded as a high-profit year for Inco, bringing in over $750 million.

In the last half century, the majority of production in Sudbury has been controlled by two principal companies; although the ownership has changed from Inco to Vale and Falconbridge to Glencore, the production of metal continues, and new discoveries are made as a result of exploration. The mines at Sudbury now exploit deep ore bodies having the deepest operation in North America at over 2400 m in Creighton Mine. As the shallower ore deposits were located and mined, the emphasis increasingly moved to the discovery of new ore bodies at greater depths.

The total historic production of nickel from Sudbury is shown in Fig. 1.5, where the production output of Sudbury (in kt nickel in ore) is shown together with the average yearly nickel and copper grade. The plot also shows the evolution of ownership of the principal deposits as the companies changed names and ownership (Tables 1.1 and 1.2). The principal historic events that had a major impact on the amount of nickel production coming out of Sudbury are also shown, and it is easy to see the major impacts of labor unrest, economic recessions, both the first and second world wars, and the cold war on nickel production. The effects of local and global events on the production profile explain the sudden departures in production from the steady increase through the 20th century that reflected the global increase in demand for nickel in the production of stainless steel, alloys, electroplating, and

Table 1.2 The Sudbury Ore Deposits by Location, Size, and Geological Environment

Name	Deposit	Discovery Year	Production Years	Estimated Ni in Ore Produced (kt Ni) up to 2015 (*/+ = See Footnote)ᵃ	Reserves + Resources Mmt ore	Ni Grade wt%	Cu Grade wt%	Pt + Pd + Au g/t	Date of Last Formal 43-101 Compliant Reserve and Resource Information Published	Source or Information on Reserves and Resources	Present Ownership	Current Producer/Historic Producer	Geological Environment	Specific References (See Also: Royal Ontario Nickel Commission, 1917; Dressler (1984 a–c); Ames et al., 2008)
Very large ore systems	Copper Cliff Mine Complex	1884	1886–2015	1100	63.51	1.23	1.17		Dec. 31, 2005	North Mine, South Mine, and Kelly Lake; Inco and Falconbridge (2006)	Vale	Producing mine complex with 2 shafts and historic workings	Offset deposit	Souch et al. (1969); Cochrane (1984); Farrow and Lightfoot (2002)
	Coleman-McCreedy East Mine	1889	1970–2015	500	25.92	1.59	3.05		Dec. 31, 2005	Inco and Falconbridge (2006)	Vale	Producing mine	Contact-footwall deposit	Moore and Nikolic (1994); Coats and Snajdr (1984)
	Levack-McCreedy West Mine	1889	1914–2015 (intermittent)	>1300b	16.45	1.52	1.23		Dec. 31, 2008	Levack and McCreedy West; Farrow et al. (2008a,b)	KGHM	Producing mine complexes with multiple shafts and historic mine workings	Contact-footwall deposit	Souch et al. (1969); Farrow et al. (2008a,b)
	Creighton Mine Complex	1886	1901–2015	2500	44.70	1.28	2.00		Dec. 31, 2005	Creighton 3 plus Creighton 9; Inco and Falconbridge (2006)	Vale	Producing mine	Contact-footwall deposit	Souch et al. (1969)
	Garson Mine	1891	1908–2015	900	15.49	1.67	1.36		Dec. 31, 2005	Inco and Falconbridge (2006)	Vale	Producing mine	Modified contact	Souch et al. (1969)
	Falconbridge Mine	1899	1932–88	c							Glencore	Historic producer	Modified contact	Owen and Coats (1984)
	Frood-Stobie Mine Complex	1884	1923–2015	3400	38.45	0.73	0.69		Dec. 31, 2005	Inco and Falconbridge (2006)	Vale	Producing Mine Complex	Breccia Belt Deposit	Zurbrigg (1957); Souch et al. (1969)
	Murray (-Elsie)	1883	1889–1971 (intermittent)	290	16.91	1.08	0.79		Dec. 31, 2005	Inco and Falconbridge (2006)	Vale	Historic producer	Contact deposit	Souch et al. (1969)

Name	Deposit	Discovery Year	Production Years	Estimated Ni in Ore Produced (kt Ni) up to 2015 (*/+ = See Footnote)[a]	Reserves + Resources Mmt ore	Ni Grade wt%	Cu Grade wt%	Pt + Pd + Au g/t	Date of Last Formal 43-101 Compliant Reserve and Resource Information Published	Source or Information on Reserves and Resources	Present Ownership	Current Producer/ Historic Producer	Geological Environment	Specific References (See Also: Royal Ontario Nickel Commission, 1917; Dressler (1984 a–c); Ames et al., 2008)
Large ore systems	Victor	1890	1960 (shallow deposit)	3	20.03	1,68	2.13		Dec. 31, 2005	Inco and Falconbridge (2006)	Vale	Project awaiting development	Contact-footwall deposit	Morrison et al. (1994); Jago et al. (1994)
	Craig-Onaping	1950	1958–2015	c	3.80	1.56	0.40		Dec. 31, 2005	Inco and Falconbridge (2006)	Glencore	Producing Mine	Contact-footwall deposit	
	Strathcona-Fraser Mine	1890	1968–2015	c	3.50	1.09	2.55		Dec. 31, 2005	Inco and Falconbridge (2006)	Glencore	Producing Mine	Contact-footwall deposit	Li and Naldrett (1992)
	Nickel Rim Mine	2001	2005–2015	c	15.00	1.78	4.02		Dec. 31, 2005	Nickel Rim and Nickel Rim Depth; Inco and Falconbridge (2006)	Glencore	Producing Mine	Contact-footwall deposit	McLean et al. (2005)
	Crean Hill	1885	1906–2000 (intermittent)	220					Dec. 31, 2005		Vale	Historic producer	Modified Contact	Gibson et al. (2010)
	Totten-Worthington Mine	1885	1885–2015 (intermittent)	25	9.12	1.39	2.00		Dec. 31, 2005	Inco and Falconbridge (2006)	Vale	Producing mine	Offset deposit	Lightfoot and Farrow (2002); Farrow and Lightfoot (2002); Murphy and Spray (2002)
	Lockerby Mine	1888	1971–2015 (intermittent)	c	12,71	0.74	0.49		Dec. 30, 2012	Lockerby Depth zone; www.fnimining.com	First Nickel Inc.	Producing mine	Contact deposit	Clow et al. (2005)
	Little Stobie	1885	1902–1997 (intermittent)	145							Vale	Historic producer	Contact-footwall deposit	Davis (1984)

(Continued)

Table 1.2 The Sudbury Ore Deposits by Location, Size, and Geological Environment *(cont.)*

Name	Deposit	Discovery Year	Production Years	Estimated Ni in Ore Produced (kt Ni) up to 2015 (*/+ = See Footnote)[a]	Reserves + Resources Mmt ore	Ni Grade wt%	Cu Grade wt%	Pt + Pd + Au g/t	Date of Last Formal 43-101 Compliant Reserve and Resource Information Published	Source or Information on Reserves and Resources	Present Ownership	Current Producer/ Historic Producer	Geological Environment	Specific References (See Also: Royal Ontario Nickel Commission, 1917; Dressler (1984 a–c); Ames et al., 2008)
Discrete deposits	Milnet			2	0.29	0.65	0.70	1.65	2002	Bailey (2011, 2013)	Wallbridge	Historic producer	Offset deposit	Bailey (2013)
	Onaping Depth			0	25.78	2.28	1.00		Dec. 31, 2005	Onaping Depth and Dowling North; Inco and Falconbridge (2006)	Vale, Glencore	Undeveloped	Contact deposit	
	Sheppard (Davis)	1890	1892–93	0.01							Vale	Historic producer	Contact deposit	
	Gertrude and Gertrude West	1884	2000–13	22	0.37	1.03	0.35		Dec. 31, 2005	Inco and Falconbridge (2006)	Vale	Historic open pit mines	Contact deposit	
	Ellen	1886	1961–71; 2011–15	15	0.58	1.18	0.72		Dec. 31, 2005	Inco and Falconbridge (2006)	Vale	Open pit and ramp	Contact deposit	
	Whistle	1897	1992–2010	45	6.68	0.75	0.21		Dec. 31, 2005	Whistle Mine; Farrow et al. (2008a,b)	KGMH	Historic producer	Contact-footwall deposit	Lightfoot et al. (1997a–c); Farrow et al. (2008a,b)
	McConell			0	0.84	0.85	2.33		Dec. 31, 2005	Inco and Falconbridge (2006)	Vale	Undeveloped	Modified contact deposit	
	Podolsky	2002	2008–13	b	2.24	0.37	4.23		Dec. 31, 2008	Farrow et al. (2008a,b)	KGMH	Historic producer	Contact-footwall deposit	Ames and Farrow (2007); Farrow et al. (2008a,b)
	Norman-Norman West		None	0	7.60	1.60	1.30		Dec. 31, 2005	Inco and Falconbridge (2006)	Vale, Glencore	Undeveloped	Contact deposit	
	Capre	1912	None	0	2.17	1.14	0.26		Dec. 31, 2005	Inco and Falconbridge (2006)	Vale	Undeveloped	Contact-footwall deposit	Stewart and Lightfoot (2010)

Name	Deposit	Discovery Year	Production Years	Estimated Ni in Ore Produced (kt Ni) up to 2015 (*/+ = See Footnote)ᵃ	Reserves + Resources Mmt ore	Ni Grade wt%	Cu Grade wt%	Pt + Pd + Au g/t	Date of Last Formal 43-101 Compliant Reserve and Resource Information Published	Source or Information on Reserves and Resources	Present Ownership	Current Producer/ Historic Producer	Geological Environment	Specific References (See Also; Royal Ontario Nickel Commission, 1917; Dressler (1984 a–c); Ames et al., 2008)
Discrete deposits (cont.)	Joe Lake (WD16)	1926	None	0							Vale	Undeveloped	Contact deposit	
	Nickel Lake (WD150 and 155)	1926	None	0	2.36	1.02	0.50		Dec. 31, 2005	Inco and Falconbridge (2006)	Vale	Undeveloped	Offset deposit	
	Trillabelle	1891	none	0							Vale	Undeveloped	Contact deposit	
	Chicago Mine	1890	1892–97	~0.5							Vale	Historic producer	Modified Contact	
	Sultana	1891	1892	~0.01	0.30	1.20	0.30		Dec. 31, 2005	Inco and Falconbridge (2006)	Vale	Historic producer	Contact deposit	
	Cryderman	1892	None	0							Vale	Undeveloped	Modified Contact	
	Lindsley	1928	1953–57, 1971–2008	c							Glencore	Historic producer	Contact-footwall deposit	Binney et al. (1994)
	Blezard	1885	1889–95	~1	3.31	1.41	0.97		Dec. 31, 2005	Inco and Falconbridge (2006)	Vale	Historic producer	Contact deposit	
	Vermilion	1887	1896–1916 (intermittent)	~0.1							Vale	Historic producer	Offset deposit	
	Ella			0							Vale	Undeveloped	Contact deposit	
	McKim	1883		0							Glencore	Historic producer	Contact deposit	
	Tam O'Shanter	1893	None	5							Vale	Undeveloped	Contact deposit	

(Continued)

Table 1.2 The Sudbury Ore Deposits by Location, Size, and Geological Environment (cont.)

Name	Deposit	Discovery Year	Production Years	Estimated Ni in Ore Produced (kt Ni) up to 2015 (*/+ = See Footnote)ᵃ	Reserves + Resources Mmt ore	Ni Grade wt%	Cu Grade wt%	Pt + Pd + Au g/t	Date of Last Formal 43-101 Compliant Reserve and Resource Information Published	Source or Information on Reserves and Resources	Present Ownership	Current Producer/ Historic Producer	Geological Environment	Specific References (See Also: Royal Ontario Nickel Commission, 1917; Dressler (1984 a-c); Ames et al., 2008)
Discrete deposits (cont.)	MacLennan	1907	1965–73	22							Vale	Historic producer	Offset deposit	
	Victoria	1886	1899–1923	20	0.53	1.41	1.23		Dec. 31, 20058	Farrow et al. (2008a,b)	KGHM	Historic producer	Contact deposit	Farrow et al. (2008a,b)
	New Victoria	2010	None	0	12.50	2.20	2.30	8.50	Jun.3, 2011	Farrow et al. (2008a,b)	KGHM	New Project in development	Offset deposit	Farrow and Morrison (2012)
	Kirkwood	1892	1900–69 (intermittent)	20	2.37	1.24	0.84		Dec. 31, 2008	Farrow et al. (2008a)	KGHM	Historic producer	Offset deposit	Farrow et al. (2008a,b)
	Morrison	2005	2009–15	b							KGHM	Producer	Footwall deposit	
	Broken Hammer	2004	2014–15	0	0.26	0.10	0.88	9.82	Sep. 18, 2013	Bailey (2011, 2013)	Wall-bridge	Open pit	Footwall deposit	
Significant showings	Hess		None	0							Vale	Undeveloped	Offset deposit	
	Manchester		None	0							Vale	Undeveloped	Offset deposit	
	McIntyre	1898	None	0	0.30	1.04	0.77		Dec. 31, 2005	Inco and Falconbridge (2006)	Vale	Undeveloped	Offset deposit	
	Flett	? 1929	None	0							Vale	Undeveloped	Contact deposit	
	North Range Shaft	1898	None	0							Vale	Undeveloped	Contact deposit	

ᵃReported reserves and resources come from the last public domain complete reports for 43-101 compliant reserves and resources. Companies do not always report resources. No correction has been applied for mined material.
ᵇProduction from the Levack Mine Complex prior to acquisition by FNX; smaller producers including KGHM (previously FNX and Quadra–FNX), and First Nickel produced 95 kt Ni from all Sudbury Basin properties to 2010 (Mudd, 2010).
ᶜProduction from Glencore Mines is not broken out by mine; total of 1824 kt Ni in ore produced is from Mudd (2010); the majority of the production came from the Falconbridge Deposit.

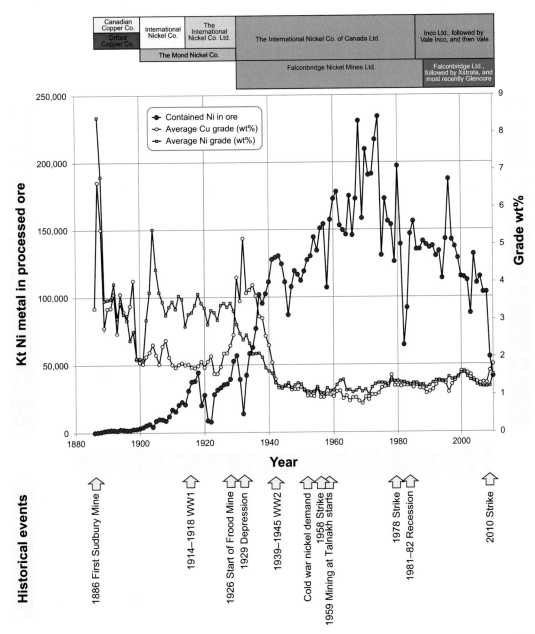

FIGURE 1.4 Synthesis of Sudbury Nickel Production, Nickel Grade, and Copper Grade From Discovery to the End of 2014

The data are sourced from Mudd (2010) with extrapolation to the end of 2014. The principal mining companies and major historical events controlling nickel supply and price are shown.

other industrial uses for nickel metal. The discovery of the Voisey's Bay Deposit in Labrador in 1994 (Naldrett et al. 1996a,b; McNish, 1999) and the subsequent development of a mine at the deposit in 2005 led to the production of nickel metal in Sudbury from concentrates shipped from Voisey's Bay to Sudbury for processing. The nickel in these shipped concentrates are not included with the Sudbury total, so the actual production of metal from Sudbury ore has dropped in the 21st century in response to the changing economics and availability of alternative high-grade sulfide concentrate feeds.

The historic production of nickel expressed as a ratio for Vale's seven principal mine complexes and other smaller mines is shown in Fig. 1.5A on the basis of the average production of ore over 5-year intervals. The deposits at Copper Cliff, Frood-Stobie, Levack-Coleman, Garson, Murray, Creighton, and Crean Hill have contributed the lions share to Vale's historic nickel production with more than 90% of the metal coming from these seven major deposits. Of these deposits, the Frood-Stobie Deposit has produced the largest contribution, followed by Creighton Mine Complex. Over the last 30 years, an increasing contribution of nickel has come from deposits discovered in the north part of the Sudbury Basin at Levack and Coleman. Fig. 1.5B shows the production on the basis of the principal style of mineralization which relates to the position of the ore bodies relative to the base of the SIC (i.e. ore bodies hosted at the base of the SIC are contact type, those in the footwall beneath the contact ore bodies are footwall type, those hosted in the Offset Dykes are offset type, and the remainder are a variant of the offset type referred to as the breccia belt type).

The discovery of new ore bodies at Sudbury has resulted from two approaches. One group of discoveries has been made as a result of following the known surface ore deposits to depth by carefully drilling at the edges of the ore bodies and along the trends of known mineralization; in the cases of the very large deposits such as Creighton, Frood-Stobie, Murray, Garson, Falconbridge, and Levack, this approach has been successful and in many cases has maintained a supply of new ores that has pushed the closure of the mine complex well into the future. Another group of deposits have been discovered as a result of exploration work where geologists have identified the possibility of mineralization that is not obviously attached to known ore bodies, and does not extend to surface; these discoveries are often challenging and harder to make, but the use of a combination of geological information, geophysical methods, and geochemistry ensures that the company can minimize both expenditure (to pay for expensive drilling) and risk (failure to find an economic ore body). Exploration success of this type yields new discoveries which, under the right economic conditions, can be mined. This exploration success is important because it shows the value of better datasets, new ideas and approaches, and careful exploration. Examples of this type of new discovery include the Kelly Lake, Onaping Depth, McCreedy East 153, Podolsky, Capre, Victor, Nickel Rim South, and Victoria Deposits. These discoveries together with the un-mined deposits and as yet unfound deposits are the future of the Sudbury Camp.

POSITION OF THE SUDBURY CAMP AS A FUTURE PRODUCER OF NICKEL

Increasing depth of mine operations coupled with the associated cost of mine development mean that the next generation of ore deposit discoveries at Sudbury need to be high-grade; the size and grade of the deposit will need to support the capital and operating costs of the mines and processing facilities. Clever approaches to finding more ore underpin this success, and the exploration geologist is a key player in growing the future development of Sudbury through the discovery of new ore deposits as well as incremental, yet strategic growth in the size of the mined deposits.

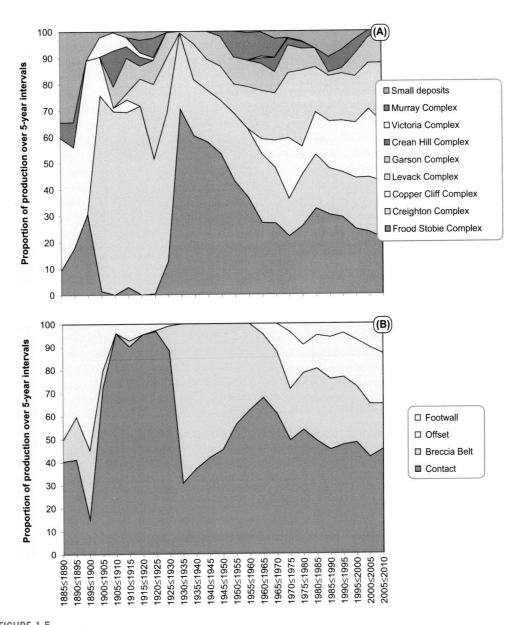

FIGURE 1.5

(A) Normalized historic production by mine complex based on data from Vale. (B) Normalized historic production from the four principal deposit types described in the text.

Continued pressure on sulfide nickel production comes from global developments. One of these has been the increased global production of nickel from near-surface laterite nickel deposits (Dalvi et al., 2004; Mudd, 2009) using well-tested technologies such as pyrometallurgy and Caron furnaces as well as new technologies such as high-pressure acid leaching of ores. The migration from sulfide sourced nickel to laterite nickel was initially driven by the large laterite resources which could be developed using inexpensive hydroelectric energy to power ferronickel smelters (PT Vale Indonesia operates the Soroako laterite mines and ferronickel plant in Sulawesi, Indonesia; Dalvi et al., 2004). More recently, there has been a transition to low-pressure and high-pressure leaching using hydrometallurgy plants that have been very expensive to develop and slow to ramp-up to full production (Mudd, 2009). Over the past 10 years, there has been a transition toward processing of nickel laterite ores shipped directly from Indonesia and the Philippines, principally to blast furnaces and electric arc furnaces in China. This shift in mining and processing activity was driven mainly by the enormous growth of the Chinese economy in the last 15 years, and their demand for nickel to produce low quality stainless steel (Lennon, 2007). Although many of the electric-arc furnace plants in China remain in operation, the Indonesian Government has now banned direct ore shipments, as it wishes to encourage companies to process ores within Indonesia and develop its industry and economy. This has led to an increase in the amount of directly shipped laterite production from the Philippines.

The position of sulfide nickel as a major component of future production is secured by the fact that laterites contain very little value-added metal (other than cobalt), whereas sulfide deposits continue to generate value from copper and by-product metals such as platinum, palladium, gold, and silver.

THE GEOLOGY OF SUDBURY IN A NUTSHELL

The history of the developments in science that have been triggered or supported by the geology of the Sudbury Structure is an important context to understanding the origin of the ore deposits. In order to explain these ideas, it is first necessary to briefly introduce the geology of the Sudbury Structure. In a nutshell, the Sudbury Structure comprises three principal rock associations which are shown in broad stratigraphic context in Figs. 1.6 and 1.7 together with the presently accepted interpretation of these rocks (based on Grieve, 1994), namely:

1. The 1.85 Ma SIC is a differentiated magmatic body comprising noritic, gabbroic, and granophyric rocks of the "Main Mass" (Fig. 1.7). The base of the SIC is irregular and the depressions are typically occupied by inclusion-rich mineralized norites and SLGRBX which have a quartz-feldspar matrix which comprise the Sublayer. The Main Mass is surrounded by a number of radial and concentric dykes composed of QD which often contain magmatic breccias in association with mineralization; these are termed the Offset Dykes; they tend to be discontinuous because of late faulting or due to the primary emplacement process which generates physical discontinuities, breaks, and jogs in the dyke; this is the origin of the term "Offset Dyke" at Sudbury (Figs. 1.6 and 1.7).
2. The surrounding and underlying country rocks are Archean and Proterozoic in age, and are heavily brecciated and disrupted by the Sudbury impact event. These rocks contain pseudotachylite veins, termed Sudbury Breccia (SUBX), that appear to be derived by shock-induced comminution and partial melting of the host.

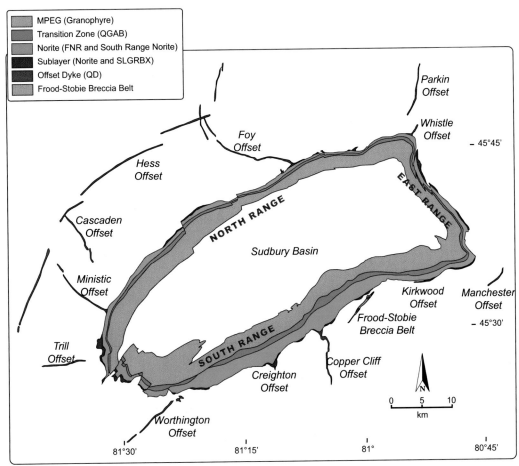

Legend:
- MPEG (Granophyre)
- Transition Zone (QGAB)
- Norite (FNR and South Range Norite)
- Sublayer (Norite and SLGRBX)
- Offset Dyke (QD)
- Frood-Stobie Breccia Belt

FIGURE 1.6 Geological Map of the Sudbury Structure Highlighting Location Information, Terminology, and Nomenclature

3. A sequence of breccias which cap the SIC comprise the Onaping Formation, and these rocks grade into progressively upward into deep-water sedimentary rocks of the Onwatin and Chelmsford Formation (Fig. 1.7).

The Sudbury ore deposits broadly group into four types which are discussed in detail in Chapter 4. They include the following:

1. Contact deposits located at the lower contact of the Main Mass in association with physical traps and structures.
2. Footwall deposits which often occur in the country rocks below the contact deposits or the eroded remnants of contact deposits.
3. Offset deposits occurring in the radial Offset Dykes.
4. Deposits associated with wide domains of SUBX and partially melted to recrystallized country rocks.

TERMINOLOGY USED TO DESCRIBE THE GEOLOGY OF THE SUDBURY STRUCTURE

The literature on the SIC and the associated ore deposits extends back to the early 20th century with classic papers describing the elliptical shape of the complex, and breaking out the rock units of the igneous sequence. It is necessary to introduce this terminology at an early stage so that it is familiar to the reader. The nomenclature at Sudbury was established long before the classification scheme of Streckeisen (1967), and so caution is required in interpreting specific rock names. The names of the rock types (QD norite, etc.) do not match the modern definitions, but it would be inappropriate to change the nomenclature of Sudbury geology after over 100 years of common usage.

The term "Sudbury Structure" refers to all of the rock formations created by the 1.85 Ga Sudbury impact event (Table 1.3). The SIC is an elliptical-shaped, differentiated igneous body comprising noritic, gabbroic, and granophyric rocks which comprise the "Main Mass" of the SIC (Figs. 1.7 and 1.8). The SIC is described in two sectors which are broadly the South Range, and the North and East Range (Fig. 1.6) where "range" refers to the elevated topographic expression within the SIC.

THE IGNEOUS ROCKS

The lower part of the Main Mass of the SIC is a sequence of noritic rocks which are termed Quartz Norite overlain by Brown or Green Norite in the South Range and Mafic Norite overlain by FNR in the North and East Ranges. Although these norites record quite different metamorphic histories (greenschist to amphibolite facies), they also have hallmarks of a different differentiation history. The norite stratigraphy is overlain by the QGAB which contains an upper unit of hornblende gabbro with acicular needs of amphibole that is termed the Crows Foot Granophyre. The Transition Zone is overlain by the MPEG, which is also called the Granophyre.

A discontinuous unit of norite (and locally granite) with abundant mafic and ultramafic inclusions is developed at the base of the SIC, and localized in a series of troughs and discrete depressions in the base of the SIC which are termed embayment structures (Figs. 1.7 and 1.8). The basal contact of the SIC with the underlying country rocks is marked by a discontinuous group of magmatic breccias which contain magmatic Ni–Cu–PGE sulfide ores. This unit is termed the Sublayer, and the depressions at the base of the Main Mass which contain the Sublayer are termed embayments; groups of embayments aligned along a trend are termed troughs (Figs. 1.7 and 1.8). The ore deposits that reside in these depressions at the base of the SIC are termed contact ores. Each embayment or trough is normally the locus for the development of a single ore deposit, and the names of many of the ore deposits are assigned to both to the ore body and/or mine complex, and the associated host rocks; for example, the Creighton Deposit is developed in association with the Creighton embayment structure (Fig. 1.6).

Igneous rocks of the SIC are also present as radial and concentric dykes which are termed Offset Dykes. They are composed largely of QD and magmatic breccias, and the ore deposits associated with these dykes are termed Offset Deposits; examples include the Copper Cliff Deposits contained in the Copper Cliff Offset (Fig. 1.7). The radial Offset Dykes are named clockwise from the South, the Copper Cliff Offset, the Creighton Offset, the Worthington Offset, the Trill Offset, the Ministic Offset, the Cascaden Offset, the Foy Offset, and the Parkin Offset (Fig. 1.7). The concentric Offsets Dykes are the Hess to the north of the SIC and the Manchester to the south. Small discontinuous segregations of QD also occur as "pods" within the South Range Breccia Belt which hosts the Frood-Stobie Deposit (Fig. 1.7).

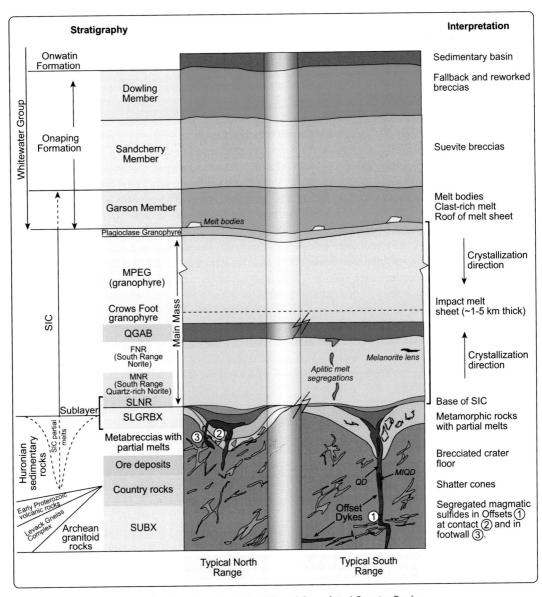

FIGURE 1.7 Simplified Geological Stratigraphy of the SIC and Associated Country Rocks

Modified after Grieve (1994) and Lightfoot et al. (1997b).

THE TARGET COUNTRY ROCKS

Extensive modification of the country rocks by brecciation and partial melting has given rise to enormous complexity in the terminology of the country rocks adjacent to the SIC. Within the embayment structures, magmatic breccias are developed, but beneath these embayments, the breccias include megabreccias (where large blocks of country rock are separated by a granitic matrix, SUBX, and sometimes magmatic sulfides; Fig. 1.8 and Morrison, 1984). With increasing distance away from the SIC, the extent of melting and brecciation declines. Proximal to the SIC, rocks which have been intensely brecciated in situ and partially melted are termed metabreccias (Fig. 1.8).

The rocks beneath the SIC are termed the Footwall, and they comprise country rocks which include Archean granitoid rocks and Archean gneisses (Levack Gneiss Complex) to the north and a more diverse group of metavolcanic and metasedimentary rocks to the south (the Huronian Supergroup). The country rocks have been intruded by gabbros and granitoid rocks. The gabbros belong to both Early Proterozoic East Bull Lake Type intrusions and the Nipissing Diabase (James et al., 2002a,b; Lightfoot et al., 1993a,b); the granites comprise the Creighton and Murray Plutons (Dutch, 1979). Close to the base of the SIC, the country rocks are extensively thermally metamorphosed, recrystallized, and partially melted. A plethora of inadequate and uninformative names are used in the literature to describe these rocks, the most challenging of which is the term "metabreccia" which is broadly used to describe country rocks adjacent to the SIC that exhibit metamorphic and partial melt textures (Figs. 1.7 and 1.8). In most cases it pays to name the rock type and qualify the name by describing the extent of thermal recrystallization and melting. The country rocks of the SIC are also subject to a complex terminology, with amphibolitized mafic intrusive rocks termed "Sudbury Gabbro" and "Gray Gabbro."

Extensive brecciation of the country rocks by the impact event produced the development of pseudotachylite breccias in the footwall of the SIC (Fig. 1.8); these rocks are collectively termed SUBX (French, 1998).

The rocks overlying the Main Mass are a sequence of fallback and reworked breccias which may contain melt bodies, and are collectively termed the Onaping Formation. The Onaping Formation grades upward into the Onwatin and then the Chelmsford Formations, which collectively comprise the Whitewater Group (Fig. 1.7).

Ores at the basal contact of the SIC and associated with the SLNR and SLGRBX are termed Contact Ore Bodies, whereas those in the immediate footwall are termed Footwall Ore Bodies. And likewise, the ore bodies in the Offset Dykes are referred to as Offset Deposits. The descriptions of specific ore types have also grown on a historic basis and the terminology is provided in Table 1.3 as it is commonly used today.

SUDBURY GEOLOGY AT THE CENTER OF IMPORTANT EARTH SCIENCE DEBATES

The Sudbury story is not only rooted in its economic context of ore deposit discovery but has also played a fundamental part in scientific debates in geosciences. The scientific debates about Sudbury follow the great historic geological debates such as the ideas about whether change is a uniform slow process (Uniformitariansim; Hutton, 1795; Lyell, 1830) or a sudden short-lived violent event (Catastrophism; Cuvier, 1813). The manifestation of this debate is still active; for example, in the context of the relative roles of asteroid impact versus flood basalt eruption, there is still no consensus regarding

Table 1.3 Summary of Key Thematic Development in Debates on Sudbury Geology

	Fundamental Concept	The Nature of the Debate	Brief Synopsis	Model 1	Model 1 References	Model 2	Model 2 References	Present Understanding	General References
Fundamental igneous petrology	Origin of the Main Mass of the SIC	Petrology and geochemistry models for emplacement and differentiation of the SIC	The relative importance of differentiation versus emplacement of compositionally different magmas derived from different sources was debated as a fundamental topic in igneous petrology; the SIC played a key part in this debate. On the one hand, one group of scientists considered it to be the product of in situ differentiation; on the other hand another group considered it to be the product of multiple different magmas	In situ crystal differentiation of one batch of magma	Harker (1926); Bowen (1925, 1928)	In situ crystal differentiation of two or more different batches of magma possibly derived from different sources	Knight (1917); Phemister (1926)	One melt sheet source with evidence of in-situ crystal differentiation from the bas-up and top-down	Rayleigh (1896)
	Differentiation history of the Main Mass	Source of compositional diversity in the Main Mass	Although it is widely believed that the impact melt sheet is derived by differentiation of a single parental magma, the mechsanisms by which the melt differentiated is not fully understood. The exact extent to which the melt sheet has assimilated country rocks, and the mechanism by which this happened without significant changes in incompatible element ratios remains incompletely understood	The melt sheet evolved by crystallization of one magma type which was density-layered, and it appears likely that the QGAB records some of the last melts to crystallize	Kuo and Crocket (1979); Golightly (1994); Lightfoot et al. (2001); Therriault et al. (2002); Lightfoot and Zotov, 2006	The variety of rocks in the Main Mass was produced by liquid immiscibility.	Bain (1925); Zieg and Marsh (2005); Farrow and Lightfoot (2002)	One parental melt which underwent early density separation, and then crystallization from base-up and the top-down	Lightfoot et al. (2001)

(Continued)

Table 1.3 Summary of Key Thematic Development in Debates on Sudbury Geology *(cont.)*

Fundamental igneous petrology (cont.)

Fundamental Concept	The Nature of the Debate	Brief Synopsis	Model 1	Model 1 References	Model 2	Model 2 References	Present Understanding	General References
Estimates of bulk composition of parental magmas	The best approach to establishing the composition of the primary magma response for the SIC; relative importance of chilled margins, bulk estimates of average silicate rocks, and influence of assimilation of crust (Naldrett, 2004)	Finer grained rocks at the base of the SIC were initially used as estimates of magma composition; with no addition of melt to the primary magma, this margin should be equivalent to the bulk magma and represent the average composition of the SIC if it was a closed system. Assimilation of crustal rocks would decouple these estimates of primary magma composition	Estimates for starting melt composition can be made based on most rapidly cooled rocks Chilled norite, QD and/or glass shards from the Onaping breccias)	Bain (1925); Lightfoot et al. (1997a); Ames et al. (2002)	Composition of the bulk melt is modified by crustal assimilation	Bowen (1925); Keays and Lightfoot (2004)	The bulk composition of the starting melt is likely recorded in the composition of glass shards from the Onaping Formation and in quenched QD from the Offset Dykes	Lightfoot and Farrow (2002)
Requirement for mantle-derived magma	Traditionally it has been believed that magmatic Ni-Cu-PGE sulfide deposits form from ultramafic-mafic magmas rather than intermediate to evolved magmas. (Naldrett, 2004)	In one group of models a major hidden layered complex developed at depth beneath the SIC is the source of many of the exotic mafic and ultramafic inclusions in the Sublayer. Geochemical investigations indicate that the lithophile element composition of the SIC is swamped by a melt of broadly upper crustal composition, with the possibility of up to 20% mantle-derived magma, but no proof for a contribution from the mantl	No contribution of mantle melt is recognized or required to form the SIC and the associated ore deposits	Keays and Lightfoot (2004)	A contribution of mafic-ultramafic magma is required to provide the Ni, Cu, and PGE that are contained in the ore deposits	Naldrett (1984a-d); Naldrett et al. (1986)	It is now widely believed that no mantle contribution is required to form the SIC, and all of the metal and sulfur in the ores can be derived from the melt sheet (Lightfoot et al., 2001; Keays and Lightfoot, 2004)	Lightfoot et al. (1997a, 2001)

Formation of the Sudbury ore deposits

Fundamental Concept	The Nature of the Debate	Brief Synopsis	Model 1	Model 1 References	Model 2	Model 2 References	Present Understanding	General References
Triggers to sulfide saturation	Role of crustal assimilation as a trigger to sulfide saturation of the Sudbury magma	It has long been debated whether crustal sulfur must be added to a magma to trigger sulfide saturation. In one group of models, addition of crustal silica is called on to lower the solubility of sulfur and trigger the segregation of an immiscible sulfide; in a second group of models, the sulfur is considered as being added from the crust. The debate over silicification versus cooling of crustally-derived melts is still debated	Silicification of the magma triggers the formation of immiscible sulfides	Naldrett (1984a–d); Rao et al. (1985)	Addition of crustal sulfur is required to form immiscible sulfides	Ripley et al. (2015)	A variety of opinions, but it is now widely believed that silicification alone will not be enough to form economic concentrations of metals	Keays and Lightfoot (2009)
Formation of immiscible magmatic sulfide	Evidence for in situ segregation and accumulation of sulfides	The association of Ni–Cu sulfide mineralization with diorite breccia at the base of the SIC indicated a magmatic origin of dense immiscible sulfide. The dense sulfides were also injected into the footwall of the SIC to form Offset type deposits. The idea also gathered acceptance as the principals in ore genesis are very similar to those used to separate metals in a Bessemer Furnace, and they are supported by experimental studies	Saturation of the silicate magma in sulfide and gravitational settling of dense sulfide to the lower contact of the SIC	Bell (1891a,b); Coleman (1905a–c, 1913a–d); Gregory (1925); Bateman (1917); Hawley (1962)	Hydrothermal concentration of sulfide at the base of the SIC	Molnar et al. (2001); Farrow and Watkinson (1996)	It is now accepted that the sulfide ores has a primary magmatic origin from the SIC	Bessemer (1905); Kullerud (1963)

(Continued)

Table 1.3 Summary of Key Thematic Development in Debates on Sudbury Geology *(cont.)*

	Fundamental Concept	The Nature of the Debate	Brief Synopsis	Model 1	Model 1 References	Model 2	Model 2 References	Present Understanding	General References
Formation of the Sudbury ore deposits *(cont.)*	Mechanism by which compositional diversity is produced in Sudbury ores	Relative roles of sulfide differentiation, hydrothermal fluids, and deformation in controlling ore composition	The ores of the SIC range in composition from pyrrhotite–pentlandite-rich through pyrrhotite-rich chalcopyrite–pentlandite-rich to chalcopyrite–pentlandite, and chalcopyrite–pentlandite-millerite, to millerite–bornite-rich variants. Ore bodies are also modified by syn-magmatic and postmagmatic deformation	Composition variation produced by hydrothermal processes	Dickson, (1903); Wandke and Hoffman (1924)	Compositional variation produced by magmatic differentiation of sulfides	Keays and Crocket (1970); Naldrett et al. (1979, 1982); Farrow and Lightfoot (2002); Naldrett et al. (1994a,b, 1999)	The primary variations are produced by sulfide fractionation; local effects at the edge of highly fractionated sulfides involve volatile activity and some remobilization of Cu and PGE. Ore bodies are sometimes modified by deformation and sulfide kinesis	Abel et al. (1979), Naldrett et al. (1979, 1982); Hoffman (1978); Ames et al. (2008), Farrow and Lightfoot (2002)
Impact origin of the Sudbury Structure	Relative roles of endogenic and impact processes	Impact versus volcanic origin of the Sudbury Structure	The debate over the impact origin of the SIC developed in a series of steps: 1. Evidence for an endogenic explosive magmatic event; 2. The development of the impact hypothesis; 3. The progressive reconciliation of melt sheet formation, syn-cratering deformation, and ore genesis with the impact hypothesis; 4. General acceptance that the Sudbury Structure has an impact origin, but has been modified by deformation	The earliest documentation of evidence for the impact origin of the Sudbury Structure was followed by a sequence of research investigations that progressively eroded the endogenic model.	Muir (1984)	The recognition that evidence of impact tectonics, melt sheet and dyke genesis, and regional influence of these processes has developed.	Grieve et al. (1991a,b), Grieve (1994), French (1998), and Addison et al. (2005, 2010)	The debate over the relative roles of impact processes versus endogenic processes is now largely resolved in favour of impact genesis of the Sudbury Structure	Dietz (1964); Addison et al. (2005, 2010); Farrow and Lightfoot (2002)

Impact origin of the Sudbury Structure (cont.)

Fundamental Concept	The Nature of the Debate	Brief Synopsis	Model 1	Model 1 References	Model 2	Model 2 References	Present Understanding	General References
Genesis of pseudtachylite breccia matrix	In situ comminution of target rocks versus local melting and/or injection of melt sheet contribution	The relative roles of target rocks and melt sheet components in the genesis of pseudotachylite matrix	In-situ genesis of pseudotachylite by shock-induced compression and cataclasis.	Dressler (1984 a-c); French (1998); LaFrance and Kamber (2010)	Drainage of initially super-heated impact melt into tension fractures in the crater floor to form pseudotachy-lite matrix	Riller et al. (2010)	The compositional variations in SUBX matrix often require a large contribution of local country rocks in their genesis	Dressler (1984 a-c); O'Callaghan et al. (2015); Roussel et al. (2003)
Sequence of events in the formation of the SIC	The exact sequence of events that lead to the geological relationships in the Sudbury Structure	The relative timing of formation of the ore deposits, the SIC and associated dykes, and the breccias above and beneath the SIC	Late injection of the Sublayer and associated ores	Pattison (1979)	Formation of the ore deposits in response to settling and accumulation toward the base of a superheated melt sheet	Lightfoot and Farrow (2002); Keays and Lightfoot (2004)	Geological relationships now provide firm evidence for a sequence of events at Sudbury that can be understood in the context of impact processes coupled with syn-impact and postimpact deformation events	Dressler (1984 a-c); Peredery and Morrison (1984a,b); Lightfoot and Farrow (2002)
Deep structure of the SIC	Large folded impact structure versus a smaller less heavily deformed crater	Configuration of the SIC at depth is known from deep drilling and Lithoprobe seismic traverses; below a depth of ~3 km, the relationships are less certain	Large impact structure deformed by thrusting from south	Milkereit et al. (1994a,b)	Smaller folded impact structure	Card and Jackson (1995)	The original crater diameter is now believed to have been ~150-200 km; there is no evidence for thrusting of the South Range over the North Range	Gibson (2003); Lightfoot et al. (2001)

(Continued)

Table 1.3 Summary of Key Thematic Development in Debates on Sudbury Geology *(cont.)*

	Fundamental Concept	The Nature of the Debate	Brief Synopsis	Model 1	Model 1 References	Model 2	Model 2 References	Present Understanding	General References
Impact origin of the Sudbury Structure (cont.)	Availability of superheat	Was the Sudbury magma superheated?	The debate over availability of superheat is important to the formation of the Sudbury ore deposits as prtracted cooling of the melt would allow for very efficient sulfide settling before silicate crystallization	Melt sheet superheated to ~1700°C	Zieg and Marsh (2005); Lightfoot et al. (2001); Keays and Lightfoot (2004)	Melt was not superheated	Latypov et al. (2010)	Evidence for superheat includes the thick thermal aureole, the extent of footwall melting, and the efficiency of sulfide segregation	Watts (2010)
Deformation of the SIC	Deformation history	To what extent has the Sudbury Structure been folded by deformation and what role does this play in controlling the location of mineralization?	Ductile and brittle deformation history of the SIC and the importance of deformation in the control of ore deposits: relative roles of syn-magmatic and postmagmatic events	Postmagmatic deformation. Examples include: Tri-shear model for the South Range deformation zone; Thrusting and flexural-slip during regional buckling of the SIC	Lenauer (2012); Lenauer and Riller (2012a,b); Mukwakwami et al. (2012, 2014a,b)	Syn-magmatic deformation: Crater-floor modification by terrace collapse accompanied crater re-adjustment and resulted in the formation of Footwall and Breccia Belt style ore deposits	Spray (1997); Scott and Spray (2000)	The relative roles of syn-magmatic and postmagmatic deformation are key in controlling the geometry of the SIC and the associated ore deposits; these structures largely modify pre-existing magmagic ore deposits	Lightfoot (2015); Mukwakwami et al. (2012, 2014a,b); Scott and Spray (2000)

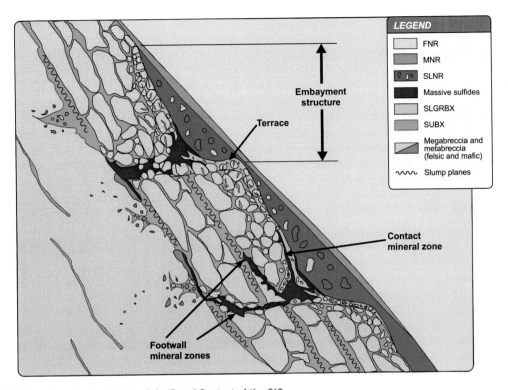

FIGURE 1.8 Simplified Geology of the Basal Contact of the SIC

Extensively modified after Morrison (1984).

the exact trigger to mass extinction events (Wignall, 2001; Courtillot, 1994; Morgan et al., 2006; Keller et al., 2009).

The historic importance of Sudbury to debates on petrology, geochemistry, formation of ore deposits, structure and tectonics, and impact events is worth consideration. Sudbury is actually at the heart of many of the great debates and revolutions in Earth Sciences; these are broken out into four main groups, namely: (1) the origin of the igneous rocks of the SIC; (2) the origin of the ore deposits; (3) the nature of the catastrophic event which created the Sudbury Structure; and (4) the deformation history and deep structural configuration of the SIC.

FUNDAMENTAL IGNEOUS PETROLOGY

Early debates on the origin and evolution of magmas dwelt on the processes by which a diversity in rock compositions could be produced by in situ magmatic processes such as crystal fractionation and crystal settling (Bowen, 1928) versus the emplacement of diverse magmas generated at depth in several different events (Harker, 1926; Table 1.4). The discovery and identification of the noritic rocks comprising the SIC dates back to Bell (1891a–d), and the first reported evidence for crystal fractionation processes appears in the work of Walker (1897) where he suggested that the compositional diversity in the norites, gabbros, and granophyric rocks of the SIC could be generated by crystal fractionation. The debate was started, and Harker (1926) stepped in to suggest that whereas there was some evidence for crystal fractionation, this process could not explain the broad differences in chemistry, so he proposed a model in which two or more pulses of magma were injected into their final resting place. The important role of crustal assimilation by the Sudbury magma was developed by Bain (1925), and the relative contributions of mantle versus crustal melts have been debated on the basis of geochemistry by numerous authors using modern datasets (Naldrett et al 1984a,b; Faggart et al., 1985a,b; Lightfoot et al., 1997a,b, 2001), where the challenge has been an increasing realization that there may be no mantle contribution at all to the original magma. The debate also evolved to include open system processes (Table 1.4), and papers by Phemister (1926), Naldrett et al. (1994a,b) and Chai and Eckstrand (1996) report further evidence for both fractionation and multiple emplacements of magmas. The debate started by Bell (1891a–d), Walker (1897) and Harker (1926) continues to this day, but it is now underpinned by a new hypothesis that brings impact-generated crustal melts into the discussion (Grieve, 1994). More about it will be discussed later in the chapter, but let us look at some of the other historic debates at Sudbury.

ORIGIN OF THE ORE DEPOSITS

Another major debate triggered by the Sudbury ore deposits was the question of genesis of the mineralization (Table 1.4). The earliest descriptions of the ores showed them to be associated with the noritic rocks at the base or outer edge of the SIC (Barlow, 1904; Coleman 1905a-c, 1913a-d), and this was accompanied by a debate on the relative roles of magmatic versus hydrothermal process (Bell, 1891a,b; Dickson, 1903; Wandke and Hoffman, 1924) as well as in situ versus injection of the sulfides into their final resting place (Howe, 1914; Bateman 1917). The idea of gravity segregation of dense sulfide melts from the magmas of the SIC first appeared in Gregory (1925), and it is this model that has dominated through the literature (Hawley, 1962; Souch et al., 1969), even though there has been considerable debate about whether this process happened in situ (Lightfoot et al., 2001; Keays and Lightfoot, 2004) or deeper-level chambers that fed magma into the SIC (Naldrett, 1984a-d). Notwithstanding, there has been a vociferous and enduring effort to explain the formation of the ores from hydrothermal fluids (Dickson, 1903; Wandke and Hoffman, 1924; Watkinson 1994); however, many proponents of the hydrothermal processes are moving toward the importance of ore body modification rather than primary genesis (Farrow and Lightfoot, 2002; Mukwakwami, 2012). The debate remains important to both science and exploration at Sudbury, but the importance of magmatic processes in the primary genesis of the ore systems at Sudbury is now generally recognized.

Yet another debate triggered by Sudbury geology is the origin of the enormous range in the compositions of the ore bodies and the parts of individual ores. Mining company geologists who discovered ore deposits wrote some of the key papers on their geology. The debate arose from detailed studies undertaken on ore bodies as a result of mining activity (Zurbrigg, 1957; Hawley, 1965; Souch et al., 1969), and the mechanisms by which this diversity in ore composition could be generated were extensively debated. One group of models dwelt on a role for hydrothermal processes (Watkinson, 1994); another group of models recognized diversity due to the segregation and accumulation of sulfide from the magma in response to the efficiency of the process and the volumes of sulfide melt relative to silicate magma (Kullerud, 1963; Naldrett and Kullerud, 1967); Naldrett et al., 1994a,b, 1999). Yet another group of models strove to understand the compositional diversity in ores ranging from Ni-rich to Cu-rich, and they recognized the important role of inherent fractionation of the sulfide magma through the formation of monosulfide solid solutions as well as crystallization of sulfide minerals (Keays and Crocket, 1970; Naldrett et al., 1979, 1982, 1994a,b, 1999). The diversity in ore body compositions and the exploration methods used in the search for new ores are very much underpinned by the understanding of these processes of sulfide evolution.

Table 1.4 Sudbury Nomenclature and Terminology

Group	Term	Synonymous Term (in Brackets) and Abbreviations	Explanation or Definition	References
The host rocks of the Sudbury Structure	Sudbury Structure		Also called Sudbury Basin and Sudbury Irruptive, is the sum total of all of the rocks created or modified by the 1.85Ga impact event	Grieve (1994)
	Sudbury Basin		The area enclosed by the SIC and including the Whitewater Group	Giblin (1984a,b)
	Sudbury Event	Impact event	The bolide impact event which produced the Sudbury Structure; sometimes termed an astrobleme	Modified after Giblin (1984a,b)
	SUBX	Pseudotachylite breccia	SUBX is a pseudotachylite; it is a glassy or very fine-grained rock that is composed of an extremely fine-grained or glassy matrix that often contains inclusions of wall-rock fragments. Pseudotachylite characteristically occurs in veins; is dark in color; and is glassy in appearance. It often has the appearance of the basaltic glass. Typically, the glass has been completely devitrified into very fine-grained material with radial and concentric clusters of crystals. It occasionally contains crystals with quench textures that began to crystallize from the melt. It can be formed in fault zones, but the extensive outcrops of SUBX associated with the Sudbury Structure are believed to have formed by impact melting. It is widely thought to have formed by frictional effects within the crater floor and below the crater during the initial compression phase of the impact and the subsequent formation of the central uplift. The most extensive examples of impact related pseudotachylites come from impact structures that have been deeply eroded to expose the floor of the crater, such as Vredefort crater, South Africa and the Sudbury Structure, Canada	French (1998); Grieve (1994); Spray (1995)
	Shatter cone		Conical shape structures that radiates from the *apex* of the cones repeating cone-on-cone at a variety of scales in the same outcrop. In finer-grained rocks such as sany siltstones, they form an easily recognizable "horsetail" pattern with thin grooves (*striae*) on the cone surface. Coarser grained rocks tend to yield less well-developed shatter cones, which may be difficult to distinguish from other geological formations such as slickensides. Geologists have various theories of what causes shatter cones to form, including compression by a shock wave as it passes through the rock or tension as the rocks rebound after the pressure subsides	French (1998); Guy-Bray et al. (1966)

(Continued)

Table 1.4 Sudbury Nomenclature and Terminology *(cont.)*

Group	Term	Synonymous Term (in Brackets) and Abbreviations	Explanation or Definition	References
The host rocks of the Sudbury Structure *(cont.)*	South Range Breccia Belt	Breccia Belt	A discontinuous belt where SUBX, metabreccia, and melt bodies of QD are developed in the South Range. The unit is continuous in the area of the Frood and Stobie Mines, and extends toward the contact of the SIC to the east. Toward the west, the belt is more discontinuous and largelky comprises SUBX and metabreccia. The belt has some of the attributes of a major fault zone, but it also has aspects of a SUBX Belt in which a magma with the composition of QD was introduced from above	Spray (1997)
	Metabreccias	Diatexite and metatexite	A name given to a group of rocks which occur beneath the Sublayer. The rocks are invariably fragmental, partially melted, and/or thermally metamorphosed. I apply this term strictly to describe a footwall rock that has been brecciated, melted, and then variably metamorphosed by the thermal effect of the melt sheet and postmagmatic regional metam orphism; the term can apply to a wide range of rock types, and typically the extent of brecciation and thermal metamorphism declines away from the SIC. The term is not used to describe rocks that have mobilized into their final resting place as a fragment-charged melt (Lafrance et al., 2014), the terms SLGRBX and Lecocratic Breccia are used to describe these rock units (Pattison, 1979)	Farrow (1995)
	Megabreccias		Large blocks of country rock which are clast supported, but separated by a matrix of SLGRBX or metabreccia	Pattison (1979); Morrison (1984); Dressler (1984 a-c)
	Trough	(Embayment)	Continuous physical depressions at the base of the SIC which contain SLNR and SLGRBX, and often controls the distribution of contact ore bodies	Morrison (1984); Pattison (1979); Lightfoot et al. (1997a)
	Embayment	(Trough)	Isolated physical depressions at the base of the SIC, or deeper segments of trough structures; embayments often contain SLNR and SLGRBX, and mineralization is often concentrated in association with embayment features	Morrison (1984); Pattison (1979); Lightfoot et al. (1997a)

Group	Term	Synonymous Term (in Brackets) and Abbreviations	Explanation or Definition	References
The host rocks of the Sudbury Structure (cont.)	Sudbury Gabbro	Metagabbro; amphibolite; Nipissing Diabase	Sudbury Gabbro is a term used to describe a group of amphibolites and metagabbros developed to the south of the Sudbury Structure. The term refers to a group of rocks that have been traditionally grouped with the 2.2 Ga Nipissing Diabase, but they are unusually mafic in comparison. The term falsely implies a genetic link to the SIC, but this is not the case; these rocks are older and often form inclusions in the SIC	Lightfoot and Farrow (2002)
	Levack Gneiss Complex	(Feldspathic Gneiss Complex)	A belt of high-grade metamorphic gneisses adjacent to the base of the SIC in the North Range	Giblin (1984a,b)
	Sudburite	Hornfels	A local name for melanocratic volcanic rocks containing hyersthene and augite phenocrysts in an equigranular groundmass of pyroxene, biotite and plagioclase	Coleman (1914a,b); Le Maitre (2002)
Rocks of the SIC	SIC	(Sudbury Irruptive, Nickel Irruptive); SIC	A differentiated sheet of igneous rock with associated sulfide deposits and Offset Dykes	Lightfoot et al. (1997a); Giblin (1984a,b)
	Offset Dykes (radial and concentric)		Radial and concentric dykes of QD which are located in the country rock below and outboard of the SIC; the term "offset" refers to primary discontinuities that are lateral steps created by changes in host rock or physical windows where the magma does not form a continuous dyke. The rock type is dominantly QD (± inclusions) with local development of Footwall Breccia radial Offset Dykes are connected to the base of the SIC and extend away from the contact a moderate to high angle. Concentric Offset Dykes wrap around the SIC	Giblin (1984a,b); Grant and Bite (1984); Cochrane (1984); Lightfoot et al. (1997b)
	Main Mass (of the SIC)		This term refers to igneous rocks of the SIC that rest above the Sublayer. It comprises the Mafic Norite, FNR, QGAB, and Granophyre in the North Range, and the Quartz-rich Norite, Brown and Green Norites, QGAB and MPEG in the South Range. The term excludes the rocks comprising the Offsets Dykes and Sublayer	Giblin (1984a,b); Naldrett and Hewins (1984); Naldrett et al. (1984a,b); Lightfoot et al. (2001)
	QGAB	(Transition Zone; Oxide-rich Gabbro); TZQG	The Transition Zone is a unit of QGAB developed at the interface between the noritic and granophyric rocks. It is composed of QGAB	Naldrett and Hewins (1984); Naldrett et al. (1984a,b); Lightfoot et al. (2001)

(Continued)

Table 1.4 Sudbury Nomenclature and Terminology *(cont.)*

Group	Term	Synonymous Term (in Brackets) and Abbreviations	Explanation or Definition	References
Rocks of the SIC *(cont.)*	South Range		The southern segment of the SIC between the Cameron Creek Fault in the west and the flexure of the SIC in Falconbridge Township	Giblin (1984a,b)
	North Range		The northen part of the SIC extending from the Ceameron Lake Fault to the flexure structure in Norman Township	Modified after Giblin (1984a,:)
	East Range		A portion of the SIC, often considered as a part of the North Range, with which it ahs similar geology. It extends from flexure of the SIC in Norman Township southward to the faulted flexure in Falconbridge Township	Giblin (1984a,b)
	SLNR	(Basic Norite; Hanging wall Breccia; Inclusion Norite; Inclusion Basic Norite; Xenolithic Norite)	A fine-medium grained norite with inclusions comprising dominantly melanocratic and ultramafic rocks. Occurs at the base of the Main Mass and tends to be restricted to embayments and troughs. Also termed igneous Sublayer. The Offset Dykes are no longer considered to be part of the Sublayer. There are several variants of Sublayer described in Chapter 3	Pattison (1979); Giblin (1984a,b); and Farrow and Lightfoot (20C2)
	SLGRBX	(Footwall Breccia; Late SLGRBX; Contact Breccia; Gray Breccia)	A term used to describe a unit of fragment-laden rock with a granitic matrix. The matrix contains both melt contributions and xenocrysts and footwall and exotic xenoliths so it is best considered a xenomelt. The matrix is composed of plagioclase, quartz, pyroxene, amphibole, biotite, and gran ophyric intergrowths of K-feldspar and quartz. It is an important rock type in the North, West, and East Ranges, but is rarely developed in the South Range. It occurs beneathy the SLNR or is intercalated with the SLNR. In some classifications it is placed in the footwall (Pattison, 1979), but beacause it is a xenomelt and an igneous component of many embayments and troughs, this book classifies it is a part of the Sublayer	Pattison (1979); Giblin (1984a,b); and Farrow and Lightfoot (2002)
	Mafic Norite	(Biotite Norite; Black Norite; Dark Norite; Femic Norite; Lineated Gray Norite; Poikilitic Norite)	A discontinuous unit of basal Main Mass melanorite developed in the North, West, and East Range at the base of the Main Mass	Naldrett et al. (1984a,b); Giblin (1984a,b)
	FNR	(Gray Norite)	A continuous unit of weakly differentiated noritic rock that rests above the Mafic Norite in the Main Mass of the North, East and West Ranges	Naldrett et al (1984a,b)
	Plagioclase-rich granophyre		The uppermost rocks of the MPEG Unit	Giblin (1984a,b)

Group	Term	Synonymous Term (in Brackets) and Abbreviations	Explanation or Definition	References
Rocks of the SIC (cont.)	MPEG	(Granophyre)	A continuous unit of weakly differentiated quart-K-feldspar with lesser plagioclase and amphibole. It is often weakly vesicular and occurs in all three Ranges of the SIC	Naldrett et al. (1984a,b)
	Quartz-rich norite		A unit of weakly differentiated green, brown or black norite that develops toward the base of the norite in the South Range. The basal part of this unit is characterized by large grains of blue quartz	Giblin (1984a,b)
	South Range Norite	(Brown Norite; Green Norite)	The norites developed above the Quartz-rich Norite and below the QGAB in the South Range. These norites may be green, brown or black in color	Giblin (1984a,b); Naldrett and Hewins (1984); Naldrett et al. (1984a,b); Lightfoot and Zotov (2006)
	Mineralized Inclusion-bearing Quartz Diorite (MIQD)	(QD breccia); inclusion quartz diorite; MIQD	A variety of mineralized and inclusion-bearing rock found in radial and concentric dykes and discontinuous patches in SUBX; related to the SIC. Now considered to pre-date the Sublayer. The matrix of this rock is equivalent to a quartz mononite in the Streckeisen (1976) classification of igneous rocks	Modified after Giblin (1984a,b); Wood and Spray (1998); Lightfoot et al. (1997a)
	QD	QD	A variety of rock found in radial and concentric dykes and discontinuous patches in SUBX; related to the SIC. Now considered to pre-date the Sublayer. This rock is equivalent to a quartz mononite in the Bas and Streckeisen (1991) classification of igneous rocks	Modified after Giblin (1984a,b); Wood and Spray (1998); Lightfoot et al. (1997a)
Definitions relating to Sudbury ore deposits	Contact ore body	Contact ore	The portion of the mineral reserve and/or resource that occurs at the lower contact of the SIC	Lightfoot et al. (1997b)
	Massive Sulfide	MASU	An ore type consisting almost entirely of sulfide minerals	Giblin (1984a,b)
	Contorted schist inclusion sulfide		An ore type characterized by contorted fragments of schist and/or inclusions of quartz	Giblin (1984a,b)

(Continued)

Table 1.4 Sudbury Nomenclature and Terminology *(cont.)*

Group	Term	Synonymous Term (in Brackets) and Abbreviations	Explanation or Definition	References
Definitions relating to Sudbury ore deposits *(cont.)*	Inclusion massive sulfide	IMS	An ore type characterized by the presence of angular fragments of footwall rocks in massive sulfide	Giblin (1984a,b)
	Fragmental sulfide		An ore type characterized by the presence of subangular fragments of sulfide which comprise part of the fragment population; typically occurs with blebby to disseminated sulfide	Giblin (1984a,b)
	Gabbro-peridotite inclusion sulfide	GPIS	An ore type characterized by inclusions of mafic and ultramafic rocks with sulfide > >norite	Giblin (1984a,b)
	Ragged disseminated sulfide		An ore type characterized by small, close-packed inclusions in a matrix of norite and sulfide. The inclusions are typically mafic in composition, and the sulfides often exhibit a cusp-like or inter-fragmental shape	Giblin (1984a,b)
	Interstitial sulfide		An ore type characterized by sulfides which fill the interstices between euhedral plagioclase and pyroxene	Giblin (1984a,b)
	Footwall ore body		The portion of the mineral reserve and/or resource that occurs below the lower contact of the SIC in country rocks	Farrow and Lightfoot (2002)
	Narrow vein sulfide		Veins of massive sulfide developed in the footwall of the SIC; these veins often have sharp walls	Farrow et al. (2005)
	Stringer sulfide		An ore type in which narrow veins are filled with sulfide minerals	Giblin (1984a,b)
	Low-sulfide, high-precious metals mineralization	LSHPM	Rocks or ores containing with low sulfide content and potentially economic concentrations of the PGE	White (2012)

Group	Term	Synonymous Term (in Brackets) and Abbreviations	Explanation or Definition	References
Definitions relating to Sudbury ore deposits (*cont.*)	Reserves		An "Ore Reserve" is the economically mineable part of a Measured and/or Indicated Mineral Resource. It includes diluting materials and allowances for losses, which may occur when the material is mined or extracted and is defined by studies at Pre-Feasibility or Feasibility level as appropriate that include application of Modifying Factors. Such studies demonstrate that, at the time of reporting, extraction could reasonably be justified	JORC (2012)
	Resources		A "Mineral Resource" is a concentration or occurrence of solid material of economic interest in or on the Earth's crust in such form, grade (or quality), and quantity that there are reasonable prospects for eventual economic extraction. The location, quantity, grade (or quality), continuity and other geological characteristics of a Mineral Resource are known, estimated or interpreted from specific geological evidence and knowledge, including sampling. Mineral Resources are sub-divided	JORC (2012)
	Exploration target		An Exploration Target is a statement or estimate of the exploration potential of a mineral deposit in a defined geological setting where the statement or estimate, quoted as a range of tons and a range of grade (or quality), relates to mineralization for which there has been insufficient exploration to estimate a Mineral Resource	JORC (2012)

In any normal geoscience debate, the aforementioned list would be a compelling testimony to the historic legacy of geoscience research, but the Sudbury Structure has been the subject of yet more controversy.

THE CATASTROPHY THAT CREATED THE SUDBURY ROCKS

Another group of debates has grown up around the origin of the very complex rocks at the base of the SIC and in the immediate footwall. Sudbury is home to some of the most extensive and complex examples of breccias developed in association with a differentiated igneous complex. These breccias were found to be associated with the ore deposits (Speers, 1956, 1957; Wilson, 1956a,b), and their origin was initially considered to be a product of explosive volatile-rich magmatic activity. The breccias broadly break out into pseudotachylite breccias developed around the SIC (French, 1998) and both magmatic and country rock breccias developed at the base of the SIC (Pattison, 1979) and in association with radial and concentric dykes (Grant and Bite, 1984; Lightfoot and Farrow, 2002). The origin of these different groups has been reconciled in the context of an explosive magmatic event triggered by an asteroid impact (Grieve, 1994).

In 1964, Dietz proposed the idea that the Sudbury Structure was the product of the impact of a large Cu–Ni-rich iron asteroid. His work suggested that a shallow crater on the scale of 50 km across and 3 km deep was excavated, and the ores were splash-emplaced contributions from the asteroid. The heat generated in the crust in response to impact was considered to be the driving force behind the generation, upwelling, and emplacement of magma from depth. The overlying rocks were considered to be crater infill sedimentary rocks. Dietz had established the basis for the impact revolution at Sudbury, although his view that the metals in the ores come from the asteroid are likely not correct. The idea caught on quite slowly, and was not embraced by some workers who considered it rather ad hoc and convenient to explain some very complex rocks with such a simple model that side-steps the importance of endogenic models (Muir, 1984). The supporting evidence for the impact model arrived in bits and pieces as workers investigated the origin of shatter cones (Guy-Bray et al., 1966), the development of shock metamorphism in the country rocks (French, 1968a,b), and then the compilation of evidence started to underpin the theory and link it to the ore deposits (Morrison, 1984; Peredery and Morrison, 1984a,b).

A more holistic model appeared as result of studies of other terrestrial and extraterrestrial impact craters, and the fact that Sudbury has many of the hallmarks of an impact event was formalized in papers by Grieve et al. (1991a,b), Grieve (1994), and Golightly (1994). The debate over the role for meteorite impact versus volcanogenic processes was also supported by studies of the breccias overlying the SIC (Muir and Peredery, 1984; Roussel, 1984a,b), and detailed studies identified microdiamonds and fullerenes produced by the impact event (Heymann et al., 1999; Becker et al., 1994a,b). The recognition that the radial and concentric dykes of the SIC were derived from impact melts was discussed by Grant and Bite (1984), and the origin of the whole SIC as a crustal melt sheet was proposed by a number of authors (Grieve et al., 1991a,b; Grieve, 1994; Lightfoot et al., 1997c; 2001; Theriault et al., 2002; Keays and Lightfoot, 2004). Recent research has linked the 1.85 Ga Sudbury event to U-Pb zircon ages for materials present in what are considered to be 1.85 Ga distal ejecta horizons associated with the 1.85 Ga impact event (Addison et al., 2005). These models of epicontinental impact would trigger tsunami events and formation of chaotic distal deposits may relate to the demise in formation of Archean banded-iron formations (Slack and Cannon, 2009; Cannon et al., 2010). The impact origin of Sudbury is now firmly established, but the exact process is one that remains opaque in the details of what happened in a catastrophic event that imparted vast amounts of kinetic energy into potential energy in a few seconds – the physical and chemical effect of striking the surface of the planet with a multi-km diameter bolide remain difficult to conceive, let alone model. These conditions are unlike those encountered in traditional relatively low-temperature intrusive and volcanic events.

THE DEEP STRUCTURE AND DEFORMATION HISTORY OF THE SIC

The opportunity to understand the primary configuration of the Sudbury Structure by undertaking a reconstruction of the effects of subsequent deformation are also a critical part of deciphering the processes responsible for the present geometry of the Sudbury Structure and the associated ore deposits (Riller et al., 1999). The extent of deformation that accompanied the post-impact re-adjustment of the impact crater versus the deformation that postdated the formation of the Sudbury Structure and represents a response to far-field tectonics is not always clearly defined and understood. Geological models that describe the Sudbury Structure at depth utilize seismic reflection data acquired along traverse lines across the Sudbury Structure (the Lithoprobe Project, a Canadian Geoscience Program summarized in Boerner et al. 1994a,b). These public domain data together with drilling undertaken by mining companies close to the lines have generated multiple different interpretations for the crustal structure in the Sudbury Region (Milkereit et al., 1994a,b; Card and Jackson, 1995). There has also been an extensive debate on the degree of folding and flexing of the SIC (Cowan and Schwerdtner, 1994), and the deformation that gave rise to some of the major structures which cut through and control ore bodies (Lenauer and Riller, 2012a,b; Mukwakwami et al., 2012). Three-dimensional reconstructions of Sudbury which respect the various structures are rarely shown in the public domain (Gibson, 2003), but the geometry of the contacts and faults are now modeled routinely by mining companies to depths of up to 3 km around the Sudbury Basin using three dimensional imaging software. Regional airborne geophysical datasets can also be integrated to generate tests of models using potential field data (McGrath and Broome, 1994a,b; Hearst et al., 1994a,b). Much of this information underpins the ongoing exploration efforts by companies working to find new mineral deposits at Sudbury and hence remains confidential.

ATTEMPTS TO RECONCILE IMPACT MODELS WITH IGNEOUS PROCESSES

The holy grail of opportunities offered by all of these strands of theory, thesis, and dispute, is the reconciliation of the ore-forming process with the impact event that produced the Sudury Structure. Aspects of the morphology and genesis of the rocks which contain the ores have been described by Grant and Bite (1984), Morrison (1984), and Golightly (1994) in the context of the impact model, and the sequence of events and processes have been documented in more recent papers devoted to detailed geochemical studies of the SIC where a sequence of events in the formation of the ore deposits and their host rocks was outlined (Farrow and Lightfoot, 2002; Lightfoot and Farrow, 2002; Lightfoot et al., 1997a-c; Keays and Lightfoot, 2004).

The evolution of theories and hypotheses for the formation of the Sudbury Structure is in itself a classic example of polarization around models that involve longer lived planetary processes versus sudden catastrophic events triggered from outside. On the one hand, more traditional models call on heat supplied from within the Earth resulting in unusually explosive magmatic processes to generate both the breccias and the igneous rocks of the SIC through melting of the crust by mantle-derived magmas (Muir, 1984). On the other hand, a group of models require an exogenic contribution of energy from an asteroid, which produced the extreme brecciation of the country rock and the igneous rocks of the SIC by crustal melting (Table 1.4). The gradual transition of ideas from purely endogenic to impact models came slowly with the recognition that all of the features of the Sudbury Structure can be reconciled with an impact event and impact-triggered magmatic activity. Some models claim middle ground in this debate, and invoke both an impact event as well as contributions of mantle-derived magma that have been extensively contaminated (Chai and Eckstrand, 1996). Although there have been many detailed geochemical studies of the SIC, there remains no overwhelming evidence for a mantle contribution the magma (Lightfoot et al., 1997b). Notwithstanding the important details of these debates, the weight of evidence now favors an impact event, and an origin of the SIC by melting of the crust to produce the rocks of the SIC (Table 1.4).

Into this context, the ore deposits must also be placed, and a second and equally important set of debates have polarized around the relative roles of magmatic and hydrothermal processes in the genesis of the ore deposits. On the one hand, some have sought to explain all of the sulfide ore deposits as a product of magmatic processes involving the saturation of silicate magmas in sulfur, the segregation of immiscible sulfide, and the accumulation of these dense sulfide melts to form ore bodies at the outer margin of the SIC. A second set of models seeks to explain the presence of mineralization below the base of the SIC, and calls on evidence for hydrothermal remobilization of metals, and deposition of ores in favorable settings beneath the SIC. The weight of evidence now indicates that the vast majority (>95%) of the sulfide ore deposits are formed by magmatic processes, and the extent of hydrothermal modification is restricted to the margins of ore bodies and structures.

THE IMPACT PROCESS AND THE FORMATION OF THE SULFIDE ORES

The origin of the igneous rocks and their role in ore formation has been a matter of some debate (Keays and Lightfoot, 2004). On the one hand, these efforts have demonstrated that the igneous portion of the complex was produced by crustal melting to produce the differentiated rocks of the SIC, the QD of radial and concentric dykes, and comminution and partial melting to produce the pseudotachylite breccias which abound in the country rocks surrounding the SIC. The bulk chemical composition of the rocks comprising the SIC and the Offset Dykes approximate the average composition of the crustal target rocks, with no conclusive evidence for, or requirement for a contribution from the lithospheric or asthenospheric mantle-derived magma. On the other hand, the rocks which comprise the base of the complex are a series of magmatic breccias which are contained in a discontinuous unit termed the Sublayer which is associated with both ore deposits and mineral occurrences at the lower margin of the SIC and the QD Offset Dykes that also contain magmatic breccias. Moreover, these rocks contain the world's second largest historic production plus current reserves of magmatic Ni–Cu–PGE sulfide, and have generated over 11 Mmt of nickel production since their discovery in 1883. The host rocks containing many of the ores are mafic to intermediate in composition (norites and QD), and the ores are directly associated with a population of exotic inclusions of ultramafic-mafic composition. The rock types comprising the SIC are ultramafic, mafic to intermediate and evolved in composition. All have the same U-Pb age as the SIC, and so have the rocks which contain fresh cumulus olivine, orthopyroxene and chrome spinel which were generated at the time of the SIC. Under normal circumstances of magmatic differentiation of crustal melts, it would be unlikely to generate rocks with ultramafic compositions, or trigger the segregation of immiscible magmatic sulfides that were enriched in Ni, Cu, and PGE. One of the great enigmas and riddles to be solved at

Sudbury is to reconcile the process of impact melt generation with the formation of ultramafic rocks and Ni–Cu–PGE-enriched sulfide ores.

WHERE DO ALL THE METALS AND SULFUR IN THE ORES COME FROM?

The origin of the metals at Sudbury has long been considered as the product of segregation of immiscible sulfide from the magmas that produced the igneous rocks (Bell, 1891a,b), and the descriptions of the formation of magmatic sulfides has become a key part of the Sudbury literature (Naldrett, 1984a-d; Naldrett, 2004). The actual recognition that the ore-forming process produced a depletion of nickel in the noritic rocks was recognized by Golightly (1994) grounded on observations made earlier of Ni-depletion in basaltic rocks at Noril'sk (Lightfoot et al., 1990) and the importance of these data to the formation of the Noril'sk ores (Naldrett et al., 1992; Naldrett and Lightfoot, 1993). Lightfoot et al. (2001) provided the first detailed chemostratigraphy of metals through the SIC and showed that the development of very low nickel and copper abundances in the norites reflected the segregation and removal of immiscible sulfide. Detailed studies of other traverses (Lightfoot and Zotov, 2006) and the abundances of precious metals (Keays and Lightfoot, 2004) provided further support for depletion of the silicate melt in metals through segregation and accumulation of magmatic sulfide at the base of the SIC.

The wealth of previous work on Sudbury, the excellent opportunity to study the Structure using three dimensional datasets from the mining industry and remote sensing plus geophysical datasets, and the opportunity to more completely present the geological relationships between ores and country rocks are at the very focus of this book which attempts to provide a grand unified theory for the enigmatic Sudbury Structure. To get to this position, the first step is to explain the principal rock associations and modern nomenclature so there is a clear context for the geological information presented in this text.

MAGMATIC NICKEL SULFIDES – AN INTRODUCTION AND OVERVIEW

The ore deposits at Sudbury are classic examples of the magmatic style of Ni–Cu–PGE sulfide ore deposits described in a textbook by Naldrett (2004); they are part of a group of deposits which form by the segregation and accumulation of a Ni–Cu-rich immiscible sulfide melt from a silicate magma. Despite their magmatic associations, the Sudbury Deposits have typically been allocated to a unique group of astrobleme-associated deposits linked to a crustal melt-sheet (Grieve, 1994; Naldrett, 2004); no other global examples of this type of deposit are known outside of Sudbury. Notwithstanding this unique classification, many of the same processes control the development of other magmatic sulfide ore deposits, and the actual classification scheme that breaks Sudbury out into a different group gives the misleading impression that the ore deposits share no common features with other magmatic sulfides, which is evidently not the case from both a geological and genetic point of view (Lightfoot, 2007; Lightfoot and Evans-Lamswood, 2015).

In order to best understand the Sudbury ore deposits, it is necessary to have a basic understanding of magmatic sulfide ore deposits formed from mantle-derived mafic and ultramafic magmas. The relatively young, undeformed and unmetamorphosed deposits at Noril'sk provide some of the best-preserved examples of magmatic Ni–Cu–PGE-sulfide ore deposits (Naldrett et al., 1992, 1995), and much has been learnt about the processes responsible for the formation of magmatic sulfides from the very clear examples at Noril'sk (Naldrett et al., 1992, 1995; Naldrett and Lightfoot, 1993). The formation of magmatic sulfide ores can be viewed as the result of a rather unique set of conditions that produce economic concentrations of metals. Tony Naldrett refers to this process as a pathway "From the Mantle to the Bank" in his 2010 paper on magmatic sulfide ore deposits (Naldrett, 2010a,b). The stages can be broken out according to depth into seven stages, namely:

1. Partial melting of the mantle to produce a magma which contains normal concentrations of metals, the proportions of which are governed by the composition of the magma source and the degree of melting (Keays, 1995; Barnes and Lightfoot, 2005; Fig. 1.9A).

2. Magmas may pond close to the base of the crust (Cox, 1980) and then migrate towards the surface along conduits created by transtensional cross-linking structures within major deep-seated, crustal-scale structures such as rift zones and continental sutures (Fig. 1.9A,B; Lightfoot and Evans-Lamswood, 2015).

3. Sulfide saturation of the silicate magma results in the formation of an immiscible sulfide melt within the silicate melt; base and precious metals will concentrate into this melt because of their high sulfide melt/silicate melt partition coefficients (Naldrett, 2004; Fig. 1.10A).

4. Enrichment of the sulfide melt in base and precious metals is controlled by the ratio of volume of silicate magma to sulfide magma (Campbell and Naldrett, 1979). If the ratio is very low, the base metal content of the sulfide will be low, and if the ratio is very high then the metal contents of the sulfides produced by equilibration of the sulfide magma with the silicate magma will be very high (Naldrett, 2004; Fig. 1.10A).

5. Economic concentrations of mineralization can form if the dense sulfides settle and accumulate in a physical trap where the country rocks form a "container" in which the sulfides are localized by pooling before the magma crystallizes (Lightfoot, 2007; Barnes and Lightfoot, 2005; Fig. 1.10B,C). In some cases, it appears likely that the magmatic sulfides are transported through the crust from the point at which they form to the point at which they form an ore body (Tang, 1992; Lightfoot et al., 2012).

6. Modification of the primary magmatic sulfide can take place during cooling, where a monosulfide solid solution is formed (which exsolves down-termperature to form pyrrhotite and pentlandite), and a residual Cu-rich and Ni-bearing sulfide melt is expelled to form what is often termed a fractionated sulfide melt. This process produces a range in mineralogy of the ores from pyrrhotite–pentlandite-rich ores to chalcopyrite–pentlandite, and bornite–millerite mineralization (Naldrett, 2004)

7. Postmagmatic modification of the ores can include hydrothermal modification, metamorphism, and structural displacement. These processes can entirely re-form and/or remobilize the ore body by partial melting and deformation (Lightfoot et al., 2011b), or remobilize part of the mineralization, and change the metal concentration in the sulfide (Farrow and Watkinson, 1997).

The ore deposits at Sudbury have a classic magmatic association of pyrrhotite–chalcopyrite–pentlandite, and occur at the base of the SIC, so their magmatic origin has long been recognized (Gregory, 1925; Souch et al., 1969; Hawley, 1962), although there has been some debate regarding the extent of hydro-thermal modification (Farrow and Watkinson, 1997). What has always been less clear is the role of partial melting and magma transport in their genesis in the context of the impact origin of the SIC.

The Sudbury ore deposits are the type examples of magmatic sulfide ores, and other discoveries such as Noril'sk and Jinchuan carry evidence in the literature of the important impact of the description of Sudbury ores in the recognition of the rocks that gave rise to the discoveries. Even though the Sudbury Deposits are uniquely associated with an impact structure, many of the processes recognized in

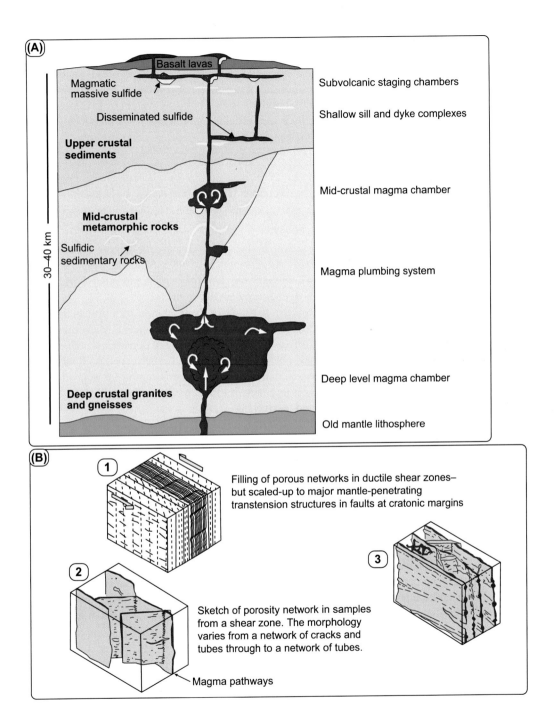

FIGURE 1.9

(A) The traditional models for nickel sulfide ore formation involve a path "from the mantle to the bank."

(B) Pathways through the crust are established in strike-slip transtensional fault structures.

Based on Geraud et al., 1995, Lightfoot et al. (2012).

mafic and ultramafic-hosted and associated deposits such as Thompson, Kambalda, Raglan, Pechenga, Noril'sk and Jinchuan (Naldrett, 2004; Barnes and Lightfoot, 2005) are similar to those historically recognized to be important in the control of the compositional diversity in Sudbury ores (Lightfoot, 2007).

The ore deposits at Sudbury are perhaps not the best example of processes involving a path from "The Mantle to the Bank" (Naldrett, 2010a,b), but there is an increasing weight of evidence to suggest that they are good example of ores formed by a very efficient natural smelter which localizes immiscible magmatic sulfides from the "crust to the bank."

PROCESS CONTROLS IN THE FORMATION OF MAGMATIC SULFIDE ORE DEPOSITS

Naldrett described what have come to be recognized as process controls (Naldrett, 2010a,b) in the formation of what Barnes et al. (2015) refer to as "mineral systems." It is worth looking at the elements of these processes and the key roles that they play (Lightfoot et al., 1997c, Naldrett, 1999); this provides a critical context in understanding how close Sudbury ore deposits compare to other world examples, and how the exact nature of the magmatic heat engines were different.

A snapshot of process controls is listed previously where the evolution of a potentially pregnant magma capable of forming an ore system is tracked from the mantle to a deposit. Many things can "go wrong" on the way from the mantle to the surface, in the sense that an ore deposit may never form. Although not all of the elements of this process model are entirely agreed, a perfect constellation of events is required to generate a world-class ore system that underpin mine production for many decades. We will look at the history of a mafic magma as it migrates from the mantle to the surface, and show how it generates associated magmatic sulfide ore deposits (Fig. 1.10A).

MELTING OF THE MANTLE

Starting in the mantle, the first step involves melting of lherzolite, peridotite, or garnet peridotite. The exact causes of melting remain hotly debated by scientists, but the ones that most commonly matter in the context of nickel are melting events triggered by deep mantle plumes (Pirajno, 2000; Ernst and Bleeker, 2010). These are unusual domains of upwelling heat in the mantle, likely produced at the core–mantle interface, and localized or focused by the geodynamic impact of plate tectonics. Not all nickel sulfide ore deposits are linked to plume events, but many link to hot mantle in areas of plume activity or unusually hot mantle created in areas of crustal thickening (eg, mantle sources modified by subduction zones can be tapped along deep-penetrating strike-slip structures; Lightfoot and Evans-Lamswood, 2015). A key feature of the magmas that ultimately form ore deposits is that they are undersaturated sulfides when they leave the mantle (Keays, 1995). This generally means that they form by high degrees of partial melting.

MIGRATION OF THE MAGMAS THROUGH THE CRUST, AND MAGMATIC DIFFERENTIATION

Melts generated by plume events or anomalous heat are able to pond together and migrate from depth by virtue of their lesser density. They may gather in deep subcrustal chambers (Cox, 1980), or pass into the crust to form giant differentiated intrusions with a thickness of 10 km or more [eg, the Bushveld

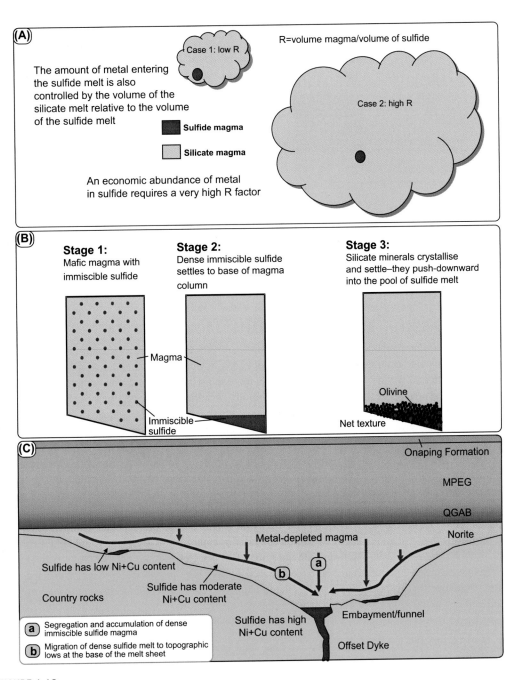

FIGURE 1.10

(A) Significance of magma–sulfide ratio as a control on the amount of metal in the sulfide liquid; (B) role of gravitational settling (the billiard ball model); (C) effect of slope and localization of ore deposits in "containers."

(B) After Naldrett (2004); (C) after Ames et al. (2002) and Ripley et al. (2015).

Complex in South Africa (Naldrett et al., 2009) the Stillwater Complex in Montana (Keays et al., 2010), and the Kiglapait Complex in Labrador, Canada (Morse, 1969)]. Not all magmas end up in giant differentiated intrusions, although many that do are able to form world class deposits of magmatic precious metals such as the Merensky Reef of the Bushveld Complex (Naldrett et al., 2009), whereas others are inherently devoid of metals and so they appear incapable of forming metal-enriched reef horizons (eg, the Kiglapait Intrusion; Lightfoot et al., 2011a). If the metals are not present in the magma when it leaves the mantle, then there is one less ingredient that makes it viable for the melts of the source to form a magmatic sulfide ore deposit.

Mantle-derived magmas tend to loose heat as they migrate adiabatically from depth toward the surface (Fig. 1.9A); in some circumstances, the magma can entirely crystallize without any process happening to localize the metals and form a potential ore system. Vast regions of the planet are covered in basaltic rocks that arise as the magmas flood out at surface to form flood basalts in large igneous provinces (Ernst, 2014). These magmas are often erupted as highly fractionated basaltic magmas that have lost olivine, pyroxene, and plagioclase at depth, to crystallize as common basalt at surface (examples are the Deccan Trap in India, the Columbia River in USA, and many other large igneous provinces as described in Ernst (2014). It takes a unique set of conditions if this common magmatic differentiation process is to be upset and result in concentration of the metals found in the ores.

TRIGGERS TO SULFUR SATURATION OF MAFIC MAGMAS

The key trigger to the formation of a magmatic sulfide is a process by which an immiscible sulfide melt forms from the silicate magma due to the presence of excess sulfur which cannot be accommodated in the silicate melt structure (Irvine, 1975). Other controls include temperature and iron content of the magma (Keays, 1995 and references therein). The exact solubility of sulfur in mafic magmas has been the subject of experimental and theoretical studies, and it appears likely that silica-poor magmas dissolve more sulfur than silica-rich magmas, and that a protracted period of cooling of the magma before crystallization can provide ample opportunity for magmatic sulfides to segregate (Li and Ripley, 2005) so either the addition of sulfur from the crust or the reduction of sulfur solubility is required to trigger the segregation of an immiscible sulfide magma. The relative roles of the two processes are hotly debated, but it appears empirically more likely that sulfide segregation occurs in response to addition of crustal sulfur than as a result of changes in pressure, temperature, oxygen fugacity or silicate content of the magma (Keays and Lightfoot, 2007; Ripley and Li, 2013). Indeed, many magmatic sulfide ore deposits reside in association with intrusions that rest in packages of sulfidic metasedimentary rocks (Raglan, Kambalda, Pechenga, Noril'sk; Naldrett, 2004), although some do not (Jinchuan; Naldrett, 2004). The debate over the relative roles of silicification versus derivation of S from crustal sources is an important topic which is discussed at greater length later in this book. For now, it is the fundamental control of S solubility in the magma that is considered key to the saturation process which ultimately controls the ore-forming ability of the magma.

THE IMPORTANCE OF GRAVITY-ASSISTED SETTLING OF DENSE MAGMATIC SULFIDE, AND PATHWAYS THROUGH THE CRUST

For the sake of the discussion, let us assume as of now that the parental magmas had achieved sulfur saturation. These conditions have immediate physical implications. Silicate magmas typically have densities of 2.9–3.1 g/cc whereas sulfide magmas have a density of 4–5g/cc. This large difference in density within a magmatic system provides the driving force required for the segregation of the dense sulfide melt from the silicate melt; the sulfides will tend to localize toward the base of the magma column or magma chamber unless some physical process continues to carry the immiscible sulfide upward in association with less dense entrained crystals (de Bremond d'Arsa et al., 2001) or by some tectonic mechanism that allows melt to pass through a network of crustal chambers (Lightfoot and Evans-Lamswood, 2015). The separation of sulfide is a very important stage in the formation of an economic ore body versus a subeconomic or uneconomic mineral zone. The pathways between the mantle and the crust along which the magmas and entrained sulfides can pass are often controlled by space created in cross-linking structures within strike-slip fault zones inboard of cratonic margins (Lightfoot and Evans-Lamswood, 2015; Fig. 1.10B).

CONTROLS ON THE COMPOSITION OF SULFIDE MAGMAS

Sulfide melt separation from the silicate magma is a key step, and the formation of a magmatic sulfide in a silicate magma has the effect of upsetting the partitioning behavior of elements such as Ni, Cu, Co, and Pt, Pd, and Au which would normally be governed by partitioning between silicate magmas and silicate minerals (Keays, 1995; Rollinson, 1993). The Nernst partition coefficient is the key control, and this coefficient is a measure of how readily metals enter the magma versus a crystallizing mineral or immiscible sulfide (Rollinson, 1993). This is because nickel partitions into olivine with a partition coefficient of about 6–29 (Rollinson, 1993), but it more readily partitions into a sulfide melt with a partition coefficient of 500–1500; the removal of a dense immiscible sulfide will deplete the magma in nickel far more readily than the crystallization and setting of less dense olivine. Likewise, copper and PGE do not easily partition into silicate minerals, but they readily enter magmatic sulfide, with copper entering sulfide with a partition coefficient of about 2000, Pt with a coefficient of about 20,000, and Pd with a coefficient of almost 100,000 (Peach et al., 1990). In a simple closed system, with a small volume of magmatic sulfide, the degree of enrichment of metals in the sulfide magma will be Pd > Pt > Cu > Ni based on the partition coefficient (Keays, 1995).

The segregation of large amounts of immiscible sulfide from tiny volumes of silicate magma will result in quite a different outcome from a process of segregation of smaller amounts of sulfide from a much larger volume of silicate magma. This distinction is the basis of the "R-factor" equation which predicts the concentration of a metal species in the sulfide melt relative to the silicate melt as a function of the partition coefficient of the element (Campbell and Naldrett, 1979; Fig. 1.10A). Very few magmatic systems are closed, and this has led to further refinement of the equation to allow for open system magmatic processes with the introduction of the "N factor" (Naldrett et al. 1996a,b). In all of these models, the fundamental control is the relative volume of sulfide magma to silicate magma, and the concentration of the metals in the starting magma. If the initial magma is devoid of metals, then it will take an enormous volume of silicate magma relative to sulfide magma to generate a sulfide magma with elevated base and precious metal concentration levels. Examples of deposits believed to have formed with low R factor are the Moxie–Khatadin mineralization in Maine (Naldrett et al., 1984a,b) and the Voisey's Bay South mineralization in Labrador (Kerr, 2012; Li et al., 2001); examples of ore deposits formed in magmatic systems with especially high R factor include the Noril'sk-Talnakh Deposits in Siberia, Russia (Naldrett et al., 1992, 1996). Examples of deposits formed from magmas with very low precious metals concentrations include Voisey's Bay in Labrador (Lightfoot et al., 2012), where the low abundance levels reflect the abundance of precious metals relative to nickel and copper in the source region from where the magmas were generated.

SULFIDE LOCALIZATION AND CONCENTRATION MECHANISMS IN "CONTAINER ROCKS"

Once a dense, metal-rich, immiscible, sulfide melt forms, it must be localized in one place in order to form an ore deposit. This is a critical stage in which the dense sulfide is localized in physical depressions in intrusions or at breaks in the magma conduit. The degree to which massive sulfide can localize is controlled by the primary process of concentration where the velocity of flow of the magma drops (Naldrett et al. 1996a,b describe this as a principal control in the formation of the ore deposits at Voisey's Bay). It is also controlled by the extent to which the overlying silicate cumulate minerals press down onto the sulfide; this is termed the "billiard ball model" by Naldrett (2004), and it is an especially important control in rocks that forms by accumulation of silicate phases such as olivine and orthopyroxene to form adcumulate-textured rocks (Naldrett, 1973, 2004; Fig. 1.10B). At Sudbury, only a few of the rocks develop textures indicative of this process, but this may be a function of the efficiency of sulfide segregation and accumulation from the overlying magma.

Magmatic sulfide may be trapped in the silicate magma during relatively rapid cooling, and this generates the disseminated and blebby sulfide textures that are commonly recognized; in these cases the efficiency of sulfide segregation has been arrested by the presence of crystals or rock inclusions in the magma which prevent efficient gravity-aided segregation of the sulfide. These styles of mineralization can be ore-grade where the volume of the sulfide and the metal content of 100% sulfide is large (Noril'sk; Naldrett et al. 1996a,b and Mt Keith, Barnes et al., 2011), but most deposits of this type are sub-economic to uneconomic (e.g. Franchuk et al., 2015), and they may have to wait until the right market conditions allow for their development.

The controls on the localization of sulfide are often structural, whereby the sulfide magma is localized and concentrated in physical depressions or into structures that are active during magmatism (Fig. 1.9B; a good example of this is the Voisey's Bay Deposit in Labrador where the geometry of the host intrusion plays an important part in controlling the distribution of mineralization; Naldrett et al. 1996a,b; Lightfoot et al., 2012). Although there are many good examples of situations in which sulfides settle out of the magmas and accumulate at the lowest point down the slope, there are other examples where the quantities of sulfide or the lack of crustal sulfur sources appear to require the emplacement of magmatic sulfide. This is clearly the case at Voisey's Bay (Lightfoot et al., 2012), but it also appears increasingly likely that other deposits such as Jinchuan and Noril'sk were formed by the injection of sulfide-laden magmas (Tang, 1992; Lightfoot and Evans-Lamswood, 2015). Sudbury offers some very special insights to the controls on sulfide localization by both the effect of early gravitational segregation of dense sulfide melt toward the base of the SIC and subsequent lateral transport down-slope toward topographic lows in the melt sheet where there are embayments, troughs and Offset Dykes (Fig. 1.10C). The details of the supporting geological evidence and the controls on localization of ore deposits will be treated in Chapters 2 and 4. Not all of the Sudbury ore deposits are free of structural modification, but it is this primary geometric control of the uneven topography of the country rocks at the base of the melt sheet which leads to the term "container rocks" being applied to magmatic sulfide ore deposits (Lightfoot et al., 2012).

The economic viability of a nickel sulfide ore deposit depends on the grade and tonage as well as the presence of other elements of commercial importance such as copper and the PGE. A useful index in the evaluation of mineralization is the metal concentration in sulfide. The metal concentration in the sulfide component of the rock is termed its metal tenor (nickel tenor is expressed as $[Ni]_{100}$ where the amount of nickel in sulfide is calculated based on the stoichiometry of the sulfide minerals that comprise the ore; Kerr, 2003). Metal tenor is a useful indicator of compositional diversity in simple magmatic sulfide ores which comprise an assemblage of the principal sulfide minerals pyrrhotite [$Fe_{(1-x)}S\{x = 0\text{-}0.17\}$], pentlandite [$(Fe,Ni)_9S_8$] and chalcopyrite ($CuFeS_2$). If the sulfide assemblage is more complex and contains minerals such as pyrite (FeS_2), millerite (NiS), bornite (Cu_5FeS_4) or less commonly arsenides, tellurides, or sellenides, then the calculation of metal tenor must fully respect the complex mineral assemblage if it is to be reported (Kerr, 2003; Chapter 4).

The extent to which economic grades of mineralization are achieved in the system depends strongly on the efficiency of concentration of the sulfide melt in one location. This is illustrated in Fig. 11.1C where the sulfides segregate from a column of magma and accumulate toward the base of a chamber and then move laterally along the sloped floor of the magma chamber to reach a topographic low where they accumulate (Ames et al., 2002; Ripley et al., 2015).

SULFIDE DIFFERENTIATION

During cooling of the magmatic sulfide liquid, the crystallization is controlled by a combination of sulfide mineral phases and solid solution series. The cooling sulfur-poor Ni–Cu–Fe sulfide melt possibly undergoes early crystallization of pentlandite (Sugaki and Kitakaze, 1998), but is generally believed to crystallize a monosulfide Fe–Ni–S solid solution (mss), which then exsolves to pyrrhotite and pentlandite on further cooling (Naldrett, 2004). This process extracts iron, nickel, and sulfur from the magma leaving behind a sulfur- and iron-poor melt with very elevated platinum, palladium and copper, with some residual nickel. Cooling of the sulfide magma generates a solid mss which can accumulate to form an ore that is moderately rich in nickel, but typically has lower abundance of copper and PGE (depending on the proportion of trapped copper-rich liquid and the amount of Cu in solid-solution in the mss); the products of mss crystallization often comprises pyrrhotite crystals with flame pentlandite exsolved along the crystal cleavage (Hawley, 1962). The liquid remaining is able to crystallize around the Mss to form loop-textures (as described in the Noril'sk and Voisey's Bay deposits; Lightfoot and Zotov, 2014; Lightfoot et al., 2012) or these liquids can be expelled from the Mss to form discrete ore stringers which then crystallize to form copper-rich deposits. The process of sulfide melt fractionation is an important feature of the Noril'sk and Sudbury ore deposits (Keays and Crocket, 1970; Naldrett et al., 1994a,b, 1996a,b). Further crystallization of the copper-rich residual liquid produces an intermediate solid solution (iss), which exsolves to an assemblage of pentlandite and chalcopyrite (Naldrett, 2004). Any remaining fractionated liquid will then be very rich in nickel and copper and is able to crystallize to form assemblages of chalcopyrite–pentlandite–millerite, millerite–bornite, and bornite–chalcopyrite. The final stages of sulfide fractionation are not completely understood, but the textural relationships between sulfide minerals in veins provide important information about the sequence of events that is discussed in Chapter 5.

The very last portions of sulfide melt are often rich in silver and precious metals, and they can also contain carbonate and fluid components. These small-volume hydrous melts are able to generate styles of mineralization of subeconomic interest that occupy the distal fringes of ore bodies, and they are often referred to as low-sulfide-high-precious-metal (LSHPM) styles of mineralization (Farrow et al., 2005).

The slow pace of sulfide differentiation coupled with the instability of magma chambers and melt sheets can result in the migration of the more fractionated sulfide melts away from the earlier formed sulfide cumulates; in the case of the Noril'sk Deposits, the Cu-rich breccias in the hanging wall of the

Kharaelakh Intrusion are probably the product of remobilization of Cu-rich magma as the chamber adjusts in response to tectonic activity along the Noril'sk-Kharaelakh Fault (Lightfoot and Zotov, 2014). In the case of the Sudbury footwall deposits, this process likely happens in response to syn-magmatic and postmagmatic crater floor re-adjustment (Lightfoot, 2015).

POST MAGMATIC DEFORMATION, METAMORPHISM, AND HYDROTHERMAL MODIFICATION OF SULFIDE MINERALIZATION

Postmagmatic deformation can have a major additional influence on the composition and distribution of massive sulfide mineralization. Deformation zones which intersect mineral zones provide opportunity for sulfides to move from areas of compression to areas of extension. The sulfides are able to flow in a plastic form at relatively low pressures and temperatures, and they will tend to occupy dilatant zones in zones of rifting and transtension (Fig. 1.9B; Lightfoot and Evans-Lamswood, 2015). The mobilization of sulfides along with fragments of country rock generates rocks that resemble sulfide breccias, and the process of mobilization is termed sulfide kinesis. This process can result in the partial to entire detachment of sulfides from the parental intrusion, and at very high pressures, partial melting of the sulfides can result in fractionation of mss from pentlandite, resulting in a compositional change in the sulfide (eg, the Thompson Deposit in Manitoba, Canada; Lightfoot et al., 2011a,b).

The hydrothermal genesis of Ni sulfide ore deposits is recognized in deposits such as Avebury (Keays and Jowitt, 2013), Enterprise in Zambia (Capitrant et al., 2015), and the GT34 iron-oxide-nickel deposit in the Carajas Belt in Brasil (Siepierski, 2008). These deposits are possibly linked to mafic-ultramafic precursor rocks, but their genesis likely follows processes active in hydrothermal systems that cross redox boundaries, so the reader is referred to studies of the Zambia-Congo Copper Belt (McGowan et al., 2003; Capitrant et al., 2015) and the Carajas iron-oxide copper-gold (IOCG) systems (Monteiro et al., 2008; Grainger et al., 2007) as a primary source of insight into these process models.

In summary, many of the elements of process models for the formation of magmatic Ni–Cu–PGE sulfides from mafic and ultramafic magas can be reproduced in a crustal system if the melt becomes sulfide saturated and if there are sufficient volume of metal-undepleted magma to form ore deposit. This is one of the main themes which this book describes, but first it is worth providing some background to the impact origin of the Sudbury Structure, and the formation of the igneous rocks and magmatic sulfides from a crustal melt sheet.

EVIDENCE FOR IMPACT ORIGIN OF SUDBURY

Most models for the origin of the Sudbury Structure call on a major impact event, impact-induced igneous activity and protracted tectonic re-adjustment. Although a very few scientists consider the evidence for impact genesis of the SIC to be equivocal, the vast majority now accept the evidence that the Sudbury Structure is the product of a large 1.85 Ga impact event which gave rise to a complex multi-ring impact structure containing a melt sheet. There is a large literature on this topic. A summary of some key facets of the impact model set a useful context, and it is also worth considering aspects of the previously endorsed endogenic and hybrid-impact-igneous models for the origin of the SIC and associated ore deposits.

The evidence for an impact origin of the Sudbury Structure and the SIC break out into two levels; one group of features is compelling and has evolved since the pioneering work of the early advocates of impact (Dietz, 1964; French 1967, 1968a,b, 1970, 1998; Dence, 1972; Guy-Bray et al., 1966; Peredery, 1972a,b) to include an impact model for not just the target rocks but also the SIC and the associated magmatic sulfide ore deposits (Morrison, 1984; Morrison et al., 1994; Grieve et al., 1991a,b; Grieve, 1994; Keays and Lightfoot, 2004). The other evidence are considered more supportive and taken with the main evidence provides an overwhelming case for an impact origin (Grieve et al., 1991a,b; Grieve, 1994; Scott and Spray, 2000; Mossman et al., 2003; Masaitis et al., 1997; Slack and Cannon, 2009; Addison et al., 2005). The objective of this book is not to take apart this evidence piece by piece and test the impact hypothesis; but to underpin the science of how the whole system evolved as a consequence of this process, so the reader is referred to the supporting literature (Table 1.5).

COMPELLING EVIDENCE FOR AN IMPACT EVENT

1. *Morphology of the Sudbury Structure*: The elliptical shape of the SIC and the Sudbury Basin is a product of multiphase deformation during NW–SE crustal shortening of an original circular shape. The deep structure of the SIC is consistent with an original circular shape, with an original dimension of at least 80 km. The SIC melt may have originally extended beyond the present outcrop of the SIC and covered an outer ring structure where Offset Dykes were formed (Ames et al., 2002; Ripley et al., 2015; Table 1.5).

2. *Pseudotachylite breccias (SUBX)*: The SUBX is considered to be the product of high-pressure frictional melting of the country rocks as well as partial melting (French, 1998; Table 1.5). Although there remains some level of debate as to whether the pseudotachylite breccia matrix is derived from local frictional melting of country rocks (Lafrance and Kamber, 2010) or from an influx of melt derived from the proto-melt sheet (Riller et al., 2010).

3. *Shatter cones*: Dietz (1964) suggested that the shatter cones in the basement rocks surrounding the Sudbury Structure are the product of a high-pressure shock event (Table 1.5). An example of shatter cones developed in the McKim Formation in the footwall of the SIC is shown in Fig. 1.11.

4. *Evidence of high-pressure planar deformation*: Dence (1972) recognized planar deformation features in quartz in the Archean footwall rocks of the SIC, and kink banding is reported in feldspar and zircon (French, 1998).

5. *Metamorphism and brecciation of footwall rocks*: The development of a wide metamorphic halo in country rock is recorded in breccias, megabreccias, and partial melting of country rocks (Dressler, 1984 a-c); Boast and Spray, 2003; Jørgensen et al., 2013, 2014). The extent of this event provides support for a high-energy catastrophic event (the impact event) which was followed by a thermal metamorphic process where the heat source was the SIC melt sheet.

6. *Distal ejecta blanket and impact layer*: The distal effect of Sudbury event included giant tsunamis which created debris layers, and distal fall-out blankets which contain glassy shards with zircons formed at 1.85 Ga (Addison et al., 2005; Slack and Cannon, 2009). These features are recognized up to 1000 km from the Sudbury Structure, and point to a globally significant geological event.

Table 1.5 Summary of Main Evidence for Asteroid Impact at Sudbury in the Formation of the SIC; Supporting Evidence and Further Reading

Group	Geological Feature	Interpretation	Description of Evidence – the Main Points	References
Regional geology	Shape of the Sudbury Structure	Elliptical shape of the Sudbury Structure is due to the compression of a primary circular crater; the primary form of the Sudbury Structure is considered an impact-generated crater	Reconstruction of the primary geometry is supported by the removal of the effects of faulting and thrusting of the South Range toward and possibly over the North Range	Golightly (1994)
	Basal topography of the SIC	Impact topography with excavated troughs and embayment structures	Basal morphology including overall slopes of the basal topography and discrete depressions and fractures are inconsistent with emplacement of melt from beneath	Morrison (1984); Ames et al. (2002); Ripley et al. (2015)
	Marginal collapse and ring structure morphology	Main impact crater occupied by the SIC, and the outer margins represent collapse zones outside which the concentric Offset Dykes represent an outer ring of a multi-ring structure which possibly are the remnants of a distal impact melt	Morphology of the outer contacts of the SIC show evidence for slumping and marginal collapse in the Frood-Stobie Breccia Belt and the footwall of the Levack embayments	Scott and Spray (1999); Spray et al. (2004); Lightfoot (2015)
SIC	Main Mass	Differentiated impact melt sheet	Elliptical shape at locus of disruption; chemical and isotopic composition of bulk upper crust	Lightfoot et al. (1997a, 2001)
	Offset Dykes – radial and concentric	Radial and concentric impact structures filled with melt	QD with the composition of the average impact melt. This melt was injected in several phases along radiating and concentric structures produced by impact event	Grant and Bite (1984)
	SLNR	Basal noritic melt sheet rock which is heavily contaminated by country rocks and contains abundant fragments	Igneous-textured norite matrix with inclusions of mafic and ultramafic material that are heavily recrystallized and sometimes partially melted.	Pattison (1979); Morrison (1984)
	SLGRBX	Partially melted target rocks containing clasts which are heavily recrystallized and partially melted	Matrix is often sub-igneous to igneous textured, and the fragments range from felsic through to ultramafic; some inclusions of noritic rock may come from the basal part of the SIC, and so this unit is sometimes termed the "late SLGRBX"	Pattison (1979); Morrison (1984)

Group	Geological Feature	Interpretation	Description of Evidence – the Main Points	References
SIC (cont.)	Melt segregations and patches in country rocks	Partial melts as enclaves within heavily recrystallized and brecciated country rocks produced by incomplete melting by heat transfer from the melt sheet	Stringers and segregations of quartz-feldspar between fragments and within country rocks; commonly develops in more felsic country rocks	Lightfoot et al. (1997a)
	SUBX matrix	Derived from host rocks by local catacaustic milling and/or frictional melting processes; the matrix is often a recrystallized igneous melt	Glassy to sub-igneous textured fine-grained rock containing abundant local country rock inclusions	French (1998); Spray (1995); Reimold (1995); Roussel et al. (2003); Riller et al. (2010); Lafrance et al. (2014).
	Onaping Formation impact melts	A unit of aplitic to granophyric composition produced by melting of material which fell onto the top of the melt sheet	Discontinuous unit of melt developed at top of the Basal Member, and often having transitional contact relationships with the Onaping Formation breccias; these enigmatic rocks may be melt bodies generated within the overlying breccia	Grieve (1994)
	Magmatic sulfide ores	The formation of the ores in response to the segregation and accumulation of immiscible sulfide from the melt sheet is consistent with derivation from a crustal target	The composition of the sulfides in the Offset Dykes Sublayer, and basal Main Mass are consistent with in situ segregation and concentration of metals from the melt sheet	Naldrett et al. (1999); Lightfoot et al. (2001); Keays and Lightfoot (2004)
Breccias	Megabreccias	Broken-up target rocks; large blocks of country rock separated SLGRBX, partial melts, and SUBX	Impact-generated basal megaclast breccia	Morrison (1984)
	Hinge rocks	Large fragments of country rock that are rotated parallel to the contact and aligned with the crater walls by the impact event	Commonly dyke fragments rotated parallel to the walls of embayment structures and separating the igneous rocks of the Sublayer from the recrystallized country rocks; hinge rocks are often partially melted and heavily recrystallized	Lightfoot et al. (1997c)
	Onaping Basal Formation breccia	Fragment laden material represents a combination of fallback and material transported into the crater after impact by ground surge	Basal Onaping Member in gradational contact with the upper plagicoalse-bearing granophyre with both allochthonous and autochthanous components	Grieve (1994); Muir and Peredery (1984); Grieve et al. (1991a,b)

(Continued)

Table 1.5 Summary of Main Evidence for Asteroid Impact at Sudbury in the Formation of the SIC; Supporting Evidence and Further Reading (cont.)

Group	Geological Feature	Interpretation	Description of Evidence – the Main Points	References
Breccias (cont.)	Shatter cones	A unique indicator of shock deformation at high pressures in target rocks	Distinctive curved, striated fractures that form a cone shape in a variety of target rock types. The surface of the shatter cones form a "nested" texture with positive and negative features defining the cone. Striations branch along the surface of the cone forming a distinctive radial pattern	French (1998)
	Chaotic country rock assemblages	Chaotic translocation of country rocks along contacts and bedding planes	Discontinuous stratigraphy in target rocks produced by rotation and transportation of target rocks; boundaries are often SUBX zones or zones of incipient SUBX development; most easily recognized in Huronian Formation metasedimentary rocks with will developed rhythmic cycles of fine and coarse-grained sedimentary rocks	Dressler (1984 a-c)
	Recrystallized rock fragments in Sublayer and Offset Dykes	Recrystallization and melting of inclusions	Strong recrystallization and sometimes partial melting of fragments in norite magma produces entire recrystallization and melting with border textures that are often gradational	Lightfoot et al. (1997a)
	Distal ejecta blanket and impact layer	An 1850 Ma distal ejecta blanket and impact layer recognized in the Paleoproterozoic iron range of Michigan and Minesota	Distal effect of Sudbury event included giant tsunamis which created debris layers, and distal fall-out blankets which contain glassy shards with zircons formed at 1.85Ga. These observations triggered debate on the demise of banded iron formation at c. 1.85 Ga as a result of ocean mixing	Addison et al. (2005); Slack and Cannon (2009).

Group	Geological Feature	Interpretation	Description of Evidence – the Main Points	References
Mineralogy	Planar deformation features in minerals	High pressure shock deformation of planar features in minerals	Mineral fracture planes in quartz and higher temperature polymorphs and feldspars; fractures are also observed in zircon	Dence (1972); French (1998)
	High pressure polymorphs of minerals	Quartz pseodomorphs after primary tridymite	Recognized in the upper MPEG at the upper interface of the SIC with the Onaping Formation	Stevenson (1963)
	Shock metamorphic textures in country rocks	Extensive metamorphism of country rocks represent a metamorphic halo around a superheated melt sheet	Mineral paragenesis consistent with metamorphic zonation; textures are indicative of very high temperature and rapid cooling	Boast and Spray (2003); Jørgensen et al. (2013, 2014); French (1998)
	Diaplectic glass	Shock-generated glass (maskelynite stage)	Textures described in Onaping Formation	Peredery (1972a,b)
	Metamorphic halo	Extensive metamorphic halo below the SIC; thickness of halo broadly relates to the vertical extent of melt sheet rocks	Metabreccia consisting of Pyroxene hornfels	Boast and Spray (2003); Jørgensen et al. (2014); French (1998)
	Impact glass shards	Impact glass in the Onaping Formation produced by quenching of primary impact melt	Shards of glassy composition in the lower part of the Onaping Formation	Peredery (1972a,b); Ames et al (1998)
	Kink banding in mica	High pressure deformation produces kink banding in micas within country rocks	Kink-banding is at high angle to original cleavage (foliation) and associated quartz and K-feldspar show shock-deformation	French (1998)
Geochemistry	Crustal signatures—major, trace element, and isotope ratios	The average composition of the SIC approximates to average upper crustal target rocks in the Sudbury Region	Major and trace elements and Sr-Nd-Pb isotope systematics all are consistent with an overwhelming target rock contribution to the Sudbury melt sheet; although the mixing models are not unique, least squares mixing models using the major target rocks are consistent with these models	Grieve et al. (1991a,b); Grieve (1994)
	No geochemical requirement for mantle-derived magmas	Compositions of most mafic and ultramafic rocks in the SIC have either genetic links to the melt sheet (eg, Mafic Norite), or compositions unlike normal mantle-derived magmas (eg, olivine melanorites)	Extensive datasets for mafic-ultramafic inclusions in the Sublayer provides no direct confirmation of a mantle contribution; most can be formed by in situ recrystallization or melting of mafic-ultramafic target rocks and incorporation into the Sublayer	Lightfoot et al. (1997a,c)

(Continued)

Table 1.5 Summary of Main Evidence for Asteroid Impact at Sudbury in the Formation of the SIC; Supporting Evidence and Further Reading (cont.)

Group	Geological Feature	Interpretation	Description of Evidence – the Main Points	References
Geochemistry (cont.)	Impact diamonds	Microdiamonds discovered in the carbonaceous member of the Onaping Formation	Polycrystalline agregates 0.1-0.6 mm in diameter with graphite; carbon isotope ratio signature is consistent with derivation from Onaping Formation carbon	Masaitis et al. (1997)
	Fullererenes (buckyballs) in the Whitewater Formation	C_{60} and C_{70} ring structure of fullerenes in the Onaping Formation breccias	Generally considered the product of reformation of organic carbon as a product of impact events	Mossman et al. (2003)
	Re-Os isotope composition of the Sudbury sulfides	An overwhelming contribution (>90%) of crustal Os is required to explain the composition of the Sudbury ores	Highly radiogenic osmium reservoir required to form Os in the sulfide ores; a mantle contribution of Os would be swamped by the crustal Os in the ores.	Walker et al. (1991a-d); Dickin et al. (1992)
	S isotope data on mineralized rocks	Local heterogeneity in the S isotope composition of the Sublayer, Offset Dykes and Main Mass is consistent with derivation by impact melting and local assimilation of underlying country rocks	Systematic variations in S isotope composition recorded around the SIC provide evidence for relative homogeneity of the melt sheet, but some local assimilation of crustal S from the adjacent target rocks	Ripley et al. (2015); Ames et al. (2008)
	Iridium anomaly	Iridium anomaly from impact bolide	Iridium anomalies have been detected in proximal ejecta within the Onaping Formation	Mungall et al. (2004); Ames et al. (2005); Ames and Farrow (2007)

FIGURE 1.11 Shatter-Cones Developed in the South Range Footwall of the SIC in Coniston at Location NAD27-17N, 514018E 5147588N

SUPPORTING EVIDENCE FOR AN IMPACT EVENT

1. *Onaping Formation fallback breccias with glass shards and melt bodies*: These rocks are considered to be the product of fallback of fragments launched into the air above the crater, and the subsequent reworking of this material by partial melting and influx of water (Ames et al., 2002). The glass shards and melt bodies were interpreted by Peredery (1972a,b) to be produced by impact melting.

2. *Magmatic breccias at the base of the SIC*: The Sublayer Embayment rocks are considered to be a clast-laden unit developed at the base of the melt sheet in response to both local erosion of the fragmented country rocks by the magma, and accumulation of mafic–ultramafic inclusions (Keays and Lightfoot, 2004).

3. *Bulk composition of the SIC*: The rocks of the Main Mass and Offset Dykes have compositions consistent with a dominant contribution from the continental crust. Although the door is still open to the possibility of very small contributions of mantle-derived magma, there is no reason to call on endogenic contributions of magma. The support for a crustal origin for the Main Mass and the Offset Dykes rests in whole-rock geochemistry (Lightfoot et al., 1997a, 2001), and the isotope geochemistry of the Main Mass (Faggart et al., 1985a,b; Naldrett et al., 1986; Deutsch et al., 1989; Dickin et al., 1999; Cohen et al., 2000; Prevec et al., 2000; Darling et al., 2010a,b, 2012). A crustal origin for the ore deposits is supported by Re-Os and S isotope ratio studies (Walker et al., 1994; Ripley et al., 2015).

4. *Iridium anomaly*: Iridium anomalies have been detected in proximal ejecta within the Onaping Formation (Mungall et al., 2004; Ames et al., 2005; Ames and Farrow, 2007).

5. *Diamonds*: Polycrystalline aggregates (0.1–0.6 mm in diameter) of diamond with graphite have been detected in the Onaping Formation (Masaitis et al., 1997). The formation of this high-pressure polymorph is consistent with impact conditions.

6. *Complex hydrocarbon molecules*: C_{60} and C_{70} ring-structure fullerenes have been found in the Onaping Fomation (Mossman et al., 2003). These fullerenes are considered to represent the product of reformation of organic carbon at high pressure.

7. *Mineral psueodomorphs*: After high-pressure tridymite (Stevenson, 1963) and diaplectic glass, produced by shock melting of minerals (French, 1998).

The development of complex impact structures such as those observed in Sudbury is considered to be a product of rapid initial deformation and slower subsequent re-adjustment of the crust (see French, 1998; Grieve, 1994; Osinski and Pierazzo, 2010). This process is captured in a series of cross sections in Fig. 1.12A–E which record the principal stages in the development of a large impact crater, and the terminology associated with these stages. The process of formation involves the initial development of a large transient crater by the excavation of target rocks; a shock wave passing through the target rocks results in crustal melting to form a melt sheet and shock frictional melting of the country rock to form pseudotachylite and crater floor breccias (Fig. 1.12A,B). Initial development of the crater is rapidly followed by uplift at the center of the crater to produce a central uplift (Fig. 1.12C,D). Some minutes after the impact, the peripheral part of the crater collapses and the original melt and fallback debris are draped over the crater (Fig. 1.12C). The walls of the crater start to collapse inward along faults, and finally the crater takes on the morphology of a basin with a floor covered by the melt sheet and fallback; the melt sheet undergoes assimilation and cooling (Fig. 1.12D). The final crater form is shown in Fig. 1.12E and an inset shows detail of the topography of the melt sheet where it is connected to a radial dyke. The details of this model and how it is refined in the case of the Sudbury Structure is discussed in Chapter 5, but the model provides a context which provides the reader with a conceptual model for crater development.

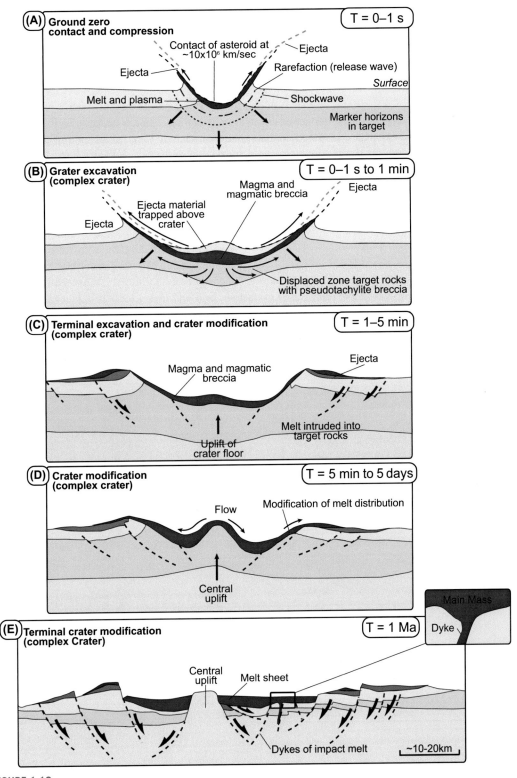

FIGURE 1.12

(A–E) Simple model for impact crater development in terrestrial structure scale to the Sudbury event.

Modified after Grieve (1994), French (1998), and Osinski and Pierazzo (2010).

SOME FEATURES NOT RESOLVED BY THE SUDBURY IMPACT MODEL

The evidence summarized here for impact at Sudbury is by no means an exhaustive list, but the reader can see a wealth of support for an impact origin. Notwithstanding, some aspects of Sudbury Geology remain unusually coincidental, and some scientists have found it hard to reconcile them with the impact hypothesis. One observation that has always been a thorn in the side of the Sudbury impact community is the serendipitous location of the Sudbury Structure at a major suture between the Proterozoic and Archean which contains an abundance of mafic volcanic and intrusive rocks (Muir, 1984). There are no other known 1.85 Ga magmatic events within this continental margin other than the Spanish River carbonatite which has a similar but very imprecise Rb–Sr age of 1838 ± 95 Ma (Fig. 1.13; Sage, 1987), so the Sudbury event was an almost unique expression of magmatic activity at 1.85Ga. Although there are both magnetic and gravity anomalies associated with the Sudbury Structure, these features are intrinsic to the melt sheet and associated crustal rocks. The development of a second gravity high to the east of Sudbury (Fig. 1.13) is likely the product of a hidden rift complex containing thick Archean banded iron formations which crop out in the Temagami Greenstone Belt and near Emerald Lake east of Sudbury (Fig. 1.13). These rocks coincide with the possible development of a hidden graben structure trending between Sudbury and Temagami (Milkereit and Wu, 1996). The coincidence of the SIC with especially mafic crust has been argued to be an unusual coincidence, but the fact remains that coincidences such

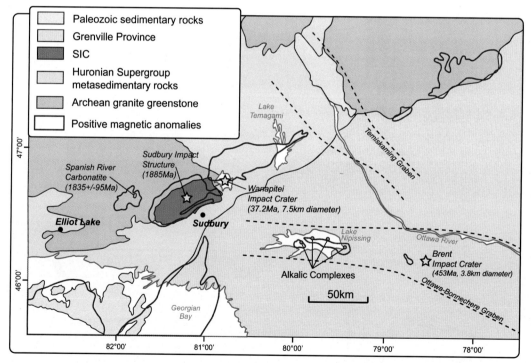

FIGURE 1.13 Regional Setting of the Sudbury Structure in the Context of Potential Field Data and Tectonic Corridors

Modified after Muir (1994); Easton et al. (2010); and Sage (1996).

as this do happen, and there is no evidence whatsoever to explain this coincidence through endogenic processes. Moreover, the presence of mafic crust in the region was a bonus from the perspective of target rocks which are enriched in metals that are found in the Sudbury ore deposits, whether as proto-ores (Pattison, 1979) or simply the target rocks for crustal melting (Vogel et al., 1998a-b, 1999).

SYNTHESIS OF THE APPROACH USED IN THIS BOOK

The objective of this book is to assemble a current understanding of the geology of the SIC, and the sequence of events that gave rise to the observed rocks and the ore deposits. The objectives can be set out in a concise list; namely:

1. What was the sequence of events at Sudbury which resulted in the formation of the SIC and the associated ore deposits?
2. Where did all of the metal and sulfur in the ores come from, and what processes allowed such efficient concentration of these elements?
3. What processes resulted in the formation of such a wide range of ore types at Sudbury, and what controls the location of the ore bodies?
4. What was the relative importance of syn-magmatic tectonic activity versus postmagmatic deformation in controlling the morphology of the Sudbury Structure and the associated ore bodies?
5. How can this information be used to help inform a new generation of geologists and their models in a way that encourages fundamental advances in understanding the geology of the Sudbury Structure?
6. What critical knowledge can be taken away from Sudbury to understand the mechanism of formation of magmatic sulfide ore deposits, and the best tools to apply in order to uncover the next generation of discoveries?

This book examines examples of different ore deposits around the SIC and classifies them into five groups; namely (1) basal contact-related massive nickel-copper sulfide mineralization; (2) footwall-hosted and contact-related copper-nickel-PGE sulfide mineralization; (3) Offset Dyke-associated Ni–Cu–PGE sulfide mineralization; (4) ores hosted in QD melt bodies in the pseudotachylite breccia belt surrounding the SIC; (5) sulfide mineralization modified by deformation and/or hydrothermal activity. The case studies presented are used to show how many of the deposits have evolved in similar ways, and how the observations can be used to better understand the magmatic class of nickel sulfide ore deposits.

An understanding of the geology of the SIC and the associated ores has helped to inform geological, geochemical, and geophysical methods in the exploration for new ore bodies and extensions to known ore deposits. This book shows examples of deposits located by the application of this geoscience understanding, and how the information can be used to flag more prospective versus less prospective environments for exploration.

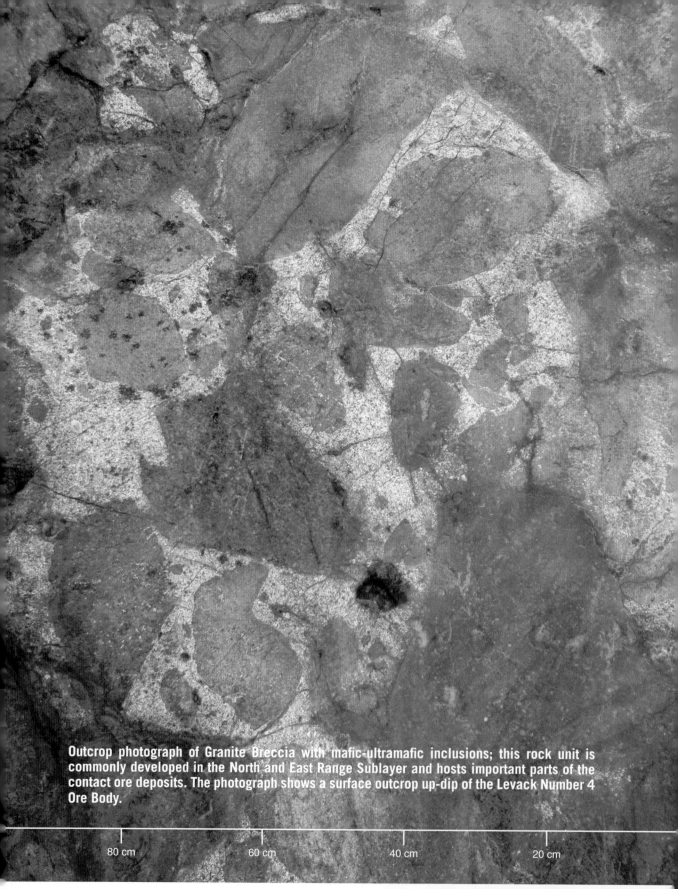

Outcrop photograph of Granite Breccia with mafic-ultramafic inclusions; this rock unit is commonly developed in the North and East Range Sublayer and hosts important parts of the contact ore deposits. The photograph shows a surface outcrop up-dip of the Levack Number 4 Ore Body.

80 cm   60 cm   40 cm   20 cm

A SYNTHESIS OF THE GEOLOGY OF THE SUDBURY STRUCTURE

INTRODUCTION

In the 1880s, the hunt for mineral wealth in the Sudbury Region was driven by gossans and the near-surface sulfide mineralization with few preconceived ideas about why the ore deposits were there, what processes created them; and most importantly, how to find more of them. The exploration approach evolved with a growing understanding of the importance of both the magmatic breccias and the geological environments that contain the ore deposits (eg, Zurbrigg, 1957; Souch et al., 1969). A progressive shift took place from a simple two-dimensional view of the mineralization and the host rocks, to a three-dimensional understanding of how these surface manifestations of mineralization extend to depth along mineralized trends, and then how these characteristics could be used to identify mineralized zones that are blind to the surface.

The Creighton Deposit is a good example of how understanding has evolved between the surface discovery of mineralization in 1886 to the current understanding of a deposit that extends to a depth of more than 3 km. Mining of this deposit commenced with an open pit and then moved to progressively more advanced methods of underground mining that required a commensurate growth in the understanding of the trajectory of the ore zone and the quality of the mineral based on diamond drilling (Boldt, 1967). The exploration process evolved to include detailed exploration drilling, systematic description of the drill core, three-dimensional modeling of the ores and the host rocks, and detection of mineralization using new geological models and new geophysical and geochemical tools (eg, Lightfoot, 2007; King, 2007; McDowell et al., 2007). The Creighton ore deposit is now mined to a depth of 2.5 km and explored to a depth of 3 km. The rich ore bodies discovered in the process of exploration have triggered the development of mine technologies to extract these ores in a profitable way.

Mapping of the surface and subsurface geology of the Sudbury Structure provides an important basis on which in which to understand the sequence of events in the evolution of the Sudbury Structure and its ore deposits. A series of progressively more detailed surface geological maps have been created over the past 125 years (Bell, 1891a–d; Collins, 1937a,b; Cooke, 1946; Card, 1965, 1968; Dressler, 1984a; Ames et al., 2005). The more recent maps have benefited from some knowledge of the three-dimensional geology of the Sudbury Structure arising from exploration and geophysical input provided by mining companies. It is the development of methods for the three-dimensional visualization of geological relationships in diamond drill core anchored by maps created at surface and underground that has triggered the most recently advances in the knowledge of Sudbury geology that are the subject of this chapter.

This chapter is designed to provide a basic framework of the regional and local geology of the Sudbury Structure. The regional geological setting of the target rocks is described so that the reader can understand the relevance of impact-generated melts and breccias to the distribution of ore deposits. The geology of the SIC and the associated breccias are described with special emphasis on the

Nickel Sulfide Ores and Impact Melts. http://dx.doi.org/10.1016/B978-0-12-804050-8.00002-X

environments in which magmatic sulfide ore deposits were formed. The postimpact evolution of the Sudbury Region is reviewed and the kinematics and chronology of the deformation events influencing the present-day geometry of the Sudbury Structure is highlighted in the context of postimpact modification of the ore bodies. The chapter closes with a timeline for the regional evolution from 2.8 Ga to the present, and a possible timeline for the evolution of the Sudbury Structure between the impact event and ~100 Ma postimpact. This sets the scene for Chapter 3 where the petrology and geochemistry of the SIC is provided to support a refined model for the S-saturation history and evolution of the impact melts.

AN IMPACT EVENT AT THE SOUTHERN MARGIN OF THE SUPERIOR PROVINCE CRATON

It has long been recognized that cratonic margins provided a locus of development of nickel sulfide ore deposits (eg, Maier and Groves, 2011). This is illustrated very clearly in North America, where the Proterozoic-aged Thompson, Raglan, Duluth, and Sudbury nickel sulfide ore deposits are located at the margin of the Superior Province Craton (Fig. 2.1) whereas the Proterozoic-aged Voisey's Bay deposit is located at the edge of the Nain Craton (eg, Lightfoot and Evans-Lamswood, 2015).

The linkage between nickel deposits and cratonic margins was first recognized by the geologists working for mining companies such as Inco and Western Mining whose exploration strategies over the last ~25 years was based on searching for mineralization in these settings. The importance of cratonic margins has more recently has been recognized in the literature (eg, Maier and Groves, 2011; Begg et al., 2011). Fig. 2.1 shows the regional gradient of the gravity response to illustrate that the Superior and Nain Province craton margins are also the location of large amounts of dense mafic-ultramafic rock which coincide with the belts that contain the nickel sulfide ore deposits. In the absence of a good understanding of Sudbury geology and the overwhelming evidence for a meteorite impact origin (Table 1.4), it would be quite natural to include the complex and its associated ores as a product of a Proterozoic mantle event at the cratonic margin. However, a search for mantle contributions to the igneous rocks of the Sudbury Structure has failed to demonstrate any convincing evidence or requirement for contributions of mantle-derived magmas in the formation of the Sudbury Structure and the associated ores (eg, Lightfoot et al., 1997b).

Although the craton margin model is a useful broad guide to prospective belts for nickel, it is the more detailed structure of the crust on a belt scale that is required for effective exploration. Strike-slip fault zones likely provided the pathways of mantle-derived magmas from depth to near surface. These pathways controlled the development of transtensional space in which the small intrusions with nickel sulfide ore deposits were emplaced (Lightfoot and Evans-Lamswood, 2015). Sudbury appears to be an anomaly in the context of current models that link the development of nickel sulfide ore deposits to Proterozoic-aged mafic-ultramafic rocks at the boundaries major cratons.

The Sudbury Structure was formed astride the southern margin of the Archean-aged Superior Province Craton and the Proterozoic Southern Province. The Archean comprises both granitoid rocks and mafic-felsic gneisses, whereas the bimodal mafic and felsic volcanic rocks and sedimentary rocks of the Southern Province formed at a rifted cratonic margin. These rocks are host to at least two important mafic intrusive events that pre-date the Sudbury Structure. Of all the possible terrains that the asteroid could have serendipitously hit, it was a target of diverse and unusual crust at the southern margin of the Superior Province Craton that provided the framework for the formation of the Sudbury Structure.

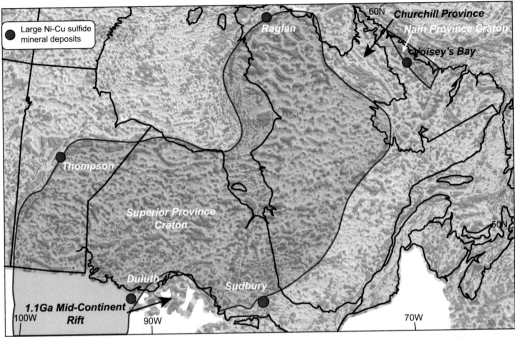

FIGURE 2.1 The Boundary of the Superior and Nain Province Cratons (after Maier and Groves, 2011) Superimposed in the Analytic Signal of the Gravity Response Derived From Public-domain Data (Geological Survey of Canada)

The location of some of the largest nickel deposits is shown (red circles), and they coincide with the margin of the Superior and Nain Province Cratons.

The geology of the target area at the time of the Sudbury impact event is important; the diversity in rock types and structure was likely a control on the primary shape of the Sudbury Structure, the thickness of the SIC and the extent and geometry of the Offset Dykes, as well as the diversity in breccia types. The presence of a target stratigraphy of mafic volcanic and intrusive rocks at the point of impact coupled with many small examples of magmatic Ni–Cu–PGE-rich sulfides in association with the Paleoproterozoic intrusions are significant, as they may have enhanced the metal and sulfur contribution from the target rocks to the melt sheet (Keays and Lightfoot, 2004), and thereby contributed to the formation of economic ore deposits.

On the one hand there appears to be overwhelming evidence for an impact origin of the Sudbury Structure; on the other hand, the impact happened in just the right place to involve target rocks of especially mafic-ultramafic composition (and possibly with preexisting mineralization).

The topography of the Sudbury Region at the time of impact is not fully understood. Whether the target topography was a mountain range or a more mature subdued topography adjacent to a continental shelf has an important impact on the development of the impact crater and the potential role of saline hydrothermal fluids derived from seawater. Notwithstanding, the far-field effects of the Sudbury impact event is recorded in distal tsunami and/or ejects deposits created by the impact event (Cannon et al., 2010). The weight of evidence indicates that the Sudbury impact event occurred at the edge of the Superior and Southern Provinces in an area directly adjacent to the continental shelf.

Through the chapter, I use a combination of surface geological data and information gathered from exploration and three-dimensional visualization to address the following questions:

1. How much diversity exists in the geology of the country rocks that formed the target of the Sudbury impact event, and what is the significance of the geological setting with respect to melt sheet and ore formation?

2. What are the characteristics of the wide diversity in breccias in the Sudbury Structure, how were these breccias produced, and which types of breccia occur in association with the ore deposits?

3. What is the importance of structure in controlling the geometry of the Sudbury Structure and the distribution of the ore deposits? Were preexisting structures re-activated in response to the impact event, which structures were active during crater modification, and what effects did far-field magmatic and tectonic events have on the evolution of the Sudbury Structure?

4. What specific processes happened in response to by the Sudbury impact event and what role did they play in the localization of the magmatic sulfides into ore deposits?

5. Is it possible to establish a framework timeline for the regional evolution of the crust in the area of the Sudbury Structure, and what were the specific timelines of events responsible for the impact genesis of the SIC and the associated ore deposits?

AN OVERVIEW OF THE REGIONAL GEOLOGICAL SETTING OF THE SUDBURY STRUCTURE

As described in Chapter 1, the Sudbury Structure comprises three principal rock groups, namely: (1) The magmatic rocks of the Main Mass of the SIC, the magmatic breccias at the lower and outer margin of the SIC (the Sublayer and Offsets) and the melt contributions in breccias overlying the SIC (the Onaping Formation). (2) Metasedimentary rocks of the Onwatin and Chelmsford Formations which overlie the SIC. (3) The brecciated country rocks of the impact target developed within a ~100 km radius of the SIC, but dominantly within less than 1 kilometers of the base of the SIC. It is the regional geological setting of this group of rocks that is discussed in this section.

The Sudbury Structure is positioned north of the Grenville Front, and conspicuously astride the Archean and Proterozoic terrains of Central Ontario (Fig. 2.2). The Proterozoic rocks to the south of the Sudbury Structure record the evolution of a continental rift magmatic and sedimentary belt formed between ~2.45 and 2.2 Ga, termed the Southern Province. The geophysical manifestation of this event is recorded along a major corridor extending ESE to WNW over ~400 km (Fig. 1.13). The igneous rocks of the Southern Province comprise a package of volcanic rocks including basalt and rhyolite, comagmatic differentiated anorthosite-gabbro intrusions of the East Bull Lake Group and granitoid intrusions such as the Murray and Creighton plutons, and a suite of gabbroic to melagabbroic intrusions of the Nipissing and Sudbury Gabbro events (Fig. 2.2). Some of the mafic intrusive rocks of the Southern Province are mineralized with disseminations of magmatic Ni–Cu–PGE sulfide mineralization, but the concentrations of these sulfides are small and highly disseminated when compared to the deposits associated with the SIC.

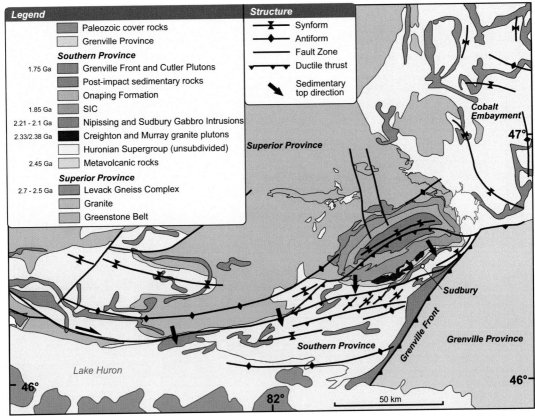

FIGURE 2.2 Regional Geological Setting of the Sudbury Structure Showing the Location of the Archean and Proterozoic Country Rocks and the Younger Grenville Province

Modified after Riller (2005).

The timing of deformation of the Southern Province to form dominantly east- to northeast-trending fold axes is not firmly established. Much of the 2.23 Ga Nipissing Diabase (Corfu and Andrews, 1986; Lightfoot and Noble, 1992) and the Sudbury Gabbro is folded by the ENE trending structures, and so major deformation may have occurred after the emplacement of these gabbroic rocks (Card et al., 1972). The 2400–2140 Ma Blezardian Orogeny (Stockwell, 1982; Riller and Schwerdtner, 1997; Raharimahefa et al., 2014) and the early stage of the 1900–1700 Ma Penokean Orogeny (Bennett et al., 1991) likely created the major structural fabrics in the Southern Province before the 1.85 Ga Sudbury event. The exact timing of the Sudbury event relative to deformation and metamorphism is also not firmly established, but it is known that metasedimentary and metavolcanic rock fragments with staurolite are contained within the SIC. On these grounds, it is believed that impact occurred during a possible hiatus between early and late Penokean Orogenic events.

At the time of impact, the Sudbury Region was located at the margin of a Paleoproterozoic sea in which a thick assemblage of sedimentary rocks including shales and iron formations had formed prior to, and after the Sudbury event (Fig. 2.3A, B). These sedimentary rocks contain an ejecta layer

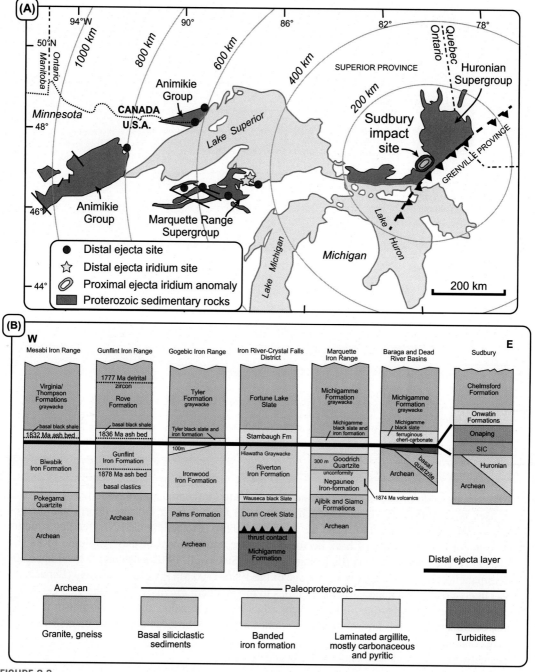

FIGURE 2.3

(A) Distribution of the distal ejecta sites of Sudbury age in Michigan and Minnesota and location of Proterozoic sedimentary basins; (B) Correlation of stratigraphy in the iron ranges of the Lake Superior Region and the location of the distal ejecta layer.

(A) After Cannon et al. (2012); (B) After Cannon et al. (2012); Addison et al. (2005); Fralick et al. (2002).

comprising chaotic flow breccias and glass shards (Addison et al., 2005, 2010; Cannon et al., 2012). These rocks are now present up to 1000 km from the Sudbury Structure in northwestern Ontario, Michigan, and Minnesota. They have been attributed to a continental scale tsunami created by the Sudbury impact event.

The central basin of the Sudbury Structure comprises breccias and sedimentary rocks deposited in progressively deeper water. There is also a growing weight of evidence to suggest that the magmatic ore deposits have at least some hydrothermal alteration overprint produced by halogen-rich fluids possibly derived from seawater (eg, Hanley, 2002; Ames and Farrow, 2007). It would therefore appear plausible that the Sudbury Structure formed at a continental margin, but the there is no evidence of pre-1.85 Ga unconsolidated sedimentary material like that shown in Fig. 2.3B entrained in the magmatic breccias or beneath the lower contact of the SIC, so the impact event is unlikely to have occurred outboard of the continental margin.

The Sudbury Region rapidly developed into a basin shortly after the impact event, and it is likely that the deep water sedimentary rocks preserved in the core of the Sudbury Structure were originally developed over a much wider area between ~1.85–1.82 Ga (Long, 2004, 2009; Ames et al., 2008).

The 1.85 Ga Sudbury Structure was buckled to generate the broad elliptical shape of the Sudbury Structure with tight fold hinges to the south and a gentler buckling to the west, north and east of the SIC (Thomson, 1957a; Zolnai et al., 1984; Lenauer and Riller, 2012a,b). The oval shape of the Sudbury Basin was created by reverse faulting and thrusting such that the southern part of the Sudbury Structure was transported over the northern part (Burrows and Rickaby, 1930; Shanks and Schwerdtner, 1991a,b). The rocks of the Sudbury Structure were also affected by several other families of faults and folding (Roussel et al., 2002). The deformation history of the SIC was protracted, yet the Sudbury Basin retains a high degree of integrity of the outer contact as a single elliptical geological feature which was presumably a response to the competent behavior of the Main Mass during faulting and folding. The deformation events which produced the present configuration of the SIC likely included major thrusting during the Penokean and Mazatzal orogenies, and a minor contribution to deformation and metamorphism to the south of the Sudbury Structure from the Yavapai and Grenville orogenic events (Riller, 2005; Raharimahefa et al., 2014). A more detailed account of the structural geology of the region is given later in this chapter.

The Sudbury Structure is cut by two major regional dyke series, namely: (1) 1.23 Ga NW–SE-trending olivine diabase and diabase dykes (Dudas et al., 1994; Shellnutt and MacRae, 2011) which are locally metamorphosed by the Grenvillian event (Bethune and Davidson, 1997); (2) W–E-trending diabase dykes that cut the southern margin of the SIC broadly parallel with the major W–E structures; these dykes are locally termed "trap dykes" and they are cut by the olivine diabase (Cooke, 1946).

GEOLOGY OF THE SUDBURY STRUCTURE

The geological framework of the Sudbury Structure is shown in Fig. 2.4A together with a simplified stratigraphic section shown in Fig. 2.4B; these diagrams are designed to provide a guide to the reader as the rocks are described according to the following stratigraphic groupings: (1) The country rocks of the Sudbury Region that predate the formation of the SIC; (2) The SIC (the Main Mass, the Offset Dykes, and the Sublayer); (3) The breccias in the footwall of the SIC; (4) The Whitewater Formation.

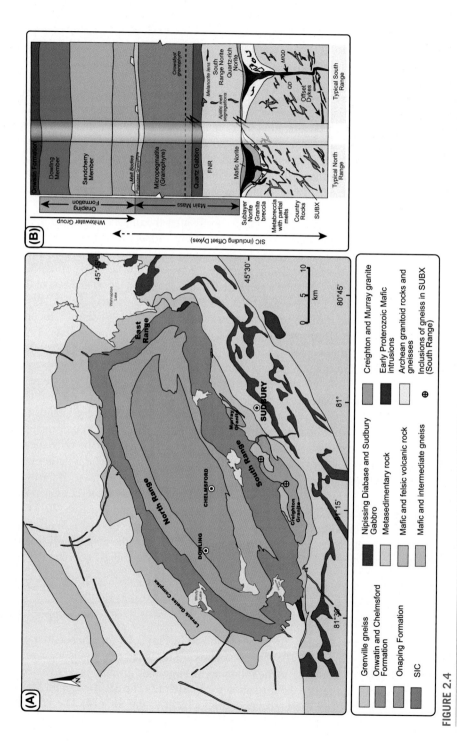

FIGURE 2.4

(A) Geology of the country rocks surrounding the SIC. (B) Summary of the stratigraphy of the Sudbury Structure as a framework for the nomenclature and sequence of topics discussed in Chapter 2.

(A) Modified after Dressler (1984a) and Ames et al. (2005).

THE TARGET COUNTRY ROCKS

The country rocks can be broadly broken out into Archean rocks to the north and east, and Paleoproterozoic rocks to the south of the SIC (Fig. 2.4A). The rocks to the north broadly comprise an older series of granitoid intrusions that are typically separated from the SIC by a group of mafic, intermediate, and felsic gneisses which belong to the Levack Gneiss Complex (James and Golightly, 2009; Fig. 2.4A). The Levack Gneiss Complex includes mafic enclaves, heavily deformed gabbroic intrusions (eg, the Joe Lake Gabbro), and much smaller ultramafic bodies (Moore et al., 1993, 1994, 1995). Many of these rocks may have been exhumed to high crustal levels prior to the Sudbury impact event (Prevec et al., 2005; Siddorn and Halls, 2002).

The Paleoproterozoic sedimentary and volcanic rocks of the Southern Province to the south of the SIC comprise a package of tholeiitic basaltic rocks which comprise the Elsie Mountain Formation and rhyolite flows and tuff beds which comprise the Copper Cliff Formation (Fig. 2.4A). These are separated by the highly brecciated, deformed and metamorphosed Stobie Formation that comprises metavolcanic and pelitic sedimentary rocks with large enclaves of basalt, amphibolite and gabbro typically partially enveloped in pseudotachylite breccia. The overlying Huronian Supergroup sedimentary rocks are less strongly deformed and show a broad progression from shallow epicontinental to deepwater sedimentary rocks with three periods of diamictite formation related to glaciation events.

The broad stratigraphy of the target rocks is described below where the rocks have been assigned to three major groups, namely: (1) Archean granitoid rocks and gneisses; (2) Felsic and mafic volcanic rocks of the Huronian Supergroup, and; (3) Sedimentary rocks, also of the Huronian Supergroup.

ARCHEAN GRANITOID ROCKS AND GNEISSES OF THE SUPERIOR PROVINCE

Archean rocks are located northwest of the Sudbury Basin (Fig. 2.4A). They are part of a Neoarchean greenstone-granite terrane termed the Abitibi Subprovince. In the Sudbury area the Archean rocks comprise:

1. Highly deformed basalt and andesite flows, tholeiitic and calc-alkaline tuff breccias and sulfide-bearing volcanogenic metasedimentary rocks of the Benny Greenstone Belt (Card and Jackson, 1995);
2. The Cartier batholith is part of the Algoma suite of large batholiths in the southern part of the Superior Province (Card, 1979). These intrusions were emplaced at high crustal levels as extensive ~5 km thick sheets during the late stages of the Kenoran Orogeny (2642 Ma, Meldrum et al., 1997). The Cartier batholith consists of coarse-grained leucogranite with local rapakivi and subporphyritic textures (Card and Wodicka, 2009).
3. The Levack Gneiss Complex that forms a dome-shaped collar up to 5 km wide, around the North and East Ranges of the SIC (Fig. 2.4A). The gneisses comprise tonalite-granodiorite orthogneiss with abundant inclusions, boudins and layers of mafic to diorite gneiss, diorite, gabbro, anorthosite, amphibolite, iron formation and pyroxenite. A typical example of this gneiss in the footwall of Coleman Mine is shown in Fig. 2.5A where the pink gneiss is cut by Sudbury Breccia and sulfide veins. The Levack Gneiss Complex was metamorphosed to granulite facies and then retrograde amphibolite facies metamorphism, a regional greenschist facies metamorphism were overprinted (James et al., 1992a,b; James and Dressler, 1992.). High-grade metamorphism occurred at 6.0–8.5 kb and 750–900°C, corresponding to depths of 20–30 km

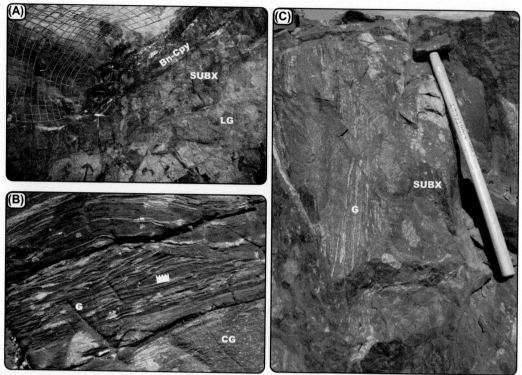

FIGURE 2.5

(A) Levack Gneiss (LG) cut by Sudbury Breccia (SUBX) and a bornite-chalcopyrite stringer vein (Bn-Cpy) in the footwall at Coleman Mine. (B) Enclave of gneiss (G) locally surrounded by an envelope of Sudbury Breccia (not shown) within a porphyritic phase of the Creighton Granite (CG), approximately 700 m south of the Creighton Trough; the extent of deformation in this gneiss raft is quite different when compared to Huronian-aged metasedimentary pendants. (C) Block of orthogneiss (G) within Sudbury Breccia (SUBX); U–Pb zircon ages indicate an Archean contribution and the extent of deformation is atypical of Huronian metasedimentary blocks; the debate remains open regarding the source of the gneiss fragments shown in (B) and (C).

(James et al., 1992a,b). Subsequent retrograde metamorphism involved the breakdown of garnet to plagioclase and biotite at 1.5–3 kb at 550°C; the timing of uplift may predate the Sudbury event (Prevec et al., 2005). Krogh et al. (1984) obtained a U-Pb zircon age for tonalite gneiss of 2711 Ma and a younger age of 2647 Ma for a mobilized felsic enclave.

The Levack Gneiss Complex consists of contorted and boudinaged lenses of mafic gneiss within a broader domains of felsic gneiss. The contact zones between mafic gneiss and felsic gneiss are often exploited by psuedotachylite veins generated by the impact event, and these contacts tend to play an important role in controlling the localization of footwall mineralization.

The Levack Gneiss Complex contains metagabbro and amphibolite gneiss enclaves; although there are no convincing examples of fresh ultramafic rock in the distal footwall of the SIC, there are altered and deformed websterites and wherlites in the footwall of the North Range of the SIC at Fraser Mine (the Fraser Intrusion; Moore et al., 1993, 1994, 1995) and bodies of fresh olivine melanorite to lherzolite in the immediate footwall of the McCreedy West Deposit (Moore et al., 1995) that may be produced by the Sudbury event from melanocratic target rocks.

Enclaves and bodies of metagabbro that are collectively grouped as "Gray Gabbro" occur in the breccias to the south of the SIC, and they possibly comprise fragments of an Archean protolith, but they are more likely Early Proterozoic mafic target rock that have been exhumed by impact processes into the breccia belt developed in the footwall of the South Range. These rocks are typically fine to medium grained gabbros with variable degrees of metamorphism and partial melting. These rocks do not contain primary magmatic sulfide ores, but they do contain anomalous concentrations of sulfide mineralization associated with structures and they are the dominant footwall rock type at Crean Hill, Little Stobie and Blezard (eg, Gibson et al., 2010).

The development of the Levack Gneiss Complex outside of the North and East Range of the Sudbury Structure is restricted to controversial examples of highly contorted mafic gneiss enclaves in the Sudbury Breccia (SUBX). These rocks may be an exhumed fragment of the Levack Gneiss Complex (Petrus et al., 2012; Fig. 2.5B,C) or fragments of Huronian metasedimentary rock that have been extensively deformed and melted (Bleeker et al., 2015). They are located in domains of SUBX to the south of Creighton Mine and south of the Tam O'Shanter contact deposit (Fig. 2.4A). Metasedimentary enclaves developed near to these gneises show weak deformation and clear graded bedding typical of the Huronian Supergroup metasedimentary rocks; in contrast, the enclaves of gneiss do not record clear evidence of a Huronian provenance.

Between Sudbury and Temagami is a broad WSW–ENE-trending domain with a strong positive gravity and magnetic response (Fig. 1.13 and Muir, 1984); this anomaly has often been considered as a possible twin to the Sudbury Structure. The interpretation of this feature is still debated. Various interpretations exist for these rocks, namely: (1) The anomalies are considered to be a stratigraphy of hidden rift-related mafic volcanic rocks (Milkereit and Wu, 1996); (2) The anomalies are interpreted to represent a series of hidden mafic-ultramafic intrusions (with dubious linkages to the Sudbury event; Muir, 1984), and (3) The anomalies represent segments of an Archean greenstone belt that contain iron formation and mafic intrusions; this may be an extension of the Temagami and Emerald Lake Greenstone Belts. Although greenstones of this type are not developed in the immediate Sudbury Region, the trend of this belt extends towards Sudbury and it is possible that these rock types were present in the Sudbury Region at the time of the impact event.

HURONIAN SUPERGROUP METAVOLCANIC AND METASEDIMENTARY ROCKS

The 2.45–2.22 Ga Huronian Supergroup comprises a thick package of sandstones, mudstones, carbonates and conglomerates with lesser mafic and felsic volcanic rocks that were regionally metamorphosed to greenschist facies and locally metamorphosed to amphibolite facies.

The stratigraphy of the Huronian Supergroup is summarized in Fig. 2.6A (after Long et al., 1999; Young et al., 2001, and Long, 2004), and the distribution of the main rock groups in part of the Southern Province is shown in Fig. 2.2, and in the immediate footprint of the south part of the Sudbury Structure in Fig. 2.7.

A thick package of volcanic rocks and deep-water sedimentary rocks were formed within a transtensional basin produced by left-lateral movement along the paleo-Murray Fault system (Long and Lloyd, 1983; Long, 2004). These rocks are allocated to the Elliot Lake Group, and they have been subdivided into the Elsie Mountain, Stobie and Copper Cliff Formations (Fig. 2.6A, Fig. 2.7).

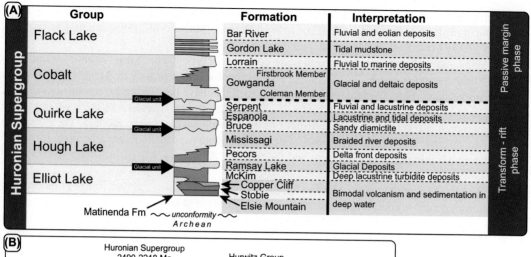

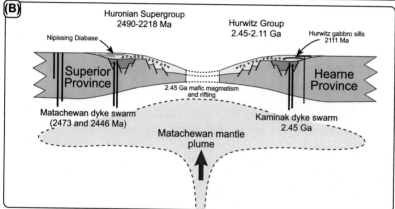

FIGURE 2.6

(A) Lithostratigraphy and tectonic history of the Huronian Supergroup. The interpretation of the units follows the description in Young et al. (2001). (B) Tectonic setting of the Huronian Supergroup at the southern-rifted margin of the Superior Craton.

(A) Based on Long (2004). (B) After Bleeker (2004).

The Elsie Mountain Formation consists of a sequence of massive and pillowed flows of tholeiitic basaltic rocks (Figs. 2.6A and 2.8A). Flow-top breccias and minor volcaniclastic rock units occur between 20–80 m thick basalt flows. Laminated mudstone intervals (2–3 m thick) indicated a deep-water setting. The Elsie Mountain Formation basalts have been genetically linked to mafic intrusions of the East Bull Lake Type (Vogel et al., 1998a,b) and appear to represent an enriched type ocean-ridge basaltic magma produced by a far-field mantle plume event (Figs. 2.6A and 2.6B).

The contact of the Elsie Mountain Formation with overlying volcanic and volcaniclastic rocks in the Stobie Formation is transitional and is placed by Innes (1978a) where interflow sedimentary rocks

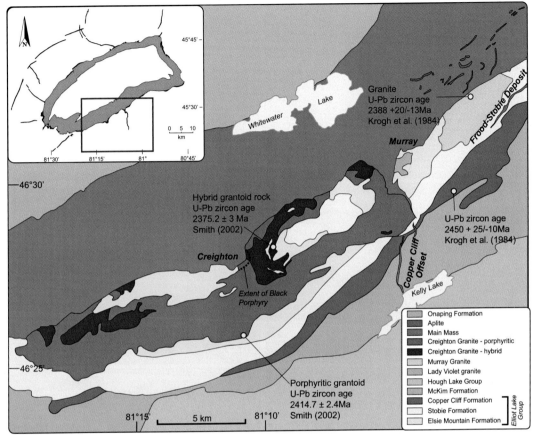

FIGURE 2.7 Simplified Geology of the Footwall to the South of the SIC Showing the Principal Huronian Supergroup Formations, the Position of the Main Mass, and the Location of the Creighton and Murray granites

The location of aplitic melts which cross cut the SIC above the Murray pluton are shown from Dressler (1984a), but they are known to extend to the west above the Creighton Pluton.

Based on Ames et al. (2005); Krogh et al. (1984), and Smith (2002).

make up more than 15% of the section, or where individual flows are less than 20 m thick. The Stobie Formation consists predominantly of felsic to mafic metavolcanic rocks, pyroclastic breccias and pelitic metasedimentary rocks (Figs. 2.6A and 2.8B). Individual flows are often massive and develop flow-top breccias and pillows. Siltstone and muddy-sandstone occur as interflow sediments in the middle and upper parts of the formation.

The Elsie Mountain Formation is overlain by the Copper Cliff Formation which consists of quartz-feldspar crystal tuff and felsic lithic tuff and tuff breccia with minor flow-banded rhyolites (Card et al., 1977; Figs. 2.6A and 2.8C). These felsic volcanic rocks are possibly syngenetic with the Creighton and Murray Granites.

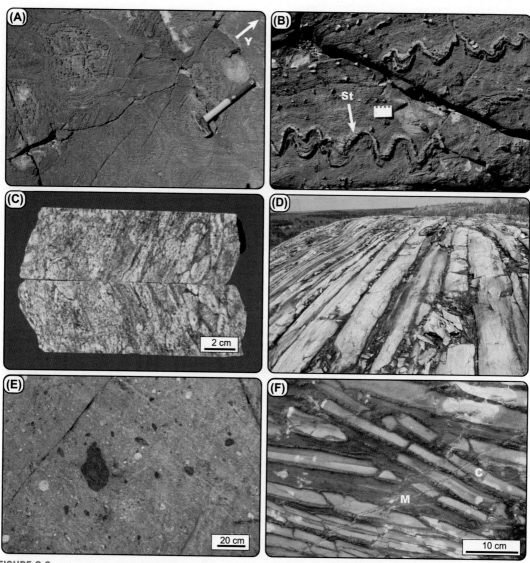

FIGURE 2.8

(A) Pillow basalts in the Stobie Formation, ~250 m from the contact at the Murrray Deposit, Sudbury; direction of younging (Y) is shown. As shown in the photograph, the pillows have competent cores that are strongly recrystallized with the development of amphibole porphyroblasts (this type of rock is quite common in the immediate footwall of the SIC, and it is termed Sudburite). (B) Stobie Formation folded pelitic metasedimentary rocks with bands containing staurolite crystals (St) replaced by chlorite; west of Frood Mine in the Breccia Belt. (C) Flow banding and lapilli tuff of the Copper Cliff Rhyolite, Copper Cliff. (D) Turbidites of the Mississigi Formation south of Highway 17 in the Algonquin area of Sudbury; note the fine bands of silt have acted as slip-planes with development of pseudotachylite between the competent beds or arkose. (E) Locally layered diamictite of the Ramsay Lake Formation; south of Kelly Lake. (F) Espanola Formation calcareous siltstone and limestone (C) within a mudstone matrix (M), Parkin Township.

Collectively, the Elsie Mountain, Stobie and Copper Cliff Formations are located proximal to the SIC, and they have been extensively modified by syn-impact brecciation, and postimpact thermal metamorphism. This modification overprints an earlier phase of thermal metamorphism and deformation related to regional scale orogenesis. The extent of this deformation, irrespective of the effects of the impact event at 1.85 Ga appears to be more extreme than that found in the remainder overlying Huronian Supergroup Formations, and this hints at a regional deformation event between the formation of the Copper Cliff and McKim Formations.

In the Sudbury Region, the Copper Cliff Formation is overlain by the McKim Formation (Figs. 2.6A, 2.7). Card et al. (1977) mapped mudstones, muddy sandstones, argillites, and thin-bedded sandstone, and proposed that many of these units were deposited as turbidites. The Elliot Lake Group unconformably overlies the Copper Cliff Formation to the west of Sudbury; the rock types are dominantly arkose and conglomerate that accumulated on alluvial fans.

The Hough Lake Group overlies the Elliot Lake Group, and comprises the Ramsey Lake, Pecors and Mississagi Formations (Figs. 2.6A, 2.8D). The Ramsay Lake Formation comprises a thick sequence of diamictites that were deposited as a glacial till (Long, 2009; Figs. 2.6A and 2.8E). The Pecors Formation rests conformably on the Ramsay Lake Formation, and is characterized by thinly laminated mudstones and muddy-sandstones of deltaic origin. The Mississagi Formation is between 1600 and 3400 m thick to the south of the SIC. It is characterized by medium- to coarse-grained sandstone of arkosic to subarkosic composition, with abundant planar and trough cross-stratification. The formation is predominantly fluvial in origin, and was deposited from shallow braided rivers that flowed from a series of tributary basins in the Cobalt Embayment to the east.

The Quirke Lake Group consists of the Bruce, Espanola and Serpent formations exposed in the Sudbury District. The Bruce Formation is a diamictite with thin beds of sandstone and mudstone. The Espanola Formation is poorly developed in the Sudbury Region, and consists of limestone, siltstone, and calcareous siltstone (Figs. 2.6A and 2.8A, F). The Serpent Formation is developed west and south of Sudbury and consists of fine-grained sandstone. The Quirke Lake Group comprises sedimentary rocks formed in a shallow water epicontinental setting.

The Cobalt Group is dominantly developed east of Sudbury, and comprises the Gowganda and Lorrain Formations; this group rests unconformably on the Archean basement (Fig. 2.6A). The Gowganda Formation consists of a heterogeneous sequence of diamictite, sandstone, siltstone and mudstone. The Lorrain Formation consists of planar and trough cross-stratified fine- to coarse-grained sandstone with local pebble horizons. This group is largely formed in a fluvial setting.

The overall stratigraphic succession of the Huronian Supergroup as shown in Fig. 2.6A is consistent with early rift-related magmatism developed at the southern margin of the Superior Province (Fig. 2.6B; Long, 2009). This initial rifting event was followed by subsidence and deep water sedimentation, followed by uplift accompanied by the formation of shallow water epicontinental sedimentary rocks. The uplift continued, and deltaic and fluvial sedimentation was interrupted by three periods in which glacial sedimentary rocks were deposited (Fig. 2.6A).

In the area of the Sudbury impact event, the target rocks comprised principally felsic and mafic volcanic rocks of the Elsie Mountain, Stobie and Copper Cliff Formations, and fine-grained sandstones and siltstones that belong to the Stobie, McKim, Ramsey Lake, Pecors and Mississagi Formations. These rocks comprise the dominant Huronian Supergroup rocks that underlie the Subury Igneous Complex in the South Range (Fig. 2.7), and also important potential crustal contributions to the melt sheet.

MAFIC AND ULTRAMAFIC INTRUSIVE ROCKS IN THE SUDBURY REGION

The mafic and ultramafic intrusions in the country rocks surrounding the SIC are important in the sense that they comprise the most primitive known target rocks from which the SIC and its ores were formed, and they represent some of the principal inclusion types found in the magmatic breccias of the SIC (Lightfoot et al., 1997a). The distribution of the principal rock groups shown in Fig. 2.2 and comprise the East Bull Lake, Nipissing Diabase and Sudbury Gabbro type intrusions.

Early Proterozoic East Bull Lake Type Intrusions

Intrusions of the ~2480 Ma East Bull Lake intrusive suite occur as a ENE-trending belt along the boundary of the Archean Superior and Proterozoic Southern Province in the Sudbury Region (Fig. 2.9A). The suite is part of a bimodal magmatic event recorded in the volcanic rocks of the Huronian Supergroup, and they appear to have formed from an intracontinental mantle plume rifting event (Fig. 2.6B) similar in age to those recorded in Finland and Wyoming that were once contiguous with the Proterozoic rocks of Ontario (Bleeker and Ernst, 2006). It has been proposed that a single plume head to the south of the Sudbury Region resulted in the development of the East Bull Lake Suite and the contemporaneous Matachewan dyke swarm (Fig. 2.9B; Bleeker and Ernst, 2006).

The East Bull, Agnew, and River Valley Intrusions crystallized from broadly similar tholeiitic low-Ti, high-Al, PGE-rich parent magmas of a boninitic affinity that originated by differentiation in deeper hidden magma chambers (eg, Vogel et al., 1998a,b, 1999; Peck et al.,1993; Easton et al., 2010, Ciborowski, 2013). The primary magmas were likely second-stage melts derived from a sublithospheric depleted mantle source that was stabilized in the Archean, modified by a Neoarchean subduction and accretion events, and then tapped along cratonic margin structures during rifting. The magmas crystallized a large amount of plagioclase feldspar to produce a pronounced Fe-enrichment trend. Subordinate olivine and orthopyroxene crystallization in the lower parts of the stratigraphy (Fig. 2.10) is typically followed by thick sections of layered leucogabbronorite and gabbronorite, and ferrogabbro and ferrosyenite in the uppermost layers. The margins of the intrusions are commonly brecciated and a marginal inclusion-rich gabbronorite is often developed with disseminated, blebby and interstitial magmatic chalcopyrite–pyrrhotite–pentlandite mineralization (1–5% sulfide, $\leq$10 g/t Pt + Pd) within 5–50 m of the contact of the intrusions (Fig. 2.10).

The Paleoproterozoic mafic to anorthositic intrusions located in the immediate footwall of the Sudbury Basin are the Chicago Mine anorthosite, the Drury Township gabbro-anorthosite, the Norduna anorthosite, and the Fraser wehrlite. The larger Drury and Norduna Intrusions are considerably more feldspathic than those developed in the East Bull Lake and Shakespeare-Dunlop Intrusions (Prevec and Baadsgaard, 2005), but the Fraser Intrusion (which occurs as an envlave in the Levack Gneiss Complex in the footwall of Fraser Mine) is considerably more primitive. Large bodies of olivine melanorite are also present at Levack and McCreedy West Mines. They are developed within the Sublayer Granite Breccia. These rocks are fresh, and likely comprise part of the ultramafic inclusion suite of the SIC, and are considered to be part of the SIC that may be formed from primitive target rocks.

A group of medium to coarse-grained amphibolitized pyroxenites and medium-grained gray-colored gabbros occur in the South Range Breccia Belt to the west and East of the Frood-Stobie Deposit (Fig. 2.11). The deposits are hosted in discontinuous segments of QD and metabreccia (see Table 1.3) developed in the metavolcanic and metasedimentary stratigraphy of the Elsie Mountain Group and between large bodies of megacrystic amphibolite and metagabbro (Fig. 2.11).

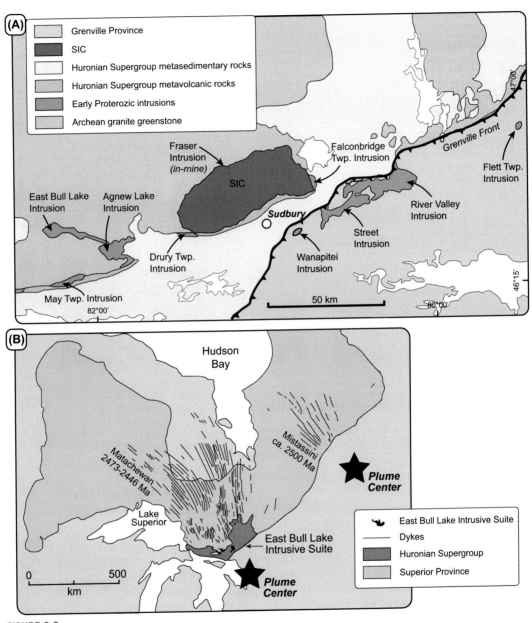

FIGURE 2.9

(A) Geological map showing the location of Early Proterozoic Intrusions in the Sudbury Region. (B) Position of the Early Proterozoic Intrusions of the Sudbury Region in relation to the Matachewan diabase dykes (~2490–2450 Ma), and the position of the inferred plume head responsible for this event.

(A) After Easton et al. (2004); (B) after Lamothe (2007).

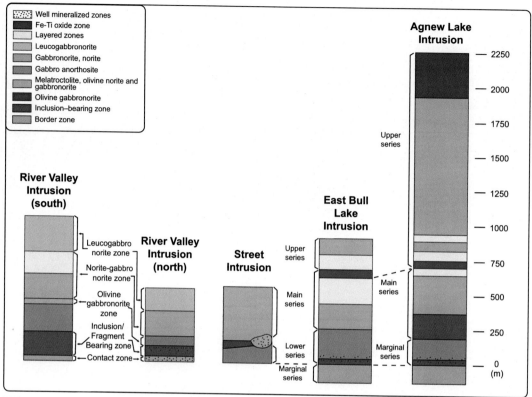

FIGURE 2.10 Stratigraphy of Mafic Rock Types Developed Within Different Intrusions Belonging to the East Bull Lake Type

After James et al. (2002a,b) and Easton et al. (2004, 2010).

Megacrystic amphibolite and metagabbro comprise a suite of medium to coarse-grained igneous rock within the Breccia Belt. The distribution of these rocks is shown in Fig. 2.11, and they are collectively grouped and termed the Frood Intrusion (Fig. 2.11). A typical example from the eastern part of the Breccia Belt is shown in Fig. 2.12A,B where 1–5 cm megacrysts of amphibole, possibly after pyroxene are contained in a fine-grained chlorite-rich matrix. The Frood Intrusion is typically enveloped and sometimes cut by SUBX (Fig. 2.12C), but some contacts of fine-grained amphibolite with metasedimentary rock are possibly indicative of primary intrusive contacts.

These rocks have received no attention beyond the work of Vale and preliminary geological maps produced by the Ontario Geological Survey (Johns, 1996). They represent an important footwall and hanging-wall rock type of the Frood-Stobie Deposit, and they extent for almost 10 km from west to east along the Breccia Belt. The Frood Intrusion possibly extends as far west as Ethel Lake near Victoria Mine (Lightfoot et al., 1997c); further work is in progress to establish whether these intrusions are part of the Paleoproterozoic suite of East Bull Lake Type Intrusions.

Approximately syn-magmatic with the East Bull Lake Group is the Matachewan Diabase Dyke swarm which is sometimes developed as rafts and fragments of plagioclase-porphyritic diabase in the

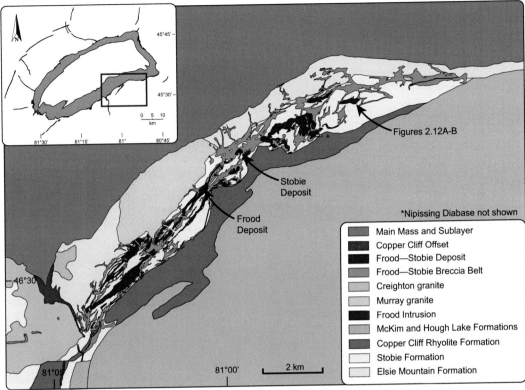

FIGURE 2.11 Geological Map of the Frood-Stobie Breccia Belt East of the Copper Cliff Offset

The map shows the extensive development of SUBX, and the presence of target rock types of poorly characterized mafic affinity that bound and crop out along strike from the Frood-Stobie Deposits. A particular group of medium- and coarse-grained amphibolites and metagabbros are collectively grouped as members of the Frood Intrusion; several of these bodies contain very coarse-grained megacrystic amphibole (Fig. 2.12A,B).

Sublayer of the SIC in the North and East Ranges, and transported blocks occur in South Range SUBX hosted in the Creighton Granite (Fig. 2.12D). These dykes appear to be quite common in the Archean basement, but they appear to be rarer in the Southern Province, perhaps because they were syngenetic with the mafic volcanic rocks of the Elsie Mountain Formation.

The Nipissing Diabase

The Nipissing Diabase crops out largely in the Southern Province between Sault Ste Marie in the west and Cobalt in the east, and it is part of a large igneous province that includes the Seneterre dyke swarm that is linked to a plume head in Ungava (Fig. 2.13A,B; Jambor, 1971; Ernst, 2014). It is a very important intrusive rock type developed in the South Range of the Sudbury Structure.

Most of the intrusions are undulating sills that are generally conformable with the stratigraphy of the Huronian Supergroup (Fig. 2.14A; Hriskevich, 1968; Jambor, 1971); there are also less common dykes (Lightfoot and Naldrett, 1996).

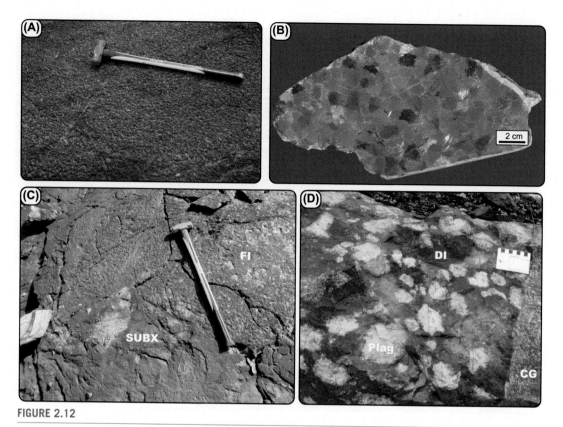

FIGURE 2.12

(A) Megacrystic amphibolite body in the Frood-Stobie Breccia Belt. The location of this body is shown in Fig. 2.11. (B) Slab of megacrystic amphibolite showing the coarse-grain size and locally recrystallized appearance of the rocks comprising the "Frood Intrusion." (C) Inclusions of megacrystic amphibolite derived from the Frood Intrusion in SUBX of the Frood-Stobie Belt. (D) Inclusion of plagioclase glomeroporphyritic diabase (DI) in Creighton Granite (CG); this rock type possibly represents a fragment of a Matachewan dyke that may have been exhumed from depth during emplacement of the Creighton Granite.

U–Pb geochronology has yielded crystallization ages of 2.2 Ga (2219 Ma, Corfu and Andrews, 1986; 2212 Ma, Conrod, 1988; and 2210 Ma, Noble and Lightfoot, 1992). Despite this narrow range of U–Pb ages, Buchan et al. (1989) has suggested that the emplacement of the Nipissing Diabase may have been a protracted event which recorded a range in paleomagnetic poles based on well-constrained baked contact tests. The Nipissing Diabase is approximately the same age as the Ungava magmatic event (2219+ −1 Ma), and may be related to the Senneterre dyke swarm (2216+ /−8/4 Ma) (Fig. 2.13A; Ernst, 2014).

The rock types are typically gabbroic, but the magmatic event has historically been named based on the less-common development of finer-grained diabase. In the Cobalt and Temagami Regions to the east of Sudbury, the sills tend to be well differentiated (Fig. 2.14A,B). In these areas, the Nipissing Diabase is less strongly deformed than the intrusions of the Sudbury Region, and so they serve as better type examples illustrating the diversity in rock types.

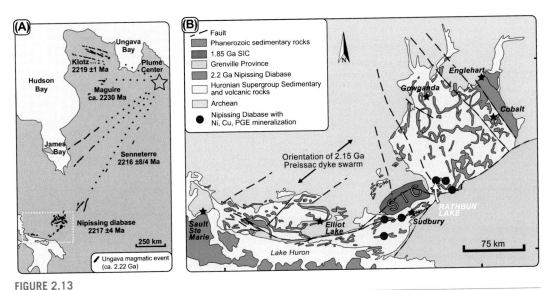

FIGURE 2.13

(A) Location of Nipissing Magmatic Province within the wider Senneterre large igneous province; (B) map showing the distribution of Nipissing Diabase and the Huronian Supergroup in the Southern Province of Ontario and some of the principal outcrops of magmatic Ni–Cu–PGE sulfide mineralization.

(A) After Ernst (2014); (B) after Jambor, 1971 and Lightfoot and Naldrett (1989).

The contact zones of Nipissing Diabase sills tend to comprise fine-grained chilled gabbro over an interval of about 50 cm, with the gabbro becoming progressively more coarse-grained over 5–10 m into a hypersthene-bearing gabbro. The main body of the intrusion largely consists of hypersthene gabbro, which grades upwards into leucogabbro and locally develops varied-texture gabbro where plagioclase, amphibole, and augite crystals vary in grain size from a few millimeters to grain sizes of several centimeters. Within the hypersthene gabbro there are occasionally coarse-grained patches of gabbro (Fig. 2.15A); these variable-texture rocks develop very long single crystals of augite that have a dendritic appearance in outcrop (Fig. 2.15B), and these rocks often show complex textures of medium- and very coarse-grained gabbro (Fig. 2.15C). These rocks have traditionally been interpreted to indicate rapid cooling, but the location of these rocks within the intrusion in areas of extensive vari-textured gabbro are more indicative of slow cooling under volatile-rich conditions.

The geological relationships in the roof zones of the intrusions are complex; partial melts of Lorrain Formation sedimentary rocks are developed in situ by the progressive disintegration and melting of the roof rocks; this is illustrated in the Kerns Intrusion west of New Liskeard in the Cobalt Embayment (Figs. 2.15D,E; Lightfoot and Naldrett, 1989), and the Obabika Intrusion to the east of Sudbury (Fig. 2.15F; Lightfoot and Naldrett, 1989). In the Sudbury Region, the more extreme granophyric products of differentiation are less commonly developed, but hypersthene-rich gabbro, leucogabbro, and varied-textured gabbro are common.

The Nipissing Diabase intrusions show evidence that the parental magma underwent assimilation of adjacent metasedimentary rocks coupled to fractional crystallization of hypersthene, plagioclase, augite, and apatite, which explains the extreme range of rock types developed within individual intrusions as a

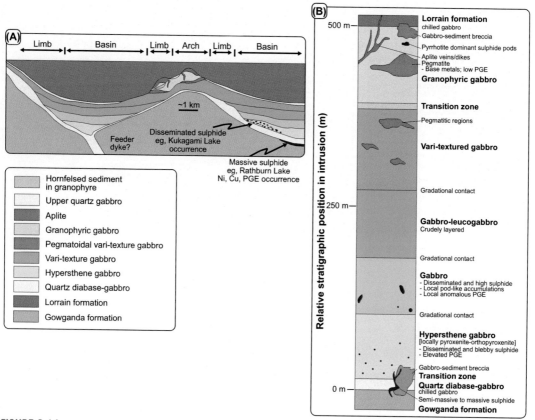

FIGURE 2.14

(A) Geological cross section of a differentiated Nipissing Diabase sill showing the development of a feeder dyke (after Lightfoot and Naldrett, 1989; Hriskevich, 1968). (B) Typical section through a differentiated Nipissing Diabase intrusion showing the diversity in rock types and the location of magmatic sulfide mineralization (after Lightfoot and Naldrett, 1989).

FIGURE 2.15   ▶

(A) Variable-textured Nipissing Diabase from the sheet hosting the Copper Cliff Offset north of Kelly Lake; the black coloration in outcrop is due to a patina of oxidation produced by fallout from early 19th Century smelting operations in the Sudbury Region. (B) Quench-textured augite patch in variable-textured Cobalt area. (C) Variable-textured gabbro, Nipissing Diabase, Kerns Township. (D) Aplites (Ap) generated by partial melting of metasedimentary rock (Msed) caught in the diorite. (E) Fragments of metasedimentary rock (Msed) arrested within an aplite vein (Ap) derived by partial melting of the Lorrain Formation sediments in the roof zone of a Nipissing Diabase sill (Obabika Township; Lightfoot and Naldrett, 1996); (F) Fragments of aplite entrained in granophyric quartz diorite, Obabika Township; (G) Blebby magmatic sulfide in gabbro; Wanapitei Intrusion near the Rathburn Lake Ni–Cu–PGE sulfide occurrence (Lightfoot and Naldrett, 1996). The blebs of sulfide are differentiated into lower pyrrhotite–pentlandite and upper chalcopyrite domains much like those found in the Insizwa and Noril'sk Intrusions (Lightfoot et al., 1984; Lightfoot and Zotov, 2014).

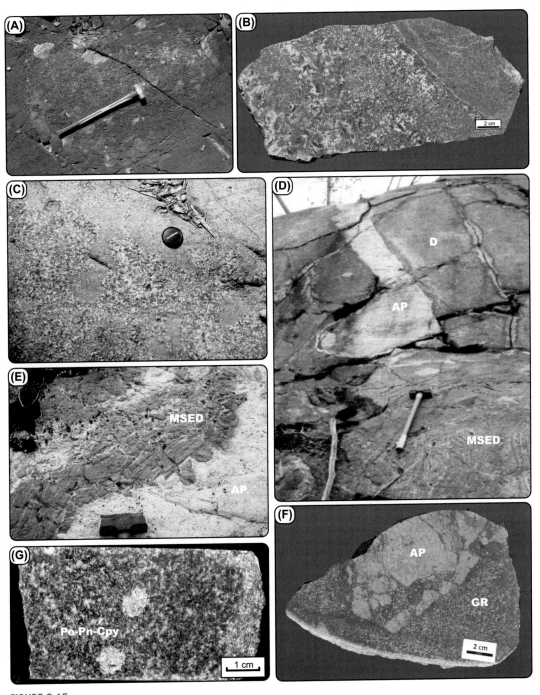

consequence of progressive assimilation of the roof sediments (Lightfoot and Naldrett, 1989). Some portions of the sills are weakly enriched in cumulus orthopyroxene (typically in the basins, Fig. 2.14A), whereas less-common evolved rocks are typically developed above arches that are developed laterally towards the edge of the sheet (Fig. 2.14A).

A comparison of the composition of contact gabbros provides evidence of a very uniform parental magma throughout the magmatic province. Despite the paleomagnetic evidence (Buchan et al., 1989), it appears that the intrusions were produced from a single parental magma type which was derived from a low-Ti mantle source perhaps modified by Archean and early Proterozoic mantle metasomatism (Lightfoot et al., 1993a).

Card and Pattison (1973) showed that mineralization associated with Nipissing Diabase varies in type and style across the Southern Province, namely: (1) Co-Ag-(PGE-Ni) sulfides and sulf-arsenides occur in quartz-carbonate veins which cut the Nipissing Diabase in the Cobalt Region; (2) Disseminated magmatic Cu-Ni-Co-(PGE) sulfides in the lower parts of Nipissing Diabase sills are present between Massey and River Valley in the Sudbury Region; and (3) Hydrothermal Cu sulfides develop in association with quartz-carbonate veins between Blind River and Sault Ste. Marie (Fig. 2.13A).

The Nipissing Diabase is known to contain marginally economic resources and subeconomic occurrences of magmatic Ni–Cu–PGE sulfide mineralization. These include the 2217 Ma Shakespeare Deposit to the west of Sudbury (11.8 million tons @ 0.33%Ni, 0.35% Cu, 0.87 g/t Au + Pt + Pd; Sproule et al., 2007; Dasti, 2014), Ni–Cu–PGE-sulfide occurrences in Kukagami and Janes Townships to the east of Sudbury (Thompson and Card, 1963; Lightfoot and Naldrett, 1996), the Carson Lake Intrusions in the Lake Panache area, and the Rathbun Lake showing in the Wanapitei Intrusion (Fig. 2.13A; Dressler, 1982a; Lightfoot and Naldrett, 1996). Lightfoot and Naldrett (1989) noted that this variation in type and style of mineralization appears unrelated to differences in lithology, degree of metamorphism, or level of intrusion of the Nipissing Diabase. Sulfide occurrences in the Nipissing Diabase are clearly of magmatic origin (Fig. 2.15G; Lightfoot and Naldrett, 1989, 1996; Lightfoot et al., 1993a; Sproule et al., 2007), but there is evidence that the massive sulfides at Rathbun Lake are either hydrothermal in origin (Finn et al. 1982; Rowell and Edgar, 1986), or created by modification of magmatic sulfides. The parental magma of the Nipissing Diabase contained ~100 ppm Ni, and there is no evidence of metal-depleted signatures that would be indicative of major segregation and removal of magmatic sulfide during crystallization of the magmas (Lightfoot and Naldrett, 1996). This is consistent with the fact that economic concentrations of high grade Ni–Cu–(PGE) sulfide mineralization have not been discovered in association with the Nipissing Diabase intrusions.

Sudbury Gabbro Intrusions

The "Sudbury Gabbro" comprises a medium-grained metagabbro or amphibole that crops out in an area south and southwest of the SIC (Fig. 2.16; Card, 1965, 1968). This domain of country rocks has experienced several phases of deformation, and extensive brecciation in response to the Sudbury Impact event. Peredery (1991) considers these rocks to be metamorphosed equivalents of the Nipissing Diabase. Lightfoot and Farrow (2002) show that the Sudbury Gabbro are amphibolites and metagabbros with up to18 wt% MgO; this is considerably more primitive than the most Mg-rich Nipissing Diabases (<12% MgO; Lightfoot et al., 1993a). Attempted to date the Sudbury Gabbro using U–Pb techniques on accessory minerals have not yet met with success (Chamberlain, Pers. Comm., 2011).

The Sudbury Gabbro contains subeconomic concentrations of disseminated and patchy Ni–Cu–PGE sulfide mineralization (Fig. 2.16). Many of the occurrences are developed up to 50 km WSW of Sudbury,

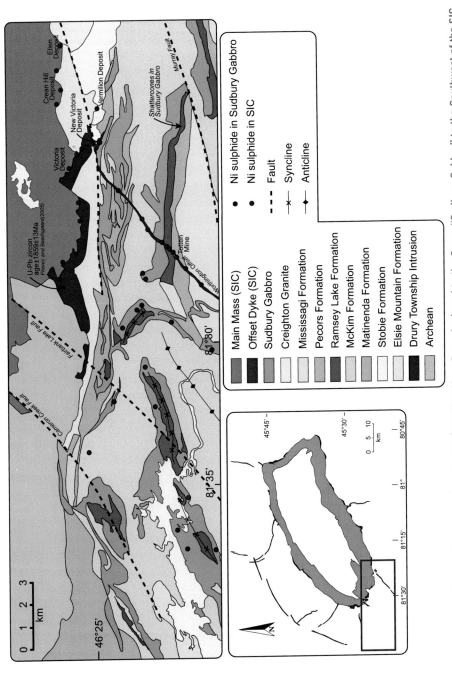

FIGURE 2.16 Map Showing the Distribution of Intrusions That Have Been Assigned to the Group "Sudbury Gabbro" to the Southwest of the SIC

Many of these bodies comprise significantly more mafic rock types than those typically found in the Nipissing Diabase (see Lightfoot et al., 1997c).

After Card (1965, 1968).

and the textures are typical of magmatic sulfides formed from the parental magma of the Sudbury Gabbro intrusions. Closer in to Sudbury, the Sudbury Gabbro has been more strongly affected by the Sudbury event, and the sulfide mineralization is localized into segregations by deformation and metamorphism.

The Sudbury Gabbro is an important inclusion type and host-rock to the ore deposits at Totten Mine on the Worthington Offset, and at Kelly Lake in the southern part of the Copper Cliff Offset. The origin of these inclusions and the mechanism by which the Sudbury Gabbro blocks were introduced into the mineralized sequence of rocks in the Offset Dyke is an important question that will be discussed further in Chapters 3 and 4.

FELSIC INTRUSIVE ROCKS IN THE SUDBURY REGION

The 2.33 Ga Creighton (Frarey et al., 1982) and 2.47 Ma Murray (Krogh et al., 1996) granitoid plutons cut the Huronian Supergroup footwall rocks in the South Range of the SIC (Fig. 2.7). The plutons intrude volcanic and sedimentary rocks of the Elsie Mountain and Stobie Formations and enclose roof pendants of both Formations, as well as rafts of basement gneiss and fragments of plagioclase glomeroporphyritic Matachewan diabase (Fig. 2.12C). The plutons were emplaced during the Blezardian Orogeny (2400–2200 Ma, Riller and Schwerdtner, 1997) at which time the rocks of the Huronian Supergroup were deformed under amphibolite-facies metamorphic conditions (Card et al., 1972; Card, 1978). Chemical similarities between the Creighton and Murray plutons and the Copper Cliff Formation, together with similar U–Pb ages, are consistent with a comagmatic relationship between these rocks as suggested by Card (1978).

The Creighton pluton is composed of melanocratic quartz monzonite with less than or equal to 20% mafic minerals, leucocratic granite consisting of equigranular quartz with recrystallized plagioclase, perthite and minor biotite, and porphyritic quartz monzonite comprising orthoclase phenocrysts in a coarse-grained matrix of quartz, plagioclase, perthite and biotite (Riller, 2009). In the field, the Creighton Granite is broken out into porphyritic and hybrid-felsic variants (Fig. 2.7).

The Murray pluton consists largely of equigranular granite. The pluton can be divided into southeast and northwest structural domains based on the development of primary magmatic versus postmagmatic fabrics produced by the Sudbury thermal event (Riller et al., 1996). There are very clear differences in grain size, color, inclusion content, and diversity in rock types between the Creighton and the Murray granites, and the different U–Pb ages (Fig. 2.7) are consistent with two different magmatic events that gave rise to these bodies.

Syntectonic emplacement of granitoid magma is supported by: (1) concordance of magmatic and metamorphic foliation within the pluton; (2) continuity of foliations across intrusive contacts; and (3) general obliquity between foliation surfaces and pluton contacts (Riller et al., 2006).

The heat source of the Creighton and Murray granite plutons is believed to be the Paleoproterozoic Matachewan plume event where granitic magmas were generated by melting of lower crust. The lower crustal melts rose to high levels in the crust and ponded as sheets close to the Archean-Proterozoic boundary where they possibly gave rise to the rhyolite flows and tuffs of the Copper Cliff Formation.

The Creighton Granite forms the immediate footwall to parts of the Creighton Deposit, and it develops a black color in close proximity to the base of the SIC and the associated mineral zones; this rock is termed the "Black Porphyry" and appears only to be developed in the footwall of the Creighton Embayment. This rock type is likely developed within the metamorphic halo the SIC, and it likely reflects recrystallization and possibly partial melting.

GEOLOGY OF THE SIC

The rocks that comprise the igneous component of the Sudbury Structure are grouped together as the SIC. They comprise the rocks shown in the simplified map and simplified stratigraphic section shown in Fig. 2.17A:

1. A group of largely inclusion-free differentiated noritic, gabbroic, and granophyric rocks that collectively comprise the Main Mass of the SIC (eg, Naldrett and Hewins, 1984; Figs. 1.7 and 1.8, Table 1.3).
2. A group of inclusion-bearing and generally mineralized rocks at the base of the SIC that occupy local topographic depressions termed troughs and embayments; these rocks generally have either an igneous-textured noritic or a xenocryst- and fragment-laden granitic matrix, and the two variations are grouped as a unit termed the Sublayer, and referred to as Sublayer Norite (SLNR) and Granite Breccia (GRBX).
3. The radial and concentric dykes of Quartz Diorite (QD) and inclusion QD and discontinuous pods of QD and inclusion-bearing QD in SUBX (Grant and Bite, 1984); these are termed the Offset Dykes; they tend to be discontinuous because of either late faulting or due to primary discontinuities, buds and swells, steps, or branches (Grant and Bite, 1984).
4. More controversially, various country rocks that have been partially transformed into melts by the Sudbury event as a result of high pressure and/or temperature; the igneous component of these rocks are the matrix of the SUBX as well as leucosomes in the metamorphic aureole of the SIC that are developed in a variety of rock types, but most commonly within ~250 m of the contact. Traditionally, these rocks have not been grouped as part of the SIC, but the weight of evidence suggests that they comprise a melt contribution derived from of the country rocks that formed leucosomes and melt-filled tension fractures so they broadly represent igneous rocks created by impact and melting of the crust.

THE MAIN MASS

The Main Mass of the SIC comprises igneous-textured rocks that crop out as an elliptical (~30*60 km) differentiated body of norite, overlain by gabbro, and capped by granophyre (Fig. 2.17A). The historic term 'Main Mass' is not especially informative, but it refers to the stratigraphy of relatively inclusion-free igneous rocks beneath the Onaping Formation, and above the Sublayer.

The Main Mass is the folded and faulted remnant of a sheet with a subhorizontal upper contact that varied in thickness from ~1.5 to 5 km largely depending on the topography of the base of the melt sheet, and had a lateral extent of at least 80 km in diameter (Golightly, 1994), and possibly further if the melts extended beyond the crater into the impact-related ring structure surrounding it.

The dip of the basal contact of the SIC is known from both detailed outcrop reconstructions based on the intersection of the basal contact with topographic contours in areas of moderate to good exposure (Dreuse et al., 2010; Fig. 2.17A), but it is also established based on a much more accurate reconstruction based on information gathered over more than 100 years of mining activity and exploration drilling. Examples of the configuration of the basal contact of the SIC are shown as a series of sections in Fig. 2.17B–K derived from a three-dimensional models of the Sudbury Structure developed by Gibson (2003) using surface and drill core data acquired by Vale, together with the results of the regional Lithoprobe seismic survey transects and gravity surveys completed across the Sudbury Basin (eg, Milkereit et al., 1994a,b).

The dip of the contact zone between the Main Mass and the Sublayer is broadly similar to that of the contact between the Sublayer and the country rocks; the dips shown on sections in Fig. 2.17 are

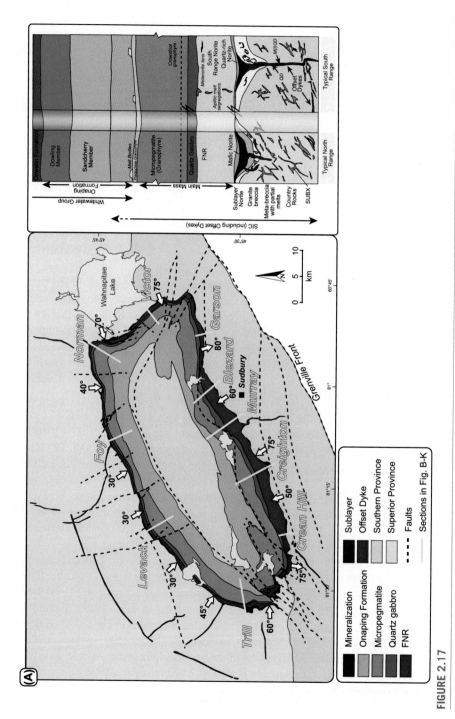

FIGURE 2.17

(A) Map of SIC Main Mass lithologies, with dip of basal contact, structural boundary of N and S Range, and names used in text. Individual sections depict the geometry of the main units of the SIC at various locations around the SIC based on a three dimensional model constrained by drill and mine data for the Sudbury Basin reported by Gibson (2003). The stratigraphy of the Main Mass discussed in this chapter is highlighted in the inset section. (B–K) The locations of the sections based on surface mapping and drilling are cross-referenced to the map shown in Fig. 2.17A; they are shown in clockwise arrangement around the Sudbury Basin.

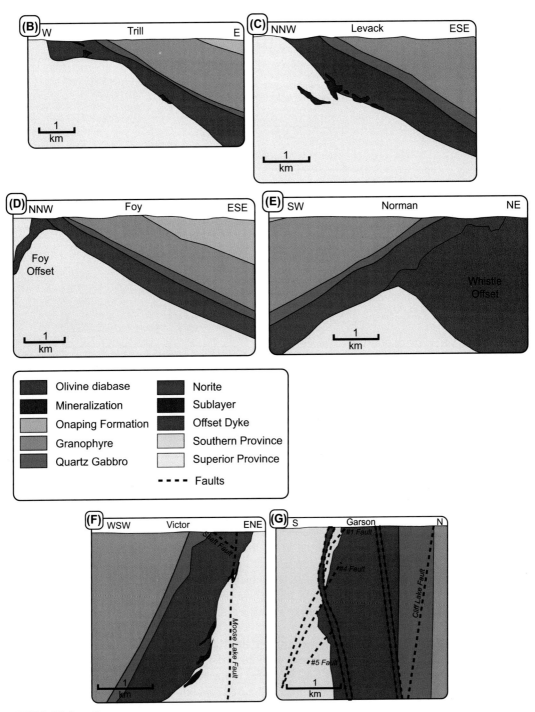

FIGURE 2.17 (*cont.*)

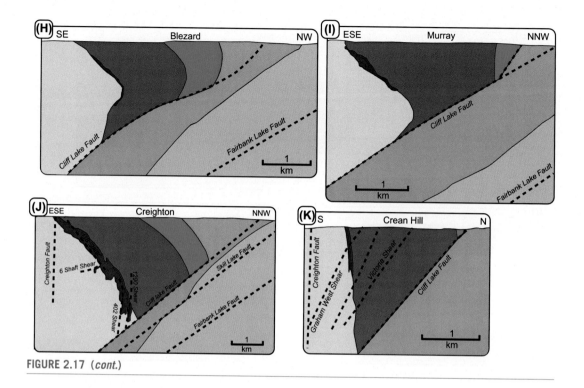

FIGURE 2.17 (*cont.*)

based on the lower contact of the Sublayer or Main Mass depending on that unit is in contact with the country rocks. There are clearly local undulations due to changes in the scale of embayments and troughs, but the contact surface provides a very good indication of the broad regional dip of the basal contact of the SIC.

The base of the Main Mass in the North Range dips inwards towards the center of the Sudbury Basin at an angle of 30–45°; the East Range contact is much steeper (~75–90°, and it is locally overturned). The dip of the basal contact of the SIC in the South Range varies from west to east along the southern margin, and also with depth. For example, at Creighton Mine, the contact dips 45° NNW to a depth of about 1.5 km, and then steepens to near vertical between 2 and 3 km depth extent (Fig. 2.17). To the west of Creighton Mine, the near vertical lower contact extends to the surface at Crean Hill; to the east, the inflexion becomes progressively shallower from Murray through Blezard and Lindlsey, and at Garson and Falconbridge, the contact is nearly vertical (Fig. 2.17). The geometry of the basal contact in the South Range is a response to both regional-scale faulting and tilting that is described later in this chapter, as well as primary variations in the topography of the basal contact of the melt sheet.

The ratio of norite:gabbro:granophyres in the Main Mass for structurally unmodified profiles is almost always ~25:15:60 irrespective of the overall thickness of the SIC. There is one segment where this rule does not hold fast; the Main Mass in the area of the Windy Lake-Onaping area has a ratio of norite:gabbro:granophyre that is closer to 15:25:60 based on both the surface geology (Ames et al., 2005) and the record from deep drill holes (Lightfoot et al., 2002). The unusually thick Quartz Gabbro narrows away from Windy Lake towards the SW and ENE.

The North and East Ranges of the Main Mass comprises rocks that are different when compared to the South Range (Fig. 1.8). The principal differences in stratigraphy between the rocks are in both the sequence of primary lithologies (Fig. 1.8), and the degree of metamorphism (and deformation) of the norite. The line demarcating the Main Mass rocks of the North and East Ranges from those of the South Range extends along the Cameron Creek Fault, broadly through the northern part of the Sudbury Basin, and then through the Norduna Fault in the east. This fault is the northernmost member of the east–west fault system that extends through the Sudbury Basin and cross cuts the Whitewater Group and the SIC. These rocks exhibit a strong ductile deformation fabric that has traditionally been termed the South Range Shear Zone (Fig. 2.17).

The boundary between the North and East Ranges and the South Range is broadly defined by fundamental changes in the geology of the rocks that comprise the SIC and the footwall. Important changes include: (1) Granite Breccia is developed as an important Sublayer rock type in the North and East Ranges, but it is largely absence in the South Range. (2) A discontinuous unit of Mafic Norite (MNR) is developed at the base of the Main Mass in the North and East Ranges, but a unit of more evolved Quartz-rich Norite is developed at the base of the Main Mass in the South Range (Fig. 2.18). (3) The

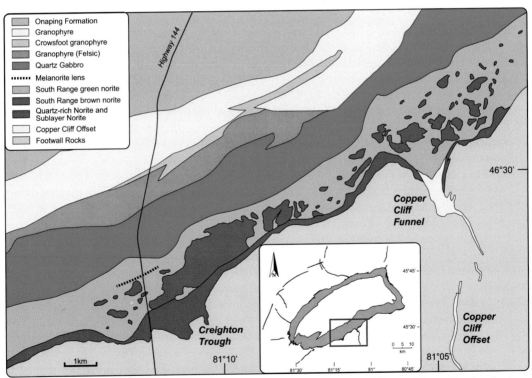

FIGURE 2.18 Detailed Geology in South Range of the SIC Between Gertrude and Murray Mines, Showing the Distribution of Brown (Fresher) and Green (More Altered) South Range Norite, the Quartz Gabbro, and the Position of the Crowfoot-Textured Granophyre in the Granophyre Unit

The location of a lens of melanorite is shown approximately two-thirds of the way up through the stratigraphy of the noritic rocks (after Lightfoot and Zotov, 2006).

footwall rocks comprise dominantly Archean granites and gneisses in the North and East Ranges, but they comprise metavolcanic, metasedimentary and intrusive mafic and felsic rocks in the South Range. The fault boundary that demarcates the South Range from the North and East Ranges is sharp, and there are no evidence of lateral gradation between the rock types in the South Range and those developed in the North and East Ranges.

The Geology of the North and East Range of the Main Mass

In this book the description of the Main Mass commences with the rocks developed at the base of the stratigraphy and works upwards through the units shown in Fig. 1.7 that is an important reference framework.

The most basal Main Mass norites of the North and East Range comprise a discontinuous (0–50 m thick) lower unit of ophitic-textured melanorite with 30–60% orthopyroxene, large plagioclase plates containing cumulus orthopyroxene, and abundant phlogopite mica that is termed the Mafic Norite (MNR) unit (Fig. 2.19A). The MNR varies in thickness from up to 100 m over embayment and trough structures to a very thin or absent unit where there is no Sublayer. Although there is local recrystalliza- tion of orthopyroxene to amphibole, the MNR is largely fresh or exhibits a small to moderate degree of alteration. The MNR often contains weak disseminated sulfide mineralization as well as irregular xenocrysts and/or xenoliths of anorthosite.

The MNR is in sharp or gradational contact (over 1–10 m) with an overlying sequence of norite that contains ~20% orthopyroxene; this unit is termed the Felsic Norite (FNR) (Fig. 1.7). The FNR is a weakly differentiated hypidiomorphic equigranular textured rock that contains no inclusions (Fig. 2.19B), but contains trace to disseminated sulfide mineralization that changes from dominantly pyrrhotite at the base to trace pyrite within ~100 m above the base of the unit. The lower two-thirds of the FNR contains orthopyroxene, but the upper third is devoid of hypersthene and contains only augite, so it would challenge the traditional definition of a norite (Fig. 2.19C). As described in Chapter 1, the historic terminology of the SIC has generated some examples of classification oxymorons. The term "Felsic Norite (FNR)" is one of them; the upper part of the stratigraphy is essentially gabbroic based on modern classification schemes (eg, Streickeisen, 1967).

Although it would be logical to draw the boundary between the FNR and the Quartz Gabbro at the point where orthopyroxene disappears, this is not always easy when the pyroxenes are replaced by amphibole, and so the entry of cumulus magnetite that can be flagged with a simple hand-held magnet or a magnetic susceptibility meter is a more convenient approach in establishing the boundary; this is the approach that is widely used by geologists in the mining companies working in the Sudbury Region.

The Quartz Gabbro stratigraphy of the Main Mass is locally termed the "Transition Zone" as the boundaries with the underlying noritic rocks and the overlying granophyric rocks is gradational over several meters. The Quartz Gabbro varies in grain size, mineralogy, and potential field properties as the proportions of mafic minerals declines from base to top (two examples of an oxide-rich mafic Quartz Gabbro from the base of the unit and a more evolved Quartz Gabbro from the upper part of the unit are shown in Fig. 2.19D,E).

The transition from the Quartz Gabbro Unit into the Granophyre Unit (sometimes termed the Micropegmatite) is more gradual than the transition from FNR into Quartz Gabbro; it is marked by the absence of magnetite and a marked increase in the proportion of quartz and K-feldspar as pyroxene dis- appears and the plagioclase content declines to ~25%. The quartz gabbro is transitional into the low- ermost granophyres over several tens of meters, and the rocks are often highly evolved in composition

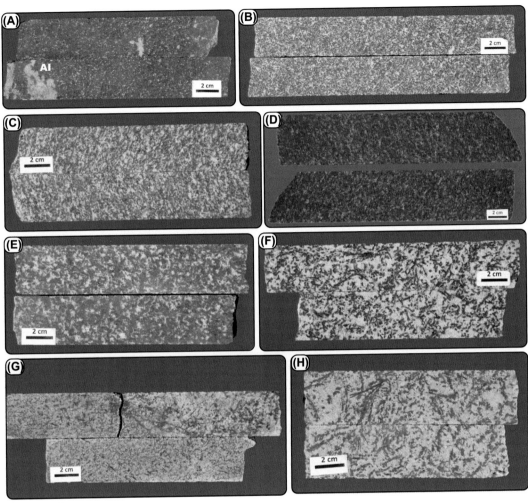

FIGURE 2.19

(A) North Range MNR from borehole 700180 (3133 ft.) showing anorthositic clots in a biotite-rich coarse-grained norite. (B) North Range basal FNR from borehole 528480 (2942 m. depth). (C) North Range upper FNR from borehole 528480 (depth: 2295 m. depth). (D) North Range lower Quartz Gabbro from borehole 528480 (2105 m. depth). (E) North Range Upper Quartz Gabbro transitional into basal Granophyre from borehole 528480 (1641 m. depth). (F) Transitional quartz gabbro to granophyre from borehole 528480 (1307 m. depth). (G) Basal Granophyre of the North Range (bore hole 528480 1153 m.). (H) Crowsfoot Granophyre in borehole 528480 (947 m. depth) in the North Range; this horizon comprises 1–50 cm blades of amphibole in quartz and K-feldspar. (I) Plagioclase-rich Granophyre from the roof of the Main Mass in the North Range at Onaping. (J) Plagioclase lamination and trace disseminated sulfide in a melanorite lens developed towards the top of the South Range norite stratigraphy, Highway 144 (Lightfoot and Zotov, 2006). (K) Mafic inclusion (MI) in Quartz Gabbro (QG) beneath the Granophyre in the South Range Main Mass, Highway 144. (L) Crowsfoot-textured granophyre in basal Granophyre, Highway 144. (M) Basal South Range quartz-rich norite containing feldspar xenocrysts derived from an inclusion of Creighton granite derived from the footwall of the SIC. (N) Contaminated norite (NR) with feldspar xenocrysts (FX) and fragments of mafic metavolcanic rock (MV) derived from the country rocks; this unit cross cuts the South Range Norite. (O) Aplite vein (AP) cutting through the basal Main Mass quartz-rich norite north of the Murray granite pluton.

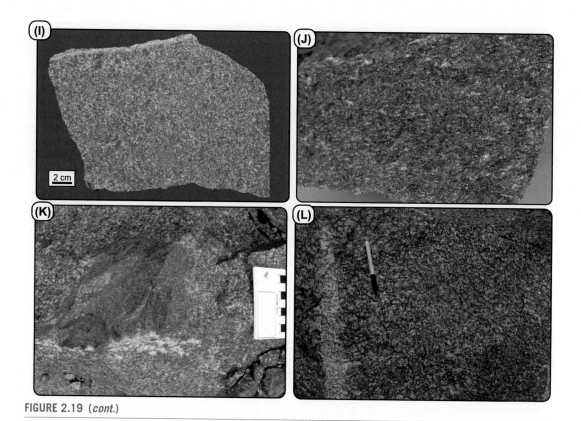

FIGURE 2.19 (*cont.*)

(Fig. 2.19F,G). A layer is locally developed near the base of the Granophyre that contains 1–50 cm-long blades of amphibole contained within a granophyric matrix (Fig. 2.19H) towards the base of the Granophyre; this rock type forms a semicontinuous stratigraphic horizon low in the Granophyre around the North Range that is termed the "Crowsfoot Granophyre," and resembles similar rocks developed in the Granophyre Unit of the South Range.

The Granophyre unit comprises a weakly differentiated and heavily altered series of rocks comprising quartz and K-feldspar in micrographic intergrowths with variations in the proportions of tabular plagioclase crystals and small quantities of amphibole (Fig. 2.18G). The texture, grain size, and proportions of mafic to felsic minerals tend to vary through the stratigraphy. The uppermost Granophyre contains abundant plagioclase feldspar (~35%; Fig. 2.19I), and this unit is often transitional into the melt bodies developed towards the base of the overlying Onaping Formation (Ames et al., 2008).

The Geology of the South Range of the Main Mass

In the South Range, the lowermost Main Mass norites comprise a variably altered sequence of rocks. The lowermost Quartz-rich Norite often contains blue quartz (<20%) with magmatic mica and amphibole and disseminated to trace sulfide mineralization. The Quartz-rich Norite grades into overlying South Range Norite over a distance of about 100 m as the amount of quartz, phlogopite mica, and primary amphibole declines.

FIGURE 2.19 (*cont.*)

Irregular bodies of fresh brown-colored norite termed Brown Norite occur as islands within altered green-colored norite that is termed Green Norite (Fig. 2.18). The Brown and Green Norite domains can be mapped within the lower part of the South Range Norite. Their distribution does not relate to stratigraphic position or the development of local fault structures.

The South Range Norite exhibits a cryptic increase in the proportion of ferromagnesian minerals upwards towards a locally-developed horizon of fresh black-colored melanoractic norite with a strong subhorizontal lamination of plagioclase and local development of phlogopite mica and blebby magmatic sulfide (Fig. 2.19J; Lightfoot and Zotov, 2006). This locally developed horizon forms a discontinuous layer in the immediate vicinity of Highway 144 (~1.5 km west to east), and shows elevated Cr and Ni concentrations. If the South Range was tilted by more than 45 degrees after formation, then the primary plagioclase lamination fabric should dip steeply towards the North; this relationship is inconsistent with the generally accepted idea that the basal contact of the South Range was originally subhorizontal, and then tilted by regional deformation.

Above the melanorite layer, the South Range Norite shows a cryptic decline in the proportions of ferromagnesian minerals in rocks that are similar to the green type of South Range Norite.

The transition from South Range Norite into the overlying Quartz Gabbro is marked by increased magnetic susceptibility over 1–10 m. The rocks above this transition contain fresh augite (Naldrett and Hewins, 1984) or amphibole after primary augite, and they contain occasional small (1–10 cm) fine-grained mafic inclusions (Fig. 2.19K). The central part of the Quartz Gabbro shows the local development of a Crowsfoot texture (Fig. 2.19L), which resembles that found in the lower part of the North and South Range Granophyre unit, but the rock is typically more mafic in composition (eg, compare the Crowsfoot texture in the Quartz Gabbro shown in Fig. 2.19L to that developed in the North Range Granophyre Unit shown in 2.19H).

The Granophyre Unit of the South Range Main Mass shows a cryptic and cyclical upwards increase in ferromagnesian mineral content, and the proportion of plagioclase can reach ~40% immediately below the upper boundary of the Main Mass. Both the Quartz Gabbro and Granophyre are commonly altered, and often heavily sheared by multiple splays of the South Range Shear Zone.

Felsic Enclaves within the Norite Stratigraphy of the Main Mass in the South Range

The presence of enclaves of aplite as pods, segregations, and dykes within the South Range norites is illustrated in Dressler (1984a) where he refers to them as "remobilized granitoid rocks." Their distribution in part of the Main Mass of the South Range is shown in Fig. 2.7, but these segregations are common in the norite stratigraphy above both the footwall granite stratigraphy of the Murray and Creighton plutons.

Near to the base of the Main Mass adjacent to and along-strike from embayment structures, the aplite dykes and segregations carry inclusions of footwall rocks (mafic volcanic rocks and granite fragments) and xenocrysts of feldspar derived from partial melting of footwall granitoid rocks. The quartz-rich norite of the Main Mass is sometimes extremely contaminated and contained feldspar xenocrysts that appear to be derived by local assimilation of granitoid rock derived from the Creighton Granite (Fig. 2.19M). The quartz-rich norite of the Main Mass also contains felsic segregations and mafic inclusions derived from the footwall (Fig. 2.19N).

Aplite segregations and dykes are commonly developed in the stratigraphy of the South Range Norite and Quartz Gabbro. These segregations rarely contain inclusions or feldspar xenocrysts, and they tend to form anastomosing dykes and veins (Fig. 2.19O) that appear to originate by partial melting of footwall granites and injection of the granitic melts into the overlying stratigraphy leaving the inclusions at the base.

Synthesis of the Geology of the Main Mass

The objective of this summary is to set the context for detailed petrological and geochemical data presented in Chapter 3, and to highlight some of the observations that are not reconciled with a simple uniform process of differentiation throughout the melt sheet.

1. The MNR is developed above North Range Sublayer, whereas a more evolved quartz-rich norite is developed above the South Range Sublayer. More details about this rock type will appear in Chapter 3, but the reader's attention is drawn to the fact that the most basal Main Mass norites of the North Range are more primitive when compared to those developed in the South Range. The process of differentiation of the North and South Range melt sheet appears to vary across the Sudbury Basin.

2. The FNR of the Main Mass in the North Range is generally fresh to weakly altered and become more differentiated from the base upwards. In contrast the South Range Main Mass consists of both fresh Brown Norite and altered Green Norite, and these rocks show a broad upwards reversal from more leucocratic to more melanocratic compositions over the lower two-thirds of the stratigraphy, followed by a return to more leucocratic compositions in the upper one-thirds of the stratigraphy. The extent of alteration and the sequence of noritic rock types developed in the South Range is different when compared to that of the North Range. This difference appears to require a different process of crystallization in the two spatially related segments of the melt sheet.

3. A discontinuous lens of fresh black-colored melanocratic norite is located ~two-thirds of the way above the base of the Main Mass stratigraphy of the South Range. This lens of norite contains fresh plagioclase and abundant orthopyroxene with trace to disseminated magmatic sulfide and phlogopite mica. The mafic composition and subhorizontal lamination of plagioclase feldspar in this unit appear to be inconsistent with traditional models of in situ fractional crystallization prior to deformation.

4. The Quartz Gabbros of the North and South Range are quite similar in petrology; small mafic inclusions are developed towards the top of the unit. These rocks appear to represent a transition zone in petrology between the underlying norite and the overlying granophyre, and this is why the literature often refers to this unit as the Transition Zone Quartz Gabbro. Some of the rocks within the Quartz Gabbro are derived by quenching; others contain small fine-grained mafic inclusions, and there are dykes and segregations of aplite hosted by a more melanocratic matrix. Upwards through the stratigraphy, there is a very broad change in composition from melanocratic to leucocratic compositions.

5. Granophyres from the North Range and the South Range are heavily altered and exceptionally heterogeneous in mineralogy and grain size; they exhibit a broad upwards trend towards a more mafic composition; there are local variations in the ratio of felsic/mafic minerals, and the stratigraphic position of the "Crowsfoot-textured" granophyre appears not to occur along one continuous stratigraphic horizon.

THE OFFSET DYKES

The term "Offset Dyke" is applied to radial and concentric dykes of the SIC which are shown in Fig. 2.20A; the name reflects the development of primary discontinuities created during emplacement and subsequent postmagmatic faulting. The dykes are composed of what has historically been termed quartz diorite (QD) and mineralized inclusion-bearing quartz diorite (MIQD), although strictly speaking many of the dykes are composed of quartz monzodiorite based on modern classification schemes (Streckeisen, 1967; Wood and Spray, 1998). The dykes are located in the country rocks beyond and/ or below the basal contact of the Main Mass of the SIC. The Offset Dykes comprise three styles,

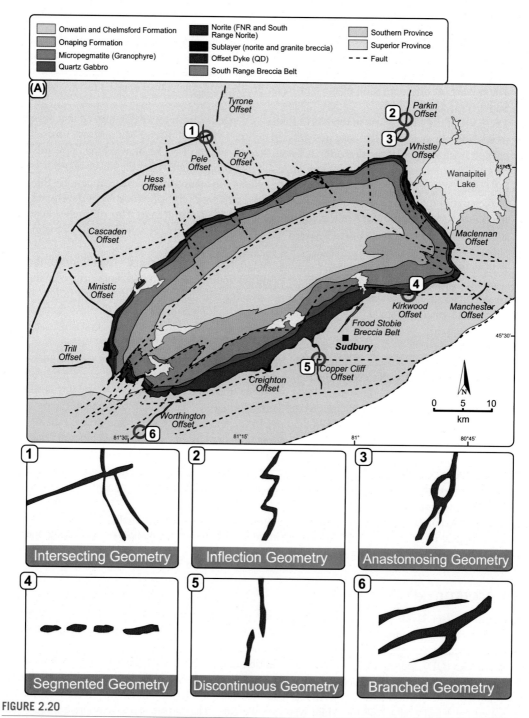

FIGURE 2.20

(A) Location and nomenclature of the Offset Dykes of the SIC; inset diagrams (1–6) show the simplified characteristics of the geology in locations where the Dykes have discontinuities or breaks (modified after Grant and Bite, 1984 and Smith et al., 2013). The location of the two offsets documented in detail in Figs. 2.21 and 2.23 are shown.

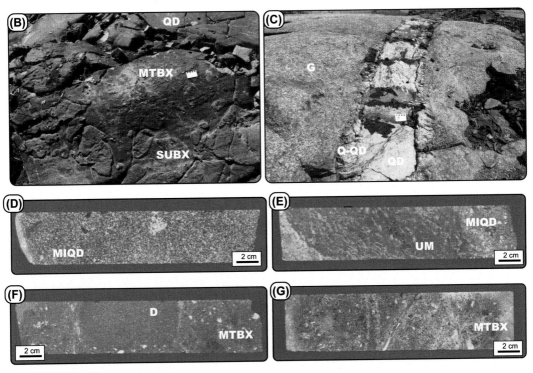

FIGURE 2.20 (*cont.*)

(B) Photograph showing geological relationships in the contact zone of one of the discontinuous segments of QD ~1.4 km to the east of Stobie Mine; QD, quartz diorite; MTBX, metabreccia developed from SUBX; (C) Quenched glass (Q-QD) and quench-textured QD from an apophysis of the Trill Offset cutting Archean granitoid rocks (G) (see Golightly et al., 2010). (D) Inclusion QD (MIQD) from an Offset Dyke in the keel of the Whistle-Norman trough (borehole 1259870, 4806 ft.). (E) Ultramafic inclusion in mineralized inclusion-bearing QD (MIQD) in the keel of the Whistle-Norman trough (borehole 1259870, 4975 ft.); the inclusion QD has a less clearly defined igneous matrix and this sample is transitional into a metabreccia. (F) Diabase inclusion (D) in metabreccia from an Offset Dyke in the keel of the Whistle–Norman trough (borehole 1259870, 4911 ft.). (G) Metabreccia (MB) from an Offset Dyke in the keel of the Whistle–Norman trough (borehole 1259870, 4846 ft.). (H) UM in the Ministic Offset inclusion-bearing quartz diorite (MIQD) matrix; note the contacts are sharp on two sides and diffuse on the third side (A-UM) with a tail of dark QD on one margin which appears to be QD that has been compositionally modified by assimilation of the ultramafic inclusion. (I) Inclusion of QD in MIQD and metabreccia. (J) Contact of the QD of the Frood Deposit with metabreccia formed from the footwall; the sample illustrates a sharp contact between igneous- and metamorphic-textured silicate matrix, but sulfide mineralization is developed in both rock types.

namely: (1) Radial dykes which are commonly contiguous with the Main Mass, and connected along troughs (eg, Creighton) and funnels (eg, Copper Cliff; Fig. 2.20A); (2) Concentric Offset Dykes which extend around the Main Mass and have no known connection to the Main Mass at the present level of erosion (eg, Machester and Hess; Fig. 2.20A), and (3) Discontinuous segmented concentric and radial Offset Dykes that consists of discontinuous pods of QD hosted by SUBX and metabreccia (eg, Frood-Stobie, Kirkwood, and Maclennan; Fig. 2.20A).

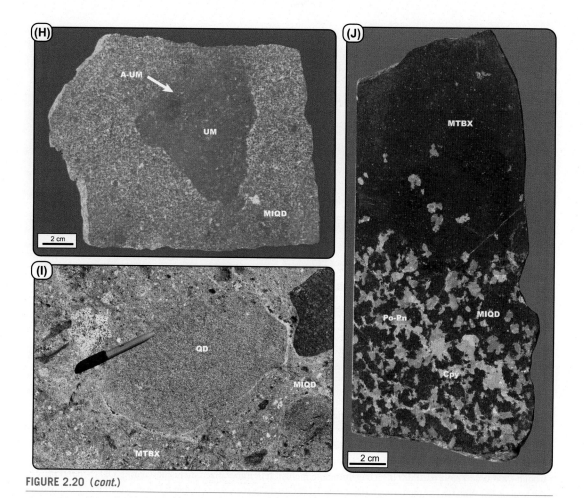

FIGURE 2.20 (*cont.*)

The Offset Dykes appear to represent injections of melt into country rocks during the earlier stages of melt sheet formation (eg, Lightfoot et al., 1997a), so their detailed geological relationships provides a foundation on which to understand the early evolution of the melt sheet. Moreover, the Offset Dykes host some of the largest and richest ore deposits within the SIC (Fig. 1.5B and Table 1.1B). For example, the Frood-Stobie Deposit is hosted in segments of inclusion QD in a belt of metamorphosed and partially melted country rocks and SUBX in the South Range Breccia Belt; the more continuous Copper Cliff radial Offset Dyke contains the Copper Cliff Mine Complex, and the radial Worthington Offset contains the Totten deposit (Table 2.1).

A detailed description of many of the dykes appears in Grant and Bite (1984), and more recent papers documenting information for each of the dykes are shown in Table 2.1. In this section attention is paid to the comparative geology of the Offsets and examples that underpin the geological relationships that constrain models of formation.

The Radial Offset Dykes

The radial Offset Dykes are the Copper Cliff, Creighton, Worthington, Trill, Ministic, Foy, and Parkin dykes (Grant and Bite, 1984; Lightfoot et al., 1997b; Smith et al., 2013; Fig. 2.20A; Table 2.1). They extend for a distance between a few hundred meters to ~20 km away from the outer contact of the Main Mass at the present level of erosion. The relationship between the dykes and the base of the SIC is typically one of gradation over a few meters from either the basal quartz-rich norite or inclusion-rich norites of the Sublayer into the underlying QD.

Radial offsets connect to the Main Mass where there are depressions (troughs and funnels) along the lower contact of the SIC. Not all troughs have associated Offset Dykes; for example the Trill trough has no direct connection to an Offset Dyke, but it may be structurally detached from the Trill Offset. In contrast, the Creighton trough has a short discontinuous dyke developed that is dextrally displaced from the keel of the trough and appears to terminate north of the Creighton Fault. Some of the longest and most continuous offsets link into the Main Mass through funnel-like depressions in the Main Mass (eg, Copper Cliff and Foy; Table 2.1).

Discontinuities in the radial Offset Dykes are due to both primary and secondary processes. For example, the Creighton Fault breaks the Copper Cliff Offset into two segments with a dextral displacement of ~700 m. Primary synmagmatic emplacement is related to anisotropy in the composition of the country rocks which controls pathways along which magmas were injected; this results in discontinuities, branching, changes in orientation, and/or widening of the dyke (Grant and Bite, 1984). The morphology of these breaks is quite variable, and examples of some of the common types of discontinuities are illustrated in Fig. 2.20A and the detailed insets (1–6) show specific examples of the different configurations.

The exact geological relationships at the distal termination of the radial Offset Dykes are poorly understood, but there is a general tendency for the dyke to decrease in width and inclusion content, and become more weakly mineralized with distance away from the Main Mass (Table 2.1).

Sulfide-poor and inclusion-free QD tends to occur at the margin of the dyke where the rock texture changes from a quench-textured fine-grained rock to a medium-grained rock with increasing distance from the margin of the dyke towards the center. The QD sometimes contains trace magmatic-textured Ni–Cu–PGE sulfide mineralization, but is normally barren or enriched in sulfides along fractures by postmagmatic transportation of metals from the deposit into the surrounding rocks (eg, Lightfoot and Farrow, 2002). The contact part of the QD sometimes contains fragments of adjacent country rocks that have spawled-off into the magma during emplacement of the dyke (Lightfoot and Farrow, 2002).

The inclusion-bearing unit of the dyke often, but not always, occurs in the center of the dyke and contains subeconomic to economic concentrations of magmatic sulfide mineralization. There are often sharp contacts between QD and MIQD that are indicative of two phases of emplacement (Lightfoot et al., 1997c). The composition and size of the inclusions varies between the different Offset Dykes. The inclusions are dominantly derived from local country rocks (eg, Copper Cliff Offset contains inclusions of metavolcanic rock and metagabbro, and the Worthington Offset contains 5 cm–10 m-sized amphibolite inclusions and small (<1 cm) fragments of mafic metavolcanic rock). Less commonly, the inclusion population comprises both local country rocks and exotic inclusions (eg, the Ministic Offset where fresh biotite-rich olivine melanorite inclusions occur with fragments of gabbro and mafic gneiss; Lightfoot et al., 1997c; Farrell et al., 1993, 1995; Farrell, 1997).

Table 2.1 Comparison of the Geology of the Sudbury Offset Dykes

Name of Offset Dyke	North or South Range	Type	Dip of Dyke	Country Rocks	Known Mineralization Related to Offset	Mines (A—Active; P—Past Producer)	Connection to Main Mass	Inflexions and Discontinuities	Funnel/ Embayment Connection to Main Mass	Splays
Hess	North	Concentric	75° N to vertical	Archean granitoid rocks	Occurrence	None	No	None known	No connection to Main Mass	None known
MacLennan	North	Discontinuous	Unclear	Archean granitoid	Small deposit	McLennan (P)	Not known	None known	No connection to Main Mass	None known
Foy-(Pele)	North	Pseudoradial	Vertical	Levack Gneiss and Archean granitoid rocks	Occurrence	Foy Offset (P)	Yes, via Foy funnel-shaped embayment	None known	Foy funnel ~750 m wide and 1 km deep contains SLNR and dyke develops 3 km to NW	None known
Trill	North	Radial	Unknown	Levack Gneiss	Occurrence	None	No connection known at this time	None known	No funnel known, but trough is very large	Well-developed bifurca-tions
Ministic	North	Radial	Subvertical	Levack Gneiss	Occurrence	None	Yes, via small embayment	None known	Funnel is quite narrow	None known
Cascaden	North	Radial	Unknown	Levack Gneiss	None known	None	No connection known	None known	No connection to Main Mass	None known
Whistle	North	Radial with discontinuous segments	Vertical	Archean granitoid, Levack Gneiss	Occurrence	Whistle-Podolsky (P)	Yes via Whistle embayment	None known	Wide deformed embayment, but no extensive funnel	None known
Parkin	North	Radial	Vertical	Archean granitoids, Levack Gneiss, Huronian metasedimentary rocks	Small deposit	Milnet (P)	Possibly a fault-displaced segment of the Whitle Offset	Milnet inflexion	No connection to Main Mass	Well-developed bifurca-tions in 2 locations
Manchester	South	Concentric with local discontinuous segments	Subvertical	Serpent Formation Metasedimenatry rock	Occurrence	None	No	None known	No connection to Main Mass	None known
Stobie East	South	Discontinuous	Subvertical	Stobie Formation metavolcanic and metasedimentary rocks	Occurrence	None	No	None known	No connection to Main Mass	None known

Podiform Geometry	Matrix Rock Types	Chareristics of Terminations	Total Length (a) (km)	Width, Range (m)	Depth Extent (b) (m)	Contacts	Brecciation of Host Rocks	MIQD With Ultramafic Inclusions	QD/ MIQD Ratio	References
No	MG QD; FG chilled QD contacts	N/A	>20	10–30	>0.5	Sharp	Local SUBX	Locally developed	~10	Wood and Spray (1998); Lightfoot et al. (1997c)
Not known	Amphibole-biotite and quartz-rich QD		<1	?	?	Not known	SUBX	Not known	Not known	Grant and Bite (1984)
No	SLNR in funnel; QD at >3 km from Main Mass	Not known	>30	500 (funnel) to 50 (distal)	>1	Often diffuse into GRBX	Extensive GRBX	Locally developed	~10	Tuchsherer and Spray (2002); Lightfoot et al. (1997c); Farrow et al. (2008a,b); Bailey (2011)
No	MG QD; FG chilled QD contacts	Not known	~3	1–5	>0.5	Sharp	Extensive GRBX	Not known	~2	Péntek (2013); Jago (2007); Golightly et al. (2010)
No	MG QD; FG chilled QD contacts	Not known	>11	10–>31	>~1	Sharp	Extensive GRBX	Locally developed	~5	Farrell (1997); Grant and Bite (1984); Smith et al. (2013)
Not known	Not known	Not known	~5	10–20	N/A	Not known	Not known	Not known	Not known	Smith et al. (2013); Bailey (2011)
Podiform QD in MTBX	Amphibole-biotite QD and MG–CG chilled QD	Faulted - originally contiguous with Parkin (?)	1.5	350 (near embayment) to 10–30	>1	Often diffuse into GRBX	Extensive GRBX	Locally developed	~0.5	Lightfoot et al. (1997c); Murphy and Spray (2002); Bailey (2011)
No	Amphibole-biotite QD	Disappears in meta-breccia	12	10–50	>1	Often diffuse into GRBX	Extensive GRBX	Locally developed	~4	Lightfoot et al. (1997c); Murphy and Spray (2002)
Main segment is continuous; western end has many <3 m QD pods in SUBX	MG granophyric QD; FG chilled QD contacts	Narrows into metabreccia and SUBX	6	10–20; pods smaller	>0.5	Sharp and chilled	Massive SUBX	Rare	>10	Lightfoot et al. (1997c); Grant and Bite (1984)
Podiform in SUBX or Granite Breccia	Amphibole-biotite QD	Narrows into metabreccia and SUBX	~1	~1–7	>1	Diffuse into SUBX	Extensive metabreccia and SUBX	No	~10	Grant and Bite (1984)

(Continued)

Table 2.1 Comparison of the geology of the Sudbury Offset Dykes (*cont.*)

Name of Offset Dyke	North or South Range	Type	Dip of Dyke	Country Rocks	Known Mineralization Related to Offset	Mines (A—Active; P—Past Producer)	Connection to Main Mass	Inflexions and Discontinuities	Funnel/Embayment Connection to Main Mass	Splays
Vermilion	South	Discontinuous	Subvertical	Amphibolite and gabbro (Sudbury Gabbro)	Small deposit	Vermilion (p)	No	N/A	N/A	N/A
Frood-Stobie	South	Discontinuous concentric	80° to NNW	Stobie Formation metavolcanic and metasedimentary rocks	Very large deposit	Frood-Stobie (A)	No	None known	No connection to Main Mass	None known
Kirkwood-McConnell	South	Discontinuous concentric	Subvertical	Elsie Mountain Formation metavolcanic rocks	Occurrence	Kirkwood (P)	Possible faulted linkage to Kirkwood Embayment	None known	No funnel	None known
Copper Cliff	South	Radial	Vertical with local flat domains	Huronian Supergroup metavolcanic and metasedimentary rocks, and gabbro intrusions	Large deposit	Copper Cliff (A)	Via Clarabelle funnel	860–865 discontinuity; 120 discontinuity	Clarabelle funnel ~1.5 km deep and up to 1 km wide contains Quartz-rich Norite and QD>>MIQD	Local splays adjacent to discontinuities
Creighton	South	Radial	Vertical	Creighton Granite	Very large deposit	Creighton (A)	Connected to Creighton embayment; heavily dissected by faults	None known	Termination of Creighton embayment has aspects of a narrowing funnel cut by major W–E structures	None known
Worthington	South	Radial	Vertical to subvertical	Huronian Supergroup metavolcanic and metasedimentary rocks, and gabbro intrusions	Large deposit	Totten (A)	Possibly, but heavily dissected by faults	Gap developed between Worthington #1 and #2	Creighton Fault detaches this Offset from Ethel Lake Victoria	Local splays adjacent to discontinuities and as offshoots at Totten #1
Ethel Lake-Victoria	South	Sheet	Flat to steep	Amphibolite.	Large deposit	Victoria (F)	Yes, via narrow heavily deformed QD–SUBX	N/A	Plexus of QD sheets developed adjacent to Victoria Embayment	N/A

Podiform Geometry	Matrix Rock Types	Chareristics of Terminations	Total Length (a) (km)	Width, Range (m)	Depth Extent (b) (m)	Contacts	Brecciation of Host Rocks	MIQD With Ultramafic Inclusions	QD/ MIQD Ratio	References
Chaotic eliptical pods of QD In massive SUBX	MG QD; FG chilled QD contacts	N/A	0.25	1–20	>0.5	Sharp	Massive SUBX	Sudbury Gabbro	~1	Grant and Bite (1984); Szentpeteri et al. (2002)
Podiform in SUBX or gabbro	Amphibole-biotite QD	Narrows into MTBX and SUBX	3	0–300	2.5 (Frood); 2 (Stobie)	Diffuse into SUBX	Extensive MTBX and SUBX	Very extensive	~0.5	Hawley (1962); Scott and Spray (2000); Lightfoot et al. (1997c); Hattie (2010)
Podiform development of QD+MIQD in SUBX and MTBX	Amphibole-biotite QD	Narrows into MTBX and SUBX	~2	<60	>0.5	Diffuse into SUBX	Extensive MTBX and SUBX	Locally developed	~1	Grant and Bite (1984); Farrow et al. (2011)
No	Amphibole-biotite, pyroxene-QD, and FG QD at contacts	Faulted termination	19 (heavily faulted)	1500 (at funnel) to 25–75 (proximal dyke), to 10 (distal dyke)	>2.75	Sharp	Extensive SUBX	Extensively developed	~2	Cochrane (1984); Grant and Bite (1984); Lightfoot et al. (1997c); Scott and Benn (2002); Magyarosi et al. (2002); Carter et al. (2005); Rickard and Watkinson (2001); Huminicki et al. (2005)
No	Amphibole-biotite and quartz-rich QD	Faulted termination	<1 (heavily faulted)		~1.5	Sharp	Black Porphyry host with some SUBX	Local	~0.5	Lightfoot et al. (1997c); Grant and Bite (1984)
No	Amphibole-biotite and quartz-rich QD; local spherulitic QD at contacts	Not known	>15	10–50	>2	Sharp	Local SUBX	Extensive	~2	Lightfoot et al. (1997c); Lightfoot and Farrow (2002); Grant and Bite (1984); Hecht et al. (2008); Lausen (1930)
No	MG-CG QD	N/A	1	200	>0.5	Sharp	Local SUBX	Not developed	~100	Lightfoot et al. (1997c); Lightfoot and Farrow (2002); Farrow et al. (2011)

Summary of the geological characteristics of the Offset Dykes. Much of the data is derived from the classic geological and petrological synthesis in Grant and Bite (1984) and maps from the published report of Lightfoot et al. (1997a). An increasing number of studies have extended the geological footprints of the Offset Dykes through careful mapping (eg, Wood and Spray, 1998; Tuchscherer and Spray, 2002 or through regional exploration that has resulted in the discovery of significant new dykes and extensions of known dykes (eg, Smith et al., 2013; Péntek, 2013).
Notes: (a), Measured along length of Offset at surface; (b), where known from diamond drilling; SUBX, Sudbury Breccia; MG, medium grained; FG, fine grained; QD, Quartz Diorite; MIQD, mineralized inclusion QD

The Concentric Offset Dykes

The concentric Offset Dykes form a discontinuous ring that wraps around the North Range Main Mass at a distance of ~15–20 km from the present erosional base of the SIC (Fig. 2.20A). The most continuous example of a concentric dyke is called the Hess Offset (Lightfoot et al., 1997c; Wood and Spray, 1998; Smith et al., 2013). In the South Range footwall there is no single continuous concentric QD dyke. The most continuous concentric South Range segment is the Manchester Dyke that approaches 7 km in length (Lightfoot et al., 1997c), but the western end of the dyke becomes discontinuous and segmented (Lightfoot et al., 1997c). The Manchester Offset is developed in an exceptionally wide domain (>250 m) of SUBX that has a very high ratio of matrix to inclusions (Lightfoot et al., 1997c).

The Discontinuous Segmented Offset Dykes

Several examples exist of bodies of inclusion QD that are enveloped in SUBX. However, the texture and mineralogy of these QDs is very similar to that developed in the Offset Dykes (Grant and Bite, 1984). The largest and best-known example of a segmented QD body contains the Frood-Stobie Deposit (Zurbrigg et al., 1957; Souch et al., 1969; Hattie, 2010) that is discussed later in this chapter and in Chapter 4.

A number of smaller discontinuous 2–30 m wide patches of QD occur in SUBX and metabreccia. Examples of discontinuous segmented Offset Dykes include a segment of a weakly mineralized QD dyke located east of the Stobie Deposit, discontinuous segments of weakly mineralized QD and strongly mineralized inclusion QD west of Kirkwood Mine, irregular patches of MIQD which host the Vermillion and New Victoria Deposits, and patches of QD in extensive domains of SUBX at the western termination of the Manchester Offset (Lightfoot et al., 1997c; Fig. 2.20A). Weak disseminated sulfide mineralization and local concentrations of semimassive sulfide in QD and inclusion-rich massive sulfide are commonly developed in the QD segments.

The segmented offsets are somewhat puzzling; in some cases the linkage of material to the SIC is no longer evident because of erosion, but in other cases the QD bodies appear to be almost entirely surrounded by SUBX. The extent to which the magmas were emplaced versus generated in situ in the breccia belts is an important question which I will return to examine in more detail later in the book.

Comparative Geology of the Offset Dykes

Table 2.1 and Fig. 2.20A summarizes the main features of the geology of the offsets and the characteristics of the discontinuities. Fig. 2.20A–J show examples of the geological relationships at the contacts of the Offset Dykes and within the QD. The following summary draws on the geology of different dykes to establish some of the common features that are important to understanding the associated ore deposits, and the mechanism of formation.

The Contact of the Offset Dykes with the Country Rocks

The contact relationships in a short segment of an Offset Dyke immediately east of the Stobie Deposit are illustrated in Fig. 2.20B. This photograph shows the contact between the QD and metabreccia is quite sharp (1–5 cm), and then there is a transition from metabreccia into SUBX over a distance of about 2 m. In the Frood-Stobie Deposit, the contact between the QD and metabreccia is very sharp as illustrated by the sample shown in Fig. 2.20J; in this sample, disseminated sulfide mineralization occurs in both the inclusion-bearing QD and in the country rock metabreccia.

The Frood-Stobie QD is a clear example of an Offset Dyke that comprises a mineralized inclusion-bearing QD that is in moderately sharp contact with subigneous-textured metabreccia that grades into SUBX (Table 2.1). The metabreccia often contain heavily melted and embayed country rock inclusions, which appear to require initial formation as SUBX, and then an additional phase of metamorphism and partial melting related to the injection of the QD magma.

Not all Offset Dykes have sharp contacts with the country rocks. In the Whistle, Parkin, and Foy Offset Dykes, the QD is often transitional through metabreccia into the surrounding SUBX, gneisses and granitoid rocks over a distance of tens of centimeters to meters (eg, Lightfoot et al., 1997c). The contact relationships of most of the Offset Dykes with the country rocks tend to become sharper at distances of more than 1 km from the base of the SIC, and although there are often domains of metabreccia at the margins, the dykes normally have knife sharp contacts with the country rocks and SUBX. In the case of the Trill Offset, a quenched QD is in very sharp contact with Archean granite, and develops both glassy quenched margins which grade into fine-grained QD with lathy plagioclase and pyroxene (Fig. 2.20C; Golightly et al., 2010).

Thickness and Continuity of the Offset Dykes

There is a general tendency for the long radial offsets to have a 500–1500 m wide funnel at the point where the Offset Dyke connects to the Main Mass; this is most evident for the Foy, Copper Cliff, and Whistle Offset dykes. The relationships of the Worthington dyke to the base of the SIC at Victoria are complex due to both primary configuration of the connection (several sheets and branches of QD are developed), and the effect of postemplacement deformation. Although the radial Offset Dykes tend to be wider in the funnel domains, the more distal dyke illustrates no clear relationship between the thickness of the Offset Dyke and the distance away from the base of the SIC.

There is a very small difference in the width of the Creighton Offset dyke over a radial distance of less than 750 m (typically 3–5 m wide). The Foy Offset extends for over ~20 km and it is typically 15–30 m wide. The Copper Cliff Offset dyke averages 40 m in width according to Cochrane (1984), but GoCAD models of the three-dimensional form of the walls of the Offset indicate that the average thickness is ~30 m, with the thicker parts of the dyke typically relating to the location of the magmatic sulfide ore bodies. The Copper Cliff Offset also exhibits primary discontinuities where the QD is absent. The more distal concentric Manchester and Hess Offset Dykes are mapped over considerable distances, with variations in thickness of 12–30 m and 15–30 m, respectively (Grant and Bite, 1984; Lightfoot et al., 1997c).

In contrast to the radial and concentric Offset Dykes, the discontinuous segmented melt bodies in the South Range Breccia Belt occur closer in to the Sublayer and vary in thickness from ~500 m at Frood-Stobie to a more typical range of 1–30 m in the Kirkwood and Stobie East segments (Fig. 2.20A). As a general rule, there appears to be no clear relationship between radial offset length and width, but concentric offsets developed close to the SIC exhibit much greater variation in thickness and continuity relative to those found at greater distances from the SIC.

Branching of the Offset Dykes

Branching refers to the common tendency of the Offset Dyke to break into two or more segments that extend a few tens or hundreds of meters away from one another before either merging again, or through termination of one of the branches. Examples of branching is documented in the Parkin Offset dyke where Tuchscherer and Spray, (2002) describe a large enclave of Levack Gneiss and metabreccia that separates two branches of the Offset (Fig. 2.20A); Lightfoot et al. (1997c) also show locations along the Parkin Offset where the dyke is locally branched within a wide zone of brecciated country rocks that are likely metabreccias (Fig. 2.20A). Examples of branching are also reported at locations in the Worthington Offset dyke (Lightfoot and Farrow, 2002), and there are segments of the Foy Offset that appear to follow two different parallel trends close to the junction of the Foy and Hess Offsets (Smith et al., 2013). The Trill Offset also develops several narrow subparallel sheets which form a network along the strike of the dyke (Golightly et al., 2010); these narrow distal QD veins have glassy margins in contact with granitic country rocks and develop glassy quench-textured QD in the core (Fig. 2.20C).

Controlling Influence of Country Rock Contacts and SUBX on the Localization of the Offset Dykes

Primary discontinuities that are unrelated to postemplacement deformation are found in both radial and concentric Offset Dykes. These discontinuities appear to be similar to the windows and breaks commonly found in dyke swarms. The controls on development of the breaks are broadly related to the way the country rocks break in response to dyke propagation (Mathieu et al., 2008). Discontinuities in the Offset Dykes occur where there are geological contacts that are normally associated with the development of SUBX where there is a strong competency contrast.

Two important examples of discontinuities that influence the localization of ore deposits in the Copper Cliff Offset are described by Cochrane (1984) and Mourre (2000). Discontinuities tend to be localized where the dyke cross cuts the stratigraphy of the country rocks; they are less common in Offset dykes that cut through more homogeneous country rocks. The Parkin Offset illustrates this relationship very clearly at Milnet where the QD has a sinistral inflection where it passes through carbonate-rich metasedimentary rocks (Fig. 2.20A; Lightfoot et al., 1997c). In locations where Offset Dykes pass through metabreccia, the Offset Dykes tend to break into discontinuous segments and in some cases local patches of QD which grade into the surrounding metabreccia; this relationship is developed in the QDs at the keel of the Whistle offset in Norman Township. Fig. 2.20D–G, I illustrates examples of inclusion-bearing QD with ultramafic inclusions that grade into metabreccia such that it is often very difficult to distinguish the two rock types without the examination of a thin section.

Diversity in Inclusion Type and Content in Offset Dykes

Offset dykes contain inclusion populations of five major types, namely: (1) Inclusions of local country incorporated into the margin of the dyke as a response to spawling of the dyke walls by the magma during injection; these inclusions often form trails which can be tracked back to the contact where the fragmentation and incorporation of the material is evident in outcrop (eg, Lightfoot and Farrow, 2002). (2) Inclusions that comprise nebulous segregations of quartz-feldspar that appear to be the incompletely digested fragments from country rock felsic gneiss; these inclusions often develop in unmineralized or weakly mineralized hypersthene QD adjacent to the contacts in North Range offsets such as Parkin and Foy (eg, Tuchscherer and Spray, 2002). (3) Amphibole-biotite QD or amphibole (after hypersthene) QD with small inclusions of mafic metavolcanic rock that comprise one of the main rock types in the MIQD unit (eg, Scott and Benn, 2002). (4) Amphibole-biotite QD, amphibole (after hypersthene) QD, and semi-massive sulfide that carries large inclusions of mafic-ultramafic rock that are sometimes of local derivation, but sometimes have no obvious source in the nearby country rocks (eg, Lightfoot and Farrow, 2002). (5) MIQD through to massive sulfide that contains inclusions of QD that appear to be derived from the inclusion free portion of the offset. The development of economic ore deposits is more typically associated with types 3–5 rather than types 1–2.

Textural and Petrological Diversity in the Matrix of the Offset Dyke

The margins of many North Range and South Range Offset Dykes have a fine-grained matrix containing acicular crystals of feldspar that have a quench-texture (Grant and Bite, 1984). The contact type

QD grades into a medium-grained hypersthene QD (near to the Main Mass) or amphibole-biotite QD over 1–3 m, and the medium-grained QD can contain local coarse-grained patches of pyroxene or amphibole-biotite QD. The entry of inclusions and significant amounts of sulfide demarcates the QD from the inclusion-bearing QD. The matrix of the inclusion-bearing QD is commonly either pyroxene QD (near to the Main Mass) or hornblende-biotite QD, which is similar to the QD. Hypersthene QD is more common in the North Range than the South Range (Table 2.1).

The distal parts of the concentric Offset Dykes such as the Manchester Offset are comprised of a more quartz-rich to granophyric QD when compared to the more common radial Offset Dykes, and granophyric QD is also recorded in the Maclennan Offset (Grant and Bite, 1984); similar granophyric QDs are also reported in some of the newly discovered dykes in the NW Range (Smith et al., 2013). The economically mineralized offsets tend to comprise amphibole-biotite and hypersthene QD, and some segments of the Offset Dykes can comprise massive sulfide and inclusion-bearing massive sulfide with no development of QD. The granophyric QD offsets however do contain examples of subeconomic Ni–Cu sulfide mineralization.

Alignment and Reaction of Inclusions with Magma

Examples of aligned inclusions within the central part of the Worthington Offset are recorded by Lightfoot and Farrow (2002), and this may reflect flow-alignment of inclusions. Rare examples of inclusions which have sharp contacts on one side and nebulous contacts on the other side appear to indicate that the inclusions moved through the QD magma leaving a trail of contamination in the wake; Fig. 2.20H shows one example of an ultramafic inclusion within inclusion QD from the funnel of the Ministic Offset. This unoriented sample is remarkable in demonstrating that more primitive ultramafic inclusions can be assimilated in what was presumably a superheated melt that had the composition of a QD. This information also offers a potential approach by which flow direction can be identified in Offset Dykes.

Metamorphic grade and deformation of the Offset Dykes

The metamorphic grade of Offset Dykes in the South Range is typically higher (amphibolite facies) than that of the Offset Dykes in the North Range (greenschist facies). The development of amphibole-biotite QD in the central parts of South Range Offset Dykes may be related to a weak planar deformation fabric defined by mafic minerals and small inclusions in the core of the dyke (Table 2.1).

Detailed Geological Relationships: Case Studies of the Worthington and Copper Cliff Offset Dykes

To illustrate the diversity in geology of the Offset Dykes, and their linkages to the ore deposits described later in this book, this section reports examples from the Worthington and Copper Cliff dykes. These two examples are chosen for case studies as they illustrate the geology of the environment that hosts economic ore deposits. Similar geological relationships can be found in other more weakly mineralized Offset Dykes, and some of these locations provide important information that help establish the reasons why some offsets contain ore deposits and others do not. The key characteristics of the other dykes are shown in Table 2.1 where references are provided to further reading.

The Worthington Offset Dyke

The Worthington Offset is one of the best exposed of all the dykes, and the relationships at surface and depth provide the basis for a three-dimensional model for the Offset Dyke and associated ore deposits. Fig. 2.21A,H show the location of the Offset Dyke in relation to the country rocks.

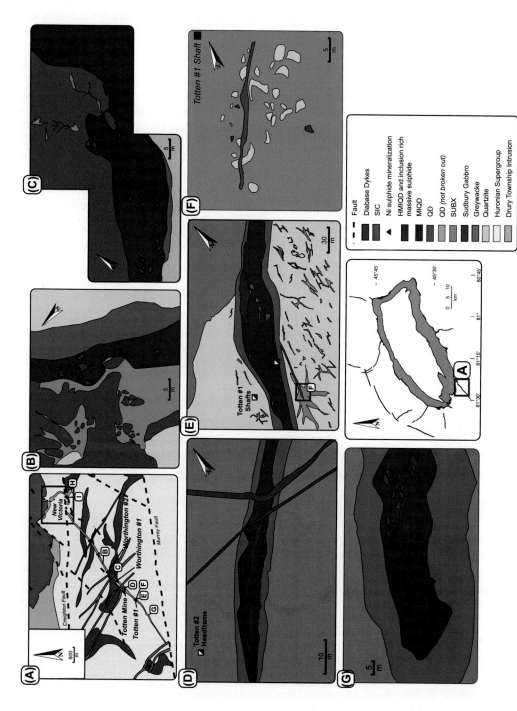

FIGURE 2.21 Geological Relationships in the Worthington Offset Dyke Summarized from Lightfoot et al. (1997c) Lightfoot and Farrow (2002), and Mapping of the Outcrop on Which the Totten Mine Complex Now Rests

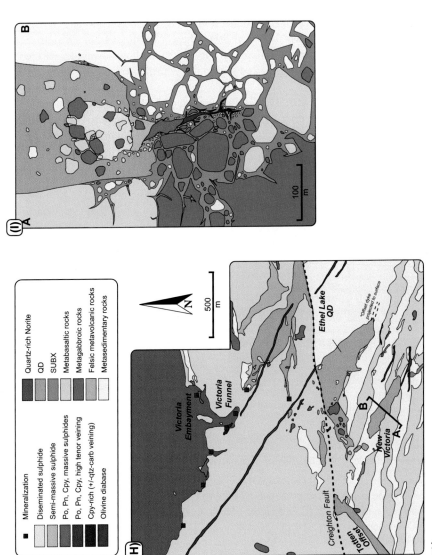

FIGURE 2.21 *(cont.)*

(A) Local geology of the Worthington Offset; (B) Geology of the Worthington Offset where it passes between metasedimentary rocks and Sudbury Gabbro. (C) Geology of the Worthington Offset where it cuts through Sudbury Gabbro; (D) Geology of the surface outcrop above the Totten #2 Deposit. (E) Geology of QD Offset Dyke developed southwest of the Totten #1 Deposit, where there is a well-developed zonation from quench-textured QD, through medium-coarse-grained QD, and then into progressively more heavily mineralized inclusion-rich QD (MIQD through VMIQD) with large amphibolite inclusions. (F) Detailed relationships in the narrow offshoot dyke of QD above the Totten #1 Deposit. (G) Geology of Worthington Offset where a zone of weakly mineralized inclusion QD is developed ~0.5–0.75 km along-strike from the Totten #1 Deposit. (H) Geology Map of the Ethel Lake QD sheet and the surface manifestation of the New Victoria Deposit. (I) Model showing geological relationships in the new Victoria Deposit in cross-section

(I) After Lightfoot et al. (1997c) and Farrow et al., (2008a,b); (J) after Farrow et al., (2008a,b).

The Offset comprises two distinctive domains located north and south of the Creighton Fault; the domain to the south is a continuous radial Offset Dyke hich cuts through metasedimentary rocks of the Huronian Supergroup and mafic intrusive rocks; the domain to the north comprises a combination of sheet-like bodies of QD (the Ethel Lake intrusion; Lightfoot et al., 1997c), discontinuous segments of QD in SUBX (eg, the new Victoria Deposit which is named to distinguish it from the original Victoria Deposit located cloe to the base of the SIC), and a narrow connection through metabreccias and SUBX to the heavily faulted Victoria Embayment shown in Fig. 2.21H. These two sections of the Offset Dyke are termed the Totten and Victoria segments, respectively. Fig. 2.22A–J shows examples of the geological relationships in the host rocks, at the contacts and within the Worthington and Victoria segments of the Offset Dyke that are discussed below.

The Totten Segment of the Worthington Offset Dyke. The Totten segment of the Worthington Offset extends southwest of the Creighton Fault for more than 15 km, and varies in width from ~10–70 m, with an average width of 20 m. The Offset Dyke exhibits a local variation in dip that is typically less than

FIGURE 2.22 Outcrop Photographs of Geological Relationships in the Worthington Offset

(A) Inclusions of amphibolite derived from the Sudbury Gabbro (SG) in a matrix comprising dominantly MIQD to the northeast of the Worthington #2 Deposit. (B) Contact of the quench-textured QD (QQD) with baked Sudbury Breccia (SUBX) containing quartzite fragments (Q) at Totten Number 1 Shaft. (C) Contact between QD and MIQD located northwest of the historic Worthington #1 Mine. (D) QD inclusion in MIQD located northwest of the historic Worthington Mine. (E) Inclusion of QD in MIQD of the Totten Deposit (sample number MU611336). (F) MIQD with fragments (SG) of amphibolite derived from the Sudbury Gabbro country rocks. (G) Narrow QD apophysis of the main dyke cutting a zone of Sudbury Breccia (SUBX) in Mississigi Formation quartzite at Totten #1 Mine. (H) Coarse-grained quartz diorite (CGQD) patches within medium-grained amphibole-biotite QD from the QD intrusion north of Ethel Lake. (I) Outcrop of Sudbury Breccia (SUBX) in quartzite (Q) with discontinuous segments of QD to the south of Ethel Lake. (J) Shatter cone fabric developed in amphibolite of the Sudbury Gabbro to the east of the Totten Deposit.

(A) See Lightfoot et al. (1997c) for detailed geological maps; (H) see Lightfoot et al., 1997c for detailed geological maps.

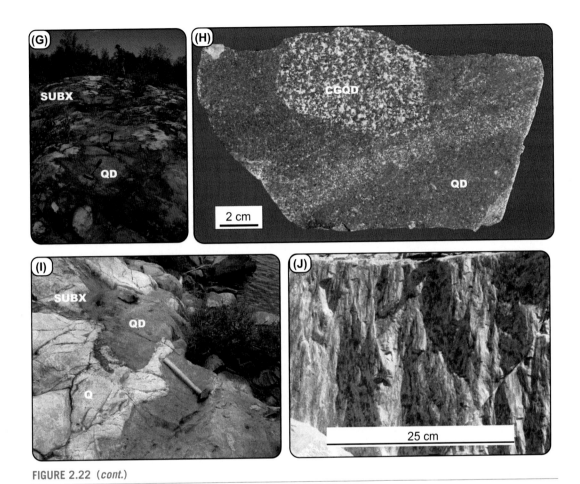

FIGURE 2.22 (*cont.*)

15° from vertical. The Offset passes through metavolcanic and metasedimentary rocks of the Huronian Supergroup and sheets of Sudbury Gabbro.

Sections of the Offset that cut through the Sudbury Gabbro intersect country rocks that comprise metagabbro and amphibolite. In these sections of the Offset Dyke, fragments of Sudbury Gabbro (0.25–10 m in diameter) occur within the contact zone of the Offset Dyke and mineralized inclusion-bearing QD has impregnated the amphibolite of the Sudbury Gabbro Intrusion to form a magmatic mega-breccia. The country rocks and inclusions both comprise similar rock types, and this is the basis for suggesting that the amphibolite and metagabbro inclusions are derived from the adjacent country rocks (Lightfoot and Farrow, 2002). An example of this geological relationship is shown in Fig 2.21B,C in a segment between the Worthington #1 and #2 Deposits. In this location, an amphibolite-rich inclusion population is present within the Offset Dyke and the adjacent country rock is Sudbury Gabbro with both SUBX and small intrusion of QD. These segments of Offset tend to be located within the center of the more massive amphibolite core of the Sudbury Gabbro; these amphibolites are heavily fractured with the development of horse-tail structures that resemble shatter cones (Fig. 2.22J). In the central part of the Sudbury Gabbro sill between Worthington #1 and #2 Deposits, the Offset Dyke branches (Fig. 2.21A) and consists largely of QD and inclusion-bearing QD with small mafic inclusions.

SUBX tends to be concentrated towards the contacts of the heavily shock-modified Sudbury Gabbro. It appears possible that the intersection of these breccia zones and shattered amphibolites provided pathways along which the QD magma could invade the amphibolite and metagabbro and incorporate fragments; these relationships are important as there is a strong spatial linkage between the localization of economic zones of mineralization and the two contacts of the Sudbury Gabbro as well as an association of Sudbury Gabbro inclusions with the economically mineralized parts of the Offset (Lightfoot and Farrow, 2002).

The contact between QD and SUBX is normally very sharp with quench-textured QD adjacent to baked SUBX (Fig. 2.22B). The margin of the offset is composed of contact type QD, and the grain-size rapidly increases away from the contact into amphibole-biotite QD. Fig. 2.21D shows an example of the relationship near to Totten Mine where the incorporation of locally derived fragments of metagrey-wacke of the Mississagi Formation into the marginal portion of the QD is developed. This type of QD is not grouped with the MIQD as the fragment population is clearly derived from the immediate wall rocks rather than injected with QD that carried immiscible magmatic sulfide and fragments with both local and exotic origins.

The Worthington Offset Dyke cross cuts and locally invades the metasedimentary rocks of the Huronian Supergroup. The degree of invasion into the country rocks by QD is controlled by the com-position and extent of brecciation of the country rock, namely: (1) QD impregnates behind large blocks of shaley metasedimentary rock that were spawled-off the walls during emplacement of the dyke, (2) The development of SUBX in the quartzites can produce changes in the geometry of the contact and local apopyhses of QD penetrate into these breccia zones (Figs. 2.21E,F, 2.22G).

Within the Offset Dyke, the contacts between QD and the inclusion-bearing QD can be either sharp or gradational. An excellent example of a sharp contact (<2 cm wide) is developed to the northeast of the Worthington Mine (Figs. 2.21D, 2.22C), where sulfide- and inclusion-free medium-grained QD of the offset margin is cut by inclusion-rich sulfide-bearing QD of the offset core. The inclusion QD contains subangular 10–50 cm sized fragments of sulfide- and inclusion-free QD, which appear to be derived from an earlier injection of QD that presumably solidified against the walls prior to the injection of the second pulse of sulfide-laden magma (Fig. 2.22D,E).

The mineralized inclusion QD comprises a continuum of rock types from a weakly mineralized QD with 1–10% small (1–10 mm) inclusions of mafic volcanic rock; this rock name is abbreviated in the text and diagrams to MIQD to reflect the composition of the silicate matrix, and the requirement for both trace to 5% sulfide mineralization and the development of fewer small inclusions. The MIQD forms a continuum into a more heavily mineralized variety (5–25% sulfide) with both small inclu-sions of unmineralized inclusion-free QD (~5% inclusions, 1–50 cm in size) and larger inclusions of amphibolite and metagabbro (5–20% inclusions; 10 cm-2 m in size; Fig. 2.22F); this group of rocks is typically very mineralized, and the rock type is abbreviated in the text and diagrams to highly mineral-ized inclusion QD (HMIQD).

As the proportion of sulfide minerals increases at the expense of the QD matrix, the proportion of inclusions tends to increase, and some segments of the Totten #2 deposit comprise 100% sulfide min-eralization as the host with a either a clast-supported matrix or floating inclusions of amphibolite and metagabbro (this is discussed in more detail in Chapter 4).

The Victoria Segment of the Worthington Offset Dyke. The geology of the Victoria segment of the Worthington Offset is shown in Fig. 2.21H (after Lightfoot et al., 1997c and Farrow et al., 2011) and as a simplified cross section through the concentric portion of the segmented offset that hosts the new

Victoria Deposit (Fig. 2.21I; Farrow et al., 2011). The heavily faulted Victoria embayment (and the original Victoria Deposit) is connected to the QD offset through a thin discontinuous zone of SUBX, metabreccia, and QD to a body of QD termed the Ethel Lake Sheet that cuts through the Elsie Mountain Formation as a sheet-like body with segments of concentric discontinuous dyke within zones of SUBX and metabreccia (Fig. 2.21H). The QD sheet comprises a weakly mineralized inclusion-free amphibole-biotite QD with local patches of coarse-grained QD (Fig. 2.22H).

An outcrop on the southwest shore of Ethel Lake consists of discontinuous segments of weakly mineralized inclusion-free amphibole-biotite QD in SUBX and metabreccia developed in a metaquartzite of the Mississigi Formation (Fig. 2.22I; Lightfoot et al., 1997c). This outcrop is a surface manifestation of a concentric belt of breccia and QD developed adjacent to a metagabbro and metaquartzite that contains the new Victoria Deposit (Fig. 2.21H,I; Farrow et al., 2011).

Synthesis of the Geological Relationships in the Worthington Offset Dyke

The geology of the Totten segment of the Worthington Offset provides a basis to understand the relationships between the Offset Dyke and the country rocks, and the internal structure of the offset.

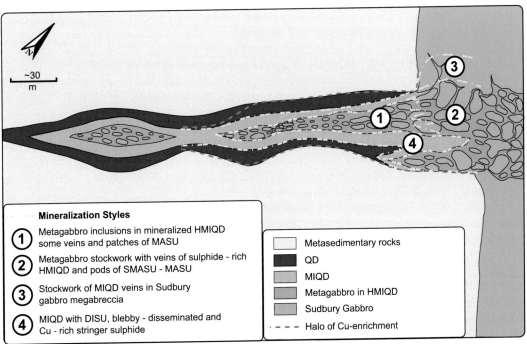

FIGURE 2.23 Geological Plan Showing a Summary of the Relationships Between the Offset Dyke and the Country rocks, and the Styles of Mineralization Developed in the Worthington Offset at Totten Mine

The Worthington Offset Dyke cuts through metasedimentary rocks and amphibolite intrusions. The inclusion-poor QD is chilled against the country rocks and contains inclusions that were spawled-off from the country rocks. The central part of the dyke consists of heavily mineralized inclusion quartz diorite (HMIQD). The MIQD grades into a megabreccia at the point where the Offset Dyke passes from metasedimentary rock into amphibolite. Styles of mineralization developed in the dyke are shown.

Based on Lightfoot and Farrow (2002).

Fig. 2.23 provides a simple summary of these relationships as found in the Totten Deposit that provide a framework for the description of the Deposit provided in Chapter 4 (Lightfoot and Farrow, 2002):

1. There is a close relationship between the location at which the Offset crosses the southern contact of a Sudbury Gabbro intrusion and the development of a megabreccia of Sudbury Gabbro and the QD and massive sulfide that contains inclusions derived from the Sudbury Gabbro country rock.
2. Where the QD is in contact with metasedimentary rocks, the dyke comprises a marginal gabbro that is in sharp contact with the MIQD; inclusions of QD from the margin are contained within the MIQD.
3. Economic mineral zones tend to be associated with the presence of inclusions of amphibolite derived from the Sudbury Gabbro. As the proportion of amphibolite inclusions declines, so too does the thickness of the mineral zone and the proportion of sulfide to QD located in the matrix between the inclusions.

The Copper Cliff Offset Dyke

The radial Copper Cliff Dyke is the most heavily mineralized Offset of the SIC and it is also well exposed and extensively explored. Fig. 2.24A shows the location of the Offset in relation to the different types of country rock. A west-facing long section between Kelly Lake and the Funnel is shown in Fig. 2.24B with the projected distribution of ore bodies, structures, and discontinuities shown in the plane of the dyke.

The Offset comprises two distinctive segments that cut from North to South through Huronian Supergroup metavolcanic and metasedimentary rocks, Creighton Granite and mafic intrusive rocks (Nipissing Diabase and/or Sudbury Gabbro). The domain to the north comprises the funnel connection to the Main Mass (Fig. 2.24A); the domain to the south of the funnel comprises the major portion of the Offset Dyke and its associated economic wealth. The Offset Dyke is cut by subvertical east–west faults and SW–NE faults that dip at ~45° towards the NW; these structures are post magmatic in age. A second group of discontinuities in the dyke are primary in origin and produced by syn-magmatic displacement (Cochrane, 1984). The ore deposits of the Copper Cliff Offset represent one of the largest resources of Ni–Cu–PGE-rich sulfide mineralization in the Sudbury Basin (Fig. 1.5B and Table 1.1B).
The Copper Cliff Funnel Segment of the Offset Dyke. The Copper Cliff Offset is connected to the Main Mass via a funnel-shaped depression that is composed of quartz-rich South Range Norite (Fig. 2.24A); amphibole-biotite QD is developed in the lower part of the Main Mass in the funnel and in the transition between the funnel and the dyke (Lightfoot et al., 1997c; Capes, 2001).

The funnel structure trends SSE of the base of the Main Mass and narrows from 1500 m to ~750 m, into the area of Clarabelle Pit where two NNE-trending structures provided space for QD and sulfide mineralization to penetrate through Stobie Formation metavolcanic and metasedimentary rocks along a zone of SUBX towards the NNE; these apophyses are locally referred to as "horns" (Fig. 2.24A).
The Copper Cliff Offset Dyke. To the south of the funnel, the offset narrows into a 20–70 m wide dyke that passes close to the contact of the Creighton Granite and the Elsie Mountain Formation with a sharp subvertical contact quenched QD in contact with weakly baked Creighton Granite (Fig. 2.25A). The center of the offset is composed of MIQD with an inclusion population dominated by fine-grained fragments of mafic volcanic rock (Grant and Bite, 1984; Lightfoot et al., 1997c). The contact between the QD and MIQD is diffuse and indicative of a shorter interval between the injections of the two melts than that proposed for the Worthington Offset Dyke.

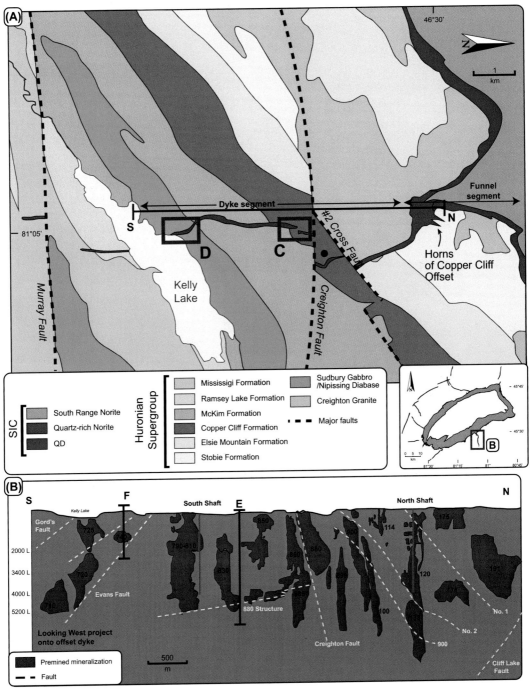

FIGURE 2.24

(A) Geology of the Copper Cliff Offset Dyke showing the major faults, discontinuities, and stratigraphy of the host rocks, and the numbering of ore bodies (historically mined and unmined mineral zones are depicted). (B) North–south long section showing the projected position of the main ore bodies.

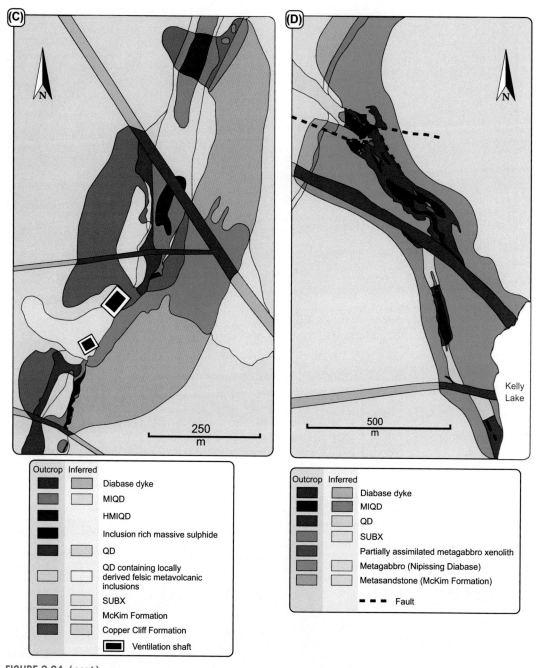

FIGURE 2.24 (*cont.*)

(C) Detailed geological map of primary discontinuity in the Copper Cliff Offset Dyke south of the Creighton fault between the 865 and 860 ore bodies. (D) Geological map of the Copper Cliff Offset to the north of Kelly Lake. (E) North-facing cross section showing the geology to the north of Kelly Lake. (F) North-facing section of the Copper Cliff Offset Dyke in the northern part of the 740 mineral zone of the Kelly Lake Ore Body (OB).

(B) After Farrow and Lightfoot (2002); (C) after Mourre (2000); Lightfoot et al. (1997c); (E) after Lightfoot and Evans-Lamswood (2015) and based on Mytny, Pers. Comm. (2010).

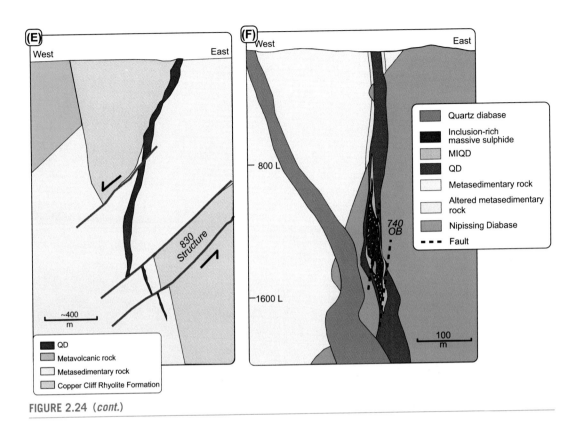

FIGURE 2.24 (*cont.*)

The dyke is broken by the Number 2 Cross Fault which dips at 45 degrees to the NW (Fig. 2.24A, B); this postmagmatic structure is a brittle-ductile fault which offsets the southern extension of the dyke ~500 m to the west. The offset continues to the south beneath the Town of Copper Cliff, and it is then cut by the Creighton Fault with a dextral offset of 700 m to the west (Fig. 2.24A, B). More details about these structures is provided later in this chapter where the Structural Geology of the SIC is presented.

An outcrop of the offset developed between the Copper Cliff Formation rhyolites and the McKim Formation greywacke exhibits a dip of ~80° WNW (Fig. 2.24B). This outcrop illustrates a primary discontinuity in the Offset Dyke that is coincident with in a thick zone of SUBX developed near to the contact of the rhyolite and the metasedimentary rocks (Mourre, 2000; Fig. 2.24C). The dyke is composed of a quenched QD in contact with the Copper Cliff Rhyolite (Fig. 2.25C); there are many fragments of rhyolite in the dyke on the western contact and Copper Cliff Rhyolite and McKim Formation metasedimentary rock at the eastern contact. At this location, the texture of the QD is coarser-grained towards the center of the Offset where MIQD, HMIQD, and semimassive sulfide with inclusions are developed (Fig. 2.24C). The heavily mineralized rocks of the Offset Dyke are the surface expression of the 860 Ore Body that is mined at depth.

The MIQD contains small (0.1–5 cm) inclusions of metavolcanic rock in an amphibole-biotite QD that develops a weak foliation parallel to the dyke contacts; this foliation is also locally found at the

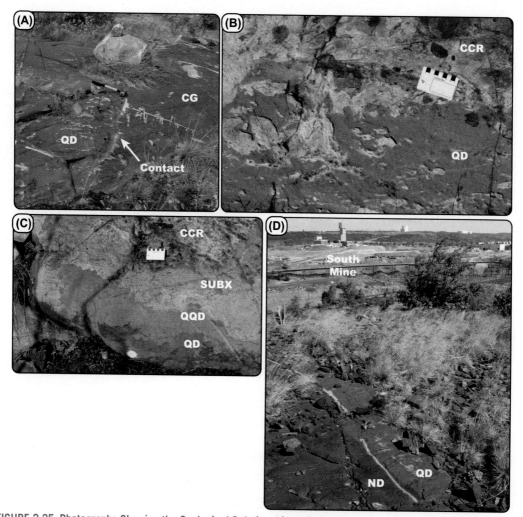

FIGURE 2.25 Photographs Showing the Geological Relationships Within the Copper Cliff Offset and With the Surrounding Country Rocks

(A) Contact of the inclusion-free unmineralized QD with the Creighton Granite (CG) on the eastern flank of the Copper Cliff Offset to the south of the northernmost headframe; (B) Contact zone of QD of the Copper Cliff Offset with the Copper Cliff Rhyolite to the south of the Creighton Fault; abundant fragments of rhyolite are entrained within the marginal QD up to 1 m from the contact; (C) Southern termination of the quenched QD (QQD) and QD of the Copper Cliff Offset Dyke in Sudbury Breccia (SUBX) developed in country rocks of the Copper Cliff Rhyolite Formation and metasedimentary rocks of the McKim Formation south of the 865 Ore Body. (D) Contact of QD of the Copper Cliff Offset Dyke with Nipissing Diabase (ND) to the north of Kelly Lake; the Nipissing is heavily baked, brecciated, and contains pods of pegmatoidal gabbro that may be primary or originate through recrystallization.

contact, but it is not as pervasive in the weakly mineralized and unmineralized QD. The Offset Dyke narrows over a distance of ~300 m from ~25 m in width down to a few centimeters in width in a zone of extensive SUBX (Figs. 2.24C and 2.25C). The QD at the termination develops a quenched texture against the SUBX and rhyolite. The SUBX has previously been described to contain 10 cm–2 m diameter pod-like segments described as QD (Grant and Bite, 1984). More recent petrographic and geochemical studies indicate that these rocks are highly recrystallized metasedimentary rock inclusions (Mourre, 2000).

To the south of the discontinuity, the Offset widens over a distance to ~70 m, and develops another zone of semimassive sulfide with inclusions and HMIQD on the eastern side that grades through MIQD into amphibole-biotite QD at the western contact.

Further to the South, the Offset Dyke is exposed where it cross cuts a body of Nipissing Diabase that is hosted by the McKim Formation on the north shore of Kelly Lake (Figs. 2.24D and 2.25D). This outcrop shows the surface expression of the dyke as a NNW–SSE trending subvertical intrusion with a segment that has a NW–SE trend. The adjacent gabbroic and metasedimentary country rocks are extensively partially melted, brecciated, and recrystallized. The dyke broadly decreases in width from ~30 m in the north (where it is cut by a NW–SE fault) to ~15 m in the south. The contacts of the dyke with country rocks are sharp and irregular; the fine-grained quenched QD at the margin grades into medium-grained amphibole-biotite QD and then into a central part of MIQD. The inclusion population in the dyke comprises small fragments of mafic volcanic rocks.

The southernmost part of the Copper Cliff Offset crops out to the south of Kelly Lake where there is a very simple transition from contact QD through medium-grained amphibole-biotite QD to a core of MIQD with fragments (0.2–5 cm in diameter) of mafic volcanic rock. The offset is terminated to the south in a zone of extensive SUBX cut by the Murray Fault. Small segments of Offset have been mapped south and west of Kelly Lake, and they are largely composed of amphibole-biotite QD (Lightfoot et al., 1997c).

The geology of the Copper Cliff Offset at depth often comprises a relatively simple continuous dyke that varies in width up to 50 m; there is no evidence of a decrease in the width of the dyke with depth. Small deviations from a vertical configuration are controlled by the composition of the host rock, but more significant deviations occur where there are low-angle primary discontinuities such as the one associated with the 865 Ore Body as well as subvertical faults that displace and sometimes thicken the width of mineralization in the dyke (Fig. 2.24E). In locations where there are changes in the composition of the country rocks and extensive development of SUBX, the offset narrows and breaks up into discontinuous segments of QD over a distance of less than 100 m vertical extent, before continuing again at depth.

The vertical distribution of mineralization in the 740 Ore Body is illustrated in section view in Fig. 2.24F; in some segments the Offset is composed largely of QD and in other sections there are wider domains if MIQD that are associated with the mineral zones. The relationship between mineralization and the contact zone of the Nipissing Diabase intrusion is not fully established (Fig. 2.24F), and in this respect the relationship is different to the one established between the Totten mineral zone and the Sudbury Gabbro (Fig. 2.23).

Synthesis of Geological Relationships in the Offset Dykes

1. Offset Dyke can be grouped into three principle types, namely: (a) radial; (b) concentric, and (c) discontinuous and segmented dykes.

2. The radial Copper Cliff Offset Dyke is endowed with large ore deposits; the Worthington Offset Dyke contains important ore deposits, but other radial Offset Dykes such as Foy and Ministic appear to contain subeconomic mineral concentrations.

3. The continuous segments of the concentric Offset Dykes contain no known economic mineralization. In contrast, the discontinuous segmented Offset Dykes enveloped in wide zones of metabreccia and SUBX, contain major ore deposits such as the Frood-Stobie Deposit.

4. Offset Dykes have primary discontinuities along their length and at depth that often occur where the dyke cuts through the contacts between different country rocks that coincide with zones of SUBX.

5. Proximal to the SIC, the discontinuous Offset Dykes commonly have sharp to gradational contacts with metabreccia and/or SUBX. Further away from the SIC, the margins of the Offsets are often rapidly cooled against the country rocks and/or SUBX.

6. The Offset Dykes record two or more injections of magma from the melt sheet. The MIQD was injected after the QD, and the MIQD contains inclusions of the QD that was initially injected into the Offset Dyke. These relationships provide critical constraints on the timing of sulfide saturation in the melt sheet (documented in more detail in Chapters 3 and 4).

7. The economically mineralized segments of the Offset Dykes contain abundant mafic and ultramafic inclusions, and few inclusions of metasedimentary or granitic country rocks.

8. Economic mineralization in the Offset Dykes tends to be associated with MIQD, and comprises HMIQD, and semimassive to massive sulfide with varying proportions of mafic-ultramafic inclusions. These inclusions are sometimes sourced from nearby country rocks (eg, in the Totten Offset; Fig. 2.23). In contrast, many of the mafic-ultramafic inclusions of the Ministic and Frood-Stobie Offset Dykes have an exotic source.

THE GEOLOGY OF THE SUBLAYER

The basal contact of the SIC is characterized by the development of embayment and trough-like depressions that contain an igneous-textured magmatic breccia with a noritic or granitic matrix. These magmatic breccias are collectively termed the Sublayer, and they are the principal host of mineralization at the lower contact of the SIC, and the source of well over half the mineral wealth produced from the Sudbury Structure (Chapter 1).

The literature on Sudbury Geology has variously included the Sublayer as part of the SIC (Lightfoot et al., 1997b), as a separate phase of the SIC (Pattison, 1979), and as partially melted footwall (eg, Dressler, 1984c). In some publications, the Offset Dykes are also included with the Sublayer, but this book does not include the Offset Dykes with the Sublayer.

The Offset Dykes and the Sublayer are considered to be part of the SIC in this book, but the Sublayer is broken-out from the Main Mass because it is a magmatic breccia.

The volume of Sublayer developed at the base of the SIC is broadly related to the size of the trough or embayment. Small to large localized depressions at the base of the SIC are termed 'Embayments' (eg, the Nickel Rim and North Range Shaft embayment structures) whereas connected groups of embayments along trends are collectively termed 'Troughs' (eg, the Creighton and Coleman-Levack trough structures).

The location of many of the embayment and trough structures developed around the margin of the SIC is shown in Fig. 2.26A, and the four diagrams shown as insets to Fig. 2.26A illustrate the shape of the embayments and troughs projected onto the base of the SIC

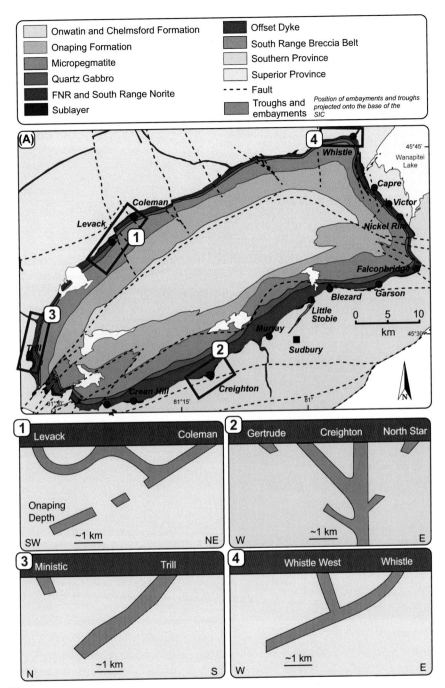

FIGURE 2.26

(A) Distribution of the Sublayer and names of the main embayment and trough features described in the text. Inset diagrams 1–4 show the geometry of the trough and embayment structures projected onto the base of the SIC; these diagrams illustrate the diversity of morphologies found in these settings on plans drawn on the surface of the SIC at different locations around the SIC. The inset diagrams (1–4) summarizes the terminology used to describe the Offset Dykes. The exact three dimensional morphology of the embayments and troughs is established based on detailed surface geology, drilling, and underground mapping, so this morphology to a depth of less than or equal to 3 km is now well established.

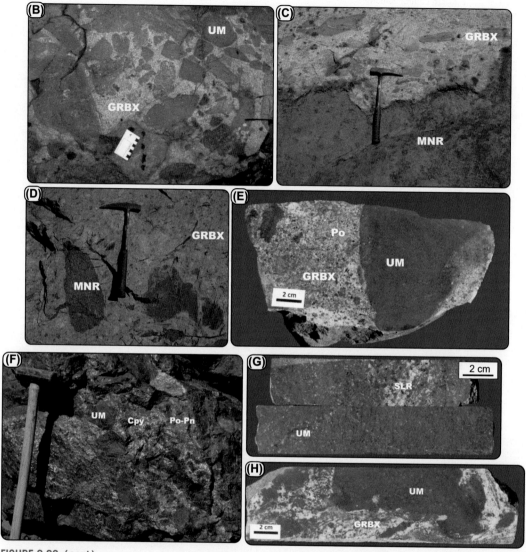

FIGURE 2.26 (*cont.*)

(B) Photograph of Sublayer from the Levack area; rounded mafic and ultramafic (UM) inclusions are separated by a Granite Breccia matrix. (C) Contact between MNR and Granite Breccia at Coleman Mine. (D) Inclusions of MNR in Granite Breccia at Coleman Mine. (E) Photograph of a slab of Granite Breccia with fragmental pyrrhotite (Po) showing an ultramafic inclusion in a granitic matrix and development of both interstitial and fragmental sulfide, Longvack Mine. (F) Typical mineralized inclusion-rich Sublayer Norite from Gertrude West Mine; UM, ultramafic inclusion, Cpy, chalcopyrite; Po–Pn, pyrrhotite–pentlandite. (G) Ultramafic inclusion in Sublayer Norite from the Trill trough. (H) Ultramafic inclusion in Sublayer Granite Breccia, Trill Embayment (drill core 60049, depth 6208 ft).

The Sublayer Norite and Granite Breccia

Embayments generally contain an upper unit of variably mineralized Sublayer Norite (SLNR) that contains both locally derived and exotic fragments of mafic-ultramafic rock, and a lower unit of variably mineralized Granite Breccia. The Granite Breccia has a matrix of granitic composition consisting of a partially melting felsic assemblage of minerals that contains locally derived (Fig. 2.26D) and exotic mafic-ultramafic inclusions (Figs. 2.26B, E, F–H) as well as fragments of locally derived granite-gneiss (Figs. 1.9 and 2.26C–E).

The SLNR is present in embayment and trough structures in both the North and South Ranges, but the Granite Breccia is principally developed in the North Range. The geological relationships between SLNR and Granite Breccia are most often like those illustrated in Fig. 1.9, but several cases exist where the Granite Breccia cross cuts the SLNR and extend upwards into the MNR and sometimes into the overlying FNR. Inclusions that are texturally and compositionally similar to the MNR are sometimes enveloped in Granite Breccia (Fig. 2.26D; Lightfoot et al., 1997c).

The Granite Breccia Sublayer is endowed with disseminated, inclusion-rich, and massive Ni–Cu–PGE sulfide mineralization. Very large troughs such as Creighton contain world-class ore deposits, whereas very small embayments such as Nickel Rim can also contain significant ore deposits (McLean et al., 2005). Large troughs such as Trill contain very small amounts of known mineralization (Fig. 1.1). Although there is no direct relationship between the scale of the embayment and the quantity of economic mineralization, embayments and troughs do play an important part in controlling the distribution of the mineralization at the base of the SIC.

Laterally away from embayments and troughs, the Sublayer decrease substantially in thickness, and although Granite Breccia is sometimes developed at the contact, SLNR tends to be absent outside of embayments and troughs. The sporadic distribution of Sublayer outside of the embayments and troughs is the reason for the distribution shown in Fig. 2.26A. Distally away from the embayment and trough structures, FNR of the Main Mass occurs in direct contact with the country rocks that rest beneath the SIC.

The SLNR, Granite Breccia, and inclusion-rich massive sulfide in embayment and trough structures contain inclusions of both local country rocks and apparently exotic fragments (eg, olivine melanorite, peridotite; Rae, 1975; Scribbins, 1978; Farrell, 1997; Lightfoot et al., 1997b,c). Ultramafic inclusions are present in the Sublayer where there are both economic ore deposits and subeconomic occurrences; an example from the Trill Embayment is shown in Fig. 2.26G where a serpentinized ultramafic is developed in SLNR with disseminated sulfide. Country rock fragments in the SLNR are commonly mafic in composition and resemble the adjacent country rock. The correlation between mineralization and the presence of mafic-ultramafic inclusions is a strong relationship; it is very rare to find extensive mineralization in association with inclusion-poor SLR. However, the size of the trough or embayment structure containing the Sublayer is not a good guide to mineral exploration potential.

Primary Geometry of the Troughs and Embayments and Variations in Thickness of the Sublayer

The topography of the lower contact of the SIC in areas of less extensive deformation is likely a good indicator of the topography of the base of the melt sheet. The basal topography was probably a major control on the thickness of the melt sheet much as other melt sheets such as Manicouagan exhibit an enormous range in thickness that relate to the topography of the basal contact zone (Spray and Thompson, 2008).

The predeformation topographic variation at the base of the melt sheet is local and regional in scale; viz: 1) A regional crater floor topography that broadly maps out the thickness of the original melt sheet on the scale of kilometers. 2) Local scale discrete troughs and embayments that vary in depth and extent on the scale of hundreds of meters to more than 4 km. It is the local topography that provides the "container morphology" of the Sublayer and represents a principal control on the distribution of contact ore deposits, whereas the regional topography of the crater floor may play a role in the differentiation of the melt sheet and the localization of dense magmatic sulfides towards the troughs and embayments.

The exact geometry of the base of the SIC is established based on exploration and geoscience data collected during surface mapping, drilling, and underground mapping in mines. In many cases, the maps showing the distribution of the embayment and trough features at the base of the SIC are constrained by tens of thousands of individual drill holes and by mapping in multiple levels, so this geometry is factually established. The Sublayer varies in thickness up to ~500 m around the edge of the SIC. Some deep troughs are characterized by extensive development of thick Sublayer (eg, Creighton and Trill have sections of Sublayer approaching 500 m in thickness), whereas other troughs are more discontinuous and comprise a series of embayment structures that are along the trend of a trough (eg, Levack-Colman where the Sublayer is typically less than 200 m thick except in some of the very localized pockets described in Chapter 4).

Fig. 2.26A shows the distribution of some of the important trough and embayment structures of the SIC, and the four insets to this diagram show the shape of the trough and embayment structures projected onto a long section of the base of the SIC. In this figure, the trough and embayment structures are shown as trends projected onto the base of the SIC as a map. At an empirical level, it is possible to break out these features into four groups based on the variation in dimensions (length, width, depth), degree of continuity, the extent to which troughs branch, and the degree of syn- or postmagmatic deformation. In reference to Fig. 2.26A and Table 2.2, the classification of trough and embayment structures is broken out as follows:

1. Simple embayments that represent localized discrete depressions in the footwall of the SIC (eg, Nickel Rim embayment).
2. Troughs that contain two or more embayments along a single broad trend (eg, Trill, Murray).
3. Complex intersecting troughs that define broad patterns on the base of the SIC; the exact geometries of these troughs include both intersecting trends that define the shapes of the letters "V", "W", and "Y" on the base of the SIC (eg, Levack-Coleman, Whistle-Norman), and trends that define multiple branches that join up with a single main corridor (eg, the Creighton-Gertrude system).
4. Troughs that have been heavily deformed during postmagmatic tectonic events that have significantly changed the geometry, but where the primary shape of the embayment and trough can still be recognized (eg, Gertrude, Falconbridge, Crean Hill, Victoria).

Case study—the Geology of the Whistle Embayment

The Sublayer mineralization in the Whistle Mine embayment (Fig. 2.27A) was mined as an open pit operation by Inco during 1988–1991 and 1994–1997; the mine production was 5.71 million tons of ore with grades of 0.33% Cu and 0.95% Ni. The mine was relatively unimportant in the context of the nickel produced at Sudbury during this period, but the open pit mine operation provided an opportunity to understand the geological relationships in the Sublayer embayment at a level of detail that is often difficult in underground workings, drill core, or two-dimensional surface exposures.

After closure of Whistle Mine, the waste rock piles that once covered the termination of the Offset were removed and used as back-fill for the open pit; the distal portions of the Offset were cleared of debris and overburden to provide a resource for exploration and future investigation of the geology of the footwall of the embayment. Following divestment of the property by Inco, a junior mining company (FNX) followed up on a weak geophysical anomaly and previously identified intercepts of Cu-rich mineralization in the footwall that were considered too small to represent a target of opportunity to Inco; FNX identified two Cu-rich mineral zones in the footwall; one of these zones became the Podolsky Mine that was developed in 2006–2012 (Farrow et al., 2008a,b).

For these reasons, and despite the small size of the mineral system, the Whistle trough was chosen to showcase some of the important geological relationships in the Sublayer as a case study.

The Whistle embayment, located at the northeastern margin of the SIC (Fig. 2.27A), is comprised of a 1 km deep depression at the base of the Main Mass that contains a thick package of Sublayer that extend downwards into a zone of metabreccias containing discontinuous segments of QD and MIQD (Fig. 2.27A); the embayment is developed above Archaean granitoids and felsic gneisses that contain local enclaves of mafic gneiss, metapyroxenite, and diabase (Fig. 2.27A). The NE–SW section in Fig. 2.27C shows the position of the base of the SIC, the location of the Whistle embayment and contact ore deposit, and the position of the footwall mineralization known as the Podolsky Deposit.

The only published descriptions of the embayment before mining commenced was made by Pattison (1979); after mining commenced at Whistle, Lightfoot et al. (1997a,c) and Farrell (1997) documented the geology of the embayment as mining operations progressed (Fig. 2.27B). After mining was completed at Whistle, the area beneath the waste rock pile to the NE of the embayment was removed and the outcrop was mapped by Murphy and Spray (2002), Bygnes (2011), and Lafrance et al. (2014), which is summarized in Fig. 2.27A.

The geological relationships between Main Mass rocks and the Sublayer at Whistle Mine are consistent with a gradation between the MNR and FNR over 1–5 cm, and a more complex diffuse and irregular contact between inclusion-bearing SLNR and MNR over less than 1 m. Through the contact zone, the inclusion population changes from anorthositic segregations in poikilitic-textured biotite-rich MNR into SLNR with plagioclase porphyritic and fine-grained diabase inclusions and melanorite segregations that texturally and mineralogically resemble the MNR. These melanorites are hosted in finer-grained subophitic SLNR (Lightfoot et al., 1997c).

The Sublayer Norite of the Whistle embayment

The SLNR is the dominant rock type in the Whistle embayment (Fig. 2.27B). It consists fine to medium-grained norite with less than 20% cumulate hypersthene, hosting patches of coarse-grained patches of norite and melanorite; these patches have both sharp and diffuse (1–10 mm) contacts with the surrounding SLNR (Fig. 2.28A,B). Diabase inclusions with abundant magnetite are distributed throughout the Sublayer, and these inclusions have relatively sharp contacts (1–5 mm) with the surrounding SLNR (Fig. 2.28C). Disseminations of sulfide occur interstitially within the SLNR, and as blotchy segregations that are often associated with feldspathic leucosomes. The SLNR matrix varies from melanocratic through to leucocratic with increasing depth in the embayment (Fig. 2.27B).

The economically important mineral zone at Whistle comprised inclusion-rich sulfides that range from a clast-supported inclusion-rich sulfide (Fig. 2:28D) through to semimassive sulfide with (Po + Pn) > >Cpy that contains floating inclusions (Fig. 2.28E,F show end-members of this

Table 2.2 Summary of geological features of main troughs and embayments

Embayment Name	Location	Type	Footwall rocks	Contact morphology	Post-SIC fault sets
Murray	South Range	Trough	Creighton Granite and Stobie Formation	Surface—60°NNW; 1500 m vertical; 2000 m 80°SSW	Thrust
Copper Cliff	South Range	Funnel	Creighton Granite and Stobie Formation	75°NNW	Minor
Tam O'Shanter	South Range	Embayment	Creighton Granite and Stobie Formation	70°NNW	Minor
North Star	South Range	Embayment	Creighton Granite and Elsie Mountain Formation	65°NNW	Minor
Creighton	South Range	Troughs	Creighton Granite, Elsie Mountain Formation and Gabbro	Surface—55°NNW; 1500 m vertical; >2000 m 85°SSW	Thrust; W–E shear sets
Gertrude	South Range	Embayment	Creighton Granite, Elsie Mountain Formation and Gabbro	65°NNW	Minor
Gertrude West	South Range	Embayment	Creighton Granite	65°NNW	Minor
Lockerby	South Range	Embayment	Elsie Mountain Formation	Subvertical	W–E shears
Crean Hill	South Range	Trough	Elsie Mountain Formation	Subvertical	W-E shears; thrusting
Victoria	South Range	Funnel	Elsie Mountain Formation and Drury Township metagabbro	Subvertical	W–E shears; thrusting
Sultana	North Range	Embayment	Archean granitoid rocks	70°E	Bounded by SRSZ
Trill	North Range	Trough	Levack Gneiss	55°E	Minor
Levack–McCreedy West–Strathcona	North Range	Troughs	Levack Gneiss - mafic and felsic	50°SSE	N–S faults; minor thrusting (?)
Coleman	North Range	Troughs	Levack Gneiss—mafic and felsic	60°SSE	N–S faults; minor thrusting (?)
North Star	North Range	Embayment	Levack Gneiss	50°S	Minor

Mineralization style	Scale of ore system kt contained Ni (*)	Style of depression	Volume of Sublayer (~km3)	Dominant sublayer host type (+)	Scientific papers and technical reports
Contact and footwall types	Large	Trough	0.5–<2	100% SLNR	Souch et al. (1969)
Contact type	Small (very large in offset)	Funnel	0.5–<2	100% SLNR	Capes (2001)
Contact type	Small	Embayment	≤0.05	100% SLNR	
Contact type	Small	Trough	≤0.05	100% SLNR	
Contact and footwall types	Very large	Trough	0.5–<2	100% SLNR	Boldt (1967); Souch et al. (1969); Dare et al. (2010a,b; 2014); Lightfoot (2015)
Contact type	Small	Trough	0.05–<0.25	100% SLNR	Dare et al. (2010a,b)
Contact type	Small	Trough	0.05–<0.25	100% SLNR	
Contact and footwall types	Medium	Trough	na	100% SLNR	Clow et al. (2005)
Contact and footwall types	Large	Trough	0.25–0<.5	100% SLNR	Gibson et al. (2010)
Structurally modified contact type	Medium	Funnel	0.05–<0.25	100% SLNR	
Contact type	Small	Embayment	≤0.05	75% SLNR; 25%GRBX	
Contact type	Small	Trough	0.5–<2	60%SLNR; 40%GRBX	Livelybrooks et al. (1996)
Contact and footwall types	Large	Trough with aligned embayments	0.25–0 < 0.5	50% SLNR; 50% GRBX	Farrow and Lightfoot (2002); Coats and Snajdr (1984); Morrison (1984); Nelles (2012); Legault et al. (2003); Farrow et al. (2005, 2008a,b); Hanley (2002); Li and Naldrett (1993a,b); Naldrett et al. (1994a, 1999)
Contact and footwall types	Large	Trough with aligned embayments	0.5–<2	50% SLNR; 50% GRBX	Gregory (2005); Stout (2009); Moore and Nikolic (1994); Jago et al. (1994); Morrison et al. (1994)
Contact	Occurrence	Embayment	0.5–<3	60%SLNR; 40%GRBX	

(Continued)

Table 2.2 Summary of geological features of main troughs and embayments (*cont.*)

Embayment Name	Location	Type	Footwall rocks	Contact morphology	Post-SIC fault sets
Whistle-Norman	North Range	Embayment	Levack Gneiss—mafic and felsic	45°SW	Minor
Ella	North Range	Embayment	Levack Gneiss—mafic and felsic	Subvertical	N–S fault repetition
Capre	North Range	Embayment	Levack Gneiss—mafic and felsic	Subvertical	N–S fault repetition
Victor	North Range	Trough	Levack Gneiss—mafic and felsic	80°W	N–S fault repetition
Nickel Rim	North Range	Embayment	Levack Gneiss—mafic and felsic	80°W	Minor
Falconbridge-Cryderman	South Range	Tectonized embayment	Stobie and McKim Formation	Subvertical	W–E shears
Garson	South Range	Tectonized embayment	Elsie Mountain and McKim Formations	Subvertical	W–E shears
Kirkwood	South Range	Embayment	Stobie Formation	Subvertical	W–E shears
Blezard-Lindsley	South Range	Trough	Murray Granite and Stobie Formation	60°NNW; 1500 m vertical	Thrust
Little Stobie	South Range	Trough	Murray Granite and Stobie Formation	Subvertical	W–E shears
McKim	South Range	Trough	Murray Granite and Stobie Formation	80°NNW	W–E shears

Mineralization style	Scale of ore system kt contained Ni (*)	Style of depression	Volume of Sublayer (~km3)	Dominant sublayer host type (+)	Scientific papers and technical reports
Contact and footwall types	Medium	Trough with aligned embayments	0.25–0 < 0.5	90% SLNR; 10% GRBX	Lightfoot et al. (1997a,c); Farrell (1997); Carter et al. (2005); Bygnes (2011); Lafrance et al. (2014); Farrow et al. (2005)
Structurally modified contact type	Small	Embayment	0.25–0 <0.5	50% SLNR; 50% GRBX	
Contact and footwall types	Medium	Embayment	0.25–0 < 0.5	50% SLNR; 50% GRBX	Stewart and Lightfoot (2010)
Contact and footwall types	Large	Trough	0.25–0 < 0.5	50% SLNR; 50% GRBX	Morrison et al. (1994); Jago et al. (1994); Lightfoot (2015)
Contact and footwall types	Large	Embayment	≤0.05	40% SLNR; 60% GRBX	McLean et al. (2005)
Structurally modified contact type	Very large	Embayment	0.05–<0.25	100% SLNR	Lockhead (1955); Owen and Coats (1984)
Structurally modified contact type	Large	Trough	0.25–0 < 0.5	100% SLNR	Mukwakwami (2012); Mukwakwami et al. (2012, 2014a,b); LeFort (2012)
Structurally modified contact type	Small	Embayment	≤0.05	100% SLNR	Thomson et al (1985); Farrow et al. (2008a,b)
Contact and footwall types	Medium	Trough	0.05–<0.25	100% SLNR	Binney et al. (1994); Bailey et al. (2004, 2006); Molnár et al. (1997)
Contact and footwall types	Medium	Trough	0.05–<0.25	100% SLNR	Davis (1984); Molnár et al. (1997); Stewart and Lightfoot (2010)
Contact and footwall types	Medium	Trough	0.05–<0.25	100% SLNR	Clarke and Potapoff (1959a,b)

*Summary of the main features of the different trough and embayment structures that host the contact ore deposits at Sudbury. The table shows a summary of sources for more detailed geological information on the troughs and embayments that host the contact ore deposits. Abbreviations: *Very large, ≥1000 kt Ni contained; large, 500–<1000 kt Ni contained; medium, 250–<500 kt Ni contained; small, <250 kt Ni contained; SLNR, sublayer norite; GRBX, Granite Breccia*

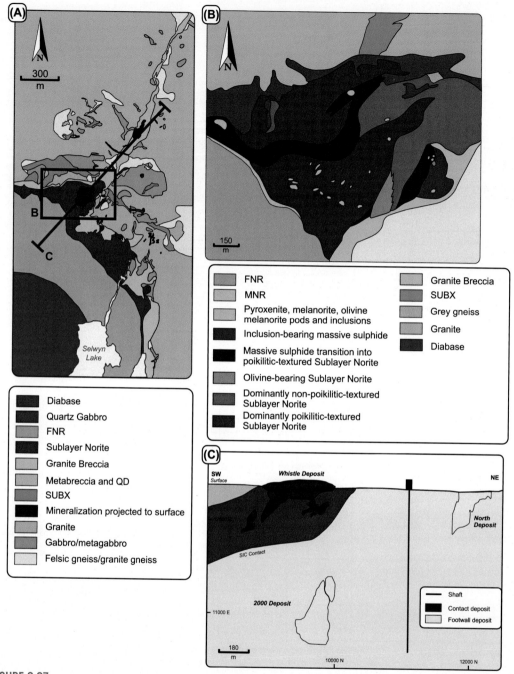

FIGURE 2.27

(A) Geological map of the Whistle area after rehabilitation of the Whistle Open Pit and exposure of outcrops beneath the rock waste pile. (B) Geological map of the Whistle embayment prior to the exposure of the contact relationships under a rock waste pile. (C) SW–NE-facing long section showing position of the Whistle embayment and the location of two footwall mineral zones, one of which is the Podolsky Deposit.

(A) Based on Meyn (1970), Lightfoot et al. (1997c), and Farrow et al. (2008b); (B) modified after Lightfoot et al., 1997c; (C) based on Farrow et al. (2008b).

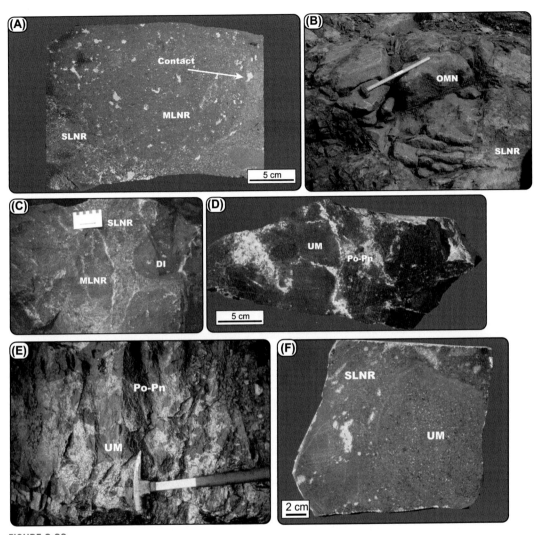

FIGURE 2.28

(A) Inclusion of melanorite (MLNR) in SLNR from the Whistle Mine embayment structure. (B) Inclusion of olivine melanorite (OMN) in SLNR at the surface NE termination of the Sublayer as it grades into leucocratic norite, Whistle Mine; (C) SLNR containing inclusions of diabase (Di) and melanorite (MLNR), Whistle Mine; (D) Ultramafic (UM) inclusions in a matrix of pyrrhotite and pentlandite (Po–Pn), Whistle Mine; (E) Ultramafic inclusions floating in massive sulfides (Po–Pn) of the Sublayer, Whistle Mine. (F) Inclusion of altered olivine melanorite (OMN) enclosed in SLNR Whistle Mine. Note the black rim around the inclusion, and the gradational contact. (G) GRBX transitioning into leuconorite (LNR) from the base of the Whistle embayment structure (H) MIQD of the Whistle Offset in contact with metabreccia (MTBX), Whistle Mine. (I) Mineralized melanorite inclusion (MLNR) in Sublayer Norite (SLNR) with blebby disseminated pyrrhotite–pentlandite (Po–Pn), Whistle Mine; (J) Ultramafic inclusion undergoing incipient melting and recrystallization with the surrounding Sublayer Norite (SLNR), Whistle Mine.

(A) Lightfoot et al. (1997b,c).

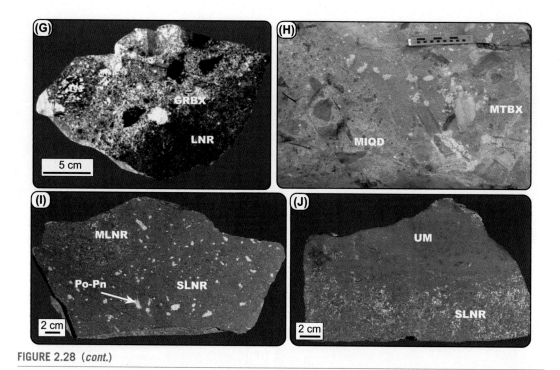

FIGURE 2.28 (*cont.*)

mineralization type). The inclusion types are dominantly diabase, norite, and melanorite; the noritic inclusions are commonly more altered when they are hosted in sulfide. The inclusion-rich sulfide grades into sulfide-rich Sublayer Norite that contains a similar range of inclusion types.

Inclusion-bearing massive sulfides of the Whistle embayment are dominantly pyrrhotite-rich sulfides with occasional 1 mm–1 cm equant pyrite porphyroblasts (up to 5 modal percent of the sulfide) and granular pentlandite. The sulfide ore zone occurred in a single zone (1–150 m wide) that roughly follows the northwestern and lower margin of the embayment from beneath the Main Mass norites outwards towards the radial Offset (Fig. 2.27A). The inclusion-rich massive sulfide zone transitions into sulfide-rich Sublayer Norite over 1–5 m at the edges of the ore body.

Geological relationships between inclusions and Sublayer Norite

The inclusion population of the Sublayer Norite comprises the following rock types: (1) abundant inclusions of poikilitic-textured melanorite (which are commonly texturally and mineralogically very similar to the MNR), olivine melanorite, and alteration products of these noritic rocks; (2) common plagioclase porphyritic to nonporphyritic magnetic oxide-rich diabase inclusions (termed the Diabase inclusions) from 2 mm up to 2 m across; (3) rare (<1% of all inclusions) altered anorthositic, troctolitic, and gabbroic inclusions that commonly have a strained fabric.

The inclusions of melanorite range in size from less than 1 cm to over 10 m in diameter, and they have gradational (<0.5 cm thick) to sharp contacts. The inclusions often contain disseminations and blebs of sulfide mineralization and small inclusions of diabase (Fig. 2:28A). In some cases the ultramafic inclusions appear to be recrystallized (Fig. 2.28F) and sometimes partially melted and recrystallized at the margins (Fig. 2.28J). Although not commonly developed, the presence of diabase inclusions within the melanorite inclusions of the Sublayer indicates that diabase was incorporated into the Sublayer magma before the melanorite inclusions were developed (Fig. 2.28C).

Some of the inclusions contain abundant olivine (5–25%), mica (2–15%), sulfide (1–5%), and apatite (0–0.5%); these rocks are termed olivine melanorites although their elevated phlogopite and apatite contents are unusual when compared to similar rock types developed in other maficintrusions. The olivine melanorite inclusions are normally coarse-grained in texture and the inclusions are sometimes zoned from one side to the other (phlogopite-rich olivine melanorite through to phlogopite-rich leuconorite). These zoned inclusions reside within the Sublayer Norite.

Diabase inclusions make up 5–10% of the inclusion population in the Sublayer at Whistle. The diabase is texturally and mineralogically very similar to heavily recrystallized 1–30 m bodies diabase developed at the margin of the embayment within the Granite Breccia and metabreccia. The diabase inclusions tend to have quite sharp contacts in hand sample, but the relationships in thin section are much more complex, and much of the Sublayer Norite at Whistle contains xenocrysts or xenoliths of variably assimilated diabase as described in Chapter 3.

Granite Breccia and Metabreccia

The footwall of the Whistle Embayment consists dominantly pink porphyritic-textured granitoid rocks. Locally, small areas of banded gray gneiss are developed. Banded and strongly contorted segregations of foliated amphibolite occur within the gneiss. The northern wall of the Whistle open pit mine consists of granitoid rocks cut by a 100 m-wide vertical zone of strong SUBX that trends due north from the embayment wall (Lightfoot et al., 1997a). This zone hosts large fragments of diabase, inclusions of granular amphibolite, and fragments of the Footwall granitoid rock. Lightfoot et al. (1997a,c) reported the presence of chalcopyrite mineralization in the SUBX, and the development of irregular 1–20 mm blebs and 1–30 cm veins of chalcopyrite in the surrounding granitoid rocks.

The contact between the footwall granitoid rocks and the Sublayer Norite is a transitional zone of Granite Breccia that contains inclusions of diabase, altered metapyroxenite, and granitoid rocks; this breccia contains 1–25% pyrrhotite–pentlandite ± chalcopyrite as disseminations, segregations, and inclusions (Fig. 2.28G). The Granite Breccia contains examples of inclusions of norite that are texturally and mineralogically similar to the SLNR and blocks of orthopyroxene-rich melanorite and altered melanorite that are texturally and mineralogically similar to those found in the SLNR. The GRBX is best developed along the keel of the Offset directly beneath and adjacent to the sulfide zone (Fig. 2.27A,B).

At the contact between the footwall granitoids and the embayment are large sheet-like bodies of diabase (up to 20 m × 50 m in dimensions) that are texturally and mineralogically similar to the diabase inclusions from within the Sublayer. The diabase bodies are heavily brecciated, recrystallized, and partially melted; they contain anhedral clusters of feldspar phenocrysts, and are they are cut by late veins of quartz that merge into the adjacent GRBX.

The GRBX and metabreccia are located at the margins of the embayment and within the footwall (Fig. 2.27A). The breccia consists of subangular to angular fragments of granitoid rock, phenocrysts presumably derived from the granitoid rocks, fragments of diabase, and fragments of highly deformed amphibolite (Figure 2.28G). The matrix is a fine-grained metamorphic-textured fragmental rock (see Pattison, 1979). The Footwall Breccia contains 1–25% sulfide, and tends to be chalcopyrite-rich (>75% of the sulfide). The disseminated to blebby and heavily disseminated sulfides in the Footwall Breccia are chalcopyrite-rich, but fragmental sulfides are pyrrhotite-rich. The SLNR fragments contain inclusions of the diabase and blebby disseminated sulfide and whispy feldspathic leucosomes (1–5 cm × 10–50 cm).

Transition from the Whistle Embayment to the Whistle Offset

The distal portion of the Whistle embayment is composed of inclusion-bearing leucocratic norite (historically termed inclusion-bearing basic norite), Granite Breccia, and metabreccia with discontinuous segments of coarse-grained QD that locally shows a coarse-grained quenched texture. The Whistle Offset trends in a northeasterly direction as a discontinuous series of lenses of quench-textured QD, QD, and MIQD within metabreccia and Granite Breccia (eg, Lafrance et al., 2014; Fig. 2:27A)

At the location where the Embayment and the Offset Dyke merge, the Sublayer changes gradually from a mineralized inclusion-bearing leucocratic norite with up to 25% inclusions to QD.

The QD appears to terminate at surface within 2 km of the Main Mass in a breccia zone that contains many fragments of diabase and granitoid rocks. The Whistle Offset is mainly hosted by Granite Breccia, metabreccia, and SUBX. Although there is some debate, a postmagmatic sinistral fault zone may displace the Whistle Offset Dyke from the Parkin Offset Dyke. There is some geological evidence to support the continuation of the QD along strike beyond this structure (www.northamericannickel.com). If this evidence is shown to be correct, this opens up the opportunity to discover new segments of both the Parkin and Whistle Offset Dykes.

Synthesis of geological relationships at Whistle

1. The contact relationships between the SLNR and the MNR of the Main Mass are transitional rather than intrusive; there is no evidence of upwards injection of SLNR through the Main Mass. Some of the melanorite inclusions in the Sublayer resemble the MNR but these inclusions are part of a suite that range from norite through melanorite to olivine melanorite, and they appear not to originate from the MNR. The geological relationship of the melanorite inclusions to the MNR is not fully understood. In some locations, the malenorite inclusions appear to originate from the MNR unit (eg, Levack), but in other locations such as Whistle, they appear to be an intrinsic part of the Sublayer assemblage.
2. SLNR contains an inclusion population dominated by norite, melanorite and olivine melanorite (~70%) and plagioclase porphyritic diabase (~30%). The diabase inclusions have moderately sharp contacts with SLNR, but the melanocratic inclusions have contacts that range from diffuse for noritic fragments to sharp and baked for some of the olivine melanorite inclusions. The melanorite inclusions contain disseminated to blebby sulfide and small diabase inclusions. Texturally and mineralogically, the MNR resembles the segregations of melanorite enclosed in SLNR, but geochemical data indicate that the MNR at Whistle Mine is compositionally different when compared to the melanorite segregations (see Chapter 3 for details), but similar to the Main Mass FNR (Lightfoot et al., 1997b).
3. The mineral zone at Whistle occupies the keel of the Offset and comprises clast-supported sulfide as well as semimassive sulfide with inclusions that do not touch. The mineral zone grades into SLNR with heavy disseminated sulfide mineralization. The inclusion population in the massive sulfide resembles that of the SLNR, but the inclusions are often more heavily altered.
4. There is a narrow zone of Granite Breccia at the base of the embayment, and this is typically adjacent to the mineral zone.
5. The embayment is connected through Granite Breccia and Metabreccia into discontinuous segments of coarse-grained and quench-textured QD that forms the radial offset.

MAGMATIC AND METAMORPHIC-TEXTURED BRECCIAS IN THE FOOTWALL OF THE SIC

The Sudbury impact event produced two types of breccias in the footwall of the SIC, namely: (1) Pseudotachylite breccias, termed SUBX, formed by cataclasis of country rocks during the first few seconds of the impact process, and (2) Metamorphic breccias, termed metabreccias, that were formed by metamorphic recrystallization, partial melting, and disruption of the target rocks and pseudotachylite breccias in response to heating by the melt sheet and postimpact re-adjustment of the impact crater (Fig. 2.4B).

SUBX is widely considered to be a product of shock melting (Spray, 1995) and fragmentation of the target rocks (Reimold, 1993, 1995) related to the shock process of the impact event. These rocks contrast with breccias produced by postimpact melting and metamorphism related to transfer of heat from the SIC to the country rocks. The formation of the SUBX predates the formation of the igneous rocks of the Offset Dykes, Sublayer, and Main Mass because the Offset Dykes cross cut and are quenched against SUBX. Moreover, partially melted SUBX fragments occur in the Granite Breccia of the Sublayer.

Metabreccias are considered to be rocks produced by recrystallization and partial melting of the country rocks adjacent to the melt sheet; this process involved the transfer of heat from the melt sheet to the target rocks. The rocks produced by contact metamorphism and melting are difficult to relate back to the country rock protolith, and terms such as "Sudburite" (Thomson, 1935a-d) have been used to describe partially melted metamorphic-textured rocks of primary igneous derivation consisting of plagioclase and pyroxene exhibiting a hornfels texture (Chapter 1). The diversity in target rock types has resulted in the production of a wide range of geological relationships in the Sudbury Breccias and metabreccias.

The development of SUBX is often localized where there are changes in country rock types, and SUBX tends to be weakly developed within mafic country rocks and localized into wide fine-grained veins in more homogenous target rocks such as granites, rhyolites, and quartzitic metasedimentary rocks.

The response of target rocks to heating is also widely different; for example, mafic volcanic rocks and gabbros close-in to the SIC tend to be recrystallized to form Sudburite, and fine-grained hornfels. Melting in these rocks tends to be restricted to leucosomes. In contrast, felsic igneous rocks and granitoids tend to melt to produce leucosomes and aplitic veins. The diversity in textures produced by recrystallization and melting of the target rocks is largely a result of the proximity to the melt sheet and the diversity in target rock types.

Classification of the Breccias

Breccias in the SIC footwall are complex; the geological relationships have defied almost 100 years of description, and only recently have detailed petrological studies started to unpick the details of mineral chemistry that will help provide a unified scheme for the classification of these rocks (eg, Jørgensen et al., 2013, 2014). Notwithstanding, mining companies engaged in exploration have found some very ingenious ways to classify the rocks in a meaningful way. Table 2.3 provides a synthesis of some of the key parameters in the classification of footwall materials that have been brecciated by the impact event. It is designed to cross-traditional boundaries between pseudotachylite type breccias as well as metabreccias, and it may also be used for Granite Breccia of the Sublayer. The scheme was originally devised for use by exploration companies; the scheme broke out the breccias based on the composition of the matrix (basic, intermediate, and siliceous), the composition of the fragments (basic, intermediate,

Table 2.3 Classification Criteria Used to Group Breccia Types at Sudbury

Composition of Matrix	Composition of Fragments	Texture of Matrix
1—Basic matrix (quartz<5%)	A—Basic igneous rock fragments	(0—Igneous)
2—Intermediate matrix (5%<quartz<15%)	B—Intermediate igneous rock fragments	1—Subigneous texture (visible plagioclase laths)
3—Siliceous matrix (quartz >15%)	C—Granodioritic Fragments	2—Medium grained Porphyroblastic (mafic mineral porphyroblasts)
	D—Granitoid rock fragments	3—Fine grained porphyroblastic (mafic mineral porphyroblasts)
	E—Metasedimentary rock fragments	4—Recrystallized (sugary texture to matrix)
		5—Aphanitic (dark color; no mineral grains)

A classification scheme for breccias in the footwall of the SIC suitable for rocks that have a metamorphic matrix and exhibit low degrees of partial melting. The application of this scheme works best for SUBX and metabreccia, but it has also been applied to Sublayer Granite Breccia.

granitic, sedimentary, volcanic) and the texture of the matrix (subigneous, porphyroblastic, recrystallized, or aphanitic) (Andy Bite, 1997, Pers. Comm.). The main parameters used in this classification are the following:

1. Composition of the matrix: (a) basic; (b) intermediate, and (c) acidic.
2. Dominant fragment type: (a) gabbroic; (b) gabbroic-dioritic; (c) diorite-granituc; (d) granitic; (e) sedimentary.
3. Texture of the matrix: (a) subigneous; (b) medium-grained porphyroblastic; (c) fine-grained porphyroblastic;
4. Recrystallized.
5. Aphanitic.

The following parameters are also important and represent a way to fit the rock description into the broader context of the host rocks:

1. *Partial melt textures*: This parameter describes the extent of development of whispy partial melt segregations, pegmatoidal veins, or melt patches
2. *Contact and structural control on breccia*: This parameter links breccia to the controlling influence of preexisting lithological contacts or structures in the target rocks.
3. *Source of fragments*: The fragments within a breccia can be derived from the immediate target rocks or they can be transported as exotic fragments.
4. *Scale of brecciation*: Extensive development of large (5–100 m diameter) blocks of country rock that are surrounded by breccias are termed megabreccias.

The Impact Breccia: Sudbury Brecciia (Pseudotachylite Breccia)

Pseudotachylite breccias are common in the footwall of the SIC for distances in excess of 50 km away from the SIC and locally as much as 150 km (Fig. 2.29; after Dressler, 1984b). The breccias comprise a matrix of psuedotachylite that was formed instantaneously at the time of the impact event (French,

1998). SUBX often occurs as partially melted enclaves in Granite Breccia (Fig. 2.30E), and it is cross cut by the Offset Dykes (Fig. 2.22B, G). These relationships confirm that the rocks of the SIC were formed after the SUBX.

The Sudbury Breccia is normally yellowish-gray through dark-gray to black in color (Fig. 2.30A–G). For example, SUBX developed in mafic volcanic and intrusive rocks tends to be very dark gray to black in color. Breccias developed intermediate gneisses tend to be dark gray, and those developed in metasedimentary rocks can be yellowish-gray to dark gray in color, and those developed in granitoid rocks are often gray to dark-gray. A range of breccia types developed in various different country rocks are shown in Fig. 2.30A–G, and their petrography and geochemistry is presented in Chapter 3.

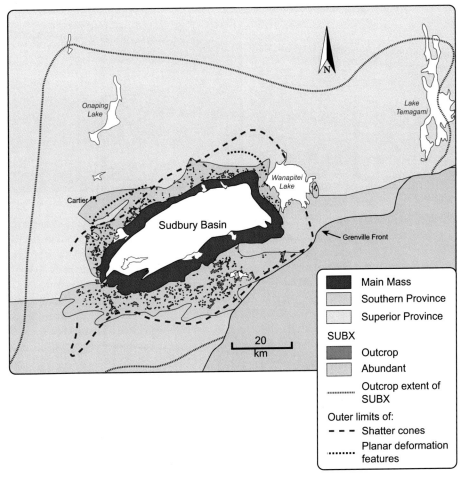

FIGURE 2.29

Distribution of SUBX around the Sudbury Structure; the map shows the location of poorly mapped discontinuous belts of which are developed in the footwall of the SIC.

Modified after Ames et al. (2005).

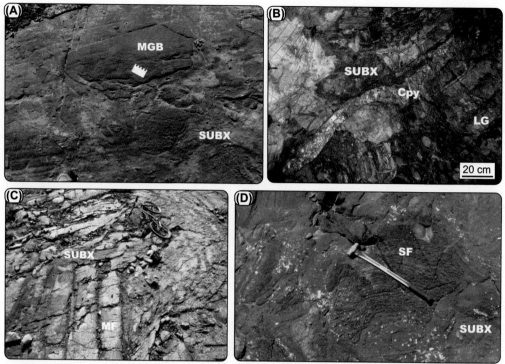

FIGURE 2.30

(A) Typical Sudbury Breccia (SUBX) with melagabbro inclusions (MGB) in the Frood–Stobie Belt are characterized by partial melting, flow fabric, rounded inclusions from multiple sources, and extensive development. (B) Example of Sudbury Breccia (SUBX) from Coleman Mine showing the development of matrix between clasts of Levack Gneiss (LG); these breccias are host to chalcopyrite (Cpy) that is one of the principal sulfide minerals of the footwall deposits at Coleman Mine. (C) Development of Sudbury Breccia (SUBX) in well-bedded Mississigi Formation (MF) quartzites; local heterogeneity disrupts planar shearing along stratigraphic boundaries, and results in extensive dislocation of the stratigraphy and local development of SUBX. (D) SUBX developed in Stobie Formation metasedimentary rocks (SF) adjacent to the east margin of the Copper Cliff Offset Dyke. (E) Partially assimilated fragment of SUBX in Granite Breccia (GRBX) showing the timing relationship between the two breccia types (from the Trill trough). (F) SUBX developed in fine grained MG at the western end of the Murray Pluton; (G) Coarse-grained porphyritic Creighton Granite containing SUBX that occupies transtensional veins in the central part of the Creighton Pluton.

Classification and Description of the SUBX

A review of the classification schemes for SUBXs is given by French (1998) and Roussel et al. (2003). Roussel et al. (2003) propose a three-fold classification, based on the nature of the matrix, namely: (1) Clastic SUBX consisting of rounded clasts in a matrix that develops flow features; (2) Pseudotachylite Sudbury Breccias that resemble those of the Vredefort Structure (Shand, 1916; Reimold and Gibson, 2006) and are characterized by a black aphanitic matrix described as a "cryptocrystalline massive, microigneous textured" rock with a grain size of ~10 μm. (3) Microcrystalline SUBX that occurs along the base of the SIC within ~500 m of the basal contact, and exhibits both metamorphic

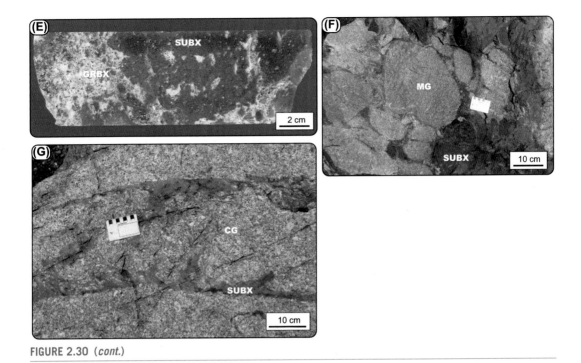

FIGURE 2.30 (*cont.*)

and partial melt textures. This scheme is reasonably well aligned with aspects of the scheme provided in Table 2.3.

This book broadly follows the subdivision used by Roussel et al. (2003). The principal subdivision in this text follows the scheme in Table 2.3, and describes the extent to which the matrix is melt-textured or contains leucosomes, namely: (1) A medium-grained matrix with biotite porphyroblasts and extensive development of flood quartz; these rocks are locally termed "hot," and they sometimes show a flow lamination and/or interstitial to blebby magmatic Ni–Cu sulfides if the rock is located near to an ore deposit (eg, Frood and Stobie Mines; Fig. 2.30A). (2) A medium-grained matrix which comprises quartz and feldspar as well as some biotite mica; the matrix often shows localized development of flood quartz enveloping microlaths of plagioclase feldspar, and the rock is often flow banded with a larger diversity in fragment types typically derived from the immediate country rocks. (3) A fine-grained matrix that comprises quartz-feldspar with a weakly developed flow fabric; the matrix surrounds rounded inclusions of local country rocks.

The gradation between these three types of breccia very broadly relates to proximity to rocks that comprise the SIC. The first type of breccia is developed adjacent to the metabreccia that surrounds the Sublayer embayments and segments of the Offset Dykes located close to the SIC. The second type of breccia is developed at some distance from the SIC, and the third type is developed in the footwall beyond the thermal halo of the SIC.

Protolith Controls on the Development of SUBX

Fig. 2.30C illustrates the local development of SUBX within a stratigraphy of arkose with shale beds from the Mississigi Formation. Local heterogeneities in the Mississigi Formation stratigraphy are caused by mafic plugs (Nipissing Diabase or Sudbury Gabbro). These mafic plugs provide a vortex where the stratigraphy of the metasedimentary rocks is heavily disturbed and where fragments of the country rocks can be rotated into right angle configurations relative to the regional stratigraphy. These loci of deformation are also the centers of development of SUBX. Much of the deformation along strike in the stratigraphy is taken up by motion along less competent boundaries between the Bouma cycles, and it is likely that these boundaries have deformed by slip motion that may not have generated significant amounts of SUBX.

SUBX is also commonly developed where the stratigraphy of the country rocks is highly variable (eg, the Stobie Formation; Fig. 2.30D) or where there are multiple types of metasedimentary-, metavol-canic, and intrusive rocks (eg, the South Range Breccia Belt; Fig. 2.30A). The fragment population in the breccia tends to be more variable in size and rock type. The extent of melting and metamorphism of the SUBX tends to increase towards both embayment structures and the segments of concentric and radial offset dykes that are located proximal to the SIC.

Relationship between Sudbury Breccia and Mineralization at Sudbury

SUBX is an important host to the Footwall ore deposits at Sudbury, and it represents an important component of the country rocks that contain the Sublayer embayments and troughs as well as the Offset Dykes. The North Range footwall ore deposits at McCreedy West and McCreedy East, and the East Range footwall deposits at Nickel Rim, Victor and Capre are developed in Archean-aged Levack Gneiss that contains abundant SUBX. These breccias occur as arcuate-shaped belts in the footwall of the SIC (Lightfoot, 2015) and Chapter 4). They typically consist of a fine-grained dark gray to black matrix that contains abundant subrounded to rounded fragments of the adjacent gneiss. The matrix exhibits local recrystallization in samples located close to the base of the SIC, but the SUBX that hosts the footwall ore deposits is not heavily recrystallized or partially melted, and in many ways resembles the breccias developed at greater distance from the base of the SIC. The localization of these breccia belts is often linked to changes in the composition of the footwall gneisses (eg, between mafic and fel-sic gneisses or between felsic gneisses and enclaves of deformed mafic intrusions). Displacement along of minor structures that cut through these breccias is an important local control on the development of the space occupied by the footwall sulfide mineralization (eg, Figs. 2.30B and 2.31A).

Examination of the petrology of the breccia matrix and the fragment population does not provide a direct indicator of mineral potential or proximity to footwall sulfide mineralization. Samples directly adjacent to sulfide veins are metamorphosed, altered, and locally contain disseminated sulfides, and the effect of this enrichment is amplified near to the terminations of veins at the periphery of the ore body where low-sulfide mineralization enriched in precious metals is sometimes developed. The use of textural and inclusion relationships as guide to the development of footwall mineralization is not straightforward outside of the envelope where disseminated and stringer sulfide mineralization is developed.

In the South Range, SUBX comprises an important component of the country rocks that contain the Sublayer embayments and Offset Dykes. In these settings, the SUBX often exhibits a greater variety in fragment types because the country rocks are commensurately more varied in composition. The SUBX matrix and the country rocks are often heavily recrystallized, partially melted, and contain fel-sic leucosomes developed by partial melting. Some of the extreme members of this rock group are the metabreccias described below, which are often developed not only from the country rock host but also the SUBX. The SUBXs located at greater distance from the Sublayer and Offset Dykes tend not to be as extensively overprinted by the development of metabreccia, and they exhibit a coarse-grained matrix with porphyroblasts of mica and amphibole and local area of partial melting. These types of SUBX frequently form envelopes around the metabreccias which in-turn surround the embayment and the off-set structures that are developed close-in to the base of the SIC. The types of SUBX are locally termed "hot," and they do encourage a careful evaluation of mineral potential, but they tend not to be the host rock types that contain large ore deposits. Frood-Stobie is an excellent example where the vast majority of the ores were either inclusion-rich massive sulfides related to MIQD (Chapter 4), whereas the imme-diate host rocks are mineralized metabreccias developed after either the country rocks (Fig. 2.20J) or SUBX. Further away from the metabreccias, the SUBX is still "hot" (eg, Fig. 2.30A), but these hot

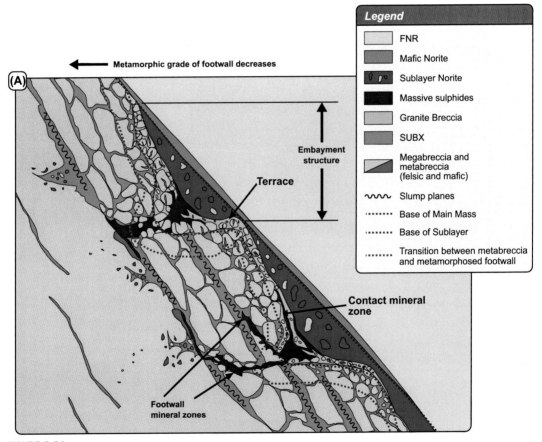

FIGURE 2.31

(A) Summary of the geological relationships in metabreccias developed in the footwall of North Range Sublayer adjacent to embayment structures. Metabreccia is typically developed beneath the Granite Breccia. (B) Map showing the extensive development of metabreccias in the footwall of the SIC between the Blezard, Little Stobie and Frood Stobie Deposits. Extensive recrystallization and melting textures are developed in these mafic footwall rocks and they define a broad domain between the contact and the eastern termination of the Stobie Deposit.

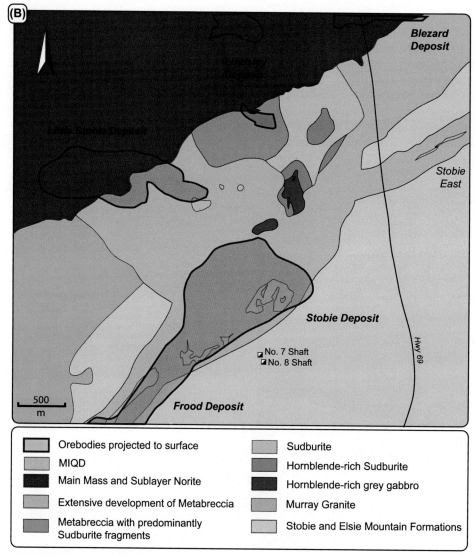

FIGURE 2.31 (*cont.*)

breccias are quite common in the broader belt which contains Frood Stobie (eg, Fig. 2.30D), and they are so common that they do not provide an obvious indicator of mineral potential.

SUBX is also an important host rock to the discontinuities developed in Offset Dykes. The localization of SUBX is normally controlled by contrasting country rock composition, so it is a combination of the presence of contacts between different country rocks and the localization of SUBX that controlled the development of discontinuities in the Offset Dykes. The SUBX developed adjacent to the dykes and in these discontinuities is often anomalously rich in sulfide mineralization, and the texture of the breccias matrix tends to be "hot." These features are important in understanding the changing geometry of the dykes, and can aid in the identification of mineralization within the dyke, but they are not a direct indicator of proximity to economic ore deposits.

SUBXs developed proximal to the Sublayer in areas where the melt sheet was thickest, show evidence of thermal metamorphism and melting. Although there has been a debate regarding whether SUBX records textural and metamorphic evidence of nearby ore systems (eg, Lightfoot et al., 1997c), it appears more likely that the variation relate to proximity to melt rocks of the SIC. In the sense that Lightfoot et al. (2001) show an empirical relationship between mineral potential and thickness of the melt sheet, it appears likely that the degree of melting of breccia has a serendipitous link to mineral potential.

Origin of the Sudbury Breccia

Much recent debate has focused attention on whether the matrix of impact-related pseudotachylite breccias are derived from nearby country rocks through cataclasis and partial melting (eg, LaFrance et al., 2008; LaFrance and Kamber, 2010; O'Callaghan et al., 2015; Reimold et al., 2015) or whether they contain contributions from an impact melt that was introduced into the breccia package (Riller et al., 2010). This debate is discussed in more detail in Chapter 3 where the geochemical evidence is presented to support the generation of the pseudotachylite breccias within the country rocks with no requirement for a contribution from the melt sheet.

Thermal Metamorphism of the Footwall: the Metabreccias

The country rocks in the immediate footwall of the SIC underlying the Sublayer are often fragmental, extensively recrystallized and partially melted by the Sudbury event (Fig. 2.31A). These rocks are collectively termed metabreccias, and they are described in terms of the degree of melting, nature of the country rock fragments, and nature of the rock fragments using the scheme shown in Table 2.3. The scale of development of metabreccias varies with distance away from the base of the Sublayer and the QD Offset Dykes.

The range in textural variants in the metabreccias is enormous, and descriptions of a rock as a "metabreccia" are normally qualified with detailed descriptions of petrology, mineralogy, and possible protolith (many of the features described in Table 2.3 apply to the description of metabreccia). The range of rock types extend from rocks showing evidence of extensive melting to rocks which are thermally metamorphosed.

Melt-textured rocks are variants on the Sublayer, Granite Breccia, and MIQD of the Offset Dykes, where the rock comprises a matrix of partially melted granite-feldspar with relict xneocryst derived from the target rock. These rocks often show evidence of a flow-texture in the leucocratic matrix. Their principal distribution of metabreccia is proximal to Sublayer, Granite Breccia, and the Offset Dykes.

Melting of country rock gneiss is also found less than 500 m into the footwall, and comprises wispy-textured leucosomes in a matrix of grains and clasts of country rock. Further away from the SIC, the rocks showing evidence of incipient melting give way to rocks that have partially recrystallized to form more leucocratic patches of quartz and feldspar, but without significant brecciation or development of leucosomes.

The range in geological relationships, degree of melting and metamorphism in the country rocks beneath an embayment structure is illustrated in Fig. 2.31A based on the geology of the Levack-Coleman area. Dressler (1984a) and Farrow (1995) reports evidence for a thermal metamorphic aureole in the

North Range footwall, but Boast and Spray (2006) suggest that northward thrusting of the SIC along the base of the Sublayer produced discontinuities in the metamorphic halo. In detail, the degree of thrusting appears not to have influenced the overall geometry of the Sublayer at the base of the North Range, although there is evidence for the development of small contact-subparallel structures (Coats and Snajdr, 1984). The metabreccias developed in the footwall of the embayments and troughs are a very good indicator of proximity to Sublayer (Fig. 2.32A, B), but there are excellent examples of metabreccias developed in association with unmineralized and weakly mineralized embayments, so the linkage appears to relate to the availability of heat, not to the presence of sulfide mineralization.

The metabreccias are sometimes weakly mineralized and they can contain the margins and distal extents of ore bodies (eg, Frood), but they are not important standalone hosts to major ore deposits at Sudbury. Metabreccia locally controls the distribution of some S-poor styles of mineralization enriched in Cu and PGE (Stewart and Lightfoot, 2010; Gibson et al., 2010). Opinions are mixed with respect to the importance of thermal recrystallization of country rocks in the context of mineral potential. In this author's experience, the range of recrystallized rocks described above can be found in both the footwalls of major mineralized trough and embayment structures as well as along segments of barren contact. The extent of the recrystallization in the North Range appears to be more closely related to the thickness of the SIC.

Studies of the rocks beneath the South Range of the SIC confirm the development of partial melt textures, thermal metamorphism, and recrystallization of the country rocks in response to heat provided by the melt sheet (Jørgensen et al., 2013, 2014). These thermal signatures outline a metamorphic halo approaching 500 m in thickness (Jørgensen et al., 2013, 2014), and these rocks show an extensive development of metabreccia and specific variations such as Sudburite within a broad footwall footprint between Little Stobie, Blezard and Stobie Deposits that is illustrated in Fig. 2.31B. The extent of thermal metamorphism of mafic volcanic and intrusive rocks in the footwall of this footprint is extreme. Fine-grained rocks exhibiting brecciation and containing weakly mineralized leucosomes and porphyroblasts of amphibole (Fig. 2.32A,B) are common features of the metabreccias developed in the Elsie Mountain Formation.

THE WHITEWATER GROUP: SUBSIDENCE, MAGMATISM, AND SEDIMENTATION FOLLOWING THE IMPACT EVENT

The rocks within the Sudbury Basin structure on top of the Main Mass of the SIC comprise the Onaping, Vermilion, Onwatin, and Chelmsford Formations; these Formations collectively make up the Whitewater Group (Fig. 2.33). The Onaping Formation is a ~1.5 km thick package of breccia consisting of altered glass and country rock fragments that show extensive hydrothermally alteration. The narrow (5–50 m thick) Vermilion Formation consists of carbonate, argillite and chert; this unit hosts the Zn–Pb–Cu volcanogenic massive sulfide mineralization of the Errington and Vermilion Deposits. The overlying Onwatin Formation is 600–1100 m thick and consists of carbonaceous and conductive argillite with minor greywacke units. The more than 850 m thick Chelmsford Formation consists of greywacke and siltstone that comprise proximal type turbidite facies.

The Onaping Formation
The Onaping Formation has historically been subdivided into the Gray, Green and Black Members on the basis of color that reflects the abundance of carbon in the rock (Peredery, 1972b; Avermann, 1999).

FIGURE 2.32

(A) Example of metabreccia developed in a gabbroic host rock ~500 m south of the contact to the east of the Little Stobie Deposit. This rock shows extensive recrystallization, development of amphibole porphyroblasts (P) and lecosome (L) partial melt textures, as well as extensive recrystallization to two pyroxene hornfels (H). This type of mafic rock in the South Range footwall has been termed Sudburite; it occurs in the metamorphic halo of the SIC. (B) Example of Sudburite developed in the Elsie Mountain Formation in the footwall of the Murray Deposit. The fine-grained hornfelsed metavolcanic rock contains porphyoblasts of amphibole and partial melt segregations associated with weak brecciation.

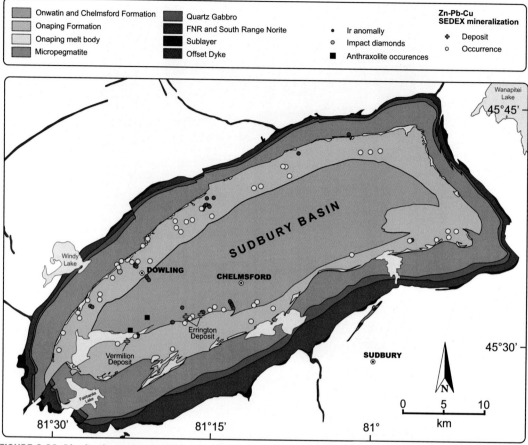

FIGURE 2.33 Distribution of the Onaping, Onwatin, and Chelmsford Formations of the Whitewater Group

The image shows the distribution of melt bodies mapped in the basal Onaping Formation, and highlights the distribution of showings and deposits of Cu–Pb–Zn mineralization. The map also shows the location of anthraxolite showings, iridium anomalies, and impact diamonds.

Based on Ames and Farrow (2007).

However, the color variations in the Onaping Formation cross cuts stratigraphic contacts and patches with different coloration can be developed throughout the Onaping Formation, so the color variations are best attributed to postdepositional alteration rather than primary diversity in lithology (Gibbins, 1994, 1997; Ames et al., 1997, 1998, 2002). A more recent subdivision breaks the Onaping Formation out into members based on the stratigraphy and volcanology of the rocks; these are the Garson, Sandcherry and Dowling Members (Figs. 2.33 and 2.34A).

The Garson Member is well exposed above the Main Mass in the southeastern part of the South Range with type sections developed near to Garson Lake (Figs. 2.33 and 2.34A). The Garson Member consists of a thick package of bedded to fragmental rock with 6–50 m clasts of quartzite (20–85% by volume) derived from the Huronian Supergroup (Ames et al., 2009). The Garson Member also contains

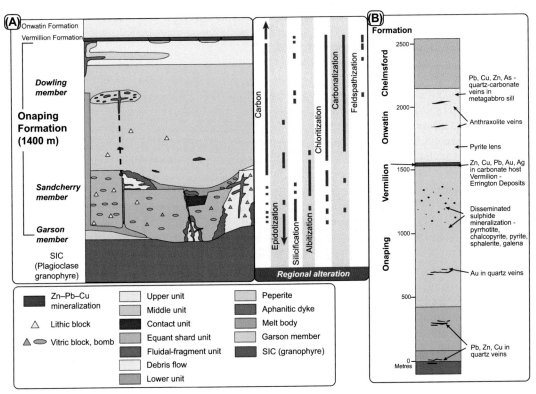

FIGURE 2.34

(A) The geological relationships in the Onaping Formation; this simplified section illustrates the terminology, thegeological relationships, their volcanological interpretation, and the extent of hydrothermal alteration. (B) Stratigraphy of the Whitewater Group and distribution of the main styles of mineralization.

(A) After Ames et al. (2002), Roussel et al. (2002); (B) after Roussel et al. (2002).

less than or equal to 15% vitric andesite lapilli and less than or equal to 5% vitric andesite bombs and blocks.

The Sandcherry Member is 300–500 m thick and characterized by more than 60 modal % vitric fragments that are equant, blocky, or fluidal in morphology (Fig. 2.34A). The units are massive to poorly bedded, they contain less than 35 modal % matrix, and they are rich in blocks and volcanic bombs. Units that consist of equant pseudomorphed glass shards are interpreted to represent fallback breccia. Andesitic hyaloclastite and peperite breccia cuts through the poorly bedded Sandcherry Member (Fig. 2.34A).

The Dowling Member comprises ~75% of the stratigraphic thickness of the Onaping Formation. This Member contains abundant carbon throughout the sequence in rocks that contain 25–40% lenticular glass fragments in a fine-grained matrix (Muir and Peredery, 1984). The base of the Dowling Member has small-scale faults associated with the intrusion of vitric aphanitic dykes and locally discontinuous debris flows and mass flows that comprise the Lower Unit (Fig. 2.34A; Gibbins, 1994;

Ames, 1999). The overlying rocks of the Dowling Member comprise a succession of increasingly fine-grained rocks dominated by tuffs. Minor reworking is found principally in fine-grained andesitic rocks that are interpreted to represent fallback deposits. The top of the Dowling Member shows minor reworking and redeposition of the tuff.

The lower 500 m of the Onaping Formation is cut by numerous inclusion-rich sills and dykes of granophyre, andesite and QD. The matrix of these rocks is similar to that of the underlying Granophyre of the SIC. This package of igneous-textured breccias has previously been termed "melt bodies," "basal intrusion," and "Basal Member" (Peredery, 1972a,b; Muir and Peredery, 1984; Anders et al., 2013), but they reside stratigraphically in the Sandcherry and Dowling Members (Fig. 2.34A; Ames et al., 2002). The melt bodies developed towards the base of the Onaping Formation tend to be heterogeneous in composition, and contain partially digested xenoliths that have created conditions of local contamination as a result of melting of the clasts (Fig. 2.35A,B).

The top of the Onaping Formation grades into carbonate and argillaceous rocks of the overlying Vermilion and Onwatin Formations. Aphanitic andesitic dykes reach 15 m in width, but are more commonly 0.5–1 m wide with flow-banding, spheroidal textures, and amygdules. Locally these dykes extend high into the Dowling stratigraphy and are sometimes enveloped in peperite (Fig. 2.34A). The dykes which cross cut higher into the Onaping Formation tend to follow structures, and they appear to be generated by the expulsion of melt from the underlying magmatic breccias.

The entire stratigraphy of the Onaping Formation contains disseminated sulfide as discrete fragments, grains within rock and glass fragments and disseminations in the melt bodies (Fig. 2.35A,B). The sulfide content rarely exceed 10% and is generally approximately less than 1% by volume, comprising principally pyrrhotite with 0.28% Ni (Desborough and Larson, 1970) and lesser pentlandite, chalcopyrite, pyrite, marcasite, sphalerite and galena.

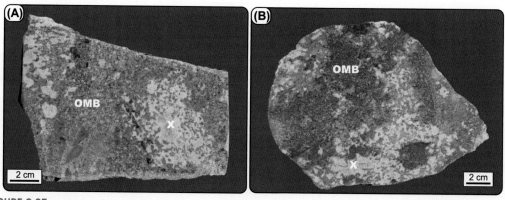

FIGURE 2.35

(A) Example of fragment-laden melt body from the basal Onaping Formation, NW of High Falls. The melt body has a felsic matrix (OMB) and appears to be part intrusive and part extrusive with flow-lineated decomposed fragments (X), a wide range of country rock types common in the Sudbury Region, vesicular structures, and the development of irregular blebs of sulfide dominated by pyrrhotite that carries small amounts of nickel. (B) Similar example of melt body with strong assimilation of fragments (X). These rocks are interpreted as impact generated melts that have been emplaced from the melt sheet into the overlying fallback breccias (eg, Peredery, 1991).

Mechanism of Formation of the Onaping

Published studies of the Onaping Formation cover a broad range of genetic models that include those with an emphasis on the volcanic origin (Bonney, 1888; Burrows and Rickaby, 1930; Thomson, 1957a; Williams, 1956; Stevenson, 1972; Muir, 1984), those requiring impact-triggered volcanism (Dietz, 1964; Thomson, 1969; Muir, 1983), models of fallback and reworking of material from meteorite impact (French, 1968a,b; Dence, 1972; Peredery, 1972a,b; Peredery and Morrison, 1984b), and interaction of melt derived from the SIC with water (Gibbins, 1994, 1997).

Evaluation of the facies changes in the Onaping Formation support two stages of evolution. In the first stage, the Garson Member was formed during postimpact slumping of the crater walls onto the top of the crater floor assemblage (Ames and Gibson, 1995) that is now considered to be the melt sheet. This was followed by the deposition of a fallback of glass shards to form the Sandcherry Member (Peredery, 1972b). The Sandcherry Member was cross cut by melts injected from beneath which were likely derived from the melt sheet.

The overlying Dowling Formation records an initial period of crater instability that resulted in mass flow deposits being formed adjacent to fault structures. The main sequence of Dowling Formation tuffs were deposited as fallout from the ash cloud column created by the impact event. The faults acted as conduits along which melts and fluids migrated and localized the development of peperites. Introduction of seawater and minor reworking of the top of the Dowling Formation marked the introduction of extensive fluids that mixed with magmatic fluids and initiated extensive hydrothermal alteration that gave rise to the conditions required to form the sea-floor volcanogenic massive sulfide mineralization developed in the overlying Vermilion Formation.

The fragmental sulfides in the Onaping Formation indicate the presence of quite abundant crustal sulfide in the target rocks; the observation that some of these sulfides in the fall back contains nickel sulfide is consistent with a target rock that contained at least some magmatic sulfide.

In summary, the Onaping Formation is presently considered to represent a package of rocks formed in response to the conditions created by the collapse of material onto the top of the melt sheet within the crater. The processes involved in the genesis of these rocks include fall-back of material onto the melt sheet, collapse of the crater walls, development of fractures through which contributions of magma from the melt sheet could escape, and extensive hydrothermal alteration which may be related to the influx of water into the central basin of the crater (Ames et al., 1998, 2002).

Perhaps the most important rocks in the context of this book are the melt bodies that contain the least number of partially digested xenoliths. These rocks exhibit minimal alteration, and their value rests in their use together with vitric fragments as an indicator of the primary composition of the melt sheet (Ames et al., 2002 and Chapter 3).

The Vermilion Formation

The Vermilion Formation conformably overlies the Onaping Formation, and it is conformably overlain by the Onwatin Formation (Figs. 2.33 and 2.34A,B). The Vermilion is the narrow ($\sim$<25 m thick) stratigraphic unit which hosts volcanogenic massive sulfide mineralization, and the description of this Formation is heavily influenced by data from a restricted part of the Sudbury Basin where drilling has been undertaken. Stoness (1994) divided the Vermilion Formation into a Lower Carbonate Member, a Grey Argillite Member, and an Upper Carbonate Member. Each of these members varies in composition depending on the distance away from the Zn–Pb–Cu sulfide mineralization (Stoness et al., 1993; Stoness, 1994). The proximal facies are local mounds of gray to pink, laminated, colloform carbonate material. The distal facies comprise dark brown, finely laminated, carbon-rich silty carbonate rocks formed as turbidites.

Mineralization was originally discovered near to the Vermilion River in 1897, and the Errington and Vermilion Deposits were located nearby in 1929 with reserves plus resources of 2.44 M tons at 5.11% Zn, 1.49% Cu, 1.37% Pb, 1.1 g/t Au, and 66.2 g/t Ag (based on data reported in Roussel, 2009). The sulfide minerals comprise pyrite > sphalerite > chalcopyrite > galena > marcasite > pyrrhotite in several strata-bound sulfide lenses in tight fold structures hosted in the lower carbonate-rich member of the Vermilion Formation (Roussel et al., 2009).

Burrows and Rickaby (1930) suggested that the mineralization was derived from hydrothermal solutions emanating from the SIC. The recognition that the mineralization was localized by both structure and stratigraphy of the host rocks led to the suggestion that the deposits were formed by hydrothermal activity localized along structures (Martin, 1957; Thomson, 1957a). The recognition that the stratigraphic host rocks comprise argillite, banded carbonate with sulfide-mineral layers, chert and black carbonaceous pyritic shale was established by Roussel (1984b, 1984c) and related to a sedimentary-exhalative process of formation (Sangster, 1970).

The Onwatin Formation

The ~600–1410 m thick Onwatin Formation consists mainly of carbonaceous and pyritic, massive to laminated argillite and siltstone with minor greywacke; it has been described in detail (Coleman, 1905a,b,c; Burrows and Rickaby, 1930; Thomson, 1957a,b; Sadler, 1958; Beales and Lozej; 1975; Arengi, 1977; Roussel, 1984b; Roussel et al., 2002; Stoness, 1994). The carbon content ranges from 0.26–4.05% (Arengi, 1977), and local anthraxolite veins have developed in response to thermal metamorphism (Fig. 2.33; Burrows and Rickaby, 1930). The rock types are considered to be deep water pelagic sediments formed in a restricted basin under stagnant anoxic conditions (Roussel, 1984b).

The Chelmsford Formation

The more than 900 m thick Chelmsford Formation is located in the center of the Sudbury Basin. The contact with the Onwatin Formation is a transition from argillite and minor greywacke through to greywacke and minor argillite. The Chelmsford Formation is interpreted as a proximal turbidite sequence (Burrows and Rickaby, 1930; Thomson, 1957b; Williams 1956; Cantin and Walker, 1972; Roussel, 1972, 1984a; Roussel et al., 2002). Beds in the formation range in thickness from 0.03 to 5.2 m with an average thickness of 1.23 m (Roussel, 1972), and they contain examples of ideal "Bouma" sequences (Bouma, 1962).

Sedimentary structures in the Chelmsford Formation include concretions, channel infill structures, current and ripple marks, convoluted laminations, load structures, and flute casts and groove marks (Cantin and Walker, 1972; Roussel, 1972). Highly equivocal evidence of Paleoproterozoic biological activity were reported as patterns on bedding planes and vertical pipe-like structures (Ames et al., 2009).

Significance of the Whitewater Group to the evolution of the SIC

The detailed stratigraphy of the Whitewater Group provides important information that underpins models of crater development and melt sheet evolution. Some of the important points developed later in this book include:

1. Best estimates of melt sheet composition may be anchored by the compositions of glass shards and andesitic melt bodies in the lower part of the Onaping Formation (Ames et al., 2002).

2. The value of compositional data for the glass shards and intrusive andesitic rocks is important in understanding the timing of sulfur saturation in the Sudbury melt sheet. In most cases they appear to contain normal metal inventories, so their formation predates sulfide segregation.
3. The basal breccias of the Onaping Formation locally contain clasts of sulfide that may be derived from the target rocks; the fact that some of these sulfide fragments have elevated Ni and Cu abundance may indicate that the target rocks contained preexisting sulfides. It is not clear whether a correlation exists between these nickeliferous sulfides and the most prospective segments of the basal contact of the SIC.
4. The stratigraphic development of the sequence of rocks in the Onaping Formation provides information about the evolution of the crater after excavation, and provides a record of the volcanic interaction between the melt sheet and the breccias developed above it.
5. The basin infill rocks provide evidence for evolution of the SIC under progressively deep water conditions, and this process may well link to the evidence for hydrothermal modification of large parts of the SIC during and after crystallization (Ripley et al., 2015).
6. The recognition of sedimentary-exhalative types of mineralization at the base of the Onwatin Formation is consistent with the development of extensive hydrothermal alteration systems that locally concentrated metals in association with carbonate rocks of the Vermilion Formation.

REGIONAL MAGMATIC EVENTS THAT POSTDATE THE SIC

The Sudbury region was affected by mantle plume-triggered magmatic events that resulted in the formation of carbonatites, dykes, and intrusions in the Sudbury Region. These events are briefly summarized below:

Sage (1987) summarizes the geology of the Spanish River Carbonatite Complex that is developed to the west of Sudbury (Fig. 1.13). This Complex has recently been mined for phosphate. The complex consists of a silico-carbonatite, sovite, ijolite, pyroxenite, and cancrinite-nepheline syenite enveloped in fenitized Archean quartz monzonite. The complex has been dated by whole-rock rubidium-strontium isotopic techniques at 1838 ± 95 Ma (Bell and Blenkinsop, 1980). Although this age is almost the same as that of the Sudbury event (Sage, 1987), a more robust U–Pb age is required to establish whether this intrusion was emplaced at the same time as the impact event.

A number of mafic intrusions are developed to the SSW of Sudbury. Card (1968) reports the geology of a small olivine amphibolite pluton that cuts Huronian Supergroup-aged rocks near to Espanola. This pluton may be comagmatic with a number of associated lamprophyre dykes. Whole-rock samples with their associated biotite separates yields ages around 1.5 Ga (Rb–Sr) that appear to be similar in age to the Croker Island Complex in the North Channel of Lake Huron (van Schmus et al., 1975; Card, 1965).

Lumbers (1975) mapped two types of diabase dykes in the Southern Province southeast of Sudbury, which he referred to simply as 'olivine diabase' and 'diabase'; these later became known as the Sudbury and Grenville swarms, respectively (Fahrig and West, 1986).

Olivine diabase dikes are confined to the Southern Province and they are part of a major northwest-trending swarm that intrudes rocks of both the Southern and Superior Provinces but these dykes do not extend across the Grenville Front Tectonic Zone into the Grenville Province. Shellnutt and MacRae (2011) suggest that the ~1.23 Ga evolved alkaline olivine diabase dykes were emplaced during the break-up of the Mesoproterozoic Columbia supercontinent, and record contributions from a lithospheric mantle source which was modified by earlier subduction events.

A number of 3–30 m wide west-east fine-grained quartz diabase dykes cuts through the southern part of the Sudbury; these are locally termed "Trap Dykes" by mine geologists. These may be part of an east–west trending dyke set developed in both the Southern Province and the Grenville Province over a lateral distance of ~600 km and dated at 595 Ma by Kamo et al., (1995).

STRUCTURAL GEOLOGY OF THE SUDBURY REGION

Much critical research and debate has been expended in understanding the tectonics of the Sudbury Region (eg, Riller et al., 1999; Roussel et al., 1997, 2002), but little of the work fully aligns an understanding of the regional tectonic framework (Table 2.4 and references provided therin) with the kinematics of local structures that control the geology of the Sudbury Ore Bodies (Table 2.5 and references provided therein).

Structures have provided an important controlling influence on the shape of the Sudbury Structure and the distribution of the rock types and ore deposits.

1. Structures that predate the impact event. There is an incomplete understanding of the role of preexisting structures as they relate to the morphology of the basal contact of the SIC, the distribution of radial and concentric offsets (and the associated ore deposits), the localization of embayments and troughs (and the associated ore deposits), and the development of belts of SUBX (Spray et al., 2004; Golightly, 1994).

2. Structures that were either reactivated or created at the time of impact or during the isostatic recovery of the lithosphere immediately after the Sudbury impact event (ie, <~1 Ma). These include the primary structures that controlled the development of the Offset Dykes, the faults developed along the crater walls that may have controlled slumping and readjustment of the crater floor and melt sheet (eg, Spray, 1997), and or arcuate zones of SUBX immediately adjacent to the base of the melt sheet near to the ore deposits at Coleman and Victor that provided slump planes along which mineralization was injected into the footwall (Lightfoot, 2015).

3. Structures that have modified the Sudbury Structure after the impact event as a result of regional far-field plate tectonic forces that affected the early stabilization of the Sudbury Structure, and drove the regional deformation events, including, but not restricted to Penokean, Mazatzal, and Grenville orogenic events. These structures control the gross morphology of the Sudbury Structure and they often intersect and dissect ore bodies (eg, Coleman-Lavack; Lightfoot, 2015) or control the development of remobilized sulfides (eg, the Garson Deposit; Mukwakwami et al., 2012, 2014a,b)

In the next section, I look at these three stages of deformation as a framework in which to understand the Sudbury Structure. I will leave it in the hands of ongoing and future studies to continue to contribute necessary detail to the chronology of the deformation events and their relative importance in controlling the geometry of the Sudbury Structure and the ore deposits. This is a segment of Sudbury Geology that will benefit enormously from the detailed ability to use U–Pb geochronology to unpick deformation events by dating of igneous events (eg, Raharimahefa et al., 2014).

Importance of structures predating the Sudbury Event

At the southern margin of the Superior Craton, the earliest major deformation event recognized in the Sudbury Region appears to have uplifted the lower crust, and introduced the Levack Gneiss Complex into the target area of the impact (Table 2.4A; James et al., 1992a,b). The exhumation of granulite facies

Table 2.4 Summary of Major Deformation Events in the Sudbury Region

Event	Age (Ga)	Structural Outcome	Influence on Sudbury Structure	References (tectonics)	Influence on Ore Bodies	References (ore deposits)	Metamorphic Conditions	Reference (Metamorphic Effects)
Wanapitei Impact event	0.004	Impact east of Sudbury to produce Wanapitei structure	Possible deformation of olivine diabase dykes and East Range, but the deformation of the East Range is probably not influenced significantly	L'Heureux et al. (2005); Clark et al. (2012)	None	na	na	na
Grenville Orogeny	1.235–0.98	Grenville Front located 8–16 km SE of Sudbury; contains rocks of Southern Province	Thermal overprint in Southern Province, with exception of Creighton and Murray faults, no evidence that Grenville-aged structures reactivate structures in Southern Province or deform/offset 1.27 Ga dykes	Brocoum and Dalziel (1974)	No significant influence			
Olivine diabase dykes	1.24	Development of dyke swarm in response to mantle plume	Thermal overprint on SIC; deformation is limited to the process of space creation	Krogh et al. (1987)	Some dykes cross cut near to ore bodies and cause local redictribution of stratigraphy	na	na	na
Mazatzal Orogeny (part of the Hudsonian Event)	~1.65	Continent–continet collisiion with far-field effect largely expressed as thrusting	Thrusting of South Range and creation of South Range Shear Zone	Piercey et al. (2007)	Major structures offset the geometry of the base of the SIC and produce changes in orientation of mineral zones	Bailey et al. (2004); Lightfoot (2015)	Mylenite formation in shears	na
Yavapai Orogeny	1.74–1.70	Deformation largely recorded to the south of Sudbury	Unknown	Raharimahefa et al. (2014)	None known	na	na	na
Late Penokean Orogeny	1.85–1.80	Deformation of SIC by ductile folding of the Sudbury Structure and early imbrication of the contact on the South Range and East Range	Folding of basin and development of fault-repetitions at base of South Range and East Range Main Mass; possible activation of fault-thrust transfer corridors	Siddorn and Lee (2005); Mukwakwami et al. (2012); Cowan et al. (1999); Roussel et al. (1997); Bailey et al. (2004); Piercey et al. (2007); Young et al (2001)	Offset of ore bodies and sulfide kinesis in mylonite zones; detachment of contact and footwall deposits	Lightfoot (2015)	Amphibolite to lower greenschist facies	Piercey et al. (2007); Magyarosi (1998)
Sudbury crater modification and readjustment	1.85 Ga (duration measured in tens of thousandsto hundreds of thousands of years)	Readjustment of crust to impact event	Crater wall readjustment; development of Whitewater Formation during regional subsidence	This text and references	Migration of fractionated sulfide melts into space created by structural modification of the crater walls (slumping); creation of VMS mineralization in Onwatin Formation	This text and references	na	

(Continued)

Table 2.4 Summary of Major Deformation Events in the Sudbury Region (*cont.*)

Event	Age (Ga)	Structural Outcome	Influence on Sudbury Structure	References (tectonics)	Influence on Ore Bodies	References (ore deposits)	Metamorphic Conditions	Reference (Metamorphic Effects)
Sudbury Impact and creation of melt sheet	1.85 Ga (duration likely measured in hours and days)	Development of radial and concentric offset dykes, establishment of breccia belts around SIC; development of initial crater floor geometry; reactivation of older fault structures	Additional exhumation of crater floor (Levack Gneiss)	This text and references	Creation of magmatic sulfide ore deposits in Offsets, contact, and Sublayer	This text and references	Footwall contact metamorphism - creation of pyroxene hornfels transitional to amphibole hornfels, and then plagioclase recrystallization	Dressler (1984a,b)
Instantaneous impact, crater excavation, and pseudotachylite event - the Sudbury Impact	1.85 Ga (duration a few minutes)	Psuedotachylite development; crater excavation; depression of crust	Fracture pathways and crustal melting along preexisting structures and contacts; excavation of crater and creation of "hings rocks" on crater walls; initial conditions for melt sheet to form from combination of melts and plasma.	This text and references	Creation of cbreccia belts that control major ore systems; creation of crater ftopoghraphy	This text and references	Shock metamorphic conditions	French (1998) and references cited
Early Penokean Orogeny	1.89–1.75	Distal arc accretion; uplift of Sudbury Region with preimpact craton margin deformation and metamorphism followed by erosion	Folding and metamorphism predate SIC in a back-arc setting where the preimpact topography was an epicontental margin; SUBX and Offsets contain chloritized staurolites from the Stobie Formation - confirmation that the Penokean event predated the Sudbury event; development of W–E system (eg. Creighton, Murray, and associated brindging structures) as transcurrent faults; initiation of N–S fault family (eg, Sandcheery Creek, Fecunis, and Bob's Lake Faults) parallel to Kapuskasing structure	Young et al. (2001); Roussel et al. (1997)	Created target rock geometry; preimpact structures and country rock boundaries possibly controlled offsets and troughs as well as geometry of base of melt sheet	Golightly (1994); Keays and Lightfoot (2004); Ripley et al. (2015)	Amphibolite facies (staurolite formed in Stobie Formation)	Dressler (1984a,b)
Nipissing Diabase (and Sudbury Gabbro?) magmatic event	2.2	Formation of Nipissing Diabase (and possibly Sudbury Gabbro) in response to far field plume event	Development of a dominantly mafic footwall to the South Range with evidence of extensive gabbroic rocks	Card et al. (1972); Lightfoot and Naldrett (1989); Lightfoot et al., 1993a; Noble and Lightfoot (1992); Corfu and Andrews (1986)	Important inclusion type in contact and offsite ore deposits and control on location of mineral shoots	Lightfoot and Farrow (2002)	Local contact metamorphism of Huronian Supergroup	na

Event	Age	Deformation	References	Controlling influence	References	Metamorphism	References
Blezardian Orogeny	2.4–2.2	Transpressive deformation of southern margin of Archean Craton	Possible generation of and migration of magmas through transpressive pathways and deformation of the granitoid intrusions	Stockwell, 1982; Riller and Schwerdtner (1997); Raharimahefa et al. (2014)	Created target rock geometry; preimpact structures and country rock boundaries possibly controlled offsets and troughs as well as geometry of base of melt sheet	Golightly (1994); Keays and Lightfoot (2004); Ripley et al. (2015)	Not documented	na
Crustal extension (rifting) and East Bull Lake magmatic event	2.4	Development of continental margin rift volcanic rocks of Stobie Formation and progressively deep-water sedimentation; emplacement of EBLI suite of rocks along major structures developed in-board of craton margin	Development of a dominantly mafic footwall to the South Range with evidence of extensive EBLI type magmatic activity	Roussel and Long (1998); Long (2004); Bleeker (2004); Young et al. (2001)	Source of important inclusion population and mafic host rocks that contributed to the melt sheet	James et al. (2002a,b); Easton et al. (2010)	na	na
Levack exhumation	2.65	Amphibolite facies metamorphism of granulites indicative of exhumation from 21–28 km depth to 5–11 km depth	Unexpected coincidence of pre-Sudbury uplift event with the outline of the northern margin of the SIC	James et al., 1992a,b; Prevec et al., 2005	None apparent	na	Amphibolite facies at 6–11 km depth	James et al. (1992a,b)
Kenoran Orogeny	2.71	Formation of granulites in the roots of the Superior Craton	Important country rock type in the North Range	Krogh et al. (1984)	Contacts between mafic and felsic gneiss controls development of SUBX and Footwall mineralization	Lightfoot (2015)	Granulite facies at 21–28 km depth	James et al. (1992a,b)

Principal deformation events, fold, and fault sets, and metamorphism, and their controlling influence on the Sudbury Structure. na, not applicable

Table 2.5 Summary of Major Fault Sets in the Sudbury Region

Importance	Fault Set Name	Location	Examples of Faults	Length	Principal Strike	Dip	Brittle/ Ductile	Type of Motion	Pre-SIC	Syn-SIC	Post-SIC Displacement	Timing of Last Major Displacement	Effect on Ore Deposits	Reference to Detailed Studies
Regional fault sets	Creighton–Murray–Garson set	South Range	Creighton	>500 km	W–E	75°N–90°N	Mostly brittle	Normal (south-side down) during Huronian sedimentatiom; reverse south-side up during Penokean; Neoproterozoic strike-slip (south-side up); steep north dip indicates that it was a growth fault	Yes	Possible	~700 m	After SIC	Displaces Copper Cliff Offset	Cochrane (1984); Lockhead (1955); Owen and Coats (1984)
		South Range	Murray	>500 km	W–E	70°S	Mostly brittle	Normal (south-side down) during Huronian sedimentation; reverse south-side up during Penokean; Neoproterozoic strike-slip (south-side up)	Yes	Possible	~2000 m	Penokean through grenville	No known effect	Card (1978); Sims et al. (1980); Cochrane (1991)
		South Range	Garson	>10 km	W–E	70°S	Mostly brittle	Last motion was normal-sinistral oblique slip	Unknown	Possible	Not known	After SIC	Controls Garson Ore Bodies	Cochrane (1991)
Regional fault sets	Fecunis Lake	North Range	Sandcherry Lake	~250 km	NNW–SSE	95°W	Mostly brittle	Sinistral	Possible	Unknown	Up to 600 m	Penokean through Mazatzal	No known effect	Roussel et al. (2002)
		North Range	Fecunis Lake	300 km	NNW–ESE	82°E	Mostly brittle	Sinistral	Possible	Unknown	600 m horizontal, 200 m vertical	Penokean through Mazatzal	Offsets contact and footwall mineralization	Coats and Snajdr (1984)
		North Range	Bob's Lake	3 km branch	NNW–SSE	85°E	Mostly brittle	Sinistral	Possible	Unknown	100 m horizontal	Penokean through Mazatzal	Offsets contact and footwall mineralization	Coats and Snajdr (1984)
		North Range	Rapid River (Michaud)	~10 km	NNW–SSE	80°E	Mostly brittle	Sinistral	Possible	Unknown	Up to 700-m	Penokean through Mazatzal	No known effect	

Regional fault sets	Set	Region	Fault	Length	Orientation	Dip	Deformation style	Kinematics			Displacement	Age	Significance	Reference
Regional fault sets	Waddell–Amy set	East Range	Waddell lake	>10 km	N–S curved	Vertical	Mostly brittle	Sinistral	Unknown	Unknown	Unknown	Penokean through Mazatzal	Offsets contact mineralization	Clark et al. (2012)
		East Range	Selwyn Lake	~5 km	N–S curved	Vertical	Brittle	Sinistral	Unknown	Unknown	Unknown	Penokean through Mazatzal	Offsets contact mineralization	Clark et al. (2012)
		East Range	Amy Lake	>10 km	N–S curved	Vertical	Mostly brittle	Sinistral	Unknown	Unknown	~1 km	Penokean through Mazatzal	Offsets contact and footwall ore bodies at Capre	Clark et al. (2012)
		South Range and Basin Sudbury	Cliff Lake Fault	>50 km	W–E curved	45°S	Anastomosing ductile and brittle	Reverse shear zone that transported the south part of the Sudbury Basin over the North Part; the faults steepen from the center of the Sudbury Basin towards the West and East	Not likely	Not likely	0.5–1.5 km	Mazatzal	Depth extent opf south range or bodies controlled by this fault	Gibson (2003)
Regional fault sets	Vermilion Lake Family and South Range Shear Zone	South Range and Basin Sudbury	South Range Shear Zone	>70 km	W–E anastomosing	45°S	Anastomosing ductile	Reverse shear zone that transported the south part of the Sudbury Basin over the North Part; the faults steepen from the center of the Sudbury Basin towards the West and East	Not likely	Not likely	Total ~6.5 km (excluding the lowermost major shear - the Cliff Lake Fault Zone)	Mazatzal	Depth extent opf south range or bodies controlled by this shear zone	Shanks and Schwerdtner (1991); Burrows and Rickaby (1930); Thomson, (1957a); Ames et al. (2005); Stevenson (1961); Roussel (2009); Bailey et al. (2004); Santimano and Riller (2012)

(Continued)

Table 2.5 Summary of Major Fault Sets in the Sudbury Region (*cont.*)

Importance	Fault Set Name	Location	Examples of Faults	Length	Principal Strike	Dip	Brittle/Ductile	Type of Motion	Pre-SIC	Syn-SIC	Post-SIC Displacement	Timing of Last Major Displacement	Effect on Ore Deposits	Reference to Detailed Studies
Regional fault sets	South Range and Sudbury Basin	South Range and Basin Sudbury	Fairbanks Lake Fault	>50 km	WSW–ENE	45°S	Anastomosing ductile and brittle	Reverse fault that transported the south part of the Sudbury Basin over the North Part; the faults steepen from the center of the Sudbury Basin towards the West and East	Not likely	Not likely	Unknown	Mazatzal	No known effect	Thomson, (1957a); Dressler (1984c), Ames et al. (2005).
		South Range and Basin Sudbury	Chicago Fault	~10 km	WSW–ENE	45°S	Ductile and brittle	Reverse fault that transported the south part of the Sudbury Basin over the North Part; the faults steepen from the center of the Sudbury Basin towards the West and East	Not likely	Not likely	Unknown	Mazatzal	Modifies contact mineralization at Chicago Mine	Thomson, (1957a); Dressler (1984c), Ames et al. (2005).
		Sudbury Basin and East Range	Airport Fault	~40 km	WNW–ESE	70°NE	Ductile and brittle	Reverse fault that transported the south part of the Sudbury Basin over the North Part; the faults steepen from the center of the Sudbury Basin towards the West and East	Not likely	Not likely	~1 km	Mazatzal	None known	Dressler et al. (1991)
		Sudbury Basin and East Range	Bailey Corners Fault	~40 km	WNW–ESE	65°SW	Ductile and brittle	Reverse fault that transported the south part of the Sudbury Basin over the North Part; the faults steepen from the center of the Sudbury Basin towards the West and East	Not likely	Not likely	~0.2 km	Mazatzal	None known	Roussel et al. (1997)
		South Range and Sudbury Basin	Vermillion Lake Fault	>40 km	W–E curved	35°SE	Brittle	Reverse fault that transported the south part of the Sudbury Basin over the North Part; the faults steepen from the center of the Sudbury Basin towards the West and East	Not likely	Not likely	150 m dip slip	Mazatzal	No known effect	Thomson, (1957a); Dressler (1984c), Ames et al. (2005).

Regional fault sets	South Range and Sudbury Basin	South Range and Basin Sudbury	Cameron Creek Fault	>50 km	W–E curved	45°S at center of basin; 60–75°S at western exist from SIC	Brittle	Reverse fault that transported the south part of the Sudbury Basin over the North Part; the faults steepen from the center of the Sudbury Basin towards the West and East	Not likely	Not likely	2800 m slip and 2680 m throw	Mazatzal	No known effect	Thomson, (1957a); Dressler (1984c), Ames et al. (2005); Shanks (1991)
Mine scale fault sets	Six Shaft Shear	Creighton Mine	Six Shaft Shear	~500 m; possibly 1.5 km	Highly variable	Variable	Ductile	Dextral with dip-slip component	No	Yes	Unknown	Syn-SIC	Controls remobilized contact ores and 126 footwall ore body	Lightfoot (2015)
	701 Fault Set	Creighton Mine	701 Fault	~1 km	W–E	Vertical	Ductile	Unknown	No	Yes	Unknown	Syn-SIC	Controls Creighton Deep footwall ore boidies	Lightfoot (2015)
	Garson shears	Garson Mine	#4 Shear	~1 km	W–E	Vertical	Ductile	Layer parallel north-dipping reverse shears formed at the base of the SIC are offset by reverse south-dipping South Range Shear Zone	No	Possible	Unknown	Late Penokean through Mazatzal	Sublayer mineralization is remobilized into shears in the footwall and Main Mass	Davidson (1984); Cochrane (1984); Lockhead (1955); Siddom and Lee (2005); Siddom and Ham (2006); Mukwakwami (2012)
	Cross faults	Copper Cliff Mine	#2 Cross Fault	~1 km	WSW–ENE	60°NW	Ductile	Right lateral	No	Not likely	20 m horizontal	Late Penokean through Mazatzal	Displaces vertical plunge of ore bodies	Cochrane (1984)
	Gords Fault	Copper Cliff Mine	Gords Fault	>1 km	WNW–ESE	45°SSW	Brittle	Unknown	No	Not likely	Unknown	Late Penokean through Mazatzal	No known effect	
Mine scale fault sets	Falconbridge Faults	Falconbridge Mine	Falconbridge Main	2 km	W–E	85°N	Brittle and ductile	Unknown	No	Not likely	Unknown	Late Penokean through Mazatzal	Sublayer mineral remobilized along fault	Lockhead (1955); Owen and Coats (1984)
	Thrust-transfer set	Strathcona	#2 Hangingwall Fault	~2 km	45°	South	Unknown	Possible thrust	No	Not likely	Unknown	Mazatzal	Some stacking possible	Coats and Snajdr (1984); Boast and Spray (2006)

rocks involved uplift from lower crust to mid crustal levels; the rocks were retrogressed to amphibolite facies in response to this uplift. The mafic and felsic Levack Gneisses form the footwall of the SIC in the North and East Ranges, and the controlling boundaries between mafic and felsic gneisses provided a structural framework that played an important part in controlling the development of SUBX and footwall mineralization (eg, Lightfoot, 2015).

Prior to the Sudbury event two major structural trends were present, namely: (1) the NNW-striking Onaping Fault System which may relate to and control the distribution of the 2.47 Ga Matachewan dyke swarm, and (2) the ENE-striking Murray–Creighton Fault System, which acted as a major Paleoproterozoic suture zone along which bimodal mafic-felsic volcanic rocks were erupted and the Huronian sedimentary basin was developed between 2.45–2.2 Ga (Spray et al., 2004). Crustal extension accompanied by basaltic magmatism (Elsie Mountain Formation) and development of East Bull Lake Suite Intrusions was a key manifestation of the rift event. These mafic rocks comprise an important part of the target in the Southern Margin, and one that has received scant attention in the academic literature despite the presence of abundant large bodies of altered and locally recrystallized coarse-grained mafic intrusions in the South Range Breccia Belt that have collectively been termed the Frood Intrusion (Fig. 2.11).

Transpressive deformation during the Blezardian Orogeny deformed the southern margin of the Superior Craton at about 2.4–2.2 Ga (Riller and Schwerdtner, 1997). The rift sequence rocks of the Huronian were deformed, and granitoid magmas may have been generated at depth and injected along structures generated by a transpressive magmatic event (Table 2.4A and references cited therein) to form the Creighton and Murray Plutons. These rocks may be contemporaneous with the formation of rhyolite flows and pyroclastic rocks (the Copper Cliff Formation), and possibly explain why the degree of deformation exhibited by the Stobie and Elsie Mountain Formations is more severe than the deformation found in metasedimentary rocks located higher in the stratigraphy. The contacts between granitoid rocks and the stratigraphy of the Huronian sequence provided the foundation for the development of SUBX zones as well as the controls on geometry of the basal contact (embayments and troughs) and the geometry of the discontinuities in the Offset Dykes in the South Range (eg, Grant and Bite, 1984; Golightly, 1994).

At 2.2 Ga, the Nipissing Diabase was emplaced (Table 2.4A); it has not yet been confirmed whether the Sudbury Gabbro is comagmatic with this event, but it appears possible that it is a more primitive manifestation of the same event localized to the SW of the Sudbury Structure (Lightfoot and Farrow, 2002). The Nipissing Diabase, like the Early Proteroizoic mafic intrusions, provided important mafic crustal contributions to the future melt sheet, and possibly metals and sulfur too. The Sudbury Gabbro is one of the principal controls on the localization and inclusion content of the deposits along the Worthington Offset. The contact between gabbroic rocks and metasedimentary rocks appear to have been exploited by the QD magma during readjustment and injection of MIQD that formed the ore bodies.

Early Penokean deformation occurred between 1.89–1.75 Ga, and is considered to be the product of a quite distal arc accretion event (Table 2.4A). The deformation event involved NNW-directed reverse faulting, uplift, and transpression at mainly greenschist facies grade. The deformation event produced folding of the Huronian stratigraphy and broad flexures in the Nipissing Diabase sills; it resulted in staurolite facies metamorphism as recorded in shales in the Stobie Formation (Fig. 2.8B), and triggered faulting along the major east–west and north–south fault sets. This produced an important structural framework that controlled aspects of the geology of the 1.85 Ga impact structure (Table 2.4A). Metamorphism prior to the development of the Sudbury impact melts is clearly established based on the presence of retrograde chlorite after primary staurolite that occurs as deformed layers cut by SUBX and QD.

The Structural Effect of the Sudbury Event

The 1.85 Ga impact event has been described in great detail above, but it is worth emphasizing the break-out of processes between early impact and excavation, melt sheet development (injection, sulfide saturation, and differentiation), and crater-wall readjustment. These processes cannot be resolved with U–Pb geochronology, but the timelines in French (1998) and Grieve (1994) for impact cratering provide a useful framework, coupled with the geological relationships between different parts of the SIC (Table 2.4A).

The details of the topography of the surface immediately before impact are not easily unpicked. There is good evidence for the development of tsunamai deposits at 1.85 Ga, which may be indicative of earthquake activity or possibly epicontinental impact, but there are no signs of sedimentary rocks slightly older than the 1.85 Ga impact, so it appears unlikely that the impact occurred in an area substantially covered in water, but it is possible that the subsidence of the area after cratering provided the conditions necessary for the development of the volcanic breccias and hydrothermal alteration by saline waters in the Onaping Formation, and progressively deeper water basin-infill conditions for the remainder of the Whitewater Group (Table 2.4A).

The geometry of the embayment and funnel structures on the crater floor possibly reflect impact-scouring of the crater floor by materials ejected from the crater (Morrison, 1984; Golightly, 1994), but they may also reflect zones of weakness in the crust where the melt sheet was better able to erode the crater floor. These zones of weakness include the boundaries between different types of country rocks that appear to have acted as a locus for the development of some of the embayments and troughs (eg, Copper Cliff and Little Stobie). Despite what appears to be some level of regularity in distribution of troughs and embayments and the primary control of lithology contacts on their distribution, there is no strong correlation between the development of troughs and embayments and the presence of preexisting crustal fault structures. For example, the largest embayment structures at Trill and Creighton both occupy regions with complex nonisotropic crust where there are no clearly developed ancient structures with a radial trend. The same is true for the Levack, Victor-Capre, Copper Cliff, and Murray trough-embayment structures.

The structurally complex troughs and embayments at the SW and SE end of the SIC are heavily modified by post-SIC deformation, and so it is difficult to establish with any certainty whether there is a pre-SIC control, but given the weight of evidence from other embayments, it appears unlikely. On these grounds it appears that the impact event created the conditions of embayment and trough formation in response to anisotropy in country rocks and the fundamental physical processes governing the formation of impact craters (Grieve et al., 1993; Osinski and Pierazzo, 2010) rather than preexisting fault structures.

The extent to which preexisting structures guided the development of the Offset Dykes and embayment structures has been debated in the literature (Golightly, 1994; Spray et al., 2004), but there is very little evidence to indicate that either the radial or concentric dykes were emplaced into long-lived structures that were reactivated at the time of the Sudbury event. In most cases, the development of these offsets and breccia belts tend to respect a simple radial and concentric distribution, with localization in areas where rocks contrast in composition and/or rheology. The Parkin, Worthington, and Copper Cliff Offset Dykes exhibit discontinuities that also link to preexisting contacts rather than fault structures. In the case of the Copper Cliff Offset, the role of country rocks appears to be important in not only the development of the discontinuities, but also controlling the orientation of the dyke and the nature of the adjacent country rocks. A feature recognized at surface and in the Copper Cliff Mine Complex is the development of transtensional structures along the length of the dyke that may be related to primary impact and melt injection into crustal fractures that exhibit strike-slip displacement (Fig. 2.24D, E).

Another group of structures were created in response to slip along zones of extensive SUBX; in the case of the Frood-Stobie breccia belt, Spray (1997) has proposed that the development of the discontinuous zones of QD in wide zones of metabreccia and SUBX is a response to movement along listric superfault structures (Fig. 2.36D). Unfortunately, there is no geological evidence for normal

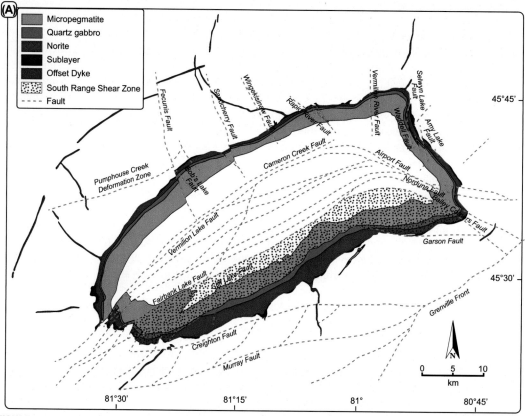

FIGURE 2.36

(A) Structural geology of the Sudbury Region showing the important fault sets that control the geometry of the SIC (see Table 2.4 and references listed). (B) Location of the Cliff Lake Fault that represents the southernmost south-dipping fault structure in the South Range Shear Zone. This diagram shows a section through the South Range in the area of the Blezard Deposit, and it highlights the effect of reverse faulting along the Cliff Lake Fault Zone. (C) A model for the development of the Frood–Stobie Offset Dyke requires the development of open space along a subvertical pre- and syn-imapct strike-slip fault structure that passes through the South Range Breccia Belt from west to east. The inset iIllustration shows the effect open space as a controlling mechanism for the development of the Frood and Stobie Ore Bodies, and the map shows the geometry of the host rocks of the Frood and Stobie QD and their possible emplacement into zones of transtension in the South Range Breccia Belt. (D) Possible model for the development of structural space in the South Range Breccia Belt; in one model the space is considered to be the product of a super-fault developed in response to crater readjustment.

(C) Based on Lightfoot and Evans-Lamswood (2015); (B) based on Spray (1997).

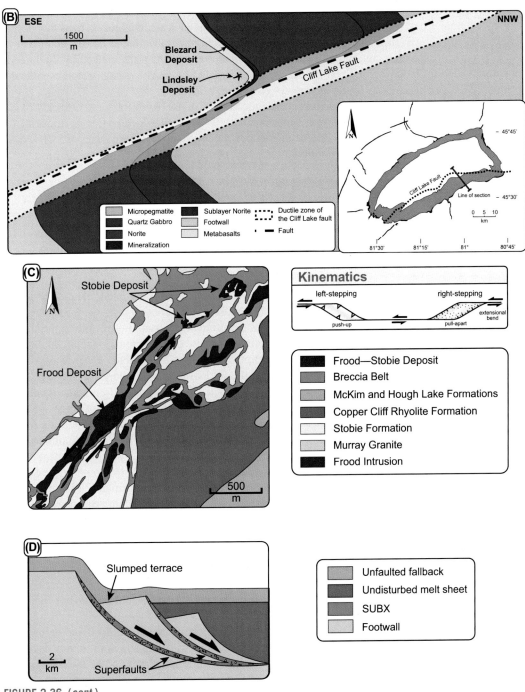

FIGURE 2.36 (*cont.*)

faulting, and the fault planes would have been active when the SUBX was developed as there is limited evidence of major belt-parallel fault structures. On a much smaller scale, the breccia belts that develop adjacent to embayment structures are associated with footwall mineralization at Levack and Coleman. An alternative explanation considers the Frood and Stobie ore deposits to be created by the injection of sulfide-laden inclusion-rich QD into footwall spaces created by transtensional tectonics guided by zones of SUBX between large blocks of country rock (Fig. 2.36D).

The importance of crater wall structures is also recognized in the North Range at Levack-Coleman and the East Range at Victor-Capre (eg, Lightfoot, 2015), where the localization of footwall ores appears to be controlled by the development of late open space structures within zones of SUBX developed between mafic and felsic gneiss of the Levack Complex (Stewart and Lightfoot, 2010).

Structural evolution of the Sudbury Region after the 1.85 Ga impact event

Late Penokean deformation lasted from ~1.85–1.8 Ga, and resulted in ductile folding of the SIC and possibly imbrication of the SIC in the South and East Ranges (Table 2.4). This deformation event resulted in the overall configuration of the Sudbury Basin as a folded melt sheet, and it produced repetitions of the basal contact which both displaced ore bodies and resulted in remobilization into the footwall by a process of translocation of sulfide and fragments termed "sulfide kinesis" (Table 2.4; Rogerio Monteiro, Pers. Comm. 2012).

After the Sudbury Event, a NW directed far-field stress regime resulted in ductile deformation of the central and southern parts of the Sudbury Basin, and this created both uplift and shortening of the Sudbury Structure (Cowan, 1996). This shortening was certainly important in controlling the shape of the Sudbury Structure, but it appears not to be the principal cause of the primary diversity in thickness of the SIC as exhibited in the North Range. The elliptical shape of the Sudbury Basin was produced by this phase of deformation together with many of the fold structures developed in the Whitewater Group (Roussel, 2009). The basin was also deformed by a NE–SW oriented compressional stress field which created of the Sudbury Basin in a northeast-southwest direction; these forces produced buckling and warping of the Sudbury Structure with synclinal fold axes defined by the Main Mass in the NE and SW corner, and an antiform developed in the East Range (the West Bay Anticline, Clark et al., 2012).

The Yavapai Orogeny occurred at 1.74–1.7 Ga, and modified the rocks to the south of the Sudbury Structure, but apparently produced negligible deformation of the Sudbury Structure (Table 2.4). The 1.65 Ga Mazatzal Orogeny appears to have had a greater influence on the Sudbury Structure, and there is increasing evidence to indicate that much of the south over north thrusting of the footwall and Main Mass occurred as a response to the far field effect of this orogeny (Table 2.4). The Mazatzal event was the last orogeny to have a major influence on the geology of the Sudbury Structure.

Injection of 1.24 Ga olivine diabase dykes in the Sudbury Region occurred in response to far field mantle plume activity and radial uplift; the dykes cut through the Sudbury Structure and occasionally produce weak displacement of contacts and local modification of ore bodies, but they do not control the formation of ore deposits (Table 2.4).

The Grenville Front marks the northwesterly limit of penetrative metamorphism and deformation produced by the Grenville Orogeny. This Front and truncates most of the structural trends developed in the adjacent Provinces, although locally some dykes and mafic intrusions can be traced across the Grenville Front. It is now generally accepted that the Grenville Orogeny involved northwest-directed thrusting and imbrication of the crust, presumably as a result of terminal collision at about 1100 Ma

(Easton, 1992). The northwest-directed thrusting may have modified the Sudbury Structure, but there is no evidence of major dislocation of the 1.23 olivine diabase dykes, and there is no known record of Grenville structures that penetrate and control the geometry of the Sudbury Structure.

Post-Grenville dyke development is largely restricted to the South Range where west–east diabase dykes termed Trap Dykes cut the Main Mass and the footwall rocks, and often pass through and modify ore bodies. The exact age of this event is not certain, and it may predate the Grenville Orogeny.

The last major deformation event to influence Sudbury was the ~37 Ma Wanapitei Impact Structure on the eastern side of the Sudbury Structure. This crater excavation event appears to have contributed to uplift in the East Range, but the influence is small in relation to the folding and faulting that occurred during the late Penokean, and it is not considered to have a major role in the configuration of the SIC or the ore deposits of the East Range (Table 2.4).

To the south of Sudbury, the Southern Province is on-lapped by Phanerozoic sedimentary rocks that obscure much of the uppermost stratigraphy of the Huronian Super-group on Manitoulin Island. The sedimentology of these rocks is a remarkable record of shallow basin created by subsidence beneath Lake Huron and the immediately surrounding area (Johnson et al., 1992).

Major Fault Groups of the Sudbury Region

Faults are identified based on surface geology maps, satellite imagery, detailed airborne magnetic and gravity surveys; confidence on the distribution of many of the faults in areas of vegetation and soil cover is enhanced by high resolution LiDAR surveys of the Sudbury Basin. In near-mine and mine settings, the detailed geometry of fault structures is tracked in diamond drilling and underground mapping; some mines also have an array of seismometers to monitor ground control conditions related to fault structures (eg, Eneva and Young, 1993).

Five major regional-scale fault groups cut the Sudbury Structure; a series of other local-scale faults structures are also recognized. The regional scale faults provide a first order control on the shape of the Sudbury Structure, but no less important are the local–scale structures which are often known from detailed near-mine and in-mine geological studies because of the roles they play in controlling and modifying the distribution of the ore deposits. The regional structures together with some of the near-mine and in-mine faults are summarized in Table 2.4 and arelocated in Fig. 2.36A. The exact relative timing of development of the fault sets in the context of the regional setting described above is not always known, but best estimates of the timing and kinematics of displacement are summarized in Table 2.4.

The Creighton–Murray Fault Set

These extensive brittle-ductile dextral faults extend in a broad west–east direction in the south part of the Sudbury Structure (Fig. 2.36A). The Creighton fault shows a displacement of ~700 m after the Sudbury event; the displacement on the Murray Fault is thought to be ~2000 m based on sediment-gabbro contacts. The faults are connected by a series of shorter ENE–WSW faults that may represent conjugate fault sets. The displacement of the Copper Cliff Offset by the Creighton Fault is a clear example of dextral kinematics, but there is surface evidence to suggest that the Creighton fault may have subsequently moved in a sinistral form to displace the tension gashes in the earlier damage zone (Fig. 2.37A). The Creighton Fault produces a major displacement of the Creighton dyke that disrupts the continuity of ore bodies in the dyke.

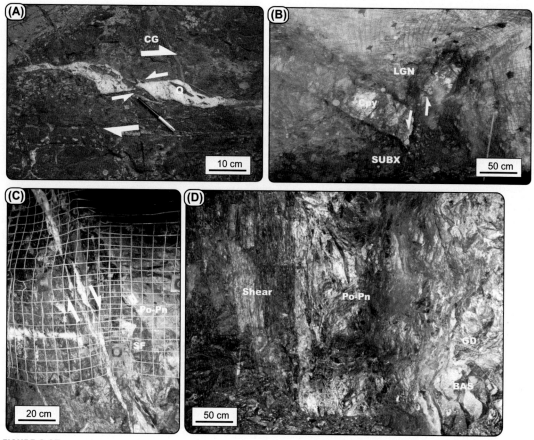

FIGURE 2.37

(A) Quartz (Q) veins in Creighton Granite adjacent to the Creighton Fault show sinistral transtension; this information coupled with the dextral displacement of the Copper Cliff and Worthington Offsets is indicative of at least 2 different motions on the fault. (B) Displacement of the 170 ore body by a splay structure related to the Bob's Lake Fault in the Coleman Mine; LGN, Levack Gneiss; SUBX, Sudbury Breccia; Cpy, Chalcopyrite. (C) The Number 2 cross-structure produces displacement of ore bodies in the north part of the Copper Cliff Offset; Po–Pn, pyrrhotite–pentlandite, SF, Stobie Formation. (D) The Number 4 Shear at Garson Mine controls the location of inclusion massive sulfide, and the footwall contact of this structure is often a locus for development of gersdorffite (Po–Pn, pyrrhotite–pentlandite; GD, gersdorffite; BAS, Elsie Mountain Formation metabasalt).

The Sandcherry–Fecunis Fault Set

These extensive brittle sinistral faults extend broadly north–south through the north Range of the Sudbury Structure (Fig. 2.36A); they offset the North Range, and there are indications that they continue in the South Range, but they have significantly less displacement in the South Range (<50 m). The largest faults are the Sandcherry Creek and Fecunis Faults with the displacement on the Fecunis fault is 600 m horizontal and 200 m vertical. The N–S faults in the Levack–Coleman mine areas are members of this family, and many of them together with their splays produce offsets between ore bodies or complexity within ore bodies. An example is the Bob's Lake Fault that separates the 170 ore body from the 170 Lower Ore Body (Fig. 2.37B). The Sandcherry–Fecunis Fault set is developed around much of the North Range, and the distribution of trajectories of the faults has a weak tendency to rotate from WNW–ESE to N–S from west to east through the North Range (Fig. 2.36A).

The Waddell–Amy Lake Fault set

These moderately extensive brittle-ductile faults cut the stratigraphy of the East Range SIC subparallel to the contact. The faults have a weakly curved geometry, and they create multiple repetitions of the basal stratigraphy of the SIC (Fig. 2.36A) by as much as 2000 m. Examples of the largest faults in this set, from north to south, are the Eatlots Lake, Waddell Lake, and the Amy Lake Faults. These faults are very important in controlling the geometry of the Sublayer and Footwall, and they can detach contact ores from the original overlying Main Mass stratigraphy (eg, the Capre contact deposits) and potentially detach footwall deposits from the contact transition zones that fed them. Although small faults are recognized in other deposits such as Victor and Nickel Rim, these deposits appear to be located in corridors between the major faults (Fig. 2.36A).

The Cameron Creek and Cliff Lake Faults, and the South Range Shear Zone

These major faults cut through the Sudbury Structure from west to east, but their geometry changes from broadly NW–SE through W–E to SW–NE from west to east (Fig. 2.36A). The faults are dominantly ductile, and they manifest as mylonite zones in the South Range Shear Zone and the Cliff Lake Fault (Fig. 2.36B). The faults plunge broadly towards the south at ~45 degrees at the center of the Sudbury Basin and their dip increases to near-vertical at the western and eastern margins of the Sudbury Basin; they are dominantly south-side up. The amount of displacement on individual faults can be large (up to 3 km), and taken together as a group, they are the principal structures that changed the geometry of the Sudbury Structure from a primary concentric structure to a basin with ~50% shortening along a broad WSW–ESE axis. This structure enters the Main Mass at the SW end of the SIC near Victoria, and exits the Main Mass towards the South East in the area of Garson and Falconbridge. The shear zone is a mylonite over several hundred meters width, and breaks into different fault zones as it traverses through the Main Mass. These structures produce extreme thinning of norite stratigraphy at the western and eastern ends of the South Range, and influence the thickness of the Quartz Gabbro in the central part of the South Range. Although the structures do disrupt the continuity of parts of the stratigraphy to the west and east, the continuous changes in chemical composition through the Main Mass at Creighton (Lightfoot and Zotov, 2006) and in other South Range profiles between the Vermilion and Blezard Deposits point to a limited influence on the stratigraphic continuity of the noritic rocks of the Main Mass rocks.

This fault set produced major offsets at the lower contact of the SIC in the South Range as revealed by exploration drilling. The Cliff Lake Fault is the lowermost structure comprising the South Range Shear Zone that cuts the Main Mass; it has been traced from west to east across the basin and it develops an undulatory geometry with a wide range in displacement as shown in Fig. 2.36B. This fault controls the depth extent of the South Range Ore Bodies, and various splays modify the shape of the ore bodies. Deformation of the South Range during shortening appears to have generated both a regional response in terms of the South Range Shear Zone, as well as deformation along the contacts of the base of the SIC with the footwall as observed at Victoria, Crean Hill, Creighton, and Garson Mines where segments of mineralization and/or noritic rocks are displaced into the footwall along these structures (Fig. 2.36B). Part of this deformation is related to anisotropy in the rocks of the Main Mass and the footwall, but these structures are also likely a minor (yet economically important) response to regional deformation. Part of this deformation appears to be the N–S buckling of the SIC. However, the development of these local features may also be a response to the development of a south-over-north thrusting that allows for the footwall rocks of the South Range to have approximately their original orientation. In this mode, the base of the Main Mass is required to represent a shear zone that was progressively rotated to produce contact angles of ~45–60° in the shallow part of Creighton Mine, and 90° in the deep part of Creighton Mine and at Garson and Crean Hill. The faults developed during regional shortening are largely south-dipping, but a number of conjugate faults with a north dip appear to have been created; examples include the north-dipping faults along the Copper Cliff Offset (Figs. 2.24A,B and 2.37C).

The effect of shear zones that intersect major mineralized embayment structures is illustrated by the Garson Deposit, where the economic mineral zones are contained in both the Sublayer trough structure and the late shear zones that cross cut the SIC and remobilize sulfides into the footwall and hangingwall (Fig. 2.37D and more detailed discussion in Chapter 4).

Several studies have been made to reconstruct the original geometry of the SIC shortly after the impact event. Reconstructions include efforts on a range of scales. On the scale of the Sudbury Basin, Cowan and Schwerdtner (1994) provide evidence for a fold-origin for the Sudbury Basin. On the scale of the South Range, Lenauer (2012) and Lennauer and Riller (2012a,b) attempt to reconcile the steep to vertical basal contact of the SIC with the vertical geometry of the ore bodies in the Copper Cliff Offset that appear to have a primary configuration. They present a tri-shear model that explains this geometry, but it does not explain the relationships found in contact ore deposits between Crean Hill in the west and Garson in the East. Mukwakwami et al. (2012, 2014a,b) provide an analysis of the Garson Deposit, and show that buckling followed by thrusting explains the development of mineralization along shear zones that penetrate into both the footwall and the Main Mass. More recent work by Papapaplova et al. (2015) shows the potential value of U–Pb accessory mineral geochronology using in situ methods to examine slickenside structures. Age dates for the Cliff Lake Fault, Six Shaft Shear, and structures controlling the footwall mineral zones at Creighton are reported in Papapaplova et al. (2015).

The Sudbury Structure at Depth

Much of the SIC has been explored with surface and underground mapping and drilling, and as the focus of work by mining companies has been on the environments favorable to Ni–Cu–PGE sulfide ore deposits, the majority of this work has focused on mineralized trough-embayment structures, their immediate footwall, and Offset Dykes. Therefore, the exploration of the Main Mass above the contact in these areas is quite complete. Between the trough structures and embayments there is less drilling, but enough to have a very good control on the geometry of the units that comprise the SIC.

As a general statement, it is true to say that the level of information above 3 km depth is enough to build an excellent three-dimensional model of the Sudbury Basin. A combination of potential field data from airborne magnetic and gravity surveys, LiDAR data, and Lithoprobe seismic data and potential field datasets have been used to refine the models presented by Gibson (2003).

The structure below 3 km has been very difficult to anchor, despite a conference meeting dedicated to a better understanding the geology beneath the center of the Sudbury Basin (http://www-icdp.icdp-online.org/front_content.php?idcat=719). Two models have been tabled in the published literature for the 3D structure of the Sudbury Basin, and they provide a useful starting point.

Milkereit et al. (1992, 1994a,b) tabled a model that respected the key reflectors from the Lithoprobe seismic program, and they suggested that the structure was heavily thrust faulted such that the South Range was lifted well above and past the center of the original impact structure. This model respects shallow data and many (but not all) reflectors, but it has the difficult problem of accounting for the near elliptical outline of the base of the Main Mass where several tens of kilometers of total displacement would be required on the faults at the center of the Sudbury Basin and significantly less displacement at the western and eastern margins. The more continuous configuration of the lower contact and the units of the SIC require much less deformation than the model proposed by Milkereit et al. (op cit).

In a second model that does not use drill core data, Card and Jackson (1995) proposed a model that respects the reflections in the Lithoprobe program as well as drilling above 3 km, and they propose that the Sudbury Structure is not as heavily deformed as Milkereit et al. (1992) propose. In the Card and Jackson (1995) model, the base of the center of the SIC (assuming no peak uplift structure) is located at 5 km depth.

These models have been tested with refined surface gravity and magnetic data; McGrath and Broome (1994a,b) support the Milkereit model with a gravity interpretation along the lithoprobe transects, and Hearst et al. (1994a) propose a similar geometry to Milkereit et al. (op cit). Both models are compared and contrasted in Lightfoot et al. (2002) and Pattison (2009).

Gibson (2003) refined the model using a three dimensional reconstruction based on geological data and potential field data; the revised model is shown in modified form in Fig. 2.38. This model provides what is a reasonable compromise between the previous models whilst respecting drill hole data. This model represents the current view on the geometry of the structure at depth, but it is still sensitive to the fact that two seismic panels from different locations in the Sudbury Basin have been merged to produce one hypothetical section.

This model has the advantage that it is consistent with fault kinematics in the South Range, and allows for less extreme offsets between units of the SIC on the southwestern and eastern southern margins where the fault structures exit the Sudbury Basin; in this way, the broadly continuous shape of the SIC is retained by motion along a series of faults and shears that have a curved and corrugated morphology and exhibit extensive branching (Fig. 2.36A,B).

A tri-shear configuration has been used by Lenauer (2012) to explain the geometric relationship between the steep basal contact of the SIC, and the vertical geometry of mineral zones in the Copper Cliff Offset Dyke. The ore bodies are distributed in a broadly vertical configuration that is indicative of emplacement of the mineralization from above (eg, Farrow and Lightfoot, 2002). The role of this tri-shear feature in the context of the geometry of the basal contact of the SIC and the South Range shear zone remains incompletely understood, and the relationship of any tri-shear deformation mechanism to the development of the South Range shear zone is not yet demonstrated or reconciled with the deep geometry of the Sudbury Structure.

GEOCHRONOLOGY OF THE SUDBURY STRUCTURE

The SIC has been the subject of numerous geochronology studies. By far the most useful have been the U–Pb studies of accessory minerals such as zircon, baddeleyite, monazite, and other accessory minerals. The earliest systematic work by Krogh et al. (1982, 1984) established beyond reasonable doubt the 1.85 Ga age of the SIC, and this remains a benchmark age for the impact-generated rocks. Since then, many other studies have dedicated efforts to establishing the contemporaneous nature of the Offset

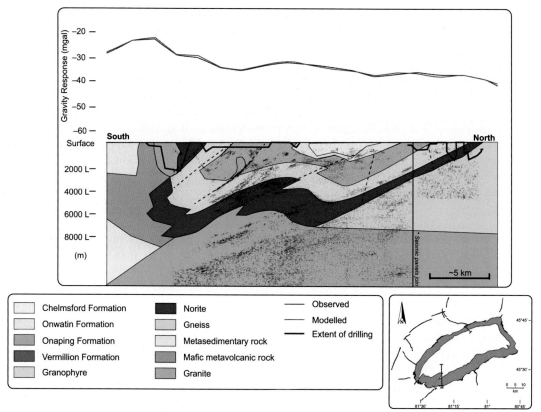

FIGURES 2.38 Geological Model Illustrating the Deep Structure of the Sudbury Structure Along the Two Profiles Shown in the Inset Map of the SIC

The Lithoprobe seismic reflection image is are shown as the backdrop to the section. The construction of the model primarily respected drill core data to the depth extent of drilling; faults were included and respected the known strike and dip orientation data. Seismic data were used to guide the extensions of faults and primary lithology contacts at depth. Gravity data was respected by two dimensional forward modeling of the three dimensional geology using average density data. A high-density shallow unit in the Vermilion Formation is required by the gravity data. The model shows the surface gravity data and the modeled data along the profile based on the geological relationships depicted in the model. The reader's attention is brought to the fact that these composite Lithoprobe panels are located in two different parts of the Sudbury Basin and juxtaposed as shown in the model.

Dykes (Corfu and Lightfoot, 1997), the age of the Sublayer and mafic-ultramafic inclusions contained in this unit (Corfu and Lightfoot, 1997), and the ages of a small subset of the target rocks and overlying Onaping Formation breccias (Table 2.6 and references therein).

The target rocks beneath the SIC have also been dated with varying success as the effect of impact-induced recrystallization of target rocks has resulted in the original ages being reset to that of the impact event; for example, some of the country rocks immersed in, or immediately adjacent to the Sublayer and Offsets have yielded similar ages to the SIC (eg, Corfu and Lightfoot, 1997; Prevec and Baaddsgaard, 2005; Ostermann et al., 1996b).

Table 2.6 U–Pb Geochronology Data for the Sudbury Region Sorted by Age

Rock Type	Rock Unit	Location	Sample Number	UTM E Zone 17, NAD 27	UTM N Zone 17, NAD 27	Mineral Dated	Age (Ma)	Error (+/-Ma)	References
Basal breccia fluidal glass	Breccia	Lower Onaping Formation	na	470447	5160084	Z	2719	4	Krogh et al. (1984)
Gray Gabbro	Gabbro	Podolsky Deposit	11AV-74	na	na	Z	2714	52	Kontak et al. (2015)
Granite xenolith	Breccia	Lower Onaping Formation	TK82-14	470447	5160084	Z	2711	7	Krogh et al. (1984)
Basal breccia fluidal glass	Breccia	Lower Onaping Formation	na	470447	5160084	Z	2708	8	Krogh et al. (1984)
Gabbro	Footwall	Falconbridge Twp.	SPA89-08a	517730	5162135	Z	2700	na	Prevec and Baadsgaard (2005)
Basal breccia fluidal glass	Breccia	Lower Onaping Formation	na	470447	5160084	Z	2695	3	Krogh et al. (1984)
Basal breccia fluidal glass	Breccia	Lower Onaping Formation	na	470447	5160084	Z	2686	2	Krogh et al. (1984)
Basal breccia fluidal glass	Breccia	Lower Onaping Formation	na	470447	5160084	Z	2679	16	Krogh et al. (1984)
Granodiorite	Footwall	Levack Gneiss Complex	CLA-93-175	462880	5165007	Z	2668	3.2	In Ames et al. (2008)
Granodiorite	Footwall	Levack Gneiss Complex	CLA-93-336	459467	5157403	Z	2668	5	In Ames et al. (2008)
Tonalite diatexite	Footwall	Levack Gneiss Complex	CLA-93-167A	465734	5163376	Z	2661	2	In Ames et al. (2008)
Foliated metagabbro	Footwall	Joe Lake Intrusion	BNB-12-094	497330	5178808	Z	>2660		Bleeker et al. (2015)
Monzodiorite	Footwall	Levack Gneiss Complex	CLA-93-065	507000	5180393	Z	2657	5	In Ames et al. (2008)
Gabbro	Footwall	Joe Lake Intrusion	na	497330	5178808	Z	2657	9	Bleeker et al. (2013)
Granite	Footwall	Cartier granite	BNB-12-095	497323	5178800	Z	2648	9	Bleeker et al. (2015)
Pegmatoid	Footwall	Levack Gneiss Complex	TK82-10	463040	5165729	Z	2647	2	Krogh et al. (1984)
Mafic gneiss	Footwall	Levack Gneiss Complex	CLA-93-339	460167	5157468	Z	2647	1	In Ames et al. (2008)
Diatexite mobilizate	Footwall	Levack Gneiss Complex	CLA-93-032B	513452	5172021	Z	2646	2	In Ames et al. (2008)
Granitic leucosome	Footwall	Levack Gneiss Complex	CLA-93-009A	515167	5168835	Z	2645	7.4	In Ames et al. (2008)
Massive granite	Footwall	Levack Gneiss Complex	CLA-93-027	516111	5173126	Z	2644	3	In Ames et al. (2008)
Granite	Footwall	Cartier granite	CLA-93-186	459702	5167882	Z	2642	1	Meldrum et al. (1997)
Dioritic to monzonitic gneiss	Footwall	Levack Gneiss Complex	Lev-1	490049	5175874	Z	2635	5	Osterman et al. (1996)
Dioritic to monzonitic gneiss	Footwall	Levack Gneiss Complex	Lev-2	490070	5175875	Z	2635	5	Osterman et al. (1996)
QD lobe in peperite	Breccia	Upper Onaping Formation	AV-162D	469188	5151337	Z	2628.6	3.2	Ames et al. (1998)

(Continued)

Table 2.6 U–Pb Geochronology Data for the Sudbury Region Sorted by Age (cont.)

Rock Type	Rock Unit	Location	Sample Number	UTM E Zone 17, NAD 27	UTM N Zone 17, NAD 27	Mineral Dated	Age (Ma)	Error (+/-Ma)	References
QD lobe in peperite	Breccia	Upper Onaping Formation	AV-162C	469188	5151337	Z	2616	5.7	Ames et al. (1998)
QD lobe in peperite	Breccia	Upper Onaping Formation	AV-162B	469188	5151337	Z	2602.8	3.3	Ames et al. (1998)
Geniss inclusion in SUBX	Footwall	South of Tam O'Shanter, South Range	na	490092	5147097	Z	2600	na	Petrus et al. (2012)
Geniss inclusion in SUBX	Footwall	South of Tam O'Shanter, South Range	TOGN	490092	5147097	Z	2519	2	Lightfoot (unpublished data)
Pyroxenite dyke	Footwall	Cuts Cartier Granite	BNB-13-087A	518496	5161906	Z	2507	4	Bleeker et al. (2015)
Granite	Footwall	Murray pluton	na	497422	5152998	Z	2477	9	Krogh et al. (1984)
Gabbro enclave in Creighton Granite (Matachewan)	Footwall	Creighton pluton	BNB-13-093	485966	5144629	Z,B	2476	7	Bleeker et al. (2015)
Biotite-hornblende granite	Footwall	Street Twp.	C3	528870	5154060	Z	2475	25,15	Corfu and Easton (2001)
Gabbro	Distal footwall	East Bull Lake Intrusion	na	na	na	Z	2475	2	Easton et al. (2010)
OPX-hornblendite	Footwall	East Bull Lake intrusive suite	C4	524997	5153551	Z	2468	5	Corfu and Easton (2001)
Orthopyroxene hornblendite bodies	Distal footwall	Sudbury Region	na	na	na	Z	2468	5	Corfu and Easton (2001)
Rhyolite	Footwall	Copper Cliff Formation	BNB-12-064	488430	5141367	Z	2465	14	Bleeker et al. (2015)
Granite	Footwall	Creighton pluton	BNB-12-058	490094	5147118	Z	2464	35	Bleeker et al. (2015)
Granite	Footwall	Murray pluton	BNB-12-059	496735	5150992	Z	2460	6	Bleeker et al. (2015)
Rhyolite	Footwall	Copper Cliff Formation	na	na	na	Z	2455	3	Bleeker et al. (2013)
Rhyolite	Footwall	Copper Cliff Formation	C81-19	496330	5148472	Z	2450	25,10	Krogh et al. (1984)
Migmatitic granite	Footwall	Grenville Front	96DM 70	523110	5151775	Z	2446	7	In Ames et al. (2008)
Gabbro	Footwall	Falconbridge Twp.	SPA89-08a	517730	5162135	Z	2441	3	Prevec and Baadsgaard (2005)
Granite	Footwall	Creighton pluton	na	na	na	Z	2437	2	Bleeker et al. (2013)
Granite	Footwall	Creighton pluton	BNB-12-060A	486067	5141778	Z	2433	4	Bleeker et al. (2015)
Granite	Footwall	Murray Pluton	na	na	na	Z	2429	2	Bleeker et al. (2013)
Rhyolite	Footwall	Copper Cliff Formation	BNB-12-098	489311	5141660	Z	2426	3	Bleeker et al. (2015)

Rock Type	Rock Unit	Location	Sample Number	UTM E Zone 17, NAD 27	UTM N Zone 17, NAD 27	Mineral Dated	Age (Ma)	Error (+/-Ma)	References
Porphyritic granite	Footwall	Creighton pluton	LH98-63	485979	5141839	Z	2415	5	Smith (2002)
Granite	Footwall	Creighton pluton	MS99-50	487759	5146465	Z	2376.3	2.3	Smith (2002)
Granite Dyke	Footwall	Creighton Pluton	285TR	na	na	Z	2344	47	Raharimahefa et al. (2014)
Granite Dyke	Footwall	Creighton Pluton	319TR	na	na	Z	2343	17	Raharimahefa et al. (2014)
Granite	Footwall	Creighton pluton	na	487501	5144302	Z	2333	33,22	Frarey et al. (1982)
QD lobe in peperite	Breccia	Upper Onaping Formation	AV-162A	469188	5151337	Z	2296.9	5.8	Ames et al. (1998)
Nipissing Gabbro	Distal footwall	Castle Mine, Cobalt Embayment	Wallrock 5001 vein	na	na	B	2219.4	3.6,3.5	Corfu and Andrews (1986)
Nipissing Gabbro	Distal footwall	Kerns Township	Kerns (N1)	585116	5273713	B	2217	2,4	Noble and Lightfoot (1992)
Nipissing Gabbro	Distal footwall	Shakespeare Intrusion	na	na	na	Z	2217		Sproule et al. (2007); Sutcliffe et al. (2002).
Mafic dyke	Distal footwall	Cutting Huronian Supergroup	BNB-12-022	358234	5128148	B	2116	5	Bleeker et al. (2015)
Nipissing Gabbro	Distal Footwall		BNB-13-048A	502367	5142640	Z,B	2215	1	Bleeker et al. (2015)
Nipissing Gabbro	Distal footwall	Hudson Township	Triangle Mt (N3)	584642	5269869	B	2209.6	3.5	Noble and Lightfoot (1992)
Mafic dyke	Distal footwall	Cutting Huronian Supergroup	BNB-12-028B	331657	5135079	B	2105	5	Bleeker et al. (2015)
Mafic dyke	Distal footwall	Cutting Huronian Supergroup	BNB-12-025	353644	5122629	B	1930		Bleeker et al. (2015)
Impactite	Distal ejecta	Thunder Bay Region	na	664785	5329200	Z	1878	1	Fralick et al. (2002)
Gabbro	Footwall	Drury Twp. intrusion	Spa-88-56a	464000	5141600	Z	1859	13	Prevec and Baadsgaard (2005)
Diabase dyke	Dyke	Strathcona Deep copper mine	JF-95-65	473718	5168688	Z	1858.7	7.6	Fedorowich et al. (2006)
QD	QD	Foy Offset	Foy-3	489437	5176737	Z+B	1854	4,3	Osterman et al. (1996)
QD	QD	Foy Offset	Foy-2	489774	5176473	Z+B	1853	4,3	Osterman et al. (1996)
Contact QD	QD	Foy Offset	Foy-4	489818	5176477	Z+B	1852	4,3	Osterman et al. (1996)
QD	QD	Foy Offset	Foy-1	490206	5176070	Z+B	1852	4,3	Osterman et al. (1996)
Granophyre	Main Mass	North Range	TK82-15	470758	5161395	B	1850.5	3	Krogh et al. (1984)
Sheared norite	Main Mass	South Range	BL-7	500500	5156060	Z	1850.1	11.2	In Ames et al. (2008)

(Continued)

Table 2.6 U–Pb Geochronology Data for the Sudbury Region Sorted by Age (cont.)

Rock Type	Rock Unit	Location	Sample Number	UTM E Zone 17, NAD 27	UTM N Zone 17, NAD 27	Mineral Dated	Age (Ma)	Error (+/-Ma)	References
MNR	Main Mass	North Range	na	473492	5168775	Z	1850	3.4,2.4	Krogh et al. (1984)
MNR	Main Mass	South Range	na	501047	5156892	Z	1850	1.3	Krogh et al. (1984)
Granite dyke	Main Mass	South Range	S81-8	495998	5155562	Z	1850	1	Krogh et al. (1984)
Granite	Footwall	Murray pluton	na	497422	5152998	T	1850	3	Krogh et al. (1984)
QD	Offset	Copper Cliff offset	C-94-12	494584	5148091	Z	1849.8	2	Corfu and Lightfoot (1997)
Melanorite lens/layer	Main Mass	South Range	BNB-13-0007	484226	5146795	Z	1849.7	0.2	Bleeker et al. (2015); Lightfoot and Zotov (2006)
FNR	Main Mass	North Range	na	466935	5162551	Z	1849.6	3.4,3	Krogh et al. (1982)
FNR	Main Mass	North Range	na	466935	5162551	Z	1849.53	0.21	Davis (2008)
Black Norite	Main Mass	South Range	Average of 6	na	na	Z	1849.53	0.21	Davis (2008)
MNR	Main Mass	South Range	na	501047	5156892	Z	1849.4	1.9,1.8	Krogh et al. (1982)
MNR	Main Mass	South Range	na	501047	5156892	Z	1849.11	0.19	Davis (2008)
FNR	Main Mass	South Range	Average of 5	N/A	na	Z	1849.11	0.19	Davis (2008)
QD	Offset	Hess Offset	BNB-13-088A	449037	5160679	B	1849.1	0.9	Bleeker et al. (2015)
Olivine melanorite pod in sublayer	Inclusion	Whistle Embayment	C-94-2	509086	5179520	Z+B	1849.1	1.1	Corfu and Lightfoot (1997)
FNR	Main Mass	North Range	na	466935	5162551	Z	1848.9	4,2.7	Krogh et al. (1984)
Diabase	Inclusion	Whistle Embayment	C-94-9	509086	5179520	Z	1848.7	1.1	Corfu and Lightfoot (1997)
QD	Offset	Pele Offset	BNB-13-090B	477219	5178116	B	1848.5	0.8	Bleeker et al. (2015)
Metapyroxenite inclusion in sublayer	Inclusion	Whistle Embayment	C-94-7	509086	5179520	Z	1848.4	1.4	Corfu and Lightfoot (1997)
Albitized Sandcherry member breccia	Altered breccia	Lower Onaping Formation	AV-J8	498828	5174642	T	1848.4	1	Ames et al. (1998)
Melanorite pod in sublayer	Inclusion	Whistle Embayment	C-94-5	509086	5179520	Z/B	1848.3	1	Corfu and Lightfoot (1997)
Sublayer Norite matrix	Sublayer	Whistle Embayment	C-94-8	509086	5179520	Z+B	1848.1	1.8	Corfu and Lightfoot (1997)
Aplite	Main Mass	Dyke, cutting SIC	02-AV-852	500672	5157455	Z	1848	4	Ames et al. (2008)
Granite xenolith	Breccia	Lower Onaping Formation	TK82-14	470447	5160084	Z	1836	14	Krogh et al. (1984)

Rock Type	Rock Unit	Location	Sample Number	UTM E Zone 17, NAD 27	UTM N Zone 17, NAD 27	Mineral Dated	Age (Ma)	Error (+/-Ma)	References
Diabase dyke	Dyke	Strathcona Deep copper mine	JF-95-65	473718	5168688	Z	1833.8	6.6	Fedorowich et al. (2006)
Diabase dyke	Dyke	Strathcona Deep copper mine	JF-95-65	473718	5168688	Z	1832.2	7.1	Fedorowich et al. (2006)
Diabase dyke	Dyke	Strathcona Deep copper mine	JF-95-65	473718	5168688	Z	1818.8	9.4	Fedorowich et al. (2006)
Diabase dyke	Dyke	Strathcona Deep copper mine	JF-95-65	473718	5168688	Z	1817.7	5.7	Fedorowich et al. (2006)
Diabase dyke	Dyke	Strathcona Deep copper mine	JF-95-65	473718	5168688	Z	1815.9	25.2	Fedorowich et al. (2006)
Sheared norite	Main Mass	South Range	BL-7	500500	5156060	Z	1815	15	Bailey et al. (2004)
Biotite granodiorite	Footwall	Eden Lake suite	92DM 247	501375	5138150	T	1749	12,8	Davidson and van Breemen (1994)
Granite dyke	Footwall	Wanapitei Complex	92DM 169	513250	5145525	Z	1747	12,6	In Ames et al. (2008)
Norite	Footwall	Wanapitei Complex	SPA-89-31	515250	5144765	Z	1747	6,5	Prevec (1993)
Biotite-hornblende granite	Footwall	Street Twp.	C3	528870	5154060	T	1720	36,19	Corfu and Easton (2001)
Albitized metasedimentary rock	Footwall	Huronian Supergroup	SC-5 Scadding mine	526624	5163735	M	1701	3.6	Schandl et al. (1994)
Albite metasomatism	Regional meta-somatic event	Sudbury	na	na	na	M	1701	4	Schandl et al. (1994)
Albitized metasedimentary rock	Footwall	Huronian Supergroup	na	520934	5170771	M	1699	3.6	Schandl et al. (1994)
Sheared norite	Main Mass	South Range	BL-7	500500	5156060	Z	1658	68	Bailey et al. (2004)
OPX-hornblendite	Footwall	East Bull Lake intrusive suite	C4	524997	5153551	Z	1471	10	Corfu and Easton (2001)
Biotite granite	Footwall	Chief Lake complex	92DM 192a	503000	5138175	Z	1464	2,1	Davidson and van Breemen (1994)
Olivine diabase dyke	Plume-LIP	Sudbury dyke swarm	na	na	na	Z	1238	4	Krogh et al. (1987)
OPX-hornblendite	Footwall	East Bull Lake intrusive suite	C4	524997	5153551	Z	1052	19	Corfu and Easton (2001)
Granitic leucosome	Footwall	Street Twp.	C2	528870	5154060	Z	995	3	Corfu and Easton (2001)
Pegmatite vein	Footwall	Street Twp.	C1	528870	5154060	Z	989	2	Corfu and Easton (2001)
Biotite-hornblende granite	Footwall	Street Twp.	C3	528870	5154060	T	987	9	Corfu and Easton (2001)
Granitic leucosome	Footwall	Street Twp.	C2	528870	5154060	Z	987	3	Corfu and Easton (2001)
Gabbro	Footwall	Falconbridge Twp.	SPA89-08a	517730	5162135	Z	986	30	Prevec and Baadsgaard (2005)
Biotite-hornblende granite	Footwall	Street Twp.	C3	528870	5154060	A	932	6	Corfu and Easton (2001)
Diabase dyke	Dyke	Fraser-Strathcona mine	JF-95-65	473718	5168688	Z	506	4	Fedorowich et al. (2006)

Summary of U–Pb geochronology data for the Sudbury Structure and associated country rocks (modified and updated after Ames et al., 2008 with references to individual studies cited in the table). Z, Zircon; B, Baddeleyite; M, Monazite; T, Titanite; nr, exact location not reported; na, not available.

The geochronology record of the Sudbury Region is far from complete with many South Range footwall rock types remaining undated, and few constraints being placed on the longevity of crater readjustment of the Sudbury Structure despite the development of new methods that offers the opportunity to eventually resolve this question (Davis, 2008).

Isotopic studies also provide an approach to establishing model ages; the importance of Proterozoic mafic target rocks as a source for the Os in sulfide ores is an example of the overwhelming evidence for the need for old mafic crust in the formation of the melt sheet (Cohen et al., 2000), and the highly radiogenic Os in many of the ore deposits (Dickin et al., 1992; Walker et al., 1991a–d, 1994) is aligned with both radiogenic isotope ratios of Sr and Nd (Faggart, 1984; Faggart et al., 1985c; Naldrett et al., 1985, 1986; Prevec et al., 2000; Dickin et al., 1999) and the requirement for ancient contributions of crustal Pb (Dickin et al., 1996; Darling et al., 2010a,b, 2012).

A SEQUENCE OF EVENTS AT SUDBURY

The summary of events in Figs. 2.39 and 2.40 are anchored by the geological relationships presented in this chapter. The resolution is shown at two scales; on one scale (Fig. 2.39), the entire history of the Sudbury Region from 2.8 to present is shown in the context of regional tectonic setting and magmatic events. In a second and more detailed image (Fig. 2.40), the approximate sequence of events is shown in the context of an approximate timeline for the Sudbury event after impact. The timescale of Fig. 2.40 is logarithmic so that the more immediate processes associated with impact on the scale of seconds and minutes can be reconciled with the evolution of the crater and melt sheet on the scale of days and years, and with the crystallization of the melts and readjustment of the crater walls on the scale of tens of years to tens of thousands of years. This requires a minimum of 16 logarithmic cycles of time to document the processes, and the exact timing in this sequence will surely change as increased understanding of impact tectonics and melt sheets.

The following geological constraints are considered to underpin the model shown in Fig. 2.40:

1. The Sudbury impact event involves an asteroid with dimensions large enough to create an ~120–200 km diameter crater with two or more rings. The impact process translated kinetic into potential energy; a high P–T plasma created from target rocks was produced; SUBX was created beneath the transient crater—largely in situ, but with some contribution possibly from the impact plasma and/or melt; ejecta material and products of tsunami events were distributed up to 1000 km from the center of impact. This impact process lasted no more than a tiny fraction of a second and produced the shocked mineralogy of the country rocks, shatter cones, and the SUBX (Fig. 2.40). The development of the transient crater took only a few minutes.

2. The transient crater stabilized; footwall rocks started to relax; melt concentrates accumulated in lows in transient crater; regional ring structures may also have once contained melt outboard of the main crater. The process of melt formation was likely initiated on the time scale of minutes, although this process continued with the assimilation and incorporation of country rocks for a considerable time due to the superheat available from the melt sheet (Fig. 2.40).

3. The QD of the Offset Dykes was injected along pathways of least resistance exploiting pseudotachylite breccias and lithological contacts. Some dykes followed continuous pathways; others followed more discontinuous paths or locally penetrated into the SUBX. More fractionated magmas outside of the edge of the main crater were injected into structures to form concentric Offset Dykes (eg, Manchester). Contact relationships between QD and SUBX

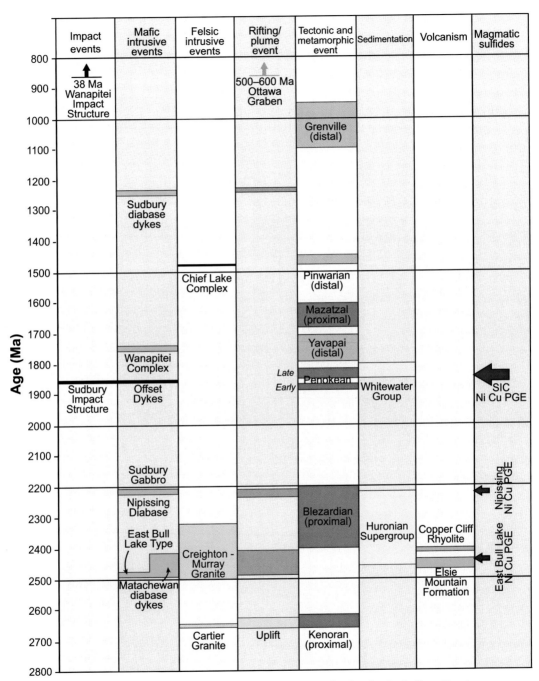

FIGURE 2.39 Illustration to Show Possible Sequence of Events as a Timeline for the Sudbury Structure

The diagram is based on geological information and U–Pb geochronology (the principal references are given in Tables 2.4 and 2.6).

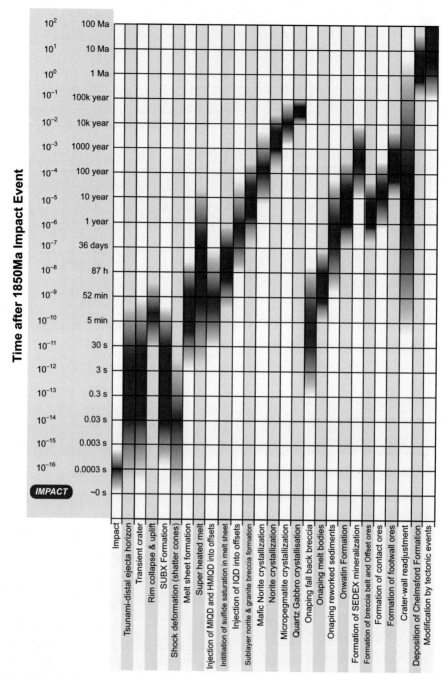

FIGURE 2.40 An Illustration to Show a Possible Timeline of Events in the Sudbury Structure Immediately After the 1.85 Ga Event Horizon

The vertical scale is a logarithmic representation of time after impact, and the diagram is used to summarize some of the key geological relationships pointing to the timing and evolution of the Sudbury Structure; exact event timelines remain very uncertain as indicated by the bars, but the diagram provides some indication of the relative timing and possible duration of events. The diagram was constructed to respect the scale of the impact process (Melosh, 1989, 1995; Melosh and Ivanov, 1999), timeline and scale of impact (Grieve, 1987, 1994; Grieve and Cintala, 1992; French, 1998), and utilizes information from surface testing of thermonuclear devices (Short, 2006) and observations from the Earth Impact Database (http://www.unb.ca/fredericton/science/research/passc/). The sequence of events and detailed aspects of this time sequence are discussed after the presentation of petrology, geochemistry, and ore body characteristics in Chapters 3 and 4.

indicate that this happened after cooling and consolidation of the matrix of the SUBX. The composition of the inclusion poor QD offers a potential indication of the average early melt sheet composition. The process of initial injection of the QD likely happened within the a few days of impact (Fig. 2.40).

4. The melt sheet assimilated country rocks and incorporated additional fragments from the target; the felsic fragments tended to melt more readily, but the mafic fragments provide a record of the development of these magmatic breccias. The melt sheet achieved sulfur saturation and ore formation commenced with the injection of sulfide-laden MIQD into the early formed offsets. The sulfide-saturated MIQD was injected after the QD and likely within a period of months (Fig. 2.40)

5. The formation of Sublayer as a balanced process involving settling of inclusions and sulfide and thermal erosion of underlying country rocks. It clearly postdated the formation of SUBX as there are digested inclusions of pseudotachylite in the GRBX of the Sublayer. Depending on melting conditions of country rocks, different phases of Sublayer develop; major assimilation of local crust is recorded in diversity of inclusion population between different embayment and trough structures. Sulfide melt accumulated towards the base of the melt sheet and flowed downwards into troughs and embayments. This process likely happened over a period of years (Fig. 2.40)

6. The melt sheet achieved a liquidus temperature with the removal of excess heat. The melt then commenced crystallization and fractionation, and it's composition was perhaps modified by assimilation of fallback material or material washed back into the crater depression after impact melt sheet commenced crystallization. This process likely took place on the scale of tens of thousands of years (Fig. 2.40).

7. Crystallization of the sulfide magma took place by progressive fractionation to produce contact and footwall ore types over a timescale of thousands of years, which happened within a more protracted period of crater stabilization (Fig. 2.40).

8. The Onaping Formation is a fallback breccia that contains glass shards; melt bodies derived from the melt sheet cross cut the Sandcherry Member. Relaxation of the crater is accompanies by development of the Sudbury Basin; inflow of saline waters add key ingredient to create hydrothermal modification of the rocks crystallized from the melt sheet, and the underlying country rocks. The processes of formation of the Onaping Formation were varied and ranged from fall back over a timescale of days, emplacement of magma from the melt sheet over a time period of years, and full development of the Onaping Formation stratigraphy over tens of years (Fig. 2.40). The development of the Cu–Pb–Zn mineralization in the Vermilion Formation likely followed the early formation of the Onaping.

9. Complete sedimentation of the Whitewater Group was likely achieved in a time frame of tens of millions of years (Fig. 2.40).

10. The Sudbury Structure was modified by several orogenic events that created the existing shape of the Sudbury Basin and modified the distribution of mineralization (Fig. 2.39).

A more detailed understanding of the sequence of events and the evolution of the Sudbury melt sheet in response to differentiation requires a more complete documentation of the petrology and geochemistry of the silicate rocks (provided in Chapter 3). This chapter has provided a foundation on which the geology of the ore deposits can be described, and their diversity in composition can then be related to the sulfide saturation, segregation, and differentiation history of the melt sheet (provided in Chapter 4).

Thin section between crossed polars of ultramafic inclusion in the Sublayer, from the Murray Mine embayment. This image depicts the unusual textures of an olivine pyroxenite; the olivines form grain clusters comprising a single olivine domain with different extinction angles and olivine grains within olivine; the surrounding intercumulus orthopyroxene is heavily fractured with undulose extinction. The sample was provided by Sherri Digout and photographed by Ben Vandenburg

PETROLOGY AND GEOCHEMISTRY OF THE SUDBURY IGNEOUS COMPLEX

INTRODUCTION AND OBJECTIVES

The objective of this chapter is to synthesize information on the petrology of the rocks that comprise the SIC, the associated breccias, and some of the country rocks. The petrology is integrated with whole-rock geochemical data to establish the relationships within and between the igneous rock groups that comprise the SIC. The extent to which these relationships can be used to understand the mechanism of formation of the SIC, the sulfide saturation history of the magma, and the process of formation of the ore deposits is then discussed and considered in the context of an impact model for the genesis of the Sudbury Structure.

The data presented for the Main Mass are used to investigate the variations within North and South Range traverses that are produced by crystallization and differentiation of the melt sheet, and to look at the variations along strike within the noritic rocks. The data provided in this chapter are designed to form a basis on which the following questions can be addressed:

1. Does a single parental magma type represent the average composition of the Main Mass of the melt sheet, and can it explain for the diversity in rock types within and between the South Range and North Range stratigraphy?
2. Is there any evidence that the melt sheet evolved in discrete segments, or was it well-mixed across the floor of the crater?
3. What mechanisms of crystallization and differentiation controlled the composition of the Main Mass?
4. Where do the Ni, Cu, and PGE in the magmatic sulfide ores originate, and what role did sulfide segregation from the melt sheet play in their formation?
5. If the weight of evidence points to segregation of metals from the melt sheet, was the melt sheet a superheated "perfect smelter" from which immiscible sulfide melts were concentrated towards the base unhindered by silicate minerals and inclusions?

The data presented for the Offset Dykes are designed to portray the compositional diversity within and between the dykes. These data provide a basis on which to understand the principal rock type hosting the ore deposits in the Offset Dykes. The key questions addressed are:

1. What petrological and geochemical evidence do the Offset Dykes provide for a sequence of events in the early evolution of the melt sheet?
2. How does the composition of the Offset Dykes help to constrain the composition of the parental magma?

Nickel Sulfide Ores and Impact Melts. http://dx.doi.org/10.1016/B978-0-12-804050-8.00003-1

3. What is the origin of the mafic and ultramafic inclusions contained in the mineralized inclusion-bearing quartz diorite (MIQD) of the Offset Dykes?

4. At what stage in the evolution of the Offset Dykes were the ore deposits formed?

A synthesis of petrological and geochemical data for the noritic part of the Sublayer is presented for rocks from within individual troughs and embayments. A comparison of the noritic matrix of the Sublayer from different embayments is presented together with the geochemistry of the entrained inclusions. These data underpin an understanding of the Sublayer as a principal rock type hosting the ores at the base of the SIC. The key questions addressed are:

1. What processes explain the diversity in the composition of the noritic matrix of the Sublayer within and between trough and embayment structures?

2. Where do the mafic and ultramafic inclusions contained in the Sublayer come from?

3. Is an impact-triggered mantle contribution of magma required to form the ore deposits or is the composition of the crustal target rocks consistent with metallogenesis?

Finally, the chapter addresses the compositional diversity in footwall breccias produced by cataclasis, melting, and thermal metamorphism. Samples of Sudbury Breccia (SUBX) from different hosts are used to investigate the geochemical relationship between host rocks and matrix. Petrological relationships in country rocks that have been extensively metamorphosed and partially melted are also examined. The objective of this exercise is to address the following questions:

1. Does the matrix of SUBX contain contributions from the melt sheet or was it generated entirely from local country rocks?

2. Can the petrology and geochemistry of SUBX provide information about proximity to the melt sheet and metal-endowment of the SIC?

3. What is the post-impact influence of thermal metamorphism on the rocks beneath the melt sheet?

Satisfying answers to these questions are a foundation to an understanding of the magmatic sulfide ore deposits in one of the world's largest mining camps and they are at the nexus of discussion about catastrophic impact in the study of planetary science.

THE MAIN MASS

The quantity and quality of mineralization at the base of the SIC varies around the lower contact as described in Chapter 2. Some contact segments have troughs and embayments with major ore deposits (eg, the Creighton Deposit), whereas other troughs and embayments contain subeconomic ore deposits (eg, the Trillabelle Deposits). Moreover, there are large segments of contact and long segments of Offset Dykes that have minor mineral occurrences but no economic concentrations of mineralization. The most central question is why there is such a diversity in mineral potential along the lower contact of the SIC. In particular, does the overlying Main Mass provide a record of the process responsible for the formation of the ore deposits?

The petrological and geochemical stratigraphy of surface and drill core traverses through the North and South Range Main Mass provide a way to understand the differentiation history of the melt sheet (eg, Lightfoot et al., 2001; Keays and Lightfoot, 2004; Lightfoot and Zotov, 2006). The noritic rocks located above ore deposits and along strike where there are subeconomic concentrations of mineralization provide an opportunity to examine the relationship between the segregation of sulfide melt from the silicate melt sheet and the development of ore bodies at the base of the SIC.

The melt sheet has been deformed by folding and faulting. Deformation has undoubtedly controlled aspects of the thickness and geometry of the SIC (eg, Riller, 2005; Santimano and Riller, 2012; Lenauer and Riller, 2012a,b), but it has certainly not obscured variations in melt sheet thickness created by crater floor topography. The traverses chosen for this text are not entirely devoid of structural complexity, but the deformation is limited to foliation and/or fault-related interruptions that can be identified in the traverses (Fig. 2.17). The presence of structure does not undermine the technical validity of using samples from largely intact stratigraphic sections through the Main Mass to establish chemostratigraphic variations.

A PETROLOGICAL AND GEOCHEMICAL TRAVERSE THROUGH THE NORTH RANGE

As discussed in Chapter 2, the Main Mass rocks of the North Range consists of the Mafic Norite (MNR), Felsic Norite (FNR), Quartz Gabbro (QGAB), and Granophyre (GRAN; also termed Micropegmatite) Units; these rocks crop out in the contiguous footprint of the western part of the SIC between Trillabelle in the west and Victor-Nickel Rim in the east of the SIC (Fig. 2.26A).

The most comprehensive studies of the Main Mass in the North Range include the petrological studies of Hewins (1971) and Naldrett and Hewins (1984), and the geochemical studies of Lightfoot et al. (1997c), Lightfoot et al. (2001), Keays and Lightfoot (2004), and Theriault et al. (2002). This section draws heavily on the geochemistry data presented in Lightfoot et al. (1997c,d, 2001) and Keays and Lightfoot (2004) for the East Range at Capreol and Victor-Nickel Rim (Fig. 3.1A,B). It also uses petrographic data from surface traverses acquired at the western end of the North Range (Trillabelle Township; Fig. 3.1A).The location of these sites and other drill core traverses discussed later in the text are shown in Fig. 3.1A–B.

The variations in major and trace elements as well as in physical properties (density and magnetic susceptibility) can be understood in the context of the petrology. Variations in incompatible trace element ratios (shown on spidergrams) can be used to constrain variations that may be due to open system mixing or contamination processes. The variation in siderophile and chalcophile element abundance provides a record of the sulfide saturation history of the magma and the accumulation of magmatic sulfide.

Petrology and mineralogy

The Main Mass rocks of the North Range were modified under lower greenschist facies metamorphic conditions, and the textures of the rocks are therefore easier to interpret in a magmatic context than the Main Mass rocks of the South Range that were metamorphosed and deformed under lower amphibolite to upper greenschist facies conditions (eg, Fleet et al., 1987; James and Golightly, 2009; Roussel, 2009). The main petrological features of the Main Mass in the North Range are shown in Table 3.1.

The modal mineralogy determined by point counting on surface samples from a traverse through the Main Mass to the east of the Trillabelle trough is shown in Fig. 3.2A,B. The modal proportions are shown as 10-point moving averages of modal mineral content based on stratigraphic position and point counting of ~500 points in 309 samples from the stratigraphy. The modal mineralogy is plotted as a function of distance from the outer contact of the Main Mass; there is no surface outcrop of the MNR along the traverse, so the section commences close to the base of the FNR, passes through the QGAB, and terminates at the top of the GRAN (Figs. 1.7 and 3.1A).

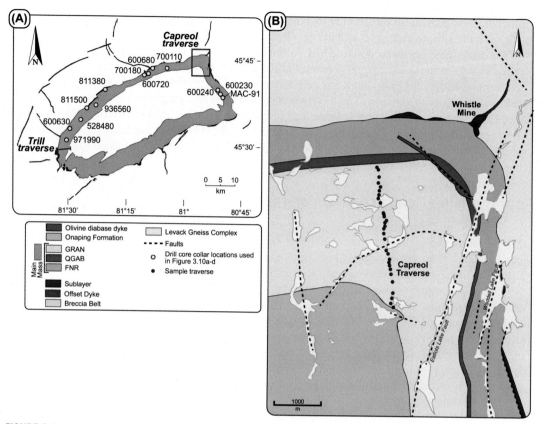

FIGURE 3.1 Location of the North Range Main Mass Sample Traverses

(A) Geological map of the SIC showing the location of the surface traverse through the Main Mass in the northeastern corner of the North Range and the collar of drill core MAC-91 (at Victor-Nickel Rim). The location is shown of the Trillabelle surface traverse. The collars of drill holes from which samples were taken to investigate the lateral variation in geochemistry of the Main Mass are shown. (B) Geological map of the traverse NE of Capreol showing sample locations in the GRAN Unit of the Main Mass. Sample locations details for the Capreol traverse are available in Lightfoot et al. (1997c,d).

The basal Main Mass rock in the North Range is a discontinuous unit of MNR which contains up to 50% cumulus orthopyroxene that is poikilitically enclosed in plagioclase (the abundance of plagioclase varies inversely with the orthopyroxene content). Cumulus chrome spinel is commonly present, and interstitial phlogopite, micrographic quartz + K-feldspar intergrowths and magmatic sulfide are commonly developed towards the base of the FNR and into the MNR (Fig. 3.3A,B). The MNR transitions into the FNR over vertical distance of ~1–10 m, and the texture changes from ophitic through subophitic to hypidiomorphic granular texture which is characteristic of the FNR where both orthopyroxene and plagioclase are cumulus minerals (Fig. 3.3C).

Table 3.1 Summary of Petrographic Features of the Igneous Rocks that Comprise the SIC and Sources of Additional Information on Petrology and Mineralogy

Unit of the Main Mass and Offset Dykes of the SIC	Rock Type Forming stratigraphic Unit	Colour	Grain Size and Texture	Equant (Cumulus) Minerals	Matrix (Intercumulus) Minerals	Sulfide and Important Accessory Minerals	Common Features and Contacts	References
South Range GRAN	Plagioclase-rich granophyre	Grey	Medium-grained	Elongate to tabular plagioclase laths (<25%)	Amphibole, biotite, quartz+K-feldspar graphic intergrowths	Trace Py	Uniform fine-grained grey upper unit of the GRAN	Peredery and Naldrett (1975)
	Granophyre (GRAN) (Micropegmatite)	Dark pink colour with green patches	Medium-grained to coarse-grained	Plagioclase (25%–35%)	Amphibole, biotite, quartz+K-feldspar graphic intergrowths	Trace Py	Variable in texture and possibly comprising 2 or more subunits	Naldrett et al. (1970); Naldrett and Hewins (1984)
South Range QGAB	Quartz Gabbro (QGAB)	Grey with local segregations of pink granophyre and local fine-grained mafic inclusions	Coarse-grained; occasional inclusions; local development of quench texture (like the crowsfoot-textured granophyre)	45%–50% plagioclase, 5%–30% amphibole; 0%–5% titanomagnetite	Clinopyroxene and quartz+K-feldspar graphic intergrowths	Trace Py. Up 0.5% apatite; accessory titanomagnetite and sphene	Grey with conspicuous amphibole	Naldrett et al. (1970); Naldrett and Hewins (1984)
South Range Norite	Melanorite (Black Norite)	Black	Black with laminated plagioclase	50%–60% plagioclase, 25%–35% orthopyroxene	Augite, phlogopite	1% Po and trace Cpy+Pn; Cr spinel(?)	Lens enclosed in Green Norite on Highway 144	Lightfoot and Zotov (2006); Baird (2007); Walker (2014); Lesher et al. (2014); Strongman (2016)

(Continued)

Table 3.1 Summary of Petrographic Features of the Igneous Rocks that Comprise the SIC and Sources of Additional Information on Petrology and Mineralogy (cont.)

Unit of the Main Mass and Offset Dykes of the SIC	Rock Type Forming stratigraphic Unit	Colour	Grain Size and Texture	Equant (Cumulus) Minerals	Matrix (Intercumulus) Minerals	Sulfide and Important Accessory Minerals	Common Features and Contacts	References
South Range Norite	Green Norite	Dark green	Medium-grained with strong replacement of pyroxenes by amphibole	50%–65% zoned dusty plagioclase; 10%–20% amphibole (hornblende) after primary orthopyroxene	<10% quartz, amphibole	Trace Py	Thick unit between Quartzose Norite and QGAB; irregular patches/enclaves of Brown Norite and a lens of Melanorite (Black Norite)	Naldrett et al. (1970); Naldrett and Hewins (1984)
	Brown Norite (enclaves in Green Norite)	Black to black-green and typically unmetamorphosed	Medium-grained hypidiomorphic granular to subophitic	50%–65% plagioclase, 20%–25% orthopyroxene	Augite, phlogopite	Trace Py	Enclaves within Green Norite	Naldrett et al. (1970); Naldrett and Hewins (1984)
	Quartzose Norite	Slightly lighter than Green Norite	Medium grained hypidiomorphic granular	55%–60% Plagioclase	10%–15% Quartz, amphibole, phlogopite	Trace Py	Lesser blue Quartz; gradation contacts with Quartz-rich and Green Norite	Naldrett et al. (1970); Naldrett and Hewins (1984)
	Quartz-rich Norite	Black and white (resembled coarse-grained QD)	Medium grained, subpoikilitic to hypidiomorphic granular	50%–55% Plagioclase	15%–25% Quartz (sometimes graphitically intergrown with feldspar, biotite (partially after augite), amphibole and phlogopite	0%–2% Po, trace Pn+Cpy	Blue flood Quartz; gradaitional into SLNR and Quartzose Norite	Naldrett et al. (1970); Naldrett and Hewins (1984)

Region	Unit	Colour	Texture	Mineralogy	Intergrowths	Sulfides	Description	References
North Range GRAN	Plagioclase-rich GRAN	Grey	Medium-grained	Elongate to tabular plagioclase laths (<25%)	Amphibole, biotite, quartz+K-feldspar graphic intergrowths	Trace Py	Uniform fine-grained grey upper unit of GRAN	Peredery and Naldrett (1975); Peredery (1991)
	GRAN (Micropegmatite)	Grey and pink	Medium to coarse-grained ophitic texture with abundant phlogopite	Plagioclase (25%–35%)	Amphibole, quartz+K-feldspar intergrowths	Trace Py	Lathy plagioclase in quartz+K-feldspar matrix with patches of amphibole; local quench-textured crowfoot-textured granophyre GRAN	Naldrett et al. (1970); Naldrett and Hewins (1984); Lightfoot et al. (1997a, 2001); Theriault et al. (2002)
North Range QGAB	Quartz Gabbro (QGAB)	Pink-grey to pink-white	Medium- to coarse-grained	30%–50% plagioclase, 0%–5% titanomagnetite, 15%–20% augite; cumulus hornblende can sometimes be recognised as a primary mineral; plagioclase is often heavily saussurized	Graphic intergrowths of quartz+K-feldspar	Trace Py. Up to 0.5% apatite, accessory titanomagnetite and sphene	Grey with conspicuous amphibole; gradaition into FNR below and GRAN above; identification of this unit based on magnetic properties alone can be misleading as titanomagnetite is not ubiquitous	Naldrett et al. (1970); Naldrett and Hewins (1984); Lightfoot et al. (1997a, 2001); Theriault et al. (2002)
North Range FNR	Felsic Norite (FNR)	Grey	Hypidiomorphic-granular texture	45%–55% Plagioclase, 10%–20% orthopyroxene; cumulus augite enters towards the top of the FNR	Clinopyroxene and quartz+K-feldspar graphic intergrowths	Po+trace Cpy+Pn in lower few tens of meters; typically Py above this level	Unit is transitional into overlying Quartz Gabro and underlying MNR. The FNR becomes more evolved from base to top.	Naldrett et al. (1970); Naldrett and Hewins (1984); Lightfoot et al. (1997a, 2001); Theriault et al. (2002); Keays and Lightfoot (2004)

(Continued)

Table 3.1 Summary of Petrographic Features of the Igneous Rocks that Comprise the SIC and Sources of Additional Information on Petrology and Mineralogy (*cont.*)

Unit of the Main Mass and Offset Dykes of the SIC	Rock Type Forming stratigraphic Unit	Colour	Grain Size and Texture	Equant (Cumulus) Minerals	Matrix (Intercumulus) Minerals	Sulfide and Important Accessory Minerals	Common Features and Contacts	References
North Range MNR	Mafic Norite (MNR)	Dark grey to black with inclusions and sulfide	Medium to coarse-grained ophitic texture with abundant phlogopite	Othopyroxene (20%–45%)	Large plates of plagioclase which ophitically enclose cumulus orthopyroxene, augite, phlogopite phlogopite, Quartz-feldspar intergrowths	Po, Pn, Cpy	Distinctive phlogopite content and large plates of intercumulus plagioclase; contact with FNR and SLNR is gradational over a few meters	Naldrett et al. (1970); Naldrett and Hewins (1984); Lightfoot et al. (1997a, 2001); Theriault et al. (2002); Keays and Lightfoot (2003)
Offset Dyke	Glassy quench-textured QD	Black	Glass with some feldspar	Plagioclase	Variably altered glass	None	Known from the margin of the Trill Offset	Golightly et al. (2010)
	Quench-textured Quartz Diorite (QQD)	White-grey	Fine-grained matrix with needs reaching 5cm in length; quench texture (sometimes called "spherulitic")	Plagioclase, orthopyroxene, (amphibole sometimes pseudomorphs hypersthene)	Quartz+K-feldpsar intergrowths and quartz; phlogopite and amphibole	Trace sulfide	Quench textures exhibited by plagioclase, amphibole and sometimes orthopyroxene	Grant and Bite (1984); Lightfoot and Farrow (2002); Hecht et al. (2008); Wood and Spray (1998); Tuschscherer and Spray (2002); Golightly et al. (2010).
	Amphibole Quartz Diorite (QD)	Grey-green	Fine- to medium-grained	Plagioclase, amphibole (after orthopyroxene) - equant crystals, but no crystal accumulation	Quartz+K-feldpsar intergrowths and quartz; phlogopite and amphibole	Trace sulfide; sometimes cut by remobilized veins of sulfide from MIQD	"Salt-and-pepper" texture	Grant and Bite (1984); Lightfoot and Farrow (2002); Hecht et al. (2008); Wood and Spray (1998); Tuschscherer and Spray (2002)

		Colour	Grain size	Phenocrysts/Minerals	Groundmass	Sulfide	Texture	References
Offset Dyke	**Orthopyroxene Quartz Diorite (QD)**	White-grey	Fine- to medium-grained	Plagioclase, orthopyroxene - equant crystals, but sign sign of accumulation	Quartz+K-feldpsar intergrowths (7%–15%) and Quartz; amphibole and phlogopite	Trace sulfide; sometimes cut by remobilized veins of sulfide from MIQD	Salt-and-pepper texture	Grant and Bite (1984); Lightfoot and Farrow (2002); Hecht et al. (2008); Wood and Spray (1998); Tuschscherer and Spray (2002)
	Granophyric QD	White-grey	Fine- to medium-grained	Plagioclase	Quartz+K-feldpsar intergrowths and Quartz; phlogopite and amphibole	Trace Py	Salt-and-pepper texture	Grant and Bite (1984); Lightfoot and Farrow (2002); Hecht et al. (2008); Wood and Spray (1998); Tuschscherer and Spray (2002)
	Mineralized inclusion quartz diorite (MIQD)	Grey-green with >5% sulfide and inclusions	Fine- to medium-grained	Plagioclase, amphibole (after orthopyroxene) - equant crystals, but no sign of accumulation	Quartz+K-feldpsar intergrowths and Quartz; phlogopite and amphibole	0%–5% Po-Pn-Cpy	Disseminated to blebby sulfide contained in matrix with inclusions	Grant and Bite (1984); Lightfoot and Farrow (2002); Hecht et al. (2008); Wood and Spray (1998); Tuschscherer and Spray (2002)
	Heavily mineralized inclusion quartz diorite (HMIQD)	Grey-green with >5% sulfide and larger inclusions	Fine- to medium-grained	Plagioclase, amphibole (after orthopyroxene) - equant crystals, but no sign of accumulation	Quartz+K-feldpsar intergrowths and Quartz; phlogopite and amphibole	>5% Po-Pn-Cpy	Blebby to connected blebs and strong disseminations of sulfide in matrix with larger inclusions	Grant and Bite (1984); Lightfoot and Farrow (2002); Hecht et al. (2008); Wood and Spray (1998); Tuschscherer and Spray (2002)

(Continued)

Table 3.1 Summary of Petrographic Features of the Igneous Rocks that Comprise the SIC and Sources of Additional Information on Petrology and Mineralogy (cont.)

Unit of the Main Mass and Offset Dykes of the SIC	Rock Type Forming stratigraphic Unit	Colour	Grain Size and Texture	Equant (Cumulus) Minerals	Matrix (Intercumulus) Minerals	Sulfide and Important Accessory Minerals	Common Features and Contacts	References
Sublayer	Granite Breccia (GRBX)	Grey-white with inclusions and sulfide	Igneous-textured to subigneous textured fine-grained matrix	Laths of plagioclase enclosed in quartz	Quartz, quartz+K-feldspar, amphibole, phlogopite	Po-Pn-Cpy	Both mafic and granitic inclusions	Hebil (1978); Pattison (1979)
	Sublayer Norite (SLNR)	Dark grey to black with inclusions and sulfide	Medium grained porphyritic to subpoikilitic texture	Plagioclase (40-50%) and orthopyroxene (up to 30%)	Augite, phlogopite, sulfide, amphibole after pyroxene	Po-Pn-Cpy	Dominantly mafic inclusions	Pattison (1979); Lightfoot et al. (1997a)
	Inclusion basic norite, (also termed leucocratic norite)	Medium grey with smaller inclusions and sulfide	Medium-grained with hypidiomorphic texture	Plagioclase (~45-55%) and orthopyroxene (<25%).	Augite, phlogopite, sulfide, amphibole after pyroxene	Po-Pn-Cpy	Typically located near to boundary of SLNR and GRBX; more leucocratic variants are described as leucocratic SLNR in Lightfoot et al. (1997b).	Grant and Bite (1984)
	Hybrid-xenolithic SLNR	Dark grey to black with inclusions and sulfide	Medium grained subpoikilitic matrix with coarse-grained patches of ophitic-textured melanorite	Matrix: Plagioclase (40-50%) and orthopyroxene (up to 30%); Inclusions typically have more orthopyroxene (<=50%).	Plagioclase tends to be intercumulus in the ophitic-textured rock	Po-Pn-Cpy	Fewer fine-grained mafic inclusions; the ophitic-texture melanorite tends to be hybridized, so the texture is different when compared to SLNR	Morrison (1984)

Py, pyrite; Bn, bornite; Po, pyrrhotite; Pn, pentlandite; and Cpy, chalcopyrite

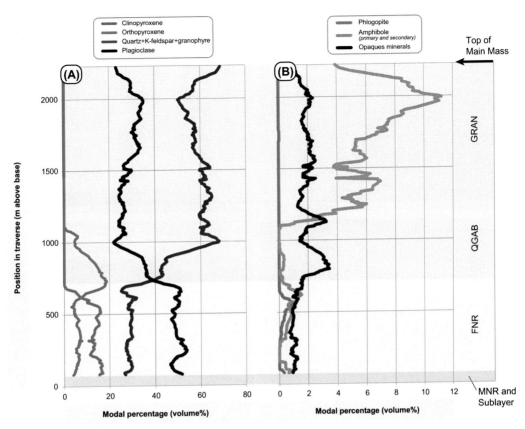

FIGURE 3.2 Modal Mineralogy from the Trillabelle Surface Traverse Presented on a Vertical Profile where Individual Data for 10 Thin Sections with 500 Point Counts per Section are Shown

The point counts reflect the primary and secondary mineralogy. (A) Variations in modal orthopyroxene, clinopyroxene, quartz + K-feldspar + granophyric intergrowths of quartz and K-feldspar, and plagioclase. (B) Phlogopite, amphibole (primary and secondary), and opaque minerals.

The FNR contains cumulus orthopyroxene and plagioclase with interstitial clinopyroxene (Fig. 3.3D). Weak to moderate alteration of pyroxenes to amphibole is commonly developed, but primary amphibole is less common.

The base of the QGAB is marked by the disappearance of orthopyroxene, and the presence of cumulus plagioclase, clinopyroxene, and amphibole; these rocks are weakly magnetic and develop intercumulus quartz + K-feldspar intergrowths (Fig. 3.3E). The QGAB locally contains abundant magnetite and apatite (Fig. 3.3F). The contact between the QGAB and GRAN Units is gradational and placed at the point where the plagioclase content decreases below 25%, clinopyroxene disappears and the proportion of quartz + K-feldspar reaches 65% (Fig. 3.2A,B).

The GRAN is strongly altered, and consists of irregular plagioclase crystals and quartz + K-feldspar intergrowths with lesser amounts of amphibole and phlogopite (Fig. 3.3G). Towards the top

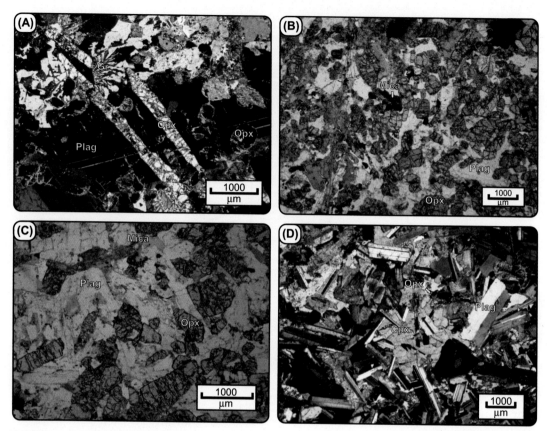

FIGURE 3.3 Petrology of the North Range Main Mass

(A) MNR in crossed polarized light (thin section C81-0251) from above the Trillabelle Embayment (drill core 60049-1 at 1855 m depth); this thin section shows elongate crystals of cumulus orthopyroxene (Opx) enclosed poikilitically in plagioclase (Plag) with interstitial quartz + K-feldspar intergrown in a graphic texture (Gr). (B) MNR in plane polarized light; from the East Range developed at Victor-Nickel Rim; the sample is from a similar drill core interval to that shown in the MNR Unit in Figure 3.4; the rock comprises ~40% modal orthopyroxene (Opx) which is poikilitically enclosed in plagioclase (Plag) with interstitial phlogopite (Mica) and sulfide (Sul). (C) Thin section of Lower FNR in plane polarized light (C79-1373); the sample comes from directly above the MNR at the Trillabelle embayment (drill core 52821-1, 1791 m depth). The orthopyroxene (Opx) and plagioclase (Plag) have a cumulus texture, and the absence of the pokilitic texture is readily recognized in thin section and hand sample as an indicator that this sample belongs to the FNR rather than the MNR. (D) Weakly altered FNR shown in cross-polarized light (thin section C79-1372) from the Main Mass stratigraphy above the Trillabelle embayment (drill core 52821-0 at 1780m depth). Orthopyroxene (Opx) is less abundant (~20%) and plagioclase (Plag) forms cumulus crystals which are pokilitically enclosed in clinopyroxene (Cpx).

FIGURE 3.3 (*cont.*)

(E) QGAB in plane polarized light from the Main Mass west of the Victor Deposit and comparable to rocks in the lower part of the QGAB in Figure 3.4; cumulus titanomagnetite (Ox) and apatite (Ap) together with weakly altered cumulus plagioclase (Plag) and amphibole (Amph) after augite occur in a matrix of quartz and intergrowths of quartz (Qz) and K-feldspar; (F) QGAB in plane polarized light in sample C79-1030 from the Main Mass above the Trillabelle embayment; the sample comes from drill core 52821-0 at 1231 m depth which is about 70 m above the top of the FNR. Clinopyroxene is the only pyroxene developed and the plagioclase is strongly altered. Primary amphibole is developed in this sample. (G) GRAN showing graphic intergrowth of quartz and K-feldspar in a surface sample from the GRAN near Capreol. (H) Plagioclase granophyre (sample C15-0183) from the top of the Main Mass west of Onaping Falls below the Onaping Formation melt bodies; sample shows abundant plagioclase feldspar in a medium-fine grained granophyre matrix with accessory amphibole and phlogopite. The samples are from the Vale thin section collection and they were photographed by Ben Vandenburg.

of the unit, the plagioclase content increases to reach 35 modal percent in the plagioclase-rich upper part of the GRAN (Fig. 3.3H; eg, Peredery and Naldrett, 1975).

Geochemical stratigraphy

Whole-rock geochemical studies of the rock types that comprise the Main Mass are simpler to establish than the matrix compositions of inclusion-bearing Sublayer and Offset rocks (see later in this chapter).

Samples acquired from the eastern part of the North Range come from locations shown in Fig. 3.1A,B; they comprise rocks from a drill core (MAC-91) and a surface traverse (Capreol Traverse). The geochemical stratigraphy is shown in Fig. 3.4A–H and the average composition of the principal rock types is given in Table 3.2. The plots comprise samples from a drill core traverse containing samples from the base of the SIC to the top of the QGAB, and a surface traverse from Capreol containing samples from the GRAN. The Capreol traverse is located in an area where the Main Mass is thinner, so the proportion of GRAN relative to FNR and QGAB is less than that which is typical of the North Range Main Mass. In the chemostratigraphic plots (Fig. 3.4A–H), the samples are spaced based on their calculated relative stratigraphic position in meters above the base of the SIC.

The lowermost rocks in the drill core traverse comprise the SLNR Unit. The noritic matrix of the Sublayer has a wide range in composition varying from noritic to melanoritic, each of which exhibits varying proportions of cumulus orthopyroxene, chrome spinel, phlogopite and magmatic sulfide content. The overlying MNR has an Mg-number exceeding 0.45, MgO > 6 wt% and low CaO (<7 wt%; Fig. 3.4A,B). The MNR also has low Sr (<450 ppm) and high Ni and Cu concentrations (>200 ppm) (Fig. 3.4A,B,G). The transition from MNR into basal FNR is recorded in the rapid decline of MgO and an increase in Sr content to ~500 ppm (Fig. 3.4B,E). Upwards through the FNR there is a decline in Mg-number, CaO, MgO, Ni, and Cu (Fig. 3.4A,B,G).

The boundary between the FNR and the QGAB is marked by an increase in CaO, P_2O_5, TiO_2, V, and Sc abundances (Fig. 3.4B,C,F). The top of the QGAB is marked by low CaO, MgO, TiO_2, P_2O_5, V, and Sc abundance levels (Fig. 3.4B,C,F).

In the GRAN (from the Capreol traverse), the TiO_2 and P_2O_5 concentrations increase towards the top of the Main Mass (Fig. 3.4C) and the abundances of MgO and CaO increase slightly through the GRAN (Fig. 3.4B).

The incompatible elements (eg, La and Y) show a very slight increase upwards through the FNR, and then a marked increase through the QGAB to reach a maximum in the GRAN (Fig. 3.4D). The concentrations of these elements falls slightly through the GRAN Unit. With the exception of some samples in the MNR, the ratios of incompatible elements such as La/Y, Th/U, and Th/Nb are relatively constant through the entire Main Mass (Fig. 3.4H). The MNR and the noritic matrix of the Sublayer have different ratios of incompatible trace elements relative to the overlying rocks of the Main Mass, and this may reflect the contribution from small inclusions and the products of assimilation of these inclusions.

A PETROLOGICAL, GEOPHYSICAL PROPERTY, AND GEOCHEMICAL TRAVERSE THROUGH THE MAIN MASS IN THE SOUTH RANGE

A case study is presented of data from a petrological traverse north of the Blezard Deposit through the noritic and gabbroic rocks below the heavily tectonized GRAN Unit which is bounded by

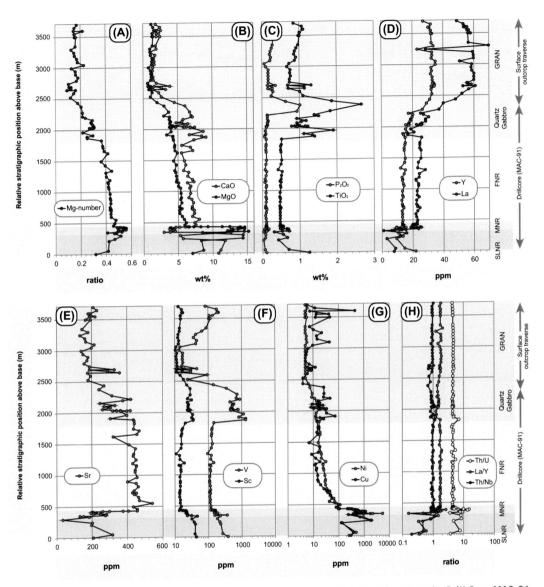

FIGURE 3.4 Whole Rock Geochemistry Variations along the Traverse which Combines Data for Drill Core MAC-91 with the Surface Traverse through the Quartz Gabbro and Granophyre at Capreol

The Main Mass of the SIC varies in thickness around the Sudbury Basin; the Main Mass is thinner in the area of the surface traverse at Capreol than the East Range; this explains why the GRAN Unit is thin relative to the thickness of the FNR and QGAB. (A) Whole-rock Mg-number versus calculated position in a normal direction to the basal contact of the SIC. (B) Variations in CaO and MgO concentration (wt%). (C) Variations in TiO_2 and P_2O_5 concentration (wt%). (D) Variations in La and Y concentration (ppm). (E) Variation in Sr concentration (ppm). (F) Variations in V and Sc concentration (ppm). (G) Variations in Ni and Cu elemental concentration (ppm). (H) Variation in incompatible element ratios; Th/U, La/Y, and Th/Nb where ratios of concentrations in ppm are shown.

After Lightfoot et al. (2001) (H) After Lightfoot et al. (1997c,d; 2001).

Table 3.2 Average Compositions of Rock Units of the North Range Main Mass

Element/oxide	Units	Method	MAC-91 Drill Core and Capreol Traverse				
			Mafic Norite (MNR)	Felsic Norite (FNR)	Quartz Gabbro (QGAB)	Grano-phyre (GRAN)	Main Mass
Rock type							
n			16	30	25	20	91
SiO_2	wt%	XRF	54.51	56.49	53.90	67.98	61.75
Al_2O_3	wt%	XRF	10.00	16.88	14.16	12.76	14.23
MnO	wt%	XRF	0.17	0.11	0.14	0.09	0.11
MgO	wt%	XRF	13.01	5.05	3.45	1.19	2.99
CaO	wt%	XRF	4.30	6.70	6.58	1.74	4.15
Na_2O	wt%	XRF	1.85	3.13	3.78	3.70	3.51
K_2O	wt%	XRF	1.22	1.43	1.38	3.48	2.44
TiO_2	wt%	XRF	0.58	0.49	1.68	0.86	0.88
P_2O_5	wt%	XRF	0.09	0.10	0.56	0.22	0.24
Fe_2O_3	wt%	XRF	11.18	7.03	11.36	6.24	7.45
LOI	wt%	XRF	1.40	1.76	1.46	1.67	1.66
Total	wt%	Calculated	98.44	99.17	98.45	99.93	99.41
FeO^a	wt%	Titration	8.11	4.61	6.96	3.24	4.39
CO_2	wt%	IR	0.08	0.06	0.06	0.07	0.06
H_2O^+	wt%	IR	0.34	0.24	0.30	0.32	0.29
H_2O^-	wt%	IR	1.63	2.03	1.99	1.54	1.77
Be	ppm	ICP-MS	1.50	1.50	3.27	2.12	2.11
Ce	ppm	ICP-MS	44.54	49.08	69.52	112	84.15
Co	ppm	ICP-OES	74.8	30.3	42.9	10.9	23.6
Cs	ppm	ICP-MS	0.83	0.99	0.49	0.41	0.61
Cu	ppm	ICP-OES	546	22.6	21.6	12.4	26.9
Dy	ppm	ICP-MS	2.26	2.51	4.19	6.05	4.56
Er	ppm	ICP-MS	1.38	1.46	2.27	3.43	2.58
Eu	ppm	ICP-MS	0.86	1.14	1.68	1.83	1.57
Gd	ppm	ICP-MS	2.73	3.07	5.24	7.68	5.74
Hf	ppm	ICP-MS	1.97	2.69	3.49	7.52	5.23
Ho	ppm	ICP-MS	0.48	0.52	0.84	1.26	0.95
La	ppm	ICP-MS	21.23	24.09	32.35	56.85	41.83
Lu	ppm	ICP-MS	0.21	0.21	0.31	0.51	0.37
Mo	ppm	ICP-OES	4.00	7.93	4.55	3.00	4.81

Element/oxide	Units	Method	MAC-91 Drill Core and Capreol Traverse				
					Quartz Gabbro (QGAB)	Grano-phyre (GRAN)	
Rock type			Mafic Norite (MNR)	Felsic Norite (FNR)			Main Mass
n			16	30	25	20	91
Nb	ppm	ICP-MS	4.74	6.06	7.84	15.97	11.30
Nd	ppm	ICP-MS	18.59	20.36	32.68	50.02	37.27
Ni	ppm	ICP-OES	555	31.1	10.8	5.12	24.2
Pr	ppm	ICP-MS	4.82	5.32	7.96	13.74	9.97
Rb	ppm	ICP-MS	40.0	46.4	49.3	93.5	70.3
Sc	ppm	ICP-OES	22.1	15.2	27.8	12.9	16.3
Sm	ppm	ICP-MS	3.32	3.64	6.05	9.15	6.81
Sr	ppm	ICP-MS	248	450	326	172	286
Ta	ppm	ICP-MS	0.27	0.37	0.48	1.01	0.71
Tb	ppm	ICP-MS	0.41	0.44	0.75	1.14	0.84
Th	ppm	ICP-MS	4.92	5.91	6.58	15.17	10.64
Tm	ppm	ICP-MS	0.22	0.22	0.33	0.51	0.38
U	ppm	ICP-MS	0.76	1.29	1.48	3.30	2.32
V	ppm	ICP-OES	158	119	606	54.4	171
Y	ppm	ICP-MS	12.49	13.07	20.81	32.96	24.33
Yb	ppm	ICP-MS	1.43	1.41	2.06	3.21	2.42
Zn	Ppm	ICP-OES	118	90.2	93.4	65.5	78.9
Zr	Ppm	XRF	77.5	111	140	276	198

Table 3.2 Average Compositions of Rock Units of the North Range Main Mass (*cont.*)

ICP-OES, Inductively-coupled plasma emission mass spectrometry; ICP-MS, Inductively-coupled plasma mass spectrometry; IR, Infrared determination; XRF, X-ray fluorescence.
The full dataset is available in Lightfoot et al. (1997c).
[a]Determination of FeO concentration; total Fe is expressed as Fe_2O_3.
Source: Lightfoot et al. (1997c,d; 2001).

the Cliff Lake Fault to the North (Figs. 2.36A,B and 3.5A,B). Geochemical variations through the South Range are highlighted based on the Creighton traverse along Highway 144 reported in Lightfoot and Zotov (2006). The variations along the Highway 144 section are considered to represent a reasonably intact section, but there is some faulting and shearing within the QGAB and GRAN.

The Blezard traverse, along which 120 samples were collected, broadly followed Highway 69 (Fig. 3.5A,B). Complete modal mineralogy acquired by point counting (500 points per thin section)

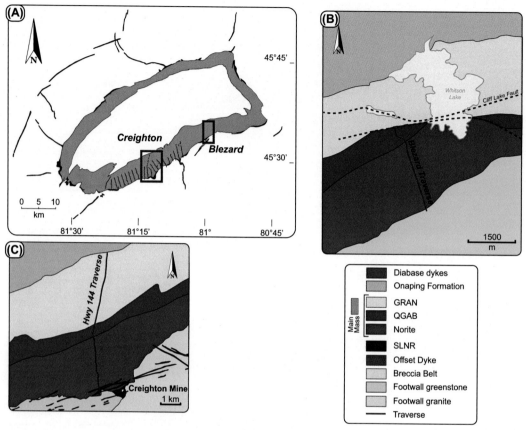

FIGURE 3.5 Location of the South Range Main Mass Sample Traverses

(A) Map of the SIC showing the location of the sample traverses at Creighton and Blezard and traverses through the South Range Main Mass. (B) Location of the Blezard surface sample traverse, and the source of weighted average modal mineralogy shown in Fig. 3.7A,B. (C) Geology map of the Creighton area showing the location of samples along the Highway 144 traverse.

After Lightfoot and Zotov (2006).

was used to establish 10 point moving average of mineral abundances to illustrate the changing mineralogy through the Main Mass to the base of the GRAN where the Cliff Lake Fault offsets the stratigraphy.

Petrological variations

The stratigraphy of the Main Mass in the South Range consists of a sequence of differentiated primary igneous rocks which have subsequently been recrystallization at lower amphibolite grade and variably altered by hydrothermal processes (dominantly in rocks located above the norite stratigraphy). Typical examples of photomicrographs of the rock types through the stratigraphy at Blezard and Creighton are shown in Fig. 3.6A–H and the locations of the traverses are shown in Fig. 3.5B–C. The variation in modal mineralogy was determined by point counting (Fig. 3.7A,B). The main petrological features of the South Range Main Mass rock units are shown in Table 3.1.

The basal Quartz-rich Norite consists of 15%–20% cumulus orthopyroxene, 50%–60% cumulus plagioclase and 10%–18% quartz. Locally, this rock type has a poikilitic texture with large patches of blue flood quartz which contains cumulus plagioclase and orthopyroxene (Fig. 3.6A). The overlying

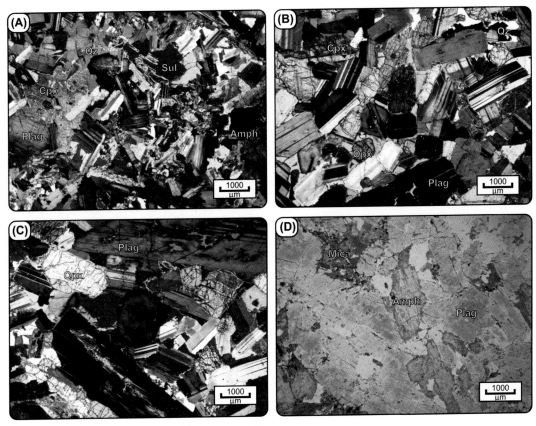

FIGURE 3.6 Petrology of the Main Mass in the South Range

(A) Basal Quartz-rich Norite in crossed polarized light (thin section 17817) from the Blezard traverse shown in Fig. 3.5B. This unaltered norite contains cumulus orthopyroxene (Opx) and intercumulus clinopyroxene (Cpx) with some (primary?) amphibole (Amph). (B) Quartz Norite in crossed polarized light (thin section 17871-1) located in the Quartz Norite of the Blezard traverse. This sample contains cumulus orthopyroxene (Opx) and plagioclase (plag) in a matrix of clinopyroxene (Cpx). (C) Fresh South Range Quartz-rich Norite in crossed polarized light; from the Blezard traverse showing cumulus orthopyroxene (Opx) and plagioclase (Plag) with lesser amounts of intercumulus augite and amphibole (thin section 17877-2). (D) Altered South Range Norite in plane polarized light (thin section 17881-2) from the Blezard traverse exhibiting alteration of primary intercmulus clinopyroxene to amphibole (Amph).

(Continued)

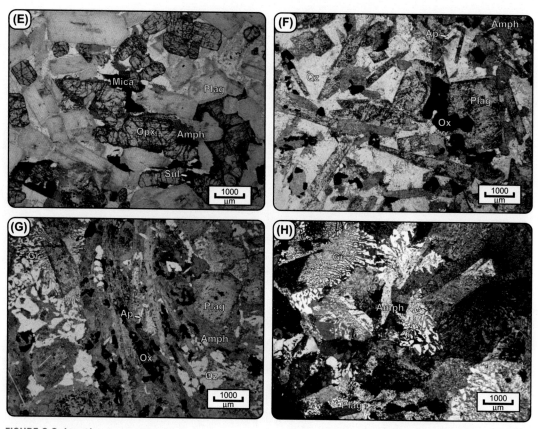

FIGURE 3.6 (*cont.*)

(E) Melanorite lens in the Quartz Norite of the Creighton traverse in plane polarized light (sample C15-0157); fresh cumulus orthopyroxene (Opx) and plagioclase (Plag) with intercumulus amphibole (Amph), biotite (Mica) and trace sulfide (Sul) is developed. (F) QGAB with primary amphibole (Amph) and cumulus titanomagnetite (Ox) in thin section C79-1000 from the Trillabelle area (bore hole 52821-0 at 1125 m). (G) Crowsfoot-textured QGAB in plane polarized light (sample C15-0196) from the upper part of the QGAB of the Creighton traverse exhibiting plumose primary amphibole laths (Amph) and apatite (Ap) separated by plagioclase (Plag), quartz (Qz), and micrographic intergrowths of quartz and K-feldspar (Gr); (H) GRAN in cross-polarized light (thin section 16320-3) from Dowling Township (bore hole 23381-0 at 616 m. Euhedral grains of albitic feldspar (Plag) are hosted in a matrix of micrographic intergrowths of quartz and K-feldspar (Gr) with subordinate amphibole (Amph) after clinopyroxene. The samples are from the Vale thin section collection and they were photographed by Ben Vandenburg.

Quartz-bearing Norite grades into South Range Norite (Fig. 3.6B,C); these rocks contain 20%–25% cumulus orthopyroxene, 55%–65% plagioclase, and 5%–10% quartz. The norites show a variable degree of alteration where the primary orthopyroxene is partly altered to amphibole (Fig. 3.6D). Irregular pods and lenses of weakly mineralized orthopyroxene-rich norite are developed at approximately 2/3rds. of the way through the norite stratigraphy, but these lenses are not as continuous or well-developed as the example identified in the Creighton traverse by Lightfoot and Zotov (2006) which is shown in Fig. 3.6E. Through the norite stratigraphy of the South Range there is a gradual upwards increase in modal cumulus orthopyroxene and plagioclase and a decrease in the proportion of intercumulus minerals such as quartz, K-feldspar and secondary amphibole (after clinopyroxene) followed by a return to rocks with lesser amounts of cumulus orthopyroxene and plagioclase (Fig. 3.7A,B).

The top of the Norite is marked by the sudden entry of opaque minerals (principally ulvospinel; Naldrett et al., 1970; Fig. 3.6F) and a decline in the proportion of orthopyroxene from 20 modal percent to zero with an increase in the clinopyroxene content (Fig. 3.7A,B). Locally, the QGAB contains fine-grained inclusions of diabase in the Creighton traverse and also a crowsfoot-textured granophyre with laths of amphibole (after clinopyroxene) illustrated in Fig. 3.6G. The content of opaque minerals and

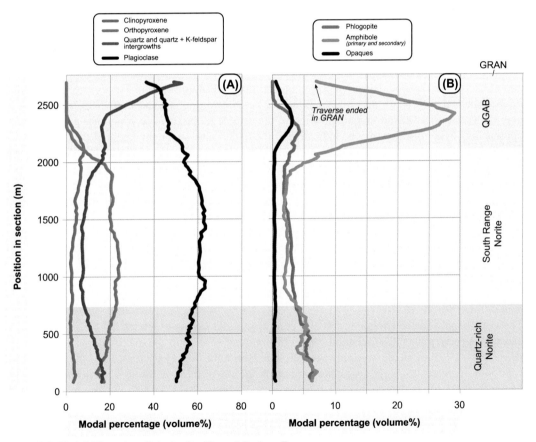

FIGURE 3.7 Modal Mineralogy Data for the Blezard Surface Traverse

(A) Variations in 10 point moving averages of 500 points per section for orthopyroxene, clinopyroxene, plagioclase, and quartz + K-feldspar + granophyric intergrowths of these minerals; (B) Variations in modal phlogopite, amphibole (primary and secondary), and opaque minerals. The traverse terminates in the lower part of the GRAN.

apatite gradually increases to reach a total of 5% in rocks containing clinopyroxene and both primary and secondary amphibole (Fig. 3.7B).

The contact between the QGAB and GRAN is gradational and placed at the point where the micrographic intergrowth of quartz-K-feldspar first exceeds modal plagioclase content. Above this level, the crowsfoot-textured granophyre contains quartz + K-feldspar, cumulus plagioclase and 2–50 cm needles of amphibole as radiating plume-like segregations. This unit is overlain by a quartz-feldspar granophyre that consists of graphic intergrowths of quartz and K-feldspar with a strong alteration overprint (Fig. 3.6H).

Mineral compositional variations through the Main Mass were reported in Pattison (1979) and Naldrett et al. (1984). The very wide range in Mg-number of orthopyroxene and clinopyroxene reported in individual samples from a traverse at Creighton reflects the compositional zonation in the minerals from individual samples; it is difficult to map out systematic variations in mineral chemistry through the stratigraphy.

Physical property and geochemical stratigraphy

Fig. 3.8A–J shows the variations in density, magnetic susceptibility, major- and trace-element geochemistry, and trace element ratio for samples from the South Range surface traverse along Highway 144 at Creighton; detailed data are reported in Lightfoot and Zotov (2006). The average compositions of the principal rock types are given in Table 3.3 based on Lightfoot and Zotov (2006). The vertical axis shows the relative stratigraphic position of the sample in the traverse which has a total thickness approaching 5100 m which may be indicative of the upper limit on the thickness of the Main Mass.

Much of the norite traverse is intact although there are local irregular aplite breccias at the base of the sequence and east-trending diabase dykes (Trap Dykes) which often exhibit slickensides along their contacts. The lower part of the QGAB is cut by a strong mylonite zone which may be a splay off the Cliff Lake Fault (Chapter 2), and this structure has produced some local disruption in the Main Mass stratigraphy. Above this level there are several fault and shear zones in the QGAB and GRAN which interrupt the continuity of the section. The thickness of the QGAB and GRAN Units relative to the Norite is less than the average expected for the SIC, and this indicates that some of the QGAB and/or GRAN stratigraphy is missing due to thrust structures, possibly associated with the Cliff Lake Fault.

The South Range Norite

The noritic rocks of the Creighton traverse comprise a unit of basal Quartz-rich Norite with slightly elevated density, magnetic susceptibility and Mg-number relative to the overlying norite stratigraphy (Fig. 3.8A–C). Through the overlying South Range Norite the density decreases and the magnetic susceptibility increases to reach a peak approximately 2/3rds. of the way through the norite stratigraphy where the background Mg-number of the norite stratigraphy also reaches a maximum. At this level, the narrow (10–50 m thick) laterally continuous (~1.5 km) lens of melanorite is developed within the stratigraphy (see Fig. 3.8E); above this level the density starts to increase, the magnetic susceptibility declines, and the Mg-number declines upwards towards the base of the QGAB (Fig. 3.8A–C).

The concentrations of incompatible trace elements (La, Y, Ba) vary in sympathy with TiO_2 (Fig. 3.8E–G), as do the concentrations of Sc and V (Fig. 3.8H), but the Sr concentration tends to vary in sympathy with Mg-number (Fig. 3.8C,G). Through the norites, there is almost no change in the ratios of incompatible elements such as Th/Nb, La/Y, and Th/U (Fig. 3.8J).

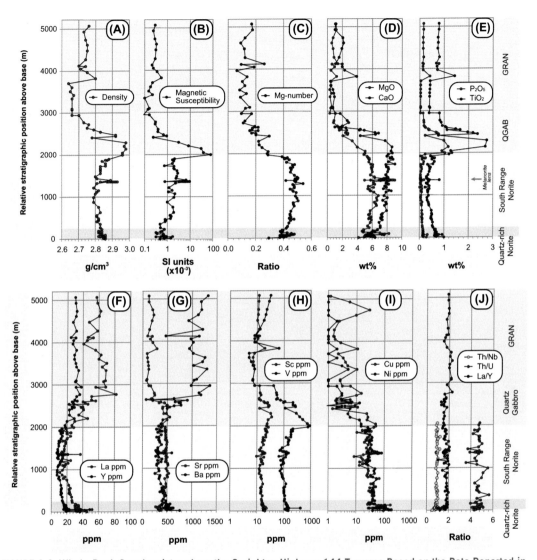

FIGURE 3.8 Whole-Rock Geochemistry along the Creighton Highway 144 Traverse Based on the Data Reported in Lightfoot and Zotov (2006) where the Vertical Stratigraphic Position is Calculated Based on the UTM Coordinates of the Sample Shown in Traverses Normal to the Lower Contact

The locations of structures and dykes are shown; these may correspond to minor discontinuities which may duplicate or remove stratigraphy. (A) Density variations (g/cm^3). (B) Magnetic susceptibility measured on hand sample slabs (SI units). (C) Mg-number (calculated based on MgO and Fe$_2$O$_3$ concentrations in wt% units). (D) MgO and CaO concentration (wt%). (E) TiO$_2$ and P$_2$O$_5$ concentration (wt%). (F) La and Y concentration (ppm). (G) Sr and Ba concentration (ppm). (H) Sc and V concentration (ppm). (I) Ni and Cu concentration (ppm); (J) La/Y, Th/Nb, and Th/U (based on elemental concentrations determined in ppm).

Based on Lightfoot and Zotov (2006).

Table 3.3 Average Compositions of Rock Units of the South Range Main Mass

Location	Units	Method	Highway 144 Traverse						
Rock Group			Quartz-ose Norite	South Range Lower Norite	Mela-norite Lens	South Range Upper Norite	Quartz Gabbro (QGAB)	Grano-phyre (GRAN)	Main Mass
n			23	24	24	15	19	19	124
SiO$_2$	wt%	XRF	58.14	56.14	55.01	55.78	59.90	70.10	62.92
Al$_2$O$_3$	wt%	XRF	15.74	17.63	15.87	17.40	14.51	12.68	14.74
MnO	wt%	XRF	0.12	0.11	0.14	0.11	0.11	0.06	0.09
MgO	wt%	XRF	5.62	6.03	8.13	5.79	2.42	0.72	3.03
CaO	wt%	XRF	6.51	7.65	7.26	8.19	4.93	1.51	4.47
Na$_2$O	wt%	XRF	2.81	2.83	2.37	2.80	3.77	3.69	3.37
K$_2$O	wt%	XRF	1.66	1.10	0.92	1.14	2.28	3.71	2.48
TiO$_2$	wt%	XRF	0.62	0.41	0.36	0.44	1.39	0.67	0.73
P$_2$O$_5$	wt%	XRF	0.14	0.08	0.06	0.09	0.48	0.12	0.18
Fe$_2$O$_3$	wt%	XRF	7.79	6.86	8.60	7.21	8.83	5.25	6.67
LOI	wt%	XRF	0.84	1.15	1.40	1.02	0.73	0.58	0.80
Total	wt%	Calculated	100.15	100.13	100.25	100.10	99.52	99.30	99.65
Ba	ppm	ICP-OES	515	381	338	356	764	1193	807
Be	ppm	ICP-MS	1.11	0.91	0.97	0.82	1.82	2.15	1.62
Co	ppm	ICP-OES	27.4	27.8	36.6	30.2	18.4	4.85	16.5
Cr	ppm	XRF	206	143	150	109	16.4	4.8	59.2
Cu	ppm	ICP-OES	55.6	29.7	35.1	23.7	15.1	6.1	17.8
La	ppm	ICP-MS	30.0	19.1	13.0	18.7	42.7	58.3	41.0
Li	ppm	ICP-OES	21.1	15.6	14.3	13.3	9.9	6.1	10.5
Mo	ppm	ICP-OES	2.0	2.1	0.9	2.0	2.0	1.5	1.8
Nb	ppm	ICP-MS	8.6	5.6	3.9	5.9	11.7	15.6	11.3
Ni	ppm	ICP-OES	57.4	39.3	55.6	31.4	7.1	1.7	17.4
Pb	ppm	ICP-MS	8.5	10.4	15.4	5.3	5.1	5.8	6.8
Rb	ppm	ICP-MS	58.4	37.9	33.5	39.0	66.7	109	75.4
Sc	ppm	ICP-OES	16.4	14.4	17.7	18.4	19.9	10.2	14.4
Sr	ppm	ICP-MS	372	466	426	464	331	142	295
V	ppm	ICP-OES	126	98	121	152	258	17	104
Y	ppm	ICP-MS	16.4	10.7	8.5	10.8	25.7	30.4	22.4
Zn	ppm	ICP-OES	70.1	63.4	86.0	58.5	62.6	46.4	56.2
Zr	ppm	XRF	116	85	66	91	182	278	188

Averages are calculated based on data in Lightfoot and Zotov (2006).
Abbreviations: see Table 3.2.
Source: Lightfoot and Zotov, 2007.

Although there is considerable variation, the lowermost Quartz-rich Norites of the Main Mass have higher Cu and Ni abundances than rocks higher in the stratigraphy due to the presence of trace sulfide minerals. The concentrations of Cu and Ni fall above this level with a further marked decline about half way through the norite stratigraphy and a small positive anomaly corresponding to the entry of the melanorite lens which contains trace disseminated sulfide minerals (Fig. 3.8I) and up to 2% phlogopite (Lightfoot and Zotov, 2006; Baird, 2007; Walker, 2014; Strongman, 2016).

The South Range Quartz Gabbro

The base of the QGAB Unit is marked by a sudden increase in both density and magnetic susceptibility that coincides with a drop in Mg-number at the top of the norites (Fig. 3.8A–C). There is a wide range in TiO_2 and P_2O_5 concentration in the QGAB (Fig. 3.8E) and also wide range in V concentration (Fig. 3.8H). The top of the QGAB Unit is marked by a decline in magnetic susceptibility, density and Mg-number to values that are lower than those of the norite.

The concentrations of incompatible elements increase through the QGAB (eg, La and Y in Fig. 3.8F), but there is no change in the ratios of the incompatible trace elements (Fig. 3.8J). The Cu and Ni concentrations fall through the QGAB Unit to background levels in the overlying GRAN (Fig. 3.8I).

The South Range Granophyre

The lower part of the GRAN straigraphy is significantly less dense than the upper part, and there is a compositional break about half-way through the GRAN below which the rocks have slightly lower CaO than those above (Fig. 3.8D). The abundances of incompatible elements like La and Y are fairly constant through the QGAB, and the ratios of these elements remain similar in the lower part of the GRAN (Fig. 3.8F,J). The concentration of Ni in the GRAN approaches the determination limits of the method employed (Table 3.3), and the variations in Cu concentration are very variable through this unit due to extensive hydrothermal alteration (Fig. 3.8I).

LATERAL VARIATIONS IN GEOCHEMISTRY, SOUTH RANGE

Samples were collected from 20 surface traverses through structurally intact Main Mass norite stratigraphy between the Crean Hill deposit in the west and the Murray deposit in the East (Fig. 3.5A) in order to investigate the lateral variations in whole-rock geochemistry of the South Range Main Mass. These profiles and data were the basis for the investigation of variations along strike in the South Range that are reported in Darling (2010a,b) and data for selected traverses are given in Appendix 3.1. The relative stratigraphic position of the rock sample is assigned based on the dip of the basal contact of the SIC and the surface position (recalculated to represent an orthogonal stratigraphic position of the sample relative to the base of the SIC). The base of the QGAB is used as a reference horizon below which the stratigraphy of the norite is "hung"; this horizon is established based on the threshold abundance of TiO_2 (<0.6 wt%) which better defines the boundary between the norite and gabbro than the hand sample petrology or magnetic properties of the rock.

Fig. 3.9A–C shows variations in Mg-number, the ratio of observed Ni concentration relative to the expected Ni concentration (Ni/Ni*), and La (ppm) in samples from the 20 traverses. The Ni/Ni* ratio is calculated from the observed whole rock Ni concentration and the expected Ni abundance calculated from the MgO concentration using an array fitted to the composition of the country rocks in the footwall of the SIC (ie, Ni* = $20.363*MgO^{0.7834}$). The data points are shown on a longitudinal section where

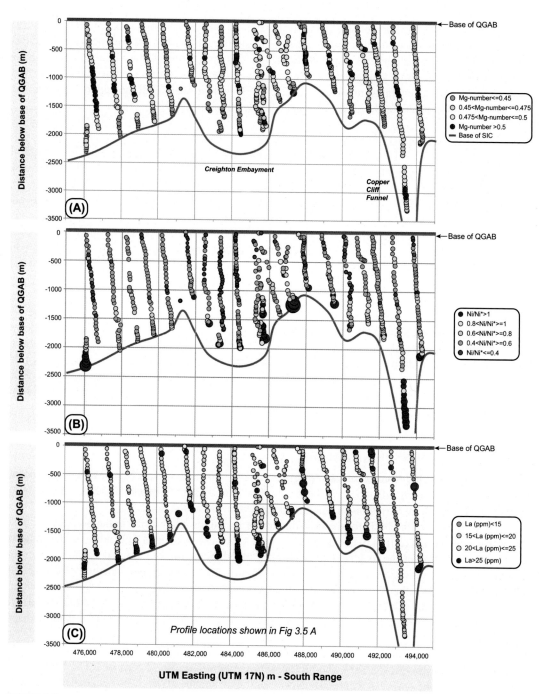

FIGURE 3.9

individual samples in the traverses are positioned based on the Easting coordinate (NAD27, UTM zone 17N), and the variations along each of the profiles are plotted relative to the base of the QGAB. The symbols are sized to elemental concentration or ratios (which is designed to pick-out the unusually high or low values), and colored by the concentration or ratio group interval to depict the lateral and vertical variations in whole-rock geochemistry. The diagram also shows the approximate projected location of the lower contact of the SIC based on drill core data; this horizon is often a short distance below the lowermost sample plotted in the traverse due to either access limitation arising from historic development of surface mine workings and/or available outcrop.

The variation in Mg-number provides an index of magmatic differentiation and it is principally controlled by addition or removal of ferromagnesian minerals and changes in the compositions of these minerals during crystallization. The profiles through the South Range stratigraphy reveal a marked tendency for the rocks to become more primitive in composition towards the center of the noritic unit (Fig. 3.9A). Some weakly mineralized samples from the Quartz-rich Norite at the base of the section have high Mg-numbers (Fig. 3.9A). All of these samples contain <1% disseminated sulfide, so the variations are not influenced by the Fe content of sulfide; rather, the variations are an intrinsic feature of the whole rock samples. In the case of the Creighton traverse documented in Fig. 3.8A, the melanocratic lens developed in this profile is not developed as a continuous unit at the level of sampling detail undertaken in the other traverses; this may be a function of lateral discontinuity or gaps in sample coverage. The variations in incompatible trace element abundances, such as La (Fig. 3.9C) are followed by other incompatible elements like Y (eg, Fig. 3.8F), and there is a general tendency for the central part of the norite stratigraphy with the highest Mg-number to have a lower La abundance than the noritic rocks developed above and below this level. Variations in Ni/Ni* highlight the mineralized samples from the base of each traverse, but they also show that all of the traverses contain norites which are depleted in Ni relative to the array of the country rocks, and some of the traverses show a more marked depletion than other traverses (Fig. 3.9B). Broadly, the variations in Ni/Ni* in the norite stratigraphy above contacts that are endowed with economic sulfide mineralization (eg, Creighton and Copper Cliff shown in Fig. 3.9A) exhibit similar levels of Ni depletion as those developed above weakly mineralized contacts (Fig. 3.9B).

The differentiation signature of the Main Mass norite is quite similar above mineralized and barren contacts in the South Range. Noritic rocks above embayment structures have elevated Ni/Ni* due to

◀ FIGURE 3.9 Chemical Variations through the Noritic Rocks of the South Range in the 20 Profiles shown in Fig. 3.5A

The relative stratigraphic position of the sample is assigned based on the dip of the base of the SIC and the thickness of the norite unit based on drill core information and surface outcrop pattern. There are no major fault structures in the stratigraphy, but there is some possibility of repetition or removal of intervals due to displacement on minor faults. The base of the QGAB is assumed to be a horizontal marker horizon along which each of the traverses is hung. The basal contact of the SIC respects the stratigraphic thickness of the noritic rocks and the base of the SIC established from diamond drilling. The data used for this plot are derived from Lightfoot and Zotov (2006). Analyses from representative traverses are reported in Appendix 3.1. (A) Variations in group interval of Mg-number. (B) Variations in group interval of Ni/Ni*. The ratio of Ni to Ni* is calculated based on the concentration of Ni in the sample divided by the expected Ni concentration based on the MgO concentration assuming a crustal model. (C) Variation in group interval of La concentration.

sulfide control, but there is no significant difference in the Ni/Ni* ratios above barren versus mineralized contacts in the unmineralized norite stratigraphy. Notwithstanding, some profiles do record more samples with strong Ni-depletion than other profiles.

Lateral variations in geochemistry, North Range

A case study of 10 drill core traverses through the North Range between Trillabelle in the west and the Nickel Rim South Mine in the east (Fig. 3.1A) is presented in Fig. 3.10A–C following a similar approach to that used for the South Range traverses. The degree of faulting and attenuation due to folding of the norite stratigraphy in the North Range is believed to be quite small relative to the primary variations in thickness of the norite stratigraphy.

Variations along strike in the North Range show the development of a systematic upwards decline in TiO_2 as the base of the QGAB is approached (Fig. 3.10A). The Mg-number also shows a systematic upwards decline through the noritic rocks (Fig. 3.10B). These signatures are consistent with progressive changes in Mg-number and TiO_2 due to the fractionation and accumulation of ferromagnesian minerals. The variations in Ni/Ni* (>1) are consistent with the presence of trace to disseminated sulfide mineralization in the lowermost noritic rocks (Fig. 3.10C), followed by a progressive upwards decline in Ni/Ni* which is due to the segregation and removal of immiscible sulfide melt from the magma to reach values as low as 0.4 (Lightfoot et al., 2001; Keays and Lightfoot, 2004).

The norite stratigraphy between the mineralized segments of contact tend to be thinner, but the rocks exhibit broadly the same upwards decline in TiO_2, Mg-number, and Ni/Ni* as in samples collected from traverses above economically mineralized segments of the basal contact. Main Mass noritic rocks developed immediately above mineralized contacts tend to have elevated Mg-number and they also have higher Ni/Ni* due to the sulfide content of the rocks. Above this level, the changes in norite geochemistry above mineralized versus barren contacts are quite similar.

Above the base of the QGAB, one particular North Range profile exhibits a remarkably thick stratigraphy of QGAB (Fig 3.10D); this traverse is located west of Windy Lake (Dowling Deep in Fig. 3.10D), and the QGAB approaches 2–3 times the thickness of that developed elsewhere in the North Range and this increased thickness is accompanied by the development of thinner FNR (Fig. 3.10D). There is no evidence that this thickening is due to structural repetition accompanying thrusting as the geochemistry is not repeated through the Dowling Deep drill core; it is a primary feature of the differentiation history of this segment of the Main Mass.

Comparison of the Main Mass in the North and South Range rock units

A comparison of the average normalized compositions of the principal rock types from the North and South Range Main Mass is shown in Fig. 3.11A where the average is weighted based of the stratigraphic interval represented by each sample.

Compositional data for the North Range Units are shown in Fig. 3.11A where the abundance levels in the major rock types are normalized against the bulk integrated composition of the Main Mass traverse of the North Range (Table 3.1). Variations in the incompatible element concentrations are controlled by the accumulation and/ or crystallization of silicate minerals from the magma. The average FNR and GRAN from the North Range traverse shows a flat pattern for the incompatible elements. The QGAB shows a sloped pattern with slight depletion in the abundance of K, Th, U, Nb, Ta, Zr, and the LREE relative to the average bulk integrated concentration of the Main Mass (Fig. 3.11A), there is a slight depletion in U, Nb, Ta, and Zr in the MNR, but it otherwise exhibits a flat profile in Fig. 3.11A.

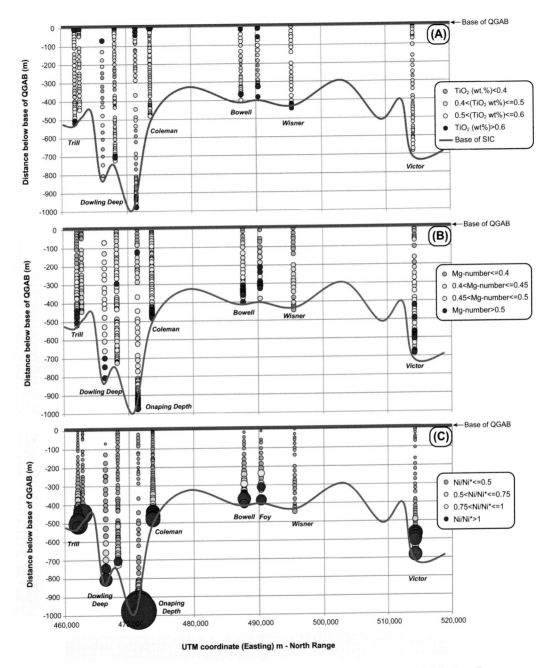

FIGURE 3.10 Chemical Variations through the Noritic Rocks of the North Range in the 10 Drill Core Traverses shown in Fig. 3.1A

The data were generated to underpin the studies by Lightfoot (2007) and Cooper (2000). The relative stratigraphic position of the sample is calculated based on the dip of the base of the SIC from the model of Gibson (2003). (A) Variation in group interval of TiO_2 (wt%). (B) Variation in group interval of Mg-number. (C) Variation in Ni/Ni* (crustal model) where the symbols are sized and colored based on the group interval of Ni/Ni*.

(Continued)

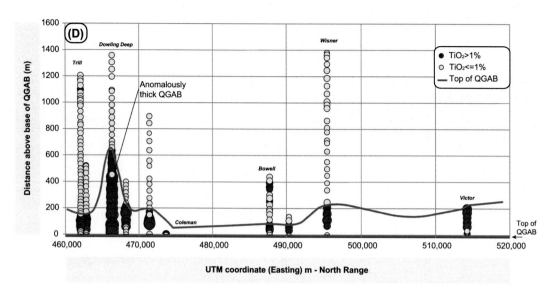

FIGURE 3.10 (*cont.*)

(D) Variation in TiO$_2$ above the base of the QGAB in drill core where the symbol is sized and colored by TiO$_2$ group interval. Note that the traverses are collared within the GRAN and so the top of the traverse does not represent the stratigraphic position of the roof of the SIC. Note the presence of anomalously thick QGAB in the Dowling Deep drill hole at Windy Lake.

As a reference point for the South Range, the average composition of the Quartz-rich Norite is also shown in Fig. 3.11A, and the profile of this rock type is flat too.

Relative to the average bulk-integrated Main Mass, the MNR, FNR, and QGAB averages are depleted, and the GRAN is enriched in incompatible trace element abundance. The patterns of all of the rock types are broadly subparallel, which is consistent with the observations of Lightfoot et al. (2001) and Keays and Lightfoot (2004) where the evidence for one parental magma type for the Main Mass was presented.

A comparison of the rocks from the principal South Range rock types is shown in Fig. 3.11B. Data points are plotted in cases where the element concentrations were determined for all samples over the same stratigraphic interval. This plot highlights the fact that the major rock types of the South Range Main Mass have patterns that are flat and enriched or depleted relative to one another. Europium exhibits a positive or negative anomaly depending on whether cumulus plagioclase is present or plagioclase has been removed from the magma. Rubidium concentrations are influenced by alteration and lower amphibolite facies metamorphism, but there may be a primary mineralogical control (possibly phlogopite) on the positive Rb anomaly in the norites and QGAB (Fig. 3.11A,B).

Implications of Main Mass petrology and geochemistry

The variations in Main Mass geochemistry are important with respect to the origin and differentiation of the magma as well as the sulfide saturation history. Several important conclusions can be drawn as discussed below:

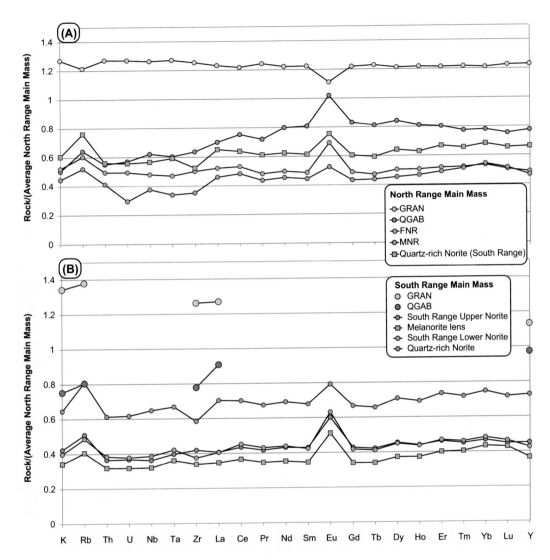

FIGURE 3.11

Concentrations of incompatible and selected compatible elements normalized to the bulk weighted average of the Main Mass in the North Range traverse (Table 3.2) based on data reported in Lightfoot et al. (2001). (A) Plots of the interval weighted average elemental abundances in North Range MNR, FNR, QGAB, and GRAN. (B) Similar plots for the South Range Quartz-rich Norite, Lower South Range Norite, the melanorite lens, Upper South Range Norite, QGAB and GRAN. Note that concentration data are not available for all samples in the QGAB and GRAN. Data source: Lightfoot and Zotov (2007).

A single parental magma composition

The weighted average incompatible elements patterns in Fig. 3.11A–B and the ratios of selected incompatible elements (Figs. 3.4H and 3.8J) are consistent with the differentiation of a single parental magma which had uniform incompatible element ratios. If there was a second magma derived from a different crustal or mantle source (Chai and Eckstrand, 1994), it would require that the second source has very similar incompatible element ratios as the rocks comprising the Main Mass of the SIC. This is unlikely given the very different incompatible element ratios of mantle and crustal reservoirs or the melts derived from these sources. Therefore, the incompatible trace element signature of the Main Mass records the trace element ratio signature of one source material.

The variations through the North Range norite stratigraphy exhibits a marked change from rocks with abundant orthopyroxene to rocks with very little orthopyroxene. In the South Range, the abundance of Orthopyroxene falls slightly above the base, and then increases, peaking in a lens of melanorite, before declining towards the base of the QGAB. Clearly the North and South Range norites evolved from a common parental magma under different physical and chemical conditions to produce the observed sequence of rocks.

Differentiation of the Main Mass

An orthopyroxene-rich unit with chrome spinel, magmatic sulfide and abundant phlogopite is developed at the base of the Main Mass where Sublayer embayments and troughs are developed in the North Range (Figs. 3.3A,B and 3.4A,B). In contrast, orthopyroxene-rich rocks are developed at the base of the Quartz-rich Norite and within the more central part of the South Range stratigraphy (the melanorite lens; Fig. 3.8C,D). The MNR of the North Range Unit is located above Sublayer which fills embayment structures. These variations attest to the ability of the SIC magma to crystallize chrome spinel and orthopyroxene despite the geochemical evidence for an overwhelming crustal origin for the melt sheet (Lightfoot et al., 1997a).

The differentiation history of the noritic stratigraphy of the North and South Range magma was clearly quite different. In the North Range a basal unit of especially melanocratic rock is developed above embayments and troughs; this discontinuous rock unit is called the MNR. It is overlain by the FNR which shows an upwards decline in MgO and Mg-number that has been attributed to fractional crystallization (Lightfoot et al., 2001). In contrast, the South Range develops a weak Mg-enrichment at the base, but shows an overall increase in Mg-number and MgO upwards through the stratigraphy before the entry of a melanorite lens (eg, Lightfoot and Zotov, 2006; Walker, 2014) about 2/3rds of the way through the stratigraphy. Above this level the norite becomes more evolved in composition.

The development of the QGAB is sudden, and it is marked by the development of ulvospinel and clinopyroxene with apatite developed at some stratigraphic levels. With increased elevation in the QGAB, the rock compositions move towards the composition of the most basal rocks of the GRAN Unit.

The compositional diversity in South Range rocks developed above the base of the QGAB is characterized in only one of the profiles that extend above the norite stratigraphy (Fig. 3.8A–J). Despite the fact that structural complexity introduced by thrust faulting along the Cliff Lake and associated faults, there is evidence to indicate a broad increase in degree of differentiation from the top to the bottom as well as indications that the magma underwent a more complex differentiation to produce two or more units within the GRAN.

The QGAB of the North Range varies in thickness; it is almost three times thicker in the area of Dowling than it is immediately along strike, and in some sections it is exceptionally thin.

The GRAN exhibits a cryptic upwards decrease in SiO_2 and incompatible trace elements coupled with an increase in MgO which is inconsistent with simple in situ fractional crystallization of a progressively evolving magma towards the roof of the melt sheet. Some of the most primitive GRAN samples are located at the top of the Main Mass, and the abundance of SiO_2 and incompatible trace elements increases downwards through the GRAN as the MgO concentration declines. These features indicate that the GRAN crystallized from the top towards the base (Lightfoot et al., 2001).

Three models have been proposed to explain the variations through the Main Mass: (1) efficient settling and accumulation of cumulus minerals to form the noritic rocks and the QGAB from a well-mixed uniform parental magma with an average composition similar to the QD Unit of the Offset Dykes (Lightfoot et al., 1997a). (2) Crystallization of the norite and granophyre units from a density stratified melt sheet (Zieg and Marsh, 2005; Keays and Lightfoot, 2004; Farrow and Lightfoot, 2002). (3) Development of a density-stratified magma produced by melting of the footwall to generate light felsic melts which float upwards in the melt sheet and dense mafic melts that tend not to localize towards the base of the melt sheet (Golightly, 1994). The geochemistry of the Offset Dykes, help to evaluate these models as they provide reasonable estimates of parental magma composition that can be used to test these alternatives.

Variations in Ni and Cu abundance through the Norite stratigraphy

The North Range and South Range traverses (Figs. 3.9A,B and 3.10B,C) through the Main Mass exhibit elevated Mg-number and Ni/Ni* (crustal model) at the base indicating a broad linkage between the development of disseminated Ni-sulfide and more primitive rock compositions. The North Range MNR is distinctly more Mg-rich than the South Range Quartz-rich Norite.

Through the norite stratigraphy there is a marked depletion of Ni and Cu from rocks with excess metal controlled by sulfide towards rocks with very low Ni and Cu relative to that expected from their MgO concentration (Figs. 3.4G and 3.8I). Both North and South Range norites of the Main Mass have Ni/Ni* < 1 and low Cu concentrations which are indicative of the removal of chalcophile elements by the equilibration of the silicate magma with sulfide melts, and density-assisted segregation of the sulfide melts.

The profiles from around the Sudbury Basin indicate that metal-depletion of the noritic rocks of the Main Mass was not restricted to domains of the melt sheet above large ore deposits. Rather, metal depletion is documented throughout the norite stratigraphy of the Main Mass irrespective of location.

THE OFFSET DYKES

The Offset Dykes of the SIC comprise radial, concentric, and discontinuous segmented bodies consisting of Units of Quartz Diorite (QD), mineralized inclusion-bearing Quartz Diorite (MIQD), and highly mineralized inclusion-bearing Quartz Diorite (HMIQD) that were emplaced into the footwall of the SIC (see Chapter 2 for details of these rock types). The literature on the petrology and geochemistry of Offset Dykes has grown since the landmark paper of Grant and Bite (1984).Comparative geochemical studies of the Sudbury Offset Dykes (Lightfoot et al., 1997b,c,d) and detailed studies of individual dykes report petrology and geochemistry for the Foy (Tuchscherer and Spray, 2002), Hess (Wood and Spray, 1998), Copper Cliff (Scott and Benn, 2002), Worthington (Lightfoot and Farrow, 2002; Hecht et al., 2008), Parkin-Whistle (Murphy and Spray, 2002; Lafrance et al., 2014), and Trillabelle Offsets (Golightly et al., 2010). New mapping studies of the footwall of the SIC have reported the discovery of previously unrecognized segments of Offset Dyke in the North Range (eg, Wood and Spray, 1998; Smith et al., 2013; Farrow et al., 2008a,b; Farrow et al., 2011).

The timing of Offset emplacement in the evolution of the Sudbury melt sheet is early, and the sulfide-free and mineralized rocks of the Offset retain a record of the early geochemical evolution of the SIC with respect to both sulfide saturation history and the extent of primary geochemical heterogeneity in the melt sheet.

The Offsets are important as they contain some of the largest known ore bodies associated with the Sudbury Structure (eg, Copper Cliff and Frood-Stobie). Quite why some of the Offset Dykes are endowed with ore deposits whereas others are weakly mineralized with no significant known economic ore deposits is an important question which can be addressed with petrology and geochemistry.

PETROLOGY OF THE QD

Detailed thin section petrology of the QD and MIQD Units is broadly consistent with a rock that comprises primary plagioclase, amphibole, and rare orthopyroxene as earlier-crystallized minerals, and later development of quartz, and quartz-K-feldspar intergrowths, and secondary amphibole. There is no evidence of cumulus enrichment in orthopyroxene and/or plagioclase, so the whole-rock composition is likely representative of liquid compositions with respect to the silicate magma.

Grant and Bite (1984) showed that pyroxene-bearing QD is rare in the Copper Cliff and Worthington Offsets, quartz-rich QD is common in the Manchester Offset, and that fresh hypersthene-bearing QD is developed in the Whistle, Foy, and Ministic Offsets in locations close to the base of the Main Mass.

The main distinction between the petrology of the QD and MIQD Units of the Offset Dykes between the North Range and South Range is in the extent of development of secondary amphibole after pyroxene (Grant and Bite, 1984) and primary amphibole. This is a function of the different grade of metamorphism in the North Range versus the South Range (e.g. James and Golightly, 2009).

Metamorphic replacement of pyroxene by amphibole creates variations in pyroxene:amphibole ratio that link to the metamorphic history of the rock; this confounds efforts to systematically look in thin section for subtle changes in mineralogy and mineral proportions along the length of concentric and radial Offsets. In one of the few studies that systematically documented variations in mineralogy along strike in an Offset Dyke, Tuchscherer and Spray (2002) provided evidence that there is a marked decline in hypersthene content in the Foy Offset away from the Main Mass over a distance of 1 km, but beyond this, the mineralogy and grain size variations are subtler and sometimes relate to the inclusion content of the MIQD with coarser-grained rocks containing fewer inclusions.

As described in Chapter 2 and Table 3.1, the Offset Dykes comprise a QD Unit that often envelopes the MIQD at the center of the Offset Dyke, or occurs as inclusions within the MIQD. The QD Unit varies in texture from the contact towards the MIQD Unit. Close to the contact with country rocks, the QD has an acicular texture of plagioclase and pyroxene (often replaced by amphibole) in a fine-grained matrix of amphibole, phlogopite, quartz, and quartz-K-feldspar intergrowths (Fig. 3.12A,B). This texture is commonly regarded as a quench, and it develops adjacent to both unbrecciated country rocks (Grant

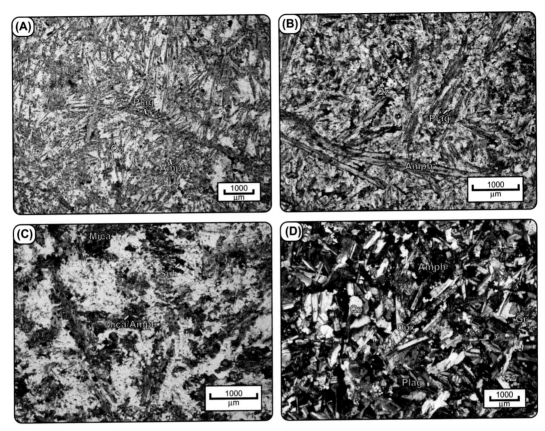

FIGURE 3.12 Petrography of the QD and Inclusions of QD

(A) Quench-textured QD in a thin section from the margin of the Manchester Offset in plane polarized light. (B) Quench-textured QD showing primary amphibole (Amph) and plagioclase (Plag) needs in a quartz, amphibole, and quartz + K-feldspar matrix froma North Range Offset Dyke in plane polarized light. (C) QD from the Copper Cliff Offset in thin section C15-0184 in plane polarized light. (D) QD from the Foy Offset shown in crossed-polarized light; sample shows fresh laths of orthopyroxene (Opx) and plagioclase (Plag) in a groundmass of amphibole, quartz, and quartz + K-feldpsar graphic intergrowths. (E) MIQD from the Worthington Offset in plane polarized light; the sample shows secondary amphibole (Amph) and primary plagioclase (Plag) with phlogopite (Mica) and magmatic sulfide (Sul). (F) MIQD from the Parkin Offset in plane polarized light with a xenocryst of pyroxene (Xen). (G) QD from the Stobie Ore Body in plane polarized light showing the development of immiscible blebs of sulfide in a matrix of plagioclase, amphibole, K-fedpsar + quartz, and phlogopite. (H) Olivine melanorite inclusion from within the MIQD of the Ministic Offset (C05-2059) in cross-polarized light; small crystals of cumulus olivine and larger clusters of olivine grains are common in this inclusion; they reside in a matrix of plagioclase, clinopyroxene, orthopyroxene, biotite, and sulfide (see Farrell, 1997 for additional information). The samples are from the Vale thin section collection and they were photographed by Ben Vandenburg.

(Continued)

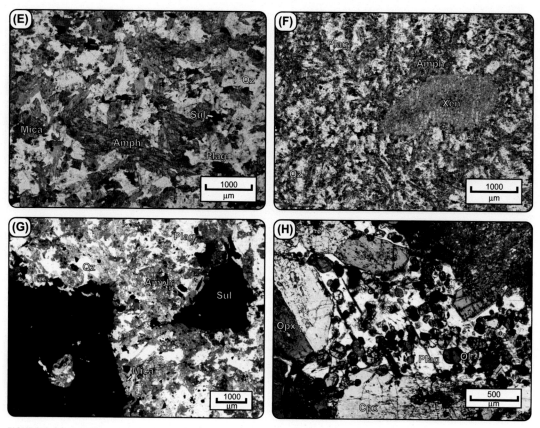

FIGURE 3.12 (*cont.*)

and Bite, 1984; Mourre, 2000) and against the matrix of SUBX (Lightfoot and Farrow, 2002). The texture of the QD changes from acicular-textured to fine-grained homogeneous QD over a distance of <1 m, but the mineralogy does not change. The fine-grained QD becomes coarse in grain size towards the center of the dyke or the contact with the MIQD. The mineralogy within an individual dyke tends to be similar, but the grain-size of plagioclase and quartz tends to increase to several millimeters. Typical examples of QD are shown in Fig. 3.12C,D.

Samples from the MIQD Unit exhibit a matrix texture similar to that found in the QD Unit (Fig. 3.12E), but with the development of inclusions (typically of mafic volcanic or intrusive rocks) that exhibit sharp contacts with the matrix (Fig. 3.12F). The MIQD Unit hosts the Frood-Stobie Deposit, and the QD envelopes immiscible blebs of sulfide and occurs as discrete segregations within the sulfide, so the textures of the rocks and ores range from immiscible sulfides hosted in MIQD through to immiscible segregations of MIQD hosted in sulfide (Fig. 3.12G; Zurbrigg, 1957). The Frood MIQD Unit contains many mafic-ultramafic inclusions. The presence of ultramafic inclusions has also been recognized in other Offset Dykes such as Parkin, Foy, and Ministic (Farrell, 1997 and Fig. 2.20H).

Fig. 3.12H shows an example of a fresh olivine melanorite from the Ministic Offset Dyke where small rounded crystals of olivine are enveloped in orthopyroxene, clinopyroxene, plagioclase, and phlogopite with interstitial magmatic sulfide minerals.

The mineralogy of the Offsets shows some variation in the degree of development of quartz and quartz + K-feldspar intergrowths. For example, the Manchester Offset Dyke is petrologically more evolved in composition than the other South Range Offset Dykes (Grant and Bite, 1984). Smith et al. (2013) have suggested that more recently discovered North Range Offset Dykes exhibit a range in degree of differentiation from pyroxene-rich to granophyric QD.

The details of the petrology of the Offset Dykes have been important in the identification of the principal hosts of the mineralization, but there is presently no strong petrological evidence that helps to distinguish the mineral potential of different Offset Dykes. Very similar rock types are developed in the QD and MIQD Units of heavily mineralized and weakly mineralized Offset Dykes. For this reason, the whole-rock geochemistry presented in this chapter and the compositions of the sulfide mineralization (see Chapter 4) provide a more instructive path in understanding the mineral potential and emplacement history of the Offset Dykes.

GEOCHEMISTRY OF THE QD AND MIQD

Representative sampling and nomenclature

This chapter uses compositional data for average samples from the QD Unit and the matrix of samples from the MIQD Unit where the larger inclusions have been removed (as described in Lightfoot et al., 1997c). There is some level of inherent uncertainty associated with sampling inclusion-bearing rocks, and it is impractical and impossible to cut rock samples to produce a matrix that is 100% devoid of fragments; in the study of Lightfoot et al. (1997c,d), the objective of the sampling was to collect a matrix sample that contained no more than 5% inclusions. The analyses from these studies are the principal data source used here.

QD and MIQD are the names applied to a group of rocks that strictly should be termed quartz monzonites (eg, Wood and Spray, 1998) based on the projection of CIPW normative compositions into the rock classification scheme proposed by Streckeisen (1967). On the plot of total alkalis versus silica (TAS; Cox et al., 1979) the Offsets fall within or near to the field of diorites (Fig. 3.13). Following Wood and Spray (1998) and Tuchscherer and Spray (2002), the term QD is retained for this rock type, and where the rocks contain mineralization and inclusions, the samples are grouped as MIQD.

Geochemical variations along strike and across the Offset Dykes

To illustrate the variations along strike and across strike in a single Offset Dyke, samples from the Worthington Offset were acquired (Fig. 3.14A; Lightfoot et al., 1997c,d), and they are grouped based on whether they represent the QD or MIQD Units of the dyke. Fig. 3.14B–D shows an example of the variations in SiO_2, MgO, and Ce/Yb versus distance from the base of the SIC (expressed as UTM Easting) for samples from the SW-trending Worthington Offset Dyke.

The data presented here and in other studies (eg, Grant and Bite, 1984; Tuchscherer and Spray, 2002), are consistent with very small changes in the composition of the QD and the matrix of the MIQD Units with distance away from the base of the SIC. In the case of the Worthington Offset Dyke, the QD and MIQD Units become slightly richer in SiO_2 and poorer in MgO with distance along the Offset, and the

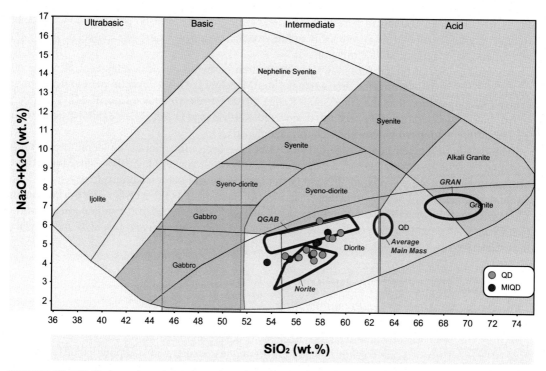

FIGURE 3.13 TAS Diagram after Cox et al. (1979) Showing the Location of the Average Composition of the QD and MIQD Units of the Offset Dykes and the Compositional Fields of North and South Range Norite, QGAB, and GRAN

Averages in Table 3.4 are based on data reported in Lightfoot et al. (1997c,d) grouped based on inclusion and sulfide content as described in the text. Note the clustering of QD and MIQD compositions close to the fields of the noritic rocks and QGAB of the Main Mass, and their position relative to the GRAN and average Main Mass.

most primitive rocks are located close to the Main Mass. There is a subtle hint that Ce/Yb increases with distance from the lower contact of the SIC.

Geochemical relationships between the Offset Dykes

The average compositions of the QD, MIQD and, in some cases the HMIQD Units of each Offset Dyke are given in Table 3.4 based on the data in Lightfoot et al. (1997c,d). Compositional data for inclusions and their possible country rock source are given in Table 3.5 after Lightfoot et al. (1997c,d).

Compositional variations between the QD and the MIQD Units of the North Range Offsets are shown normalized to average bulk integrated Main Mass of the North Range (Fig. 3.15A). This plot shows that the averages of the QD and MIQD Units from Whistle, Parkin, and Foy Offsets have almost identical patterns in terms of both relative and absolute trace element abundance levels. Relative to the average North Range Main Mass, the QD Unit of the Offset Dykes is depleted in most incompatible elements; they have similar Eu and Rb abundances, and they are enriched in Sr. It is evident from Fig. 3.15A that these three Offsets have very similar chemical composition. Relative to the average

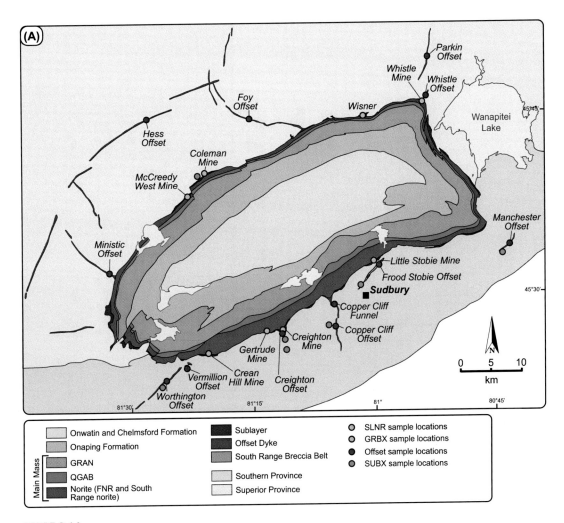

FIGURE 3.14

(A) Map of the SIC showing the location of the sample groups from the Offset Dykes, the SLNR, GRBX, and SUBX. Details of sample locations and analyses are given in Lightfoot et al. (1997c,d). (B) Variation in SiO_2 with position and rock type in the Worthington Offset (position is shown along the Offset as UTM 17N Easting); (C) Variation in MgO with position and rock type in the Worthington Offset (position is shown along the Offset as UTM 17N Easting); (D) Variation in Ce/Yb with position and rock type in the Worthington Offset (position is shown along the Offset as UTM 17N Easting).

(Continued)

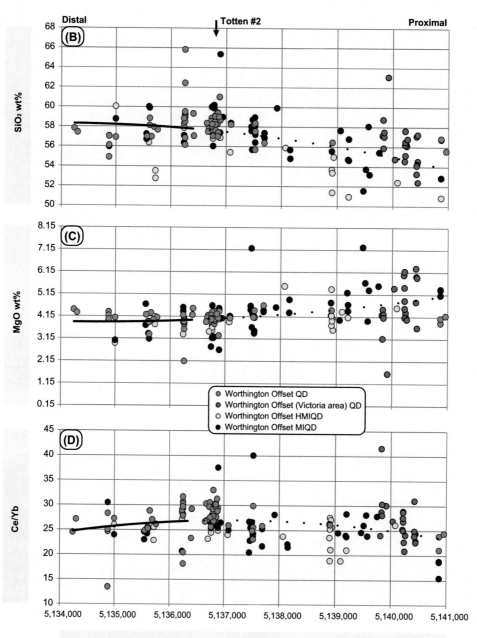

FIGURE 3.14 (*cont.*)

Table 3.4 Average Compositions of the QD and MIQD Units from the Offset Dykes

Element/oxide	Unit	Method	CCO / South Range / QD / n=13	CCO / South Range / MIQD / n=11	CCOF / South Range / QD / n=21	CCOF / South Range / QN / n=12	MO / South Range / QD / n=32	MO / South Range / MIQD / n=4	VO / South Range / QD / n=8	WO / South Range / QD / n=60	WO / South Range / MIQD / n=37	WO / South Range / HMIQD / n=25	WO / South Range / Inclusion of QD / MIQD / n=6	WO / South Range / QD / n=23	WO / South Range / Apophysis of QD / MIQD / n=2	CO / South Range / QD / n=6	FSO / South Range / MIQD / n=14	FO / North Range / QD / n=24	FO / North Range / MIQD / n=9	MNO / North Range / QD and MIQD / n=7	PO / North Range / QD / n=23	PO / North Range / MIQD / n=28
SiO_2	wt%	XRF	57.39	55.42	56.12	56.09	59.57	58.54	57.45	58.12	57.23	53.61	56.83	56.21	57.88	55.07	n.a.	58.59	57.60	58.93	58.56	57.77
Al_2O_3	wt%	XRF	15.30	14.48	15.36	15.41	14.27	14.11	14.59	15.01	14.67	13.52	14.23	15.15	15.17	14.90	n.a.	15.24	14.84	14.60	15.04	14.60
MnO	wt%	XRF	0.13	0.15	0.13	0.13	0.07	0.06	0.12	0.13	0.13	0.12	0.13	0.13	0.09	0.13	n.a.	0.11	0.11	0.12	0.11	0.12
MgO	wt%	XRF	4.06	4.37	5.00	5.08	3.89	3.86	4.28	4.05	4.44	4.06	4.31	4.78	4.01	4.42	n.a.	3.93	4.04	4.00	3.61	3.91
CaO	wt%	XRF	6.05	6.79	6.78	6.81	4.06	3.70	5.88	5.78	5.90	5.49	5.67	6.21	3.84	6.84	n.a.	5.21	5.42	4.79	4.83	4.87
Na_2O	wt%	XRF	2.55	2.68	2.78	2.79	3.68	3.80	2.40	2.71	2.75	2.51	2.70	2.54	2.39	2.89	n.a.	3.20	3.05	3.13	2.84	2.94
K_2O	wt%	XRF	2.06	1.63	1.62	1.58	2.02	1.93	1.84	1.86	1.76	1.61	2.09	1.85	3.93	1.56	n.a.	2.21	2.11	2.27	2.62	2.27
TiO_2	wt%	XRF	0.85	0.95	0.72	0.72	0.70	0.68	0.86	0.88	0.89	0.86	0.89	0.82	0.80	0.93	n.a.	0.75	0.73	0.74	0.72	0.75
P_2O_5	wt%	XRF	0.16	0.16	0.12	0.11	0.14	0.13	0.16	0.13	0.14	0.12	0.14	0.15	0.14	0.16	n.a.	0.19	0.18	0.16	0.21	0.20
Fe_2O_3	wt%	XRF	8.83	10.73	8.85	8.69	7.48	8.89	9.00	7.96	9.18	12.87	8.53	9.28	6.87	10.12	n.a.	7.80	9.37	8.20	7.41	8.60
LOI	wt%	XRF	1.32	1.03	1.04	1.06	2.98	2.59	2.03	1.19	1.12	1.06	0.75	1.76	1.55	0.60	n.a.	1.50	1.23	1.90	2.86	2.61
Total	wt%	Calculated	98.70	98.38	98.50	98.47	98.86	98.28	98.61	97.81	98.20	95.84	96.30	98.87	96.59	97.62	n.a.	98.74	98.67	98.85	98.81	98.63
FeO^n	wt%	Titration	6.13	7.57	6.27	6.17	4.68	5.15	6.50	2.78	5.39	4.69	2.83	6.56	1.55	7.37	n.a.	5.14	5.78	5.31	4.26	5.00
CO_2	wt%	Leco	0.17	0.07	0.05	0.05	1.00	1.27	0.57	0.22	0.09	0.09	0.11	0.34	0.85	0.06	n.a.	0.15	0.13	0.25	0.93	0.48
H_2O^+	wt%	Leco	0.16	0.21	0.19	0.19	0.27	0.22	0.24	0.25	0.28	0.45	0.24	1.71	n.a.	0.26	n.a.	0.20	0.23	0.23	0.31	0.27
H_2O^-	wt%	Leco	1.61	1.48	1.40	1.42	2.24	2.26	1.85	1.90	1.57	2.01	1.60	0.58	n.a.	1.24	n.a.	1.60	1.50	1.85	2.31	2.36
S	wt%	Leco	0.08	0.43	0.12	0.13	0.11	0.62	0.08	0.05	0.24	2.44	0.28	0.17	0.06	0.41	8.32	0.08	0.81	0.19	0.07	0.35
Ba	ppm	ICP-OES	519	473	512	492	448	521	654	498	530	479	595	592	604	523	458	732	676	779	803	713

(Continued)

Table 3.4 Average Compositions of the QD and MIQD Units from the Offset Dykes (cont.)

Element/ oxide	Unit	Method	South Range	South Range	South Range	South Range	South Range	South Range	South Range	South Range	South Range	South Range	South Range	South Range	South Range	South Range	South Range	North Range	North Range	North Range	North Range	North Range
Location																						
Rock group			CCO	CCO	CCOF	CCOF	MO	MO	VO	WO	WO	WO	WO	WO	WO	CO	FSO	FO	FO	MNO	PO	PO
Location and rock type			QD	MIQD	QD	QN	QD	MIQD	QD	QD	MIQD	HMIQD	Inclusion of QD	QD	Apophysis of QD	QD	MIQD	QD	MIQD	QD and MIQD	QD	MIQD
n			13	11	21	12	32	4	8	60	37	25	6	23	2	6	14	24	9	7	23	28
Ce	ppm	ICP-MS	68.31	63.15	58.72	61.09	64.45	65.01	67.59	68.41	70.16	62.16	73.50	67.23	74.01	83.02	52.70	76.36	73.30	73.48	76.17	76.20
Co	ppm	ICP-OES	30.3	52.4	36.0	34.6	24.3	45.0	32.4	30.3	39.2	121.8	41.6	37.9	25.7	43.2	388.6	26.1	67.9	32.4	25.3	38.6
Cr	ppm	XRF	179	157	243	266	162	173	184	156	189	188	122	238	162	177	140	169	196	169	144	171
Cs	ppm	ICP-MS	7.7	4.9	3.0	2.7	1.5	1.8	4.5	5.8	6.2	4.5	5.5	3.3	6.3	2.8	3.6	1.3	1.8	1.0	2.2	1.8
Cu	ppm	ICP-OES	106	639	281	246	56	1695	516	107	454	4281	437	234	29	2974	8449	66	956	297	83	854
Dy	ppm	ICP-MS	4.91	5.08	4.30	4.10	3.76	3.78	4.55	4.60	4.95	4.57	4.96	4.40	3.63	7.45	5.36	3.81	3.76	3.85	3.57	3.68
Er	ppm	ICP-MS	2.80	2.98	2.54	2.46	2.06	2.10	2.57	2.66	2.85	2.66	2.86	2.56	2.01	4.40	3.22	2.02	1.99	2.06	1.86	1.92
Eu	ppm	ICP-MS	1.49	1.48	1.39	1.31	1.32	1.39	1.45	1.45	1.47	1.29	1.51	1.42	1.55	1.67	1.28	1.52	1.45	1.49	1.48	1.52
Gd	ppm	ICP-MS	5.54	5.58	4.91	4.90	4.57	4.58	5.34	5.25	5.67	5.11	5.54	5.04	4.69	7.79	5.21	4.98	4.78	4.95	4.92	5.04
Hf	ppm	ICP-MS	3.8	4.0	2.6	2.4	3.3	3.5	4.0	3.9	4.2	3.9	4.1	3.6	3.5	2.4	3.3	4.0	4.0	3.8	4.1	3.8
Ho	ppm	ICP-MS	1.04	1.10	0.89	0.84	0.77	0.77	0.90	0.95	1.01	0.94	1.03	0.91	0.74	1.50	1.14	0.74	0.71	0.79	0.71	0.74
La	ppm	ICP-MS	33.96	31.29	28.65	29.93	31.97	32.35	33.47	33.37	34.25	30.01	35.93	33.09	36.45	40.23	25.50	37.40	36.09	36.64	38.16	37.61
Lu	ppm	ICP-MS	0.42	0.46	0.38	0.37	0.29	0.30	0.38	0.38	0.41	0.38	0.40	0.38	0.31	0.64	0.47	0.29	0.29	0.31	0.27	0.27
Mo	ppm	ICP-OES	3.6	3.6	7.3	5.7	3.9	5.5	9.0	7.5	6.5	4.6	10.0	9.6	n.a.	4.0	6.6	7.3	8.2	3.0	3.0	3.0
Nb	ppm	ICP-MS	10.4	10.3	8.9	9.4	8.5	8.4	10.3	10.3	10.8	9.6	10.4	8.6	9.9	16.8	9.0	8.1	7.9	8.8	8.4	8.4
Nd	ppm	ICP-MS	30.28	28.36	26.77	27.07	27.64	28.16	31.04	30.03	31.90	28.28	32.79	29.93	31.37	40.47	25.24	34.28	33.23	31.92	32.58	33.00
Ni	ppm	ICP-OES	89	583	228	230	70	715	135	84	396	3949	552	253	61	1381	8833	81	1023	203	77	487
Pr	ppm	ICP-MS	8.22	7.63	7.03	7.22	7.70	7.82	8.34	7.98	8.43	7.43	8.64	7.92	8.44	10.53	6.41	9.34	9.01	8.84	9.08	9.12
Rb	ppm	ICP-MS	117	69	57	63	73	79	78	76	77	67	90	70	162	73	60	80	74	84	87	76

			20.2	24.3	19.9	18.4	17.2	16.5	19.6	21.1	18.9	18.4	21.8	19.8	19.4	21.3	19.3	15.3	14.5	17.3	16.2	17.3
Sc	ppm	ICP-OES																				
Se	ppb	ICP-OES	127	1192	382	n.a.	90	2570	75	84	254	5010	n.a.	n.a.	n.a.	n.a.	15551	97	1350	360	104	562
Sm	ppm	ICP-MS	5.97	5.77	5.13	5.04	5.19	5.18	5.95	5.66	6.09	5.42	6.08	5.70	5.60	8.22	5.36	6.17	5.95	5.87	5.91	6.06
Sr	ppm	ICP-MS	281	263	345	354	196	183	235	302	293	274	313	341	245	288	182	423	369	362	417	420
Ta	ppm	ICP-MS	0.74	0.68	0.66	0.69	0.56	0.57	0.69	0.70	0.73	0.67	0.73	0.60	0.72	0.98	0.73	0.47	0.44	0.53	0.51	0.49
Tb	ppm	ICP-MS	0.86	0.88	0.74	0.73	0.69	0.69	0.81	0.80	0.86	0.79	0.85	0.77	0.67	1.22	0.84	0.70	0.68	0.72	0.69	0.71
Th	ppm	ICP-MS	8.64	8.07	6.81	7.46	8.22	8.27	9.01	8.71	9.14	8.33	8.54	8.56	9.40	9.67	7.20	8.07	8.06	8.91	8.47	8.26
Tm	ppm	ICP-MS	0.44	0.47	0.38	0.36	0.31	0.31	0.37	0.40	0.42	0.39	0.45	0.39	0.31	0.65	0.50	0.29	0.28	0.31	0.28	0.28
U	ppm	ICP-MS	2.04	1.85	1.47	1.67	2.09	2.09	2.13	2.09	2.21	2.10	2.06	1.93	3.04	2.14	2.04	1.40	1.42	1.85	1.63	1.57
V	ppm	ICP-OES	162	200	161	145	133	134	156	170	160	162	151	168	138	172	176	136	132	132	128	135
Y	ppm	ICP-MS	25.1	26.6	20.9	20.4	18.8	18.2	21.6	23.6	23.2	20.5	24.0	24.6	9.5	42.1	30.4	17.4	16.0	19.3	17.8	18.6
Yb	ppm	ICP-MS	2.80	2.99	2.49	2.49	1.94	2.00	2.47	2.52	2.74	2.56	2.65	2.51	1.93	4.24	2.97	1.88	1.85	1.97	1.75	1.78
Zn	ppm	ICP-OES	113	122	95	91	36	44	90	97	95	94	96	99	83	103	139	98	85	107	83	95
Zr	ppm	XRF	166	157	128	128	152	149	166	172	175	161	175	152	172	195	150	169	170	158	178	167

n.a., Not analyzed; CCO, Copper Cliff Offset; CCOF, Funnel of Copper Cliff Offset; MO, Manchester Offset; VO, Vermillion Offset; WO, Worthington Offset; FSO, Frrod-Stobie Offset; FO, Foy Offset; MNO, Ministic Offset; PO, Parkin Offset; QD, Quartz diorite; QN, Quartz norite; MIQD, Mineralised inclusion Quartz diorite; HMIQD, Heavily mineralised inclusion Quartz diorite.

Averages are based on data in Lightfoot et al. (1997c)

Source: Lightfoot et al., 1997c,d.

Table 3.5 Average Compositions of Inclusions from Offset Dykes and Possible Country Rocks from which they may Originate

Formation	Units	Nipissing Diabase	Nipissing Diabase	Sudbury Gabbro	Sudbury Gabbro	Sublayer	Sublayer	Sublayer	Sublayer
Rock		Gabbro (>8 wt% MgO)	Gabbro (5–<8 wt% MgO)	Amphibolite Inclusion	Amphibolite Intrusion	Ultramafic Inclusion	Ultramafic Inclusion	Granite Breccia	Granite Breccia
Location		Sudbury Region	Sudbury Region	Totten Mine	Totten Region	Gertrude Mine	McCreedy West Mine	Wisner Township	Coleman Mine
n		21	11	8	8	2	2	3	2
SiO_2	wt%	50.36	50.47	49.88	50.85	39.43	44.33	n.a.	48.86
Al_2O_3	wt%	13.29	14.66	7.72	6.80	3.41	3.28	n.a.	14.52
MnO	wt%	0.16	0.17	0.23	0.20	0.18	0.17	n.a.	0.20
MgO	wt%	9.55	7.31	14.66	16.10	34.27	25.55	n.a.	4.42
CaO	wt%	11.79	10.41	10.35	10.26	1.98	8.82	n.a.	7.75
Na_2O	wt%	1.17	1.42	0.81	0.68	0.20	0.15	n.a.	3.68
K_2O	wt%	0.69	0.68	0.78	0.21	0.17	0.37	n.a.	1.25
TiO_2	wt%	0.48	0.64	0.39	0.38	0.23	0.36	n.a.	2.35
P_2O_5	wt%	0.02	0.06	0.02	0.08	0.05	0.06	n.a.	0.50
Fe_2O_3	wt%	10.40	11.69	9.51	10.41	13.92	13.24	n.a.	15.00
LOI	wt%	1.74	2.04	1.65	2.02	5.06	3.11	n.a.	0.86
Total	wt%	97.93	97.52	n.a.	97.98	98.88	99.42	n.a.	99.36
FeO	wt%	n.a.	n.a.	2.14	8.53	7.60	7.33	n.a.	9.77
CO_2	wt%	0.26	0.39	0.24	0.17	0.44	0.31	n.a.	0.05
H_2O^+	wt%	n.a.	n.a.	n.a.	0.15	0.58	0.18	n.a.	0.28
H_2O^-	wt%	n.a.	n.a.	n.a.	3.10	5.22	3.59	n.a.	1.26
S	wt%	0.15	0.76	0.02	0.01	0.44	0.18	0.15	0.08

		36.1	62.4	1.0	4.2	30.3	76.5		2.3
Au	ppb	36.1	62.4	1.0	4.2	30.3	76.5	n.a.	2.3
Ba	ppm	143	148	123	86	90	232	761	689
Be	ppm	n.a.	n.a.	n.a.	1.50	1.50	0.44	1.33	1.69
Ce	ppm	10.46	13.61	11.12	11.01	12.66	25.16	51.45	76.69
Co	ppm	49	57	54	55	125	110	17	53
Cr	ppm	717	150	1451	2099	n.a.	3211	43	18
Cs	ppm	0.72	0.94	1.19	1.24	0.37	0.37	0.23	0.66
Cu	ppm	1309	1463	178	44	645	1207	230	91
Dy	ppm	2.09	2.64	1.81	1.71	1.13	2.11	2.67	7.11
Er	ppm	1.27	1.58	1.09	0.96	0.70	0.98	1.37	3.70
Eu	ppm	0.51	0.64	0.47	0.46	0.29	0.81	1.27	2.65
Gd	ppm	1.86	2.36	1.68	1.67	1.18	2.90	3.10	8.48
Hf	ppm	0.97	1.25	0.84	n.a.	n.a.	1.59	2.28	5.84
Ho	ppm	0.43	0.55	0.40	0.34	0.23	0.40	0.53	1.44
La	ppm	4.90	6.25	5.30	5.07	6.14	10.26	25.30	35.79
Lu	ppm	0.19	0.23	0.16	0.14	0.12	0.13	0.17	0.53
Nb	ppm	1.46	2.47	1.64	1.43	1.70	2.14	5.68	16.49
Nd	ppm	5.72	7.61	5.82	5.94	5.82	14.16	21.17	41.59
Ni	ppm	601	457	458	345	2067	637	104	46
Pb	ppm	n.a.	n.a.	4.79	n.a.	n.a.	n.a.	14.57	11.00
Pd	ppb	85.7	116.5	1.0	11.8	131.3	167.0	n.a.	2.5
Pr	ppm	1.37	1.79	1.36	1.44	1.52	3.42	6.03	10.14
Pt	ppb	51.8	68.9	n.a.	n.a.	n.a.	n.a.	n.a.	n.a.
Rb	ppm	22.4	23.4	29.8	8.6	10.5	14.4	32.5	31.0
Sc	ppm	34	29	35	36	10	27	9	24
Se	ppb	3176	2485	n.a.	46	2190	497	n.a.	136

(Continued)

Table 3.5 Average Compositions of Inclusions from Offset Dykes and Possible Country Rocks from which they may Originate (cont.)

Formation	Units	Nipissing Diabase	Nipissing Diabase	Sudbury Gabbro	Sudbury Gabbro	Sublayer	Sublayer	Sublayer	Sublayer
Rock		Gabbro (>8 wt% MgO)	Gabbro (5–<8 wt% MgO)	Amphibolite Inclusion	Amphibolite Intrusion	Ultramafic Inclusion	Ultramafic Inclusion	Granite Breccia	Granite Breccia
Location		Sudbury Region	Sudbury Region	Totten Mine	Totten Region	Gertrude Mine	McCreedy West Mine	Wisner Township	Coleman Mine
n		21	11	8	8	2	2	3	2
Sm	ppm	1.51	1.94	1.42	1.47	1.14	3.28	3.87	8.93
Sr	ppm	124	150	85	55	41	69	519	419
Ta	ppm	0.15	0.18	0.14	0.14	0.11	0.11	0.33	0.93
Tb	ppm	0.31	0.39	0.28	0.27	0.19	0.40	0.43	1.26
Th	ppm	1.28	1.61	1.23	1.36	1.51	3.77	2.63	4.80
Tm	ppm	0.18	0.23	0.17	0.14	0.11	0.13	0.20	0.56
U	ppm	0.42	0.50	0.36	0.38	0.36	0.29	0.40	1.09
V	ppm	196	191	232	237	64	110	87	220
Y	ppm	10.5	12.0	17.0	8.7	6.1	10.3	12.4	34.4
Yb	ppm	1.22	1.51	1.02	0.93	0.75	0.86	1.13	3.33
Zn	Ppm	60	60	100	83	72	137	75	142
Zr	Ppm	42	55	34	41	32	64	na	241

Analytical methods are given in Table 3.4. The composition of SLGRBX matrix is given for rocks collected from two locations.
Source: Lightfoot et al. (1997c,d). na - not analyzed.

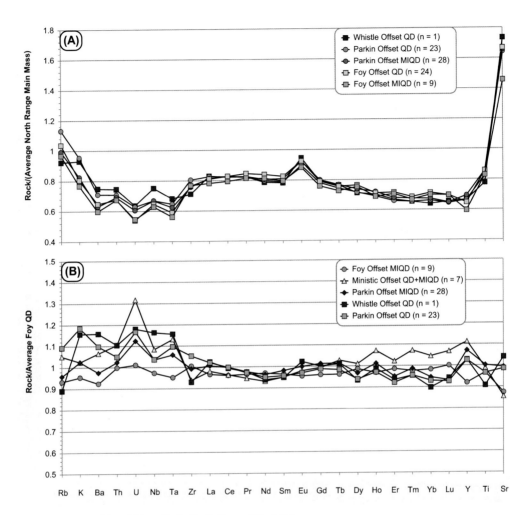

FIGURE 3.15 Compositional Diversity in North Range Offset Dykes

(A) Variations in the composition of the QD and MIQD Units based on average data for samples from the Foy, Parkin, and Whistle Offsets; in this plot, the averages are normalized to the average North Range Main Mass used in Fig. 3.11A–B. Since it is more informative to compare one Offset to another, the average composition of the QD Unit from the Foy Offset has been used as a benchmark in normalization. (B) Variations in the composition of the QD and MIQD Units based on average data for samples from the Foy, Parkin, and Whistle Offsets normalized to the average of the Foy QD Unit (Table 3.4). (C) Comparison of the average composition +/− 1 standard deviation of the QD Unit of the Parkin and Worthington Offset Dykes (normalized to the average of the Foy QD Unit). (D) Comparison of the average composition +/− 1 standard deviation of the QD and MIQD Units of the Worthington Offset (normalized to the average of the Foy QD Unit).

(Continued)

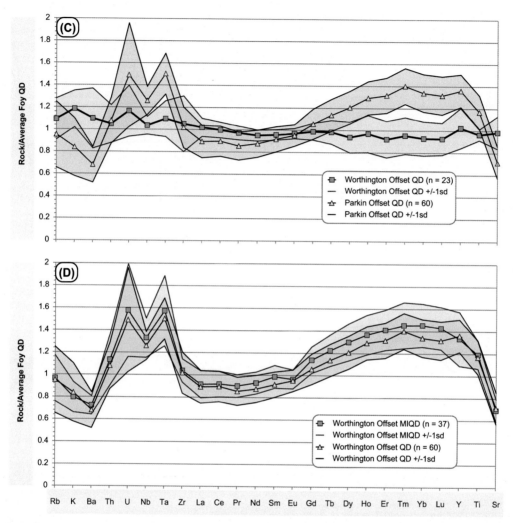

FIGURE 3.15 *(cont.)*

(E) Variations in the composition of the QD, MIQD, and basal Main Mass Quartz-rich Norite Units based on the average composition of samples from the Copper Cliff Offset and the Funnel (the average data are normalized to the composition of the average Foy QD Unit). (F) Variations in the composition of the QD, MIQD, and HMIQD Units based on the average composition of samples from the Worthington Offset and the Ethel Lake QD normalized to the composition of the average Foy QD Unit. (G) Variations in the composition of the QD and MIQD Units based on the average composition of samples from the Vermillion, Creighton, Frood-Stobie, and Manchester Offset Dykes normalized to the average composition of the Foy QD Unit. (H) A similar plot to Fig. 3.15I, but the averages are compared to the higher MgO variant of the Nipissing Diabase. Averages of rocks from the SIC and the Sudbury Region are based on data reported in Lightfoot et al. (1997c,d), Lightfoot et al. (1993a), and Lightfoot and Naldrett, (1996). (I) Comparison of the average composition of the QD Unit from the Worthington Offset, low and high-Mg variants of the Nipissing Diabase, and the Sudbury Gabbro intrusions which partially contain the Worthington Offset at Totten Mine or occur as inclusions in the semimassive sulfides.

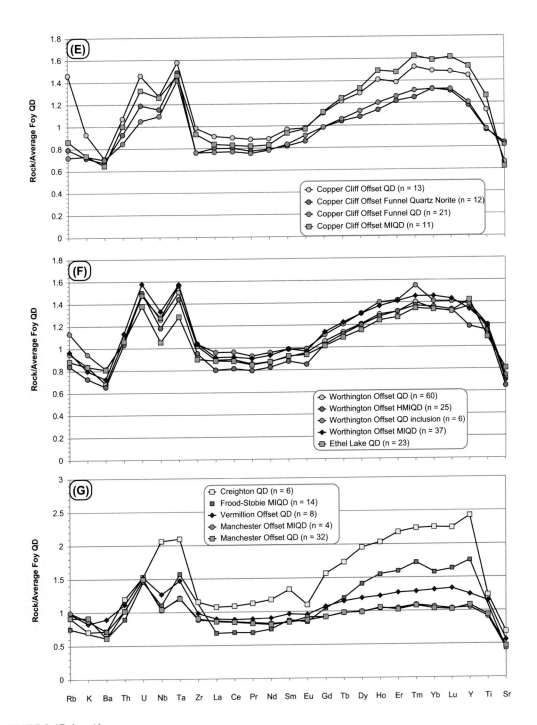

FIGURE 3.15 (*cont.*)

(*Continued*)

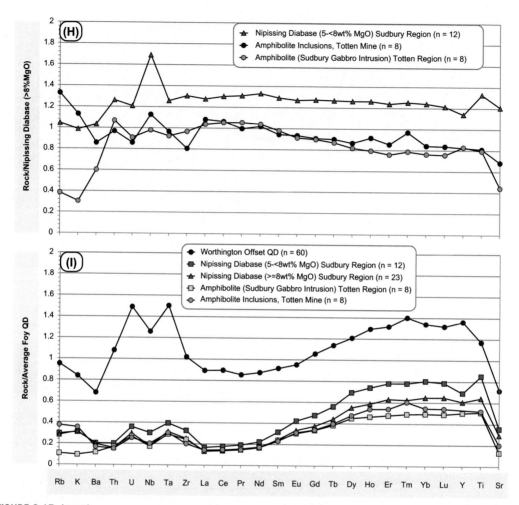

FIGURE 3.15 (*cont.*)

Main Mass of the North Range, the positive anomalies for Sr and Eu might be explained if the estimate of average Main Mass is not representative of bulk melt composition. In detail, the averages of the QD and MIQD Units from these North Range Offset Dykes exhibit elevated Rb and K, a weak negative slope in the REE, and low normalized abundances of Ba, Th, U, Nb, and Ta relative to the LREE.

To better compare the compositions of the QD and MIQD Units developed within and between Offset Dykes, the average composition of the QD Unit from the Foy Offset is chosen as a normalization factor. The average compositions of the QD and MIQD Units are compared to the average composition of the QD Unit from the Foy Offset in Fig. 3.15B. This plot shows the similarity of the patterns of the QD Unit from the Ministic, Parkin and Whistle Offset Dykes and the MIQD Unit from the Ministic,

Parkin and Foy Offset Dykes. The patterns fall close to a normalized ratio of 1, and within uncertainty, they represent rocks derived from a similar magma type without notable magmatic differentiation. These Offsets are developed in the footwall of the SIC over a lateral distance approaching 70 km strike length of the base of the SIC.

Fig. 3.15C shows the average (+/− 1 standard deviation) composition of the QD Unit from the Parkin and Worthington Offset Dykes. The plot illustrates that the samples from the two Offset Dykes have significantly different compositions. In contrast, Fig. 3.15D shows that the average composition of the QD and MIQD Units (+/− 1 standard deviation) from the Worthington Offset have almost identical patterns.

To facilitate comparison with the North Range, the average composition of the QD Unit from the Foy Offset Dyke is used to normalize the data from other Offset Dykes. Fig. 3.15E shows data for the Copper Cliff Offset; the rock groups are based on analyses of rocks from the Funnel of the Copper Cliff Offset (the Quartz-rich Norite and QD), the MIQD, and the QD Units. This plot shows a broad similarity in the shape of the pattern for each of the rock group, but the rocks from the Funnel have lower trace element abundances relative to the abundances in the QD and MIQD Units of the dyke. The lower trace element abundances may be due to the settling of cumulus plagioclase and orthopyroxene into the Funnel of the Offset Dyke. The patterns of the rocks of the Funnel and those of the dyke shown in Fig. 3.15E are similar and support a genetic linkage of the rocks to a common parental magma. Fig. 3.15E provides evidence to support the idea that the Copper Cliff Offset was derived from a different magma composition to the Foy and other North Range Offsets.

Fig. 3.15F shows normalized patterns for the QD (and inclusions of QD from within the MIQD), MIQD, and HMIQD Units of the Worthington Offset Dyke. For each rock type, the pattern resembles the rock groups from the Copper Cliff Offset Dyke. The QD and MIQD Units share similar patterns which indicate that the silicate contribution to both the mineralized and barren QD shared a common source. The similarity in the profile of the QD inclusions to the composition of the QD Unit provides strong support for the ideas that these fragments were derived from a previously solidified wall-rock consisting of QD (Lightfoot et al., 1997a).

Fig. 3.15G completes the comparison with data for the Creighton, Frood-Stobie, Vermillion, and Manchester Offsets. In this diagram the patterns for the highly mineralized Creighton and Frood-Stobie Offsets share some similarities with Copper Cliff, but the sample groups from the Vermillion and Manchester Offset Dykes show quite different slopes when compared to the other South Range Offset Dykes. In many respects, the patterns of the Vermillion and Manchester Offset Dykes are more similar to the North Range pattern.

The MIQD from the Worthington Offset contains dominantly metagabbro and amphibolite inclusions with subordinate small mafic inclusions. The geology of the Totten segment of the dyke was described in Chapter 2, where the close association between amphibolite inclusions and ore grade mineralization was described, and linked to the locus of the intersection of the Offset Dyke with an older sill of Sudbury Gabbro in the country rocks. An important question to be addressed is the source of the inclusions. Fig. 3.15I shows the compositions of average amphibolite from the inclusions within the MIQD and the patterns of country rock Sudbury Gabbro (Lightfoot et al., 1997c,d), and 2.2 Ga Nipissing Diabase (Lightfoot et al., 1993a). This plot illustrates that the inclusions from within the Offset Dyke have patterns with the same slope and abundance levels to the Sudbury Gabbro. This confirms that these inclusions were locally derived from the walls of the Dyke.

The similar shape of the patterns of the Sudbury Gabbro and the 2.2 Ga Nipissing Diabase is consistent with the Sudbury Gabbro being a more primitive cumulate rock derived from a Nipissing parental magma type. This relationship is also illustrated by normalizing the rock compositions to average primitive Nipissing Diabase (Fig. 3.15H). This plot confirms that the different mafic rocks exhibit a flat pattern, although there are unusual spikes in Rb and K that may be due to alteration.

The pattern of the average QD Unit from the Worthington Offset is also shown in Fig. 3.15H. Although the abundance levels of the incompatible elements in the average of the QD Unit are much higher than the mafic country rocks, the overall shape of the pattern has some similarity to the mafic footwall rocks, and this points to a significant contribution of this material to the Quartz Diorite.

The compositional diversity within and between Offset Dykes is illustrated in the box and whisker plot in Fig 3.16A,B. The rock groups are compared in terms of Sr concentration and Ce/Yb ratio. Within an individual Offset Dyke, the rocks comprising each group do not differ greatly in chemistry. The radial Offset Dykes of the South Range differ in their Sr concentration, but they have similar Ce/Yb to the concentric Offset Dykes (Fig. 3.16A,B). In contrast, the North Range Offset Dykes have higher Sr and Ce/Yb than the South Range Offset Dykes.

NI AND CU ABUNDANCES IN THE OFFSET DYKES

Fig. 3.16C shows the abundances of Ni and Cu normalized to that of the average North Range Main Mass in the most important rock groups that comprise the Sudbury Offset Dykes. This plot provides a way to benchmark the metal concentrations in the Offset Dykes relative to a model composition for the Main Mass.

The degree of enrichment in Ni and Cu is greater in the MIQD Unit than the QD Unit, but even the unmineralized QD averages have 3–15 times more nickel than the bulk average Main Mass. Although absolute enrichment factors can only be calculated if the bulk composition of the Main Mass is known with certainty, the plot at least provides some sense of relative enrichment of Ni and Cu in the Offset Dykes relative to a weighted average of the Main Mass composition.

The majority of samples from the QD Unit are either enriched in Ni and Cu due to their elevated sulfide content or they comprise unmineralized samples which have elevated Ni and Cu abundance levels relative to the bulk estimated composition of the Main Mass. The corollary to this observation is especially important to understanding the origin of the ore deposits. If the average composition of the QD Unit provides an index of parental magma composition, then the average Main Mass is depleted in Ni and Cu relative to the Offset Dykes, and one way to achieve this depletion is through sulfide saturation and segregation of magmatic sulfides from the Main Mass magma to form the ore deposits.

Comparison of the QD to average crust

No reliable estimates exist of the composition of average crust in the Sudbury Region at the time of the impact event. Example of averages that provide some indication of the composition of the target rocks in the Sudbury Region at 1.85 Ga include upper continental crust (Taylor and McLennan, 1985), bulk andesitic crust (Taylor and McLennan, 1985), bimodal felsic-volcanic rocks (Taylor and McLennan, 1985), and average post-Archean crust as represented by shales produced during Proterozoic weathering and depositional cycles (Taylor and McLennan, 1985). Fig. 3.17 shows the composition of various

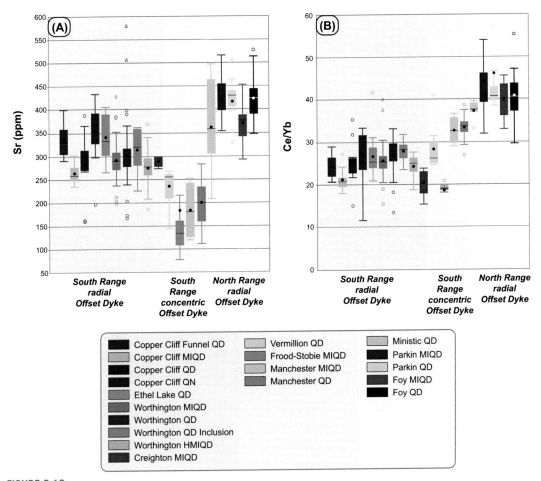

FIGURE 3.16

(A) Box and whisker plot comparing the Sr concentrations in the QD, MIQD, and HMIQD Units from within individual Offset Dykes and allowing for comparison of abundance levels in different Dykes and their geological setting in the North and South Ranges. Symbology: circle – average, line – median, central box – middle of data between Q1 and Q3, outlier circle 1.5*(Q3-Q1), triangle – far outlier further than 3*(Q3-Q1), and whiskers – extreme values that are not outliers; (B) Box and whisker plot comparing the Ce/Yb ratio of samples from the QD, MIQD, and HMIQD Units from within individual Offset Dykes and allowing for comparison of abundance levels in different dykes and their geological setting in the North and South Ranges. Symbology is explained in Fig. 3.16A. (C) Diagram showing the average concentration of Ni and Cu in the QD and MIQD Units from different Offset Dykes normalized to the bulk integrated average composition of the North Range Main Mass (Table 3.1). The diagram shows that the average compositions of mineralized samples are strongly enriched relative to the Main Mass, but all of the unmineralized QD samples are also enriched in Ni and Cu relative to the Main Mass. Data come from Lightfoot et al. (1997c,d).

(Continued)

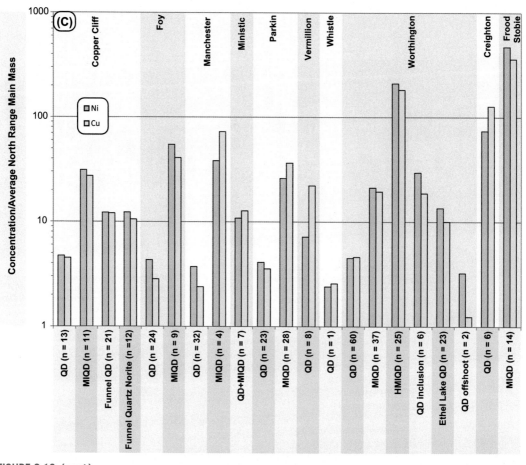

FIGURE 3.16 (*cont.*)

crustal averages and post-Archean shale normalized to the average composition of the QD Unit of the Foy Offset Dyke. The fits are all very dependent on estimates of target composition, but it is instructive that the average post-Archean shale composition has a flat profile with a ratio close to 1 which represents the best fit of the model crustal averages shown in Fig. 3.17.

To establish the exact composition of the target rocks and their relative contributions to the Sudbury melt sheet is not straightforward. Although materials can be sampled from drill core and from the surface outcrops, much of the original target is hidden beneath the deeper parts of the SIC, lost to erosion, and melted to form the SIC. The similarity of the average composition of the QD Unit to post-Archean shales encourages further efforts to establish average crustal compositions, and one approach to this discussed later in this chapter is the use of matrix compositions from SUBX which may provide more robust estimates of the bulk composition of some of the major rock formations in the target.

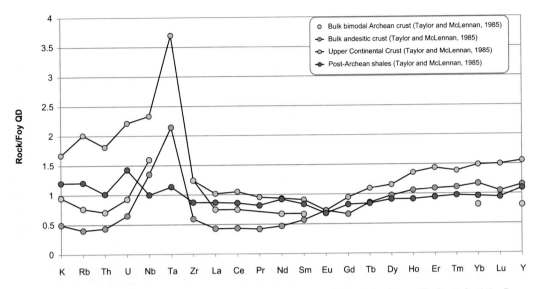

FIGURE 3.17 Composition of Average Crust Normalized to the Composition of the Quartz Diorite Unit of the Foy Offset

Published averages of Archean crust, andesitic crust, upper continental crust, and post-Archean shales (Taylor and McLennan, 1985) are shown. The pattern of post-Archean shales is quite flat and comes close to the concentrations found in average Foy QD.

SUMMARY: FEATURES OF THE OFFSET QUARTZ DIORITE

We now return to examine how the petrologic and geochemical data for the Offset Dykes help to address some fundamental questions raised in the introduction:

1. *What petrological and geochemical evidence do the Offset Dykes provide for a sequence of events in the early evolution of the melt sheet?* The geochemistry of the QD and the MIQD Units are very similar in terms of incompatible trace element abundance patterns, but they have different Mg-number, Cu, Ni, Co, and S abundance levels because the inclusion-bearing rocks contain magmatic sulfide. Although the MIQD post-dates the QD Unit, the magma giving rise to the two rock groups shows no difference in composition due to crystal differentiation or contamination. The matrix of the QD differs from that of the MIQD in sulfide content, but otherwise the abundances of major elements and the abundances and ratios of compatible and incompatible trace elements are very similar. This indicates that the two melts share a common source, and that the two rock types represent sulfide-undersaturated and sulfide-saturated examples of a common magma type. The MIQD contains inclusions of QD and is typically in contact with the QD that is often quenched against the dyke walls. These variations indicate that the MIQD was emplaced after the QD Unit was formed.

2. *How do the Offset Dykes help to constrain the composition of the parental magma?* The chemical composition of the QD Unit offers an indication of the starting composition of the melt sheet prior

to differentiation, assimilation of footwall and/or inclusions, and sulfide saturation. The silica content of the QD overlaps with that of the noritic rocks and the QGAB of the Main Mass; the GRAN and average Main Mass have much higher silica contents than the QD. The QD Unit has a reasonably constant composition within each dyke, but there are differences between North and South Range Offset Dykes which are indicative of contamination of the magma by assimilation of different country rock types.

3. *What is the origin of the mafic and ultramafic inclusions contained in the MIQD of the Offset Dykes?* The metagabbro and amphibolite inclusions from the Worthington Offset are derived from the nearby Sudbury Gabbro. Inclusions from the Ministic and Foy Offset Dykes do not have a recognized source. In all cases, the presence of mafic-ultramafic inclusions in the MIQD is related directly to the development of mineralization.

4. *At what stage in the evolution of the Offset Dykes were the ore deposits formed?* The introduction of mineralization into the Offset Dykes clearly post-dated the emplacement of the QD Unit as the MIQD contains inclusions of QD.

Four further questions can also be addressed with the data:

1. *Are there variations in composition of the Offset Dykes that relate to distance from the base of the SIC?* The petrology of the Offset Dykes is generally quite uniform with small variations in texture, degree of alteration and Quartz content. There are subtle changes in chemistry along the length of radial Offset Dykes, but these are not significantly greater than the variations across the Offset Dyke.

2. *How do the dykes vary in composition and relate to the Main Mass?* The patterns in incompatible element chemistry of North Range Offsets provide a close, but not perfect, match to the patterns of the average North Range Main Mass, but the abundance levels of the incompatible elements are lower in the average of the QD Unit than the average Main Mass. The North Range Offset Dykes all have similar ratios of the incompatible trace elements, and they are derived from a similar magma reservoir. Many of the South Range Offsets are also similar to one another in incompatible element abundance, but they are different when compared to the North Range Offset Dykes, with different degrees of relative enrichment or depletion in the incompatible elements. These variations point to an early regional heterogeneity in the incompatible trace element ratios of the primary melt.

3. *Do the Offset Dykes provide an insight to the composition of the primary magma of the melt sheet that can be used to understand the early sulfide saturation history of the melt sheet?* The information presented in Figs. 3.14 and 3.15 indicate that there was some primary heterogeneity in Offset melts at the time of their injection. All of the Offset rock types and Offsets have elevated Ni and Cu abundance levels relative to the estimate of average Main Mass, and the background abundances level of Ni and Cu in the QD Unit are indicative of a starting melt with ~70–100 ppm Ni and a similar abundance of Cu. The bulk average Main Mass is Ni- and Cu-depleted with respect to this reservoir.

4. *Is there any primary geochemical signature of sulfide-endowed versus barren Offsets or specific petrological or geochemical vectors that support exploration for ore deposits?* There are important variations in sulfide and inclusion content, and these features are visible in hand sample. There are subtle variations in chemistry along the Worthington Offset, but these variations are quite small and the linkage to mineral potential is restricted to those elements

which are present in the sulfide mineral assemblage (S, Fe, Ni, Cu, Co, and the PGE). The majority of the radial North Range dykes have different compositions when compared to the radial South Range dykes, and this variation is indicative of early heterogeneity in the melt sheet. The composition of the MIQD Unit that contains the large deposits at Frood-Stobie and Creighton are markedly different in composition when compared to other South Range Offset Dykes.

THE SUBLAYER

The petrology and geochemistry of the silicate rocks that host contact style sulfide mineralization is the subject of this section; the details of the sulfide ore deposits are provided in Chapter 4. The Sublayer consists of two types of magmatic breccia that are developed at the base of the SIC; these rocks have a noritic or a ganitic matrix, and they have been termed the Sublayer Norite (SLNR) Unit and the Sublayer Granite Breccia (GRBX) Unit. As described in Chapter 2, both rock types contain abundant inclusions of mafic and ultramafic rocks, but the GRBX typically contains abundant granite and gneiss fragments whereas the SLNR typically only contains mafic and ultramafic inclusions. Beneath the Sublayer are a complex group of metamorphic-textured country rocks which show partial melting and incipient melting (metabreccias), and their textures are very different when compared to rocks further into the footwall of the SIC.

The Whistle Mine embayment is described in detail in Chapter 2. This is a type location used to identify the petrological and chemical diversity in the noritic rock types of the Sublayer. The data for these rocks provide an unparalleled opportunity to describe the compositional variations within a single embayment, and these data can then be compared and contrasted with noritic Sublayer from other troughs and embayments in the North Range (Levack and McCreedy West Mines) and South Range (Crean Hill, Creighton, and Little Stobie Mines).

The petrology of the GRBX is briefly described, but there is sadly no systematic study of the matrix geochemistry which can be used to investigate the origin of this group of magmatic breccias. A study like this is long overdue.

The broad objectives of this section are as follows:

1. To understand the diversity in textures and geochemical variations within a single embayment structure, and the sequence of events, using Whistle Mine as a case study.
2. To compare and contrast the petrology and geochemistry of the rocks from Whistle Mine with the rocks in other trough structures where there are economic ore deposits.
3. To examine the petrological relationships between inclusions and matrix, and to understand the geochemical significance of these observations.
4. To understand some of the principal controls on the composition of the noritic matrix of the Sublayer.

PETROLOGY OF THE SUBLAYER NORITE AND THE INCLUSIONS

The SLNR matrix typically comprises cumulus orthopyroxene and plagioclase with interstitial augite, phlogopite, sulfide, and less common granophyric intergrowths of K-feldpsar and quartz. The orthopyroxene has a cumulus texture and comprises 30%–50% of the rock, and plagioclase laths comprise 30%–40% of the rock; these rocks range in composition from norites (which can be poikilitic to ophitic in texture; Fig. 3.18A,B) through to leuconorite (which is typically poikilitic in texture and is sometimes referred to as inclusion-bearing basic norite; Fig. 3.18C,D). The SLNR commonly contains inclusions, and it is not always straightforward to understand the relationship between the matrix and inclusions when there is extensive hybridization.

Diabase inclusions in the SLNR at Whistle Mine show micro- and macroscopic textures indicative of assimilation into the noritic matrix. The matrix of the SLNR often shows extreme textural

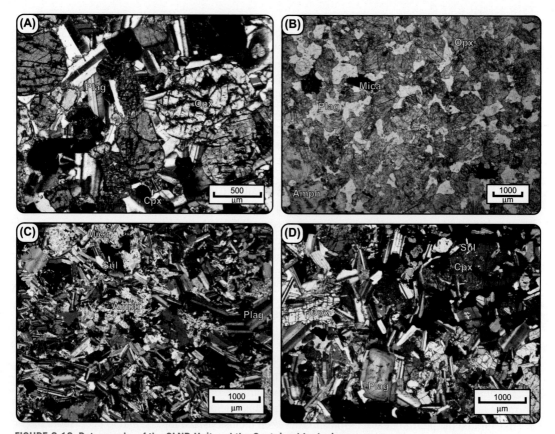

FIGURE 3.18 Petrography of the SLNR Unit and the Contained Inclusions

(A) Sample of melanorite from the Sublayer, Whistle Mine; crossed polarized light. Cumulus orthopyroxene (Opx) and plagioclase (Plag) with lesser amounts of intercumulus clinopyroxene (Cpx). (B) Sample of melanorite from the Sublayer at Creighton Mine; sample 16982 shown in crossed polarized light to illustrate the development of magmatic phlogopite. (C) Sample of norite from the Sublayer shown between crossed polars in sample 16412-2 from Whistle Mine. The fine-grained texture, abundant heavily digested inclusions, and poikilophitic texture is common in more leucocratic norite from the Sublayer. (D) Norite from the Sublayer at the boundary with the GRBX Unit at Whistle Mine; plane polarized light photograph of sample C70-2298 showing plagioclase (Plag) and clinopyroxene (Cpx) in interstitial sulfide (Sul).

heterogeneity on the scale of the thin section. These textures likely arise from the incorporation, meta-morphism, and melting of diabase inclusions in a norite matrix (Fig. 3.18E).

A common Sublayer rock type at Whistle Mine exhibits a hybridized texture consisting of a matrix of poikilitic-textured norite with inclusions and segregations of ophitic-textured norite and melanorite (Fig. 3.18F). This rock type has been termed hybrid-xenolithic norite.

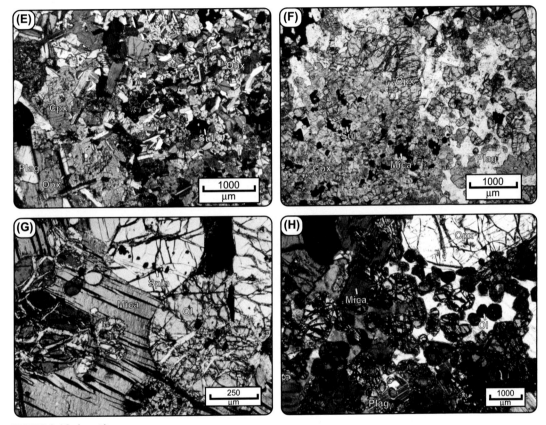

FIGURE 3.18 *(cont.)*

(E) Coarse-grained norite in contact with a with fine-grained diabase inclusion from the Sublayer at Whistle Mine; cross polarized light. The contacts between the inclusion and the matrix are often hard to identify, and very often the diabase is hybridized into the Sublayer with small residual crystals of plagioclase and pyroxene contained in the matrix of the Sublayer. (F) Norite containing a melanorite inclusion from the Sublayer at Whistle Mine; photograph is in plane polarized light. Phlogopite (Mica) and sulfide (Sul) occur in both the inclusion and the matrix. (G) Olivine phlogopite melanorite inclusion from Whistle Mine in cross-polarized light; a similar rock sample yielded an U-Pb zircon age of 1.85 Ga (Corfu and Lightfoot, 1997). Such rock types are common in the Sublayer at Sudbury, but there are no clear examples in the literature of similar rock types in terrestrial mafic-ultramafic cumulate rocks. (H) Olivine melanorite inclusion from the Sublayer at Whistle Mine between crossed polars; note the very small olivine crystals enclosed in orthopyroxene and plagioclase.

(Continued)

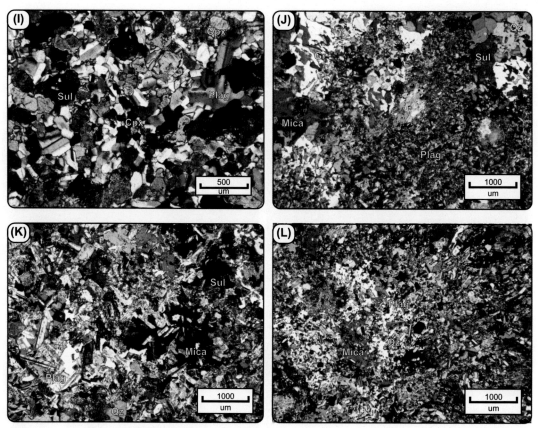

FIGURE 3.18 (*cont.*)

(I) Inclusion of diabase from Whistle Mine shown in plane polarized light to illustrate the very strong recrystallization; a similar rock was dated by Corfu and Lightfoot (1997) and yielded a U-Pb zircon age of 1.85 Ga on zircon. (J) Norite transitional into GRBX from the Sublayer at Whistle Mine (sample C75-2028 shown in crossed polarized light). (K) Norite gradational into footwall metabreccia in cross-polarised light; showing intercumulus phlogopite with sulfide and laths of plagioclase; this rock grades locally into a metamorphic-textured mafic protolith. From the Sublayer in southern Cascaden Township. (L) Quartz and phlogopite hosting small crystals of plagioclase developed in the transitional zone between norite of the Sublayer and metabreccia (often termed inclusion-basic norite); cross-polarized light; sample is from southern Cascaden Topwnship. The samples are from the Vale thin section collection and they were photographed by Ben Vandenburg.

Another group of inclusions that is common in the SLNR at Whistle Mine are olivine melanorite and altered olivine melanorite inclusions. These inclusions tend to have baked and recrystallized contacts, but the interior of the inclusions can be fresh (Fig. 3.18G,H) or weakly altered. These inclusions exhibit remarkable textures where the fresh samples contain cumulus chrome spinel enclosed within fresh rounded grains of olivine and orthopyroxene. The rounded olivine grains often reside in orthopyroxene. The minerals developed between these olivine and orthopyroxene crystals are plagioclase, augite, phlogopite and sulfide. The phlogopite commonly contains small elongate needles of apatite. These inclusions have remarkably high phlogopite content, sometimes approaching 15%, and up to 1% apatite (eg, Lightfoot et al., 1997c; Warner et al., 1998). Inclusions of olivine melanorite are common in other Sublayer troughs/embayments and Offset Dykes (Farrell, 1997; Lightfoot et al., 1997c).

The olivine melanorite inclusions are one common group of inclusions described in Rae (1975) and Scribbins (1978) and the subject of ongoing research to help establish whether these inclusions record evidence of impact shock textures (Wang et al., 2016), and if so, whether their source is related to shocking and melting of preexisting mafic-ultramafic rock formations in the basement or interaction of mantle-derived magma with the melt sheet. An example of this type of inclusion is shown in the photograph that appears at the front of this chapter.

Another common inclusion type at Sudbury is diabase. These inclusions commonly show evidence of recrystallization by thermal metamorphism (Fig. 3.18I). In many cases they contain plagioclase with a glomeroporphyritic texture, and this has led to the suggestion that they were derived from Early Proterozoic (Matachewan?) basement dykes.

PETROLOGY OF THE GRANITE BRECCIA

The GRBX is restricted in distribution to the North Range contact Sublayer and occurs either as a Unit beneath the norite within embayments or troughs, or as units which cuts through the SLNR and MNR (eg, the Victor Deposit). Variants of GRBX are developed in the South Range as granite and mafic clast-leaden rocks cross-cut the Quartz-rich Norite within a few tens of meters of the base of the SIC or as local units developed beneath the Sublayer where the country rocks are dominantly granitic. Most examples occur in the North and East Ranges, but there is one good example of this rock type east of Crean Hill Mine in the South Range (Chapter 2).

The petrology and mineralogy of the GRBX have been described and discussed in Fairbairn and Robson (1942); Greenman (1970); Hebil (1978); Pattison (1979); Coats and Snajdr (1984); Deutsch et al. (1989); Lakomy, 1988; Farrow and Watkinson (1996), and McCormick et al. (2002a,b). The GRBX consists of a matrix-supported breccia which contains fragments of mafic and felsic material varying in size up to tens of meters in length. Some of these fragments are derived from the local footwall (including SUBX); others belong to a similar population to that found in the adjacent SLNR, and yet another group has petrological and geochemical similarities to the MNR (indicating that the GRBX was mobile after crystallization of the Main Mass; Hebil, 1978; Lightfoot et al., 1997c).

The matrix of the GRBX comprises principally of plagioclase with lesser quartz, K-feldspar, epidote, chlorite, pyroxene, phlogopite, and amphibole. Sulfide minerals can reach ore grade concentrations, and comprise principally pyrrhotite, pentlandite and chalcopyrite. The texture of the matrix varies from an igneous texture adjacent to the norite (Fig. 3.18J), and it changes progressively through to a textural variety with oikocrystic quartz enclosing tabular plagioclase to a polygonal metamorphic-textured quartz and feldspar in what was likely originally an igneous-textured protolith that in-fills the breccia matrix and forms leucosomes in the underlying granite-gneiss (Figs. 3.18K–L). The variations in petrology described in McCormick et al. (2002a,b) are consistent with an early origin for the GRBX as part of the Sublayer; this is also supported by the presence of mafic-ultramafic inclusions similar to those found in the SLNR. The crystallization history was likely protracted; parts of the GRBX more distal from the melt sheet show metamorphic textures whereas those close to the norite exhibit an igneous texture.

ESTABLISHING THE MATRIX COMPOSITION OF INCLUSION-BEARING NORITIC ROCKS

The challenge in establishing the composition of the SLNR matrix is more complex than that encountered in the MIQD of the Offset Dykes. As described in the previous section, the Sublayer matrix contains not only clear examples of mafic inclusions that can be easily removed by serial cuts of rock slabs, but the matrix also exhibits the development of leucosomes, partially melted inclusions that have textures which exhibit gradational relationships with the matrix, and a hybrid textural relationship between patches of ophitic to subophitic norite within poikilitic norite. There is no simple solution to the identification of exact matrix composition as it is almost impossible to unpick the detailed geological relationships and prepare samples that reflect the end members producing this variability. The compositional data for the matrix that are used here comes from Lightfoot et al. (1997c,d) where efforts were taken to remove >90% of the identifiable inclusions in hand sample, but this study accepted that the matrix exhibits textural variations which are not controlled by the sample preparation strategy.

LITHOGEOCHEMISTRY OF THE MATRIX AND INCLUSIONS IN THE SLNR

Table 3.6 reports average compositional data for the matrix of SLNR from the Whistle, Levack, Crean Hill, Creighton, and Little Stobie Embayments using data reported in Lightfoot et al. (1997c). Table 3.6 also provides average compositions of mafic-ultramafic inclusions in the Sublayer from the Whistle and Levack Mines.

Sublayer matrix

Box and whisker plots for MgO concentration (normalized to 100% free of volatile elements) and Ce/Yb ratio (an index of the slope of the REE pattern) for the matrix of SLNR and some of the inclusion types from Whistle and other embayment and trough structures are shown in Fig. 3.19A,B. The extent of variation in samples of SLNR matrix within embayments is quite narrow relative to the variability between embayments.

   The average compositions of SLNR matricies from the Whistle and Creighton embayments normalized to the average composition of the Foy QD Unit are compared in Fig. 3.19C. The plot shows the standard deviation of multiple rock sample analyses from each location (+/− 1 sigma). The composition of the matrix from Whistle is clearly different when compared to the matrix from Creighton for the spectrum of trace element data shown in this spidergram. These variations indicate that the noritic matrix of the Sublayer evolved with a different composition in the Whistle and Creighton embayments, and this may be due to the composition of the footwall and the composition of the dominant inclusion types.

   The average compositions of the volumetrically dominant poikilitic-textured matrix of the SLNR, and the volumetrically minor leuconorite (Lightfoot et al., 1997c) which is locally developed near to the Offset Dyke which extends ENE from the embayment within heavily brecciated country rocks (Lafrance et al., 2014) are compared in Fig. 3.19D. The profile of the norite matrix exhibits a steep

Table 3.6 Average Composition of Sublayer Norite from Trough and Embayment Structures from Around the SIC

Element/ Oxide	Unit	Method	Whistle Mine	Whistle Mine	Whistle Mine	Whistle Mine	Whistle Mine	Whistle Mine	McCreedy West Mine	Crean Hill Mine	Fraser Mine	Little Stobie Mine	Creighton Mine	Creighton Mine
Location and Rock Type			Sublayer Norite	Melanorite in Sublayer Norite	Olivine Melanorite in Sublayer Norite	Leucocratic Sublayer Norite	Diabase Inclusion in Sublayer Norite	Altered ultramafic rock in Sublayer Norite	McCreedy West Sublayer Norite	Sublayer Norite	Fraser Sublayer Norite	Sublayer Norite	Sublayer Norite	Sublayer melanorite
n			53	78	22	12	13	15	32	9	13	7	35	19
SiO_2	wt%	XRF	49.83	48.67	45.53	57.38	49.03	43.41	51.79	49.74	55.03	49.73	51.66	48.39
Al_2O_3	wt%	XRF	13.20	10.32	7.11	14.60	13.97	6.30	12.99	15.11	11.03	15.84	15.05	7.54
MnO	wt%	XRF	0.18	0.18	0.20	0.13	0.21	0.16	0.17	0.16	0.16	0.16	0.15	0.21
MgO	wt%	XRF	7.88	10.62	18.48	4.45	6.33	23.06	8.43	5.93	9.89	5.93	6.64	12.90
CaO	wt%	XRF	8.41	7.31	5.90	5.38	9.89	5.21	7.37	7.49	4.88	7.42	7.25	5.62
Na_2O	wt%	XRF	2.36	1.66	0.55	3.21	2.39	0.09	2.40	2.04	2.01	2.09	2.60	1.17
K_2O	wt%	XRF	1.01	1.26	0.87	2.07	0.71	0.54	1.07	1.01	1.30	0.70	1.36	0.85
TiO_2	wt%	XRF	0.83	0.75	0.56	0.82	1.06	0.43	0.71	0.90	0.60	0.62	0.74	0.57
P_2O_5	wt%	XRF	0.18	0.24	0.16	0.24	0.13	0.13	0.17	0.18	0.12	0.20	0.15	0.06
Fe_2O_3	wt%	XRF	13.78	15.01	15.81	8.66	14.37	13.60	12.50	14.18	11.32	14.37	11.20	17.54
LOI	wt%	XRF	1.26	2.18	3.22	1.92	0.86	5.46	1.04	1.41	1.92	1.39	1.64	0.61
Total	wt%	Calc	99.01	98.21	98.39	98.84	98.94	98.37	98.69	98.15	98.27	98.45	98.42	95.46
FeO^a	wt%	Titration	8.31	8.99	9.18	5.44	8.80	5.65	7.80	8.84	7.71	8.16	7.55	12.43
CO_2	wt%	Leco	0.11	0.10	0.17	0.27	0.09	0.28	0.14	0.30	0.05	0.19	0.08	0.11
H_2O^+	wt%	Leco	0.30	0.30	0.31	0.24	0.18	0.35	0.18	0.11	0.40	0.13	0.24	0.28
H_2O^-	wt%	Leco	1.31	2.03	3.45	1.96	1.41	5.88	1.16	1.78	1.60	1.31	1.99	1.07
S	wt%	Leco	1.04	1.42	0.85	0.28	0.24	0.40	0.87	1.62	0.71	0.98	0.86	2.70
Ba	ppm	ICP-OES	479	455	362	794	333	240	475	454	418	383	440	299
Ce	ppm	ICP-MS	39.73	50.03	42.38	74.16	25.40	36.47	41.24	49.88	41.69	47.49	49.57	37.78

(Continued)

Table 3.6 Average Composition of Sublayer Norite from Trough and Embayment Structures from Around the SIC (cont.)

Element/Oxide Location and Rock Type	Unit	Method	Whistle Mine Sublayer Norite	Whistle Mine Melanorite in Sublayer Norite	Whistle Mine Olivine Melanorite in Sublayer Norite	Whistle Mine Leucocratic Sublayer Norite	Whistle Mine Diabase Inclusion in Sublayer Norite	Whistle Mine Altered ultramafic rock in Sublayer Norite	McCreedy West Mine McCreedy West Sublayer Norite	Crean Hill Mine Sublayer Norite	Fraser Mine Fraser Sublayer Norite	Little Stobie Mine Sublayer Norite	Creighton Mine Sublayer Norite	Creighton Mine Sublayer melanorite
n			53	78	22	12	13	15	32	9	13	7	35	19
Co	ppm	ICP-OES	87.7	110.7	119.8	31.9	56.8	105.7	79.8	98.5	68.9	118.4	80.9	151.1
Cr	ppm	XRF	460	803	1171	182	160	713	657	357	1352	360	465	888
Cs	ppm	ICP-MS	0.75	1.22	1.05	0.96	0.67	0.80	0.76	2.40	0.99	1.86	3.46	1.31
Cu	ppm	ICP-OES	694	814	666	1166	250	144	584	4658	419	6446	1367	12305
Dy	ppm	ICP-MS	3.33	3.32	2.52	3.71	3.53	1.99	2.98	4.21	2.41	4.05	3.58	3.41
Er	ppm	ICP-MS	1.80	1.72	1.21	1.96	2.05	1.00	1.61	2.41	1.42	2.43	2.06	2.08
Eu	ppm	ICP-MS	1.31	1.30	1.05	1.63	1.20	0.71	1.24	1.37	0.99	1.36	1.25	0.75
Gd	ppm	ICP-MS	4.07	4.50	3.62	5.25	3.55	2.85	3.76	4.66	2.95	4.32	4.14	3.54
Hf	ppm	ICP-MS	2.02	2.30	1.89	3.37	1.35	1.49	1.99	2.05	2.56	1.13	1.44	1.27
Ho	ppm	ICP-MS	0.67	0.64	0.47	0.73	0.76	0.37	0.59	0.89	0.52	0.87	0.71	0.70
La	ppm	ICP-MS	17.88	21.94	18.72	35.22	11.47	15.44	19.13	23.58	20.17	22.69	24.29	18.02
Lu	ppm	ICP-MS	0.26	0.24	0.16	0.27	0.31	0.14	0.24	0.37	0.23	0.41	0.30	0.34
Mo	ppm	ICP-OES	3.35	3.53	3.45	3.17	3.46	3.73	3.47	3.00	3.00	3.00	7.52	3.95
Nb	ppm	ICP-MS	3.62	4.67	4.30	7.76	3.21	3.37	3.83	8.76	5.18	5.33	7.39	6.68
Nd	ppm	ICP-OES	21.31	26.59	21.83	34.06	14.18	19.20	21.06	23.58	18.05	21.98	23.75	17.86
Ni	ppm	ICP-OES	836	1130	1090	169	226	809	772	1951	622	1899	1559	4446
Pr	ppm	ICP-MS	5.27	6.55	5.47	9.09	3.36	4.81	5.31	6.18	4.91	5.74	6.20	4.75
Rb	ppm	ICP-MS	26	42	33	58	19	26	29	46	39	27	58	39
Sc	ppm	ICP-OES	30	26	19	18	35	17	26	26	21	22	22	29
Se	ppb	ICP-OES	1635	2362	1843	327	539	n.a.	1304	639	870	695	n.a.	n.a.

	unit	method												
Sm	ppm	ICP-MS	4.53	5.38	4.43	6.36	3.36	3.67	4.27	4.88	3.45	4.39	4.54	3.61
Sr	ppm	ICP-MS	406	296	170	449	342	52	422	293	284	334	367	131
Ta	ppm	ICP-MS	0.20	0.25	0.26	0.43	0.20	0.18	0.21	0.54	0.33	0.37	0.45	0.46
Tb	ppm	ICP-MS	0.60	0.61	0.49	0.73	0.59	0.37	0.54	0.72	0.44	0.68	0.62	0.55
Th	ppm	ICP-MS	1.94	2.59	2.49	6.33	1.13	1.98	2.20	4.13	4.09	4.36	4.82	4.55
Tm	ppm	ICP-MS	0.26	0.25	0.17	0.28	0.31	0.14	0.23	0.37	0.22	0.40	0.29	0.31
U	ppm	ICP-MS	0.38	0.52	0.52	1.18	0.23	0.44	0.43	1.11	0.87	1.10	1.03	1.08
V	ppm	ICP-OES	227	175	123	144	303	103	197	185	152	144	171	177
Y	ppm	ICP-MS	16.05	14.91	11.55	17.49	18.00	9.23	14.30	21.97	13.45	22.28	17.24	18.59
Yb	ppm	ICP-MS	1.69	1.57	1.07	1.75	1.96	0.91	1.52	2.39	1.40	2.63	1.97	2.15
Zn	ppm	ICP-OES	121	129	147	110	129	122	121	142	117	155	107	162
Zr	ppm	XRF	87	104	89	152	68	66	89	126	100	88	102	86

n.a. - not analysed

Averages are based on data. Average compositions of inclusions from the Sublayer at Whistle and Creighton Mines.

Source: Lightfoot et al. (1997c,d).

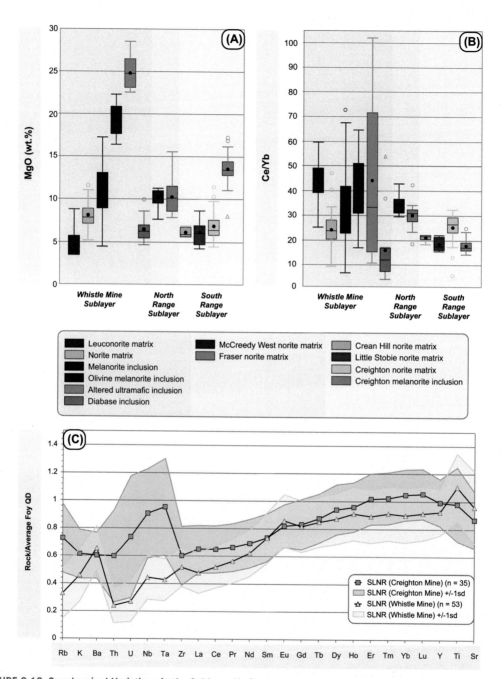

FIGURE 3.19 Geochemical Variations in the Sublayer Norite

(A) Comparsion of MgO concentrations (wt%) in rocks from the Sublayer at Whistle Mine (leucocratic norite matrix, norite matrix, melanorite inclusions, olivine melanorite inclusions, and altered ultramafic inclusions) relative to norite matrix from the McCreedy West, Fraser, Crean Hill, Creighton and Little Stobie embayments. Symbology is explained in Fig. 3.16A. (B) Similar to A, but the vertical axis is Ce/Yb ratio which is an index of the slope of the REE pattern.

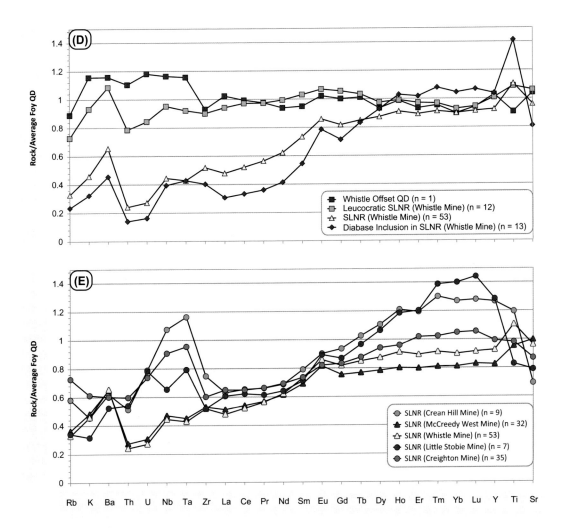

FIGURE 3.19 (*cont.*)

(C) Comparison of spidergrams for average SLNR from Whistle and Creighton where the standard deviation is shown for each group of samples. (D) Compositional variations relative to the average of the QD Unit from the Foy Offset for samples of SLNR matrix, leucocratic SLNR matrix, diabase inclusion, and QD from the Whistle embayment. (E) Compositional variations in samples of SLNR matrix from South Range Mines at Crean Hill, Creighton, and Little Stobie and North Range Mines at McCreedy West and Whistle; the average analysis of matrix are normalized to the composition of the QD Unit from the Foy Offset Dyke. (F) Detailed geochemical relationships between average analyses of SLNR matrix, and inclusions of melanorite, olivine melanorite and altered melanorite from the Whistle embayment. The average of the analyses is normalized to the average composition of the QD Unit from the Foy Offset Dyke. (G) Normalized abundances of trace elements in ultramafic inclusions, malnorite inclusions, and SLNR from the Creighton mineral system; the analysis are normalized to the average composition of the QD Unit of the Foy Offset Dyke.

After Lightfoot et al. (1997c,d).

(Continued)

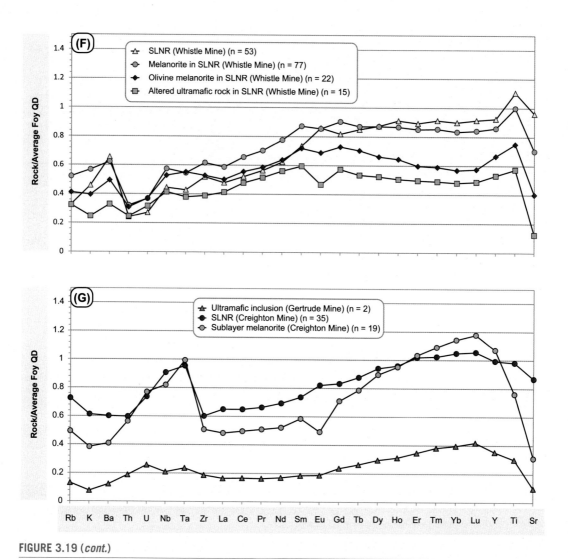

FIGURE 3.19 (*cont.*)

slope with strong depletion in the LREE over the HREE and low abundance levels of many of the LILE and HFSE. In contrast, the leuconorite matrix plots close to the flat profile of QD from the Whistle Offset Dyke with slight depletion in Th, U, Nb, and Ta, and weak enrichment in K and Ba relative to the QD; these are features of the norite (Fig. 3.19D).

The leuconorite is commonly developed towards the base of embayments and troughs, and has been regarded as a transitional rock type which develops between the SLNR and GRBX, and between SLNR and the QD of the Offset Dyke. The data from the Whistle Embayment indicate that this rock was likely formed from the Offset magma prior to the development of the SLNR. The SLNR was then formed by the assimilation of more refractory mafic inclusions.

The composition of average diabase inclusions from within the SLNR in the Whistle embayment plot close to the composition of the SLNR, but the pattern of the diabase is slightly steeper and has a very positive TiO_2 anomaly compared to the weaker anomaly of the norite matrix. The petrography of the norite matrix is indicative of substantial amounts of digestion of a diabase protolith to form the complex textures shown in Fig. 3.18D. Lightfoot et al. (1997b) suggested that the composition of the SLNR at the Whistle Mine contained an overwhelming chemical contribution from the diabase; this suggestion was later substantiated by Sm-Nd isope data reported by Prevec et al. (1997a-c).

Fig. 3.19E shows average spidergram patterns of norite matrix from the Crean Hill, McCreedy West, Little Stobie, Creighton, and Whistle Embayments. In this diagram, the patterns of norites from Whistle and McCreedy West are the most similar, but they are not identical. The patterns of SLNR matricies from the Creighton, Little Stobie, and Crean Hill Embayments are different when compared to each other and when compared to those of the North Range. The similarity in the North Range Sublayer matrix composition coupled with the common occurrence of diabase as an inclusion type is consistent with the matrix composition being largely controlled by the assimilation of diabase into the parental melt in both the Whistle and McCreedy West Embayments. In detail, each of the South Range embayments exhibits differences in slope and relative enrichment of the trace elements, so it is possible that the matrix compositions evolved in response to different inclusion compositions. Although there are large differences in the petrology and composition of the footwall, it is not known whether different embayment and trough structures are characterized by inclusions that have different compositions.

This information is very useful in helping to understand the extent of interaction between the Sublayer matrix and the mafic inclusion population. However, it is less clear whether these relationships distinguish environments where small deposits are developed (eg, Whistle) versus those hosting very large ore deposits (eg, Creighton). There is no strong evidence to suggest that the compositional variation in the matrix links to the scale of the ore deposit developed in the embayment or trough.

Mafic and ultramafic inclusions

The composition of inclusions from the Whistle Mine SLNR show a large range in MgO concentration between the groups of samples, but Ce/Yb ratio are remarkably variable within each group of samples despite the different MgO concentrations. This indicates that the inclusion population shows greater compositional diversity in incompatible trace elements than the matrix (Fig. 3.19A,B).

The compositional variations in mafic-ultramafic inclusions in the Whistle SLNR are shown in Fig. 3.19F. The pattern of the average melanorite inclusion type hosted within the SLNR is different when compared to the composition of the matrix and these two rock types differ in MgO content by ~3% (Table 3.5). Both the olivine melanorites and altered ultramafic inclusions have different patterns when compared to the SLNR, and much higher MgO concentrations (Table 3.5). These plots indicate that the poikilitic-textured norite matrix is not simply a more evolved magma type that produced earlier cumulates represented by the ophitic-textured melanorite, olivine melanorite, and altered ultramafic inclusions.

The geochemistry of average olivine melanorite and altered olivine melanorite (typically from within the inclusion-rich semimassive sulfide) from the Whistle embayment is shown in Fig. 3.19F. These rocks have slightly depleted levels of LILE and LREE relative to the melanorite inclusions, but they are strongly depleted in HREE relative to the melanorite. Fig. 3.19B shows that these inclusions have a

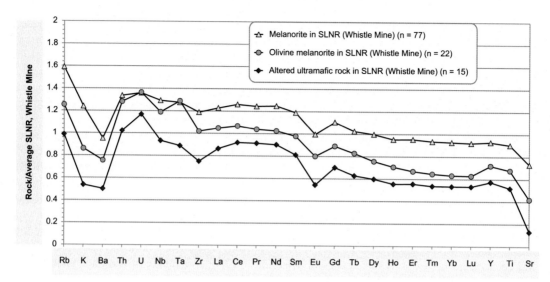

FIGURE 3.20 Composition of Melanorite, Olivine Melanorite, and Altered Ultramafic Inclusions from within Semi Massive Sulfide at Whistle

Analyses are normalized to the composition of SLNR matrix from the Whistle embayment.

After Lightfoot et al. (1997c,d).

wide range in slope of the REE patterns, so they have had a more complex crystallization history than the matrix, and are not simple cumulate products of a magma corresponding to the norite matrix.

The compositions of melanorite and ultramafic inclusions from the main Creighton trough and the Gertrude Trough which is part of the Creighton systems are shown in Fig. 3.19G. This plot shows the average composition of SLNR from Creighton. In the same way that the inclusions at Whistle have no obvious genetic link to the matrix, the same can be said of the inclusions at Creighton and the Gertrude segment of the Creighton Deposit which have patterns that are different when compared to the norite matrix.

Fig. 3.20 shows the compositions of the melanorite, olivine melanorite, and altered ultramafic rock inclusions normalized to average SLNR from the Whistle Mine. The three inclusion types have broadly similar patterns with the abundance levels of the trace elements being the highest in the most evolved inclusion types. The difference when compared to the norite matrix indicates that these inclusions are unlikely to be cognate xenoliths derived from a noritic magma undergoing crystal differentiation. Despite the range in composition within the inclusion groups, the observation that the average patterns are similar in shape lends some support to the idea that they were derived from a similar source.

GRANITE BRECCIA

Two examples of GRBX matrix compositions are shown in Fig. 3.21 where the compositions are normalized to that of the average of the QD Unit of the Foy Offset Dyke. Both of these samples were selected to represent rocks with igneous textures. The two Wisner has a flat pattern which is strongly

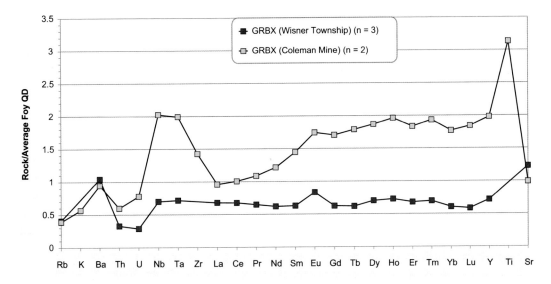

FIGURE 3.21 Average Composition of GRBX from two North Range Locations

The analyses serve to illustrate that two different trough/embayment structures have different matrix compositions for the GRBX. No systematic study of GRBX matrix from around the North and East Range has yet been completed.

depleted relative to the average composition of the QD Unit of the Foy Offset Dyke. In contrast, the GRBX from the Coleman Mine is strongly depleted in the LILE and LREE with enrichment in HFSE and Ti. Whether GRBX are derived by melting of the target crust or whether they have contributions from the melt sheet is not yet constrained by a robust database for this rock type.

GENESIS OF THE SUBLAYER

Returning now to examine the questions posed in the light of the data for the Sublayer matrix and inclusion populations:

How does the noritic matrix vary in composition within an embayment and what sequence of events produced the Sublayer? The data from the Whistle Mine are consistent with a relatively homogenous noritic matrix composition. The matrix has both a petrological and geochemical fingerprint of assimilation of diabase into the parental magma and illustrates the incorporation of a series of mafic-ultramafic inclusion types from a different source. Leucocratic SLNR is compositionally similar to the QD of the Whistle Offset and likely formed early in the development of the embayment. The relationship between the norite and GRBX of the Sublayer remains incompletely understood. The GRBX has both igneous and metamorphic textures, so it is possible that this rock type was formed initially by partial melting of the target, but this melt did not homogenize with the noritic Sublayer. Rather it cooled from the base upwards, and was then recrystallized to produce the observed metamorphic textures. The GRBX often cross-cuts the SLNR and the MNR, and inclusions of MNR as well as mafic-ultramafic inclusions like those in the Sublayer are developed in the GRBX. This indicates that the crystallization of the GRBX occurred over a protracted period of time, and solidification did not occur until after the MNR Unit had formed at the base of the Main Mass.

How does the petrology and geochemistry of the rocks from the Whistle embayment compare to that of rocks from other trough structures where there are larger ore deposits? Norites from the North and South Range embayments are different in chemistry, and each embayment is associated with a Sublayer matrix that is unique in composition depending on local controls such as the composition of the inclusions and/or the footwall rocks. There are no clear fingerprints in the silicate matrix that provide information or indicators of the scale or grade of the sulfide ore deposits associated with the Sublayer. The formation of the ore deposits happened in response to the process of segregation and accumulation of sulfide melts into pockets at the base of the SIC which have undergone a complex history of assimilation of inclusions and/or country rocks.

What are the petrological relationships between inclusions and matrix of the Sublayer? Mafic-ultramafic inclusions from the Whistle and Creighton embayments are indicative of a different source for these inclusions relative to the Sublayer and the melt sheet. The textures of the rocks point to a primary igneous origin, but the textures and chemical compositions of these inclusions are very unusual for terrestrial mafic-ultramafic rocks, and there is a paucity of basement rocks which are sufficiently primitive in composition to represent the sources of the inclusions. Further work is in progress to establish the source of these inclusions (Wang et al., 2016). They may originate by a number of mechanisms, such as: (1) Shock metamorphism of preexisting ultramafic rocks in the basement; (2) Melting and reformation of material from the asteroid; (3) Formation of cognate xenoliths derived by the early crystallization of a part of the melt sheet that was especially mafic in composition and perhaps generated by assimilation of mafic-ultramafic rocks from the basement; (4) Derivation from mafic-ultramafic inclusions derived from the basement, but partially or wholly melted in the Sublayer, but incompletely mixed into the magma; (5) Derivation from a pulse of mantle-derived magma that was introduced in response to melting triggered by the impact event.

What are the principal controls on the diversity of the SLNR matrix composition? A detailed investigation of the petrology and geochemistry of the diabase inclusion population from the Whistle Mine embayment supports the idea that these rocks have a major control on the composition of the matrix of the Sublayer.

THE SUDBURY BRECCIA
PETROLOGY OF SUDBURY BRECCIA

The matrix of SUBX comprises a groundmass of altered fine-grained aphanitic and comminuted country rock minerals. The textures also record a progressive recrystallization due to thermally metamorphosed to local anatexis characterized by the development of igneous-textured flood quartz containing small laths of plagioclase feldspar (eg, Dressler et al., 1992; Roussel et al., 2003; Morrison et al., 1994).

Examples of the two end members are illustrated in Fig. 3.22A,B. The thin section shown in Fig. 3.22A represents an example of the fine-grained aphanitic matrix which has a recrystallized vitric texture. The example in Fig. 3.22B shows the development of larger quartz grains which contain laths of plagioclase feldspar which may have been generated by in situ anatexis of the finer-grained rock shown in Fig. 3.22A. The matrix of SUBX varies between these two examples as shown, and this variability is largely related to distance from the base of the SIC or from the Offset Dykes.

FIGURE 3.22 Petrology of SUBX Samples from the Levack Area Illustrated by a Distal Sample of SUBX which has Undergone No Melting and a Sample Proximal to the SIC that is Partially Melted

(A) An example of SUBX matrix with characteristic comminution of groundmass where the individual minerals in the matrix are dominantly quartz and feldspar (C72-3202). This sample shows no sign of partial melt textures. (B) Example of SUBX showing the development of pervasive igneous plagioclase crystals and quartz in the matrix; this rock has a texture indicative of partial melting (C73-3219). The samples are from the Vale thin section collection and they were photographed by Ben Vandenburg.

The nature of pseudotachylite, and how the impact process generated the matrix has been the subject of considerable debate. In one group of models, the matrix is considered to be a comminuted clastic material derived from the adjacent country rocks (eg, Reimold et al., 2015). A second group of models proposes that the rock flour is generated by quenching of frictional melt (Maddock, 1983) produced by rapid frictional slip along faults to produce a frictional type of pseudotachylite (Spray, 1997) or during the passage of the shock wave generated by the impact event to produce shock-dominated pseudotachylite (Spray, 1998). A third group of models invoke geochemical contributions of impact melt into the breccia matrix (Riller et al., 2010).

GEOCHEMISTRY OF THE MATRIX OF SUBX

The petrology and geochemistry of the matrix of SUBX is best understood by comparing and contrasting the composition of best estimates of matrix composition with that of the principal host rock, and by comparison of the matrix of SUBX located in different types of footwall beneath the SIC. This approach was taken by O'Callaghan et al. (2015, 2016), and it examines whether the breccia matrix is a product of comminution and shock-melting of the target rocks by high-pressure impact and associated super-faults (Spray, 1997) versus injection of material derived from the impact process in the form of a melt. The data also provide an approach in establishing whether the matrix of the SUBX can be generated from local target rocks as suggested by Lafrance et al. (2014), or whether there might be a contribution of magma from the impact melt as proposed by Riller et al. (2010) for the Vredefort impact structure. At Vredefort, the pseudotachylite breccias (eg, Gibson and Reimold, 2001) have a remarkable resemblance to the SUBX but the depth of erosion of the impact structure exceeds that of Sudbury by 5–10 km.

The average composition of SUBX samples from different locations around the SIC together with the dominant host rock are given in Appendix 3.3. The whole rock samples are based on the average composition of local country rocks adjacent to the breccia zones whereas the breccia matrix is sampled to either minimize visible fragments or to allow for a visual estimate of the fragment content in samples where the matrix and fragments cannot easily be separated as described in O'Callaghan et al., 2016).

SUBX developed in the footwall of the Creighton Embament is largely hosted in granite. Although there are inclusions of diabase, gabbro and metasedimentary material in the SUBX and some development of aplites and granodiorites in the Creighton Pluton, the dominant composition of the immediate host is given in Appendix 3.3. The minor and trace element geochemistry of the breccia matrix and the host granite are compared in Fig. 3.23A using the average composition of the QD Unit of the Foy Offset Dyke as a normalization value. The average composition of the matrix and granite and the standard deviation (+/− 1 sigma) of multiple analyses are consistent with a close similarity in the composition of the matrix and the host (Fig. 3.23A). Although small amounts of other country rock types could be added to generate the matrix of the SUBX, the vast majority of the breccia matrix originates from the Creighton granitoid pluton, and there is no evidence for contributions of melt such as those represented by the QD Unit (O'Callaghan et al., 2015, 2016).

Fig. 3.23B shows the composition of a mixed rock comprising breccia matrix and 50–75 volume percent fragments of granitoid rock (location shown in Fig. 3.14A); the composition of this rock falls between that of pseudotachylite matrix and the Creighton granite. This is consistent with the greater proportion of granitic fragments in the matrix. The composition of gabbroic inclusions from the SUBX is very different from that of the granite and the matrix, so the effect of this component on the chemistry of the matrix is considered to be quite small compared to the dominant contribution from the granitoid rocks.

Fig. 3.23C repeats the analysis using the compositions of granitoid rocks (the host) and SUBX matrix from the Murray Pluton in the South Range (location shown in Fig. 3.14A). This comparison shows that the matrix composition is not significantly different when compared to that of the granite which contains the pseudotachylite. Importantly, the breccia matrix composition developed in the Murray granite (Fig. 2.29C) is not significantly different when compared to the matrix of SUBX from the Creighton granite (Fig. 2.29B). The Murray and Creighton granites are geologically and texturally quite similar as described in Chapter 2, so it is no surprise that the breccia matrix developed from this protolith is also similar.

The composition of the SUBX matrix and the Copper Cliff Rhyolite host (location shown in Fig. 3.14A) are shown in Fig. 3.23D. The correspondence between the lithophile element composition of the matrix and the host granite is very close. The fit for Ni and Cu is poor, but this likely relates to the remobilization of these metals from the proximal Copper Cliff Offset (ore bodies are ∼100 m away from the sample site) into the SUBX matrix.

The composition of SUBX matrix and the host rock quartzite from the Totten segment of the Worthington Offset (near to the Totten #1 Deposit; Fig. 3.14A) is shown in Fig. 3.23E. The patterns for the host and breccia are quite similar, and there is no evidence to suggest that they formed from different sources.

Fig. 3.23F shows an analysis for the breccias and country rocks near to the Manchester Offset (Fig. 3.14A). In this case, the breccia matrix is strongly enriched relative to the country rocks, so there is presumably a missing component that has not been incorporated into this particular model. The negative slope of the REE portion of the plot relative to the country rocks is not readily reconciled with a contribution of impact melt, but melanocratic igneous rocks like those at Creighton help to explain these variations (O'Callaghan et al., 2016). Metagabbroic rocks are developed along the trajectory of the Manchester breccia belt.

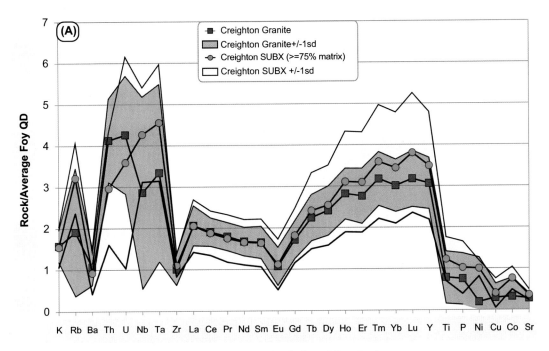

FIGURE 3.23 Compositional Diversity in Sudbury Breccia Matrix and Host Rocks

(A) Composition (+/−1 standard deviation) of SUBX matrix and host granitoid rocks from the footwall of the Creighton embayment. The compositions are normalized to the average composition of the QD Unit of the Foy Offset Dyke. The SUBX matrix and host rocks have patterns that are indicative of a large contribution of Creighton granite in the formation of the pseudotachylite matrix. (B) Comparison of the composition of matrix of SUBX from the Creighton footwall with different matrix:inclusion properties; also shown is the composition of metagabbro inclusions. (C) Comparison of the matrix of the SUBX to the composition of host granite in the Murray granitoid pluton. (D) Comparison of the matrix of the SUBX to the composition of host rhyolite in the Copper Cliff Formation. (E) Comparison of the matrix of the SUBX to the composition of host quartzite in the McKim Formation at Totten #1 Deposit. (F) Comparison of the matrix of the SUBX proximal to the Manchester Offset to the composition of host greywacke. (G) Comparison of the matrix of the SUBX to the composition of host Levack Gneiss in the Levack Gneiss Complex at Coleman Mine. Data for this study were provided by Vale and underpin the work of O'Callaghan et al. (2015). H. Variations in Ni/Ni* (the variation in MgO versus Ni depicts the power law relationship for country rocks surrounding the SIC) in the matrix of Sudbury Breccia from different locations around the SIC developed proximal to ore deposits (Creighton and Coleman) and more distal from known ore deposits; see also O'Callaghan et al. (2015, 2016). Symbology is explained in Fig. 3.16A.

(Continued)

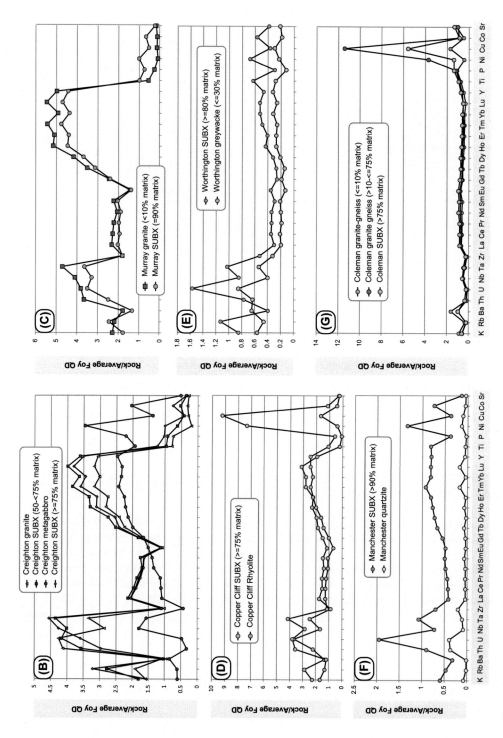

FIGURE 3.23 (cont.)

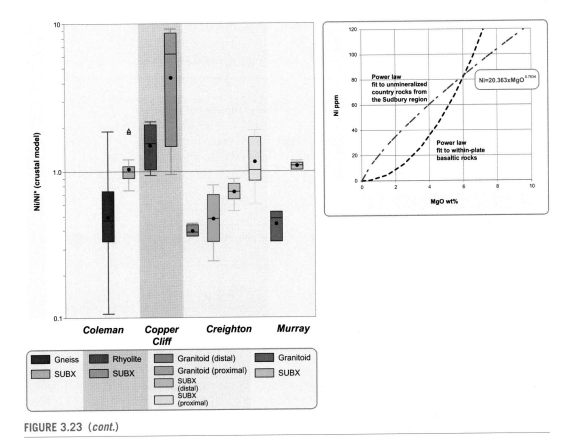

FIGURE 3.23 (*cont.*)

The compositions of SUBX matrix, mixed breccia and fragments of gneiss, and the country rock Levack Gneiss which contains it in the footwall of Coleman Mine are shown in Fig. 3.23G. The patterns of all three components are very similar and this encourages the development of models where the dominant chemical contribution to the pseudotachylite comes from the country rock.

One final analysis of breccia versus host rock composition is shown in Fig. 3.23H based on the concentration of Ni relative to expected concentrations in the target rocks (Ni/Ni*). The index relates the composition of the rock to the expected composition based on a model for the variation in MgO versus Ni in crustal rocks from the Sudbury Region. The diagram serves to illustrate the point that the breccia matrix shows a slight enrichment in Ni relative to the country rocks. In the case of the Copper Cliff samples, the proximity to the mineralized Copper Cliff Offset likely explains the strong increase in Ni in the SUBX matrix relative to the rhyolite host rock. The explanation for this may rest in hydrothermal remobilization of metals from the mineral zones into the nearby country rocks accompanying lower amphibolite metamorphism. The relationships at Coleman and Creighton support this idea, with samples from near to the ore deposits showing stronger enrichment than those developed further away.

SIGNIFICANCE OF THE PETROLOGY AND GEOCHEMISTRY OF THE SUDBURY BRECCIA MATRIX

The weight of evidence favors the in situ generation of the matrix of the SUBX by comminution and frictional melting of local country rocks. In all of these comparisons, there is no requirement for a contribution from the melt sheet (Lafrance et al., 2008; Reimold et al., 2015; O'Callaghan et al., 2015, 2016).

An investigation was carried out to establish whether the chemical composition of the matrix of SUBX exhibits any evidence for anomalous enrichment in base or precious metals which may hint at the proximity of the breccia to an ore deposit. There is evidence to indicate local enrichment in Ni/Ni* (and Cu/Y; not shown) in the breccia matrix developed proximal to mineral deposits. The elevated Ni/Ni* and Cu/Y can be detected several tens of meters away from mineral zones in the Offset and Footwall environments.

The average composition of the matrix of SUBX is useful in establishing an approximate bulk composition of the Upper Crust in the Sudbury Region, and may offer a better solution to estimating the composition of the average crust in the Sudbury Region relative to the use of global crustal averages.

METAMORPHIC AND PARTIAL MELT TEXTURES IN THE FOOTWALL OF THE SIC

As discussed in Chapter 2, many of the rocks in the footwall of the SIC exhibit recrystallization and partial melting related to processes that post-date the formation of SUBX, but predate the crystallization of the SIC. The variations in petrology of these rocks (Figs. 3.24A–F) is a manifestation of the thermal effects of the SIC, with most intensive recrystallization and partial melting developed proximal to the base of the SIC, and the more distal rocks exhibiting lower grade metamorphism and less melting (illustrated in Figs. 3.24A–E).

The diversity in grain size and composition on the hand sample scale makes whole rock geochemical approaches very difficult to apply, and future studies of these rocks will need to unpick the details on the scale of individual minerals (eg, Jørgensen et al., 2013). This group of rocks exhibits geological relationships which are consistent with both melting and hydrothermal processes as described in the

FIGURE 3.24 Petrology of the Mafic Metabreccias in the Footwall of the SIC

Variants of this rock type include fine-grained amphibole porphyroblastic mafic country rocks with local melting and breccia textures, termed Sudburite. The metabreccia is a distinctive basic fragment-rich unit commonly developed beneath the SLNR, and often developed where the country rocks have a mafic composition. (A) Example of development of flood Quartz in metabreccia from the footwall of the SIC in MacLennan Township (bore hole 600001, 1666 m, sample C75-0410). (B) Metabreccia comprising mafic fragments in a Quartz-feldspar matrix that shows incipient partial-melt texture. This sample (74-2734) comes from Capre Lake, Capreol Township. (C) Metabreccia showing development of partial melt veins and segregation in mafic fragments in the footwall of the Victor Deposit in MacLennan Township (sample C74-2709). (D) Gabbroic partial melt segregations containing residual hornfelsed diabase fragments from Wisner Township (74-3663). (E) Sudburite developed from Elsie Mountain metavolcanic rocks in the footwall of the SIC at Murray Mine (C15-0151). (F) Sudburite developed from Elsie Mountain metavolcanic rocks in the footwall of the SIC at Murray Mine (sample C15-0154). The samples are from the Vale thin section collection and they were photographed by Ben Vandenburg.

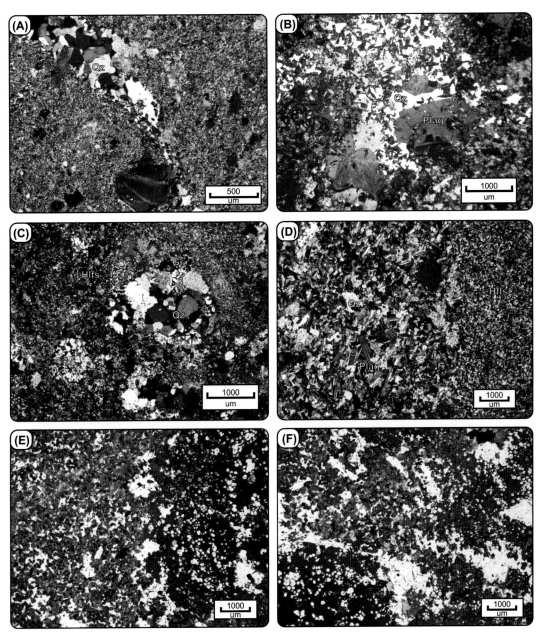

◀ FIGURE 3.24

classic work of Fairbairn and Robson (1942). The extent to which the processes resulting from melting versus those triggered by volatiles can be distinguished has been remained difficult to formulate and understand as the rocks have a very complex heritage linked to magmatic, metamorphic and deformation events in the Archean and the Proterozoic.

An interesting study by Péntek et al. (2013) claims that partial melt textures and the development of associated metabreccia provide a pathway for the transport of metal-laden fluids into the footwall. The response of the footwall to these processes will be an intrinsic function of their compositions as well as structural pathways inherited from the pre-SIC crust, accompanying crater modification, or post-dating the SIC. The context is set by examining the diversity in textures in the immediate country rocks developed around the SIC.

In the South Range footwall, which consists principally of metavolcanic and metasedimentary rocks, a rock type termed "Sudburite" (Thomson, 1935c) is developed in the footwall of the SIC where the country rocks are mafic volcanic rocks or fine-grained mafic intrusive rocks consisting of plagioclase and pyroxene with a hornfels texture (Figs. 3.24E–F shows example of Sudburite in thin section). Proximal to the base of the SIC, the mafic volcanic rocks are heavily recrystallized with the development of melt segregations containing coarse-grained amphibole set in a weakly mineralized quartz-feldspar matrix (Figs. 3.24A–B). These segregations locally develop into metabreccias with small mafic fragments entrained within either a finer-grained recrystallized country rock or enclosed within melt segregations. On a scale of $<\sim300$ m from the base of the SIC, the mafic volcanic rocks exhibit the development of strong recrystallization and development of amphibole porphyroblasts inside the finer-grained recrystallized volcanic host rock. These textures extend into the footwall of the Murray and Little Stobie embayments. The origin of Sudburite is now believed to be associated with thermal metamorphism of the footwall of the SIC by heat supplied from the melt sheet.

In the North Range, another group of metabreccias is developed in the footwall that consists principally of Levack Gneiss and granitoid igneous rocks. These breccias were derived by thermal metamorphism and melting of the country rocks in the footwall of the SIC. An important group of metabreccias originate by partial melting of the Levack Gneiss Complex in the North Range to produce both highly recrystallized and partially melted mafic and felsic rocks (Figs. 1.8 and 3.24C,D). Their description and classification into diatexites and metatexites was proposed by Farrow (1995).

The most recent petrographic and mineral chemical study of the metabreccias from the footwall of Whistle Mine is given in Lafrance et al. (2014) where these rocks are developed immediately below the keel of the Whistle embayment, and where the conditions allowed the development of a small Cu-PGE-rich ore body (the Podolsky Deposit) (eg, Farrow et al., 2008a,b, 2011). Detailed petrological, geochemical, and mineralogical studies of these rocks by Lafrance et al. (2014) highlighted the importance of early development of metabreccias related to thermal melting at the base of the embayment structure and in association with the QD of the Whistle Offset Dyke. The pathways between contact and footwall ore bodies often show extensive recrystallization and partial melting which represent members of the metabreccia group, and these melt segregations often contain disseminations and stringers of sulfide minerals. These rocks may have an important part to play in the localization of particularly high grade Ni-Cu-PGE sulfide into the footwall in deposits like Victor, Capre, McCreedy East, and Podolsky.

The response of the country rocks to thermal effects of the melt sheet are perhaps most aggressively developed in the South Range above the Creighton and Murray Intrusions. Granitoid rocks in the South Range were able to produce aplitic melts and melt breccias which cross-cut into the base of the SIC, or accumulate within the immediate footwall rocks. In some cases, these rocks have been mapped as

"Sublayer" (eg, Dressler, 1984a,b), but their field relationships indicate that they originate by partial melting of the footwall and transportation of the fragment-laden melts into cross-cutting veins in the basal Quartz-rich Norite (Fig. 2.19M,N).

MELT SEGREGATIONS, DYKES, AND VITRIC FRAGMENTS IN THE ONAPING FORMATION

The stratigraphy of the Onaping Formation contains a number of examples of rocks produced by volcanic and intrusive magmatic processes (Ames et al., 2002). Summarized in Chapter 2 and described in detail in Ames et al. (2002), these rocks comprise melt bodies at the base of the Onaping Formation that may represent intrusions of melt derived from the SIC (Fig. 3.25A,B; Peredery, 1972a,b, 1991; Anders et al., 2013, 2015), aphanitic dykes which cross-cut the stratigraphy of the Dowling and Sandcherry Members of the Onaping Formation, and vitric bombs and glass shards from the Onaping Formation.

GEOCHEMISTRY OF MELT BODIES AND GLASS SHARDS

The compositional averages of these rocks together with the average composition of the host stratigraphy of different units of the Dowling and Sandcherry Member are shown in Table 3.7 based on Ames et al. (2002).

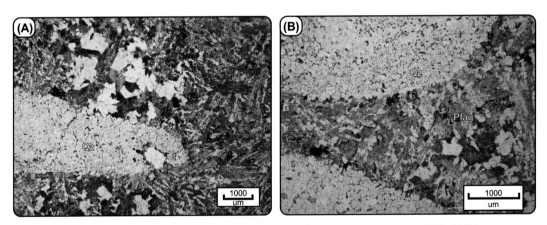

FIGURE 3.25 Petrology of Melt Bodies in the Basal Onaping melt bodies (Anders et al., 2013, 2015)

The melt bodies are irregular in shape and contain abundant fragments derived from the country rocks beneath the SIC (Archean and Proerozoic) which comprise up to 70% of the rock and often have shock metamorphic features. The melt is fine- to coarse-grained and sometimes amygdaloidal. Sample C15-0181 comes from a body in the basal Onaping Formation west of Onaping Falls in the North Range of the SIC; the location is described in Peredery (1991). (A) The matrix comprises intergrowths of quartz and K-feldspar with laths of plagioclase (Plag); lithic clasts of quartzite (Qz) have partly reacted with the matrix (plane polarized light; sample C15-0181). (B) Igneous-textured matrix penetrating between clasts of quartzite (plane polarized light; sample C15-0181). A detailed discussion of the basal Onaping melt bodies is given in Anders et al., 2013, 2015). The sample is from the Vale thin section collection and was photographed by Ben Vandenburg.

Table 3.7 Average Composition of Rocks that Represent Melt Contributions to the Onaping Formation

Unit/Rock Type	Units	Equant Shard	Fluidal Fragment	Conductive	Contact	Lower	Middle	Upper	Least Altered Victric Onaping	Vitric Blocks and Bombs	Aphanitic Dykes	Basal Intrusion
Member		Sand-cherry	Sand-cherry	Sand-cherry	Dowling	Dowling	Dowling	Dowling	Felsic Igneous rocks	Felsic Igneous rocks	Felsic Igneous rocks	Melt Body
n		12	10	4	21	7	23	8	7	6	12	5
SiO_2	wt%	64.97	64.03	65.75	62.36	61.83	60.98	62.8	61.56	62.87	62.66	64.62
TiO_2	wt%	0.51	0.54	0.49	0.56	0.51	0.49	0.45	0.61	0.57	0.61	0.57
Al_2O_3	wt%	12.15	12.75	12.03	12.21	11.81	10.87	9.64	13.59	13.48	13.58	13.56
Fe_2O_3 total	wt%	6.67	6.94	6.68	8.46	8.89	9.78	11.94	8.4	7.27	7.46	5.86
MnO	wt%	0.12	0.13	0.14	0.19	0.3	0.39	0.53	0.17	0.13	0.19	0.09
MgO	wt%	4.55	3.87	3.91	4.7	4.59	4.38	4.69	4.28	3.94	3.94	3.61
CaO	wt%	3.61	3.68	3.67	3.82	4.15	4.41	3.05	2.63	2.57	2.26	3.72
Na_2O	wt%	4.76	4.6	4.65	4.4	3.56	2.23	1.15	4.19	3.8	3.69	3.78
K_2O	wt%	1.16	2.02	1.7	1.43	2.08	1.68	1.64	2.66	3.53	3.23	2.34
P_2O_5	wt%	0.13	0.14	0.11	0.15	0.14	0.14	0.16	0.15	0.14	0.15	0.13
LOI	wt%	1.78	1.65	1.38	2.03	2.03	2.78	2.9	2.31	2.08	2.38	2.06
CO_2	wt%	0.08	0.19	0.06	0.17	0.22	1.71	1.06	0.09	0.1	0.39	0.07
C	wt%	0.15	0.1	0.3	0.14	0.41	0.47	0.4	0.1	0.1	0.11	0.1
S	wt%	0.11	0.22	0.19	0.19	0.36	0.61	0.71	0.13	0.14	0.10	0.03
Total	wt%	100.75	100.86	101.06	100.81	100.88	100.92	101.12	100.87	100.72	100.75	100.54
FeO	wt%	5.07	5.34	5.15	6.32	6.21	6.17	6.24	6.24	5.37	5.49	4.2
Ag	ppm	0.1	0.1	0.1	0.1	0.1	0.1	0.2	0.3	0.4	0.2	0.1
Ba	ppm	378	654	435	429	696	657	875	1096	1330	1512	602
Bi	ppm	0.5	0.6	0.4	0.4	0.4	0.6	0.5	n.a.	0.1	0.2	0.9
Ce	ppm	50	59.4	44.3	62.4	58.1	56.2	49	65.4	64.2	65.7	57.2
Cl	ppm	265	353	285	177	193	201	106	1	1	95	283
Co	ppm	15	20	16	19	16	18	17	23	20	21	15
Cr	ppm	93	104	78	96	80	79	71	107	103	108	107
Cs	ppm	0.18	0.35	0.2	0.24	0.32	0.41	0.51	0.39	0.54	0.45	0.26

Element	Unit											
Cu	ppm	55	154	24	62	33	66	65	51	69	120	13
Dy	ppm	2.98	3.04	2.88	3.21	2.97	2.97	2.85	3.46	3.32	3.52	3.04
Er	ppm	1.63	1.57	1.53	1.75	1.66	1.58	1.63	1.9	1.82	1.88	1.62
Eu	ppm	0.81	0.96	0.94	1.05	0.97	0.98	0.81	1.03	0.96	0.95	0.92
F	ppm	551	512	364	389	517	511	349	1	1	181	554
Ga	ppm	15.33	15.8	14.25	15.43	15.14	14.91	12.25	15.29	14.83	15.67	16.2
Gd	ppm	3.69	3.83	3.78	4	3.8	3.65	3.18	4.1	3.98	4.08	3.7
Hf	ppm	3.66	3.6	3.15	3.59	3.37	3.33	3.6	4.19	4.05	4.08	3.88
Ho	ppm	0.59	0.6	0.57	0.64	0.61	0.59	0.92	0.67	0.64	0.68	0.6
La	ppm	25.3	30.2	21.8	32.1	29.9	28.8	23.1	32.4	31.3	33.8	27.6
Lu	ppm	0.25	0.25	0.23	0.26	0.25	0.24	0.25	0.27	0.27	0.28	0.26
Nb	ppm	9.23	8.44	8.33	9.21	9.36	9.79	10.96	8.14	7.9	8.51	8
Nd	ppm	23.5	25.9	20.8	25.9	24.3	23.5	20.5	31.4	28.2	29.4	24.4
Ni	ppm	51	70	54	63	64	69	63	62	60	62	55
Pb	ppm	4	13	3	4	5	22	39	4	4	23	6
Pr	ppm	6.18	6.95	5.38	7.01	6.46	6.38	5.55	7.61	7.5	7.56	6.34
Rb	ppm	39	66	55	44	61	40	35	96	136	86	68
Sc	ppm	12	13	11	12	12	11	10	14	14	14	14
Sm	ppm	4.42	4.58	4.28	4.64	4.33	4.22	3.84	5.14	4.97	4.9	4.36
Sr	ppm	156	180	169	92	122	119	77	151	215	157	292
Ta	ppm	0.63	0.52	0.63	0.62	0.58	0.64	0.68	0.53	0.52	0.52	0.76
Tb	ppm	0.52	0.54	0.51	0.57	0.53	0.53	0.47	0.59	0.57	0.6	0.55
Th	ppm	7.76	7.52	7.1	7.67	7.46	7.06	6.71	8.17	8.38	8.13	7.96
Tl	ppm	0.18	0.33	0.27	0.24	0.38	0.28	0.33	0.57	0.77	0.57	0.37
Tm	ppm	0.26	0.25	0.25	0.27	0.25	0.25	0.25	0.28	0.27	0.29	0.27
U	ppm	4	3	3	4	3	4	4	2	3	2	2
V	ppm	109	105	103	121	119	115	121	116	108	109	87
Y	ppm	17	18	16	19	17	18	18	19	18	19	18
Yb	ppm	1.6	1.4	1.5	1.7	1.6	1.6	1.6	1.8	1.7	1.8	1.6
Zn	ppm	24	28	27	58	96	196	126	50	26	106	15
Zr	ppm	140	144	125	137	133	132	139	147	145	148	152

Averages are from Ames et al. (2002).
Source: Ames et al. (2002).

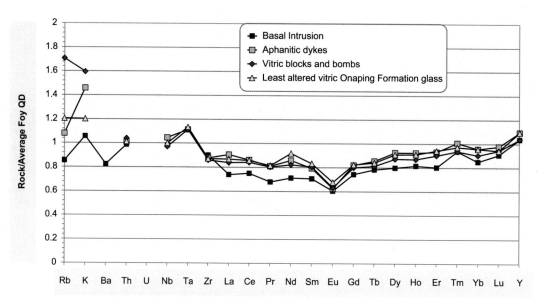

FIGURE 3.26 Compositional Data for the Basal Intrusion, Apahanitic Dykes, Vitric Blocks and Bombs, and Least Altered Vitric Fragments from the Onaping Formation

The compositions are normalised to the average composition of the QD Unit of the Foy Offset Dyke.

After Ames et al. (2002)

Minor and trace element compositional data are summarized in Fig. 3.26 where the abundances are normalized to the average composition of the QD Unit of the Foy Offset Dyke. This diagram illustrates that the various melt bodies, dykes, bombs, and vitric fragments have patterns that exhibit similar shapes (with the exception of Rb and K which are quite mobile during alteration which is extensive in the GRAN and the Onaping Formation). There is a gentle slope between the middle rare earth elements and the light rare earth elements and Y, so the patterns are not identical in shape to the average of the QD Unit of the Foy Offset Dyke. Moreover, the patterns of the melt rocks in the Onaping Formation range from slightly depleted in trace element content for the middle rare earth elements to having equivalent concentration of the heavy rare earth elements and immobile elements like Th, Nb, and Ta. The patterns are therefore slightly different when compared to those of the North Range QD Units which resemble those of the Foy Offset Dyke (Fig. 3.15B). The Onaping melts also exhibit a negative europium anomaly relative to the average composition of the QD Unit from the Foy Offset, so they have undergone removal of plagioclase.

The composition of the magmatic components of the Onaping Formation confirms that the primary magma composition of the melt sheet is close to the compositional spectrum of the average composition of the QD Units from the different Offset Dykes.

PARENT MAGMA COMPOSITION, DIFFERENTIATION, AND SULFIDE SATURATION HISTORY OF THE SUDBURY MELT SHEET
COMPOSITION OF THE PARENTAL MAGMA OF THE SUDBURY MELT SHEET

A major recurring debate in the literature concerns the identification of the composition and source of the parental magma that gave rise to the SIC. One group of models provides evidence for an overwhelming contribution from the target rocks observed in the Sudbury Region (eg, Lightfoot et al., 2001); a second group of models requires that the igneous rocks of the SIC come from both crustal and mantle sources with the noritic rocks having a mantle contribution but the GRAN being derived entirely by crustal melting (Chai and Eckstrand, 1994).

Layered on this debate is evidence that unusual mafic magma contributions are required, such as: (1) Mafic rocks with cumulus orthopyroxene comprising the basal stratigraphy of the Main Mass (the MNR), and the SLNR with the associated mafic-ultramafic inclusion populations which are often rich in olivine and chrome spinel (Lightfoot et al., 1997b; Zhou et al., 1997). (2) The development of nickel sulfide ore deposits which are normally found in association with mafic and ultramafic intrusions. Despite these relationships, the search for a primary mantle contribution has produced unconvincing evidence. Although the models allow for a contribution of 20% by volume to the SIC (Lightfoot et al., 1997a), there has been a singular lack of evidence for any material that represents the composition of a mantle source generated at 1.85Ga (eg, Cohen et al., 2000, Prevec et al., 2005).

Central to the argument for large contributions of crustal material is the overwhelming crustal trace element signature of the rocks. This is illustrated in Fig. 3.17 where the normalized spidergrams of representative crustal rocks are shown relative to the composition of the average of the QD Unit of the Foy Offset which is one of the possible estimates of the normalized pattern of the igneous component of the Sudbury Structure. The best fit comes from the average composition of Archean shales which provide a well-mixed overall estimate of typical upper crust (Taylor and McLennan, 1985). Other model crusts including average bimodal Archean greenstone, andesitic crust, and bulk Archean crust provide less tight fits, but they illustrate many of the same features as the average composition of the QD Unit.

Radiogenic Sr and Nd isotope studies of the SIC have also failed to detect mantle contributions, but the data do provide strong evidence for the development of a uniform Main Mass magma reservoir from average upper crust (Faggart et al., 1984, 1985b; Rao et al., 1984a; Prevec et al. (1997a-c); Lightfoot, unpublished data). The variability in estimates of initial Sr isotope ratio due to alteration is avoided by using the data for the Sm-Nd system where Sm and Nd are generally considered to be immobile. A synthesis of the Nd isotope data is provided in Fig. 3.27A–B where the histogram shows the variability in initial Nd isotope ratio (expressed in epsilon notation and based on the normalization factors used by Prevec et al. (1997a-c)), the diversity in initial ratios is expressed in box and whisker plots. Deviations away from the bulk composition of Main Mass and Offset rocks are recorded in the Onaping Formation and in the compositional data for inclusions in the Sublayer. In the case of the Sublayer, Prevec et al. (1997a-c) suggest that the inclusions have played a part in controlling the composition of the SLNR. The Main Mass has a Nd-isotope signature that is consistent with derivation from a crustal sources at 1.85 Ga.

The average composition of the QD Unit of the Offset Dykes has been identified as one possible index of parental magma composition. These rocks have quenched contact relationships

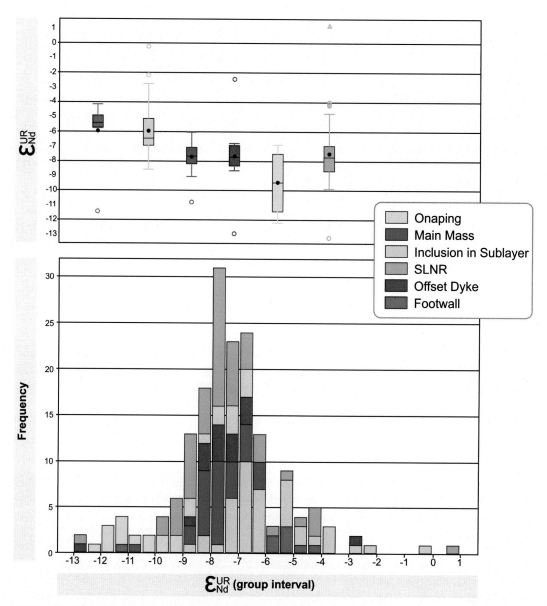

FIGURE 3.27 Compilation of Initial Neodymium Isotope Compositional Variations for Rocks from the SIC

Data sources are given in Appendix 3.2. (A) Box and whisker plot showing the compositional range in samples from the Main Mass, Sublayer, Offsets, Onaping Formation, and Footwall at 1.85 Ga. (B) Grouped frequency histogram showing the variability in initial Nd isotope ratios based on the compilation in Appendix 3.2; normalization has followed the protocols documented in Prevec et al. (2005).

with the SUBX and occur as inclusions in later batches of mineralized magma, so they were clearly generated early in the evolution of the melt and likely before the Main Mass commenced crystallization. Notwithstanding, important differences in the geochemistry of the QD Unit exist between the North and South Range Offsets. Moreover, the QD and MIQD Units of the Frood-Stobie, Creighton, and Manchester Offset Dykes are different when compared to Copper Cliff and Worthington. These differences may be explained by local variation in magma composition in the melt sheet (Lightfoot et al., 1997b), a contribution by contamination from local country rocks (Lightfoot et al., 1997b) and/or injection of the Offset Dykes at different times during the solidification of the Main Mass from magmas that were evolving in response to differentiation (eg, Smith et al., 2013). It is less easy to explain why the QD is silica-poor relative to the bulk-integrated composition of the SIC (Fig. 3.13).

The viability of models that require the Offset Dykes to be injected at different times in the development of the SIC is constrained by the absence of cross-cutting relationships between the Main Mass and the Offset Dykes as well as the geological evidence that suggests that most of the dykes were emplaced before sulfide saturation of the melt sheet (see the next section). Moreover, Offset Dykes showing a range in trace element ratios are not easily produced by closed-system differentiation of a single parental magma, so there is no clear way to generate the QD and granophyric QD comprising the QD Unit of the Offset Dykes at different stages of melt evolution recorded in the almost constant incompatible trace element ratios of the Main Mass.

There is no evidence for differential amounts of contamination within individual Offset Dykes, so it is unlikely that a QD of uniform composition would be produced by interaction of the melt with different crustal rocks during or immediately after injection of the Offsets. The most viable explanation is variability in the melt at the time of injection and prior to melt sheet homogenization. In this model, the composition of the QD Unit takes on significance as the best index of diversity in the early composition of the melt sheet.

Melts from the Onaping Formation provide independent estimates against which to test parental magma compositions. The igneous rocks of the Onaping Formation provide compositions which are consistent with an upper crustal source, but there is not a direct match with the average composition of North Range QD from the Foy Offset Dyke. It is not yet clear whether there are subtle differences in composition between melts formed in the Onaping of the North Range versus the South Range, but this may help in the future to establish whether the variability in QD composition is due to primary heterogeneity in the early melt prior to convective missing and differentiation.

CONTRASTING MAIN MASS DIFFERENTIATION TRENDS IN THE NORTH AND SOUTH RANGES

A feature which became evident from the detailed chemostratigraphic studies in Lightfoot et al. (2001) and Lightfoot and Zotov (2006) is that the variations through the North Range and South Range Main Mass are different (eg, Naldrett and Hewins (1984), just as the sequence of rock types described in previous studies of petrology indicate that the North and South Range Main Mass evolved under different conditions (Lightfoot, 2009).

To investigate the similarities and differences more closely, Fig. 3.28A,B show the variations in MgO versus TiO_2 and MgO versus Al_2O_3 where the symbols depict whether the samples come from the North Range or South Range. The trend lines (Fig. 3.28C,D) are derived from moving averages

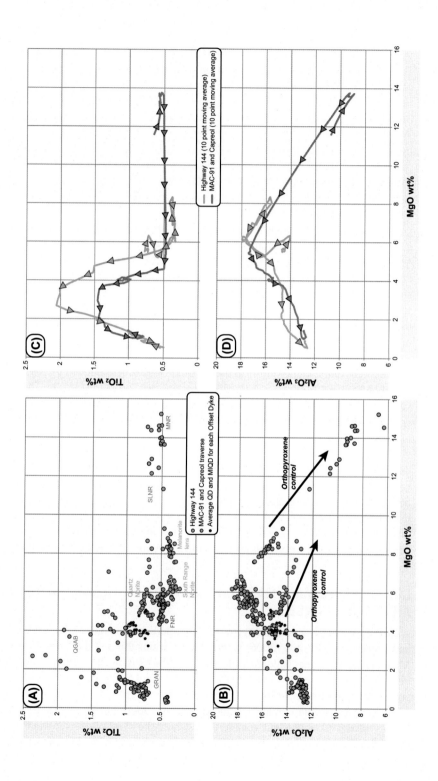

◀ **FIGURE 3.28** Comparison of the Differentiation Trends in Traverses through the South Range and North Range
Main Mass

The data points represent analyses, and the trend lines show the 10-point interval-weighted moving average of
the oxide abundance. (A and C) Variations in MgO versus TiO_2. (B and D) Variations in MgO versus Al_2O_3. The
lowermost South Range Quartz-rich Norites above the Sublayer show a progressive upwards decline in TiO_2 and
Al_2O_3 with increasing MgO after which the basal Quartz norites show a fall in TiO_2 and MgO with an increase in
Al_2O_3. Upwards through the Quartz Norite, the MgO and Al_2O_3 contents increase and the TiO_2 content decreases up
to the level at which the melanorite lens enters and there is a sudden increase in MgO and a fall in Al_2O_3 at almost
contestant TiO_2. The uppermost part of the South Range norite stratigraphy then shows a return to compositions
similar to those developed below the melanorite lens and the norites start a trend of falling MgO and Al_2O_3 with
increasing TiO_2 as they approach the base of the QGAB. In contrast, the North Range norite commences above the
Sublayer with very high MgO and low TiO_2 and Al_2O_3. These rocks rapidly progress in composition to low MgO
and high Al_2O_3 at similar MgO concentrations, and then there is a small decrease in MgO and Al_2O_3 with increasing
TiO_2 upwards towards the base of the QGAB. The trends of the South and North Range noritic sequence are quite
different. The South Range QGAB exhibits a rapid increase in TiO_2 with falling MgO, but little change in Al_2O_3, but
this decline in MgO commences at a higher MgO concentration than the North Range QGAB, and so the trends are
displaced relative to one another. The uppermost QGAB samples trend towards very low MgO, Al_2O_3, and TiO_2, and
the GRAN then commences a weak upwards increase in MgO, TiO_2, and Al_2O_3 towards the top of the Main Mass.
The rocks from the lower part of the QGAB of the North Range plots on a trend of falling MgO with increasing
TiO_2 which is displaced towards lower MgO and TiO_2 relative to the array of the South Range. The rocks from the
upper part of the QGAB from both North and South Range traverses then fall on a similar trend of falling MgO
with TiO_2.

After Lightfoot et al. (1997c,d), Lightfoot et al. (2001), and Lightfoot and Zotov (2006).

calculated upwards through the stratigraphy following the approach used in Lightfoot and Zotov (2006)
to illustrate the broader trends in chemical variation through the Main Mass.

In the North and East Range Main Mass, the most melanocratic rocks comprise the basal MNR
Unit, and the transition into the overlying FNR is rapid, followed by a regular fractionation history
towards more felsic compositions. In the South Range, the lower part of the Main Mass is a Quartz-rich
Norite which is followed by a thick package of Quartz Norite which show a progressive upwards trend
to more melanocratic compositions, reaching a maximimum about 2/3rds of the way through the stratigraphy where a lens of melanorite is developed; above this level the norites fractionate towards more
evolved compositions. These changes in petrology are marked on the trends of changing MgO, TiO_2,
and Al_2O_3 in Fig. 3.28A–D. The variations are quite complex, and they are summarized in the caption
to Fig. 3.28A–D; their interpretation is summarized below:

1. The Norite Unit: The bulk of the South Range noritic rocks show a broad upwards increase in
   MgO in the lower 2/3rds of the stratigraphy and then a decline in MgO above this level whereas
   the North Range noritic rocks show a broad upwards decline in MgO. These variations are
   consistent with orthopyroxene and plagioclase accumulation towards the base of the North
   Range norites and towards the center of the South Range norites. These data indicate that the
   mechanisms of crystallization and silicate mineral accumulation were different in the North versus
   South Range norite stratigraphy.
2. The QGAB Unit: The basal QGAB plots on an array of increasing TiO_2 consistent with control
   by titanomagnetite, but the entry of titanomagnetite commence at a lower MgO concentration

in the North Range versus the South Range. This indicates that the crystallization of the QGAB commenced from a melt that was more fractionated in composition in the North Range versus the South Range. The overlying QGAB then progressively change in composition towards the composition of the evolved basal GRAN.

3. The GRAN Unit: The differentiation sequences in the GRAN is broadly comparable in the North and South Ranges, and define a broad-upwards increase in MgO and Al_2O_3 with TiO_2.

To investigate the extent of lateral heterogeneity on the scale of the SIC (Fig. 3.29A), Fig. 3.29B–E shows variations in the interval weighted average composition of the norite as a function of the position of the sample in clockwise order around the Sudbury Basin starting at Blezard in the South Range and ending at Nickel Rim in the East Range; the numbered traverses are shown in Fig. 3.29A. The main observations from Fig. 3.29B–E are as follows:

1. The norite traverses from the South Range have a systematically higher Mg-number and slightly lower TiO_2, Ba, and Sr contents than those of the North Range. These observations indicate that there is an overall difference in the chemistry of the rocks of the norite sequence between the North and South Ranges which is presumably related to different crystallization histories and a larger overall amount of orthopyroxene and plagioclase accumulation in the noritic rocks of the South Range relative to the North Range.

2. The ratio of La/Y tends to be similar in both North and South Range norites which is consistent with the spidergram patterns in Fig. 3.11A,B. These variations are consistent with a single parental magma type for the noritic rocks of the North and South Range Main Mass.

3. The average Cu and Ni concentrations and the Ni/Ni* (crustal model) for the South Range traverses tend to be higher than for the North Range traverses. This indicates that the immiscible sulfide melt was segregated more efficiently in the melt sheet before Main Mass crystallization in the North Range relative to the South Range.

As discussed in Chapter 2, the North and South Ranges are demarcated broadly by major fault zones between Sultana and Crean Hill in the west and in the area of Norduna in the East. The North and South Range magmas evolved to produce a similar overall stratigraphic sequence of norites, QGAB, and GRAN (see Fig. 3.4A–H and 3.8A–J), and the trace element patterns of the North and South Range rocks show that the parental magma was characterized by similar ratios of the incompatible trace elements (Fig. 3.11A,B). Notwithstanding, the sequence of noritic rocks produced by differentiation of the magma are not the same in the North and South Ranges.

The explanation for the different stratigraphies of the North and South Ranges is an important question. Possible explanations together with their challenges are listed below:

1. Different parental magma compositions were generated from different crustal target rocks, and the associated melt sheets evolved independently as different domains within a discontinuous melt sheet. The present-day configuration of the SIC after regional deformation still shows the North and South Ranges juxtaposed together as one oval basin. This configuration would be unlikely if the melt sheet originally consisted of two or more discontinuous segments. Moreover, the incompatible trace element ratios of North and South Range rocks of the Main Mass are indicative of a single parental magma composition. This explanation is therefore unlikely to be correct.

2. Evolution of the melt sheet as a function of the composition of the underlying crustal target rocks or to the movement of melts of different compositions between different sectors of the impact

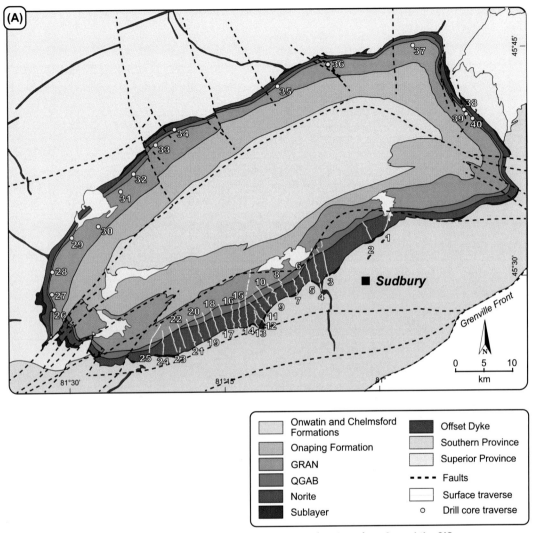

FIGURE 3.29 Compositional Variation in the Weighted Average Norites Locations Around the SIC

(A) Traverse locations and numbers depicted in Fig. 3.29B. (B) Mg-number and TiO₂ variations; (C) Ni, and Cu variations; (D) Ba and Sr variations; E. Ni/Ni* (crustal model) and La/Y variations.

After Lightfoot et al. (1997c,d), Lightfoot et al. (2001), Lightfoot and Zotov (2006), Appendix 3.1, Cooper (2000),

and unpublished data.

(Continued)

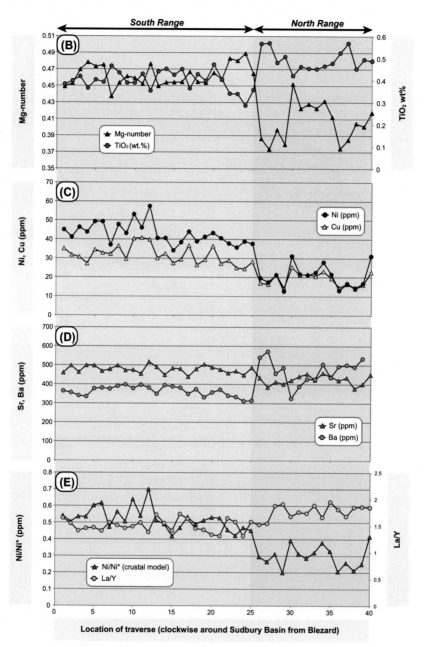

FIGURE 3.29 (*cont.*)

melt in response to post-impact lithospheric readjustment. The continuous trends through the Main Mass of the North and South Range norite, with the exception of the melanorite lens in the South Range points to gradual process of differentiation rather than mixing of different pulses of magma.

3. In a third group of models, the wide variation in thickness of the melt sheet allowed differentiation to proceed under different physical conditions in thick portions of the Main Mass in the North versus South Ranges. The sheet would differentiate in an entirely different way at the base, yet retain a common stratigraphy towards the top because of local conditions of melt convection. This explanation would work well if there was some sort of gradation between the two domains or a tendency for the thicker melt sheet of the Levack-Coleman area to develop South Range type differentiation trends, which it does not (Keays and Lightfoot, 2004).

An explanation for this difference in terms of melt sheet evolution presently remains elusive, but the data point to a separate evolution of the noritic parts of a relatively homogenous parental magma in the north versus south. The reason for the diversity in norite compositions is the subject of ongoing research (eg, Lightfoot, 2007; Lesher et al., 2014; Strongman, 2016).

A more important implication of these models relates to understanding the mineral potential of the North and South Range melt sheet.

SULFIDE SATURATION HISTORY OF THE SIC; SOURCE OF THE BASE AND PRECIOUS METALS IN THE ORE DEPOSITS

The vast majority of the mineral wealth at Sudbury was formed in the South Range based on the available information from drilling. Taking the historic data for the SIC (Chapter 1), the North Range contact, footwall, and Offset Deposits contained ~17% of the Ni produced at Sudbury whereas production from the deposits in the South Range comprise ~83% of the Ni. There is almost an order of magnitude difference in the historic mineral potential in areas where the melt sheet evolved down different liquid lines of decent.

The next step in unpicking the relationship between the Ni-Cu sulfide ore deposits and the melt sheet is to better understand the levels of metal depletion in rocks of the SIC.

Fig. 3.30A shows the variation in whole-rock MgO concentration versus Ni concentration for rock units from the North and South Ranges and from the Offsets. This diagram shows the relative position of arrays of within-plate basaltic rocks (Keays and Lightfoot, 2004), an array based on the compositions of unmineralized country rocks from the SIC target, an array of both mineralized and unmineralized country rocks from the SIC target, and a power-law relationship for the unmineralized country rocks. The power-law relationship has traditionally been used for flood basalts (Keays and Lightfoot, 2004); in the absence of arrays for crustal rocks, it has been used to illustrate that the Main Mass noritic rocks have unusually low Ni concentrations (Lightfoot et al., 2001). Irrespective of the array chosen for mantle or crustal rocks, it can be seen from Fig. 3.30A that the majority of SIC samples are depleted with respect to the composition of the rocks expected to form from the different source model compositions. The degree of depletion is actually greater if a crustal rather than a mantle model is chosen as a reference array for the calculation.

The rock groups from the Main Mass plot on a trend of increasing Ni with Cu concentration which is expected if the principal control on compositional diversity is Ni-Cu-enriched magmatic sulfide (Fig. 3.30B). In detail, there are small differences in Ni/Cu on the array of Main Mass rocks which

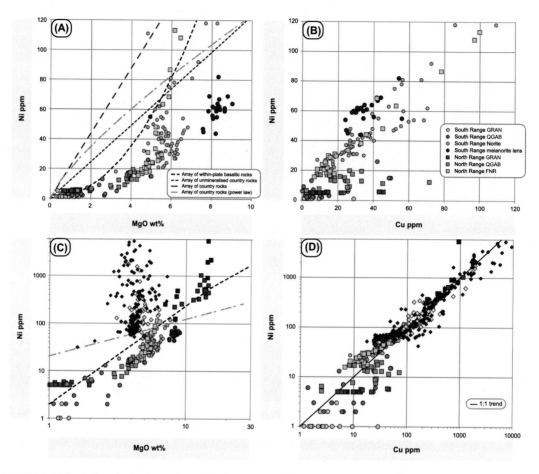

FIGURE 3.30 Variations in MgO Versus Ni and Cu Versus Ni in the Main Mass and Offsets

(A) MgO (wt%) versus Ni (ppm) showing the compositions of rocks collected along North and South Range traverses relative to the arrays of within-plate basaltic rocks (Keays and Lightfoot, 2004), country rocks from the Sudbury Region, unmineralized country rocks from the Sudbury Region, and a power law relationship for country rocks from the Sudbury Region which is used to estimate Ni*. (B) Plot of Ni (ppm) versus Cu (ppm) for the same rock series. (C) Plot of Main Mass rocks and the compositions of samples from the QD and MIQD Units of the Copper Cliff and Worthington Offset Dykes. (D) Plot of Ni (ppm) versus Cu (ppm) for the Main Mass and Worthington and Copper Cliff Offset Dykes.

After Keays and Lightfoot (2004).

reflect the stronger mobility of Cu relative to Ni under conditions of greenschist to lower amphibolite facies metamorphism. Moreover, rocks with cumulate textures like the MNR typically have elevated Ni/Cu due to the Ni content of a rock with up to 40% cumulus orthopyroxene. Notwithstanding, the vast majority of the rocks have Ni/Cu~1 (Fig. 3.30B) which is close to the bulk average composition of the Sudbury ore deposits (Naldrett, 2004).

When compared to the compositions of the Main Mass rocks from the North and South Ranges, the rocks from the Offset Dykes (QD and MIQD Units) plot above the array of the Main Mass on MgO versus Ni, yet the fall on a similar trend on the plot of Ni versus Cu (Fig. 3.30C,D). Samples from the unmineralized QD Unit has elevated Ni relative to Main Mass rocks with similar MgO concentrations, and the MIQD samples from the Offsets are displaced to higher Ni than the equivalent unmineralized QD Unit (Fig. 3.30C). The sulfide-free Quartz Diroites have normal Ni abundance (ie, chalcophile element-undepleted) relative to the power law relationship established for the target rocks in the Sudbury Region, and it may well represent the composition of this target prior to sulfide saturation. The MIQD is enriched in Ni relative to the expected abundances (based on the MgO content), and this is consistent with the observation that these rocks have compositions controlled by sulfide. Irrespective of whether the rocks are barren or mineralized, there remains a reasonably tight array of Ni versus Cu with a slope of 1:1 which is consistent with control by sulfides that have a bulk average Ni/Cu~1 which is typical of undifferentiated Sudbury mineralization (see Chapter 4).

The extent to which the noritic rocks of the Main Mass and the QD Unit of the Offset Dykes are depleted in Ni is illustrated in Fig. 3.31 where the average composition of different rock units is shown in terms of Ni/Ni* (where Ni* is estimated based on the relationships in country rocks from the

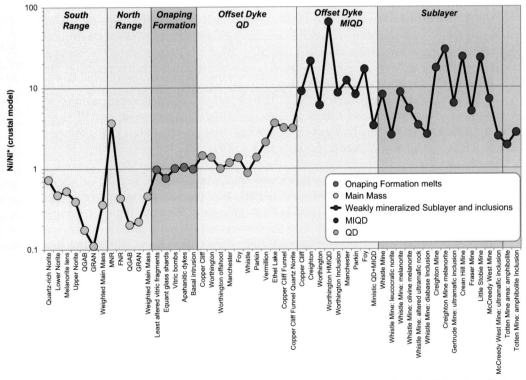

FIGURE 3.31 Variations in Ni/Ni* (Crustal Model) for Weighted Average Rock Types from the Units of the Main Mass, the Different Offset Dykes, and the Noritic Matrix of the Sublayer

After Lightfoot et al. (1997c).

Sudbury Region). Enrichment factors above 1 (Fig. 3.31, vertical axis) are indicative of sulfide control on metal enrichment whereas those below 1 are indicative of metal depletion. The crustal model allows for a somewhat more reasonable model than comparisons to average within plate flood basaltic magmas that are sulfide under-saturated (Lightfoot et al., 1997g).

This plot highlights the fact that samples of the QD Unit fall close to a ratio of 1, whereas the sulfide-free volumetrically enormous Main Mass norites (above the North Range MNR and the South Range Quartz-rich Norite) are depleted in nickel. In contrast, the average composition of the MIQD from different Offset Dykes and the average composition of Sublayer samples from different trough and embayment structures are strongly enriched in Ni together with the entrained ultramafic inclusions.

The overwhelming weight of evidence points to the derivation of the Ni, Cu, and PGE in the ore deposits from the overlying melt sheet (Lightfoot et al., 2001; Keays and Lightfoot, 2004). From the perspective of the ore deposits, the South Range contact and Offsets are endowed with an order of magnitude more known mineralization than the North Range, but from the perspective of the metal-depletion history of the Main Mass, the North Range noritic rocks are more strongly depleted in metals than the South Range. The reason for these differences may relate to the extent of convective mixing of the melt sheet and the efficiency of separation of sulfide to form the ore deposits.

SOURCE OF THE SULFUR IN THE SIC AND THE ASSOCIATED MINERALIZATION

The fact that much of the Main Mass norite is depleted in Ni, Cu, and PGE (Keays and Lightfoot, 2004) has encouraged the development of models to explain the concentration of the metals by differentiation and settling of dense immiscible sulfide (Chapter 5). What has been less clear is whether the sulfur inventory of the Main Mass supports this model, and whether different ore deposits are characterized by a range in S isotope ratios.

To investigate this problem, Ripley et al. (2015) undertook a detailed investigation of the S isotope compositions of mineralized rocks from the Sublayer, Offsets, and Footwall as well as an investigation of the sulfur isotope ratios of unmineralized rocks from the Main Mass. The position of samples analyzed from around the Sudbury Structure is shown in Fig. 3.32A.

The S-isotope composition of rocks from one traverse through the Main Mass is shown in Fig. 3.32B. In this traverse, the extremely heavy S isotope composition of the samples from the GRAN Unit is likely a response to the alteration signature of the upper 2/3rds of the SIC was influenced by hydrothermal activity (Naldrett and Hewins, 1984). There is little variation in S isotope signature through the FNR. By taking samples from the lower part of the Main Mass, and the mineralized environments related to the Sublayer and Offsets, the objective was to identify whether there is any primary isotopic variability in the sulfur through the accessible part of the lower contact (Ripley et al., 2015).

The amount of variation in S isotope ratios in the lower part of the Main Mass is quite small (Fig. 3.32C), and Ripley et al. (2015) attributed this narrow range to effective mixing within the melt sheet. In contrast, the variability associated with contact embayment structures and Offsets is wider (Fig. 3.32D) and the sulfur in the northeast corner of the SIC bracketing Whistle Mine is heavier than the sulfur from similar rocks in the North Range which in-turn tends to be heavier than that of the South Range contact and Offset environments. Ripley et al. (2015) attribute this variability to local contamination of the melt by crustal sulfur that has a range in isotope ratio compositions which offset the initial ratio of the melt sheet recorded in the overlying noritic rocks.

Despite the differences in Ni/Ni* (crustal model) between North and South Range norites, and the stronger Ni and Cu depletion of the parental magma, the available data indicate that both North

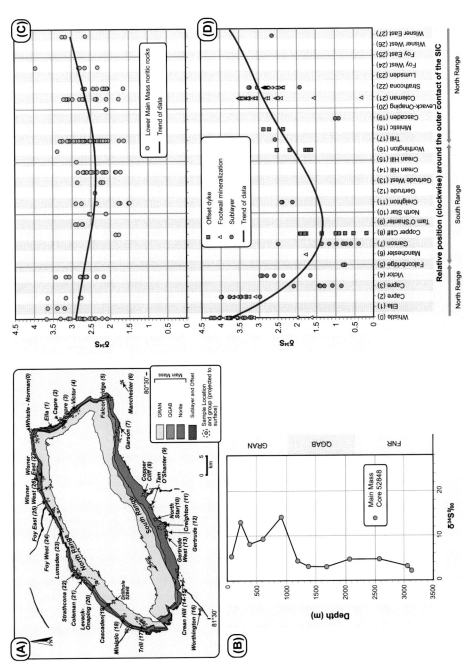

FIGURE 3.32 Variations in Sulfur Isotope Composition of Samples Collected Around the Sudbury Structure

(A) Map of sample locations and the numbering using to depict variations in a clockwise direction around the SIC. (B) Variation in a clockwise direction around the SIC of δ34S in rocks from the lower part of the Main Mass above the Sublayer; (C) Variation in S isotope ratio around the Sudbury Structure in mineralized samples from the Sublayer and associated footwall deposits. Diagram is modified.

From Ripley et al. (2015) with the permission of The Society of Economic Geologists.

and South Range norites show an upward decline in S concentration. The data for the North Range are consistent with a decline from ~800 to ~400 ppm, although the data for the South Range are not as precise, they are consistent with a similar abundance range (Keays and Lightfoot, 2004; Lightfoot and Zotov, 2006). The residual S concentrations likely reflect the S capacity of the remaining silicate magma (Keays and Lightfoot, 2004; Li and Ripley, 2005).

In summary, the sulfur isotope data are consistent with the formation of the ore deposits from sulfur that segregated as an immiscible sulfide from the melt sheet. Local variations in the contact and footwall deposits require a small amount of exchange between the country rocks and the melt sheet, and this is most pronounced in the northwestern part of the SIC. There is no evidence for major contributions of sulfur from seawater or evaporates as might be expected in a system undergoing hydrothermal modification.

SEQUENCE OF EVENTS IN THE CRYSTALLIZATION OF THE SIC

Petrological and geochemical data for the SIC coupled with the basic geological relationships described in Chapter 2 provide enough information to address a simple sequence of events in the evolution of the igneous rocks of the melt sheet. Fig. 3.33A–F is modified after Lightfoot et al., 2001, Keays and Lightfoot (2004), and Farrow and Lightfoot (2002). The model depicts the processes that occurred in a small domain of the melt sheet (Fig. 3.33A) from initial formation through to solidification (Fig. 3.33B–F). The sequence of events is based on the geological relationships, petrology and geochemistry of the silicate portions of the system. Chapter 4 will provide a more detailed view of the relationships on the scale of the ore deposit systems.

Some of the earliest rocks to form were the SUBX which comprise a matrix of ground-up country rocks with local partial melting textures; these rocks were undoubtedly formed from the local country rocks, and the geological, petrological, and geochemical evidence presented in this chapter demonstrates that the dominant host rocks of the SUBX were the principal geochemical reservoir giving rise to the breccia matrix. There is no requirement for a melt sheet contribution in the genesis of these rocks.

FIGURE 3.33 Model for the Evolution of Magmas in a Slice through the Sudbury Impact Melt Sheet

(A) Stage 1 occurred immediately post-impact and after the initial formation of the melt sheet and crystallization of the SUBX; superheated melts with a composition like that of the QD Unit were injected into the footwall of the SIC to form the Offset Dykes. These dykes typically followed weaknesses in the crust such as reactivated faults and domains of SUBX. At this stage, the melt sheet was not differentiated or sulfide-saturated. (B) Stage 2 corresponds to the saturation of the melt sheet in sulfide melts, and the segregation of the sulfide towards the base together with mafic inclusions; injection of QD with small mafic inclusions and small amounts of sulfide predated the injection of QD which was laden with sulfide and larger inclusions to form the MIQD and HMIQD Units which contains economic ore deposits in the Frood-Stobie, Creighton, Worthington, and Copper Cliff Offset Dykes. (C) Stage 3 corresponds to the formation of the Sublayer in trough and embayment structures at the base of the melt sheet; the rocks formed in these troughs contain SLNR and GRBX (mostly in the North Range). The Sublayer rocks are variably endowed with sulfide mineralization depending on the thickness and scale of concentration of metals in magmatic sulfide from the overlying and adjacent melt sheet. (D) Stage 4 shows the onset of crystallization of the melt sheet to produce the MNR; (E) Stage 5 shows the complete solidification of the melt sheet with crystallization from the base-up and top-down. (F) Cross section through the melt sheet showing the location of a small segment of the melt sheet depicted in 3.33B-F.

After Lightfoot et al., 2001, Keays and Lightfoot (2004), and Farrow and Lightfoot (2002). ▶

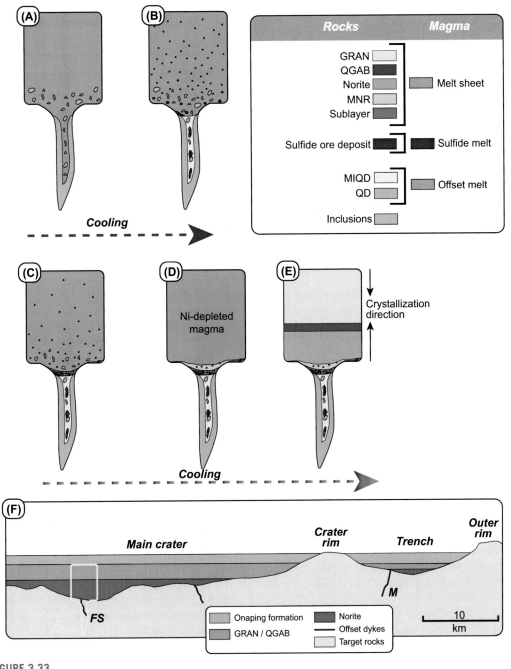

The Offset rocks of the SIC cross-cut the SUBX and show quenching against it, so the injection of magma to form the QD Unit of the Offset Dykes likely post-dated the cooling and consolidation of the SUBX.

Fig. 3.33F shows the broad structure of the impact crater after initial genesis of a melt layer immediately post-impact. The domain shown in the model (Figs. 3.33A–E) reflects the processes happening within the main melt sheet preserved as the Main Mass, and is centered on a thicker South Range part of the system localized over a funnel with a connected dyke like the Worthington Offset Dyke. Immediately post-impact the melt sheet comprised >15,000 km^3 of magma generated through melting of the target rocks (eg, Golightly, 1994).

Estimates of the bulk composition based on weighted thickness-integrated models of the SIC show that the composition of the bulk Main Mass is not easily established as there are local variations in petrology and geochemistry that reflect differentiation processes in a melt sheet that varies enormously in thickness. To overcome this problem, estimates of parental magma composition have been rooted in both the composition of the QD Unit of the Offset Dykes (Lightfoot et al., 2001) and the glassy rocks in the Onaping Formation (Ames et al., 2002). Both estimates have some uncertainty, but the homogeneity of the QD Unit in the Copper Cliff and Worthingotn Offset Dykes provides a representative indicator of the starting melt composition in the South Range whereas that of the QD Unit from Offsets in the North Range provides a slightly different initial composition which presumably records some local heterogeneity in the melt sheet when it was first formed.

The initial melt was very likely superheated (Ivanov and Deutsch, 1997; Lightfoot et al., 2001), but there remains an important debate regarding the duration of superheat in the differentiation history of the SIC (eg, Latypov et al., 2010). On the one hand, the melt generated immediately post-impact was very likely superheated by virtue of the energy imparted to the crust; on the other hand, the extent to which the heat could escape from the melt sheet rapidly depends on the extent to which it was covered with fall back material and the extent of assimilation of fragments which would reduce the excess heat in the melt.

The effectiveness of the SIC as a very perfect "sulfide smelter" depends heavily on the ability of the system to efficiently concentrate magmatic sulfides from the magma column towards the base of the SIC, and this would be optimized under superheated conditions. More completely constrained models for the physical evolution of the SIC melt sheet are clearly long required to address this question.

The first melts to be injected into the Offset Dykes resulted in the formation of the QD Unit (Fig. 3.33A), and this rock type is broadly uniform in composition between many of the larger South Range radial Offset Dykes (eg, Copper Cliff, and Worthington), but there are local variations in some of the shorter Offsets developed in the keel of troughs (eg, Creighton), as discontinuous pods in belts of breccia (eg, Frood-Stobie) or outside of the footprint of the South Range Main Mass (eg, the Manchester Offset Dyke). Notwithstanding these variations, all of the samples from the QD Unit have essential normal Ni concentrations with no evidence to suggest that the parental magmas were sulfide-saturated and hence had been depleted in the chalcophile elements. It is likely that the QD Unit was formed from a magma that was injected into the footwall of both the South and North Range prior to sulfide saturation of the melt sheet, and so these magmas originated before the formation of any of the Sudbury ore deposits (Fig. 3.33B), and their Ni concentrations provide a robust estimate of the range in average composition of the initial melt in the North and South Ranges.

Sulfide saturation of the melt sheet was likely triggered by changing physical conditions in the melt sheet as well as by the assimilation of residual fragments of country rock. During the phase of melt sheet evolution, depicted in Fig. 3.33B, the melt became sulfide-saturated and the immiscible dense magmatic sulfide melts segregated towards the base of the melt sheet together with remaining dense mafic fragments. The process of interaction of the sulfide melt with the silicate magma was likely enhanced by the high temperature of the melt sheet and the absence of crystals which would impede the settling of larger dense globules of sulfide to the base of the melt sheet. Once these sulfide melts reached the base of the sheet, they would have been guided into topographic lows at the base of the SIC and into Offset Dykes which remained open to the entry of new influxes of magma from the melt sheet. There is a minimal geochemical difference in the major and incompatible trace elements between the QD and MIQD Units, and the silicate melt giving rise to these rocks was formed from the same reservoir at a similar time. The largest difference between the two rocks types is a function of sulfide content which contributed to the enhanced base and precious metal inventory of the MIQD. There is also evidence for local assimilation of fragments in the MIQD (eg, Figure 2.20H in Chapter 2); although this has a tendency to produce a slightly wider compositional range in the MIQD Unit than the QD Unit, both rock types retain essentially indistinguishable average normalized trace element patterns. The MIQD has a sharp contact relationship with the QD Unit and contains inclusions of QD that are geochemically identical to the QD Unit of the Offset; based on these relationships, it is likely that there was a quite long time interval between QD and MIQD formation.

The next phase in melt sheet evolution is depicted in Fig. 3.33C which records the closure of the Offset systems and the formation of the Sublayer. In this time slice, the noritic rocks of the Sublayer were formed by the differentiation of the melt sheet to form cumulus orthopyroxene and plagioclase with lesser augite, phlogopite, oxide minerals, and sulfide as a matrix to mafic inclusions that were not assimilated into the melt sheet. These rocks tended to localize in trough structures where there was abundant melting of the country rocks to form breccias with a granitic matrix. Collectively the GRBX and inclusion-bearing norite were localized beneath thick melt sheets in the trough-like structures, and these environments became the principal locus for the development of sulfide mineralization. In detail, there are important differences between North and South Range Sublayer crystallization which may relate to the composition of the adjacent country rocks, but these differences are discussed in more detail in Chapter 4 where the ore systems are described.

The crystallization of the Main Mass commenced after Sublayer formation as shown in Fig. 3.33D. The differentiation process which formed the Main Mass is not completely understood; there are three principal models:

1. A highly efficient crystal differentiation mechanism that created the noritic rocks by orthopyroxene and plagioclase crystallization, and the QGAB and GRAN by crystallization of the residual melt from the base-up and the top-down (Lightfoot et al., 2001; Lavrenchuk et al., 2010).
2. In another models, the noritic and granophyric magmas are considered to be primary melts that formed by coalescence of immiscible melts and they underwent separate evolution (Zieg and Marsh, 2005).
3. A third model proposed by Golightly (1994) considers the melt sheet to form by batch melting of country rocks. The more felsic melts were less dense and rose through the melt sheet towards the roof of the melt sheet, whereas melts of more mafic crust were denser and they rose into the lower part of the melt sheet. By this process, a density-stratified magma column was generated from which melanocratic and felsic rocks were able to differentiate.

These end-member models remain the subject of ongoing research, but in both cases, the development of noritic rocks with Ni, Cu, and PGE depleted compositions (Lightfoot et al., 2001; Keays and Lightfoot, 2004) provides unequivocal evidence that the sulfide melts segregated and accumulated before the Main Mass commenced differentiation.

Complete crystallization of the Main Mass is shown in Fig. 3.33E where the broad sequence of rock types is a product of either efficient fractionation of a uniform parental magma or by the crystallization of a density-stratified melt from the base-up and the top-down.

There are several facets of Main Mass petrology and geochemistry that are incompletely understood, and these observations may help to drive the next generation of research work on the SIC, such as:

1. The North Range and South Range differentiation histories are different. The petrology and geochemistry of the South Range norites record an upward increase in orthopyroxene content and a decrease in the proportion of trapped liquid in the cumulate rocks. This reaches an apogee in the Highway 144 traverse where a melanorite lens is developed that contains magmatic sulfide minerals. In contrast, the North Range has an especially melanocratic basal unit with MgO concentrations approaching 25%; there is a sharp drop in orthopyroxene content above this melanorite, and then an upward decline in MgO with orthopyroxene content and increased proportion of trapped liquid. The reasons for these variations and their importance with respect to melt-sheet scale processes remains incompletely understood (eg, Lightfoot, 2009; Lesher et al., 2014; Strongman, 2016).

2. A second relationship which is not fully understood is the differentiation process that gave rise to the reverse differentiation profiles of the GRAN Unit. In one model, this is due to crystallization of a density stratified melt against the overlying fall-back materials at the roof of the intrusion. In a second group of models, the evolved melt is the product of expulsion of trapped liquid from the cumulate sequence, and this magma crystallized against the roof. In both models, the most evolved melts would be developed towards the base of the GRAN, and these magmas are recorded in the crowsfoot-textured granophyres.

3. The significance of the rocks developed in the upper contact unit of the SIC is described in Anders et al. (2015). The plagioclase–rich GRAN and the melt body rocks in the Onaping Formation have similar compositions to the QD Unit (Ames et al., 2002). These rocks were likely formed in response to modifications of the primary melt composition at the top of the melt sheet during phreatic volcanic evolution and alteration.

GEOCHEMICAL VARIATION, DISTRIBUTION AND THICKNESS OF THE MELT SHEET

A model for the broad-scale development of the SIC is shown in Fig. 3.34A–D based on the position of different segments of the Main Mass and different Offsets at different locations in a melt sheet that

FIGURE 3.34 Petrological and Geochemical Variations, and Melt Sheet Processes

(A) Image of the Meitner impact crater on Venus (Magellan image F-MIDR55S319.9;20; Location: 55.6°S, 321.6°E; Diameter of crater: 150 km; http://www.lpi.usra.edu/publications/slidesets/craters/slide_8.html). The 150 km diameter well-preserved multi-ring impact basin is revealed through the opaque atmosphere of Venus by the imaging radar system of the *Magellan* spacecraft. The Meitner impact structure has a flat smooth (dark) interior, two rugged circular ring structures with dark areas developed between them, and a bright irregular lobate deposit of ejecta. The crater was formed on smooth volcanic plains cut, prior to impact, by abundant parallel northeast-trending fractures. Radar image data were not acquired over the linear zone running through the eastern portion of this image mosaic. (B) Map showing one possible interpretation of the primary morphology of the Sudbury Structure immediately after impact with the location of the melt sheet, Offset Dykes, and the thickness of the melt sheet shown to illustrate the concepts in the text; (C) Section through the melt sheet showing the development of concentrations of dense immiscible magmatic sulfide ore deposits which vary in magnitude depending on the thickness and size of the melt that is scavenged. (D) Final configuration of the melt sheet showing the variation in thickness of the Main Mass and the location of large ore deposits and small mineral occurrences. Abbreviations: H, Hess Offset; Cs, Cascaden Offset; ▶

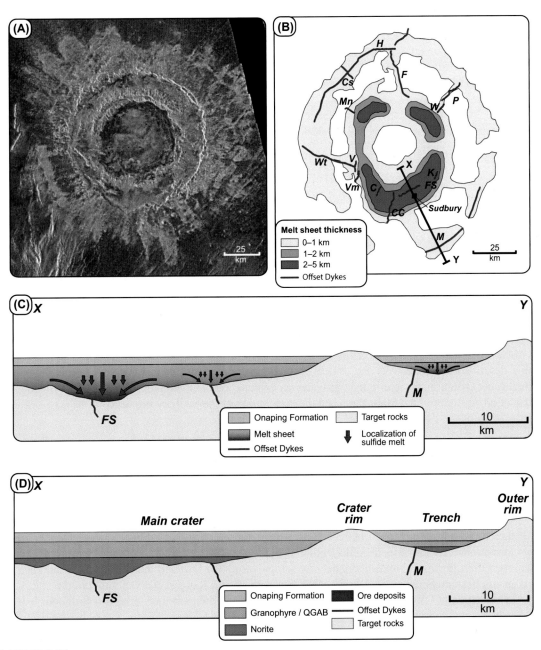

◀ **FIGURE 3.34**

Mn, Ministic Offset; Wt, Worthington Offset; V, Victoria and Ethel Lake Offset; Vm, Vermillion Offset segment; C, Creighton Offset; CC, Copper Cliff Offset; M, Manchester Offset, W, Whistle Offset, P, Parkin Offset, FS, Frood-Stobie Offset; K, Kirkwood Offset. Panels C and D depict section X–Y in map B.

(A) After Koeberl and Sharpton (2015) (D) Developed and modified after Golightly (1994), Lightfoot et al. (2002), Ames et al. (2002) and Ripley et al. (2015).

varied considerably in thickness in response to crater-floor topography. The analogue used for these models follows that depicted in Ames et al. (2005) and uses an impact structure on Venus (the Meitner impact structure) which is approximately the same size as the Sudbury Structure.

The impact crater consists of a central uplift, a surrounding melt sheet that on-laps the crater floor and varies in thickness, and an outer ring with a thinner discontinuous melt sheet (modified after Keays and Lightfoot, 2004 and Ames et al., 2002). The melt sheet possibly resembled the crater image shown in Fig. 3.34A, and it would vary in thickness as illustrated in hypothetical form in Fig. 3.34B based on a geometry and distribution that would fill the topographic lows of a crater similar to the Meitner impact structure (Fig. 3.34A).

The diversity in geochemistry through the melt sheet would reflect proximity to the center or margin of the sheet, and the style of Offset and compositional variability (with respect to sulfide content and perhaps also initial melt) would be aligned with the location of the dyke relative to the main melt sheet or the ring structure as illustrated in a slice through the impact structure shown in Fig. 3.34C (modified after Ames et al., 2002). The more mafic sections developed over thick Sublayer troughs correspond to deeper segments of the melt where larger quantities of sulfide were concentrated and where thicker intervals of mafic cumulate rocks were developed (eg, Creighton); the lateral parts of the main melt sheet may have significant embayment structures (eg, Trillabelle), but the melt sheet was thinner and there was a less well-developed thick cumulate stratigraphy represented by the noritic rocks (Fig. 3.34D).

In this model, the base of the QGAB would represent a subhorizontal marker horizon, and the stratigraphy of the rocks above this level would be produced by a common process of crystal differentiation throughout the GRAN unit.

The economic significance of this model is enormous (eg, Lightfoot, 2009), but Chapter 4 will first describe the ore deposits and critical features of the mineral systems that link to this model of melt sheet evolution. Chapter 5 then provides an explanation for the relationships between the metal-depletion signature of the Main Mass and the distribution of the Sudbury ore deposits.

SUMMARY

The following are the most important petrological and geochemical observations that help to support a model linking ore genesis to melt sheet evolution.

1. The SUBX. The matrix was formed at high pressure by comminution and frictional melting of the target rocks. The composition of the SUBX matrix was controlled by that of the target rocks, and there is no impact melt contribution. There is no evidence of sulfide-saturation accompanying the formation of frictional melts in the pseudotachylite breccias.

2. The Offset Dykes. Compositional data for samples from the QD Unit indicate small variations in composition between the Offsets of the North and South Ranges, and variability in the composition of Offsets developed in locations distal from the melt sheet or as discontinuous lenses in brecciated footwall. The composition of the average QD Unit provides a reasonable estimate of starting magma composition of the SIC, albeit that there is some primary heterogeneity in composition.

3. Timing of sulfide saturation: The QD Unit of the Offset Dykes records evidence for the exact timing of sulfide-saturation. Sulfide saturation predates differentiation of the SIC, and is timed to occur between the first phased of emplacement of melt to form the QD Unit of the Offset Dykes

and the second phase of emplacement of the sulfide-saturated magma that formed the MIQD and the Offset-hosted ore deposits.

4. Efficiency of sulfide segregation from the melt sheet. The melt sheet was possibly a "perfect smelter". Sulfide-saturation of the noritic melt and gravitational concentration of the magmatic sulfide into the "container" embayments and Offset Dykes; this process produced a stratigraphy of metal-depleted norites which records evidence of a very efficient concentration mechanism that possibly required the melt to be superheated.

5. Local development of Sublayer troughs and embayment structures: The localization of sulfide melts in depressions along the uneven base of the SIC in embayments is recorded together with a local effect on the composition of the Sublayer matrix which results from assimilation of fragments and possibly also the crater floor.

6. In situ differentiation of the melt sheet. The evolution of the North Range was different from that of the South Range. The South Range shows an upwards trend towards more mafic composition rocks that reaches an apogee in a discontinuous lens of melanorite, before trending to more evolved compositions. In the North Range the stratigraphy is very different with a basal MNR having the most melanocratic cumulate rocks in the SIC. The MNR grades into an overlying series of progressively more evolved norites where the proportion of orthopyroxene falls and the amount of trapped liquid increases with increasing elevation. The mechanism by which the Main Mass underwent differentiation is not yet fully understood, but it appears possible that the primary melt sheet comprised a layered mafic-felsic melt that underwent cooling from the base-up and top-down by crystal differentiation (see Chapter 5).

Vein of Massive Sulfide From the McCreedy East 153 Orebody; the Purple Mineral is Bornite and the Margin Consists of Millerite. The vein cuts through Levack Gneiss.

THE MINERAL SYSTEM CHARACTERISTICS OF THE SUDBURY Ni-Cu-Co-PGE SULFIDE ORE DEPOSITS

INTRODUCTION

The world-class scale of the ore deposits that comprise the Sudbury Mining Camp is summarized in Chapter 1; the ore deposits represent the second largest known concentration of magmatic sulfide mineralization after Noril'sk-Talnakh, and they have provided a fundamental source of wealth to the region and to Canada as a whole. The ore deposits at Sudbury have supported mine production for over 100 years and they are now widely accepted to have formed by the segregation, settling, and localization of dense immiscible sulfide from the impact-generated melt sheet that produced the SIC (Naldrett, 2004; Keays and Lightfoot, 2004).

The ore deposits associated with the SIC fall into two main groups, namely, (1) deposits hosted by continuous and discontinuous radial and concentric Offset Dykes, and (2) deposits associated with the Sublayer (norite and granite breccias) that are contained in troughs and embayments at the base of the Main Mass, or as vein stock-works in the underlying footwall. These geological environments contain almost all of the known Ni-Cu-Co-PGE sulfide ore deposits at Sudbury.

In the majority of cases, the mineralization at Sudbury consists of pyrrhotite, chalcopyrite, and pentlandite associated with igneous rocks of the Sublayer and Offset Dykes. Less commonly, mineralization consisting of chalcopyrite, pentlandite, cubanite, millerite, and bornite is developed in the footwall beneath the Sublayer. The sulfides range in texture from disseminated, interstitial, and blebby (in norite and grabite breccia, or Quartz Diorite (QD) as described in Chapter 2), through inclusion-rich sulfides that contain mafic and ultramafic fragments, to massive sulfide. The contact mineralization hosted by the Sublayer is often rich in Ni and Co whereas the mineralization in the adjacent footwall and within the Offset Dykes is richer in Cu and precious metals, although it retains a significant amount of Ni and Co.

The controls on mineralization at Sudbury are dominantly primary magmatic processes, but the detailed geology also reflects the controls provided by local country rocks and the effects of syn-magmatic and postmagmatic deformation and alteration. The similarities and differences between ore deposits and their host rocks offer an insight to the relative importance of primary and secondary process controls in the formation of the mineralization.

This chapter focuses on specific areas of the SIC that contain orebodies that represent the source of more than 90% of the historic metal production from the Sudbury Region. This chapter also describes deposits where mining has just started or where exploration potential exists to support a future production decision.

Nickel Sulfide Ores and Impact Melts. http://dx.doi.org/10.1016/B978-0-12-804050-8.00004-3

OBJECTIVES OF THIS CHAPTER

This chapter provides a detailed description of the ore deposits associated with the host rocks that are described in Chapters 2 and 3, and it provides a basis for genetic models that link the sulfide ores to the melts that were produced by impact processes (Chapter 5).

Geological models are provided for ore deposits associated with the Sublayer (and adjacent footwall), and the Offset Dykes. The description of each of the deposits starts with a short review of the discovery and historic production in each location. The geology of the host rocks and the mineralization is described in terms of the observed in situ mineralization in the underground rock-face and/or representative hand-samples of the mineralization. The geochemistry of the sulfide ores is presented using the ratios of the ore-forming elements that are sensitive to the proportions of the different sulfide minerals in the rock; this is the basis for a summary of the geological relationships.

The Frood-Stobie, Copper Cliff, and Totten mineral systems are described as examples found within Offset Dykes (Fig. 4.1). The review of contact and footwall mineral systems in the South Range places emphasis on the mineralization at Creighton and Garson, but descriptions are also included of the mineralization at Crean Hill, Murray, McKim, Little Stobie, Blezard, Lindsley, and Falconbridge (Fig. 4.1).

The focus of the description of the mineralization in the North and East Range deposits shown in Fig. 4.1. Levack, Coleman, Victor, and Nickel Rim South are examples of large deposits, whereas Capre, Whistle-Podolsky, and Trillabelle are smaller in scale. The North and East Range ore deposits are less deformed and metamorphosed than those located in the South Range, so they provide insight on the primary processes of formation.

The chapter is completed with a geological model that depicts the evolution of mineral systems associated with the Sublayer and immediate footwall. This model highlights the role of postimpact crater stabilization and late deformation processes. The relative roles of primary magmatic process related to sulfide differentiation and crater stabilization, and the overprint provided by regional deformation are reviewed. The importance of sulfide differentiation in the genesis of mineralization in the footwall of the South Range contact mineral systems is discussed.

Although this chapter will not address these questions directly, it will provide the information required to build a holistic model of impact, the evolution of the SIC, and the formation of the ore deposits. Specifically, it provides information that helps to address the following questions which will be addressed in Chapter 5:

1. What process controls influenced whether a contact mineral system underwent differentiation, and do contact deposits necessarily have associated footwall mineral zones?
2. Why do some parts of the basal contact of the SIC have giant ore deposits yet other parts appear to be bereft of economically significant mineralization?
3. Why do some Offset Dykes contain large ore deposits yet others contain only subeconomic concentrations of sulfide mineralization?
4. Does the known distribution of ore deposits reflect incomplete exploration, or does it reflect real spatial differences in the mineral potential of the SIC created by the impact and postimpact processes?

This chapter is designed to provide the reader with a detailed understanding of the largest and most economically important mineral systems at Sudbury, but it would not be complete without the

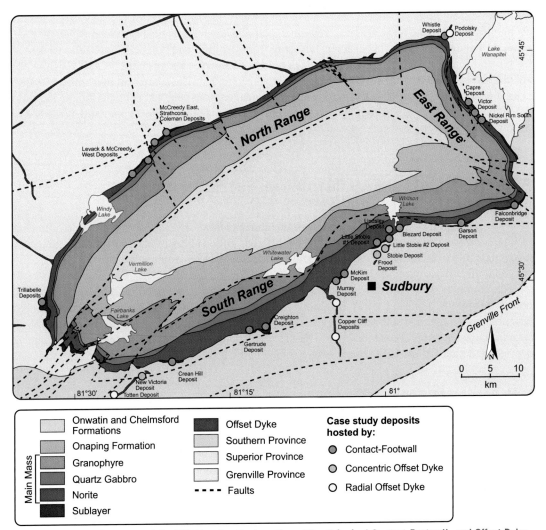

FIGURE 4.1 Geological Map of the SIC Showing the Location of the Principal Contact-Footwall- and Offset Dyke-Associated Ore Deposits (Discussed in Chapter 4)

inclusion of some examples of mineralization that are either subeconomic or have a small amount of contained metal.

THE MINERAL SYSTEM CONCEPT

Following the approach of Barnes et al. (2015), the concept of a mineral system as it relates to magmatic sulfide ore deposits is used to describe the geological environment that hosts each of the Sudbury ore deposit. The "mineral system" concept is the basis of classic studies of ore deposits, where the economically important mineral zones are a small part of a larger geological system in which they were formed (Stanton, 1972). In this chapter, the terms "ore deposit" and "orebody" are used to describe the mineral zones that are compliant with the definition of reserves and resources (CIM, 2010), or they are used in the sense that the material has previously been mined. The concept of the "mineral system" is used to describe the rocks that comprise the geological environment that contains the ore deposit or the individual orebodies. The term "mineral system" has no inherent implication that the material can be mined at a profit.

PETROLOGY AND MINERALOGY OF THE SULFIDE MINERALIZATION AT SUDBURY

This section provides a simplified overview of the nomenclature used to describe the style of mineralization, the mineralogical relationships between different sulfide minerals, the fundamental paragenetic associations between the sulfide minerals, and the variations in mineral assemblages in contact, footwall and offset types of mineralization.

The terminology used to describe the sulfide mineralization is provided in Table 4.1. This terminology covers the dominant mineralization styles at Sudbury as well as that used internationally to describe other magmatic sulfide ore deposits (Naldrett, 2004). Although there are many examples of specific and local terms used at Sudbury to describe mineralization styles on the scale of individual ore deposits and mines, these have largely been avoided in the text.

MINERALOGY OF THE SULFIDE ORES

Hawley and Stanton (1962) wrote the first systematic description of the mineralogy of the Sudbury ores in which they documented the presence of 32 different sulfide minerals. Cowan (1968) established the composition of many of these minerals with a microprobe, and Cabri and Laflamme (1976) established aspects of the compositional diversity in the precious metal minerals. The principal metallic sulfide minerals identified in the ores are shown in Table 4.2 and references are cited in the table. Many other papers and academic studies describe the sulfide mineralogy of individual samples and groups of samples (some of which are questionably representative of the overall mineral system) and other examples of which were carefully collected to represent the sulfide types and textures throughout a domain of the mineral system (Dare et al., 2010a,b, 2011, 2012, 2014). There is no single publication that describes the broader geometallurgy of the Sudbury ores, although the mining companies document these characteristics in detail in order to properly plan to mine and process the ores. Overview studies of the sulfide mineralogy are available in Naldrett (1984b, 2004), Farrow and Lightfoot (2002), and Ames et al. (2003).

Tables 4.2–4.4 provides a summary of the sulfide minerals with some information on their composition, the location where it is commonly developed or recognized, and sources for further reading.

The principal Fe-S-Ni minerals
Naldrett (1984b) reported that the most abundant sulfide minerals in contact ore types are pyrrhotite and pentlandite, with lesser pyrite (Table 4.2). The pyrrhotite and pentlandite crystallized from a

Table 4.1 Terminology Used to Describe the Diversity in Textural Relationships Between Sulfide Mineralization and the Host Silicate Rocks at Sudbury

Term Used to Describe Sulfide Texture	Definition (Common Usage)	Common Host Rocks	Examples of Mineral Systems with This Texture
Trace sulfide	Typically <1% finely diseminated sulfide in igneous rock	QD, MNR, SLNR, GRBX	Most contact and Offset-hosted mineral systems
Disseminated sulfide	1–5% sulfide occuring evenly distributed as an intercumulus mineral in a silicate rock	QD, MNR, SLNR, GRBX	Most contact and Offset-hosted mineral systems
Ragged disseminated sulfide	1–5% sulfide typically penetrated by silicate minerals and unevenly distributed in silicate rock	Sublayer melanorite, MIQD	Creighton, Whistle, Frood-Stobie
Interstitial sulfide	5–15% sulfide that fills the space between cumulate crystals in an inclusion-poor norite or QD; forms a semicontinuous or continuous network	Sublayer melanorite	Creighton, Whistle
Blebby sulfide	Round to ovate magmatic-textured sulfide blebs surrounded by inclusion-poor silicate rock; typically 5–20% sulfide.	SLNR, MIQD, HMIQD	Whistle
Coalesced blebby sulfide	Same as blebby sulfide, but blebs tend to coalesce together to form a continuous network; typically 20–40% sulfide	HMIQD	Frood-Stobie
Inclusion-bearing massive sulfide (INMS)	Inclusion or matrix-supported sulfide mineralization	Sublayer and Offset environments	Frood-Stobie, Creighton, Totten, Whistle, Victor
Contorted schist inclusion sulfide	Inclusions of locally-derived schist that has been previously deformed and now resides in the sulfide	Mineral zones that follow structures	Garson shears zones, Frood, Falconbridge
Gabbro-peridotite inclusion sulfide	Inclusion or matrix-supported mineralization such as INMS that comprises principally mafic and ultramafic inclusion types	Sublayer and Offset environments	Frood-Stobie, Creighton, Totten, Whistle, Victor
Semimassive sulfide	Sulfide with occasional floating silicate rock inclusions; 50–75% sulfide	Sublayer and Offset environments	Frood-Stobie, Creighton, Totten, Whistle, Victor
Massive sulfide	Sulfide with <25% inclusions	Sublayer and Offset environments	Frood-Stobie, Creighton, Totten, Whistle, Victor

(Continued)

Table 4.1 Terminology Used to Describe the Diversity in Textural Relationships Between Sulfide Mineralization and the Host Silicate Rocks at Sudbury (cont.)

Term Used to Describe Sulfide Texture	Definition (Common Usage)	Common Host Rocks	Examples of Mineral Systems with This Texture
Veinlet or fine stringer sulfide	Fine (<2 mm wide) connected networks of fine connected veins of sulfide in footwall rocks	Country rocks	Footwall mineral systems
Stringer sulfide	Narrow veins (2 mm–10 cm) forming connected networks of sulfide in footwall rocks	Country rocks	Footwall mineral systems
Vein sulfide	Wide (10 cm to several meters) continuous veins of massive sulfide in country rocks with sharp contacts	Country rocks	Footwall mineral systems
Low sulfide high precious metal mineralization (LSHPM)	A style of mineralization typically consisting of fine disseminations of Cu-PGE-rich mineralization in country rocks that are brecciated, metamorphosed, and partially melted; typically near a footwall mineral systems	Country rocks	Footwall mineral systems
Sulfide clast	Fragment of sulfide with subrounded to angular contacts that floats in a matrix of SLNR or Offset Dyke QD	MIQD, HMIQD, GRBX	Frood-Stobie; Levack-Coleman contact mineralization

Although there are some local terms in usage, this list can be used to describe the majority of contact, footwall, and Offset Dyke styles of mineralization.

Table 4.2 Major Rock-Forming Sulfide Minerals and the Typical Composition

Mineral Name	Empirical Formula	Fe (wt.%)	Ni (wt.%)	Co (wt.%)	Cu (wt.%)	S (wt.%)	Development	Ore Deposit Environment	References
Pyrrhotite (hexagonal)	$Fe_{(1-x)}S$	62.33	(0.1–1.4% at expense of Fe)	(0–0.15% at expense of Fe)	0	37.67	Most common sulfide in most mineral systems	Dominantly contact, offset and breccia belt	Cowan (1968); Corlett (1971)
Pyrrhotite (monoclinic)	$Fe_{(1-x)}S$	62.33	(0.1–1.4% at expense of Fe)	(0–0.15% at expense of Fe)	0	37.67	Most common sulfide in most mineral systems	Dominantly contact, offset and breccia belt	Cowan (1968); Corlett (1971)
Chalcopyrite	$CuFeS_2$	30.4	0	0	34.6	34.9	Most common copper mineral in most mineral systems	Contact, footwall, breccia belt, offset	Hawley and Stanton (1962)
Pentlandite	$(Fe, Ni)_9S_8$	32.6	34.2	(0.5–2% at expense of Ni)	0	33.2	Most common nickel mineral in most mineral systems	Contact, footwall, breccia belt, offset	Hawley and Stanton (1962)
Cubanite	$CuFe_2S_3$	41.2	0	0	23.4	33.4	Commonly exsolved in chalcopyrite or pyrrhotite	Footwall and breccia belt	Hawley and Stanton (1962); Abel et al. (1979); Cabri and Laflamme (1976); Farrow and Lightfoot (2002)
Pyrite	FeS_2	46.6	(Can reach 8% at expense of Fe)	(Can reach 1.1% at expense of Fe)	0	53.5	Common as secondary mineral; occasionally primary	Typically structurally modified	Naldrett and Kullerud (1967)
Millerite	NiS	0	64.7	0	0	35.3	Highly differentiated sulfides at vein margins	Footwall	Ames et al. (2003)
Bornite	Cu_5FeS_4	11.1	0	0	63.3	25.6	Specific parts of highly fractionated terminations of veins	Footwall	Hawley and Stanton (1962)

There is quite a wide range in Ni and Co content of pyrrhotite and pentlandite within and between the different contact ore deposits. There is no clearly established relationship between the concentration of Ni or Co in monoclinic versus hexagonal pyrrhotite.

Table 4.3 Minor Sulfide and Sulfosalt Minerals Associated with the Sulfide Ore Deposits

Mineral Name	Minor/ Trace	Formula	Ore Deposit Environment	References
Acanthite	Trace	Ag_2S	North Range footwall	Jago et al. (1994)
Altaiite	Trace	PbTe	Structurally modified sulfides and footwall sulfides	Cabri and Laflamme (1974, 1976); Dare et al. (2012)
Altaite	Trace	PbTe		Cabri and Laflamme (1976); Farrow (1995); Jago et al. (1994); Farrow and Watkinson (1997); Rickard and Watkinson (2001); Farrow and Lightfoot (2002)
Argentiferous pentlandite	Minor	$Ag(Fe, Ni)_8S_8$	Footwall	Cabri and Laflamme (1976); Farrow and Lightfoot (2002)
Bismuthinite	Trace	Bi_2S_3		Cabri and Laflamme (1976); Springer (1989); Farrow (1995); Farrow and Watkinson (1997)
Breithauptite	Trace	$Ni_7(Bi, Sb, Te)_2S_8$		
Cassiterite	Trace	SnO_2	Dominantly North Range Footwall	Farrow (1995); Jago et al. (1994)
Chalcocite	Minor	Cu_2S	North Range footwall	
Chlorargyrite	Trace	AgCl	North Range footwall	Farrow (1995)
Clausthalite	Trace	PbSe	North Range Footwall (McCreedy East 153)	Dare et al. (2012)
Cobaltite	Minor	CoAsS	Structurally modified sulfides	Cabri and Laflamme (1976); Jago et al. (1994); Stewart (2002); Carter et al. (2005); Rickard and Watkinson (2001); Farrow and Lightfoot (2002)
Covellite	Trace	CuS		Farrow and Lightfoot (2002)
Electrum	Trace	Au, Ag		Springer (1989); Money (1993); Chen (1993); Farrow (1995); Jago et al. (1994); Farrow and Watkinson (1997); Farrow and Lightfoot (2002)
Empressite	Trace	AgTe	North Range footwall	Jago et al. (1994)
Galena	Minor	PbS	Footwall and structurally modified sulfides	Hawley and Stanton (1962); Farrow (1995); Jago et al. (1994); Farrow and Watkinson (1997); Carter et al. (2005); Farrow and Lightfoot (2002)
Gersdorffite	Minor	NiAsS	Structurally modified sulfides	Cabri and Laflamme (1976); Jago et al. (1994); Stewart (2002); Carter et al. (2005); Rickard and Watkinson (2001); Farrow and Lightfoot (2002)

Table 4.3 Minor Sulfide and Sulfosalt Minerals Associated with the Sulfide Ore Deposits (*cont.*)

Mineral Name	Minor/ Trace	Formula	Ore Deposit Environment	References
Gold	Trace	Au	Footwall and Offset	Michener (1940) and Hawley and Stanton (1962)
Hauchecornite	Trace	$Ni_9Bi(Sb, Bi)S_8$	Breccia Belt	Gait and Harris (1972); Cabri et al. (1973)
Hawleyite	Trace	CdS		Farrow and Lightfoot (2002)
Hessite	Trace	Ag_2Te	Most deposit types in North and South Ranges	Hawley and Stanton (1962); Li and Naldrett (1992, 1993a); Farrow (1995); Jago et al. (1994); Farrow and Watkinson (1997); Everest (1999); Stewart (2002); Carter et al. (2005); Rickard and Watkinson (2001); Farrow and Lightfoot (2002); Dare et al. (2012)
Mackinawite	Trace	Fe_9S_8		Farrow and Lightfoot (2002)
Marcasite	Minor	FeS_2	Structurally modified sulfides	Hawley and Stanton (1962); Cabri and Laflamme (1976); Farrow and Lightfoot (2002)
Maucherite	Trace	$Ni_{11}As_9$	Structurally modified sulfides	Hawley and Stanton (1962); Cabri and Laflamme (1976); Farrow and Lightfoot (2002)
Melonite	Trace	$NiTe_2$		Cabri and Laflamme (1976); Farrow (1995); Jago et al. (1994); Farrow and Watkinson (1997); Rickard and Watkinson (2001); Farrow and Lightfoot (2002); Dare et al. (2012)
Molybdenite	Trace	MoS_2		Farrow and Lightfoot (2002)
Native Bismuth	Trace	Bi	Breccia Belt	Hawley and Stanton (1962); Farrow (1995)
Native silver	Minor	Ag	Footwall	Hawley and Stanton (1962); Farrow (1995); Jago et al. (1994); Farrow and Watkinson (1997)
Naumannite	Trace	Ag_2Se	North Range footwall	Farrow (1995)
Niccolite	Minor	NiAs	Structurally modified sulfides	Hawley and Stanton (1962)
Parkerite	Trace	$Ni_3Bi_2S_2$		Michener (1940); Hawley and Stanton (1962)
Schapbachite	Trace	$AgBiS_3$	Breccia Belt	Hawley and Stanton (1962)
Sphalerite	Minor	ZnS	Footwall and structurally modified sulfides	Hawley and Stanton (1962)
Tetradymite	Trace	Bi_2Fe_2S		Hawley and Stanton (1962)

(Continued)

Table 4.3 Minor Sulfide and Sulfosalt Minerals Associated with the Sulfide Ore Deposits *(cont.)*

Mineral Name	Minor/ Trace	Formula	Ore Deposit Environment	References
Valleriite	Minor	$(Fe, Cu, Ni)_2S_2$, $(Mg, AlCa)(OH)_2$	Footwall and Offsets	Hawley and Stanton (1962)
Violarite	Minor	$FeNi_2S_4$	Supergene oxidized pentlandite	Hawley and Stanton (1962)
Wehrlite	Trace	$(Bi, Pb), (Te, Sb)$		Cabri and Laflamme (1976); Cook et al. (2007)
Tellurbismuth-inite	Trace	Bi_2Te_2	North Range footwall (Strathcona Deposit)	Dare et al. (2012)
Talnakhite	Trace	$Cu_{18}(Fe,Ni)_{16}S_{32}$	North Range footwall (Strathcona Deposit)	
Lawrencite	Trace	$FeCl_2$	North Range footwall (Strathcona Deposit)	
Galenobis-mutite	Trace	$PbBi_2S_4$	North Range footwall (Strathcona Deposit)	
Dyscrasite	Trace	Ag, Sb	North Range footwall (Strathcona Deposit)	
Akaganeite	Trace	Beta FeOH	North Range footwall (Strathcona Deposit)	
Voynskite	Trace	$AgBiTe_2$	North Range footwall (McCreedy East 153)	Dare et al. (2012)
Tsumoite	Trace	$BiTe$	North Range footwall (McCreedy East 153)	Dare et al. (2012)
Cotunnite	Trace	$PbCl_2$	North Range footwall (McCreedy East 153)	Dare et al. (2012)
Stannopalla-dinite	Trace	Pd_3Sn_2	North Range footwall (strathcona)	

primary monosulfide solid solution (MSS) which is one of the first phases to form during the cooling of the sulfide melt (Naldrett, 2004). Two distinct forms of pyrrhotite have been identified; these are ferromagnetic monoclinic pyrrhotite and diamagnetic hexagonal pyrrhotite. Typically, the monoclinic variety is less common and occurs in three forms, namely, (1) rims to grains of hexagonal pyrrhotite; (2) coatings along fractures; and (3) surrounding the flames of pentlandite that have exsolved from hexagonal pyrrhotite. The example shown in Fig. 4.2A shows the distribution of monoclinic versus hexagonal pyrrhotite as picked out using a colloid method to stain and highlight the two types (Craig and Vaughan, 1981; Guo and Vokes, 1996).

In mineral systems that have been modified by deformation, pyrrhotite can show evidence of deformation of the crystal structure in the form of brittle fracture, slip and minor kinking, and the development of twin gliding as described in Clark and Kelly (1973) and Duuring (2003) (Fig. 4.2B,C).

Table 4.4 Precious Metal Minerals Developed in the Sulfide Ore Deposits and the Nearby Country Rocks

Mineral Name	Formula	Ore Deposit Environment	Range	References
Froodite	$PtBi_2$			Hawley and Berry (1958); Cabri and Laflamme (1976)
Geversite	$PtSb_2$	Footwall	North	Jago et al. (1994)
Hollingworthite	RhAsS	Offset	South	Cabri and Laflamme (1984); Stewart (2002); Carter et al. (2005); Farrow and Lightfoot (2002)
Hongshiite	PtCu	Unknown (concentrate)		Cabri and Laflamme (1984)
Insizwaite	$PtBi_2$	Footwall	North	Cabri and Laflamme (1976, 1984); Farrow (1995); Li and Naldrett (1992, 1993a); Farrow and Lightfoot (2002)
Irarsite	IrAsS	Offset	South	Cabri and Laflamme (1984); Rickard and Watkinson (2001)
Kotulskite	PdTe			Cabri and Laflamme (1976)
Maslovite	PtBiTe	Footwall	North	Jago et al. (1994); Farrow and Lightfoot (2002)
Merenskyite	$PdTe_2$			Cabri and Laflamme (1976)
Mertieite II	$Pd_8(Sb, As)_3$			Cabri and Laflamme (1976)
Michenerite	PdBiTe	Contact, footwall, offset and breccia belt		Cabri and Laflamme (1976); Hawley and Berry (1958)
Moncheite	$PtTe_2$	Footwall	Dominantly North	Cabri and Laflamme (1976, 1984); Springer (1989); Jago et al. (1994), Farrow (1995); Farrow and Watkinson (1997)
Niggliite	PtSn	Footwall	North	Cabri et al. (1973); Cabri and Laflamme (1976, 1984); Li and Naldrett (1992, 1993a); Farrow (1995); Jago et al. (1994); Farrow and Watkinson (1997)
Palladian melonite	$(Ni, Pd)(Te, Bi)_2$	Footwall	North	Cabri and Laflamme (1976); Farrow (1995); Farrow and Watkinson (1997)
Rhodarsenide	Rh_2As	Unknown (concentrate)		Cabri and Laflamme (1984)
Ruarsite	RuAsS	Offset	South	Carter et al. (2005)
Ruthenium	RuAsS	Unknown (concentrate)		Cabri and Laflamme (1984)
Sperrylite	$PtAs_2$	Contact, footwall, and offset	Dominantly South Range	Cabri and Laflamme (1976, 1984); Hawley and Stanton (1962); Li and Naldrett (1993a); Farrow (1995); Jago et al. (1994); Farrow and Watkinson (1997); Stewart (2002); Carter et al. (2005); Rickard and Watkinson (2001); Farrow and Lightfoot (2002)
Sudburite	PtSb			Cabri and Laflamme (1976)
Paolovite	Pd_2Sn	Footwall	North	Dare et al. (2012)
Padmaite	PdBiSe	Footwall	North	Dare et al. (2012)

The references are provided as a source of additional information.

The textural relationships between pyrrhotite, pentlandite, and chalcopyrite in mineral zones in the Offset Dykes and the Sublayer are generally typical of magmatic sulfide ore deposits (Naldrett, 2004). Pentlandite occurs as large (1–3 cm diameter) crystals, often called "eyes," within and at the margins of much larger pyrrhotite grains (Fig. 4.2D). Smaller grains of pentlandite typically occur in association with some chalcopyrite at the margins of the larger pyrrhotite crystals in a "loop texture" relationship [the term "loop texture" is used to describe similar sulfide textures in pyrrhotite-rich sulfide mineral zones at Noril'sk (Lightfoot and Zotov, 2014)]. The extent to which loops of chalcopyrite

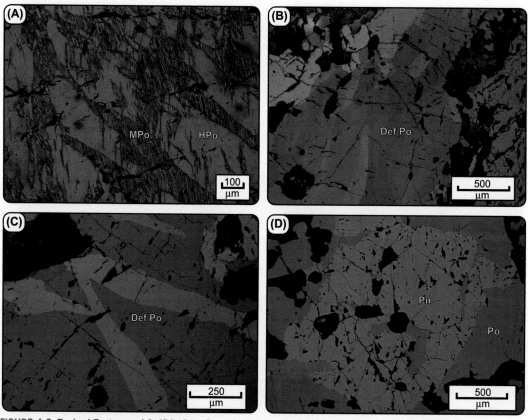

FIGURE 4.2 Typical Textures of Sulfide Ores From Sudbury

(A) Intergrowth textures of hexagonal pyrrhotite (HPo) and monoclinic pyrrhotite (MPo) shown after colloidal staining of the polished section (sample 21083357) from the Stobie mineral system. The colloid attaches to the monoclinic magnetic variant as shown in the image (Craig and Vaughan, 1981; Guo and Vokes, 1996). (B) Deformation texture of pyrrhotite (DefPo) showing the effect of low temperature deformation (Monteiro and Krstic, 2006); the sample (21100635) is from the Garson mineral system. (C) An example of planar deformation in pyrrhotite (DefPo) from the Garson mineral system (sample 21091128). (D) Granular pentlandite (Pn) hosted in pyrrhotite (sample VTDL002) is from the Copper Cliff mineral system.

plus pentlandite are developed around pyrrhotite crystals is a function of both the process of exsolution of Ni-rich and minor-Cu-rich sulfides from the monosulfide solid solution, and the degree of expulsion of trapped Fe-poor sulfide melt from between the monosulfide solid solution. In some cases, the sulfide textures comprise loops of pentlandite and chalcopyrite surrounding the pyrrhotite, providing evidence to indicate that these sulfides were formed by crystallization of monosulfide solid solution leaving a residual (Cu + Ni)-rich trapped liquid (Fig. 4.2E). There is a wide variation in the proportion of loop-textured chalcopyrite + pentlandite ± pyrite relative to pyrrhotite (Fig. 4.2E–G).

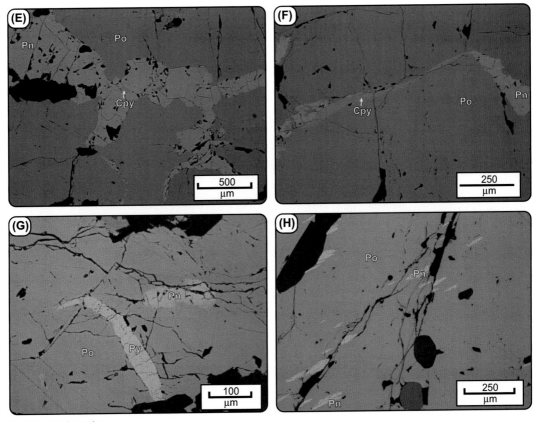

FIGURE 4.2 (*cont.*)

(E) Loop-textured matrix of pentlandite with minor chalcopyrite (Cpy) surrounding pyrrhotite (Po) (sample VTDL001) from the Copper Cliff mineral system. (F) Loops of chalcopyrite + pentlandite surrounding pyrrhotite in sample VTDL0007 from the Creighton Mine 400 Orebody. (G) Loop-textured primary pyrite at the margin of pyrrhotite and pentlandite (sample VTDL0012) from the Victor mineral system located in the contact mineral zone. (H) Flame-textured pentlandite (Pn) in pyrrhotite (Po) (sample VTDL0008) from the Creighton 400 Orebody.

Pentlandite also occurs as 1–4 μm thick, 10–100 μm long lamellae (or "flames") oriented parallel to the 0001 cleavage of pyrrhotite (Fig. 4.2H; Durazzo and Taylor, 1982). Pentlandite that occurs as very fine flames is not easily liberated from the pyrrhotite during mineral processing, so it is especially important to understand the relative proportions of flame pentlandite that is locked in pyrrhotite versus the more granular pentlandite that is easily liberated.

The formation of the commercially most-important and second-most common nickel sulfide mineral (pentlandite) in the Sudbury ores is considered to be a low temperature phenomenon (Kullurud, 1963; Kullerud et al., 1965, 1969). Some of the Ni that does not exsolve as pentlandite remains in solid solution in pyrrhotite. Corlett (1971) reported values of 0.1–1.2 wt.% Ni in pyrrhotite. Experimental

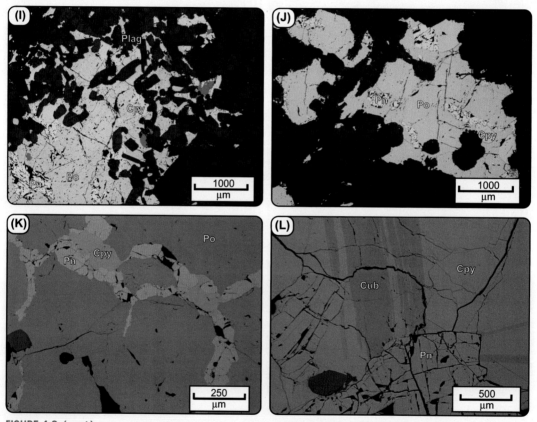

FIGURE 4.2 (cont.)

(I) Interstitial sulfide (pyrrhotite > pentlandite >= chalcopyrite) containing primary crystals of plagioclase (sample VTDL0016) from the Creighton 402 Orebody. (J) Interstitial sulfide mineralization comprising Po, pentlandite, and chalcopyrite developed between plagioclase and amphibole crystals (sample VTDL0014) from the Creighton 402 Orebody. (K) Loop-textured chalcopyrite + pentlandite between crystals of pyrrhotite from the Victor contact mineral zone (sample VTDL0022). (L) Flames of cubanite (Cub) in chalcopyrite with grains of pentlandite in chalcopyrite; this sample is from the Victor 28N footwall mineral zone (VTDL0019).

data indicates that pentlandite can form by exsolution from both monosulfide solid solution and from an intermediate solid solution (Naldrett, 2004; Ballhaus et al., 2001) that crystallize from differentiated Cu-rich sulfide melts.

The primary mechanism of concentration of Fe and Ni is the segregation of monosulfide solid solution from a sulfide melt at high temperatures (between 1000 and 1150°C; Kullerud et al., 1969). The textural form of the pentlandite is a reflection of both cooling rate and extent of metamorphic re-equilibration of the sulfide. Experimental results indicate that upper thermal stability limit of pentlandite is 610°C (Kullurud, 1963), but it may be as high as 865°C (Sugaki and Kitakaze, 1998). Pentlandite starts to revert back to monosulfide solid solution at 600°C and flames of pentlandite may be resorbed

FIGURE 4.2 (*cont.*)

(M) Millerite (Mil) and bornite (Bn) in the Victor 14N footwall mineral zone (sample VTDL0024). (N) Intergrowths of bornite and millerite with minor chalcopyrite in sulfide from the Victor 14N footwall mineral zone (sample VTDL0026). (O) Pyrrhotite grain containing a zoned crystal that consists of sperrylite (Spr) at the core, surrounded by irarsite (Ir), hollingworthite (Holl), and gersdorffite (Gdf). Sample VTDL0034 is from the Kelly Lake Orebody in the Copper Cliff Offset Dyke. (P) Chalcopyrite + pyrrhotite with an equant crystal of gersdorffite (Gdf) from the margin of a shear-hosted sulfide zone in the Garson mineral system (sample VTDL0032).

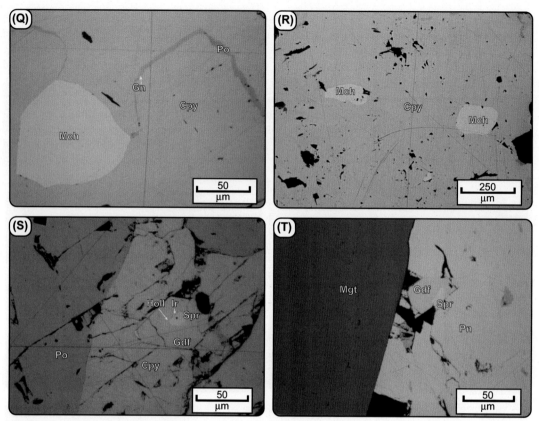

FIGURE 4.2 (*cont.*)

(Q) Crystal of moncheite (Mch) in chalcopyrite with loop-textured pyrrhotite and galena (gn). Sample VTDL0031 is from the Capre 3001 footwall mineral zone. (R) Polished sample showing the development of moncheite (Mch) in chalcopyrite from the Capre 3000 footwall mineral zone. (S) Samples consisting of pyrrhotite–chalcopyrite with a zoned grain consisting of sperrylite (Spr), irarsite (Ir), hollingworthite (Holl), and Gersdorffite (Gdf). Sample VTDL0036 is from the Kelly Lake Orebody in the Copper Cliff Offset Dyke. (T) Pentlandite in contact with magnetite (Mgt). The pentlandite contains small grains of sperrylite (Spr) surrounded by a grain of gersdorffite (Gdf). Sample VTDL0035 is from the Kelly Lake Orebody in the Copper Cliff Offset Dyke. Images 4.2A–T and petrographic descriptions were kindly provided by Ben Vandenburg.

at much lower temperatures (Naldrett et al., 1967; Craig and Vaughan, 1981). Ganular pentlandite enclosed in loop-textured chalcopyrite may have a higher upper thermal stability limit than pentlandite in monosulfide solid solution.

The textures of disseminated sulfides hosted in QD and norite are typically similar to those found in more massive sulfides, but the sulfides are typically finer-grained. Examples of pyrrhotite with lesser chalcopyrite and pentlandite formed dominantly from monosulfide solid solution are shown in Fig. 4.2I,J, and the magmatic relationships between the sulfide mineralization and the silicate minerals is illustrated by the development of poikilitic-textured plagioclase that penetrates into the sulfide (Fig. 4.2I). The disseminated sulfides show incipient development of loop-textured relationships between pyrrhotite and chalcopyrite + pentlandite (Fig. 4.2J).

There is a gradation in textures between the Fe-rich and Fe-poor sulfides, and this is recorded in contact type sulfides that grade from those dominated by pyrrhotite, through pyrrhotite separated by loops of pentlandite + chalcopyrite (Fig. 4.2K) to sulfides comprising chalcopyrite + pentlandite with lesser pyrrhotite. With a further decline in the proportion of pyrrhotite relative to chalcopyrite, the composition of this sulfide become very Fe-poor, and crystallizes as an intermediate solid solution (ISS) which consists of pentlandite grains within a matrix of chalcopyrite +/– cubanite (Table 4.2; Fig. 4.2L). Intermediate solid solution appears to stabilize between ~865–890°C (Kullerud et al., 1965; Dutrizac, 1976; Conard et al., 2007). The intermediate solid solution and residual differentiated sulfide melt typically form veins in the country rocks beneath the SIC, and they sometimes develop bornite and millerite at the vein-margins or at the terminations of the veins.

Pyrite occurs as cubic to heavily resorbed grains and as small (~1–10 mm diameter) euhedral grains at the margin of pyrrhotite. It also occurs between grains of pyrrhotite in a primary loop-textured mineral habit (Fig. 4.2G). The timing of pyrite formation is typically very late and associated with alteration and metamorphism, but the example shown in Fig. 4.2G appears to be a primary mineral formed from the sulfide melt at the edges of pyrrhotite crystals generated from the monosulfide solid solution. Naldrett and Kullerud (1967) pointed out that the composition of $Fe_{(1-x)}S$-$Ni_{(1-x)}S$ is continuous across the Fe-Ni-S system from solidus temperatures to below 300°C. They showed that the bulk composition of the Sudbury ores falls at the sulfur-rich side of the monosulfide solid solution. With cooling, the metal:sulfur ratio of monosulfide solid solution narrows (Naldrett et al., 1967) and Sudbury-type ore compositions will exsolve a small amount of pyrite. Pentlandite continues to crystallize until the monosulfide solid solution has broken down. Craig (1983) provided evidence that this will happen at very low temperatures (200–250°C). High-temperature magmatic pyrite associated with pentlandite is sometimes developed in metamorphosed magmatic nickel sulfide ore deposits (eg, South Manasan in the Thompson Nickel Belt, Franchuk et al., 2015).

Pentlandite in the Noril'sk and Thompson ore deposits is a principal host of Pd (Distler (1994); Lightfoot and Evans-Lamswood, 2015; Lightfoot, 2015). Cabri et al. (1984) established that Sudbury pentlandite contains 1.8 g/t Pd, and Dare et al. (2010a,b) determined the Pd concentrations in pentlandite in samples from the 402 Orebody in the Creighton mineral system to be ~1 g/t.

The principal Cu-Fe-S minerals

The principal Cu-Fe-S minerals are chalcopyrite, cubanite, and bornite (Table 4.2). Chalcopyrite accounts for by far the major part of the Cu in the Sudbury deposits. Cubanite occurs in many deposits in association with chalcopyrite, and is concentrated in the deeper parts of the Frood mineral system and in the Footwall Deposits at Coleman (the Strathcona and McCreedy East 153 Orebodies; Abel et al. 1979) and the Victor mineral system. Cubanite often occurs as exsolution blades in chalcopyrite (Dutrizac, 1976).

The Cu-Fe-S system exhibits large degrees of solid solution at temperatures immediately below the solidus, and the crystallization of Fe-poor sulfides generates an intermediate solid solution that has a profound effect on the composition of the residual sulfide melt that allows it to form sulfide minerals such as bornite and millerite, as well as native silver. In very Cu- and Ni-rich residual sulfide melts, the stable assemblage formed consists of chalcopyrite + cubanite + pentlandite (Fig. 4.2L), millerite + bornite ± chalcopyrite (Fig. 4.2M,N), and native silver.

Minor and trace sulfide minerals

Sphalerite and galena occur in very minor amounts, particularly in Cu-rich footwall mineral zones and as late, cross-cutting veins in association with late shear zones and faults (Table 4.3). Sphalerite and galena also occur as interstitial grains in massive sulfide and adjacent to silicate inclusions. The arsenides, niccolite and maucherite, and sulfarsenides, gersdorffite and cobaltite, commonly occur together, but are largely restricted to the South Range mineral systems that have undergone some degree of structural modification (Fig. 4.2O,P; Hawley and Stanton, 1962; Cabri and Laflamme 1976). These observations are consistent with the much higher arsenic concentrations in South Range deposits (a factor of 20 times that of the North Range; Naldrett, 1984b). Compositional ranges for cobaltite and gersdorffite overlap, with Co ranging between the two end-members of a solid solution series from 5 to 21 wt.%, and Ni from 9 to 25 wt.% (LeFort, 2012).

The platinum group minerals

Cabri and Laflamme (1976) identified or confirmed the presence of 13 platinum group minerals (PGMs) in the Sudbury ores, and since that time, many studies have been undertaken that document additional PGMs as well as new types of PGM (Farrow and Lightfoot, 2002 and Table 4.4).

Cabri and Laflamme noted that the South Range and Offset deposits have common As-bearing minerals but very few Sn-bearing minerals, whereas Sn-bearing minerals are present and As-bearing minerals are rare in the North Range mineral systems. Cabri and Laflamme (1976) also identified significant amounts of platinum group elements (PGE) in solid solution in the As-bearing minerals. Cobaltite contains from 1,200 to 7,600 ppm Pd, up to 1,900 ppm Pt, and up to 15,000 ppm Rh; gersdorffite contains from 500 to 2,000 ppm Pd, up to 2,500 ppm Pt, and up to 30,000 ppm Rh; niccolite contains up to 400 ppm of each of Pd, Pt, and Rh; and maucherite contains up to 1,000 ppm Pd, and up to 500 ppm of Pt and Rh.

Dare et al. (2010a,b) report detailed petrology, mineralogy, and laser ablation compositional data for sulfide samples from the 402 orebody located in the Creighton mineral system. This study provided evidence that the early crystallization of magmatic PGE-bearing trace sulfarsenide minerals from the sulfide melt had a far more important impact on the partitioning of PGE than any late magmatic, hydrothermal, or metamorphic processes. Dare et al. (2010a,b) argue that the arsenide mineralogy is a product of high-temperature magmatic processes. If this is correct, then the principal controls on the distribution of arsenide minerals may be primary magmatic processes rather than low temperature modification of the ores in ductile deformation zones accompanying sulfide kinesis.

Fig. 4.2Q–T provides examples of the PGM relationships in examples of Sudbury sulfides. The PGMs exhibit a complex relationship between one another and the sulfide mineral hosts. The textures are indicative of a wide range in paragenesis from early crystallization to late remobilization (Fig. 4.3). Ames et al. (2003) report detailed studies on the composition and textural relationships between PGMs, and the reader is pointed to this source for detailed information that is beyond the scope of this book.

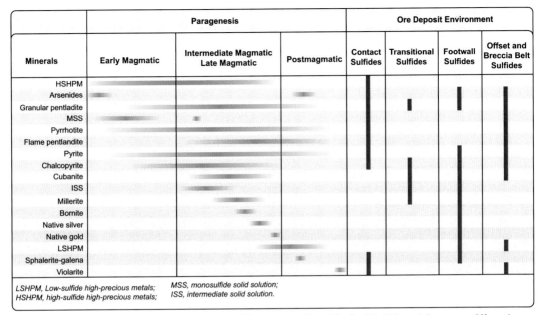

FIGURE 4.3 Generalized Paragenetic Sequence of Formation of the Principal Sulfide and Accessory Minerals Found in the Sudbury Ores Deposits

Modified after Hawley (1962) and Hawley and Stanton (1962), but updated using observations by the author of mineralogical relationships in hand sample and polished samples from contact, footwall, and offset mineral systems.

Paragenetic sequence for the crystallization of sulfide melts

Hawley (1965) was first to document a paragenetic sequence for Sudbury ores based largely on the Frood mineral system and the contact type mineral systems. The more recent discovery of more strongly fractionated footwall ore deposits in the North and South Ranges together with the development of the Offset Dyke-hosted mineral systems has provided a wealth of additional observations that underpin a modest update to the paragenetic sequence provided in Hawley's seminal contribution.

The generalized paragenetic sequence provided in Fig. 4.3 is based on textural and mineralogical relationships observed by the author in different parts of mineral systems as well as published descriptions (Hawley, 1965). In detail, all of the sulfide assemblages are easily modified by relatively low-temperature processes related to thermal metamorphism, and the exact textural relationships also vary as a function of cooling rate. The broad paragenetic sequence is anchored by mineral associations found in several different geological environments at Sudbury:

1. In Sublayer and Offset Dyke-hosted mineral systems, the early formation of monosulfide solid solution can be accompanied by the expulsion of fractionated sulfide melt. As the solid solution cools, it first forms granular pentlandite within the pyrrhotite, and then excess Cu and Ni are excluded from the monosulfide solid solution to form an assemblage of pentlandite with some chalcopyrite at the edges of the pyrrhotite grains that resembles the loop-textured Ni-rich ores described at Talnakh (Lightfoot and Zotov, 2014). Trapped Cu-rich sulfide melt that is not

removed from the sulfide forms more continuous loops. Pyrite is sometimes formed at the edges of the pyrrhotite crystals as part of the intercumulus sulfide mineral assemblage.

2. Later in the cooling of the monosulfide solid solution in the Sublayer and Offset mineral systems, exsolution of pentlandite from the pyrrhotite produces flames of pentlandite on the scale of 1–50 μm in width and up to a few centimeters in length. Under some conditions, cubanite also exsolves from the pyrrhotite, and this is textural evidence indicating that the monosulfide solid solution originally contained a small amount of Cu. The pyrrhotite is generally either monoclinic or hexagonal in crystal structure, but the exact process controls on the distribution and composition of hexagonal and monoclinic forms of pyrrhotite in Sudbury mineral systems is not fully understood.

3. Differentiation of the sulfide melt is recognized in transitional style and footwall style mineral systems in most North Range and some South Range locations. The Fe-poor sulfide melt removed from the monosulfide solid solution commenced crystallization to form an intermediate solid solution that crystallizes to dominantly chalcopyrite + pentlandite. In the transitional styles of sulfide mineralization, granular pentlandite is commonly hosted in chalcopyrite and cubanite. The sulfide assemblage has a lower Ni/Cu and lower Fe content than those found in the Sublayer and Offset Dykes (Naldrett et al., 1994a, 1999). This sulfide assemblage occurs at the base of embayment and trough structures in metabreccia. In the footwall, some of the thickest veins of massive sulfide consist of an assemblage of chalcopyrite + pentlandite with millerite developed at the edge of sulfide veins in the North Range.

4. The residual sulfide melts removed from the intermediate solid solution are very Cu-rich, moderately Ni-rich, and Fe-poor; they form the peripheral parts of footwall mineral systems that contain bornite and millerite. These are the last crystallization products of the magmatic sulfide system. This style of mineralization contains many elements that are incompatible in the principal sulfide minerals. Depending on local conditions at the periphery of the vein system, they can form native silver, gold, and carbonate vein assemblages surrounded by a weak disseminations (fringes) of low-sulfide high-precious metal mineralization associated with extensive partial melting but only localized alteration. This mineral paragenesis is often attributed to hydrothermal processes and/or the migration of fluid with elevated metal content into the adjacent partially melted country rocks to produce alteration effects (Farrow and Watkinson, 1992; Hanley and Mungall, 2003; White, 2012; Tuba et al., 2014) that are commonly superimposed on metabreccias and partially melted country rocks. This style of mineralization is a product that would be expected to form from a highly fractionated fluid- and water-rich sulfide melt during the terminal phase of differentiation of a sulfide melt (Keays and Lightfoot, 2004). Low sulfide mineralization that meets economic thresholds is typically a small part of the mineral system and typically develops adjacent to more massive offset, contact, and footwall sulfides.

The observed paragenetic sequence in the vast majority of North Range Sudbury sulfides fits into this broad scheme (Fig. 4.3), and although there is some modification of mineralization due to metamorphic re-equilibration and deformation in postmagmatic shear zones, it is the primary process of magmatic segregation and crystallization of sulfide melts that also produced the footwall mineralization in the South Range. The details matter enormously in the context of ore deposit models and process technology, so the rest of this chapter is dedicated to observations that help to unravel the timelines of ore formation at Sudbury.

GEOCHEMISTRY OF THE SUDBURY ORES

The approach to understanding the geochemistry of the Sudbury ores should be pragmatic and applied so that it can be used in exploration, but it must also provide process constraints that underpin the fundamental theories of ore formation in magmatic systems.

The geochemistry of the Sudbury ores can be understood using various approaches. In one approach, routine assay data are used to build conceptual models that can be applied to exploration. A second approach builds on the first, but involves the analysis of a much smaller group of representative ores and the host rocks for a wide range of elements that have less or no economic value but can be used to provide insights into ore-forming processes that are not readily obtained using assay data alone (Naldrett et al., 1999; Ames et al., 2014). Yet other approaches use the ratios of the metals to sulfur (Lightfoot and Farrow, 2002) and Pearce Element Ratio diagrams (Beswick, 2002) to examine the compositional diversity in the sulfide mineral systems using routine assay data. The utilization of both assay data and detailed geochemical data has a place in this book. Notwithstanding, the author's focus in this chapter has been guided by the application within the minerals industry, and so the focus is placed on the more routine geological and geochemical data that typically support exploration and mine development activities. Chapter 5 examines the observations that can be extracted from assay data in the context of more specific and detailed high quality datasets acquired on sulfide-mineralized samples from small parts of the mineral systems and orebody domains.

In this chapter, use is made of conventional element ratios that link to the diversity in mineralogy of the sulfides. Variations in the composition of sulfide can be produced by several effects and process controls that have to be taken into consideration. These include the following:

1. Inherent compositional variability that is due to the size of the sample (typically a drill core interval of 30 cm to 1 m in length) relative to the grain size of the rock (which can be on the scale of 10s of centimeters) and the local heterogeneity in the proportion of sulfides (which are occasionally on the scale of meters). These variations are due to the tendency of sulfide melts to crystallize to form coarse-grained rocks, the local separation of chalcopyrite from pyrrhotite + pentlandite (the chalcopyrite tends to wrap around inclusions, and it can be remobilized into the country rocks due to surface tension; Ballhaus et al., 2001), and segregation of a Cu-rich melt from the monosulfide solid solution (on the scale of a mine opening through to the scale of an ore deposit).
2. Compositional variation produced by expulsion of a part of the Cu-rich melt leaving behind a monosulfide solid solution and trapped Cu-rich liquid, which crystallize to pyrrhotite with lesser chalcopyrite + pentlandite. The fractionated sulfide melt is sometimes expelled into the footwall to form a domain of chalcopyrite + pentlandite + millerite, or the melt can undergo extreme fractionation to form a bornite + millerite + chalcopyrite fringe to the footwall mineral system.
3. Variations produced by postmagmatic processes accompanying deformation and metamorphism within shear zones and hydrothermal processes that sometimes overprint or obscure the relationships that were produced by primary magmatic processes.

GEOCHEMISTRY OF THE SULFIDES

Early work on the composition of the Sudbury ores by Hawley (1962) was followed by Hawley's (1965) paper on the zoning in the Frood mineral system, and discussions of Naldrett and Kullerud

(1967) and Cowan (1968) about zoning of mineralization of the mineral zones at Strathcona Mine. Keays and Crocket (1970) and Chyi and Crocket (1976) studied the distribution of PGE in mineral separates from the Strathcona Mine. Following the development of a method for analyzing PGE in bulk composite samples (Hoffman, 1978), Hoffman et al. (1979) reported on the PGE content of three Sudbury deposits. Naldrett (1981) compared the trends of major and trace element variation within the Sudbury ores with those of other Ni-Cu deposits, and Naldrett et al. (1982) documented and interpreted chalcophile and siderophile element variation within and between five different Sudbury ore deposits.

Based on detailed samples provided by Falconbridge and Inco from different deposits around the Sudbury Structure, Naldrett et al. (1994a, 1999) reported the first detailed study of the metal abundances of the different ore types; these data and the models based on them are discussed in Chapter 5.

Table 4.5 presents some of the ratios that are used to describe the case examples of mineral systems and orebody domains presented here. The table has a focus on metal ratios that are similar to those used by Naldrett et al. (1994a, 1999), but the approach and the ratios presented in Table 4.5 is designed to extract maximum value from historic data rather than smaller datasets of high quality determinations of ultra-trace precious metals such as Rh, Ir, Ru, and Os.

Naldrett (1984c) reported average Cu/(Cu + Ni) ratios for different ore samples and showed that the ratio can vary from Ni-rich end-members in contact ores and Cu-rich end-members in footwall ores. The approach used in this book follows Naldrett (1984c), in heavily utilizing the ratio of Cu/(Cu + Ni) as an index of compositional diversity and fractionation in a mineral system.

The variations in (Cu + Ni)/S is a derivative of the Cu/S and Ni/S variation diagrams used in Lightfoot and Farrow (2002); this ratio provides a guide to the metal abundance in sulfide irrespective of the presence of minerals with atypical metal and sulfur concentrations (eg, bornite, millerite, pyrite, gersdorffite, etc.). This ratio is useful as it does not make assumptions about the mineralogy of the sulfides in the way required by normative sulfide calculations. Fig. 4.4 shows a plot of Cu/(Cu + Ni) versus (Cu + Ni)/S in which the principal minerals from the Sudbury mineral systems are projected and common mineral associations are shown as tie-lines. This approach is well suited to historic databases where the analyses of the full range of PGE and other trace elements is rarely undertaken in support of commercial production.

Assay data can be plotted into Fig. 4.4, and they generally fall into the fields of contact and offset mineralization or footwall mineralization, where the distribution of assays is controlled by the proportions of the different sulfide minerals. This diagram is used to establish the compositional diversity in each mineral system in this chapter. The distribution of assays in this figure is controlled by several factors, namely:

1. *Grain size of the sulfide*: Some sulfides are relatively fine-grained (grain sizes <1 cm), so the bulk composition of a sample interval of ~1 m length of NQ-sized drill core can be representative. Sulfides can also be very coarse-grained to pegmatoidal, and the analysis is controlled by the proportions of the minerals in the sampled interval. This variation is a product of sampling, and a large number of samples are required to get a reasonable estimate of the composition of the mineralization. The net effect on Fig. 4.4 can be to produce a tendency for an assay of a coarse-grained rock to migrate toward the dominant sulfide minerals (eg, pyrrhotite + pentlandite).
2. *Sulfide differentiation*: Variations in composition that reflect the composition of the sulfide melt or cumulate as the melt undergoes differentiation from contact to footwall or within individual sulfide veins (Farrow and Lightfoot, 2002).
3. *Sulfide recrystallization and kinesis*: Recrystallization of sulfide can accompany metamorphism. Fine-grain sizes of pyrrhotite + pentlandite may reflect rapid cooling whereas coarse-grained

Table 4.5 Average Metal Tenors, Metal Ratios, and Metal/Sulfur Ratios for Sudbury Mineral Systems

Mineral System	Range	Mineral System	Sample Group	Mineral System Averages of All Samples with S ≥ 1% and (Cu+Ni) ≥ 0.3%					
				$[Ni]_{100}$ (wt.%)	$[Cu]_{100}$ (wt.%)	$[3E]_{100}$ (g/t)	Ni/Co	Cu/(Cu+Ni)	(Cu+Ni)/S
Victor	East	Contact and footwall	All samples	5.05	5.45	6.01	34.65	0.35	0.29
Victor	East	Contact	Sublayer	4.61	2.60	1.22	27.17	0.30	0.19
Victor	East	Contact	Norite host	4.41	2.43	1.31	22.69	0.31	0.18
Victor	East	Contact	GRBX host	4.80	3.20	1.52	30.24	0.32	0.21
Victor	East	Contact	Massive and semimassive	5.04	0.72	0.65	37.98	0.11	0.15
Victor	East	Footwall	Footwall	n.g.	n.g.	n.g.	115.05	0.61	0.83
Victor	East	Footwall	SUBX host	n.g.	n.g.	n.g.	125.36	0.70	1.08
Victor	East	Footwall	Metabreccia host	n.g.	n.g.	n.g.	42.03	0.40	0.38
Victor	East	Footwall	Gneiss and granitoid host	n.g.	n.g.	n.g.	112.03	0.65	0.96
Victor	East	Footwall	Massive and vein type	n.g.	n.g.	n.g.	219.39	0.80	1.03
Capre	East	Contact	All samples	4.35	2.15	5.35	28.28	0.35	0.21
Capre	East	Footwall	All samples	n.g.	n.g.	n.g.	145.32	0.79	0.61
Trill	North	Deep Sublayer embayment	All samples	4.01	2.26	3.39	18.28	0.40	0.20
Trill	North	Deep Sublayer embayment	S > 5 wt.%	3.45	1.74	2.35	18.92	0.32	0.14
Trill	North	Deep Sublayer embayment	S ≤ 5 wt.%	4.45	4.23	7.32	16.54	0.47	0.24
Whistle	North	Deep Sublayer embayment	All samples	3.06	1.45	2.38	17.61	0.32	0.13
Whistle	North	Deep Sublayer embayment	S > 5 wt.%	3.09	1.09	1.84	18.41	0.25	0.11
Whistle	North	Deep Sublayer embayment	S ≤ 5 wt.%	3.04	2.29	3.79	16.48	0.37	0.14
Blezard	South	Shallow Sublayer embayment	All samples	4.23	3.93	4.34	22.08	0.46	0.25
Blezard	South	Shallow Sublayer embayment	S > 5 wt.%	4.20	3.49	3.90	22.49	0.41	0.22
Blezard	South	Shallow Sublayer embayment	S ≤ 5 wt.%	4.39	6.13	6.49	20.42	0.55	0.29

(Continued)

Table 4.5 Average Metal Tenors, Metal Ratios, and Metal/Sulfur Ratios for Sudbury Mineral Systems (*cont.*)

Mineral System	Range	Mineral System	Sample Group	Mineral System Averages of All Samples with S ≥ 1% and (Cu+Ni) ≥ 0.3%					
				$[Ni]_{100}$ (wt.%)	$[Cu]_{100}$ (wt.%)	$[3E]_{100}$ (g/t)	Ni/Co	Cu/(Cu+Ni)	(Cu+Ni)/S
Murray	South	Shallow Sublayer embayment	All samples	3.95	2.98	1.8	20.76	0.43	0.22
Murray	South	Shallow Sublayer embayment	S > 5 wt.%	3.97	2.48	1.3	21.61	0.36	0.18
Murray	South	Shallow Sublayer embayment	S ≤ 5 wt.%	3.87	5.93	4.9	18.67	0.53	0.28
Little Stobie Number 1	South	Shallow Sublayer embayment	All samples	3.56	5.70	9.09	21.06	0.53	0.25
Little Stobie Number 1	South	Shallow Sublayer embayment	S > 5 wt.%	3.49	4.85	8.14	21.17	0.48	0.22
Little Stobie Number 1	South	Shallow Sublayer embayment	S ≤ 5 wt.%	3.64	6.54	11.90	20.84	0.58	0.27
Little Stobie Number 2	South	Footwall-hosted	All samples	3.43	5.63	10.36	19.85	0.54	0.24
Little Stobie Number 2	South	Footwall-hosted	S > 5 wt.%	3.36	4.90	9.47	20.50	0.51	0.22
Little Stobie Number 2	South	Footwall-hosted	S ≤ 5 wt.%	3.55	7.00	16.49	16.98	0.61	0.28
Levack Number 4	North	Shallow Sublayer embayment	All samples	5.30	2.16	2.2	26.86	0.30	0.23
Levack Number 4	North	Shallow Sublayer embayment	S > 5 wt.%	5.28	1.97	1.8	28.80	0.26	0.20
Levack Number 4	North	Shallow Sublayer embayment	S ≤ 5 wt.%	5.49	3.85	5.5	23.27	0.37	0.26
9 Scoop	North	Deep transitional embayment	All samples	5.12	6.01	14.8	32.82	0.43	0.44
9 Scoop	North	Deep transitional embayment	S > 5 wt.%	5.28	2.48	4.4	27.12	0.36	0.39
9 Scoop	North	Deep transitional embayment	S ≤ 5 wt.%	4.96	9.25	20.5	40.14	0.54	0.51
McCreedy East 153 and 148	North	Footwall-hosted	All samples	n.g.	n.g.	n.g.	119.31	0.83	1.22
McCreedy East 153 and 148	North	Footwall-hosted	S > 5 wt.%	n.g.	n.g.	n.g.	114.71	0.84	1.09
McCreedy East 153 and 148	North	Footwall-hosted	S ≤ 5 wt.%	n.g.	n.g.	n.g.	134.24	0.79	1.47

Creighton	South	Contact and footwall	All samples	6.72	9.45	11.52	39.89	0.51	0.44
Creighton	South	Contact and footwall	S > 5 wt.%	6.94	8.56	8.80	45.51	0.46	0.42
Creighton	South	Contact and footwall	S ≤ 5 wt.%	6.50	10.37	29.82	32.12	0.56	0.46
Gertrude-Creighton 402 Trough	South	Contact (shallow)	All samples	4.88	3.54	0.65	25.67	0.35	0.22
Gertrude-Creighton 402 Trough	South	Contact (deep)	All samples	5.94	4.99	1.33	30.76	0.40	0.29
Garson	South	Contact	All samples	4.99	5.01	19.66	28.66	0.49	0.32
Garson	South	Contact	S > 5 wt.%	5.00	4.61	16.02	32.61	0.41	0.27
Garson	South	Contact	S ≤ 5 wt.%	4.88	9.29	57.44	22.52	0.61	0.39
Copper Cliff	South	Radial Offset	All samples	5.09	7.50	16.48	25.60	0.59	0.41
Copper Cliff	South	Radial Offset	S > 5 wt.%	5.16	6.79	12.84	30.78	0.49	0.34
Copper Cliff	South	Radial Offset	S ≤ 5 wt.%	5.02	11.39	36.41	21.37	0.67	0.47
Frood	South	Discontinuous concentric Offset	All samples	3.98	4.86	7.74	24.18	0.50	0.23
Frood	South	Discontinuous concentric Offset	0–400 m depth	4.00	4.74	5.10	24.34	0.49	0.23
Frood	South	Discontinuous concentric Offset	400 m to depth	3.78	6.07	23.62	22.82	0.54	0.26
Stobie	South	Discontinuous concentric Offset	All samples	3.72	4.08	3.97	21.86	0.48	0.20
Stobie	South	Discontinuous concentric Offset	High tenor disseminated	3.88	4.14	4.62	22.15	0.47	0.21
Stobie	South	Discontinuous concentric Offset	Low tenor massive	2.81	3.57	2.66	20.44	0.49	0.17
Stobie	South	Discontinuous concentric Offset	Low tenor disseminated	2.97	4.53	5.21	19.43	0.51	0.20
Totten	South	Radial Offset	All samples	6.95	11.03	19.2	40.95	0.62	0.54
Totten	South	Radial Offset	S > 5 wt.%	7.11	10.31	15.7	44.40	0.55	0.49
Totten	South	Radial Offset	S ≤ 5 wt.%	5.74	16.59	45.0	33.96	0.71	0.59

n.g., Not given as minereal assemblage varies in S content. Calculations are based on representative assays that are unweighted. The averages are based on large numbers of analyses based on the availability of sulfur assays, S concentrations ≥ 1wt.%, and (Ni + Cu) ≥ 0.3 wt.%. The concentrations of Pt, Pd, and Au (expressed as [3E]$_{100}$ that is, Pt + Pd + Au recalculated in 100% sulfide) are provided for a representative subset of samples for which precious metal data is available.

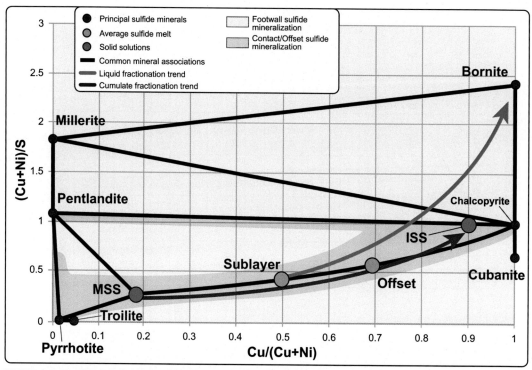

FIGURE 4.4 The Mineral Assemblage Plot Used in This Chapter Utilizes the Tie-Lines Developed Between Sulfide Minerals Expressed in Terms of Cu/(Cu + Ni) Which is Indicative of the Extent of Fractionation of Cu From Ni, and (Cu + Ni)/S Which is Indicative of the Metal Concentration in the Sulfide Component of the Rock

This plot requires available Cu, Ni, and S assays with a threshold established for the determination of ratios set at (Cu + Ni) > 0.3 wt.% and S > 1 wt.%. The positions of key sulfide minerals are based on ideal stoichiometry (Table 4.2). The effects of pyrite, arsenide minerals, and cubanite are not resolved in this diagram; the approach is based on the relative contributions of the principal sulfide minerals to contact and footwall mineral systems. MSS, monosulfide solid solution; ISS, intermediate solid solution.

textures with large eyes of pentlandite in pyrrhotite may be produced by very slow cooling. The separation of pentlandite from pyrrhotite at high temperatures and pressures in a sulfide that is undergoing deformation may result in the remobilization of sulfide which is termed "sulfide kinesis" by Monteiro (personal communication, 2006). This remobilization of sulfide is more commonly found in higher-grade metamorphic conditions where it is accompanied by the separation of pyrrhotite from pentlandite (Thompson; Lightfoot and Evans-Lamswood, 2015), but it may also occur at lower temperatures in shear zones.

The trends on Fig. 4.4 tend to be related to specific groups of minerals that occur in association with each other, specifically the following:

1. Mineralogical mixtures of pyrrhotite + chalcopyrite + pentlandite in contact ores (typically they fall along a trend of increasing (Cu + Ni)/S with increasing Cu/(Cu + Ni) (Fig. 4.4) which

reflects the mixing of chalcopyrite with pyrrhotite and pentlandite formed from monosulfide solid solution.

2. Samples falling above the array of chalcopyrite + monosulfide solid solution (Fig. 4.4) are typically mixtures of pentlandite + chalcopyrite, pentlandite + millerite + chalcopyrite, millerite + chalcopyrite and bornite + chalcopyrite ± pentlandite ± millerite. In some cases, samples are simple linear mixtures of sulfides that contain two sulfide minerals (eg, intermediate solid solution tends to recrystallize to form dominantly pentlandite + chalcopyrite), but many of the assays are products of mixing of three or more sulfide minerals.

3. Samples that fall below the monosulfide solid solution + chalcopyrite array (Fig. 4.4) contain sulfides with low metal abundances such as pyrite. The effects of other accessory sulfide and arsenide minerals can also influence where the sample plots, so it is always best to have a firm understanding of the sulfide mineralogy before interpreting the variations.

Variations in composition can be visualized by grouping the samples based on their position in Fig. 4.4, and then visualizing the data in three-dimensional models to track the compositional variation in space within the mineral system. Examples of the application include the following:

1. Visualization of changing mineral assemblages through orebodies to identify zonation of sulfide mineralogy in footwall mineral zones.
2. Identification of trends in changing sulfide mineralogy in contact mineral systems along trough structures.
3. Establishing trends in sulfide fractionation with vertical position in a mineral shoot residing in an Offset Dyke.

TRACE ELEMENT COMPOSITION OF SUDBURY SULFIDES

Cobalt

Pentlandite is the principal host for the bulk of the cobalt in the Sudbury ores (Naldrett, 1961), and there is a range in Ni/Co ratio depending on whether the pentlandite was formed in a contact or footwall environment (Lightfoot, 2015). Some cobalt exists in solid solution in pyrrhotite, and the Ni/Co ratio of pyrrhotite + pentlandite-rich sulfides varies as a function of the proportion of pyrrhotite:pentlandite. Cobalt partitions quite weakly into millerite, and so the elevated Ni/Co ratios of footwall sulfide assemblages often provide evidence that millerite is present. A small amount of Co is present in pyrite, and gersdorffite and cobaltite can contain up to 25 wt.% Co, so the variations in the Ni/Co ratio are also sensitive to the proportion of arsenide minerals.

Zinc and lead

Zinc is present in sphalerite which is sometimes found in association with chalcopyrite. Naldrett (1981) pointed out that the concentration of Zn in sulfide ores is similar to that in the host rocks, which is consistent with the experimental findings of MacLean and Shimazaki (1976) who concluded that the sulfide melt/silicate melt partition coefficient for Zn is about one. Zinc tends to be more concentrated in the Cu-rich zones of ore deposits, because Zn is concentrated into the residual sulfide liquid during fractionation of the sulfide melt (Naldrett et al., 1982).

Lead tends to occur as galena which, such as Zn, tends to be concentrated in the more fractionated sulfide melts or in association with late structures cutting through the orebodies in both the North and South Range.

Arsenic and antimony

Most sulfide mineralization from the North and East Ranges contain low As and Sb, whereas As concentrations in mineralization from the South Range tend to be quite high in samples that contain As-bearing sulfosalts such as gersdorffite, niccolite, and maucherite, minerals that are commonly observed in the South Range deposits.

Selenium

Massive sulfide ores exhibit a range in S/Se ratio depending on the degree of fractionation. Footwall deposits tend to have very high Se abundances relative to contact ores, and this may be explained by the relatively strong partitioning of sulfosalt elements such as Se, Te, Bi, and Sb into late-stage sulfide melts (Keays and Lightfoot, 2004).

Platinum group elements and gold

PGE and Au data are often averaged and expressed as concentrations in 100% sulfide, and they are then plotted by normalizing the concentrations in 100% sulfide to the average concentrations in C1 chondrite (Naldrett, 1984c). Normalization to 100% sulfide requires an understanding of the sulfide mineralogy and abundance as well as the metal concentrations in the host rock (Kerr, 2003); for mineralization with sulfide concentrations that are a combination of pyrrhotite + pentlandite + chalcopyrite, the calculation is straightforward. If millerite, bornite, gersdorffite, cobaltite, pyrite, or other sulfides are present, the calculation is influenced by the differing S contents of these minerals. For contact and offset ore types, the concentration of Pt + Pd + Au in 100% sulfide $\{[3E]_{100} \text{ (g/ton)}\}$ is indicative of the PGE concentrations in the sulfide.

PGE concentrations in sulfide tend to be higher in footwall mineral zones relative to contact zones. Platinum, Pd, and Au are enriched in the residual liquid that is mobilized into the footwall environment, whereas Ru, Os, Ir, and Rh are retained in the monosulfide solid solution in the Sublayer and Offset Dykes (Keays and Crocket (1970)). This is often expressed as a ratio of (Pt + Pd)/(Rh + Ru + Ir + Os) (Naldrett, 2004). In the context of complete orebody characterization, it is very rare to have access to data for Rh, Ru, Ir, and Os, so this approach to the understanding of ore deposits is placed in Chapter 5.

Sulfide mineralization in the footwall of the SIC is often accompanied by the development of disseminated to vein type sulfides in the country rocks that surround the orebodies. These types of mineralization are not a major contribution to Sudbury production, nor are the signatures developed at a large distance from the deposit. The style of mineralization is recognized by virtue of the high Cu and PGE concentrations relative to the sulfide content of the rock, and has been classified as a low-sulfide high-precious metal style of mineralization (LSHPM; Farrow et al., 2005). The classification of assays in this group can be made using the variations in 3E versus S, and grouping the analyses above a threshold concentration of both the 3E (eg, 5 g/t) and S concentration (eg, <2 wt.% S).

Compositional variations in mineral systems comprising pyrrhotite + chalcopyrite + pentlandite can be understood using normative sulfide mineralogy, and the proportions of pyrrhotite, pentlandite and chalcopyrite in 100% sulfide can be expressed as $[Po]_{100}$, $[Pn]_{100}$ and $[Cpy]_{100}$. In the absence of minerals that take up or dilute the Ni or Cu abundance (eg, Ni-rich or Ni-poor pyrite), the calculated normative mineralogy is a very useful indication of compositional variability in the mineral system. It should be applied with caution if the sample has low S abundance (typically >1–5 wt.% S in the sample is a reasonable threshold for meaningful results).

THE OFFSET AND BRECCIA BELT MINERAL SYSTEMS
FROOD AND STOBIE

The Frood-Stobie ore deposit was discovered in 1884 by Thomas Frood (Table 1.1) and went into major production in 1929. The history of production from Frood-Stobie is summarized in Fig. 4.5 where a timeline of major events is shown.

The deposit is famous as the largest of the Sudbury mineral systems comprising two ore deposits (The Frood and Stobie Deposits) hosted in a discontinuous Offset Dyke in the South Range Breccia Belt (Fig. 4.1). The Frood Deposit was first mentioned in Yates (1948) and was then described in Zurbrigg (1957), Hawley (1965), and most recently in a thesis study by Hattie (2010). Although small discontinuous segments of QD are developed along strike in the Breccia Belt, exploration has so far failed to identify any economically significant domains of mineralization beyond the immediate margins of the known orebodies.

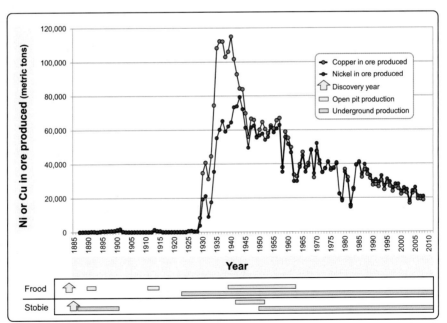

FIGURE 4.5 Plot of Time (Year) Versus Contained Ni and Cu in Ore Produced From the Frood-Stobie Mineral System

The diagram shows some of the principal events in the long history of production from one of Sudbury's most prolific mines.

Geological relationships

The Frood-Stobie ore deposit originally cropped out at surface as gossanous disseminated to semimassive sulfide in QD that is enveloped by a metabreccia developed by partial melting and metamorphism of heavily brecciated Stobie Formation country rocks. Because of mining activity, the original outcrops are long-gone, and the surface manifestation of the orebodies at depth is presently a series of large open pits and stopes that have caved to surface that are located along the ~1.5 km-long trend shown in Fig. 4.6.

The original surface geology mapping revealed that the Frood and Stobie mineral systems consisted of several thick (≤300 m), discontinuous lenses of mineralized inclusion-bearing quartz diorite (MIQD) that contain domains of semimassive inclusion-rich sulfide mineralization (Fig. 4.7A; Zurbrigg, 1957). In cross-section, the Frood mineral system rests within and at the contacts of a sheet of MIQD that dips ~70–75 degrees to the north-northwest, and extends from surface where it is 1.5 km in length to depth where it eventually narrows and disappears (Zurbrigg, 1957). In the Frood mineral system, the MIQD and mineralization progressively narrow in width and length from surface to a depth of 1.3 km as shown in the cross- and long-sections in Fig. 4.7B,C.

FIGURE 4.6 Photograph of the Stobie Pits (Foreground) and the Frood Pit in the Background

View taken along the South Range Breccia Belt from WNW to ESE showing the extensive historic development of the open pits and stopes and the Frood Mine head-frame with the Copper Cliff Smelter Complex in the background.

Photograph provided by Chris Meandro, May 2015.

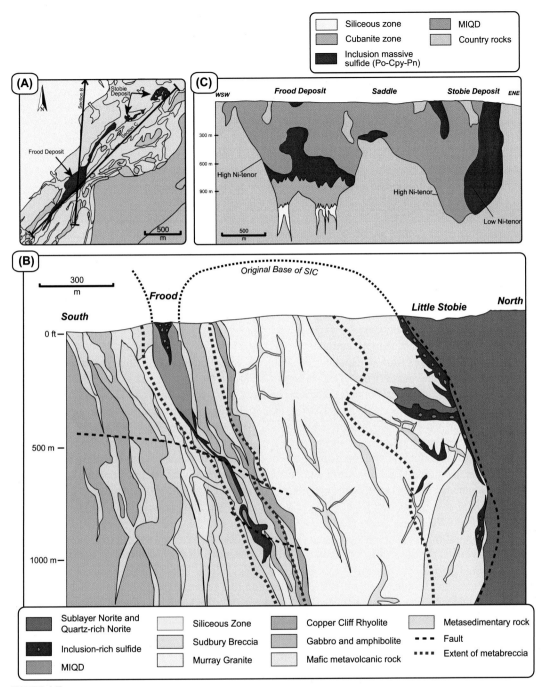

FIGURE 4.7

(A) Surface geology of the Frood-Stobie mineral system prior to open pit mining activity. (B) West-facing
geological cross-section of the Breccia Belt and the base of the SIC showing the position and characteristics of
the Frood Orebody. (C) NNW-facing geological long section of the Frood-Stobie mineral system showing the
extent of the MIQD of the Offset Dyke based on drill core descriptions. The diagram also shows the distribution of
compositionally distinctive mineralization styles in the Stobie portion of the mineral system.

(A) Modified after Zurbrigg (1957) and Freeman and Muir (1979). (B) Modified after Stewart and Lightfoot (2010); Zurbrigg (1957).

The mineralization style at Frood gradually changes with depth from semimassive sulfides with abundant mafic-ultramafic inclusions enveloped in a wide domain of heavily mineralized blebby sulfide hosted in a matrix of QD with inclusions to a zone of semimassive sulfide associated with MIQD (Zurbrigg, 1957). The near-surface sulfide types contained a greater proportion of semimassive sulfide relative to MIQD and these may have represented the deeply eroded part of a funnel that connected with the contact of the overlying melt sheet (Fig. 4.7B). The semimassive sulfide mineralization and the MIQD contain mafic fragments that were derived from the local country rocks as well as fragments of mafic-ultramafic rock that are not recognized in the adjacent country rocks.

The margins of the Frood mineral system comprise semimassive sulfide mineralization with abundant schist inclusions which are likely an indication that the margins of the Offset underwent deformation during and after emplacement. The style of mineralization is termed the "contorted schist sulfide" type and is commonly developed where orebodies are within shear zones that cut through metasedimentary and metavolcanic rocks of the Stobie Formation. This type of sulfide is formed by deformation at the margin of the sulfide ore that involves processes of infiltration of sulfide into the country rocks, incorporation, and modification of the fragments in the sulfide that is broadly similar to the mechanism described in formation of the durchbewegung type sulfides (Vokes, 1968, 1969, 1973).

Massive sulfides become more common as the MIQD narrows with depth, and the sulfide assemblage gradually changes from pyrrhotite + chalcopyrite + pentlandite to cubanite + pentlandite + pyrrhotite with increasing depth (Fig. 4.7C; Zurbrigg, 1957).

The deeper part of the Frood mineral system consists of a narrow zone of MIQD that is part of the Offset and associated fracture-filled massive Cu-PGE-rich sulfide mineralization that is termed the Silicious Zone (Hawley, 1965; Stewart and Lightfoot, 2010; Fig. 4.7C).

The Stobie mineral system has similar styles of mineralization to those described by Zurbrigg (1957) in the Frood mineral system. Some parts of the Stobie mineral system comprise semimassive sulfide mineralization (at both the north and northeastern margin of the deposit and within the center of the disseminated sulfide ore deposit) and other parts comprise MIQD with disseminated sulfide. The Stobie mineral system narrows from ~300 m wide and ~1 km long at surface to terminate at a depth of ~1 km (Fig. 4.7C).

The temporal relationships between the Offset Dyke and the country rocks are established on the basis of the contact relationships between the MIQD and mineralized metabreccias that were formed from the country rocks and Sudbury Breccia (SUBX) (Fig. 2.20J). The MIQD contains disseminations, blebs, and connected globules of sulfide mineralization that is preserved in textures that indicate the magmatic immiscibility conditions that existed between the sulfide melt and the silicate melt prior to crystallization (Zurbrigg, 1957; Fig. 4.7A). This textural relationship is consistent with the presence of co-existing sulfide and silicate melts, and the inverse relationship where globules of immiscible QD are enclosed by sulfide are found in the Frood and Stobie deposits.

The relationship between the disseminated sulfides hosted in MIQD and the massive sulfides with inclusions, but lesser amounts of silicate matrix, provides important constraints on the sequence of events at Stobie. Fig. 4.8 shows an example of a large slab of rock from Stobie Mine illustrating MIQD with blebby to disseminated sulfide mineralization, and the presence of country rock and massive sulfide inclusions. The massive sulfide inclusions have a lower proportion of pyrrhotite:pentlandite than the disseminated sulfide hosted in the MIQD, but the ratio of pyrrhotite:pentlandite of the sulfide inclusions is similar. These inclusions of massive sulfide may have been incorporated from an earlier-formed massive sulfide without complete assimilation of the sulfide fragments. The massive sulfide inclusions

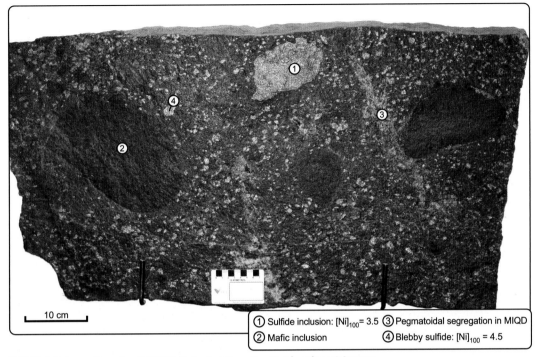

FIGURE 4.8 Large Sample of HMIQD Collected From the Stobie Mineral System

The sample contains mafic-ultramafic inclusions, feldspathic leucosomes, and both blebby disseminated pyrrhotite > chalcopyrite > pentlandite and inclusions of massive sulfide with pyrrhotite > pentlandite > chalcopyrite. The matrix sulfides have higher $[Ni]_{100}$ (~4.5) than the sulfide inclusions (~3.5). The sulfide inclusions share a similar $[Ni]_{100}$ with the lower grade massive sulfides from the eastern flank of the Stobie mineral system (see Fig. 4.6C). The sample was photographed courtesy of the Exploration Staff at Stobie Mine where it resides. More information on the massive sulfide inclusions in the disseminated sulfide is given in Hattie (2010).

share the same pyrrhotite:pentlandite ratio of inclusion-rich and massive sulfides that plunge down the eastern part of the Stobie mineral system (Fig. 2.3C).

The geological relationships in the Frood mineral system are different when compared to those of the Stobie mineral system. Unlike Stobie, Frood shows a strong vertical zonation in sulfide mineralogy with the development of cubanite and LSHPM styles of mineralization at depth. Stobie comprises domains of HMIQD and inclusion-rich to massive sulfide that were emplaced in two or more discrete magmatic events.

Styles of mineralization

The diversity in styles of economically important sulfide mineralization at Stobie and Frood are illustrated in Fig. 4.9A–H. The disseminated sulfides at Frood and Stobie comprise blebs and disseminations of sulfide consisting of pyrrhotite, chalcopyrite, and pentlandite with an inclusion-bearing quartz diorite matrix. Locally the proportion of sulfide to silicate matrix can vary from trace sulfide through to 25% sulfide (Fig. 4.9A). Larger fragments in the MIQD are often rimmed by sulfide mineralization

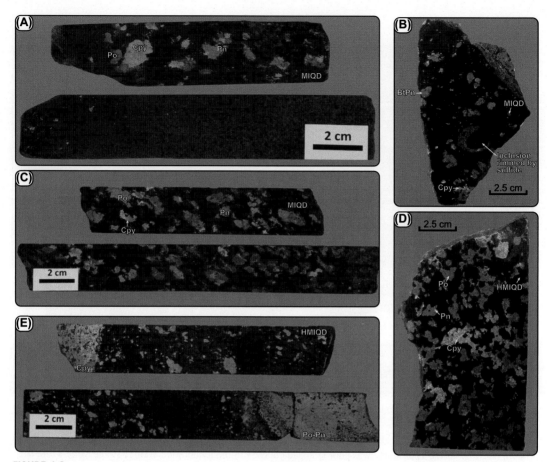

FIGURE 4.9

(A) Sample of HMIQD which grades over a few cm into MIQD. The MIQD contains large oval blebs of pyrrhotite > chalcopyrite > pentlandite with diffuse margins; in places where these blebs merge together the primary immiscible textural relationship between sulfide and silicate is preserved (borehole 117207, depth 227 m; Zurbrigg, 1957). (B) Sample of HMIQD from Stobie Mine contains small mafic inclusions which act as a locus around which sulfides tend to cling and replace the inclusion. This sample illustrates the incipient development of inclusion-rich sulfide, but the sample is grouped as a HMIQD (typical samples of HMIQD have $[Ni]_{100}$ of 4.3 wt.%, and grades 0.8 wt.% Ni, 0.7 wt.% Cu, and 0.9 g/t 3E). (C) Blebby sulfides in HMIQD from Stobie Mine; the blebs are almost, but not quite touching, so this sample does not show the coalescence of blebs which is typical of the higher grade disseminated sulfides (borehole 11207 at 216 m depth). (D) Sample showing the extreme development of coalescing blebs in HMIQD from Stobie Mine. (RX379254 has $[Ni]_{100}$ = 2.5 wt. %, 1.5 wt.% Ni, 0.9 wt.% Cu, and 1.5 g/t 3E). (E) Sample of HMIQD where the matrix contains both disseminated sulfides and larger patches of sulfide that may represent inclusions like those shown in Fig. 4.8 (borehole 117207, 81.5 m depth).

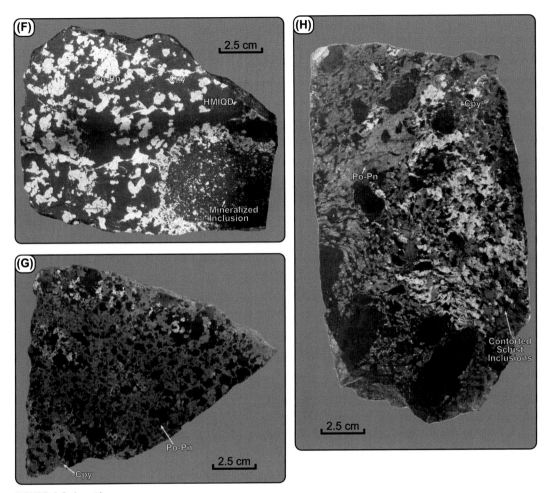

FIGURE 4.9 (*cont.*)

(F) Sample of HMIQD transitional into inclusion-rich massive sulfide. The sulfide is hosted in silicates where inclusions comprise a larger proportion of the host than QD. The sample has a low Ni tenor signature and is part of the high Ni grade, low Ni-tenor, sulfide ore shoot shown in Fig. 4.6C. (G) Inclusion-rich sulfide from the low tenor mineral zone at Stobie. (H) Semimassive sulfide from the low tenor, high-grade shoot at Stobie (RX379241 has a $[Ni]_{100}$ of 3.34 wt.%, grades 1.95 wt.% Ni, 2.86 wt.% Cu, and 1.45 g/t 3E).

along a meniscus that wraps around the inclusion (Fig. 4.9B). Disseminated sulfide commonly occurs as oval blebs which can comprise varying proportions of the different sulfide minerals (Fig. 4.9C) rather than a constant proportion of pyrrhotite + pentlandite:chalcopyrite which characterize the blebs in other magmatic sulfide systems such as Noril'sk-Talnakh and Insizwa (Lightfoot et al., 1984; Lightfoot and Zotov, 2014). As the blebs merge together, there are often discrete domains of more chalcopyrite-rich mineralization between patches with greater proportions of pyrrhotite + pentlandite (Fig. 4.9D). The

HMIQD at Stobie contains subrounded blocks of massive sulfide (Fig. 4.8) and intervals of massive sulfide that may have been undergoing partial melting and incorporation into the HMIQD (Fig. 4.9E).

The inclusion-bearing massive sulfide styles of mineralization at Stobie exhibit a range in textures comprising connected blebs of sulfide in a quartz-diorite matrix (Fig. 4.9F), clast-unsupported variety of inclusion-rich sulfide (Fig. 4.9G) and massive sulfide with a variable proportion of inclusions of mafic and ultramafic rock as well as contorted fragments of schist (Fig. 4.9H).

Geochemistry of Frood

The historic data for the Frood Deposit reported in Hawley (1965) support his suggestion that the ore deposit is zoned with respect to the sulfide mineralogy from more Ni-rich ores close to the surface toward more Cu and PGE-enriched sulfide ores at depth. Hawley (op cit) termed this variation "upside-down zoning" and he presented mineralogical observations to support his suggestion that the ores exhibit progressive differentiation with the first formed sulfides crystallizing closer to surface and the progressively more fractionated sulfides crystallizing at depth.

The legacy of historic data for the Frood mineral system offers a remarkable insight to the in situ differentiation of the sulfide melt (Hattie, 2010). Fig. 4.10A,B shows the variations in $Cu/(Cu + Ni)$ and $(Cu + Ni)$ grade with vertical position in the Frood mineral system (Fig. 4.10D) where the ratios and concentrations represent 1000 point moving averages of assays between the surface and a depth of 1200 m. The deposit shows a change from $Cu/(Cu + Ni)$ ~0.45 at surface to a ratio of 0.8 at depth whereas the concentration of $(Cu + Ni)$ in the sulfides tends to increase from 2.5 to 9% at a depth of 900 m after which it declines in the Siliceous Zone where LSHPM is developed (Fig. 4.10B). These trends support the proposal by Hawley (1965) that the ore show a progressive zonation from surface to depth. The variation has been attributed to fractionation of the sulfide magma whereby a monosulfide solid solution rich in Fe and Ni solidified close to the base of the SIC (Hawley, 1965; Farrow and Lightfoot, 2002), whereas the expelled Cu- and PGE-rich sulfide melt migrated downward through the silicate magma that occupies the Offset Dyke. Despite the limited evidence for hydrothermal alteration associated with the Cubanite Zone, another model attributes the variation in sulfide mineralogy to hydrothermal processes (Fleet, 1979).

The trends of moving averages of $Cu/(Cu + Ni)$ in the more massive sulfides [arbitrarily assumed to contain >3 wt.% $(Ni + Cu)$], sulfide mineralization with intermediate $(Cu + Ni)$ abundance levels [$1<(Ni + Cu) < 3$ wt.%], and disseminated sulfides with $0.3 \leq (Ni+Cu) < 1$ wt.% are shown in Fig. 4.10C. This plot illustrates that the vertical variation in $Cu/(Cu + Ni)$ is a feature of both low and high grade samples.

The disseminated mineralization shows a narrower range in $Cu/(Cu + Ni)$ in the sulfides in the upper part of the Frood mineral system, and the trend to depth is toward higher $Cu/(Cu + Ni)$ than the more heavily mineralized groups (Fig. 4.10C). The lowest-grade sulfides do not exhibit significant degrees of fractionation through the bulk of the upper part of the deposit, but these sulfides are more Cu-rich in the Cubanite and Siliceous Zones at depth.

The moving averages of the higher-grade samples show elevated $Cu/(Cu + Ni)$ in the Cubanite and Siliceous Zones, but they also show a less pronounced increase in $Cu/(Cu + Ni)$ with depth in the pyrrhotite + chalcopyrite + pentlandite-rich upper part of the Frood mineral system (Fig. 4.10C). The variability in $Cu/(Cu + Ni)$ with depth is greatest in the intermediate grade group which comprises dominantly HMIQD (Fig. 4.10C). On these grounds, the HMIQD may have provided the magmatic conditions required for the effective separation of monosulfide solid solution from the Cu-rich liquid in a disseminated rather than a massive sulfide melt column.

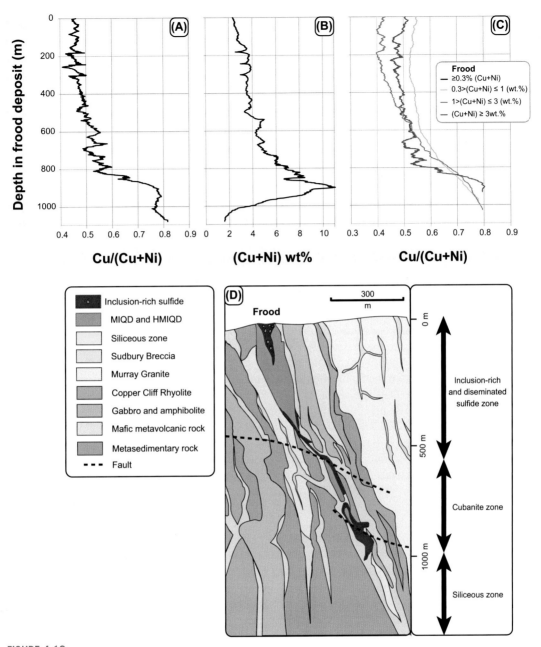

FIGURE 4.10

(A) Variation in the moving average of Cu/(Cu + Ni) with depth through the Frood mineral system. The ratio of Cu/(Cu + Ni) is based on 1000 point moving averages of assay data with increasing depth (data included in the study have S ≥1 wt.% and (Cu + Ni) ≥ 0.3 wt.%). The overall trend with depth is one of increasing Cu abundance relative to Ni. (B) Vertical change in Cu + Ni (wt.%) through the Frood mineral system, where the trend is based on 1000 point moving averages of combined grade and depth (for samples with S ≥1 wt.% and (Cu + Ni) ≥0.3 wt.%). (C) Vertical change in Cu/(Cu + Ni) through the Frood mineral system, where the changes in metal ratio [Cu/(Cu + Ni)] are based on 1000 point moving averages of metal ratio and depth; the data are grouped based on Cu + Ni grade interval. (D) Cross-section of the Frood Deposit showing the location of the zones intersected in Fig. 4.10A–C.

Geochemistry of Stobie

Unlike Frood, the Stobie mineral system comprises at least two different pulses of sulfide-laden magma (Hattie, 2010), so the treatment of the data is different when compared to Frood.

Hattie (2010) showed that the Ni concentration in the sulfide inclusions is lower than the combined blebs and disseminations contained in the HMIQD (see Fig. 4.8). So, both the high-grade sulfides and the inclusions of massive sulfide in the HMIQD have lower Ni concentrations in sulfide than those found in the disseminated sulfide.

Fig. 4.11A,B shows comparative data density plot for samples from the Frood and Stobie mineral systems for samples with >1 wt.% S and >0.3 wt.% (Cu + Ni). Fig. 4.11A shows that the vast majority of Frood data fall on a single trend of increasing Cu/(Cu + Ni) with increasing (Cu + Ni)/S. A few samples fall below this trend, and they come from a small mineral zone developed at the near-surface western periphery of the Frood Orebody.

The plot shows that the vast majority of assays from the Stobie mineral system in Fig. 4.11B cluster with a narrower range of Cu/(Cu + Ni) ratio with a data density peak of ~0.5; the spread either side is due to sulfides with different (po + pn):cpy ratios that reflects the grain-size of the sulfides and the effect of sulfide fractionation. More significantly, there are two groups of samples; one plots along an array that is displaced to lower (Cu + Ni)/S (low metal tenor) relative to the second array which is displaced to higher (Cu + Ni)/S (high metal tenor) as shown in the data density plot in in Fig. 4.11B.

Fig. 4.12 shows data for the Stobie mineral system on a data-density plot of S versus Ni concentration. As shown in Hattie (2010), this plot shows two very different clusters of analyses; the cluster with low Ni/S tends to comprise massive sulfides samples with high S concentrations approaching the stoichiometric abundance in 100 % sulfide of ~37–40 wt.% S. The second group of largely disseminated sulfides plots at higher Ni/S. Based on this plot, it is clear that two different sulfide types comprise the Stobie mineral system. A similar plot for the Frood mineral system illustrates one dominant type of mineralization which extends the upper array in Fig. 4.12, with a very small number of analyses that correspond to the array of massive sulfide mineralization style in the Stobie mineral system. The lower array in Fig. 4.11B corresponds to the lower array in Fig. 4.12, and this is consistent with the higher-grade sulfides in the Stobie mineral system having lower metal tenors in sulfide relative to the lower grade sulfides that occupy the upper array in Fig. 4.11B and the steep array in Fig. 4.12.

Comparative geochemistry of the Stobie and Frood sulfide mineralization

Both the Frood and Stobie mineral systems have average Cu/(Cu + Ni) of ~0.5 (Fig. 4.13A), but the wider variation in Cu/(Cu + Ni) is consistent with the previously discussed concept that Frood sulfides are more strongly fractionated than Stobie sulfides. Despite the fact that both deposits are truncated at the present level of erosion there is a surprising similarity in the average Cu/(Cu + Ni) ratio which is very close to the Sudbury average of 0.5 (Table 4.2).

Although the sulfides of the Frood mineral system exhibits a clear tendency to evolve toward Cu + PGE-rich compositions at depth, sulfides from Stobie illustrate the involvement of two or more different batches of sulfide magma with no clear tendency to fractionate toward more evolved compositions at depth. Despite the close proximity of these two ore deposits, they exhibit some quite different characteristics in terms of composition and emplacement history.

A detailed break out of mineralization types in the Stobie mineral system is possible based on the different trends in Figs. 4.11B and 4.12; these diagrams provide a basis to classify the assays into low and high Ni-tenor groups. The high Ni-tenor group comprises much of the disseminated sulfide

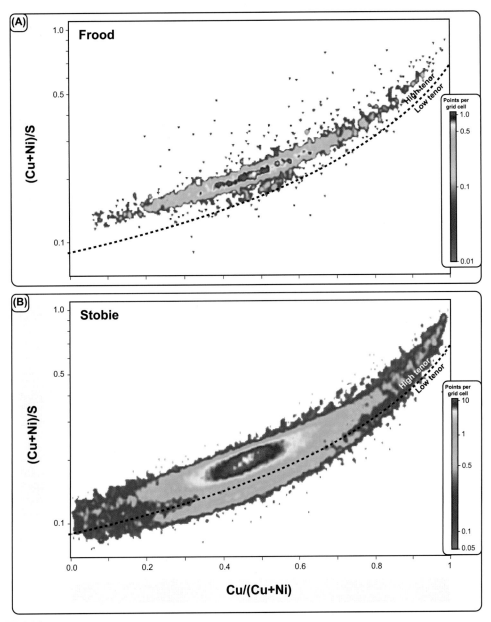

FIGURE 4.11

(A) Data density plot showing the variation in Cu/(Cu + Ni) versus (Cu + Ni)/S for samples from the Frood mineral system (samples with ≥1%S and (Cu + Ni) ≥0.3 wt.%). (B) Data density plot showing the variation in Cu/(Cu + Ni) versus (Cu + Ni)/S for samples from the Stobie mineral system (samples with ≥1%S and (Cu + Ni) ≥0.3 wt.%). The Stobie samples plot along two overlapping trends of increasing Cu/(Cu + Ni) with (Cu + Ni)/S; the array of Frood samples overlaps with the upper array of the Stobie data. High Ni tenor mineralization in the Frood system tends to be more massive whereas high Ni tenor in the Stobie mineral system tends to be associated with disseminated sulfide. The array of low tenor sulfides from Stobie is not recognized in the historic data at Frood. These observations are consistent with the formation of the Stobie mineral system from different pulse of sulfide magma injected into the Offset Dyke from the overlying melt sheet.

See Hattie (2010) for more details.

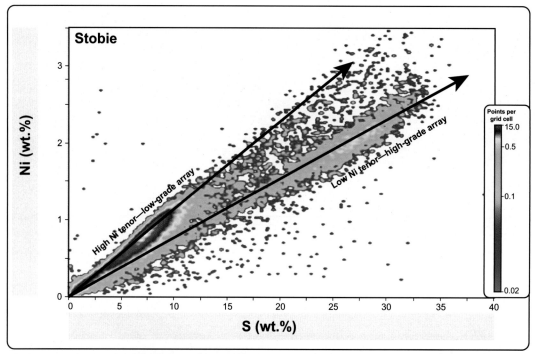

FIGURE 4.12 Data Density Distribution of Nickel and Sulfur Assays for Samples From the Stobie Mineral System

The variation in sulfur concentration versus Ni grade is shown (based on Hattie, 2010). The trends of samples with high Ni tenor and low grade as distinct from low Ni tenor and high grade are shown. The inclusions of massive sulfide enclosed in the matrix of the high Ni tenor disseminated sulfides have a composition that matches the array of the high-grade sulfide mineralization that has low Ni tenor.

mineralization whereas the low tenor group comprises the more massive higher-grade sulfide mineralization (Fig. 4.12) located on the eastern flank of the Stobie mineral systems (Fig. 4.6C). A group of analyses that correspond to more massive sulfide in the core of the disseminated part of the Stobie mineral system is illustrated in Fig. 4.12.

An example of a level plan showing the data density distribution of normative $[Pn]_{100}$ is shown in Fig. 4.14A,B for assays between the -1500 and -2000 levels with ≥ 1 wt.% S and ≥ 0.3 wt.% (Ni + Cu). The plots show that the low normative $[Pn]_{100}$ population occupies a different spatial footprint when compared to that of the high normative $[Pn]_{100}$ group. This demarcation is also associated with differences in (Cu + Ni) grade. The population with low normative $[Pn]_{100}$ is dominantly high grade mineralization and the population with high normative $[Pn]_{100}$ sulfide mineralization is dominantly low in (Cu + Ni) grade. In three-dimensional models of the Stobie mineral system, the trend of these two zones extends from depth to the surface. A small group of samples with low normative $[Pn]_{100}$ is surrounded by a shoot of high normative $[Pn]_{100}$ sulfide to the SW of the plan, and a second shoot of low normative $[Pn]_{100}$ and high-grade sulfide comprises the eastern part of the Stobie mineral system (Fig. 4.14A,B).

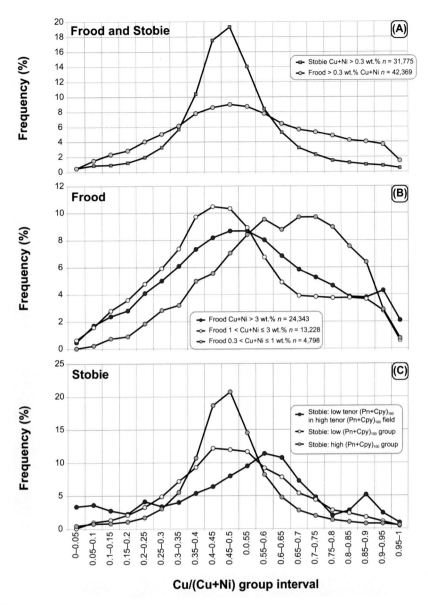

FIGURE 4.13

(A) Comparison of the percentage grouped frequency distributions of Cu/(Cu + Ni) in samples from the Frood and Stobie mineral systems based on assays of samples with (Cu + Ni) ≥ 0.3 wt.% and S ≥ 1 wt.%. Both mineral systems have a Cu/(Cu + Ni) ratio of ~0.5, but the data for Frood exhibit a broader range in Cu/(Cu + Ni) than Stobie. (B) Comparison of the percentage grouped frequency distributions of Cu/(Cu + Ni) in samples from the Frood mineral system broken out by Cu + Ni grade interval (0.3 wt.% > (Cu + Ni) < = 1 wt.%, 1 wt.% > (Cu + Ni) ≤ 3 wt.%, and (Cu + Ni) > = 3 wt.%). Samples with lower sulfide concentrations tend to have higher Cu/(Cu + Ni) than samples with intermediate and high Cu + Ni concentrations. (C) Comparison of the percentage grouped frequency distributions of Cu/(Cu + Ni) in samples from the Stobie mineral system broken out into a high Ni tenor population that tends to have low Ni grade, a low Ni tenor population from the eastern flank of the Stobie mineral system, and the low Ni tenor population from the center of the low grade mineralization (the break-out is based on normative pentlandite + chalcopyrite concentration in 100% sulfide).

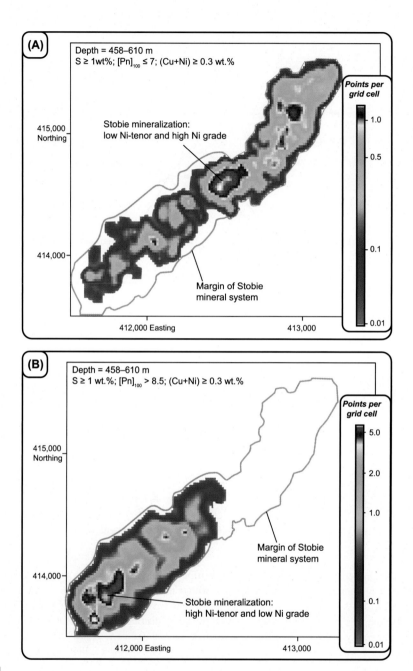

FIGURE 4.14

(A) Distribution of mineralization with low normative $[Pn]_{100}$ (≤ 7). (B) Data density plot showing the distribution of mineralization with high normative $[Pn]_{100}$ (> 8.5). Analyses were filtered to include samples with ≥ 1 wt.%S and > 0.3 wt.% (Ni + Cu) between the -1500 and -2000 level of the Stobie mineral system.

Frequency percentage distribution diagrams depicting the variation in Cu/(Cu + Ni) for the three Stobie populations based on normative $[Pn + Cpy]_{100}$ are shown in Fig. 4.13C. The Stobie samples with high $[Pn + Cpy]_{100}$ coupled with low (Cu + Ni) grade show a tight peak at Cu/(Cu + Ni) $\sim$0.5. The higher grade Stobie sulfides with low $[Pn + Cpy]_{100}$ show a broader range in compositions that also peak at Cu/(Cu + Ni) $\sim$0.5. The higher grade sulfides from within the core of the Stobie disseminated sulfide zone are displaced to higher Cu/(Cu + Ni), but they have low $[Pn + Cpy]_{100}$(Fig. 4.13C), and this is also a feature of the lower grade disseminated sulfides at Frood (Fig. 4.13B)

Important features of Frood-Stobie

The following features characterize the Frood-Stobie mineral system:

1. The Frood and Stobie deposits are the largest historic producer of Ni and Cu in the Sudbury Camp.
2. The ore deposits occur in direct association with a discontinuous QD Offset Dyke within the South Range Breccia Belt.
3. The country rocks are Stobie Formation metasedimentary and metavolcanic rocks, and a range of gabbros and coarse-grained amphibolites of the Frood Intrusion (Fig. 4.6A–C).
4. The Frood mineral system is fractionated from surface to depth as it narrows; the Stobie mineral system is not fractionated with depth.
5. The Frood mineral system comprises one main sulfide melt type that differentiated to form a pyrrhotite + chalcopyrite + pentlandite, cubanite, and PGM-enriched styles of mineralization. Stobie comprises two or more different sulfide melts that have different metal tenors; the lower tenor type at the eastern termination of the Offset Dyke was injected before the higher tenor type as it occurs as inclusions in the later (Hattie, 2010). The lower tenor domain within the core of the higher tenor disseminated sulfide has an uncertain timing relationship and origin.
6. The mineral systems are largely contained within the Frood-Stobie Offset Dyke which appears to close-off at depth. The MIQD is enveloped in a metabreccia that is developed by brecciation, thermal metamorphism and partial melting country rocks and SUBX (Fig. 4.7B).
7. The direction in which sulfide was injected into the Offset Dyke is not entirely clear as the evidence has been removed by erosion, but there is no geological evidence to connect the mineralization at depth or laterally into the base of the SIC. If the mineralization in the Breccia Belt was fed from above then this reduces the likelihood that large systems will be found with roots that were attached to the SIC prior to erosion (Fig. 4.7B).

Despite extensive efforts to locate another large mineral system in the Breccia Belt, discoveries have been much smaller in scale. By far the largest is the newly discovered Victoria Deposit within a concentric discontinuous Offset Dyke, and the much smaller Vermillion Deposit, both located toward the western end of the South Range (Fig. 2.21H,I). The Kirkwood-McConnell Deposit and other small occurrences are found in association with a discontinuous Offset Dyke at Stobie East. There may be other unrecognized concentric breccia belt rings in the footwall beneath the SIC, and this possibility will continue to intrigue exploration geologists.

TOTTEN

The Worthington Deposit was discovered by James Worthington during the construction of the Algoma Branch of the Canadian Pacific Railway. In 1885, a 20 m-deep shaft was sunk and a small quantity of mineralization was identified and mined between 1890–1894, 1907–1908, and 1913–1927. In 1927, the mine workings collapsed along with much of the surface infrastructure and the Soo Branch of the Canadian Pacific Railway line; there was no loss of life because at the time the mine was evacuated due to ground movement. The Worthington Deposit was closed as a result of this incident.

Francis Crean discovered the Totten Number 1 Deposit in 1885, and it was mined from a shallow shaft by caving methods from 1885 to 1901. Inco discovered the Totten Number 2 Deposit in 1937, and further exploration was completed in 1959–1962 with some production between 1966 and 1970, but production was terminated because the ores were of lower value when compared to other Sudbury Deposits.

Exploration for higher-grade domains of mineralization continued up to 2006 with the development of a geological model for the style of mineralization (Lightfoot and Farrow, 2002) and the application of borehole electromagnetic methods (Polzer, 2000; King, 1996, 2007) that identified conductive plates that were subsequently shown to define extensions to depth of the high-grade mineral zone.

The Totten Number 2 Deposit entered production as the newest deposit in the Sudbury Camp in 2014. Despite the fact that the Worthington Offset has contributed a relatively small component of metal production in the context of the whole Sudbury Basin, the ores at Worthington were high grade and exploration continues at depth to locate new mineral zones. Relative to the mature deposit at Frood-Stobie, the mineral zones within the Totten mineral system represent important discoveries that will help to sustain the future economic development of the Sudbury Region.

Geological relationships

In Chapters 2 and 3, much of the geological and geo chemical background information on the Totten mineral system was presented, together with a model that encapsulates the geological relationships between the mineralization, the Offset Dyke and the country rocks. The model shown in Fig. 2.23 highlights the fact that the mineral system is located within the Worthington Offset Dyke, and juxtaposed at the point where the dyke cross-cuts the boundary between the Sudbury Gabbro and metasedimentary rocks. The styles of mineralization developed in this geological setting are shown in more detail in Fig. 4.15.

The orebody at Totten consists of sulfide mineralization between clasts that float in sulfide through to mineralization supported by imbricated inclusions. The inclusion-rich style of mineralization merges gradually into MIQD as described in Chapter 2. The transition from HMIQD to MIQD is gradual and accompanied by a decline in the inclusion size and an increase in the proportion of inclusions of metavolcanic rock relative to amphibolite and metagabbro.

The high-grade sulfides of the Totten mineral system are located within a halo of MIQD that extends along strike from the deposit (Fig. 4.15). Where the country rocks are amphibolites and metagabbros, the dyke tends to be choked with inclusions of a similar type to the wall rocks, and the MIQD and semimassive sulfides occur between the inclusions (Fig. 4.15).

The timing relationships between the QD, MIQD and semimassive sulfide were discussed in Chapter 2, and constrained by the fact that the QD Unit is cross-cut by the MIQD and the MIQD contains inclusions of QD from the marginal Unit of the Offset Dyke.

Principal styles of mineralization

The styles of mineralization in the Totten mineral system are illustrated in Fig. 4.16A–E. The QD and MIQD are described in Chapter 2 (Figure 2.23). The HMIQD consists of disseminations of pyrrhotite + chalcopyrite + pentlandite as blebs with diffuse spongy margins, veinlets, and interstitial sulfides. The veinlets are often rich in chalcopyrite, and this may be a feature produced by rembolization of Cu out

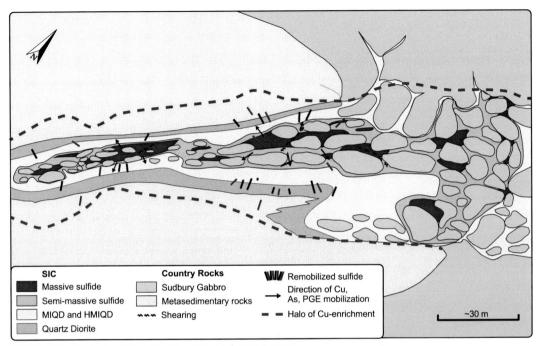

SIC
- ■ Massive sulfide
- ▨ Semi-massive sulfide
- ▢ MIQD and HMIQD
- ▢ Quartz Diorite

Country Rocks
- ▨ Sudbury Gabbro
- ▢ Metasedimentary rocks
- ∾∾∾ Shearing

- ◣◢ Remobilized sulfide
- → Direction of Cu, As, PGE mobilization
- ▬ ▬ Halo of Cu-enrichment

~30 m

FIGURE 4.15 Geological Model for the Totten Mineral System

A close association is recognized between domains of higher grade Ni-Cu-PGE sulfide mineralization and wider parts of the dyke. A linkage also exists between the best mineralization and the presence of MIQD with 25–75% inclusions. The source of the amphibolite and metagabbro inclusions is believed to rest in the nearby Sudbury Gabbro (see Chapters 2 and 3). There is a tendency for the more Ni-rich massive and semimassive sulfides to develop in the center of the mineral zone, and for chalcopyrite-rich stringer mineralization to wrap around inclusions and concentrate in rocks surrounding the MIQD. A halo of Cu-rich disseminated sulfide surrounds the main sulfide zone; this mineralization tends to have elevated 3E and arsenide mineral concentrations.

Based on Lightfoot and Farrow (2002).

of the semimassive and massive sulfide domains (Fig. 4.15). The HMIQD illustrates the transition from MIQD into inclusion-rich sulfide; the matrix between inclusions consists of both QD and semimassive sulfide (Fig. 4.16A). The transition from HMIQD to inclusion-rich massive sulfide is complete when there is no QD in the matrix as illustrated in the sample in Fig. 4.16B. As the proportion of inclusions declines, the inclusion-rich massive sulfide grades into semimassive sulfide and eventually into local patches of massive sulfide with occasional inclusions. The massive sulfide ranges in composition from domains that are entirely composed of chalcopyrite through mixtures of chalcopyrite + pyrrhotite + pentlandite to pyrrhotite + pentlandite as illustrated by Fig. 4.16C–E.

The hand samples do not capture the scale of the relationships that exist in the mine face, and so Fig. 4.17A–C shows three photographs taken of the Totten Number 2 Orebody in underground openings. The examples show a range in sulfide content with Fig. 4.17A comprising massive to semimassive sulfide with amphibolite and metagabbro inclusions floating in the sulfide matrix; the inclusions range in size from a few cm to 1.5 m in the image, but they can reach several meters in length. Fig. 4.17B,C

FIGURE 4.16 Hand Samples Showing Representative Styles of Mineralization in the Totten Mineral System

(A) HMIQD (sample RX355679 contains 3.9 wt.% Ni, 3.5 wt.% Cu, 5g/t 3E, 8 ppm As, and the sample has a nickel tenor of 8.5 wt.%). (B) Inclusion-bearing semimassive sulfide mineralization with no QD in the matrix (sample RX355677 contains 6.5 wt.% Ni, 4.4 wt.% Cu, 1 g/t 3E, 10 ppm As and the sample has a Ni tenor of 10 wt.%). (C) Massive sulfide consisting of chalcopyrite > pentlandite > pyrrhotite (sample RX355685 contains 0.8 wt.% Ni, 28.5 wt.% Cu, and 1.4 g/t 3E). (D) Massive sulfide consisting of chalcopyrite > pentlandite > pyrrhotite (sample RX355680 contains 1.1 wt.% Ni, 25% Cu, and 2 ppm 3E). (E) Massive sulfide consisting of chalcopyrite > pentlandite > pyrrhotite (sample RX355686 contains 6.5 wt.% Ni, 12.7 wt.% Cu, and 1.65 g/t 3E).

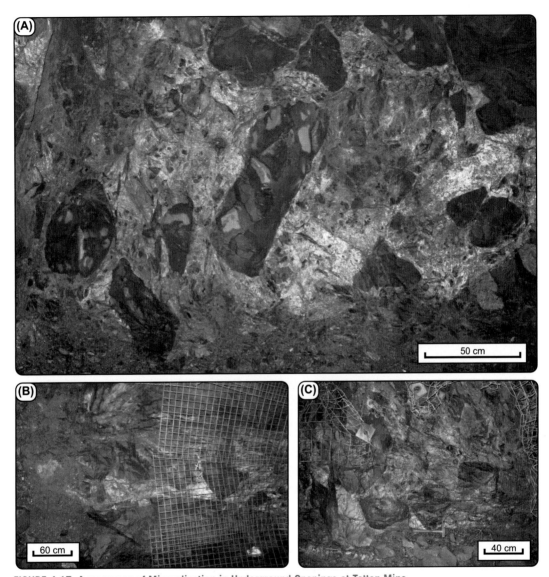

FIGURE 4.17 Appearance of Mineralization in Underground Openings at Totten Mine

(A) Semimassive sulfide in the core of the Totten Mineral system; the inclusions of black amphibolite and metagabbro are floating in a matrix of pyrrhotite > chalcopyrite > pentlandite. (B) Inclusion-rich massive sulfide, where the fragments of amphibolite and metagabbro are touching and almost support each other; the matrix is chalcopyrite > pyrrhotite > pentlandite. (C) Inclusions of amphibolite in sulfide; the mineralogy of the sulfide matrix consists of chalcopyrite > pentlandite ≥ pyrrhotite.

show variations with inclusions that almost touch one-another through to inclusion-supported sulfide matrix. It is the mineralization styles shown in Figs. 4.16 and 4.17 that form the economic heart of the Totten mineral system.

Geochemistry

Aspects of the geochemistry of the Totten mineral system were first reported in Lightfoot and Farrow (2002). Fig. 4.18A shows the grouped frequency percentage distribution of Cu/(Cu + Ni) for assays with S $\geq$ 1 wt.% and (Cu + Ni) $\geq$ 0.3 wt.%. The plot shows that the deposit is strongly enriched in Cu relative to Ni when compared to the more normal distribution of the Stobie mineral system.

The variations in frequency percent group interval distribution of Cu/(Cu + Ni) for the Totten mineral system is shown in Fig. 4.18B based on a breakout of the samples by (Cu + Ni) grade interval. Samples with >1–3 wt.% (Cu + Ni) exhibit a strongly skewed distribution with the displacement of the mode toward slightly higher Cu/(Cu + Ni) than samples with $\geq$0.3–1 wt.% (Cu + Ni). In samples with >3 wt.% (Cu + Ni), the group frequency distribution of Cu/(Cu + Ni) is weakly skewed at high Cu/(Cu + Ni) (Fig. 4.18B).

The data density plot of Cu/(Cu + Ni) versus (Cu + Ni)/S (Fig. 4.19) shows that the samples from the Totten mineral system exhibit a strong Cu-enrichment along a broad trend between pyrrhotite + pentlandite and chalcopyrite. Samples displaced above and below this array commonly contain gersdorffite-cobaltite and sometimes pyrite.

Geological model

The style of mineralization at Totten resembles that developed along much of the Worthington Offset (Lightfoot et al., 1997c). The following observations underpin a genetic model for the deposit in the context of Fig. 4.15:

1. The economically important mineralization tends to comprise much of the width of the Offset Dyke where it passes from metagabbro and amphibolite into metasedimentary rocks. Laterally away from the massive sulfide, this mineralization extends through the center of the Offset dyke where it forms an inclusion-rich massive sulfide through to MIQD.

2. The mineralization grades from disseminations in MIQD through HMIQD, inclusion-rich sulfide to semimassive and massive sulfide. The mineralization tends to comprise domains of pyrrhotite + pentlandite through to domains of chalcopyrite, and there is a tendency for Cu to be remobilized out of the domains of heavy mineralization into the envelope of MIQD. Minor elements such as arsenic are also enriched toward the margins of the massive sulfide and along strike where the massive sulfide zone narrows in width.

3. The massive sulfide was emplaced after the first phase of inclusion-free QD, and they follow a shoot which trends along the favourable contact between the Sudbury Gabbro and the metasedimentary rocks. The Sudbury Gabbro and metasedimentary rocks were strongly brecciated during the impact process to produce shatter cones in the metagabbro and SUBX. This ground conditioning was likely important as it provided a pathway for later pulses of sulfide-rich melt to enter the Offset Dyke, and it provided a source for the large blocks of metagabbro and amphibolite that were fractured and conditioned by the impact process ready to be incorporated as the melt migration down the Offset along this favourable contact.

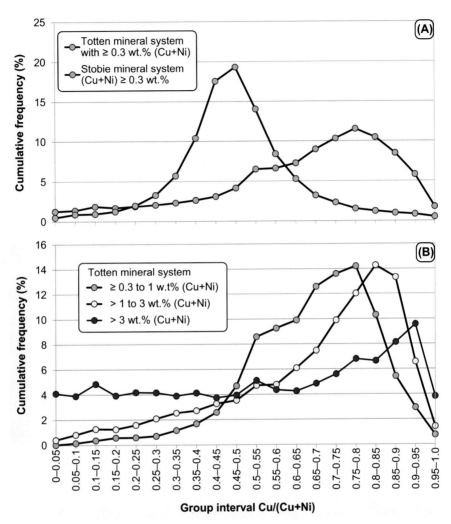

FIGURE 4.18 Geochemical Characteristics of the Totten Mineral System

(A) Comparison of the Cu/(Cu + Ni) ratios for samples from Totten with S ≥ 1 wt.%, (Cu + Ni) ≥ 0.3 wt.%) with similar data for the Stobie mineral system. Note the stronger Cu-enrichment in the Totten mineral system relative to Stobie. (B) Comparison of the variations in Cu/(Cu + Ni) in three different group intervals of (Cu + Ni) grade for samples from the Totten mineral system. The vast majority of low- and intermediate-grade samples from Totten have a skewed distribution toward Cu-rich compositions whereas the high-grade samples have a flatter less-strongly skewed distribution.

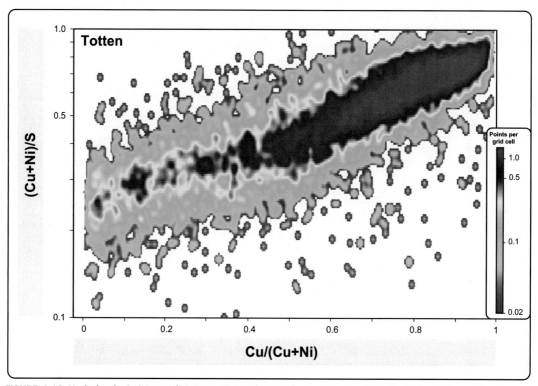

FIGURE 4.19 Variation in Cu/(Cu + Ni) Versus (Cu + Ni)/S for Assay Samples From the Totten Mineral System With ≥1 wt.% S and ≥0.3 wt.% (Cu + Ni)

The plot shows a strongly skewed distribution toward Cu-rich sulfides in all three groups, and a tendency for the more Ni-rich samples to be displaced above the extension of the trend of Cu-rich samples.

Exploration of the Totten mineral system is far from complete. New discoveries will be guided in part by the geological relationships between the Offset Dyke and the country rocks, the style of mineralization, and borehole electromagnetic anomaly targets.

COPPER CLIFF

The Copper Cliff Offset is one of the principal mine complexes that contributes to production of base and precious metals at Sudbury. The original discovery of surface and near-surface mineralization is attributed to Thomas Frood, Rinaldo McConnell and J.H. Metcalf. The development of the Copper Cliff Mine, the #1 Mine, Number 2 Mine and Evans Deposits in the late 19th and early 20th century supported mostly copper production (Fig. 4.20). The vast majority of production has occurred after 1960. The development of the Clarabelle open pit deposits in 1961–65 (Fig. 4.21) was followed by development and additional discovery of the depth extent of orebodies in the Offset Dyke that are mined from the Copper Cliff Mine Complex.

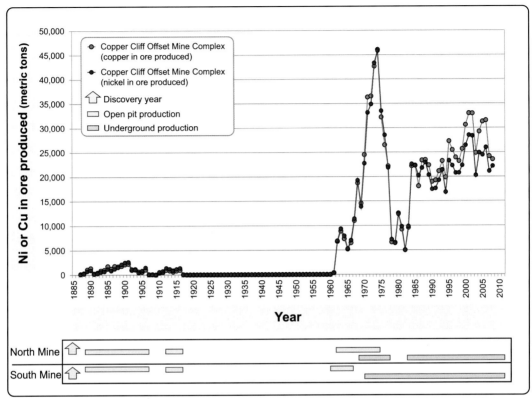

FIGURE 4.20 Graph Showing the Historic Production of Cu and Ni in Mmined Ore From the Copper Cliff Offset

Also shown are the key events in the history of mine development.

FIGURE 4.21 Photograph of the Clarabelle Open Pit From the Northeastern Margin of the Past-Producing Mine

The orebody occupies the narrow part of the keel of the Copper Cliff Funnel. The Copper Cliff Funnel consists of SLNR within a narrow rind of QD that occurs at the contact of the SIC with granite, metavolcanic and metasedimentary rock. To the east, two prominent horns of QD extend at a low angle into the country rocks on the east side of the funnel.

Geological relationships

The orebodies hosted by the Copper Cliff Offset Dyke are collectively termed the Copper Cliff mineral system. The mineralization occurs in association with a unit of MIQD which extends along much of the dyke from Clarabelle in the North to Kelly Lake in the South (Fig. 2.24A). The orebodies, shown in longitudinal section in Fig. 4.22, extend down the plunge in the Dyke to depths approaching 2500 m as a series of subvertical "ore shoots," and they also extend along cross-linking structures within the dyke between the shoots (eg, the 830 Structure in Fig. 4.22). Typically, the orebodies contain between 10% and 90% sulfide (ie, HMIQD through to massive sulfide), and the segments of the dyke between the orebodies consist of the QD and MIQD Units.

The Copper Cliff orebodies are contained in two principle settings as described in Chapter 2 (Fig. 2.24A,B). These are the Funnel and the Dyke sections of the Offset. The flank of the Funnel contains the Pump Lake and Lady Violet orebodies. The narrow southern extent of the Funnel connects to the Dyke and hosts the Clarabelle Deposit (Fig. 2.24A). The funnel consists of QD and Quartz-rich Norite and contains no definitive example of Sublayer Norite (SLNR). The margins and base of the Funnel have well-developed QD (Grant and Bite, 1984; Capes, 2001). The Pump Lake and Lady Violet Deposits consist of HMIQD and inclusion-rich sulfide in the QD at the margin of the Funnel (Fig. 4.23A).

The Clarabelle Deposit occupies the base of the funnel and a wider domain of the Offset Dyke developed below and south of the funnel. The connection between the Funnel and the Dyke plunges at about 60 degrees toward the north before being displaced to the south by the Cliff Lake Fault at a depth

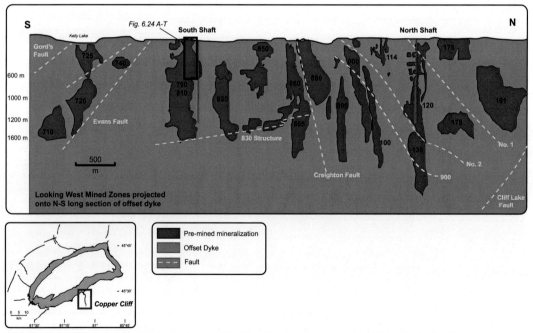

FIGURE 4.22 West-Facing Geological Long Section Showing the Orebodies and Major Faults Projected Onto the Plane of the Copper Cliff Offset

Modified after Farrow and Lightfoot (2002).

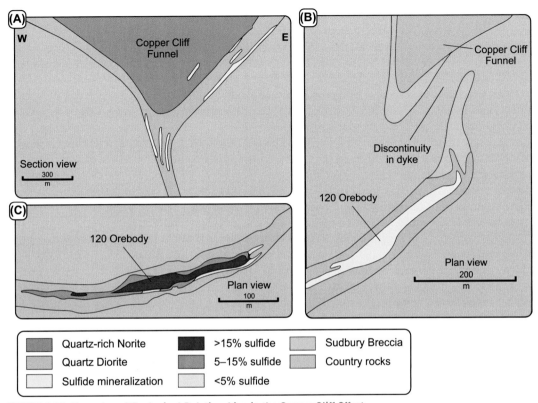

FIGURE 4.23 Examples of Geological Relationships in the Copper Cliff Offset

(A) North-facing cross-section showing the geology of the Copper Cliff Funnel and the location of the orebodies in the neck and throat of the funnel (after Cochrane, 1984). (B) Plan showing the geology of the Copper Cliff Funnel where it is partially connected at depth to the Copper Cliff Offset; the plan shows the location of the 120 ore deposit to the south of the discontinuity (after Cochrane, 1984). (C) Plan view showing geological relationships between the low, intermediate, and high grade mineralization in an economically important domain of mineralization and the Offset Dyke in the area of the 120 Orebody (after Cochrane, 1984).

Modified from publications of the Ontario Geological Survey, and used under license to Elsevier.

of >2500 m (Chapter 2 and Fig. 4.22). The contact between the funnel and the dyke is a zone of SUBX and this is an example of the many primary discontinuities in the dyke that are described in Chapter 2.

The Copper Cliff Offset Dyke hosts approximately 20 discrete orebodies that form vertical to sub-vertical continuous and discontinuous shoots that extend from surface or near-surface to depths approaching 2500 m, and many of these orebodies may extend to greater depths. The long section shown in Fig. 4.22 illustrates the distribution and names of the orebodies, the main production shafts, and the principal fault structures that break the continuity of the Offset Dyke and the ore deposits.

The orebodies are divided into two principle types by Cochrane (1984), depending on the location of the sulfides within the Offset Dyke. These are: (1) concentrations of sulfide in the center of the Offset Dyke that form elongated pipe-like orebodies where the mineralization is separated from the country rock by barren QD. The 120, 100, and 900 orebodies are examples of this type and Cochrane (1984) refers to them as "120-type orebodies" (Fig. 4.22), and (2) concentrations of sulfide developed at the margins of the Dyke. This second group of mineral zones may switch over from one side of the Offset Dyke to the other along strike, and these switchovers typically occur where the Offset Dyke is narrow and hosted in well-developed SUBX. The 880, 865, 850, 830, 810, and 800 orebodies are all examples of this type and are collectively referred to as "810-type orebodies."

The association of orebodies with primary discontinuities in the dyke was recognized by Cochrane (1984) and described in detail by Mourre (2000). The 120 orebody is located in a 40 m wide segment of the Offset Dyke immediately to the south of the Funnel, and it is separated from the Funnel by a discontinuity in the Offset Dyke (Fig. 4.23B). The orebody is contained within a 25 m wide zone of HMIQD in the center of the Offset Dyke. The zone has a strike length of 200 m and extends from near surface to almost 1 km in depth. In plan, the sulfides form a high-grade core in the center of the Offset Dyke, surrounded by a lower-grade fringe zone that tails off to the south. The mineralization consists of blebby sulfide-textured HMIQD. This mineralization type, which is typical of the Copper Cliff Offset orebodies, is described by Hawley (1965), Souch et al. (1969), and Naldrett (1961).

The 810 Orebody is located approximately 3 km to the south of the 120 Orebody and south of the Creighton Fault (Figs. 4.22 and 4.23C). The Offset Dyke is approximately 60 m wide and the sulfides are located along the eastern contact of the Offset Dyke with metasedimentary country rocks. The 810 Orebody is located near an important primary discontinuity of the dyke developed in a domain of SUBX in felsic metavolcanic and metasedimentary country rocks. A discontinuity in the Offset Dyke in the 810 Orebody is illustrated by a series of eight sections and level plans in Fig. 4.24A–T (modified after Mourre, 2000). The sections and plans illustrate the relationship of the mineral zone to the Offset Dyke, and show how the sulfides switch from one side of the dyke to the other, and how a primary discontinuity in the dyke between the 500 and 1250 Levels are located in a zone of SUBX much like the break described in Chapter 2 between the 850 and the 865 orebodies at surface (Fig. 2.24C).

Characteristics of the mineralization

The styles of mineralization in the Copper Cliff Offset are similar to those developed in the Totten mineral systems. The predominant types of mineralization are (1) MIQD in the Offset Dyke and Quartz-rich Norite in the Funnel; both contain subeconomic quantities of interstitial to ragged disseminated sulfide mineralization and inclusions (Fig. 4.25A,B). (2) HMIQD comprises a significant part of the Copper Cliff Offset Dyke (Fig. 4.25C). (3) Inclusion-rich massive sulfide has a matrix comprising different proportions of quartz diorite to sulfide (Fig. 4.25D–G). This style of mineralization grades into an inclusion-bearing sulfide type of mineralization as the proportion of sulfide/silicate in the matrix increases and the inclusions become larger (Fig. 4.25G–I). (4) The inclusion-rich sulfide grades into inclusion-bearing semimassive and massive sulfide that comprises pyrrhotite > (chalcopyrite + pentlandite), (pyrrhotite + chalcopyrite) > pentlandite, and chalcopyrite > pentlandite > pyrrhotite (Fig. 4.25J–L).

The orebodies typically occur within the Offset Dyke, but there are examples of mineral zones that show complex relationships with country rocks. One example is the impregnation of sulfide into a metastaurolite schist adjacent to the semimassive sulfide mineral zone of the Offset Dyke south of the Creighton Fault (Fig. 4.25M).

Geochemistry

The grouped-frequency variation in Cu/(Cu + Ni) for mineralization from the Copper Cliff Offset is shown in Fig. 4.26A for all samples with (Cu + Ni) $\geq$ 0.3 wt.%. The variations are described in the context of the allocation of samples to orebodies shown in Fig. 4.22. The grouped frequency distribution profiles for the mineral systems at Frood and Totten are compared to the bulk composition of mineralization from the Copper Cliff Offset in Fig. 4.26A.

Relative to the Frood mineral system, the Copper Cliff mineral system is skewed to higher Cu/(Cu + Ni), and the Totten mineral system is skewed to even higher Cu/(Cu + Ni) (Fig. 4.26A). The depth of erosion at Totten is likely quite large given it's distal location relative to the original base of the SIC, and the depth of erosion is probably larger than much of the Copper Cliff Offset. Frood-Stobie may be one of the least eroded large mineral systems associated with an Offset Dyke, and it's

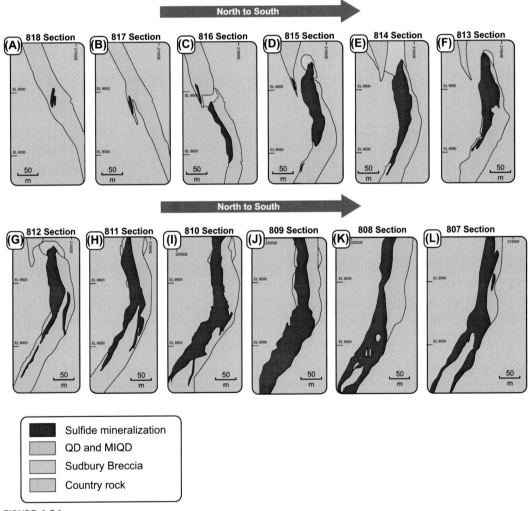

FIGURE 4.24

(A–L) A series of 12 vertical sections from north to south showing the geological relationships between the silicate- and sulfide-bearing rocks in the Copper Cliff 810 Orebody domain.

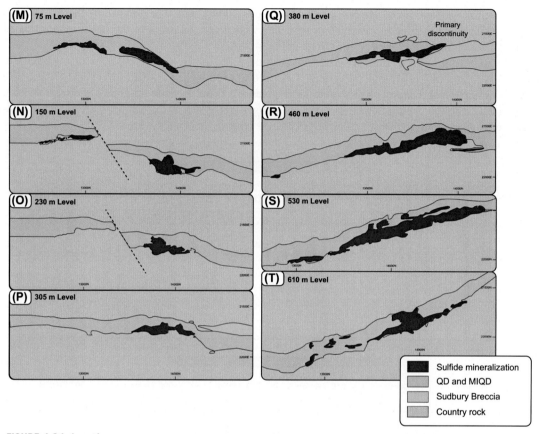

FIGURE 4.24 (*cont.*)

(M–T) A series of level plans documenting the geological relationships in the 810 Orebody with increasing depth. The primary discontinuity in the Offset is related to the development of SUBX in the country rocks and to a change in geometry of both the Offset Dyke and the associated mineralization.

Based on Mourre (2000).

relationship to the base of the SIC may be as little as 1 km distance. It is possible that the variations in Fig. 4.26A are an artifact of the different levels of erosion of the original mineral systems in different Offset Dykes as well as a function of the vertical distance to the original base of the SIC which is now removed by erosion.

A comparison of some of the individual orebody domains of the Copper Cliff mineral system relative to the system as a whole is illustrated in Fig. 4.26B–D. Orebody systems located close to the Funnel and in the proximal dyke are illustrated by the Clarabelle and 191 domains. Both the Clarabelle and 191 domains have a more symmetric pattern of Cu/(Cu + Ni) than the overall Copper Cliff mineral system, but the 191 is characterized by slightly higher Cu/(Cu + Ni) than Clarabelle. Further to the south, but still within the section of the Offset Dyke that is north of the Creighton Fault, the 120–138 and 900 Orebody domains have a Cu/(Cu + Ni) distribution pattern that is essentially indistinguishable

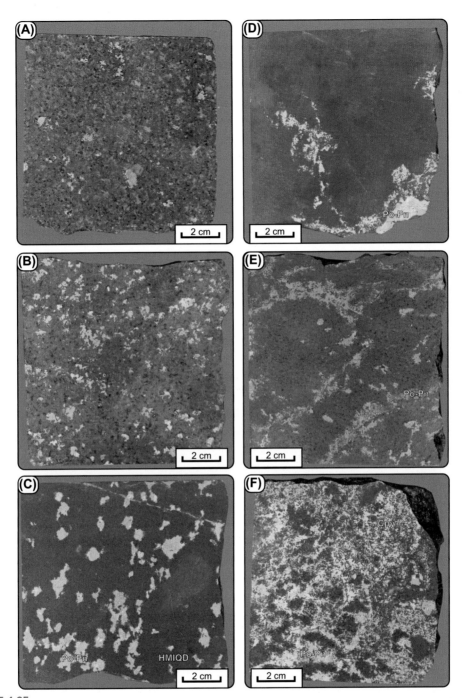

FIGURE 4.25

(A) Quartz-rich SLNR from the Clarabelle Funnel with ragged disseminated sulfide mineralization. (B) Quartz-rich SLNR with abundant inclusions and interstitial ragged disseminated sulfide mineralization from the Clarabelle Funnel. (C) HMIQD with abundant inclusions and blebby disseminated sulfide mineralization from the 830 Orebody. (D) Mafic inclusions separated by narrow segregations of sulfide mineralization, Clarabelle Deposit. (E) Inclusions of quartz-bearing norite (possibly derived from the Sublayer) separated bya zone of sulfide mineralization that appears to impregnate the margins of the inclusions; Clarabelle Deposit. (F) Inclusion-bearing massive sulfide from the Clarabelle Deposit.

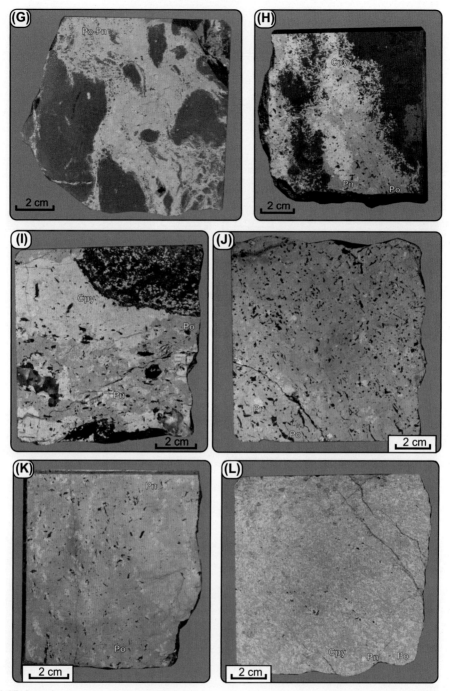

FIGURE 4.25 (*cont.*)

(G) Sulfide mineralization (pyrrhotite > pentlandite ≥ chalcopyrite) with subrounded inclusions of mafic volcanic rock, Clarabelle Deposit. (H) Inclusion-rich sulfide mineralization with chalcopyrite ≥ pyrrhotite > pentlandite from the Copper Cliff Offset; note the development of interstitial sulfide within the inclusion; this may be a replacement texture. (I) Inclusion-rich sulfide mineralization with chalcopyrite > pyrrhotite > pentlandite from the Copper Cliff Offset. (J) Massive sulfide (pyrrhotite ≫ pentlandite ≥ chalcopyrite) from the Copper Cliff Offset. (K) Massive sulfide (pyrrhotite > pentlandite ≥ chalcopyrite) from the Copper Cliff Offset. (L) Massive sulfide (chalcopyrite > pyrrhotite > pentlandite) from the Copper Cliff Offset.

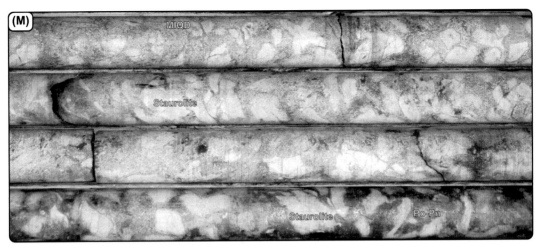

FIGURE 4.25 (*cont.*)

(M) Drill core interval (borehole 1210720, 562–565 m) from the Copper Cliff Offset south of the Creighton Fault; drill core shows the replacement of low-melting point fraction of the groundmass of a meta-staurolite schist from the Stobie Formation by pyrrhotite + pentlandite, leaving meta-staurolite crystals separated by sulfide mineralization. Scale: diameter of drill core is 3.65 cm.

from the signature of the bulk composition of the Copper Cliff mineral system (Fig. 4.26C). Continuing southward and more distal to the SIC, the mineralization located to the south of the Creighton Fault show a wide range in composition. The 790–810 Orebody domain exhibits a strongly skewed pattern toward high Cu/(Cu + Ni), but other orebody domains like the 830 have a pattern similar to that of the overall Copper Cliff mineral system.

There are significant differences in Cu/(Cu + Ni) between the different orebody domains and the ratios do not vary systematically with distance along the Offset Dyke. Although erosion has likely removed a substantial proportion of many of the orebodies, the differences in Cu/(Cu + Ni) are calculated from data collected from samples over large vertical distances in each mineral shoot. To illustrate variation in sulfide composition with depth in an ore zone system, Fig. 4.27A–C shows data for the 120–138 orebody system. These two orebodies comprise a system that has a relatively narrow range of (Cu + Ni)/S in samples evenly covering much of the range in Cu/(Cu + Ni) as shown by the data density plot of assay data in Fig. 2.47A. Although there are subtle variations in sulfide metal tenor, the analyses represents mineralization that has a composition controlled by simple mixing between an original monosulfide solid solution with a fairly constant pyrrhotite/pentlandite ratio and chalcopyrite. In order to establish whether there are systematic changes in sulfide chemistry with depth in this narrow vertical 1.6 km long orebody, 1000 point moving averages were calculated based on vertical position below the surface, and plotted as trends in Fig. 4.27B,C. From surface to depth in the orebody system there is an overall trend toward higher Cu/(Cu + Ni) with a small increase in (Cu + Ni) (Fig. 4.27B,C) that is similar to the variations described in the Frood mineral system. The trend is broken by heavy concentrations of more Cu-rich sulfides at around 300–400 m depth (dominantly due to contributions from the 138 orebody system), where the mineralization comprises an unusually high-grade lens of Cu-rich sulfide that is partially hosted in SUBX.

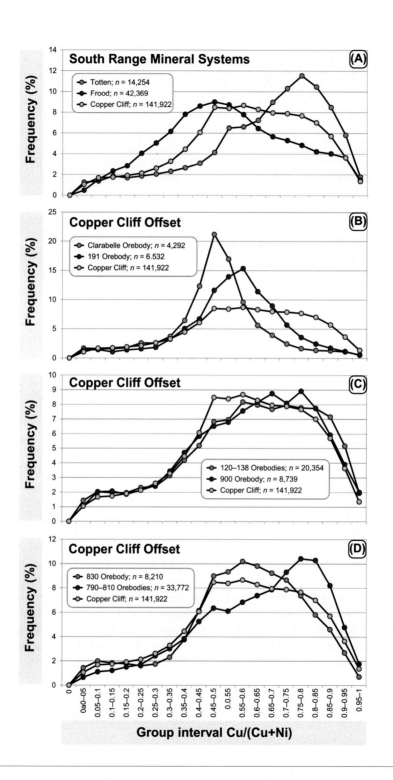

FIGURE 4.26

The very large 790–810 orebody domain exhibits a wide range in composition as illustrated by the data density plot of Cu/(Cu + Ni) versus (Cu + Ni)/S shown in Fig. 4.28A. There is not only a wide spread in Cu/(Cu + Ni) that reflects the variation in (pyrrhotite + pentlandite):chalcopyrite, but there is also a wide range in (Cu + Ni)/S throughout the range of Cu/(Cu + Ni) (Fig. 4.28A). Taking the upper and lower portions of this trend as representative of low versus high (Cu + Ni) tenor mineral styles shown in Fig. 4.28A, it is possible to investigate the distribution of these two styles of mineralization with position in a long section of the dyke. Fig. 4.28B,C show the distribution of the low and high metal tenor variants on a data density plot; this plot illustrates that the higher tenor material occupies a broad vertical shoot that is displaced to the north of the low metal tenor group which broadly wraps over the top and runs down the south side of the high metal tenor mineralization. These two groups appear to represent different domains within the mineral system. Although it is possible that these sulfides are related by fractionation within the dyke, the observation that they are separated along a considerable strike length on either side of the discontinuity described in Fig. 4.24B,C lends support to the concept that they represent different pulses of sulfide melt that were injected into the Offset at different times much like the ores at Stobie appear to belong to two or more different pulses of sulfide- and inclusion-rich magma.

The Kelly Lake orebody domain comprises a number of historic discoveries as well as a large new discovery made at the turn of the 20th Century (described in Polzer, 2000). Taken as a group, this orebody domain exhibits a very wide range in not only Cu/(Cu + Ni), but also (Cu + Ni)/S, that is indicative of at least two different sulfide tenor populations in the initial sulfide melt. These populations are distinguished in a data density plot (Fig. 4.29), and the lower of the two trends dominantly comprises mineralization from the 740 OB whereas the upper trend comprises mineralization from the 710, 720, and 725 Orebody systems that fall along a separate plunge located to the south of the 740 Orebody system (Fig. 4.22). The 710–720–725 Orebody domain contrasts with many of the other Copper Cliff

◀ **FIGURE 4.26**

(A) Comparison of the ratio of Cu/(Cu + Ni) for samples with ≥0.3% (Cu + Ni) from the Copper Cliff mineral system (including the mineralization contained in the Funnel), the average Totten mineral system, and the average Frood mineral system. Totten has the strongest asymmetry and enrichment of Cu relative to Ni, followed by Copper Cliff and then the Frood mineral system. (B) Comparison of the Clarabelle and Copper Cliff 191 Orebody domains to the overall composition of the Copper Cliff mineral system. Note the displacement of these near-contact orebody domains to lower Cu/(Cu + Ni) than the bulk of the Copper Cliff mineralization. (C) Comparison of the 120–138 and 190 Orebody domains to the overall signature of the Copper Cliff mineral system. This Offset Dyke-hosted sulfide mineralization has a signature much closer to the bulk average of the mineralization in the Copper Cliff Offset with a small displacement to higher Cu relative to the whole system which is consistent with the presence of the Clarabelle and 190 Orebody domain analyses in the calculation of bulk composition. (D) Average compositional data for the 790–810 and 830 Orebody domains illustrating the similarity of the 830 data to the average Copper Cliff mineral system, and the displacement of the 790–810 Orebody domain to very high Cu relative to Ni. The different orebody domains in the Copper Cliff Offset have undergone different degrees of fractionation, and a hypothesis worthy of further testing is whether some of the mineral domains were actually generated from different pulses of sulfide melt that were injected into the Offset.

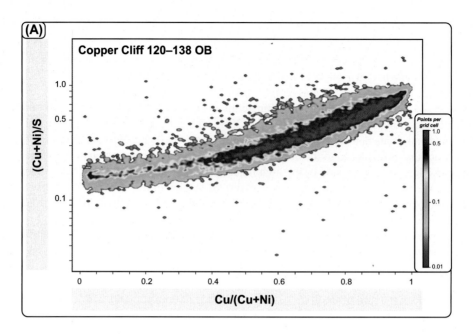

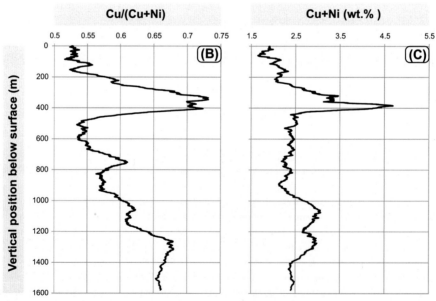

FIGURE 4.27

(A) Data density plot of Cu/(Cu + Ni) versus (Cu + Ni)/S for the 120–138 Orebody domains which form a continuous (>1.6 km long) ore shoot in the northern part of the Copper Cliff Offset. (B) Variation in Cu/(Cu + Ni) based on 1000 point moving averages of data with vertical depth in the mineral shoot. High values of Cu/(Cu + Ni) at 400 m depth are associated with the 138 Orebody domain, but the majority from the 120 Orebody domain shows a progressive increase in Cu/(Cu + Ni) with depth. (C) An analysis of the change in Cu + Ni (1000 point moving average) with vertical position in the 120–138 Orebody. There is a slight increase in Cu + Ni grade with depth, and the spike at 400 m depth is due to the 138 Orebody.

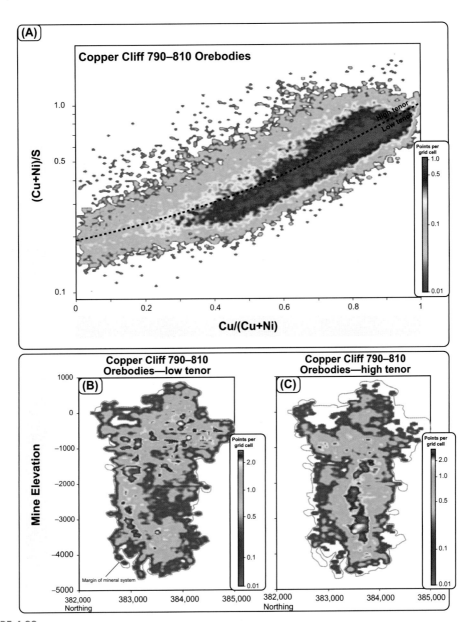

FIGURE 4.28

(A) Data density plot in the plane of the Offset Dyke showing the low and high metal tenor populations 790–810 orebodies demarcated by the curve shown in this image. (B) Data density plot showing the spatial distribution of the lower metal tenor group in the west-facing vertical plane of the Offset Dyke (ie, data below the curve on Fig. 4.28A). (C) Data density plot showing the spatial distribution of the higher metal tenor group in the west-facing vertical plane of the Offset Dyke (ie, data above the curve on Fig. 4.28A). The different Ni tenor populations appear to correspond to different parts of the mineral system, and they comprise shoots of mineralization that rake down the Offset Dyke along different trajectories. These sulfides possibly represent different influxes of sulfide melt derived from the overlying melt sheet.

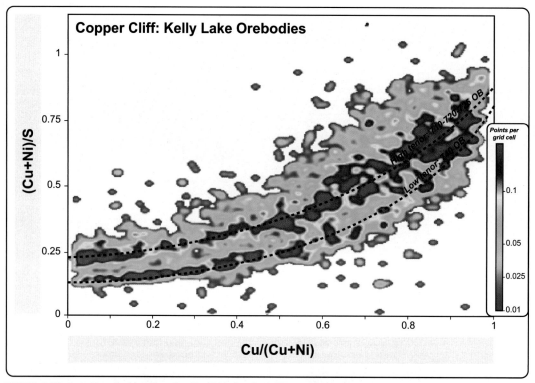

FIGURE 4.29 Data Density plot Showing the Variation in Cu/(Cu + Ni) Versus (Cu + Ni)/S for the Kelly Lake Orebody Domain From the South Part of the Copper Cliff Offset Dyke

The deposit comprises four orebodies that breakout into two overlapping trends on this diagram. The lower trend is the 940 Orebody domain, and the upper trend comprises the remaining three domains. The Kelly Lake orebody domain is another example of a mineral system that comprises two or more discrete pulses of sulfide melt derived from the base of the melt sheet at different times in its evolution.

orebodies in having an increasingly strong plunge to the south with depth (Fig. 4.22). These orebodies also represent the southernmost known major concentration of sulfide mineralization in the Offset Dyke.

Geological model

There are many geological observations that must be reconciled with a model for the formation of the mineralization in the Copper Cliff Offset; some of the important ones are listed below:

1. A large number of the mineral zones in the Offset Dyke have a vertical configuration with local flattening along gently dipping discontinuities like the flat part of the 830. The orebodies at the south end of the Offset plunge steeply to the south, whereas those at the north end plunge toward the north.
2. Economically important domains of mineralization often occur near to primary discontinuities in the dyke where there are changes in country rock type and development of extensive SUBX.

3. There is no systematic change in composition of the orebody systems from north to south along the Offset Dyke but mineralization to the south of the Creighton Fault is broadly different in composition to that developed to the north of the fault.
4. Vertically continuous orebody systems tend to be zoned from low to high Cu/(Cu + Ni) with increasing depth (eg, the 120 Orebody domain with the exception of the 138 Orebody). A tendency toward more copper-rich compositions also exists in other continuous vertical shoots that plot on tight arrays of Cu/(Cu + Ni) versus (Cu + Ni)/S. Some orebodies show evidence for two or more sulfide melt contributions that correspond to low and high metal tenor mineralization (eg, 790–810 and the 740–710–720–725 Orebody domains).
5. Structural modification of the Offset Dyke occurs along west-east structures such as the Creighton Fault, and the gentle to steep north plunging structures which offset the orebody domains. These structures offset the position of the mineral shoots, and they control their continuity at depth.

A genetic model for the Copper Cliff mineral system is shown in Fig. 4.30. The sulfide melt was injected from above where a funnel or embayment structure once existed that is now eroded-away (Farrow and Lightfoot, 2002).

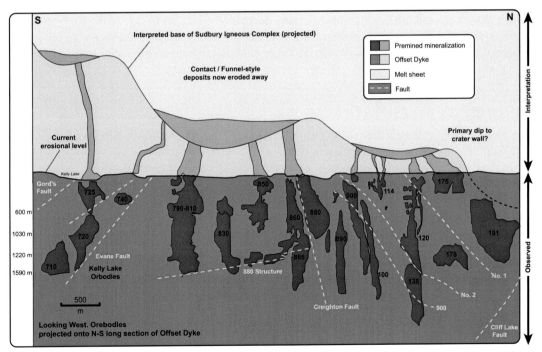

FIGURE 4.30 Geological Model for the Copper Cliff Mineral System

The present day distribution of orebodies and the level of erosion are shown. The hypothetical position of the base of the overlying melt sheet is shown as an N–S-oriented Sublayer trough and/or funnel. The mineralization was derived from the melt sheet and different batches were injected into the Offset Dyke at different times during the re-adjustment of the crater floor. Modified from Farrow and Lightfoot (2002).

A special challenge is to reconcile the geometry of the vertically plunging ore shoots with the observed dip of the base of the Main Mass and the keel of Copper Cliff Funnel. If the original base of the melt sheet was approximately flat, then the mineral zones might be expected to have an original vertical plunge. After deformation by tilting of the South Range, these ore shoots would then be expected to plunge toward the south which is clearly not the case. To solve this enigma, a deformation mechanism has been described that accommodates the geometry through a shearing mechanism within the basal Main Mass (the tri-shear model of Lenauer and Riller, 2012a,b). Although there is undoubtedly some level of deformation in the South Range, there is no local geological evidence that relates the geometry of the Clarabelle Funnel and adjacent north-dipping Main Mass to the South Range Shear Zone, and there is evidence to suggest that the more northerly orebodies in the dyke and funnel actually dip parallel to the base of the SIC.

The relationship between the Funnel and the Offset Dyke may be considered as a primary feature developed along the axis of an embayment feature along the base of the Main Mass that actually represents a primary flat depression in the melt sheet where magmatic sulfides were localized. In this scenario, at least part of the northward to overturned dip of the basal contact of the South Range would be due to the topography at the base of the melt sheet, and the effect of deformation would increase with proximity to the South Range Shear Zone.

The orebodies to the south of the Creighton Fault typically have higher Cu/(Cu + Ni) (0.62), $[Ni]_{100}$ (5.0 wt.%) and $[3E]_{100}$ (9.1 g/t) than those developed on the north side of the Creighton Fault (0.56, 4.4 and 5.0, respectively; see Farrow and Lightfoot, 2002). This may reflect a deeper level of erosion that has exposed more fractionated styles of mineralization in the orebodies to the south of the Creighton Fault.

OTHER OFFSET DYKES

The vast majority of known deposits in Offset Dykes are associated with Frood-Stobie, Copper Cliff, and Worthington-Victoria Offset Dykes.

In the East Range, small deposits exist along the Parkin Offset at Milnet Mine (Fig. 2.20A; Meyn, 1970; Murphy and Spray, 2002), and occurrences extend SSW of the Milnet occurrence for ~2 km (Bailey, 2013). In the Whistle Offset, mineralization is associated with the Whistle Offset Dyke at the Podolsky Deposit, and a small mineral zone is developed near to the surface (the North Zone; see Farrow et al., 2008a,b). Another small deposit was mined from the MacLennan Offset in the footwall of the East Range (Grant and Bite, 1984).

In the North Range, small Ni-Cu sulfide orebodies occur near to the Funnel of the Foy Offset and at a primary discontinuity in the Dyke at the Nickel Offset Deposit (Fig. 2.20A; Tuchscherer and Spray, 2002). Subeconomic concentrations have been identified in other segment of the Foy, Ministic and Hess Offset Dykes (Fig. 2.20A; Lightfoot et al., 1997c).

In the South Range, the Kirkwood Offset contains segments of mineralization located to the SW and SE of Kirkwood Mine, and a part of this mineralization comprised the historic production from Kirkwood Mine and un-mined deposits in the McConnell Offset Dyke west of Garson Mine (Fig. 2.20A; Farrow et al., 2008a,b). Small mineral zones occur in association with QD pods along strike from the Frood-Stobie mineral system (Grant and Bite, 1984).

A small Offset Dyke along the keel of the Creighton Trough contains economic concentrations of sulfide mineralization (Lightfoot et al., 1997c). In the context of the Creighton Ore System, this is a small portion of the historic production, but the orientation of this Offset provides an important control on the development of footwall mineralization in the 403 Orebody described later in this chapter.

It is clear that not all Offset Dykes are equally endowed with mineralization. In one scenario, exploration effort has simply failed to locate the ore deposits, and in a second model, the degree of mineral endowment varies between and along Offset Dykes as a function of some other processes and controls. This is a question that is given further consideration in Chapter 5.

CONTACT AND FOOTWALL MINERAL SYSTEMS

The close spatial relationship between Ni-rich contact orebodies and Cu-Ni-PGE-rich footwall mineral zones is well established in the North Range at Levack-Coleman and Victor-Nickel Rim South, but this association is less well appreciated in the South Range. A significant proportion of the South Range ores comprise Cu-Ni-PGE-rich styles of mineralization that occur in direct association with the contact deposits, but are contained in the country rocks beneath the SIC. The extent to which these mineral zones are remobilized into late structures, versus emplaced and modified by postimpact structures can be understood by examining the geology, mineralogy and geochemistry.

MINERAL SYSTEMS IN THE SOUTH RANGE

The South Range mineral systems at Creighton and Garson are discussed in some detail as examples of primary localization and emplacement of sulfide with late kinesis of sulfide into space created by postimpact deformation. These case study examples have geological analogues in the Crean Hill, Murray-McKim, Blezard-Lindsley, and Falconbridge mineral systems. Together with the Creighton and Garson Systems these deposits are the source of over half of the metal produced from the Sudbury Camp.

Creighton

Albert Salter, who was a land surveyor working for the Geological Survey of Canada, recognized the significance of a magnetic anomaly at Creighton with respect to mineral potential in 1856, but the mining rights remained unclaimed until the deposit was discovered by a prospector called Henry Ranger, in 1886 (Fig. 4.31A). Creighton became one of the few Sudbury mines to almost continuously produce ore for >100 years from an open pit and nine different shafts. The past production from Creighton Mine and the major events in the development of the mineral system are illustrated in Fig. 4.31A. Boldt (1967) provide a detailed case study of the development of the Creighton mineral system in the first half of the 20th century, and it is a very colourful history marked by the progressive development of methods to identify the continuity of mineralization at depth and to develop mining methods. Fig. 4.31B is a photograph of the early phase of underground development that extended from the base of the open pit. The deposit continues to be developed into rich ores that have been explored to a depth of 3000 m.

Geological Relationships

The geology of the Main Creighton Trough is shown in an N–S section and a level plan (5000 Level) in Fig. 4.32A,B. The mineralization is associated with the SLNR at the base of the SIC in a trough that plunges down the base of the SIC from surface to a depth of 1800 m with a dip of ~60 degrees toward the NNW. At ~1800 m depth, the contact flattens over a short interval, and then steepens to subvertical before terminating against the Cliff Lake Fault at ~3500 m depth (Fig. 4.32A).

In the upper part of the trough, the southernmost termination of the Sublayer transitions into a short Offset Dyke composed of MIQD. This is the Creighton Offset; it is a discontinuous sheet penetrating into the country rocks away from the SIC for up to 300 m as shown in Fig. 4.32A. The connection between the Offset Dyke and the keel of the embayment is close to the base of the trough, and the position of this small Offset Dyke is shown in Fig. 4.32A,B. The Offset Dyke is broken into segments by late SW-NE fault structures, although it has not yet been traced to the south of the Creighton Fault (Fig. 4.32A).

In the upper part of Creighton Mine, most of the contact mineralization is hosted either directly in the Sublayer, or within structurally detached segments of the Sublayer that rest in the footwall. The mineral zones are semicontinuous as shown in Fig. 4.32A, but there are contact parallel shears that displace mineralization and Sublayer host rocks into the footwall, in some cases locally detaching the mineral zones from the base of the SIC (Fig. 4.32B).

The surface and near-surface ore deposits comprise the open pit and sublevel caved deposits, and the 117, 118, and 125 orebodies that are located along the Main Trough (Fig. 4.32A). The configuration

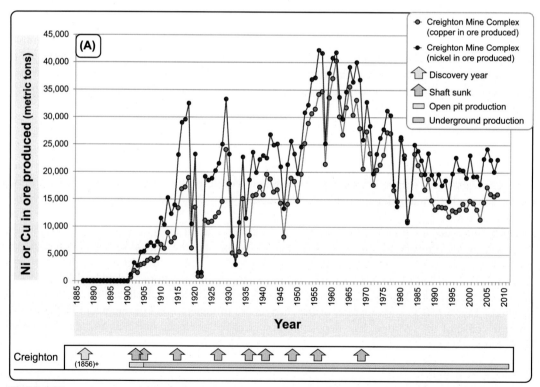

FIGURE 4.31

(A) History of production of Ni and Cu in ore from the Creighton mineral system. The major events in the development of Creighton Mine are shown.

FIGURE 4.31 (*cont.*)

(B) Photograph showing the second level of Creighton Mine in c. 1905. The image is from the Department of Mines and Northern Affairs, Ontario Government (image RG 13–30, I0004649).

(B) Used with the permission of the Archives of Ontario.

of these orebodies is illustrated in Fig. 4.33A–J using a series of level plans between 800 and 1340 m depth and two sections centered on the 125 Orebody. These plans show the juxtaposition of the ore deposit in association with Sublayer at the base of the Creighton Trough, but they also show the detachment of parts of the orebody and the Offset Dyke by dextral shear zones.

At the flattening of the contact, a curved NE-SW fault termed the Six Shaft Shear displaces the contact mineralization away from the eastern flank of the trough into the base of the trough. This

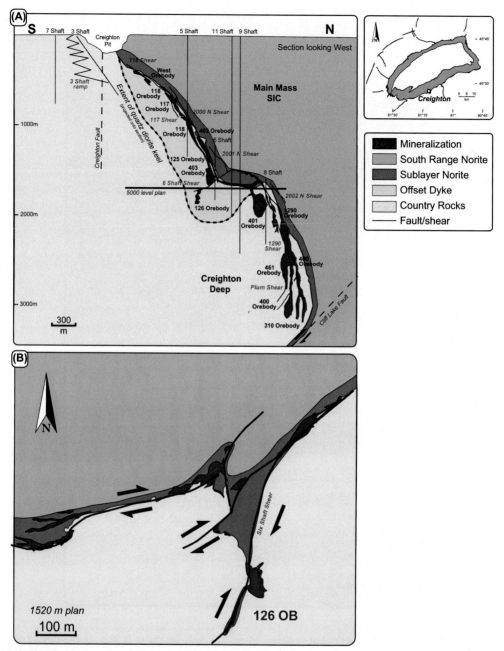

FIGURE 4.32

(A) Geological cross-section based on surface mapping, diamond drill core, and underground mapping. The section faces toward the WSW, and is aligned along the axis of the Creighton Trough. The position and names of the different orebodies, the principal fault structures, location of the Offset Dyke (projected onto the plane), and location of a plan section are shown. (B) Plan of the 1520 m Level at Creighton Mine, showing the effect of structures in detaching mineralization from the base of the SIC and the location of the Six Shaft Shear in relation to the 126 footwall style orebody.

The section and level plan were prepared by Rob Pelkey and this version is modified after Lightfoot (2015).

fault intersects the 126 Orebody which is located entirely in the footwall with no associated Sublayer (Fig. 4.32B). The 126 Orebody is a classic example of a primary magmatic sulfide, even though it is along strike and adjacent to the Six Shaft Shear, the massive sulfide does not contain fragments of sheared country rock; the mineralization appears to have been injected into a structurally created space as a magmatic sulfide rather than tectonically transported from the contact. The 126 Orebody is unusually Ni- and Cu-rich with almost double the metal tenor of the typical contact style of mineralization at

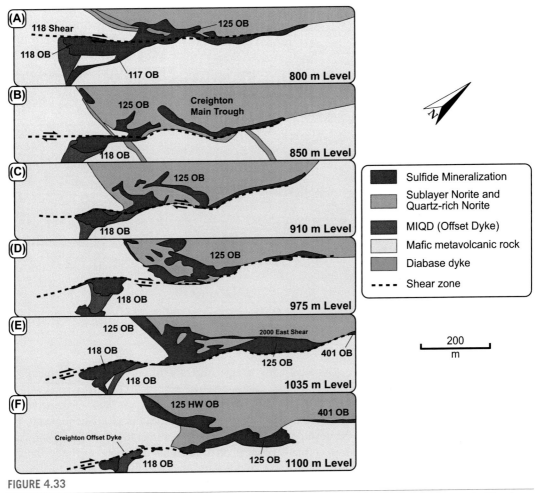

FIGURE 4.33

(A–J) A series of plan views through the upper part of Creighton Mine between 2600 and 4400 ft. levels. The plans show the progression down the contact and the development of structurally detached Sublayer mineralization as well as true footwall styles of mineralization. The position of the Offset Dyke (consisting of MIQD) is also shown. Kinematics are based on underground sections prepared by mine geology staff.

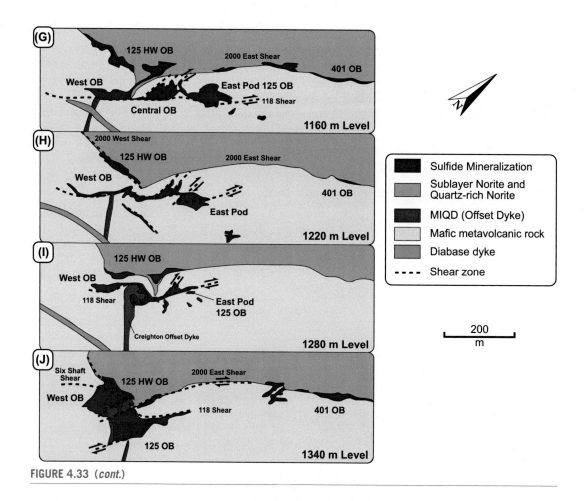

FIGURE 4.33 (*cont.*)

Creighton, so the characteristics are consistent with the injection of a differentiated sulfide melt from the contact of the SIC.

At ~1600–1800 m depth, the base of the trough flattens, and the western flank of the trough is connected to the Offset Dyke; there is also a primary lobe of SLNR that extends into the footwall. The configuration of this lobe controls the development of the upper part of the 401 Orebody as shown in Fig. 4.34A, whereas the Offset Dyke is a principal control on the location of the 403 Orebody which is largely hosted in the footwall as a series of massive sulfide veins that are almost devoid of inclusions (Fig. 4.34B–D).

With increasing depth, much of the mineralization peels away from the base of the SIC into the footwall as a series of wide veins (Figs. 4.32A and 4.34E). At depth the contact orebodies are hosted in the Sublayer, but the 400, 401, 310, and 320 Orebodies are all hosted within the footwall of the SIC in association with SUBX and contacts between a recrystallized Creighton Granite (called the Black Porphyry), metavolcanic and metasedimentary rocks. Although there are examples of structures developed parallel to the orebodies and across the orebodies (Fig. 4.34E), the walls of the sulfide veins are typically not heavily deformed and the relationship between sulfide and wall rocks indicate primary injection of the sulfide into the footwall along the fractures. This environment comprises the Creighton Depth Deposit, which will be a focus of mining activity for the next 30 years in the Sudbury Camp.

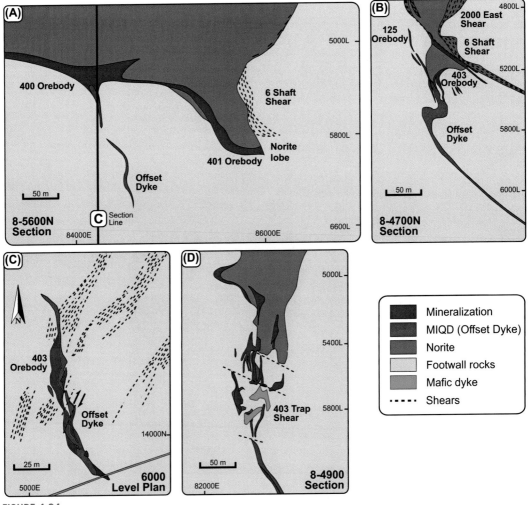

FIGURE 4.34

(A) Section showing the position of a primary protuberance of Sublayer is along strike from the connection between the Creighton Offset Dyke and the Sublayer Embayment. (B) Section showing the relationship between the 125 Contact Orebody and the 403 Footwall Orebody which is associated with the Creighton Offset and occupies sheets of massive sulfide that extend downward beneath the termination of the Offset Dyke. (C) Plan view of the 403 Orebody on 6000 Level. (D) Sectional view of the 403 Orebody.

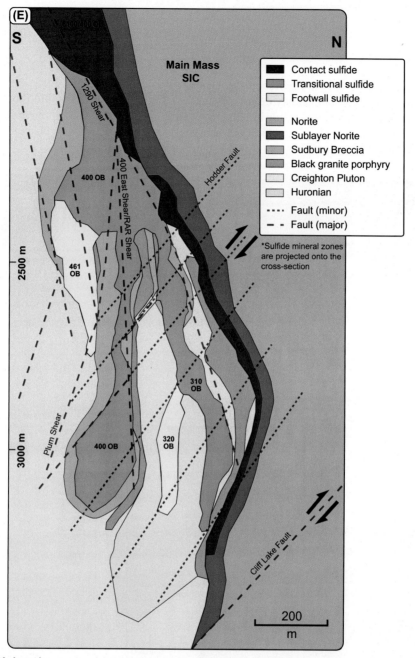

FIGURE 4.34 *(cont.)*

(E) Simplified geological model for the orebodies in the deep part of the Creighton mineral system. The diagram shows the footwall and contact orebodies projected onto a west-facing section (they are oriented in a WSW-direction so the true width is less than that shown). The sulfide mineralization varies from contact type with pyrrhotite > chalcopyrite > pentlandite through transitional type below the contact with (pyrrhotite + chalcopyrite) > pentlandite into footwall type with chalcopyrite > (pyrrhotite + pentlandite). The positions of major fault structures that cross-cut and possibly also control the primary geometry of the footwall mineralization is shown.

The section of the deep part of Creighton was prepared with assistance from Rob Pelkey and is modified after Lightfoot (2015).

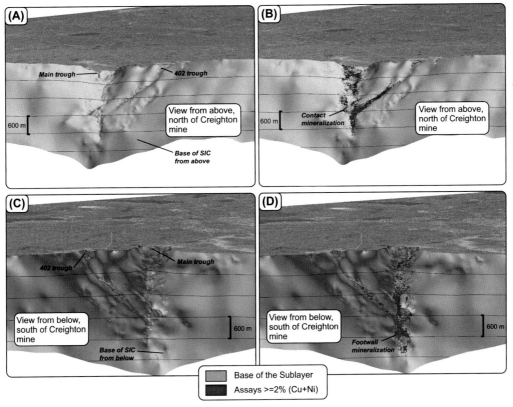

FIGURE 4.35

(A) Geological wireframe of the Creighton embayment showing the geometry of the troughs and embayments that form depressions in the lower contact. (B) The position of the contact orebodies developed in the Sublayer is shown based on assay values for Cu + Ni. Views A and B are from the NNW at an inclination of ~25 degrees downward. (C) Geological wireframe of the Creighton embayment showing the topography of the lower contact from the footwall. (D) The position of the footwall orebodies and structurally detached sulfides in the footwall developed below the Sublayer. Images C and D are viewed from the SSE at an inclination of ~25 degrees up. Images A and B depict the base of the SIC viewed from above and to the north, whereas images C and D depict the base of the SIC viewed from beneath the surface to the south of Creighton Mine. The three-dimensional model was prepared by Lisa Gibson of Vale.

The geology of the Creighton Trough is shown in images from a model prepared in three-dimensional space (Gibson, 2003; Fig. 4.35A–D). Fig. 4.35A,B shows a gray wireframe depicting the base of the SIC based on surface maps, level plans, and drill hole data. The view in Fig. 4.35A looks downward onto the contact from the north from above the surface. The surface reference image is a colour aerial photograph which extends at surface south of the contact of the SIC. Fig. 4.35B shows superimposed points that correspond to drill core samples with a grade of (Cu + Ni) exceeding 2 wt.%. The horizontal lines of Fig. 4.35A,B depict the location of 600 m spaced depth contours on the wireframe depicting the base of the SIC. The depth extent of the wireframe is not shaded beyond the projected location of the intersection of the Cliff Lake Fault with the base of the SIC. Fig. 4.35C,D depicts the base of the SIC and the location of samples with >2wt.% (Cu + Ni) in the footwall; it is viewed from below and to the south of the wireframe from beneath the surface and with an air photograph showing the surface to the north of the contact.

The Main Creighton Trough merges with a strongly mineralized western arm just above the level of the 403 Orebody (Fig. 4.35A,B). This arm is termed the Gertrude-402 Trough and extends from surface to join the Main Trough at a depth of ~1500 m below surface. In the 402 Trough, the mineralization is largely hosted by the Sublayer, but a small amount of it is hosted in the footwall (Fig. 4.35C,D). A second discontinuous trough extends parallel to the 402 from surface at Gertrude West (Fig. 4.35A,B).

A weakly defined embayment structure is present near to the surface to the east of Creighton, and contains the near-surface North Star Deposit. A connection between this embayment and the Main Creighton Trough has not yet been established.

The Main Creighton Trough extends from surface to 3000 m depth, and it contains a thick sequence of heavily mineralized Sublayer (Fig. 4.35A,B) where the majority of the sulfides are contained in the Sublayer or in structurally detached zones in the immediate footwall (Fig. 4.33A–J). In contrast, much of the mineralization developed in the deep part of the Creighton mineral system is hosted deeper in the footwall beneath a more structurally sheared and attenuated part of the primary Sublayer trough (Fig. 4.35B, D)

Characteristics of mineralization

The Creighton mineralization consists principally of different proportions of pyrrhotite, chalcopyrite, and pentlandite in disseminated, inclusion-bearing sulfide, and massive sulfide.

The disseminated sulfides are principally hosted by noritic Sublayer, and consist of blebby to interstitial sulfides (Fig. 4.36A). Intervals of inclusion-rich norite are separated from intervals of coarse-grained melanocratic norite (Fig. 4.36B; Lightfoot et al., 1997c). The disseminated style of mineralization (Fig. 4.36C) grades into inclusion-rich sulfide that contains country rock fragments in a heavily mineralized noritic Sublayer matrix (Fig. 4.36D).

With an increased proportion of inclusions the mineralization grades into inclusion-rich sulfide where the matrix is almost entirely sulfide (Fig. 4.36E,F). As the proportion of sulfide continues to increase, the inclusions float in semi-massive sulfide (Fig. 4.36G,H). The massive sulfides show a wide range in composition from pyrrhotite + pentlandite-rich with round grains of pentlandite in pyrrhotite (Fig. 4.36I) through loop textured sulfides with fine segregations of chalcopyrite + pentlandite between pyrrhotite crystals (Fig. 4.36J), to coarse-grained intergrowths with pyrrhotite + chalcopyrite > pentlandite (Fig. 4.36K) and chalcopyrite-rich ore types with coarse-grained pentlandite (Fig 4.36L).

In the deeper parts of the deposit shown in Fig. 4.37A, the mineralization consists of massive sulfide with pyrrhotite > chalcopyrite ≥ pentlandite in the contact Sublayer (Fig. 4.37B) through to

FIGURE 4.36 Examples of the Principal Styles of Sulfide Mineralization in the Creighton Mineral System   ▶

(A) Disseminated sulfide in inclusion-poor SLNR (RX355673) from the Main Creighton Trough. (B) SLNR with interstitial "ragged disseminated" sulfide; the inclusions are gray-coloured gabbro. This type of Sublayer matrix is coarser-grained than usual and occurs at the base of the 401 Orebody; the composition and texture corresponds to the Creighton melanorite as discussed in Chapter 3. (C) Disseminated sulfide in SLNR from the 402 Trough, Creighton Mine. Note the diversity in Sulfide content and the development of mafic inclusions in the SLNR. (D) Inclusion-rich noritic matrix of the Sublayer with abundant disseminated sulfide through to matrix sulfide between inclusions. This drill core comes from the 402 Trough.

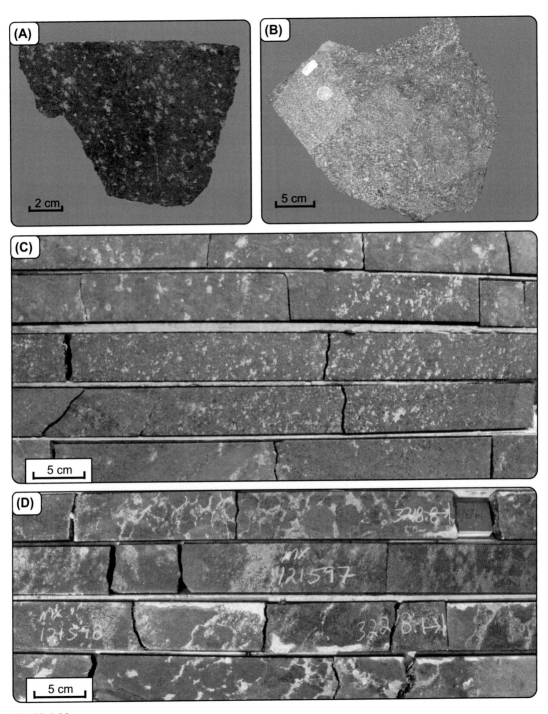

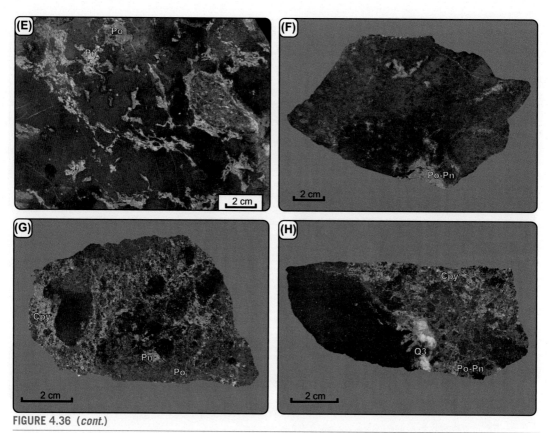

FIGURE 4.36 (*cont.*)

(E) Inclusion-rich massive sulfide from the 401 Orebody; this sample comprises noritic, gabbroic and ultramafic inclusions separated by sulfide. (F) Mafic inclusions separated by a thin rind of massive sulfide; samples comes from the upper part of the 400 Orebody at Creighton Mine (sample RX355672 has 0.87 wt.% Ni, 0.21 wt.% Cu and Ni_{100} = 4.3 wt.%). (G) Inclusion-bearing massive sulfide comprising mafic inclusions in a matrix of pyrrhotite > = chalcopyrite > pentlandite. This sample comes from the 7680 Level in the upper part of the 400 OB (sample R355669 contains 6.3 wt.% Ni, 2.85 wt.% Cu, 1.5 g/t 3E, and it has Ni_{100} = 8.5 wt.%). (H) Inclusion-bearing semimassive sulfide consisting of mafic inclusions and quartz clasts in a matrix of chalcopyrite > pyrrhotite > pentlandite from the 7810 Level of the Creighton 400 Orebody (sample RX355667 contains 5.4 wt.% Ni, 2.9 wt.% Cu, 3 g/t 3E, and has a Ni_{100} of 9.7 wt.%).

increasingly Cu-rich rich styles of vein mineralization in the footwall with chalcopyrite > pyrrhotite > pentlandite (Fig. 4.37A,C) to reach an apogee in massive sulfide veins with chalcopyrite > pyrrhotite > pentlandite. Examples of the grades of these types of mineralization are shown in Fig. 4.37A–E (Lightfoot, 2015). Close to the contact, the mineralization has a clear spatial relationship to the Sublayer, but the mineralization hosted in the footwall is typically massive sulfide with no mafic-ultramafic inclusions, sharp veins contact with the country rock, and only rare cases where noritic enclaves from the Sublayer are found in the adjacent footwall. The footwall vein mineralization at Creighton resembles that of the North Range; however, millerite or bornite are not developed, and pyrrhotite is generally present in even the most chalcopyrite + pentlandite-rich mineral zones.

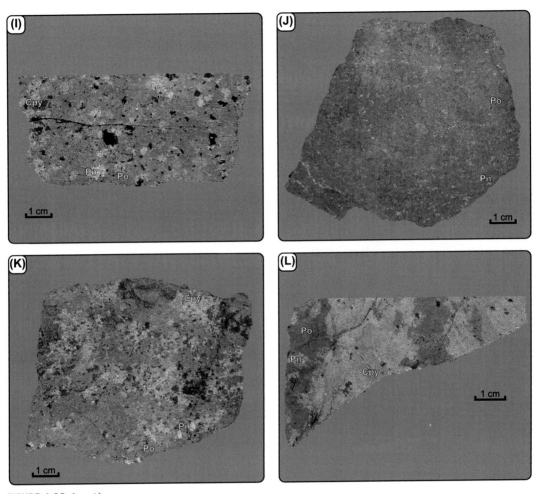

FIGURE 4.36 (*cont.*)

(I) Massive sulfide from the 7810 Level of the Creighton 400 OB showing the development of eyes of pentlandite in chalcopyrite + pyrrhotite (sample RX355668 contains 7.1 wt.% Ni, 5.6 wt.% Cu, 10.9 g/t 3E, and has a Ni_{100} of 8.4 wt.%). (J) Massive pyrrhotite + pentlandite from the Gertrude Deposit near surface along the 402 Trough (sample RX355671 contains 4.14 wt.% Ni, 0.06 wt.% Cu, 0.14 g/t 3E, and has Ni_{100} = 5 wt.%). (K) Massive pyrrhotite > chalcopyrite > pentlandite from the 7350 Level of the 400 Orebody at Creighton Mine (sample RX187425 contains 6.55 wt.% Ni, 2.2 wt.% Cu, 0.17 g/t 3E, and has a Ni_{100} of 7.8 wt.%). (L) Massive (chalcopyrite + pyrrhotite) > pentlandite from the 400 Orebody at Creighton Mine (sample RX187422 contains 2.79 wt.% Ni, 15.45 wt.% Cu, 2.3 g/t 3E, and has Ni_{100} = 3.4 wt.% which is quite low because of the elevated chalcopyrite content of this sample).

Geochemistry

The variation in the chemistry of the Creighton mineral system is illustrated in terms of Cu/(Cu + Ni) and the normative pentlandite abundance in the sulfide component of the rock, expressed as $[Pn]_{100}$. Fig. 4.38A,B shows the percentage grouped frequency distribution of Cu/(Cu + Ni) and $[Pn]_{100}$ for the Main Creighton Trough and for the 402 Trough. These plots show that the sulfide mineralization in

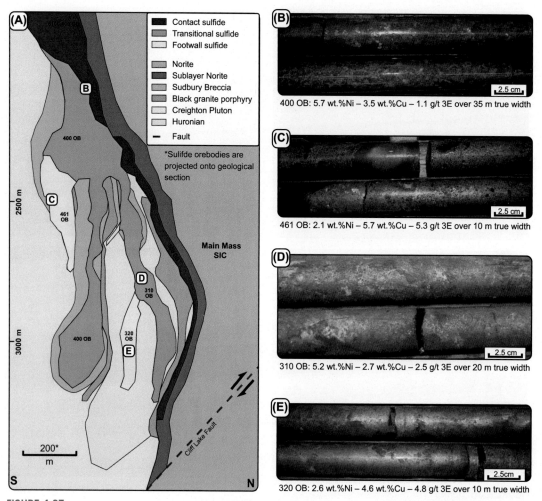

400 OB: 5.7 wt.%Ni – 3.5 wt.%Cu – 1.1 g/t 3E over 35 m true width

461 OB: 2.1 wt.%Ni – 5.7 wt.%Cu – 5.3 g/t 3E over 10 m true width

310 OB: 5.2 wt.%Ni – 2.7 wt.%Cu – 2.5 g/t 3E over 20 m true width

320 OB: 2.6 wt.%Ni – 4.6 wt.%Cu – 4.8 g/t 3E over 10 m true width

FIGURE 4.37

(A) Geological model showing the principal ore associations in the deep part of Creighton Mine from drilling between 2.4 and 3.0 km depth. The location of the ore types are shown in photographs of representative intervals of drill core. (B) Contact style mineralization from Creighton Depth; this sample comes from the contact portion of the system in a drill hole through the 400 Orebody. The mineralization consists of pyrrhotite > chalcopyrite ≥ pentlandite; in this drill core, the true thickness of massive sulfide encountered was 35 m of 5.7 wt.% Ni, 3.5 wt.% Cu, and 1.1 g/t 3E (Lightfoot, 2015). (C) Example of transitional style mineralization developed in a footwall vein in the 461 Orebody. This sample consists of chalcopyrite ≥ pyrrhotite > pentlandite. The drill hole intersected 10 m true thickness of 2.1 wt.% Ni, 5.7 wt.% Cu, and 5.3 g/t 3E (Lightfoot, 2015). (D) An example of Ni-rich mineralization that is located in the footwall in the 310 Orebody. The interval comprised 20 m true width of 5.2 wt.% Ni, 2.7% Cu and 2.5 g/t 3E (Lightfoot, 2015). (E) An example of a footwall Cu-rich interval of mineralization from the 320 Orebody that comprises 10 m true width of 2.6 wt.% Ni, 4.6 wt.% Cu, and 4.8 g/t 3E (Lightfoot, 2015).

The samples shown were photographed by Rob Pelkey who was responsible for the drill program that yielded these spectacular results.

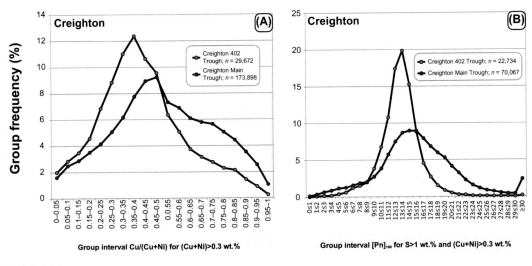

FIGURE 4.38

(A) Comparison of the percentage grouped frequency distribution of Cu/(Cu + Ni) for samples from Vale's Creighton mineral system. The profiles compare the averages of assays from the 402 Trough and the Main Creighton Trough. (B) Comparison of the percentage grouped frequency distribution of normative $[Pn]_{100}$ in samples from the 402 Trough and the Main Trough of the Creighton mineral system.

the 402 Trough tends to have a narrow range in $[Pn]_{100}$ and a skewed distribution of Cu/(Cu + Ni) to higher Ni concentrations than Cu. In contrast, the mineralization from the Main Trough at Creighton has a much wider range in $[Pn]_{100}$ and a more symmetric pattern of Cu/(Cu + Ni) (Fig. 4.38A,B). These plots indicate that the metal tenor of the sulfides in the 402 Trough is lower than that of the Main Trough, and that the sulfides from the Main Trough are skewed toward higher normative pentlandite in sulfide.

The 402 mineralization occupies a trough that extends from surface at Gertrude Mine to depth where it merges with the Main Creighton Trough. Fig. 4.39A,B shows the trends of 1000 point moving averages (calculated with increasing depth) for samples within the 402 Trough. Fig. 4.39A shows a broad overall increase in Cu/(Cu + Ni) that is interrupted as the main Creighton Trough is approached by the development of local concentrations of more Cu-rich or Ni-rich sulfide mineralization. Fig. 4.39B shows the calculated normative abundances of chalcopyrite and pentlandite in 100% sulfide ($[Cpy]_{100}$ and $[Pn]_{100}$) increase with depth along the 402 Trough. The geochemical evidence supports the idea that the mineralization becomes progressively richer in Cu relative to Ni, and the metal tenor of the sulfide increases with depth along the 402 Trough.

The variation in moving average of Cu/(Cu + Ni), $[Cpy]_{100}$, and $[Pn]_{100}$ for sulfide mineralization from the Main Creighton Trough are shown in Fig. 4.39C,D. Over a similar depth interval to the 402 Trough, the Main Trough has a low Cu/(Cu + Ni) (0.4–0.5), but the ratio is not as low as the mineralization in the 402 Trough (<0.4). The normative abundances of pentlandite and chalcopyrite in 100% sulfide are higher over the same depth interval in the Main Trough relative to the 402 Trough

(Fig. 4.39B,D). Below the point where the 402 and Main Troughs join (shown in Fig. 4.39C,D), the Cu/(Cu + Ni) ratio exceeds 0.5, and although normative pentlandite in sulfide does not increase, the amount of normative chalcopyrite in sulfide increases markedly.

The composition of the Creighton mineralization varies with the host rock type. Fig. 4.40A,C illustrates this point in the context of the percentage grouped frequency variations in Cu/(Cu + Ni) according to the rock type that hosts the mineralization. The arrays for samples hosted in the Sublayer tend to be symmetric in both the 402 Trough and the Main Trough (Fig. 4.40A,C), but the material hosted in country rock granitoids, mafic metavolcanic rocks, and metasedimentary rocks are strongly skewed in the Main Trough to high Cu/Ni and moderately skewed to high Cu/Ni in the 402 Trough

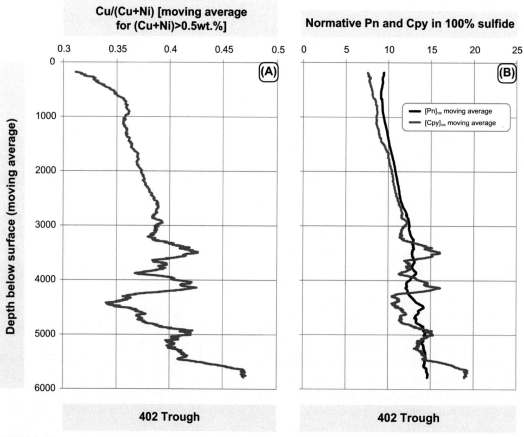

FIGURE 4.39

(A) Variation in Cu/(Cu + Ni) in the 402 Trough of the Creighton mineral system. The diagram is constructed using 1000 point moving averages of the ratio and the depth. (B) Variation in normative $[Pn]_{100}$ and $[Cpy]_{100}$ in the 402 Trough of the Creighton mineral system. The diagram is constructed using 1000 point moving averages of the ratio and the depth.

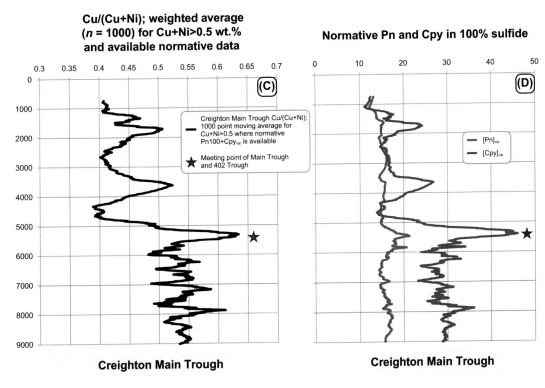

Cu/(Cu+Ni); weighted average (n = 1000) for Cu+Ni>0.5 wt.% and available normative data

Normative Pn and Cpy in 100% sulfide

Creighton Main Trough Cu/(Cu+Ni); 1000 point moving average for Cu+Ni>0.5 where normative Pn100+Cpy100 is available

★ Meeting point of Main Trough and 402 Trough

— [Pn]100
— [Cpy]100

Creighton Main Trough

Creighton Main Trough

FIGURE 4.39 (cont.)

(C) Variation in Cu/(Cu + Ni) in the Main Trough of the Creighton mineral system. The diagram is constructed using 1000 point moving averages of the ratio and the depth. (D) Variation in normative $[Pn]_{100}$ and $[Cpy]_{100}$ in the Main Trough of the Creighton mineral system. The diagram is constructed using 1000 point moving averages of the ratio and the depth.

(Fig. 4.40A,C). In contrast, the massive sulfides and inclusion-bearing sulfides have compositions that are displaced to Ni-rich compositions in both the 402 and Creighton Troughs (Fig. 4.40A,C).

The variations in $[Pn]_{100}$ are also shown by host rock type for the Main Creighton and 402 Troughs in Fig. 4.40B,D. Data from the 402 mineralization show a narrow range in $[Pn]_{100}$ for mineralization from the three different types of host rock as shown in Fig. 4.40D. Mineralization from the Main Creighton Trough has a wider range in normative $[Pn]_{100}$; the composition of the massive and inclusion-rich sulfides are displaced to higher $[Pn]_{100}$ than the sulfides hosted in either the Sublayer or the country rocks.

A data density plot showing the variations in Cu/(Cu + Ni) versus (Cu + Ni)/S is shown in Fig. 4.41A for samples of Sublayer and unclassified types of norite from the Main Creighton Trough; the analyses define in a broad cluster with a wide range in (Cu + Ni)/S, but a tendency for the majority of the samples to have a Cu/(Cu + Ni) ratio from 0.35 to 0.6. In contrast to the mineralization hosted in norite, the sulfides hosted in granitoids, metagabbros, metavolcanic and metasedimentary rocks show a clustering toward high Cu/(Cu + Ni) and (Cu + Ni)/S, and a few samples fall on a tie line between chalcopyrite

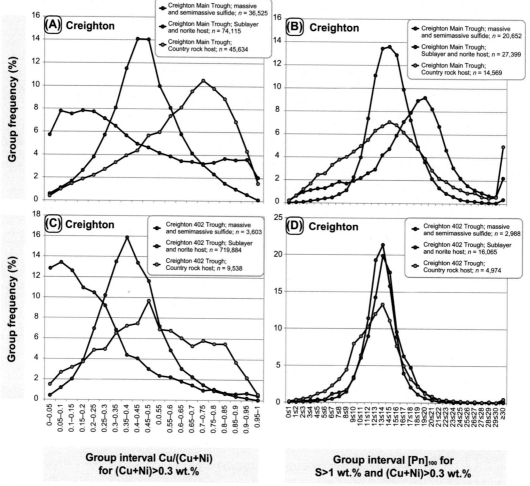

FIGURE 4.40

(A) Comparison of the percentage grouped frequency distribution of Cu/(Cu + Ni) for sulfide mineralization from the Main Trough of the Creighton mineral system. The three patterns correspond to massive and semimassive sulfides, sulfides hosted in noritic rock of the SIC, and sulfide hosted in footwall rocks. The sulfides hosted by norite have a bell-shaped curve with a Cu/Ni ratio of 1. The massive and semimassive sulfides occupy a broad range in Cu/(Cu + Ni) with a peak at Ni-rich compositions whereas the sulfides in the footwall show a strong enrichment in Cu relative to Ni. (B) A similar plot to subpart A, but the histogram class compares the percentage grouped frequency intervals of normative $[Pn]_{100}$. This plot shows that the norite-hosted sulfides have a relatively narrow range in normative $[Pn]_{100}$, the sulfides hosted in the country rocks have a similar mode, but a wider range in normative $[Pn]_{100}$, and the sulfides from the footwall are typically richer in $[Pn]_{100}$. (C) Comparison of the percentage grouped frequency distribution of normative $[Pn]_{100}$ for sulfide mineralization from the 402 Trough of the Creighton mineral system. The three patterns correspond to massive and semimassive sulfides, sulfides hosted in noritic rock of the SIC, and sulfide hosted in footwall rocks. The sulfides hosted by norite have a skewed distribution. The massive and semimassive sulfides are heavily skewed toward Ni-rich compositions whereas the sulfides in the footwall show a weak enrichment in Cu. (D) This plot shows that the norite-hosted sulfides have a relatively narrow range in normative $[Pn]_{100}$, the sulfides hosted in the country rocks have a similar mode, and a slightly wider range in normative $[Pn]_{100}$, and the sulfides from the footwall define a curve with a tail toward low $[Pn]_{100}$.

and pentlandite (Fig. 4.41B). The sulfides hosted in the country rocks are clearly quite different in composition when compared to those hosted in the norite.

The massive sulfide and inclusion-bearing semimassive sulfide exhibit both Cu-rich and Ni-rich compositions, and there are a number of samples along the chalcopyrite + pentlandite tie-line (Fig. 4.41C). At any given Cu/(Cu + Ni), there is a narrower range in (Cu + Ni)/S when compared to the norites and the country rocks. The data for samples from the MIQD of the Offset Dyke exhibit very Cu-rich compositions, and largely fall along the chalcopyrite + pentlandite tie line (Fig. 4.41D).

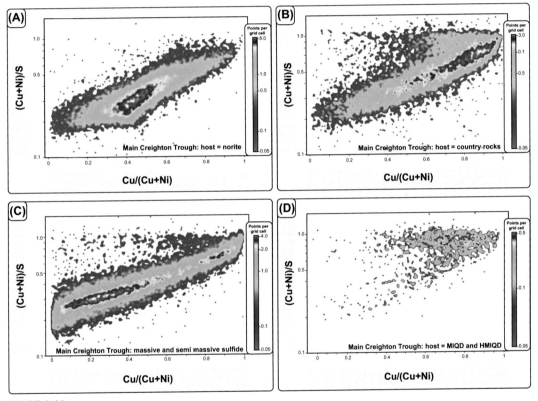

FIGURE 4.41

(A) Data density variation in Cu/(Cu + Ni) versus (Cu + Ni)/S for mineralized samples of norite from the Main Creighton Trough. The samples plot in a dense cluster with a Cu/(Cu + Ni) ratio of ~0.35–0.55, and there is a wide spread in (Cu + Ni)/S over a range of Cu/(Cu + Ni). (B) A similar data density plot for mineralized samples from the Main Creighton Trough that are hosted in country rocks shows that most of the samples are Cu-rich, and a few fall at the chalcopyrite-rich end of the pentlandite + chalcopyrite tie line. (C) Variation in composition in massive and semimassive sulfides from the Main Trough at Creighton. This sulfide type occupies a broad range in composition with a peak toward the composition of pyrrhotite + pentlandite. (D) Composition of MIQD and HMIQD from the Creighton Offset showing a typical footwall pattern of data density along a pyrrhotite + pentlandite + chalcopyrite tie line.

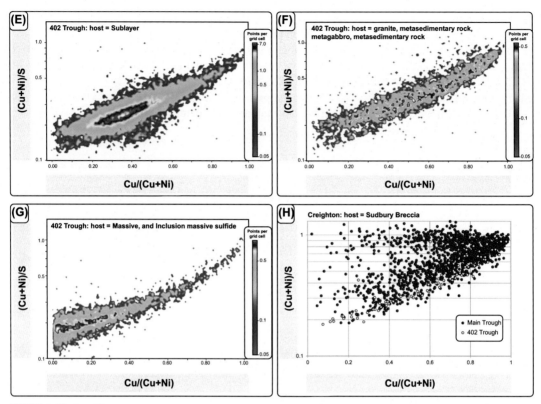

FIGURE 4.41 (*cont.*)

(E) Data density variation in Cu/(Cu + Ni) versus (Cu + Ni)/S for mineralized samples of norite from the 402 Trough. The samples plot in a dense cluster with a Cu/(Cu + Ni) ratio of ~0.25–0.50, and there is a wide spread in (Cu + Ni)/S over a range of Cu/(Cu + Ni). (F) A similar data density plot for mineralized samples from the 402 Trough that are hosted in country rocks shows that the samples plot along a broad array. (G) Variation in composition in massive and semimassive sulfides from the 402 Trough at Creighton. This sulfide type occupies a broad range in composition with a marked peak toward the composition of pyrrhotite + pentlandite. (H) Comparison of the compositions of sulfides hosted in SUBX from the Main and 402 Troughs. Samples from the Main Trough plot partly on the chalcopyrite + pentlandite tie-line whereas the small number of samples located beneath the 402 Trough fall along the array controlled by mixtures of pyrrhotite, pentlandite, and chalcopyrite.

Analogous plots for the 402 Trough are shown in Fig. 4.41E–H. The norite-hosted mineralization has a narrow range in Cu/(Cu + Ni), but there is a wide range in (Cu + Ni)/S at a given Cu/(Cu + Ni ratio (Fig. 4.41E). The mineralization hosted in the country rocks shows a very wide range in Cu/(Cu + Ni) (Fig. 4.41F). The massive and semimassive sulfides have a peak abundance at low Cu/(Cu + Ni) and a wide range in (Cu + Ni)/S at low Cu/(Cu + Ni) (Fig. 4.41G). Finally, the mineralization hosted in SUBX from the Main Trough has very high Cu/(Cu + Ni) with many samples along the chalcopy-rite + pentlandite tie-line; in contrast, samples hosted in SUBX from beneath the 402 Trough plot along the tie line between pyrrhotite + pentlandite + chalcopyrite (Fig. 4.41H).

Geological Model

Following 100 years of mining activity at Creighton, there is a very good three-dimensional understanding of the distribution of rock types, structures, ore types, ore composition, and ore mineralogy. The key details that help establish a genetic model for the Creighton ore system are summarized below:

1. The mineralization is principally controlled by physical depressions at the base of the SIC which comprise a series of troughs that rake down and merge with one another at depth on the southern flank of the SIC to generate the configuration shown in Fig. 4.35A,C.
2. The Main Creighton Trough has a discontinuous keel composed of MIQD which extends as far as 300 m into the footwall, and which controls the localization of Cu-rich mineralization.
3. With increasing depth along the 402 and Main Troughs, there is a tendency for the style of mineralization to become richer in pentlandite and chalcopyrite at the expense of pyrrhotite, and the composition of the ores change from low Cu/Ni to higher Cu/Ni. This overall change in relative abundance of Cu and Ni and metal tenor indicates that the ores become richer downward along the troughs and their compositions reach an apogee where the troughs merge to form high tenor mineralization that occupy fractures in the footwall. These tend to be related to primary injections of norite into the footwall.
4. Several different faults likely guided primary localization of fractionated sulfides into the footwall (eg, the 126, 403, 461, 400, and 320 Orebodies). Other groups of faults displaced the contact and footwall mineralization during regional deformation (eg, 125, and the upper parts of the 400 Orebody). Late structures appear to produce inflexions in the orebodies in response to the thrusting of the South Range toward the north along structures parallel to the Cliff Lake Fault which offset the contact of the SIC to the south at a depth of ∼3500 m.

Garson

The Garson Deposit was discovered by John Thomas Cryderman in 1891 and production commenced in 1908. The deposit remains an important producer of Ni and Cu in 2016 (Fig. 4.42).

This deposit differs from Creighton in the degree of deformation. A number of postmagmatic shears that comprise part of the South Range Shear Zone exit the Sudbury Basin through the Garson Embayment. Because of this relationship, Garson is an important case study demonstrating the effects of structural controls that have modified a primary magmatic sulfide ore deposit that was originally controlled by the Sublayer in a trough or embayment at the base of the melt sheet. Whether there were primary domains of fractionated sulfide mineralization in the footwall prior to deformation is not entirely certain, but much of the mineralization associated with the shear zones cut through both the footwall and hangingwall of the SIC and was therefore likely to be have been formed in association with postmagmatic structures activated by regional deformation.

Geological Relationships

The contact of the base of the SIC is overturned at Garson, and dips at approximately 75–85 degree to the south. A primary embayment structure containing SLNR is host to part of the mineralization, but this embayment is sheared along strike parallel to the contact for 1.5 km and to a depth of >2 km. The country rocks comprise metasedimentary and metabasaltic rocks of the Elsie Mountain Formation. The country rocks contain abundant SUBX, and both rock types are extensively recrystallized to form metabreccia within ∼100 m for the primary contact of the SIC. The country rocks and the rocks of

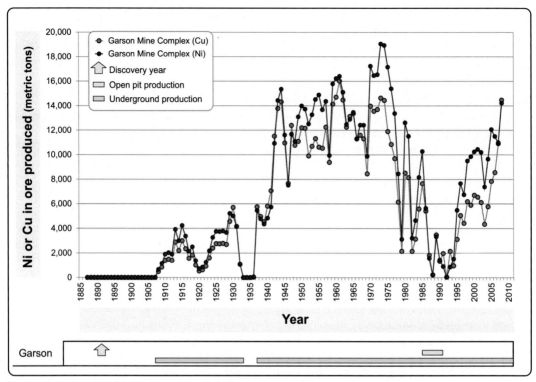

FIGURE 4.42 Plot Showing the Metal Contained in Ore Produced From the Garson Deposit

Major events in the development of the Garson mineral system are shown.

the SIC are cut by a number of postmagmatic shear zones that dip toward the south at 75–80 degrees. These shear zones contain a significant amount of mineralization that is not associated with Sublayer.

A systematic review of the Garson Ore Deposit was undertaken for Inco by Aniol and Brown (1979) and some of the most important section and level plans from Aniol and Brown (1979) are reproduced in modified form by Mukwakwami et al. (2012). Fig. 4.43A,B shows two vertical west-facing sections through the Garson mineral system that illustrate the relationships between the country rocks, the SIC, and the shear zones that cross-cut the Garson Trough. Fig. 4.43C–G illustrates surface and level plans of the relationships at the horizons shown in Fig. 4.43A,B (after Aniol and Brown, 1979). The principal deformation zone is the #1 Shear that separates the SIC and some of metabreccias developed immediately below the SIC from the underlying metavolcanic and metasedimentary rocks; this shear zone extends from the surface to a depth of >1300 m and it largely follows the base of the SLNR which occupies what was once a primary trough. The contact style of mineralization at Garson is very similar to that encountered in the Sublayer of the Creighton mineral system, but the style of mineralization developed proximal to the shear zones at Garson contains abundant country rock fragments (often comprising contorted schist fragments) and represents a zone where the primary magmatic sulfides have been modified and translocated along structures by a process of remobilization that has been termed sulfide kinesis (R. Monteiro, personal communication, 2006). The shears are commonly associated

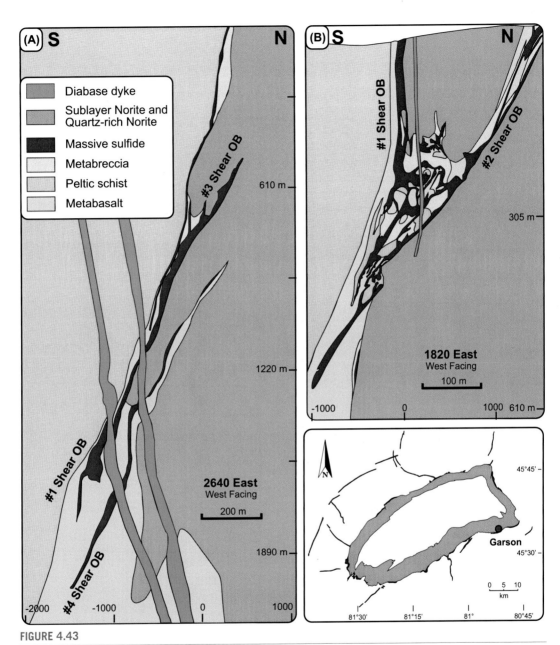

FIGURE 4.43

(A) West-facing geological cross-sections through the western part of the Garson mineral system (line 2640E) showing the relationship of sulfide mineralization to the Sublayer and to shear zones that cut through the SIC and the footwall at a low angle. (B) West-facing geological cross-sections through the shallow eastern part of the Garson mineral system (line 1820E) showing the relationship of sulfide mineralization to the Sublayer and to shear zones that cut through the SIC and the footwall at a low angle.

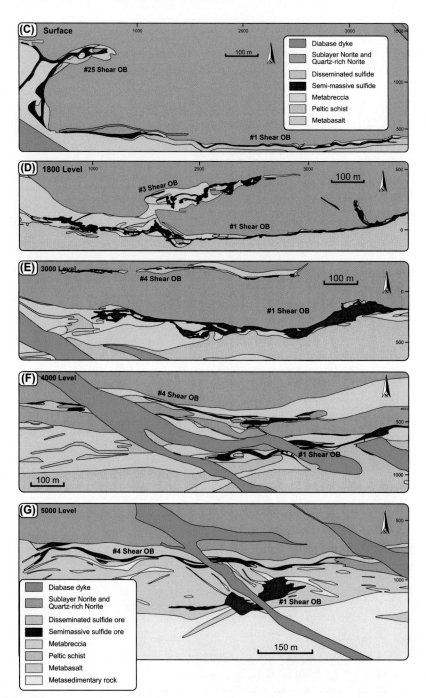

FIGURE 4.43 (*cont.*)

(C) Geological plan showing the relationships between the Main Mass, Sublayer, footwall, and shear zones at surface. (D) Geological section showing the relationships between the Main Mass, Sublayer, footwall, and shear zones on the 1800 level. (E) Geological plan showing the relationships between the Main Mass, Sublayer, footwall, and shear zones on the 3000 level. (F) Geological plan showing the relationships between the Main Mass, Sublayer, footwall, and shear zones on the 4000 level. (G) Geological plan showing the relationships between the Main Mass, Sublayer, footwall, and shear zones on the 5000 level.

All sections and plans are based on geological mapping from level plans and diamond drilling by company geologists reported in Aniol and Brown (1979) and Mukwakwami (2012).

with discontinuous marginal concentrations of nickel and cobalt arsenide minerals (gersdorffite and cobaltite; LeFort, 2012), which penetrate into both the footwall and hangingwall.

The #2 Shear zone is a splay of the #1 Shear, and both structures cross-cut and break the continuity of the base of the SIC to the north (Fig. 4.43A,B); mineralization is associated with the shear in both the hangingwall and footwall of the SIC (Fig. 4.43B). The #3 Shear zone is similar to the #2, but it does not control mineralization in the footwall. The #4 Shear zone is subparallel to the #1 Shear and cross-cuts both the SIC and footwall, and relates to the development of sulfide into both the hangingwall and footwall as shown in Fig. 4.43A.

The surface and level plans in Fig. 4.43C–G illustrate the relationship between the primary Garson Trough and the E–W oriented #2, #3, and #4 Shear zones that are associated with mineralization that appears to have been remobilized into the SIC. The mineral zones inside the SIC tend to be enveloped in norite with disseminated sulfide (Aniol and Brown, 1979) but local domains of metabreccia are developed after country rocks that appear to have been tectonically emplaced into the SIC (Mukwakwami, 2012).

Characteristics of the Mineralization

The semimassive and massive sulfide orebodies at Garson comprise both Sublayer-hosted varieties that resemble those of the Creighton mineral system, and shear-hosted mineralization. Mineralization in the shears consists of pyrrhotite + pentlandite + chalcopyrite with aligned fragments of schistose mafic volcanic and metasedimentary rock (Fig. 4.44A). The margins of the veins develop foliated inclusions that are oriented subparallel to the contact of the vein. The contacts of the veins sometimes contain less than 50 cm wide intervals of pyrrhotite + pentlandite + chalcopyrite with abundant gersdorffite and cobaltite (LeFort, 2012; Fig. 4.44B).

The principal types of mineralization at Garson consist of contorted schist sulfide, semimassive sulfide and massive sulfide. Disseminated sulfides also occur in association with the notites in the Sublayer. Semimassive sulfides in the shear zones often contain small mafic fragments (Fig. 4.44C,D), and the massive sulfides range in mineralogy from chalcopyrite > (pyrrhotite + pentlandite) through pyrrhotite > chalcopyrite > pentlandite to pyrrhotite > pentlandite ≫ chalcopyrite as illustrated in Fig. 4.44E–G.

The Garson mineral system exhibits a broad change in pyrrhotite stoichiometry from predominantly monoclinic pyrrhotite in the upper part of the deposit to hexagonal pyrrhotite in the deeper part of the deposit (Aniol and Brown; 1979; Fig. 4.45).

Geochemistry

A summary of the compositional diversity in the sulfide mineralization at Garson is presented in Figs. 4.46 and 4.47 following the approach taken for the Creighton mineral system. Variations in Cu/(Cu + Ni) and calculated normative $[Pn]_{100}$ illustrate the diversity in sulfide composition and the range in metal tenor of sulfide as a function of the sulfide content and the host rock type.

The grouped frequency variation in Cu/(Cu + Ni) for the Garson mineral system is shown in Fig. 4.46A,B. The average compositions of massive and semimassive sulfides, sulfides in the Sublayer and Quartz-rich Norite, and sulfide hosted in mafic metavolcanic rocks, sedimentary rocks, and SUBX are based on the description of the host rock. The mineralization hosted in the norite rock group has a bell-shaped curve with a peak at Cu/(Cu + Ni) = 0.5; this type of mineralization has a very low concentration of mineral phases such as gersdorffite and/or cobaltite which take in different proportions of Ni

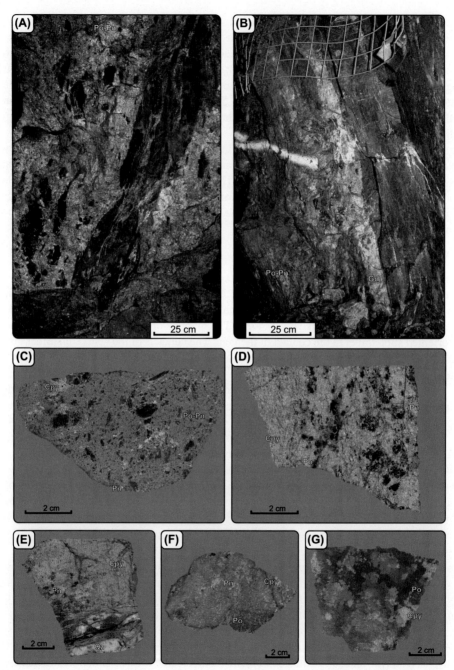

FIGURE 4.44

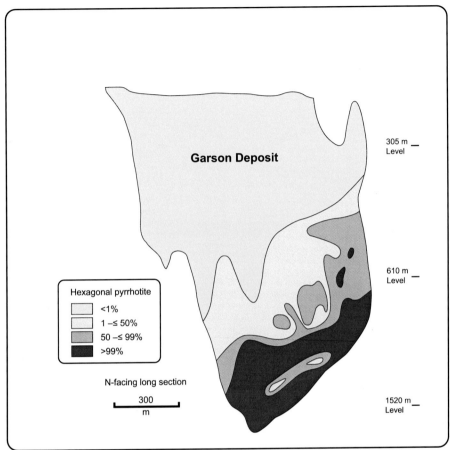

FIGURE 4.45 Simplified North-Facing Long Section Through the Garson Mineral System Showing the Broad Variation in Proportion of Hexagonal to Monoclinic Pyrrhotite. After Aniol and Brown (1979).

◀ **FIGURE 4.44**

(A) Massive sulfides with a few inclusions of mafic rock and country rock in the #1 Shear at Garson; this style of mineralization is close to the Sublayer and consists of pyrrhotite + pentlandite > chalcopyrite. (B) Massive sulfide in the Number 4 Shear at Garson Mine. This style of mineralization is hosted in footwall basaltic rocks. The contacts of the vein are heavily sheared and fragments of local country rock are incorporated into the sulfides. A marginal concentration of gersdorffite and/or cobaltite is commonly developed at the edge of the vein. (C) Massive sulfide from the main contact ore system at Garson Mine where pyrrhotite ≫ chalcopyrite > pentlandite (sample RX355666 contains 3.7 wt.% Ni, 1.8 wt.% Cu, 5.6 g/t 3E, and has a [Ni]100 of 4.8 wt.%). (D) Massive sulfide from the Garson Ramp mineral system where chalcopyrite > pyrrhotite > pentlandite (sample RX355663 contains 2.8 wt.% Ni, 9.5 wt.% Cu, 45.5 g/t 3E, and has a $[Ni]_{100}$ of 2.8 wt.%). (E) Massive sulfide from the Garson Ramp Deposit where chalcopyrite > pyrrhotite > pentlandite (sample RX355661 contains 1.3 wt.% Ni, 19.3 wt.% Cu, 4.9 g/t 3E). (F) Massive sulfide from the Garson Ramp Deposit where pentlandite > chalcopyrite > pyrrhotite (sample RX355664 contains 13.45 wt.% Ni, 7.5 wt.% Cu, 43.2 g/t 3E and has only 33 ppm As). (G) Massive sulfide from the main contact deposit at Garson Mine where pyrrhotite > chalcopyrite > pentlandite (sample RX355686 contains 0.9 wt.% Ni, 4.85 wt.% Cu, 0.35 g/t 3E, and has a $[Ni]_{100}$ of 3.4 wt.%).

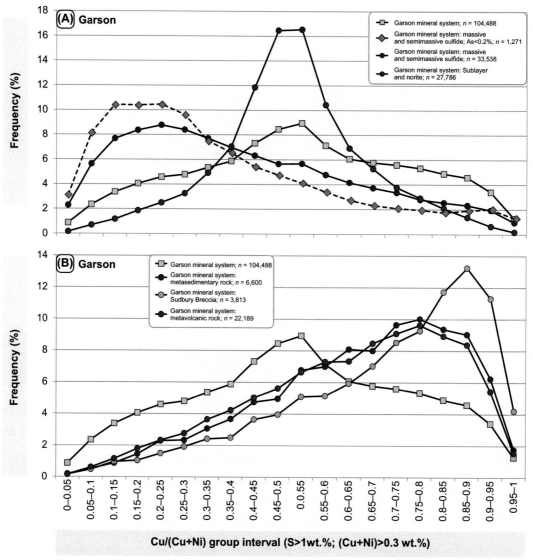

FIGURE 4.46

(A) Percentage grouped frequency distribution of Cu/(Cu + Ni) for the Garson mineral system. The diagram shows the bulk composition of the Garson mineral system as a symmetric bell curve. The composition of mineralization hosted in norite exhibits a narrower distribution with a mode of 0.5. The massive and semimassive sulfides exhibit a skewed curve with a peak at Ni-rich compositions, and this sense of skewing is exhibited by samples after a filter is applied to remove samples with elevated arsenic content (Ni is variably concentrated in the gersdorffite + cobaltite solid solution series). (B) Comparison of the grouped frequency distribution of Cu/(Cu + Ni) for the Garson mineral system to subsets of samples hosted in metasedimentary rock, metavolcanic rock, and SUBX. Note the tendency for the mineralization now hosted in shear structures to be displaced to elevated Cu/(Cu + Ni).

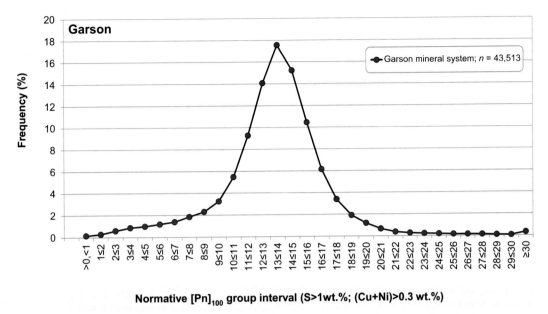

FIGURE 4.47 Percentage Grouped Frequency Distribution of Normative [Pn]$_{100}$ in Samples From the Garson Mineral System

The plot illustrates a bell-shaped curve for both the overall average composition of the mineral system and for samples with <0.2 wt.% As. The effect of gersdorffite and cobaltite on the composition of the mineralization is quite small on the scale of the whole deposit.

and Co in solid-solution (Cooper, 1976; LeFort, 2012), and so the grouped frequency distribution is not strongly influenced by minerals other than pyrrhotite, pentlandite and chalcopyrite.

The massive and semimassive sulfides are strongly depleted in Cu/(Cu + Ni), and this is also a signature of the massive and semimassive sulfides that contain < 0.2% arsenic and which do not have compositions dominantly controlled by gersdorffite and/or cobaltite. The greater abundance of Ni-rich sulfides in the massive and semimassive sulfides is complemented by the tendency of sulfide mineralization hosted in the footwall shear zones (SUBX and metabreccia, metavolcanic and metasedimentary rock) which are enriched in Cu (Fig. 4.46A,B). It appears that the sulfides in the shear zones are displaced toward Cu-rich compositions, and the bulk effect of Ni-enrichment due to gersdorffite is small in comparison to the degree of Cu-enrichment (Fig. 4.46A,B).

Fig. 4.47 shows the grouped frequency percentage distribution of calculated normative [Pn]$_{100}$ for all Garson samples that have been analysed for sulfur; the plot shows a bell-shaped distribution with a mode of 14–15% normative Pn in 100% sulfide. The contribution of Ni from gersdorffite has a negligible impact on estimates of normative [Pn]$_{100}$ as demonstrated by comparing the grouped frequency distributions of all samples versus those with <0.2 wt.% As (Fig. 4.47).

Geological Model

The primary configuration of the Garson mineral system is controlled by a classical trough/embayment structure at the base of the SIC in which SLNR, inclusion-rich massive sulfide and massive sulfide were originally developed. The trough is cut by several shears that are tangential to the contact and along which sulfide has been remobilized into the footwall and the hangingwall. The mineralization in the footwall metavolcanic, metasedimentary, and SUBX samples is enriched in Cu relative to Ni. The Cu-rich composition of the mineralization in the shears is likely due to sulfide kinesis that remobilized chalcopyrite leaving pyrrhotite + pentlandite in the contact ores. There could also be a primary Cu-rich sulfide in syn-magmatic footwall structures, but this is not resolved with the available data.

A number of models have been proposed for the origin of the mineralization in the Garson system:

1. Injection of primary sulfide melt into pre-existing faults prior to the solidification of the SIC (Lockhead, 1955, unpublished reports referenced in Mukwakwami, 2012).
2. South-over north faulting of the south-dipping mineralized trough at the base of the SIC (Binney et al., 1994; Bailey et al., 2004).
3. North-over-south thrusting of a north-dipping trough at the base of the SIC, followed by rotation and south-directed reverse faulting (Mukwakwami, 2012).

In a model that was specifically developed for the Garson mineral system, progressive folding of the competent rocks of the SIC produced back-thrusts by flexural slip between the competent Main Mass in the contact zone with the country rocks thereby producing an imbrication of the rock sequence (Mukwakwami, 2012; Mukwakwami et al., 2012). These faults were later reactivated as high-angle reverse, south-over-north shear zones during subsequent regional orogenic events (Chapter 2 and Mukwakwami et al., 2012).

The structural evolution of the deposit is illustrated in Fig. 4.48A–C following the model proposed by Mukwakwami et al. (2012). The progressive buckling of the SIC along the South Range with the development of layer-parallel north-dipping reverse fault zones is due to mechanical decoupling at the interface between the SIC and the country rocks during early deformation (D1; Fig. 4.48B). Reactivation of the D1 shear zones during the D2 event accompanies overturning of the SIC with the development of the south-over-north shear zones that contain a substantial portion of the mineralization in the Garson mineral system (Fig. 4.48C).

Fig. 4.48D–I show the sequence of events on the scale of an N–S section through the Garson mineral system. The initial geometry of the mineralization was likely controlled by a trough at the base of the SIC into which sulfides were concentrated by primary magmatic processes. The possibility that fault zones pre-dated the SIC (Lockhead, 1955) and were re-activated by the impact and crater re-adjustment may have provided opportunity for some primary injection of magmatic sulfide into these structures, and this possibly weakened the rocks so that subsequent deformation took advantage of these faults to form shear zones. Mukwakwami et al. (2012) noted that unpublished reports by Lee and Siddorn (2006) and Siddorn and Ham (2006) identified massive sulfides that often cross-cut the tectonic fabric in the country rocks, and they documented the observation that the sulfides contain abundant inclusions of contorted and foliated schist that indicates deformation prior to emplacement of the sulfide. Based on this evidence Mukwakwami et al. (2012) suggested that the Garson shears were formed at the time of impact or crater re-adjustment, and provided channels for the injection of sulfide away from the contact.

The observation that the sulfide mineralization in the shear zones is more Cu-rich (Fig. 4.46B) is consistent with the expulsion of fractionated Cu-rich melt from the contact system. However, the presence of massive sulfides in enclaves of country rock and metabreccia within the SIC along the #3 Shear is strongly indicative of post magmatic modification. It is possible that the primary footwall veins were developed in response to impact and crater re-adjustment. Notwithstanding, it appears likely that the postmagmatic translocation and kinesis of sulfide was a principal control on the geometry of the mineral zones.

Following the primary formation of the contact and possibly footwall mineralization (Fig. 4.48D,E), displacement along the #2, #3, #4, and #5 Shear zones resulted from northward-directed thrusting. This produced north-over-south imbrication of the country rock against the SIC along the principal weaknesses developed subparallel to the contact (Fig. 4.48F,G). Continued imbrication and tectonic transportation of sulfides and slices of footwall into the SIC (Fig. 4.48H) producing the geological

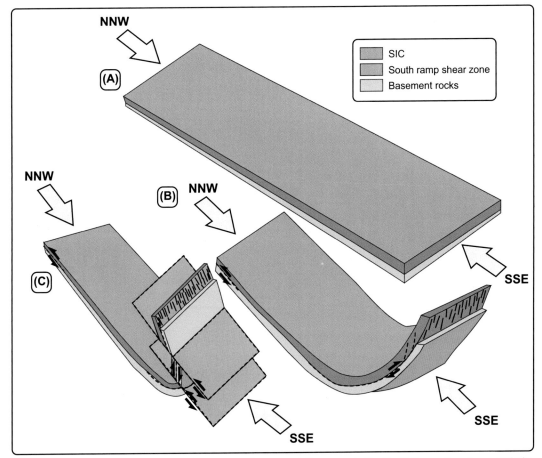

FIGURE 4.48 Schematic Diagram Showing the Development of Structures in the South Range Near to and Within the Garson Mineral System as a Response to Progressive Buckling of the South Range

(A) The SIC was formed as a melt sheet within the impact crater. (B) Layer-parallel north-dipping reverse shears were formed in the South Range in response to the anisotropy at the interface between the SIC and the footwall. (C) As deformation proceeded, the lower contact of the SIC was overturned, and motion on the fault structures was reversed.

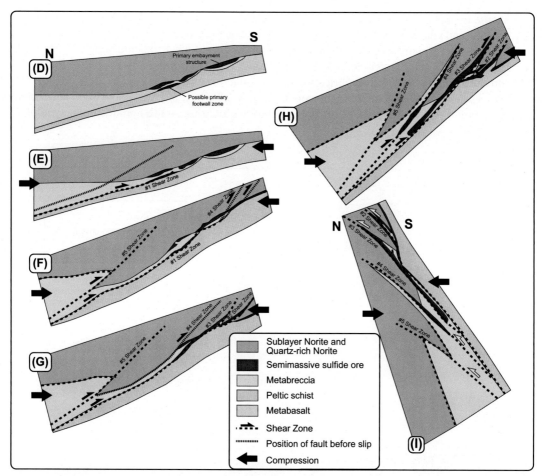

FIGURE 4.48 (*cont.*)

(D) The evolution of the Garson mineral system during NW-SE compression commenced with an initial geometry where the sulfides existed at the base of the SIC in the Sublayer, and possibly in the footwall along favourable contacts. (E) Progressive slip along the shear zones (eg, #5) occurred along north-dipping thrusts which cut through the SIC and imbricate the sequence. (F) Additional slip on other structures such as the #2 and #3 shear zones commenced. (G) Continued imbrication accompanied tectonic transport of footwall material along structures into the SIC. (H) Continued deformation resulted in metavolcanic rocks being interleaved in shears that cut the Main Mass. (I) The SIC was overturned and then the shears were reactivated.

Model and sections are modified after Mukwakwami (2012).

relationships documented in the level plans and sections (Fig. 4.43A–G). The termination of this D1 event was followed by the south-over-north slip along the shear zones (Fig. 4.48I)

The Garson mineral system retains many petrological and geochemical traits that are characteristic of contact deposits in the South Range, but it also exhibits classic examples of translocation of country rocks and kinesis of sulfides along the shear zones. The overprint of deformation coupled with the addition of elements from the immediate country rocks (eg, arsenic) may explain the unusual geo-chemical and mineralogical signature of the shear-hosted mineralization (LeFort, 2012).

Other South Range contact-footwall mineral systems

The Creighton and Garson mineral systems have contributed to metal production from the Sudbury Basin for over a century, but they are only two of seven major mineral systems in the South Range that collectively provided more than 50% of production from the Sudbury Camp. The other major mineral systems in the South Range are described in order to provide the basis for a comparison between the South Range mineral systems.

Falconbridge

The Falconbridge mineral system provided the principal source of historic production to Falconbridge Limited from an open pit and underground mines located at the eastern end of the South Range (Davidson, 1984; Lockhead, 1955; Owen and Coats, 1984). The Falconbridge Deposits were "almost" found by Thomas E. Edison when he conducted a dip needle magnetic survey of Falconbridge Township in 1901. Edison attempted to sink a shaft through overburden, but he failed and let the claims revert to the Crown (Falconbridge staff, 1959). In 1917, the orebody was discovered after a shaft was sunk through overburden into sulfide mineralization, and the claims were acquired by Falconbridge Mines Limited. This orebody was developed from shafts sunk in 1929, 1935, 1937, 1945, 1950, and 1958 (Falconbridge staff, 1959); the mineral zones were exhausted in1984 and mining was terminated on both the Falconbridge and Falconbridge East Ore Deposits. The Falconbridge Deposit was the principle producer that established Falconbridge Limited in the Sudbury Camp, and underpinned production from the Falconbridge smelter. Lateral parts of this deposit have not yet been mined (eg, Vale's Cryderman Deposit).

Blezard and Lindsley

The Blezard contact deposit is located in a primary trough at the base of the SIC, and the near surface part of the ore deposit was mined between 1889 and 1893. The depth-extent of the Blezard Deposit, which remains undeveloped, continues into the Lindsley contact and footwall deposits (Binney et al., 1994; Bailey et al., 2004). Lindsley was developed following underground exploration in 1991. The two deposits are part of a mineral system that has been modified by deformation at the lower boundary by the South Range Shear Zone.

Little Stobie

The Little Stobie mineral system consists of a contact orebody (Number 1 deposit) and a primary zone of footwall mineralization (Number 2 deposit) injected into metabreccias developed at the contact between the Murray Granite and the metavolcanic country rocks. The deposit was discovered by James Stobie in 1885, and was mined in 1902. The first shaft was sunk in 1966–67 and production commenced 1971 and terminated in 1999. This mineral system provides a good example of primary contact and footwall mineralization.

Murray and McKim

The Murray and McKim Deposits comprise both contact and footwall styles of mineralization developed in less strongly deformed rocks to the east of the Copper Cliff Funnel. The McKim deposits were mined by Falconbridge from shafts sunk in 1948 and 1957 (Falconbridge staff, 1959). The Murray

Deposit was the first to be discovered in the Sudbury Basin (1883). The patented mining claims were sold to A.H. Vivian and Co. of Swansea, Wales, who worked it from 1889–94. British America Nickel Corporation Limited purchased the property in 1912. A shaft was sunk to a depth of 700 ft. in 1914 and then re-opened for short periods in 1923 and 1924. Inco acquired the property in 1925 and sank a new shaft to mine the deposit between 1941–43 and then 1950–71, after which it was closed. The vast majority of the ores are primary contact styles of mineralization, but McKim and Murray show examples of footwall mineralization in metabreccias developed at the contact between the Murray Granite and the Elsie Mountain Formation.

Crean Hill

The Crean Hill Deposit is located at the western end of the South Range. Francis C. Crean discovered the deposit in 1885. Production commenced in 1906 and continued intermittently until 1919. After many programs of exploration, the mine produced continuously from 1964 to 1972 and 1986 to 2000. The deposit occupies a primary trough that controlled the development of the Sublayer. The deposit is cut by many faults that extend through the deposit at a low angle to the contacts. Some parts of the Crean Hill deposit that are contained in the footwall are very Cu-rich, and it is possible that they represent zones of primary footwall mineralization. Notwithstanding, there is overwhelming evidence for postmagmatic deformation at the base of the SIC and it is possible that sulfide kinesis modified the footwall mineralization. The recognition of low-sulfide mineralization with elevated 3E concentrations near surface around the Crean Hill embayment is interpreted to result from secondary remobilization of metals along zones of structural weakness into the country rocks (Gibson et al., 2010).

Primary contact and footwall mineralization

The Creighton and Garson mineral systems each exhibit the primary control of a trough structure along which magmatic sulfide mineralization was concentrated in the Sublayer. An examination of the other South Range contact-footwall mineral systems confirms that this is also a common feature of the deposits that have not been described in detail here. For example, the Murray mineral system occupies an embayment structure in the Elsie Mountain Formation metavolcanic rocks; the ores are classic examples of massive sulfides, inclusion-massive sulfides, and disseminated sulfides associated with SLNR (Fig. 4.49A; Souch et al., 1969). In the case of the McKim mineral system, the contact mineralization is related to the Sublayer at the base of the SIC that forms a smaller trough than Murray. The contact between metavolcanic rocks and the Murray Granite in the footwall of the Murray-McKim mineral system has acted as a pathway along which mineralization with (chalcopyrite + pentlandite) ≫ pyrrhotite are located. The host rocks of this style of mineralization are described as breccias developed between the Murray Granite and metasedimentary and metavolcanic rocks that are now considered to be part of the Elsie Mountain Formation (eg, Fig. 4.49B shows an example from the McKim mineral system after the description provided in Clarke and Potapoff, 1959a,b).

The Little Stobie mineral system occupies both a primary embayment structure at the base of the SIC, and breccias developed between mafic metavolcanic rocks and the Murray Granite (Fig. 4.50A–C; Davis (1984). The style of mineralization developed in the Number 2 Orebody at Little Stobie (Fig. 4.50A,C) is remarkably similar to that described in the footwall zone of the McKim mineral system (Fig. 4.49B). The Little Stobie, Murray and McKim chalcopyrite + pentlandite-rich styles of mineralization have primary magmatic associations that exploit breccias developed along the contact between granitic and metavolcanic rocks.

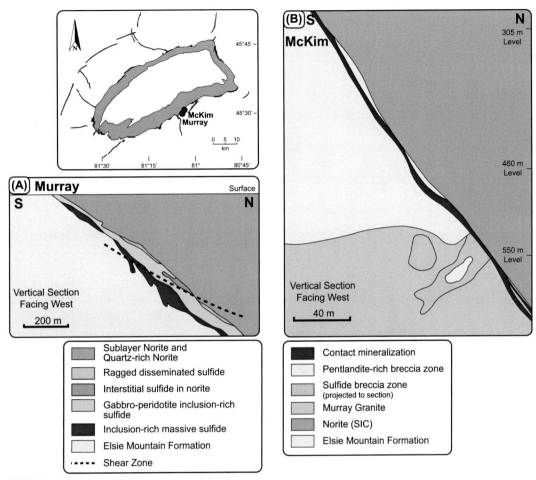

FIGURE 4.49

(A) West-facing geological cross-section of the shallow part of the Murray mineral system. (B) West-facing geological section through the McKim mineral system showing the development of both contact style mineralization and breccias comprising abundant chalcopyrite + pentlandite in a domain that rests between the Murray Granite and the Elsie Mountain Formation.

Modified from publications of the Ontario Geological Survey, and used under license to Elsevier.

The geology of the Blezard and Lindsley Deposits and the mineral system that contains them is shown in Fig. 4.51A–E. The Blezard Deposit consists of contact style mineralization associated with SLNR in a trough-like structure which rakes down the base of the SIC, and extends into the Lindsley contact mineral zone (Binney et al., 1994). The footwall mineralization in the Lindsley Deposit (Zone 4) is hosted within the Murray Granite where the basal contact of the SIC reverses in dip (Fig. 4.51A). The Murray granite forms an intrusion at depth beneath the heavily brecciated McKim Formation metavolcanic and metasedimentary rock types that crop out at surface. The

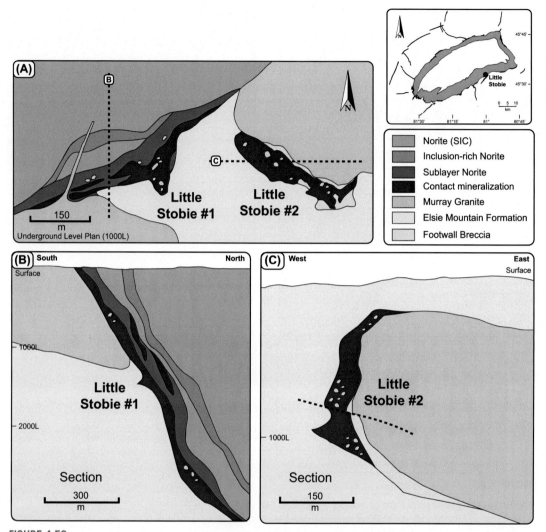

FIGURE 4.50

(A) Geological plan (1000 Level) of the Little Stobie mineral system showing the Number 1 contact orebody and the Number 2 footwall-hosted orebody. (B) Geological section through the Number 1 Orebody. (C) Geological section through the Number 2 Orebody in the Little Stobie mineral system.

Modified from publications of the Ontario Geological Survey, and used under license to Elsevier.

footwall mineral zone at Lindsley is developed within 80 m of the base of the SIC and consists of vein-like bodies of massive sulfide with locally elevated Cu and PGE concentrations (Zone 4) relative to the abundance levels in contact mineralization (Binney et al., 1994).

The Little Stobie Number 2 Orebody (Fig. 4.50A,C), the Lindsley 4 mineral zones (Fig. 4.51A), and the Murray and McKim (chalcopyrite + pentlandite)-rich breccia zones (Fig. 4.49A,B) are all part

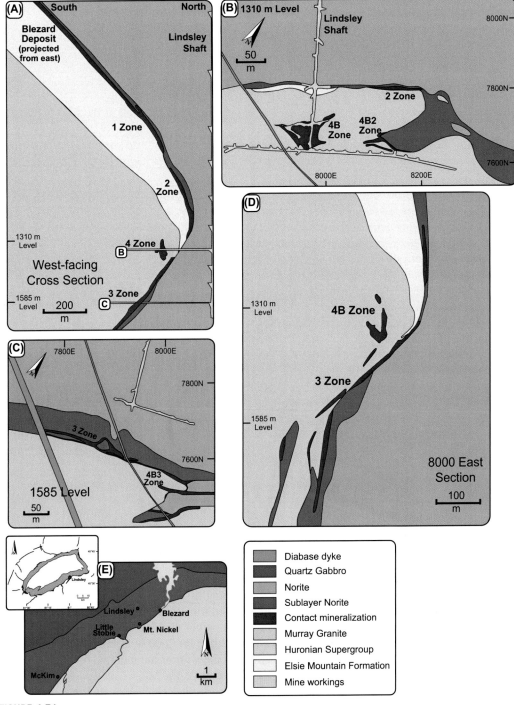

FIGURE 4.51

(A) Geological cross-section facing west through the Blezard-Lindsley mineral system. (B) Geological plan of the 1310 Level in the Lindsley Ore Deposit showing both contact and footwall styles of mineralization. (C) Geological relationships on the 1585 Level where a spur of norite containing contact mineralization occurs in the granite and the footwall mineralization is surrounded by granite. (D) Geological cross-section facing west through the Lindsley Deposit showing the relationships of mineralization Sublayer and Murray Granite. (E) Location map showing the locations of the Blezard, Lindsley and Little Stobie Deposits.

of a spectrum of mineralization styles that resemble the mineralization in the footwall of the Creighton Trough that collectively comprise the 126, 403, 461, and 320 Orebodies (Fig. 4.32A). All of these examples of mineralization are the products of fractionation of sulfide melt at the base of the SIC. Each example is possibly a primary fractionated sulfide melt emplaced into the footwall along zones of primary structural weakness created by the impact process (Lightfoot, 2015) but these mineral zones were all modified by sulfide kinesis accompanying postmagmatic deformation.

Structural modification and kinesis of primary magmatic sulfide mineralization
There is no doubt that the features of primary magmatic sulfides are modified by translocation and kinesis of sulfide during deformation. The key question is the importance of this process in respect to simply modifying a pre-existing zone of mineralization versus creating a zone of mineralization along a structure in the footwall. The Garson mineral system provides a framework for this discussion, but it is not the only deposit to show evidence of remobilization along shear zones and fault structures.

The Victoria, Crean Hill, Kirkwood, Falconbridge, Falconbridge East, and Cryderman mineral systems were created as magmatic sulfide concentrations associated with Sublayer in embayments and troughs at the base of the SIC. The similarities to the contact mineralization at Creighton include the identification of primary depressions at the base of the SIC, the development of disseminated sulfides in highly altered SLNR (Owen and Coats, 1984), the presence of mafic-ultramafic inclusions in the sulfide mineralization, and the similarity in the mineralogy and composition of the sulfide ores. However, each of these mineral systems is affected by structures that may pre-date the SIC and influence the stability of the crater walls, but the record of deformation is invariably most easily attributed to extensive deformation associated with much younger orogenic events in the South Range.

The Falconbridge West and East mineral system occur within fault zones located at the base of the SIC and in the underlying metavolcanic and metasedimentary rocks of the Elsie Mountain and Stobie Formations. The extent of deformation has largely obscured any primary trough or embayment at the base of the SIC, but the composition of the ores, their relationships with SLNR, and their localization all point to a magmatic source that has been extensively modified (Owen and Coats, 1984). The types of sulfide mineralization comprise semimassive sulfides with contorted fragments of schist in association with a shear zone that follows the contact (termed the Main Zone) at surface and to depth in the plane of the long section shown in Fig. 4.52A. Deformation has also produced a series of structures that either cross-cut the Main Zone or represent splays of this structure (Fig. 4.52A–C).

Early in the evolution of the Falconbridge mineral system, faults provided pathways for fluids that caused extensive silicification and quartz-carbonate veining associated with breccias in the path of the fault zones. The presence of fragments of norite derived from the SIC within the mineralization is indicative of deformation after crystallization of the lower part of the Main Mass.

The Main Fault and associated splays extends along strike for 2000 m at the basal contact of the SIC, extending from the Main Mass into the Sublayer to the west of the mineral system and exiting the mineral system into the country rocks in the east (Fig. 4.52B). The Main Zone fault and the contact of the SIC dips steeply north in the upper levels, is near vertical at 1200 level, and then dips steeply south to the depth of exploration (Fig. 4.52D,E). A subsidiary fault of the Main Zone is the South Fault, and where these two faults intersect with the right lateral Number 1 Flat Fault is an important control on the sulfide mineralization (Figs. 4.52 and 4.53). A group of faults termed the Tangential Shears (Owen and Coats, 1984) branch off the Main Fault as a series of shears that produce swells in the thickness of the contact mineralization (Fig. 4.53A,B). These splays vary in length, the most continuous is the

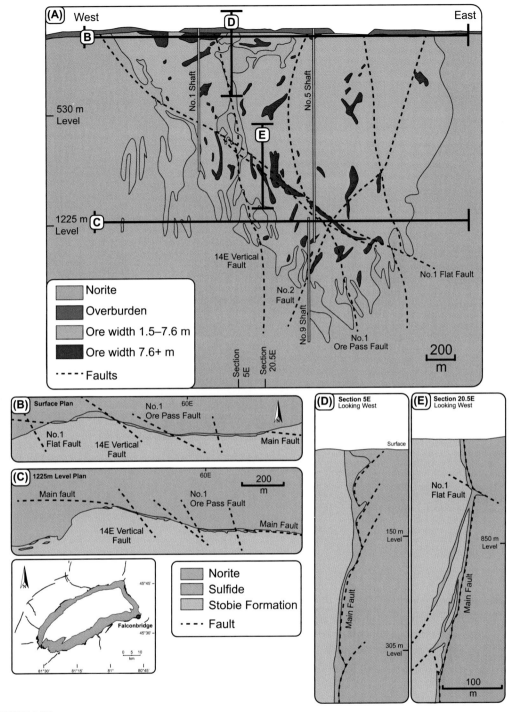

FIGURE 4.52

(A) North-facing long section through the Falconbridge mineral system showing the location of the mineral zones and principal cross-cutting fault structures. (B,C) Surface and 4025 Level plan showing the intersection of two principal fault sets and their significance in controlling the distribution and thickness of sulfide mineralization in the Falconbridge System. (D,E) West-facing sections through the Falconbridge Deposit showing geological relationships at locations shown in subpart A.

Modified from Coats and Snajdr (1984). (A–C) Modified from publications of the Ontario Geological Survey, and used under license to Elsevier.

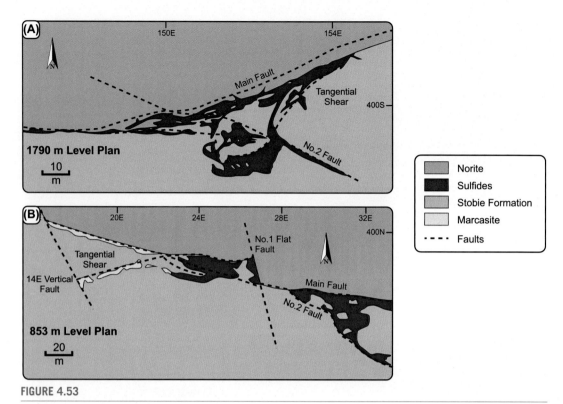

FIGURE 4.53

(A) Plan showing relationship of structure to thick mineral zones on 1790 m Level. (B) Plan showing relationship of structure to thick mineral zones on 853 m Level.

Modified from Coats and Snajdr (1984).

Number 2 Fault, and it is associated with chalcopyrite-rich sulfide mineralization in the footwall, but it also cuts into the hangingwall norites of the SIC.

A number of vertical faults cross-cut the contact mineral zone (eg, 14 East and 78 East in Fig. 4.53A,B); these faults are believed to be late as they are characterized by marcasite, carbonate, galena, and sphalerite.

The effects of structural controls in the Falconbridge mineral system are important in controlling dilations within the contact orebody, and shear-hosted mineralization in the footwall and hangingwall.

The bulk composition of Falconbridge ores is not published, but a subset of data in Owen and Coats (1984) is indicative of a heavily skewed composition toward Ni-rich ores (Cu/(Cu + Ni) = 0.22). The extent to which the faults at Falconbridge provided primary pathways for migration of Cu-rich sulfide melts into the footwall is unclear, but Owen and Coats (1984) suggest that substantial quantities of Cu-PGE-rich ores were either eroded or remained undiscovered when the mine was closed in 1984. A skewing of the composition of the contact ores toward Ni-rich compositions such as the Falconbridge mineral system is indicated by assays from the Cryderman Deposit.

Crean Hill is a further example of a deposit where there is a strong structural control that has re-distributed the primary magmatic sulfides into subvertical faults which cross-cut the lower contact

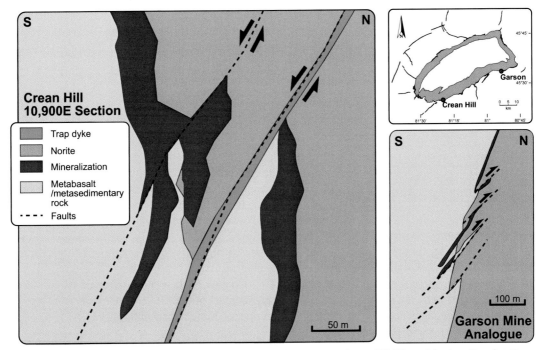

FIGURE 4.54 West-Facing Section Along Line 10,900 East Through the Crean Hill Mineral System Showing the Development of Structures That Are Remarkably Similar to the Shear Structures at Garson Mine

This section is produced from underground mapping and drilling, and was prepared by Sandy Gibson.

of the SIC, and in which mineralization is translocated into the footwall and hangingwall by kinesis (Fig. 4.54). Evidence for primary early footwall mineralization at Crean Hill is provided by Cu-rich mineralization that resides in the deep footwall zone. Much of the mineralization identified in the footwall appears to be the product of deformation and postmagmatic hydrothermal remobilization of metals into country rock structures (Gibson et al., 2010) to produce low-sulfide mineralization with elevated precious metal contents.

There are examples of South Range orebodies that possibly formed by primary injection of differentiated Cu-rich sulfide melts into the footwall during the differentiation of the SIC. The examples described from Creighton, Lindsley and Murray-McKim are the least equivocal examples, and their geological relationships are consistent with primary lithological contacts and structural controls which created space in the footwall into which the Cu-rich melts could migrate. It is less clear whether other examples of mineralization with elevated Cu ± PGMs was originally formed in the footwall and then modified by deformation, or whether the process of sulfide kinesis and translocation was able to fractionate the composition of the sulfide ores to produce enrichment of Cu in the footwall.

In the case of the sediment-hosted nickel deposits in the Thompson Deposits, there is evidence to suggest that a diversity in pyrrhotite:pentlandite ratios in the mineral zones reflects the separation of pyrrhotite from pentlandite during deformation, and that the Ni tenor of the sulfide mineralization is not related to primary magmatic processes (Lightfoot and Evans-Lamswood, 2015; Lightfoot, 2015).

Bailey et al. (2006) suggest that shearing of contact magmatic sulfides resulted in replacement of pentlandite with pyrrhotite and liberation of Ni. At Sudbury, the pressure-temperature conditions were much lower than Thompson, so separation of sulfides by partial melting accompanying deformation has not been recognized. The question of the origin of the sulfide mineralization in the footwall of the SIC in the South Range remains one that requires further investigation.

NORTH AND EAST RANGE MINERAL SYSTEMS

The vast majority of production of Ni, Cu, and PGE has come from ore deposits associated with the South Range. Of the production in the North Range, the Levack-Coleman mineral system represents the single largest source of produced metal, and the Nickel Rim South Deposit in the East Range accounts for much of the remainder of the production outside of the South Range. Smaller deposits at Capre, Whistle, Podolsky and Victor have contributed to past production, but future production may focus on the development of the Victor mineral system where large un-mined resources of contact and footwall mineralization remain.

The geology of the ore deposits in the Levack-Coleman, Victor and Nickel Rim South Mineral Systems are used as examples. Other smaller North and East Range deposits are then briefly compared to the larger systems to help establish a framework for Chapter 5 where a critical question will be addressed. Namely, why are some segments of the outer contact of the SIC are endowed with large ore deposits yet other segments have only small to subeconomic mineral concentrations?

Levack-Coleman

The Levack Deposit was the first of a cluster of deposits to be identified in the North Range. The discovery was made in 1889 by James Stobie and Rinaldo McConnell. Mond Nickel Company acquired the property in 1913 and operations began that year, with production starting in 1914. The Levack Mine was closed down during the depression of the early 1930s and reopened in 1937. The original Levack Deposit together with various satellite deposits such as McCreedy West and Morrison have been in almost continuous production under Inco and then FNX, Quadra-FNX, and KGHM.

The Coleman Deposits were discovered in 1892 by Thomas Baycroft. The land position was held by a succession of companies, with Inco and Falconbridge, then FNX, Quadra-FNX, and KGHM undertaking the development of segments of the different orebodies that comprise this Levack-Coleman mineral system. Production from these deposits has a complex history with several companies involved in mining and processing of the ores (see Table 1.2 for more details).

The styles of mineralization comprise:

1. Contact-related, embayment and trough-hosted, SLNR and Granite Breccia (GRBX) with pyrrhotite > chalcopyrite > pentlandite (eg, the Coleman Main Orebody, Fraser, and Craig; Naldrett et al., 1984b; Morrison, 1984; Morrison et al., 1994; Farrow and Lightfoot, 2002);
2. Transitional styles of mineralization contained in GRBX and metabreccia beneath deep embayments that have a spatial link to footwall-hosted mineralization. The principal sulfide minerals are chalcopyrite > pentlandite > pyrrhotite (eg, the 9 Scoop Orebody beneath the Levack Number 4 Orebody; Gibson et al., 2010);
3. Footwall type ores that comprise highly fractionated assemblages of chalcopyrite > pentlandite > millerite > bornite >> pyrrhotite in vein systems that are sometimes spatially connected to the contact ore deposits (McCreedy 153, McCreedy West, Strathcona, and the Morrison Deposits;

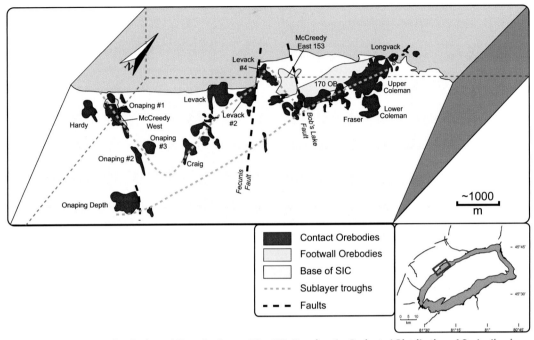

FIGURE 4.55 Long Section Projected Onto the Base of the SIC Showing the Projected Distribution of Orebodies in a Broad "W" Shape, and the Naming of the Contact and Footwall Deposits That Are Contained in the Mineral System

The diagram is after Farrow and Lightfoot (2002).

Morrison et al., 1994; Jago et al., 1994; Naldrett et al., 1994b; Watkinson, 1994; Farrow et al., 2005; Gibson et al., 2010; Stewart and Lightfoot, 2010).

4. Low-sulfide, high precious metal disseminations that envelope parts of the footwall orebodies (Farrow et al., 2005; Stewart and Lightfoot, 2010; Gibson et al., 2010; Péntek et al., 2013; Tuba et al., 2014)

The orebodies are controlled by a series of embayments and troughs that collectively describe the shape of the letter "W" on the lower contact of the SIC (Fig. 4.55). The contact mineral zones typically reside within the trough and embayment structures, the transitional style of sulfides occupy the terminations of especially deep embayments and as veins in the adjacent footwall, and the footwall mineralization typically resides in belts of SUBX developed beneath the embayments. Collectively both the contact and footwall mineralization styles have a distribution that maps out the topography of the basal contact of the SIC (Fig. 4.55).

Geological Relationships

The Geological relationships in the McCreedy West Ore Deposit are shown in Fig. 4.56A–D (after Farrow and Lightfoot, 2002; Farrow et al., 2005). The deepest part of the embayment contains GRBX overlain by inclusion-bearing norite of the Sublayer. The GRBX is thickest where the contact style mineralization connects to the footwall zone that is hosted in metabreccia, Levack Gneiss and SUBX

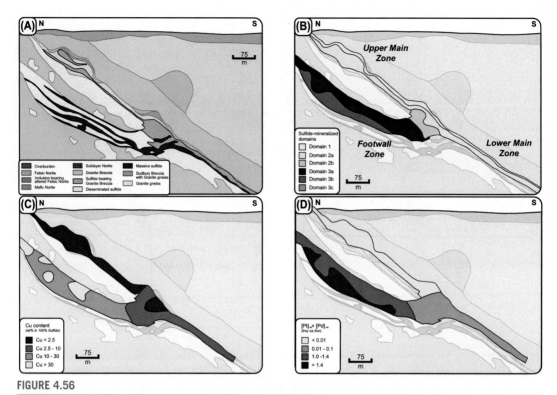

FIGURE 4.56

(A) Geological section through the McCreedy West Ore Deposit based on level plans and drill core data. This deposit is a very good example that shows the continuity between Ni-rich contact mineralization through a transitional type of chalcopyrite + pentlandite-rich mineralization into the Footwall zones of chalcopyrite + pentlandite + bornite + millerite. (B) Model section showing the distribution of mineral domain types in the McCreedy West Ore Deposit. (C) Model section showing the distribution of Cu sulfide tenor in the McCreedy West Ore Deposit. (D) Model section showing the distribution of $[Pt + Pd]_{100}$ sulfide in the McCreedy West Ore Deposit.

The relationships are shown in modified form after Farrow and Lightfoot (2002).

(Fig. 4.56A). The transition from contact to footwall is recorded in the degree of recrystallization and partial melting of the host rocks which decreases with distance from the base of the Main Mass. The SUBX zone that hosts the footwall veining is intensely recrystallized and locally contains quartz diorite "melt patches." The host rocks containing the mineralization in the McCreedy West Footwall Zone comprise equal proportions of SUBX and Levack Gneiss.

The Main Mass norites that overlie the McCreedy West embayment consists of a discontinuous unit of SLNR which is interfingered with Mafic Norite (MNR) that represents the basal stratigraphy of the Main Mass as well as hybridized types of MNR that are contained in the Sublayer. Both types of MNR consist of poikilitic-texture norite with phlogopite mica and interstitial sulfide, and it is only the development of inclusions and the hybridization textures with normal SLNR that distinguishes the Main Mass from the Sublayer. The Main Mass Felsic Norite (FNR) located directly above the transitional domain between contact and footwall is marked by the presence of abundant hornfelsed mafic inclusions (Fig. 4.56A). These fragments are not observed in the MNR and FNR outside of the connection zone between the contact and footwall mineralization.

The McCreedy West Deposit is located at the far western end of two domains of SUBX that may represent crater wall slump features; one of these horizons controls the lateral development of the McCreedy West mineral zone (Fig. 4.56A; Farrow and Lightfoot, 2002).

Sulfide mineralization in the GRBX is massive, semimassive and disseminated. The dominant sulfide mineral is monoclinic pyrrhotite, with less common pentlandite, pyrite, and chalcopyrite concentrated as granular masses along pyrrhotite grain boundaries. Magnetite grains are ubiquitous. The disseminated sulfides typically occur as subrounded blebs of sulfide (pyrrhotite ± pentlandite, pyrite, chalcopyrite) within the quartzo-feldspathic matrix of the GRBX (Farrow et al., 2005).

The Footwall vein assemblage consists dominantly of chalcopyrite with variable amounts of cubanite and pentlandite, and minor pyrrhotite, millerite, bornite and violarite. Chalcopyrite tends to enclose the other sulfide minerals. Cubanite occurs as exsolution laths within chalcopyrite. Pentlandite is rarely localized at the lower contacts of the sulfide veins as lenses or bands. Millerite dominantly occurs with quartz + sulfide ± carbonate near the terminations of veins, and in veins with sheared contacts. Bornite is restricted to the terminations of massive sulfide veins and commonly contains macroscopic inclusions of native silver (Farrow and Lightfoot, 2002; Farrow et al., 2005).

The McCreedy West Deposit is strongly zoned between contact and footwall, and is broken out into a series of domains (Fig. 4.56B) that are established based on sulfide mineralogy. The variations in mineralogy are also related to Cu and Pt + Pd tenor in 100% sulfide (Fig. 4.56C,D).

The Geological relationships in the McCreedy East 153 Orebody are shown in Fig. 4.57A,B. The sulfide mineralization is divided into three types based on geological relationships, mineralogy, and geochemistry (Farrow and Lightfoot, 2002). These comprise: A. Thick (≤24 m) massive sulfide veins dominated by chalcopyrite > pentlandite >> pyrrhotite; B. Thinner (<50 cm) veins of bornite + chalcopyrite + millerite + pentlandite that occur at the margins of the chalcopyrite + rich thick veins, and C. Narrow veins, segregations, and disseminations of bornite + millerite that occur toward the periphery of the massive sulfide veins. The center of the deposit is dominated by thick veins; the margin consists of thin veins, and there is an outer halo of the third style of mineralization. The halo of veins and disseminations extends along strike away from the orebody as well as above and below it as shown in Fig. 4.57A,B, whereas the more massive trunk veins comprise the core domain of the deposit.

The Strathcona Copper Zone is a third example of a footwall deposit in the Levack-Coleman mineral system. Stringer sulfides in the Deep Zone are connected to the contact style mineralization (Fig. 4.58A–C). The stringer mineralization consists of massive chalcopyrite enclosing a small percentage of pentlandite, and the veins are developed ~500 m below the SIC (Abel et al., 1979; Abel, 1981). The stringer and vein mineralization in the Copper Zone is entirely detached from both the contact and the Deep Zone footwall mineralization (Fig. 4.58B).

A fourth style of footwall mineralization developed in the Levack-Coleman mineral system is the footwall epidote zone at Fraser Mine. The mineralization is associated with a large body of heavily altered ultramafic rock (after primary websterite and pyroxenite; Moore et al., 1993, 1994, 1995) which forms a broad horizon in the footwall of the SIC along which the magnetite-epidote-rich style of mineralization is developed (Fig. 4.59). The unusual alteration assemblages provide one of the most convincing examples of a deposit formed by hydrothermal modification of a primary magmatic sulfide (Farrow and Watkinson, 1996).

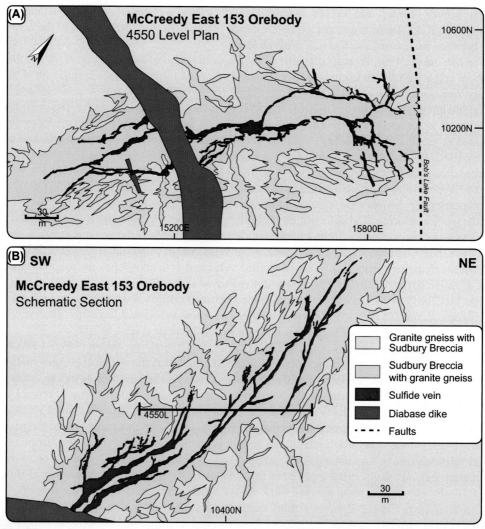

FIGURE 4.57

(A) Level plan showing the mineralization style in the McCreedy East 153 Orebody on the 4550 Level. (B) West-facing section showing the geological relationships in the stepped vein system of the McCreedy East Orebody along the section shown in subpart A. The host rocks comprising Sudbury Breccia with granite gneiss typically contain a large proportion of gneiss relative to breccia matrix (>75:25).

Diagram modified after Stout (2009) from drill core data and underground maps supplied by Vale.

Characteristics of the Mineralization

The geological relationships in the mine openings in the 153 Orebody provide spectacular examples of sulfide veins that occupy dilational space created by structures that cross-cut both the host Levack Gneiss and the SUBX. One of these examples is shown as the front image to this chapter. The veins have sharp contacts and they impart minimal alteration on the immediate country rocks (alteration

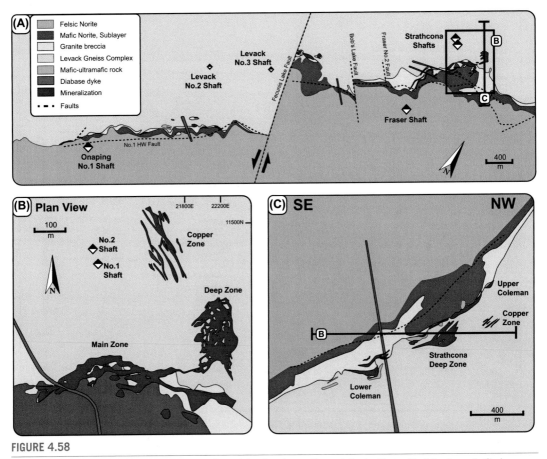

FIGURE 4.58

(A) Geological level plans showing the location of the Strathcona Deposit in the footwall of the Levack-Craig mines. (B) Detailed plan showing the geological relationships between contact and footwall ores in the Strathcona Ore Deposit. (C) North-west-facing section through the Strathcona Footwall ore deposit.

Modified from publications of the Ontario Geological Survey, and used under license to Elsevier.

zones are typically <5 cm wide). The veins vary in composition from the center to the margin from chalcopyrite + pentlandite-rich (Fig. 4.60A) through chalcopyrite + millerite + pentlandite (Fig. 4.60B), bornite + chalcopyrite + millerite (Fig. 4.60C) to an apogee in narrow veins of bornite and/or native silver associated with 1–10 cm wide carbonate veins in heavily chlorized SUBX (Fig. 4.60D).

The connection between the McCreedy East 153 footwall deposit and the contact appears to be through a series of narrow veins of chalcopyrite + pentlandite that comprise the 148 mineral zone through to the 153 Orebody. These veins connect up with the base of the Levack Number 4 Orebody through a transition zone named the "9 Scoop Orebody" (Gibson et al., 2010). The principal ores of the contact environment are magmatic assemblages of pyrrhotite + chalcopyrite + pentlandite as disseminations, inclusion-bearing massive sulfides and massive sulfides (Souch et al., 1969). The transition zone mineralization in the 9 Scoop Orebody consists of progressively higher metal tenor mineralization with depth in the embayment structure; chalcopyrite + pentlandite content eventually entirely replaces the pyrrhotite.

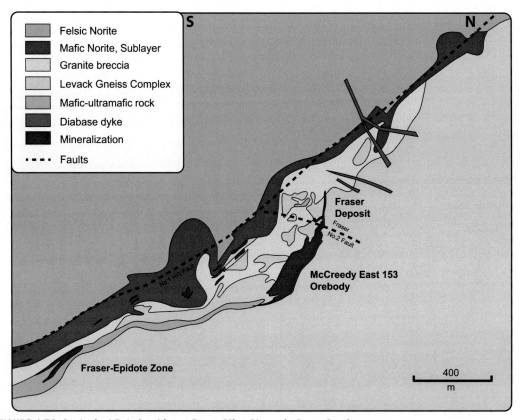

FIGURE 4.59 Geological Relationships at Fraser Mine Shown in Cross-Section

The Fraser Orebody is associated with an ultramafic footwall intrusion which appears to influence the development of magnetite-epidote styles of alteration in the footwall.

Modified from publications of the Ontario Geological Survey, and used under license to Elsevier.

The contact style disseminated sulfide in SLNR is typically blebby to interstitial (Fig. 4.61A), and the host norites are typically fresh. The inclusion-bearing massive sulfide typically consists of pyrrhotite > chalcopyrite + pentlandite (Fig. 4.61B), and this style of mineralization is contiguous with massive sulfides that consist of variable proportions of chalcopyrite, cubanite, and pentlandite (Fig. 4.61C,D). The footwall mineralization consists of chalcopyrite containing eyes of pn and intergrowths of chalcopyrite + bornite (Fig. 4.61E; in this case with the unusual development of an ultramafic inclusion in the sulfide mineralization), zoned veins consists of millerite at the margins and bornite at the core (Fig. 4.61F) and the development of chalcopyrite crystals in bornite (Fig. 4.61G).

Geochemistry

The chemical composition of the contact and footwall mineral zones at Levack and Coleman show a strong bi-modality in $Cu/(Cu + Ni)$ and $(Cu + Ni)/S$ that correspond to the compositions of mineralization in the contact and footwall, respectively. Fig. 4.62A illustrate these differences in the context

FIGURE 4.60

(A) Massive chalcopyrite > pentlandite vein with rim of millerite; McCreedy East 153 Orebody. (B) Discrete vein sets of chalcopyrite + millerite + pentlandite cutting Levack Gneiss, McCreedy East 153 Orebody.

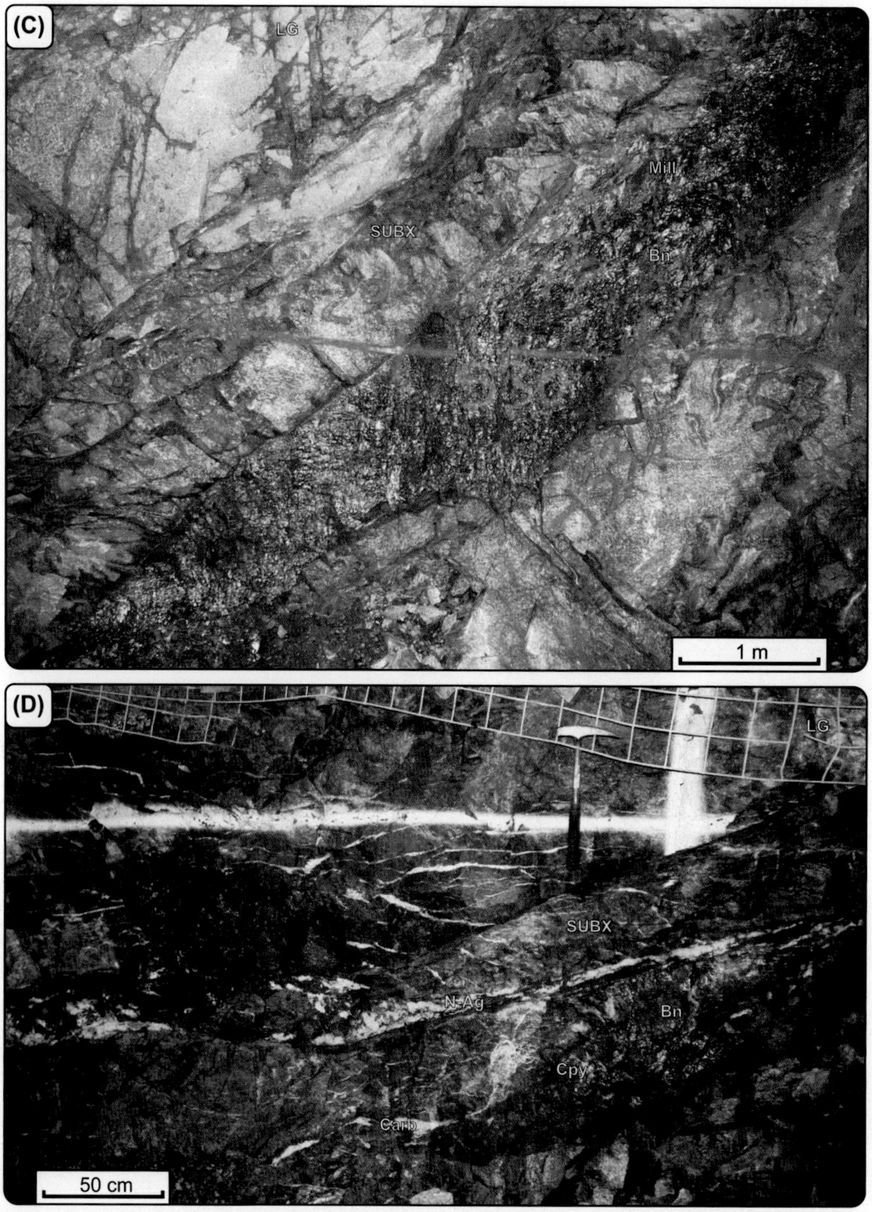

FIGURE 4.60 (*cont.*)

(C) Massive bornite within Levack Gneiss and SUBX, McCreedy East 153 Orebody. (D) Termination of bornite vein into quartz-carbonate with native silver, McCreedy East 153 Orebody.

Photographs A–B were provided by Sandy Gibson.

FIGURE 4.61

(A) Disseminated sulfide in SUBX from the McCreedy East 153 Orebody (sample RX355684 contains 0.98 wt.%Ni, 1.66 wt.%Cu, 1.9 g/t 3E, and has $[Ni]_{100}$ = 8.3 wt.%). (B) Inclusion-bearing massive sulfide from the McCreedy East 153 Orebody (sample RX355682 contains 2.5 wt.% Ni, 28 wt.% Cu, and 13.1 g/t 3E). (C) Massive chalcopyrite + pentlandite from the McCreedy East 153 Orebody (sample RX355681 contains 0.16 wt.% Ni, 32.2 wt.% Cu, and 28.5 g/t 3E). (D) Massive chalcopyrite + cubanite with pyrrhotite and pentlandite from the McCreedy East 153 Orebody (sample RX355683 contains 0.48 wt.% Ni, 32.6 wt.% Cu, and 19.3 g/t 3E). (E) Cubanite and chalcopyrite inter-grown together in the periphery of the McCreedy East 153 Orebody; the massive sulfide contains an ultramafic inclusion (Um). (F) Vein of massive millerite (Mill) grading toward the vein margin into bornite (Bn). McCreedy East 153 Orebody. (G) Massive bornite containing eyes of chalcopyrite from the periphery of the McCreedy East 153 Orebody.

Samples E and F were provided by Sandy Gibson and sample G was provided by Jason Letto.

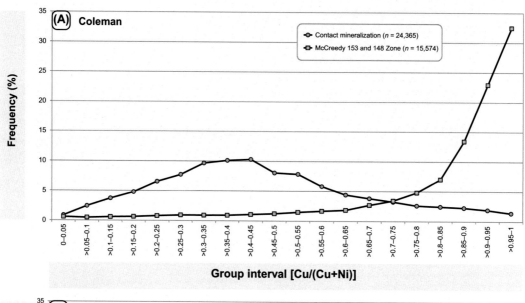

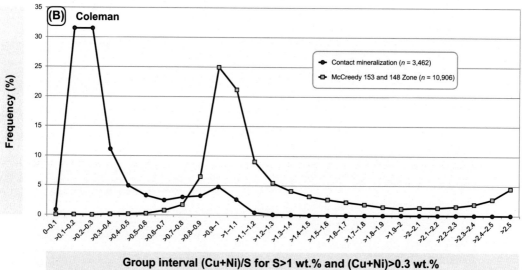

FIGURE 4.62

(A) Variations in percentage grouped frequency distribution of Cu/(Cu + Ni) for samples of contact-associated mineral system (Levack No. 4 Orebody and Nine Scoop) versus the mineralization hosted in the associated footwall mineral system (148 and 153 mineral zones). (B) Variations in percentage grouped frequency distribution of (Cu + Ni)/S for samples of contact-associated mineral system (Levack No. 4 Orebody and Nine Scoop) versus the mineralization hosted in the associated footwall mineral system (148 and 153 mineral zones).

of available data from related parts of the contact and footwall mineral zones that comprise the Levack Number 4 and 9 Scoop Deposits (parts of the Sublayer-hosted contact mineralization) and the McCreedy East 153 Deposit and the 148 Zone mineralization (that comprise the spatially connected footwall mineralization). The contact mineralization has a peak $Cu/(Cu + Ni)$ that is displaced to lower ratios than the footwall mineralization. The variations are expressed in Fig. 4.62B in terms of percentage grouped frequency distribution of $(Cu + Ni)/S$ where the contact sulfides are displaced to low concentrations of metal in sulfide, and the footwall mineralization is displaced to much higher metal concentrations in sulfide. This difference largely reflects to the very low concentrations or absence of pyrrhotite (which is the principal source of Fe) in footwall mineralization, and the high abundance of pyrrhotite in contact type mineralization.

The variations in the data available for the system as a whole are illustrated in a data density plot in Fig. 4.63 which shows the compositional array of contact mineralization connecting pyrrhotite + pentlandite with chalcopyrite, with a strong bias in the density of data toward the Ni-rich compositions. In contrast, the footwall mineralization is anchored by arrays connecting chalcopyrite + pentlandite, chalcopyrite + pentlandite + millerite, chalcopyrite + millerite, bornite + millerite, and chalcopyrite + bornite at much higher $(Cu + Ni)/S$. The pattern in Fig. 4.63 is common to many, but not all North Range

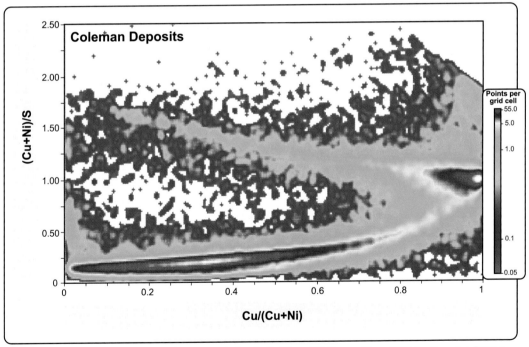

FIGURE 4.63 Data-Density Plot Showing the Variation in $Cu/(Cu + Ni)$ Versus $(Cu + Ni)/S$ for Samples From a Representative Part of Vale's Coleman Segment of the Levack-Coleman Mineral System

Samples with >1 wt.% S and >0.3 wt.% $(Cu + Ni)$ are used to generate the grided image of density distribution. The image shows the mineralogical controls on the composition of the sulfides.

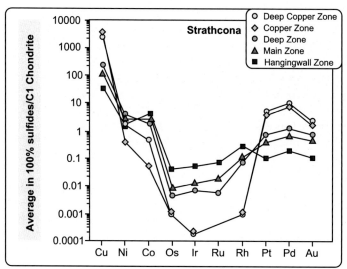

FIGURE 4.64 Variation in Chondrite-Normalized Platinum Group Element Abundances in 100% Sulfide for Samples From Strathcona

The extreme enrichment in Pt, Pd, and Au with extreme depletion in Os, Ru, Ir and Rh is a signature consistent with magmatic differentiation of the primary sulfide melt to form contact ores that have the inverse abundances of these elements (Naldrett et al., 1999).

Modified after Li et al. (1992).

footwall ore deposits, but not all deposits achieve the extreme levels of fractionation required to produce bornite (Capre and Podolsky; Stewart and Lightfoot, 2010; Farrow et al., 2005).

Extensive studies of the fractionation of precious metals between contact and footwall mineral zones have been presented in Naldrett et al. (1994b, 1999). A detailed study of Strathcona illustrates the differences that exist between contact, transitional, and footwall mineral zones from representative sampling of this portion of the Levack-Coleman mineral system. The compositions of average mineralization types from above the contact, the Main Zone, the Deep Zone, and the Footwall Stringers located in Fig. 4.58B are shown as chondrite-normalized patterns for the abundances of the elements in 100% sulfide assuming a simple sulfide stoichiometry comprising different proportions of chalcopyrite, pentlandite, and pyrrhotite (Fig. 4.64). The plots illustrate the strong enrichment of Ni, Co, and Rh in the contact mineralization, and the displacement of footwall mineral compositions to higher Cu, Pt, Pd, and Au, but very low Ni, Co, Or, Ir, Ru, and Rh (Fig. 4.64). These patterns are discussed in Naldrett et al. (2006) and shown to be a product of the fractionation of the primary sulfides by magmatic processes. A discussion of these models is given in Chapter 5.

Geological Relationships

Geological relationships in the Coleman-Levack mineral system are products of postimpact orogenesis (eg, the N–S trending Fecunis and Bob's Lake Faults; Fig. 4.55), synmagmatic structural control on footwall mineral emplacement (eg, the breccia zones associated with the footwall mineral zones in the 153, McCreedy West and Strathcona deposits), and synmagmatic localization of ore deposits along

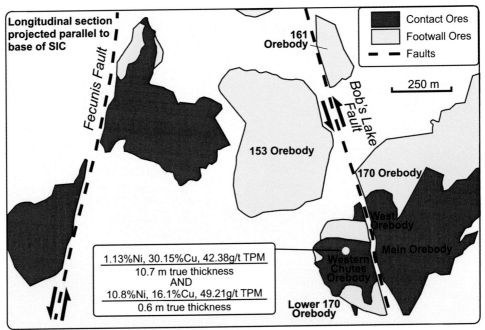

FIGURE 4.65 Geological Model for the 170 Contact and Footwall Orebodies at Coleman Mine

The displacement along the Bob's Lake Fault is responsible for the offset of the 170 Orebody. The determination of the slip direction and displacement was a key factor in the discovery of an extension to the footwall mineral zone to the west of the fault. Typical drill core results are indicated on the image from exploration work reported by Sandy Gibson.

The image is modified after Lightfoot (2015).

troughs and embayments at the base of the SIC in response to sulfide saturation, segregation, and localization of the magmatic sulfides.

The structures that offset the ore system and individual parts of orebodies are very important; the present understanding of these structures is summarized in Chapter 2 and illustrated in Fig. 4.55. The displacement on these faults is important as it creates secondary discontinuities; the sense and magnitude of displacement on these structures is important in the identification of spatially-detached zones of mineralization. The example shown in Fig. 4.65 illustrates how the kinematics of the Bob's Lake Fault helped geologists to identify a displaced down-plunge extension of the 170 Orebody that is located in the footwall of the SIC beneath the main contact deposit at Coleman.

The postimpact localization of mineralization in the footwall appears to coincide with the country rocks that are located near to the major embayment and trough structures described earlier in this chapter (Fig. 4.55A,B).

It is equally important to understand the significance of primary controls on sulfide mineralization, and this is illustrated in Fig. 4.66.

The contact sulfide mineralization associated with SLNR is rich in nickel (pyrrhotite + pentlandite > chalcopyrite) and the transitional style of contact mineralization associated GRBX and nearby

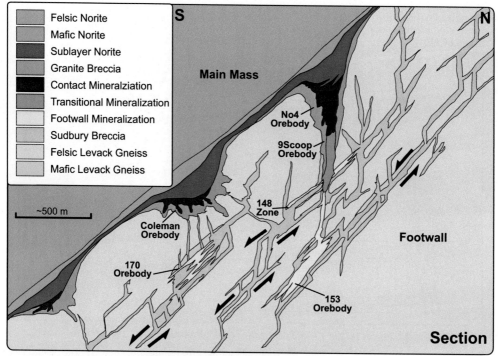

FIGURE 4.66 Geological Model in Section View Illustrating the Relationships Between Contact, Transitional, and Footwall Mineralization at Coleman

The footwall mineral zones cross-cut both Levack Gneiss and SUBX and they tend to follow the contacts between mafic and felsic gneisses.

The diagram was credited to Sandy Gibson and is published with acknowledgement in Lightfoot (2015).

footwall megabreccias is increasingly Cu- and PGE-rich (chalcopyrite + pentlandite > pyrrhotite; Souch et al., 1969) in the deeper part of the Levack Number 4 Deposit. Beneath the Levack Number 4 Deposit, the mineralization is located within heavily brecciated country rocks that contain the 9 Scoop Zone (Stewart and Lightfoot, 2010). The nearby footwall consists of Levack Gneiss and SUBX with veins consisting principally of chalcopyrite + pentlandite +/- millerite that belong to the 148 Zone (Fig. 4.66). With increasing depth along an arcuate zone of SUBX, the 148 Zone connects to the McCreedy East 153 Orebody that comprises assemblages of chalcopyrite + pentlandite, chalcopyrite + pentlandite + millerite, bornite + millerite, chalcopyrite + millerite, and native silver as described earlier in this chapter (Fig. 4.57A,B). The mineral zones are localized in association with a major concentric arc of SUBX, and they also tend to be localized close to the contacts between felsic gneiss and mafic gneiss (Fig. 4.66).

The relationship of the footwall mineral zones to SUBX and gneiss is very passive; there is minor hydrothermal alteration of the country rocks. The footwall rocks appear to be cut by syn-magmatic structures that created of space along which the fractionated magmatic sulfide mineralization could travel and localize in the footwall (Fig. 4.66). Although it is likely that hydrothermal processes contributed to the local alteration effects in the country rocks and transported metals into the economically minor domains of low sulfide-high precious metal mineralization (Farrow and Lightfoot, 2002; Farrow et al., 2005), it was the primary effect of sulfide differentiation and early emplacement of magmatic sulfide into the footwall that created the geological relationships in this mineral system (Naldrett et al., 1994b, 1999).

Victor, Nickel Rim South, and Capre

A cluster of economically significant deposit is located at or near the lower contact of the SIC in the East Range (Fig. 4.67A). The original discovery of mineralization in this area was made based on gossanous outcrops of Sublayer that still remain as the surface expressions of the Capre, Victor, and Nickel Rim South mineral systems.

The Victor Deposit was discovered in 1890 by John Leslie and George MacDonald and the domain close to surface was mined out by 1960. Extensive exploration in the last quarter of the 20th Century resulted in the discovery of the Victor Deep contact mineral zones, and the subsequent identification of associated mineralization in the footwall (Morrison et al., 1994). Exploration ceased for a number of years, but was picked up again in the first decade of the 21st century, and drill testing of geological, geochemical, and geophysical targets has identified new intervals of contact and footwall mineralization along the broad plunge of the contact mineral system (Lightfoot, 2015). The deposit remains undeveloped at this time.

The Capre contact deposit was discovered in 1912 by Tom Bennett and it remains undeveloped. Exploration work resumed in the early 21st century with a focus on the footwall environment as a target for exploration; three detached mineral zones were discovered in rapid sequence beneath the Capre contact deposit (Stewart and Lightfoot, 2010).

The near surface mineralization at Nickel Rim was identified and mined from 1953 to 1958. A deep drilling exploration program by Falconbridge resulted in the discovery of the Nickel Rim South Deposit in 2001. The Nickel Rim South Deposit was placed into production in 2010.

Geological Relationships

The geology of the Nickel Rim South mineral system is shown in Fig. 4.67B (after McLean et al., 2006). The base of the SIC has a small very localized embayment that hosts a thick sequence of SLNR and GRBX contained within a footwall comprising partially melted and metamorphosed granodiorite and felsic gneiss. The country rock consists of granodiorite gneiss with local development of mafic gneiss; SUBX is strongly developed within ~200 m of the base of the SIC. The underlying felsic gneiss with extensive development of SUBX that contains both transitional and footwall types of mineralization extend to almost 500 m below the base of the Main Mass (Fig. 4.67B). The contact mineral zone is contained in SLNR and GRBX, the transitional style of mineralization is contained in what McLean et al. (2005) show as SUBX, but these are more likely metabreccias developed after SUBX and country rock granodiorite. The footwall mineralization is hosted in SUBX and felsic gneiss.

The contact sulfides at Nickel Rim South comprise (pyrrhotite + pentlandite) > >chalcopyrite, the transitional mineralization as an elevated pentlandite + chalcopyrite content relative to pyrrhotite, and the footwall mineral zone consists of mineralization that is zoned from chalcopyrite + pentlandite at the core through chalcopyrite + pentlandite + millerite, to chalcopyrite + millerite, to chalcopyrite + bornite, and then native silver in association with fringe zones of mineralization that has a low sulfide content (Fig. 4.67B).

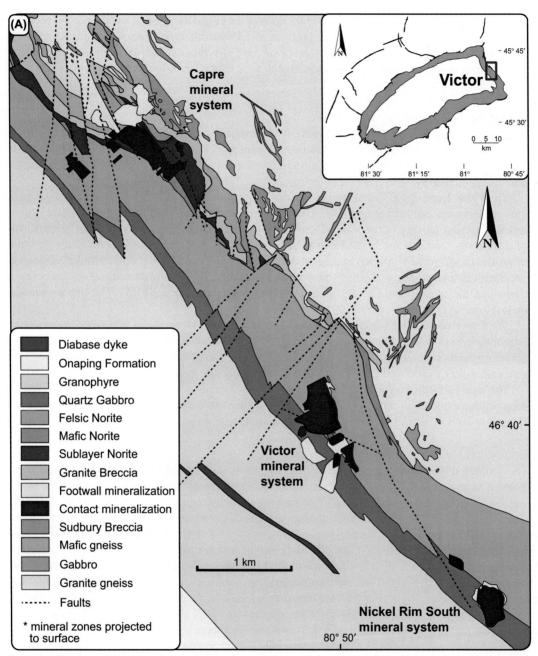

FIGURE 4.67

(A) Geological map of the East Range showing the location of the Nickel Rim South, Victor, and Capre mineral zones projected to surface.

(A) After Stewart and Lightfoot (2010).

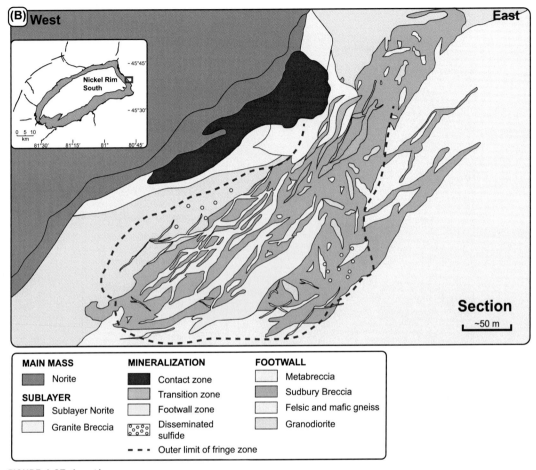

FIGURE 4.67 *(cont.)*

(B) Geological model for the Nickel Rim South Deposit located in the East Range of the SIC.

(B) Modified after McLean et al. (2005).

The Nickel Rim South discovery shares many features in common with the footwall styles of mineralization reported in Morrison et al. (1994) in the deeper parts of the adjacent Victor Deposit. The geology of the Victor part of the East Range mineral systems comprises contact, transition, and footwall style mineralization as illustrated in a synthesis of the geological relationships from Lightfoot (2015) in Fig. 4.68A. The lower part of the Main Mass of the SIC consists of MNR overlain by FNR, and the MNR is typically weakly mineralized with pyrrhotite + chalcopyrite + pentlandite up to 10 modal percent. The underlying Sublayer trough consists of both norite and GRBX, and these units have a very complex relationship that is the product of both primary intermingling of the magmatic noritic with the partially molten granitic breccias and postimpact deformation accompanying crater stabilization and orogenesis. Both the SLNR and the GRBX contain pyrrhotite + pentlandite + chalcopyrite mineralization in which the proportions of pyrrhotite + pentlandite is large in comparison to the amounts of chalcopyrite. The

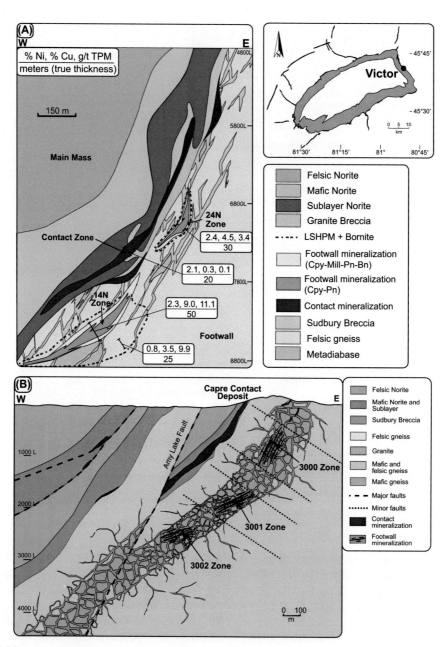

FIGURE 4.68

(A) Geological model for the Victor mineral system, showing a possible relationship between contact and footwall mineralization in section view. The model is based on drill core data, and the exact location and geometry of the connection between the contact ores and the footwall ores has not yet been fully established. Representative examples of drill results (true interval in meters) with grades or Ni, Cu, and 3E are shown. (B) Geological section showing the relationship of the Capre footwall mineral system to the contact mineralization.

(A) The original version of this diagram was prepared by Enrick Tremblay and it appears in Lightfoot (2015).

sulfides in the Sublayer consist of disseminated, inclusion-rich massive sulfide, and semimassive sulfides in which the pyrrhotite crystals often contain rims of granular pentlandite associated with chalcopyrite.

The exact locations of the transition zones between contact and footwall mineralization are not yet firmly established by drilling (Lightfoot, 2015), but there appear to be connections between the contact and the footwall mineralization which comprise veins of chalcopyrite + pentlandite that extend through zones of extensively melted and metamorphosed metabreccia derived from footwall granodiorite to felsic gneiss and SUBX. These zones connect to the footwall mineralization that consists of chalcopyrite + pentlandite that grades into chalcopyrite + pentlandite + millerite, through bornite + chalcopyrite + millerite, to bornite + chalcopyrite, and then into veins containing native silver. The apogee of development of this style of mineralization is in the deepest known footwall zones (14 North and 28 North Zones), but the mid-level footwall mineralization (24 North Zone) also contains a significant amount of chalcopyrite + pentlandite and local development of millerite (Fig. 4.68A; Lightfoot, 2015).

The Capre contact deposit is located in a faulted slice of the contact of the SIC where a deep embayment structure is developed (Fig. 4.68B). The contact mineralization consists of mineralization with pyrrhotite + pentlandite > chalcopyrite. The underlying country rocks comprise extensive metabreccias developed after felsic and intermediate gneisses, gabbro-diabase, and SUBX with localized stringers of chalcopyrite + millerite. The Capre footwall mineralization has no continuous connection to the contact mineral zones. It is located in a series of three mineral zones comprising massive vein style chalcopyrite + pentlandite and chalcopyrite + pentlandite + millerite within partially melted and metamorphosed footwall rocks (Fig. 4.68B; Stewart and Lightfoot, 2010).

Characteristics of the Sulfide Mineralization

The sulfide mineralization developed in the Victor mineral zones is illustrated in Fig. 4.69A–H where the north-facing section illustrates the geological context of the samples (Fig. 4.69A), and the photographs show the style of mineralization developed in drill core. The contact sulfides range from disseminated through inclusion-rich to the semimassive sulfides shown in Fig. 4.69B; the mineralization exhibits a similar proportions of pyrrhotite:pentlandite:chalcopyrite irrespective of texture or sulfide content, and is typical of the contact zone where the Ni:Cu ratio is 2:1 or higher.

The transition zone style of mineralization consists of intergrowths of chalcopyrite + pentlandite with lesser pyrrhotite (Fig. 4.69C) in veins that appear to connect into the footwall mineralization that consists of a core zone of chalcopyrite + cubanite + pentlandite (Fig. 4.69D) surrounded by veins of millerite + pentlandite and chalcopyrite + bornite (Fig. 4.69E,F). The peripheral part of the mineral zone consists of rare veins of chalcopyrite + bornite that show the development of quench-textured chalcopyrite laths in a matrix of bornite (Fig. 4.69G). The outermost fringe zone of mineralization consists of bornite veins (Fig 4.69H) surrounded by chalcopyrite + bornite blebs and disseminations that are hosted in metamorphosed and partially melted mafic and felsic gneiss. The host rock gneisses lack development of continuous banding and shows clear examples of felsic leucosomes created by partial melting, breccias developed by disruption, and thermal metamorphism.

The Sulfide mineralization developed at Nickel Rim South is essentially the same type as that developed at Victor (Figs. 4.67B and 4.68A). The zonation at Capre is similar in the contact and parts of the footwall, but a bornite fringe is not developed.

Geochemistry

The Nickel Rim South Deposit is described as a zoned system comprising a sequence of sulfides produced by fractionation processes that relate the Fe–Ni-rich contact mineralization to the Cu–Ni–PGE-rich

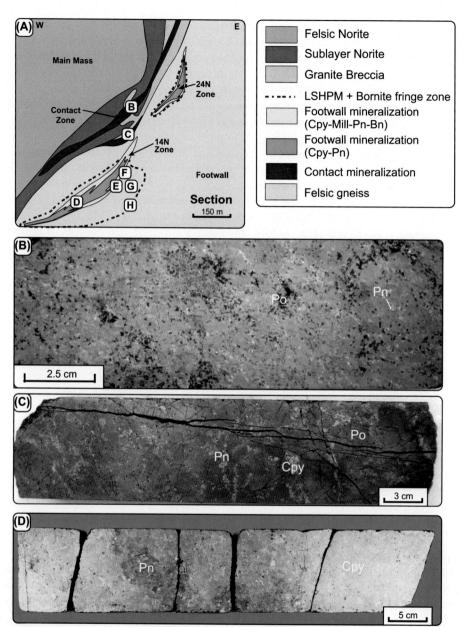

FIGURE 4.69 Diversity in Styles of Mineralization in the Victor Contact and Footwall Mineral System in Section

(A) Geological model of the mineral zones showing the location of the images. (B) Massive pyrrhotite >>
pentlandite ≥ chalcopyrite from the contact zone at Victor (2.1 wt.% Ni, 0.3 wt.% Cu, 0.1 g/t 3E over 20 m true
width). (C) Massive transitional type pyrrhotite > chalcopyrite > pentlandite from the contact zone at Victor
(2.4 wt.% Ni, 4.5 wt.% Cu, 3.4 g/t 3E over 30 m true width). (D) Massive footwall chalcopyrite + pentlandite
mineralization from the footwall zone at Victor 14N Orebody (drill core 811759; 2512.3–2514.2, which hosts the
interval shown, contains 9.1 wt.% Ni, 22.7 wt.% Cu, 46.62 g/t 3E).

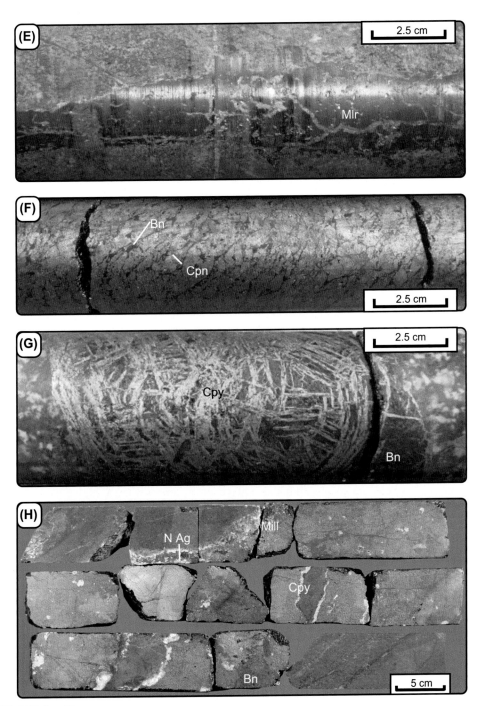

FIGURE 4.69 (*cont.*)

(E) Peripheral style of millerite > pentlandite > chalcopyrite mineralization from the 24 North Zone at Victor (0.8 wt.% Ni, 3.5 wt.% Cu, 9.9 g/t 3E over 25 m true width). (F) Peripheral massive sulfide zone at Victor with (chalcopyrite + cubanite) > bornite. (G) Vein from the peripheral zone at Victor with blades of chalcopyrite in a matrix of chalcopyrite; this texture appears to be one of rapid quenching of a sulfide magma in a narrow vein cutting partially melted Levack Gneiss. (H) Fringe zone of massive bornite veins with lesser chalcopyrite associated with carbonate veins that contain native silver (drill core 811759 at 2653 m depth; 1.59 wt.% Ni, 60.5 wt.% Cu, 5.8 g/t 3E).

footwall mineralization (Fig. 4.67B). Overall, the sequence exhibits an increase in metal tenor of sulfide with distance away from the base of the SIC. The highest grades of copper, nickel, platinum and palladium occur in the lowermost portion of the central and thickest part of the footwall mineralization (McLean et al., 2005). Ni and Cu grades increase within the transitional and upper portions of the Footwall zone, and PGE grades show a dramatic increase from the contact, through transitional zones and attain exceptionally high values within the core of the footwall zone (McLean et al., 2005). Mineralogically the higher grades of nickel and copper can be attributed directly to the changing sulfide mineral assemblage through this fractionated suite. Lower base metal contents typify areas dominated by high Fe–S mineralogy (pyrrhotite, pentlandite, chalcopyrite and cubanite), with higher base metal abundances corresponding to low Fe–S mineralogy (chalcopyrite, pentlandite, millerite and bornite).

The contact and footwall mineral zones at Victor comprise sulfide with a wide range in the proportions of sulfide minerals that mirrors the range found at Nickel Rim South. The ratio of Cu/(Cu + Ni) and (Cu + Ni)/S for the Victor mineral system is illustrated in the data density plot shown in Fig. 4.70A. Contact mineralization falls along the array linking sulfides with pyrrhotite + pentlandite to those with elevated chalcopyrite content, but the bulk of the samples on this array fall at low Cu/(Cu + Ni) relative to average Sudbury mineralization, and were likely formed by the removal of Cu-rich sulfide melts into the footwall below the embayment structure.

The footwall mineralization falls along mixing arrays displaced to high Cu/(Cu + Ni) between chalcopyrite + (pentlandite + /-millerite) and bornite + chalcopyrite + millerite (Fig. 4.70A). Relative to average Victor sulfide mineralization, the contact sulfides have very low Cu/(Cu + Ni) irrespective of whether they are hosted in SLNR or GRBX (Fig. 4.71A; Table 4.5). In contrast, sulfides hosted in the footwall gneiss or SUBX have a very high Cu/(Cu + Ni) relative to the contact mineralization (Fig. 4.71B). Massive and semimassive sulfides have compositions that peak at both Ni-rich and Cu-rich compositions, and these sulfides belong to veins in both the contact and footwall environments (Fig. 4.71B). Variations in (Cu + Ni)/S in the same rock groups are shown in Fig. 4.71C,D, where the lower metal tenor sulfides are developed in the Sublayer environment whereas higher tenor mineralization is present in the Footwall.

The compositional diversity in the contact and footwall mineral zones at Capre is shown in Fig. 4.70B. The sulfides do not show a clear transitional assemblage, and the footwall mineralization is devoid of the bornite-silver associations that are found in the footwall zones at Victor and Nickel Rim South.

Similarities and Differences between the East Range Mineral Systems

The geological, mineralogical, and geochemical relationships common to all three mineral systems are summarized below:

1. Sublayer embayment structures containing SLNR and GRBX contain the contact sulfide mineralization. The Nickel Rim embayment is a localized discrete feature of the base of the SIC. The Victor Trough consists of multiple embayment structures along a trough structure that extends from surface to depth. The Capre embayment is wide and deep, but is detached from the rest of the SIC in a fault slice

2. The contact-associated mineralization in all three mineral systems is Fe-Ni-rich with pyrrhotite + pentlandite > chalcopyrite. The Ni tenors of the (pyrrhotite + pentlandite)-rich sulfides vary between the deposits with much of the Victor mineral system having slightly lower Ni tenor than the contact system at Nickel Rim South (based on $[Ni]_{100} = 4.6$ wt.%

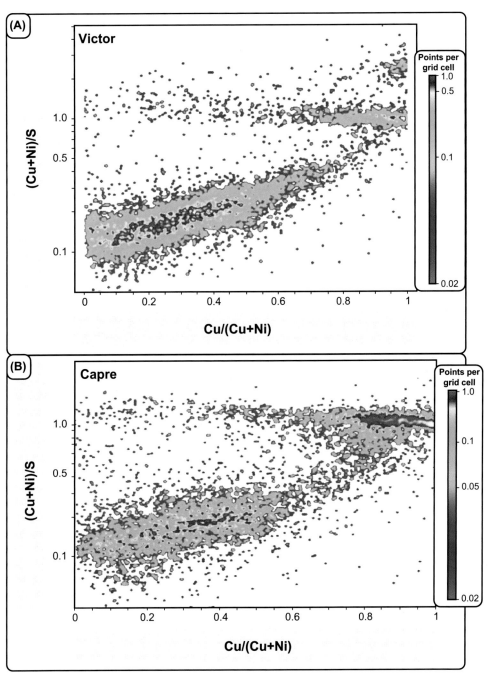

FIGURE 4.70
(A) Data density plot showing the variation in Cu/(Cu + Ni) versus (Cu + Ni)/S in Vale's Victor mineral system.
(B) Data density plot showing the variation in Cu/(Cu + Ni) versus (Cu + Ni)/S in Vale's Capre mineral system.

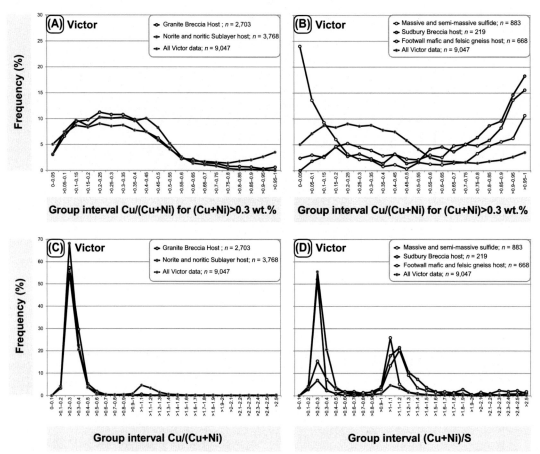

FIGURE 4.71 Percentage Grouped Frequency Histogram Showing Variations in Cu/(Cu + Ni) and (Cu + Ni)/S for Vale's Victor Mineral System, and Variations in the Contact Versus the Footwall

The data shown are for samples with S ≥ 1 wt.% and (Ni + Cu) ≥ 0.3 wt.%. (A) Variation in Cu/(Cu + Ni) for all samples from Victor mineral system, samples of mineralization hosted in SLNR and norite, and samples hosted in GRBX. (B) Variation in Cu/(Cu + Ni) for all samples from Victor mineral system, samples of mineralization hosted in footwall mafic and felsic gneiss (including mafic enclaves), and SUBX; also shown is the average for all massive and semimassive sulfides (includes both contact and footwall mineralization). (C) Variation in (Cu + Ni)/S for all samples from Victor mineral system, samples of mineralization hosted in SLNR and norite, and samples hosted in GRBX. (D) Variation in (Cu + Ni)/S for all samples from Victor mineral system, samples of mineralization hosted in footwall mafic and felsic gneiss (including mafic enclaves), and SUBX; also shown is the average for all massive and semimassive sulfides (includes both contact and footwall mineralization).

for Ni-rich contact mineralization shown in Fig. 4.70A (Table 4.5) and an average $[Ni]_{100}$ of 6.1wt.% is calculated from data presented in McLean et al., 2005). The Capre Fe-Ni-rich contact mineralization has $[Ni]_{100} = 4.4$ wt.% (Table 4.5).

3. Cu- and PGE-rich transitional styles of sulfide mineralization are developed in metabreccias after felsic and intermediate gneiss, granitoid rocks, and SUBX. Transition style mineralization is developed at Victor and Nickel Rim South. These zones are contiguous with the contact style of mineralization. The erosion level and structural complexity at Capre makes the identification of the transition zone less clear-cut, but there is a weak tendency for the near-surface mineralization to trend to higher Cu/(Cu + Ni) at the northern end of the contact, and there are intervals of narrow-vein mineralization with chalcopyrite + pentlandite + millerite in this segment of the footwall.

4. The host rocks to all three mineral zones comprise mafic, intermediate and felsic gneiss, granitoid rocks, diabase-gabbro bodies, and SUBX. The transition zone mineralization appears to be hosted by metamorphosed, recrystallized and partially melted rocks that are grouped as metabreccias. The footwall zones tend to form vein sets that cross-cut the contacts of SUBX and country rocks, so they clearly postdate the crystallization of the SUBX. There is a slight tendency for the mineralization at Capre and some of the mineralization at Victor to follow thick zones of SUBX developed between mafic, intermediate and felsic gneisses or between gneisses and diabase-gabbro bodies.

5. With distance from the contact, the Victor, Nickel Rim South and Capre deposits broadly change sulfide assemblage from pyrrhotite + pentlandite > chalcopyrite through pyrrhotite + chalcopyrite + pentlandite to chalcopyrite + pentlandite in the contact and transition zones. In the footwall, there is a change from chalcopyrite + pentlandite to chalcopyrite + pentlandite + millerite, and chalcopyrite + millerite. At Victor and Nickel Rim South, chalcopyrite + bornite + millerite, bornite + millerite, bornite + native Ag, and halo domains with high precious metal concentrations in very weakly mineralized rock (LSHPM zones) are developed at the periphery of the footwall orebody. At Capre, no bornite has been identified (Stewart and Lightfoot, 2010).

The Whistle-Podolsky mineral system

The Whistle Deposit has been studied in great detail due to the access provided by an open pit development through much of the period over which the deposit was mined (Lightfoot et al., 1997b,c; Zhou et al., 1997; Prevec et al., 2000; Farrell, 1997; Farrell et al., 1993, 1995). The geology of the Sublayer and the rocks that contain the mineral zone are documented in Chapter 2 (Figs. 2.26–2.28). The contact and footwall mineralization of the Whistle-Podolsky mineral system cropped out at surface during the mining activity, but the immediate footwall was covered by piles of waste rock until rehabilitation of the property and further exploration by FNX uncovered many of the world-class outcrops that remain to this day at Whistle. The footwall mineralization is developed in a zone of metabreccia and QD along strike from the Whistle Mine in the Whistle Offset; the geological relationship between massive sulfide and country rocks is illustrated in Fig. 4.72A (after Carter et al., 2005), and the generalized relationships in a section through the Podolsky Deposit are shown in Fig. 4.72B (after Farrow et al., 2005).

The disseminated sulfides in the SLNR have elevated (pyrrhotite + pentlandite):chalcopyrite, and they comprise weakly blebby sulfides that exhibit a spongy texture between the sulfide and the marginal silicate rock as shown in Fig. 4.73A. The disseminated sulfides grade into inclusion-rich sulfide,

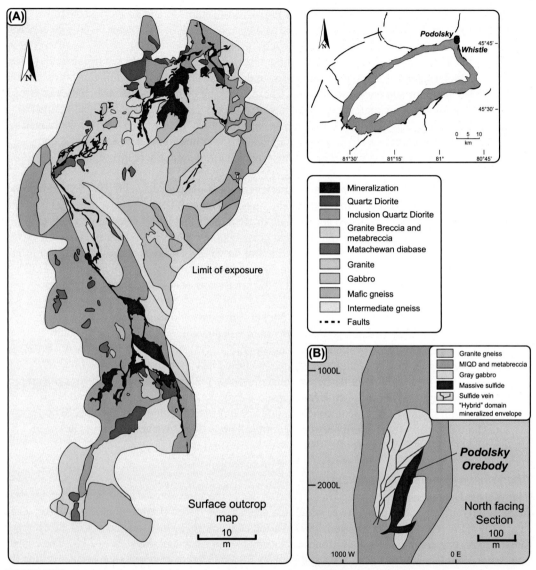

FIGURE 4.72

(A) Plan showing geological relationships in the North Zone at Whistle-Podolsky. (B) Geological section showing relationships in a near-surface footwall zone at Whistle Mine

(A) Modified after Carter et al. (2005). (B) Modified after Farrow et al. (2005).

and the inclusions often comprise fragments that appear to be derived from the Sublayer as described in Chapter 2 and 3 (Fig. 4.73B). Tantalizing developments of chalcopyrite blebs in granitoid rocks beneath the Whistle were identified during the mining activity (Fig. 4.73C; Lightfoot et al., 1997c), and drilling of the footwall identified intervals of chalcopyrite + pentlandite mineralization as veins in the footwall. Subsequent to the divestment of the property by Inco to FNX, the Podolsky Deposit

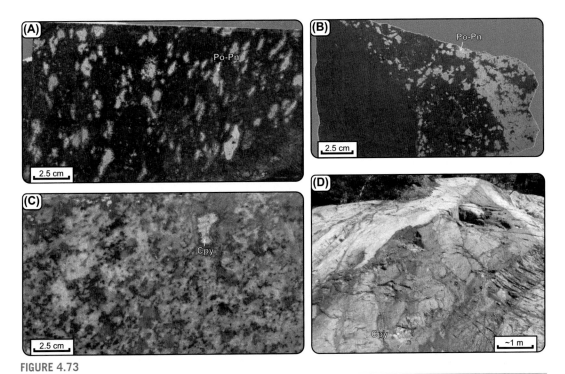

FIGURE 4.73

(A) Disseminated to blebby pyrrhotite + pentlandite mineralization hosted in SLNR from Whistle Mine. (B) Inclusion-rich massive sulfide from the contact Sublayer at Whistle Mine (sample RX355675 contains 0.65 wt.% Ni, 0.09 wt.% Cu and 0.1 g/t 3E). (C) Disseminations of chalcopyrite in Levack Gneiss directly adjacent to the Whistle Open Pit (this sample contains 0.06 wt.% Ni, 0.9 wt.% Cu, and 0.2 g/t 3E; see Lightfoot et al., 1997c). (D) Massive chalcopyrite veins along strike from Whistle Mine in the metabreccias of the footwall. The massive sulfides appear to occupy tensional features created by transtension.

was discovered beneath these early intersections of vein mineralization, and the North Zone shown in Fig. 4.72A was identified by surface prospecting along the Offset dyke and stripping of the outcrop. The resultant exposure provides excellent examples of massive chalcopyrite + pentlandite + millerite mineralization occupying arcuate veins (possibly tension gashes) between mineralized fractures in the metabreccia and QD (Fig. 4.73D).

The composition of the sulfide from the Whistle Deposit is shown as averages of the contact mineral system in Fig. 4.74A,B. The strongly skewed variations in Cu/(Cu + Ni) attest to the strong Ni-enrichment of the contact ores, and this has long been recognized as a signature of a contact deposit from which fractionated sulfide has been removed. The variations in Ni tenor of sulfide of sulfide peak at ~3.5 wt.% Ni as shown in Fig. 4.74B, and the Ni-poor composition of these sulfides relative to many of the other contact orebodies at Sudbury resulted in the development of the deposit at a time when additional feed to the Sudbury infrastructure at Inco was required.

The Trillabelle mineral system

Two small subeconomic concentrations of Ni-Cu sulfide mineralization occur in the Trillabelle embayment (Fig. 4.75). The near surface deposit is hosted in SLNR on the southern flank of the trough which

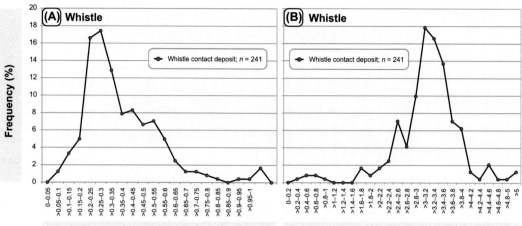

FIGURE 4.74

(A) Variations in percentage grouped frequency of Cu/(Cu + Ni) in contact mineralization from Whistle Mine.

(B) Variations in percentage grouped frequency of normative $[Pn]_{100}$ in contact mineralization from Whistle Mine.

plunges at 30 degrees toward the NE and connects to the Trillabelle Deep mineral zone. The embayment structure is at least as large as that developed at Creighton where some of the largest Sudbury orebodies are developed in association with the SLNR, but Trillabelle has no equivalent concentrations of metal in terms of either grade or tonnage. The range in Cu/(Cu + Ni) at Trillabelle is less skewed than that from the contact mineral system at Whistle, so there is no clear evidence in the observed sulfide chemistry for the segregation and removal of Cu-rich liquid from the contact ores (Fig. 4.76A). In terms of metal tenor, the mineralization at Trillabelle is quite similar to Whistle with $[Ni]_{100}$ between 2.5 and 4.5 (Fig. 4.76B), which is comparable to the lowest tenor style of mineralization at surface near to Creighton (eg, Gertrude and Gertrude West).

Trillabelle provides a good example of two small lenses of subeconomic mineralization associated with Sublayer that fills a trough that approaches the dimensions of the Main Creighton Trough. An important question addressed in Chapter 5 is why some volumetrically large domains of Sublayer in large troughs have world-class ore deposits yet others fail to pass the hurdle for a small economic mineral zone.

SYNTHESIS OF GEOLOGICAL RELATIONSHIPS IN THE NORTH RANGE MINERAL SYSTEMS

The geological relationships with increasing depth through an embayment or trough are summarized below. The sequence of rock types is summarized from the base of the Main Mass into the footwall of the SIC following the schematic relationships at Coleman (Fig. 4.66) and Victor-Nickel Rim South (Figs. 4.67B and 4.68; Chapter 2 described the geology of the host rocks in detail).

1. The Norite represents the most basal unit of the inclusion-free stratigraphy of the Main Mass. Above the embayment structures it tends to comprise a thick unit of MNR overlain by the Lower

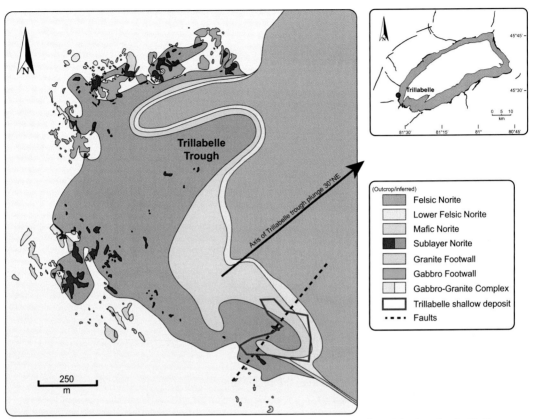

FIGURE 4.75 Outcrop Geology of the Trillabelle Embayment Showing the Location of the Low-Grade Shallow Trillabelle Deposit

There is a discontinuous unit of MNR at the base of the SIC.

FNR; both of these rock types are mineralized with disseminated (trace-10%) sulfide, and the MNR contains common anorthositic clots. This sequence of rocks thins laterally away from the embayments in areas of barren contact.

2. Noritic Sublayer occupies embayment depressions at the base of the melt sheet, and this unit forms a discontinuous inclusion- and sulfide-bearing basal contact unit of the SIC in embayments and troughs. The initial impact melt sheet had an irregular floor, and the depressions along the crater floor localized both immiscible magmatic sulfides and mafic-ultramafic inclusions. The Sublayer has a complex relationship with the MNR as shown in Fig. 4.59, and the two units evolved by a complex interaction during cooling to generate a hybrid-xenolithic unit of Sublayer that comprises a finer-grained matrix of norite with coarser-grained enclaves of broken-up MNR in addition to country rocks fragments. The hybrid-xenolithic norite grades into inclusion-bearing sulfide, and then it transitions into semimassive sulfide.

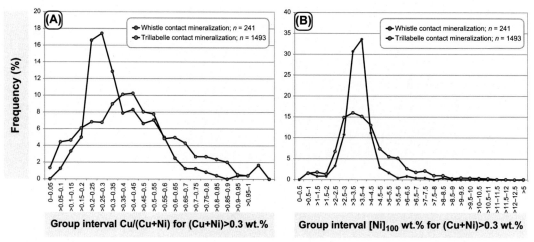

FIGURE 4.76

(A) Variation in Cu/(Cu + Ni) in the Trillabelle Deposits; the average composition of contact sulfides from the Whistle contact deposit is also shown. (B) Variation in [Ni]$_{100}$ (wt.%) in the Trillabelle Deposits; the average composition of contact sulfides from the Whistle contact deposit is shown for comparison.

3. Toward the lower margin of the Sublayer the norite tends to contain small mafic-ultramafic inclusions as well as felsic inclusions in a leucocratic norite matrix. This rock type grades over a few meters into GRBX that consists of a granitic matrix with inclusions of mafic-ultramafic rock, country rock, disseminated and inclusion-bearing massive sulfide, fragments of SUBX and often fragments of pyrrhotite + pentlandite-rich sulfide. The sulfides associated with the noritic Sublayer extend into the GRBX where they comprise inclusion-rich to inclusion-bearing massive sulfide, and they tend to become more chalcopyrite + pentlandite-rich with increasing distance above the base of the SIC.

4. The footwall that is located proximal to the base of the Sublayer in the embayment or trough consists of partially melted blocks of country rock separated by leucosomes and GRBX. With increasing distance from the base of the embayment or trough, the blocks are separated from each other by domains of GRBX, felsic partial melt segregations, and then metabreccia. With greater distance into the footwall, the megabreccia fragments merge into the country rock as the degree of partial melting and metamorphism declines. The metabreccias often carry some stringer mineralization and they appear to provide the pathways along which sulfide melts migrated from the contact into the footwall.

5. Footwall gneiss is identified where the banding and partial melt textures can be ascribed to processes that pre-date the SIC event in the ancient deep crustal Levack Gneiss Complex. These gneisses often contain arcuate domains of SUBX which can be up to a hundred meters wide, and extend beneath the base of the SIC. The breccias provided a favorable pathway along which Cu-rich sulfide melts migrated from the Sublayer into the footwall (Fig. 4.66).

PROCESS OF FORMATION OF THE SUDBURY MINERAL SYSTEMS

The generation of the contact and footwall sulfide mineralization, and it's localization in the embayment and trough structures at the base of the SIC can be explained as shown in a model depicted in Figs. 4.77A–E.

The early development of the Sublayer in a trough or embayment at the base of the melt sheet is depicted in Fig. 4.77A. The base of the SIC likely comprised a melt with the initial composition close to that of QD. This sulfide-undersaturated melt was injected after the cooling of SUBX into radial and concentric fractures created by the lithostatic relaxation of the crust after impact. These crystallized to form the Offset Dykes (Fig. 3.33A–F).

The melt sheet was probably superheated for some time after it's formation, and underwent rapid assimilation of mafic-ultramafic fragments that either remained in the melt-column or were assimilated from beneath the melt sheet (Keays and Lightfoot, 2004; Fig. 4.77A). Eventual settling and partial assimilation of these fragments and cognate xenoliths contributed to the broadly noritic composition of the lower part of the melt sheet.

The melt sheet achieved sulfur saturation early in its evolution before any silicate mineral crystallization could hinder the localization of sulfides segregating from the magma column. This saturation event also postdated the injection of melt to form the QD Unit of the Offset Dykes (not shown in Fig. 4.77A–D). The magmatic sulfides generated from the melt sheet by sulfur saturation were localized toward the base of the melt column by gravitational force to concentrate into depressions at the base of the melt sheet where they were injected as immiscible sulfides together with QD and inclusions to form the mineral systems in the Offset Dykes that underwent quite rapid cooling. In the absence of an Offset, the sulfides continued to accumulate toward the base of the melt sheet and equilibrate with mafic-ultramafic inclusions to form a xeno-melt as shown in Fig. 4.77A.

Segregation and accumulation of sulfide continued until the silicate melt reached its liquidus temperature. At this point the effect of crystallization was as important, or more important than the process of assimilation of mafic-ultramafic inclusions. The development of the basal stratigraphy of MNR was initiated at this point, but the crater remained unstable, and portions of partially crystallized MNR were incorporated whilst the noritic Sublayer melt continued to melt the underlying felsic basement rocks to initiate the formation of the GRBX Unit of the Sublayer. This complex process resulted in the redistribution of sulfide melt, with a tendency for the dense sulfides to settle toward the base of the Sublayer sequence as shown in Fig. 4.77B.

The noritic Sublayer partially melted nearby footwall rocks to form the GRBX which underwent local hybridization with the SLNR to form a leucocratic variant of the SLNR. The adjustment of the crater floor and disruption of the contact between the SLNR and the MNR resulted in the generation of hybridized norites within the Sublayer at the base of the Main Mass as shown in Fig. 4.77C. Partial melting of the footwall to form metabreccia was accompanied by the crystallization of monosulfide solid solution from the sulfide melt; the monosulfide solid solution would eventually exsolve to form the pyrrhotite + pentlandite + chalcopyrite that comprise a major component of the contact mineralization. This process was accompanied by progressive deepening of the embayment by crater subsidence and slumping together with melting of the adjacent footwall to generate pathways that would soon be exploited by residual sulfide melts.

The footwall mineralization represents a relatively small proportion of the original sulfide melt; possibly as small as 10% of the massive sulfide pooled in the trough. The fractionated sulfide melt

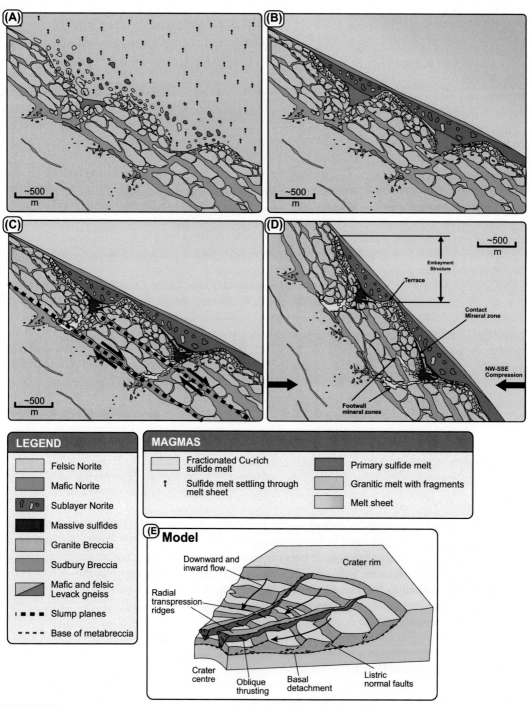

FIGURE 4.77

migrated into the country rocks along permeable zones of country rock that underwent partial melting, and from there into the underlying stratigraphy through pathways created by impact readjustment along listric fault zones that followed weaknesses in the crater wall created by contrasting country rock types and associated zones of SUBX. The residual Cu-rich melts that were expelled downward during the formation of monosulfide solid solution were transported from the contact environment along open space pathways of least resistance into the footwall, where these sulfides then crystallized to form the footwall associations shown in Fig. 4.77D.

The mineral zones developed in the footwall of the North Range contact ore deposits include McCreedy West, McCreedy East 153, Coleman 170, Strathcona, Podolsky, Capre 3000, 3001, and 3002, Victor 24N, 14N, and 28N, and Nickel Rim South. Of these systems, the extent of fractionation varies from chalcopyrite + pentlandite +/- millerite (Podolsky, Victor 24N, Strathcona) through chalcopyrite + pentlandite + millerite (Capre 3000, 3001, 3002; Coleman 170) to an apogee of chalcopyrite + pentlandite + millerite-bornite (McCreedy West, McCreedy East 153, Victor 14N and 28N, and Nickel Rim South). Different mineral systems appear to show different degrees of evolution, and there is no requirement that all footwall mineral systems evolve all the way to bornite and native silver.

The location of the footwall mineralization is spatially related to the contact mineralization, and even if minor alteration has remobilized metals, the variations are essentially the product of differentiation of a sulfide melt undergoing cooling and ongoing redistribution (Naldrett et al., 1994a, 1999). The development of mineralization in the footwall is ascribed to syn-magmatic modification of the crater wall as illustrated in Fig. 4.77D. The adjustment of the crater walls by a process of slumping has long been recognized as a potential mechanism by which space is created in the footwall adjacent to the SIC; a model that described this process is shown in Fig. 4.77E. The slump planes provide a locus of open space along which differentiated sulfide melts migrated. The overlying melt sheet modified the topography of the crater floor by thermal erosion to remove pronounced terrace features and radial trenspression ridges.

◀ FIGURE 4.77 A Geological Model in Section View for the Development of Contact Ore Deposits at Sudbury

(A) Immediately postimpact, the strongly brecciated country rocks form an intact container for the melt sheet. The melt sheet is superheated and continues to assimilate fragments of mafic country rocks. Sulfide saturation is achieved and concentrations of sulfide melt are localized in embayment and trough structures at the base of the melt sheet. Sulfide melt continues to form from the overlying magma. (B) Crystallization of the Sublayer and MNR commences, but the remainder of the melt sheet has not yet started to crystallize. The massive sulfide has ponded in the embayment structure and monosulfide solid solution begins to segregate. (C) The crater walls cave into the melt sheet along fault structures into which the fractionated Cu-PGE-rich sulfides are expelled after crystallization of the contact sulfide as a monosulfide solid solution. (D) The sulfide melt migrates to the base of the terrace. Fractionation and migration of the sulfides takes place into the footwall along zones of SUBX to create an Orebody that is zoned from chalcopyrite + pentlandite through chalcopyrite + pentlandite + millerite, chalcopyrite + bornite + millerite, bornite + millerite, bornite + chalcopyrite, to native silver associated with carbonate. (E) Illustration of the effects of slip on the crater wall that occurs as part of the crater modification process within <1 Ma of the impact event. Note that the same slip relationships are developed around the central peak-uplift structure in more complex craters, and these crater-wall collapse features will be associated with trans-tensional graben structures rather than the transpressional ridges that are shown in this figure.

Extensively modified after Morrison et al. (1994). (E) Modified from Kenkmann and Dalwigk (2000) with the permission of John Wiley and Sons.

The range in sulfide assemblages in the Victor, Nickel Rim South, and Capre mineral systems is similar to that developed in the McCreedy West, Coleman 170, McCreedy East 153, and Strathcona ore deposits. In detail, some of these mineral systems contain bornite and native silver (eg, McCreedy East 153; Victor 14N), whereas others appear not to produce such evolved sulfide types (eg, Podolsky, Capre, Strathcona). The process controls governing the compositional diversity in each mineral system were clearly quite similar, but local variations appear to have influenced the extent of differentiation of the sulfide melt. This observation is important as there are many examples of South Range mineral systems that exhibit unusually Cu-PGE-rich sulfide assemblages in footwall rocks that were possibly formed by similar processes, but under conditions where differentiation was unable to achieve the levels seen in the North Range mineral systems.

The extent of alteration of the footwall in proximity to the contained mineralization is quite minor. Extreme differentiation of the sulfide melt to form stringers of bornite and native silver in quartz-carbonate veins is possibly a product of syn- and postmagmatic processes, and it may relate to the emplacement of young dykes that cross-cut the mineral zones. The process controls may be the following:

1. Migration of the sulfide melt along crystal boundaries during low degrees of partial melting of the country rocks.
2. An increasing load of volatiles within the sulfide melt as fractionation proceeds.
3. Wide-scale alteration due to interaction of primary magmatic sulfides with externally derived hydrothermal fluids to generate new mineral systems in the footwall.
4. Redistribution of sulfide during postmagmatic deformation and the emplacement of mafic dykes.

Differentiation of sulfide melt is also recognized in mineral systems hosted by the Offset Dykes. The process is well-developed in the Frood mineral system where the ores becomes progressively more Cu-rich and PGE-rich with depth. Features of the deep parts of the Frood system are evident in the Vermilion and New Victoria mineral zones (Chapter 2). Individual orebodies along the Copper Cliff and Worthington Offsets also show enrichment in Cu/(Cu + Ni) with depth that is consistent with progressive differentiation of the sulfide melt.

ROLE OF SULFIDE KINESIS DURING DEFORMATION

As described earlier in this chapter, the term "kinesis" is used to describe the redistribution of sulfide and the fractionation of sulfide minerals during the reformation of sulfides accompanying metamorphism and deformation (Lightfoot and Evans-Lamswood, 2015). The process of deformation involves mechanical erosion of the country rocks and incorporation of inclusions from this source into the sulfide by processes of sulfide infiltration into the country rock, plucking of fragments of country rock into the sulfide, and attrition of the fragments. This process generates mineralization that has been described as "durchbewegung" (Vokes, 1968, 1969, 1973).

The mineral zones developed in the footwall of the South Range comprise examples of sulfide mineralization in the footwall with (chalcopyrite + pentlandite) > pyrrhotite, and these mineral zones tend to be associated with contact mineralization that has pyrrhotite > (chalcopyrite + pentlandite). The assemblages of rocks in the footwall vary in degree of deformation between two extremes:

1. Mineralization that appears relatively unmodified by faults or shear zones: The sulfides typically reside in metabreccia, SUBX, and country rocks. There are few examples of the styles of inclusion-rich sulfide that result from the kinesis of sulfide into shear zones and structures. Rather, the sulfides appear to occupy primary openings in the footwall that were created by crater wall

re-adjustment. Examples of these types of mineral domains include the Creighton 126,403, 461, and 320 Orebodies, the McKim and Murray pentlandite + chalcopyrite breccia zones, the Little Stobie Number 2 Zone, and the Lindsley Number 4 Zones.

2. Structurally modified zones of contact mineralization that have locally been translocated into the footwall by sulfide kinesis along faults and shear zones. The massive sulfides normally contain angular fragments of schist derived from the walls of the faults. The Garson shear zone-hosted mineralization is the best example, but the effects of kinesis of primary sulfides into structures are also evident in the Crean Hill, Frood, Falconbridge and Cryderman mineral systems where the sulfide ores have been termed "sulfide schists."

The effect of regional scale postmagmatic translocation of the entire mineral system by fault structures is important. Two examples are shown schematically in Fig. 4.78A,B based on the

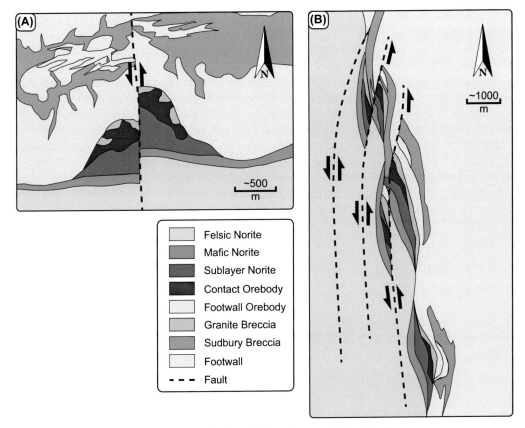

FIGURE 4.78 The Principal Controls of Late Faults and Shear Zones on Mineralization

(A) Plan showing the morphology of an embayment structure after N–S transcurrent faulting has offset contacts and mineral zones. This relationship is developed in the North Range as a result of motion on the family of faults that include Fecunis and Bob's Lake. (B) Plan illustrating the confounding effect of north-south transcurrent faulting with dip-slip motion in the East Range. The displacement of contacts and mineral zones requires a careful understanding of the direction and magnitude of displacement along faults.

Levack-Coleman and East Range mineral systems, and a third example from Garson is illustrated in Fig. 4.48A–I.

At Coleman, the Levack Number 4 Orebody resides in an embayment that is dissected by a major N–S transcurrent fault (the Fecunis Fault). This fault is one of several that offsets not only the massive contact ore deposits in the Sublayer, but also the underlying stratigraphy of country rocks that contain the footwall orebodies. By reconstruction of the displacement along these faults it is possible to identify potential continuations of zones of mineralization in the footwall (Fig. 4.78A).

Faulting can also occur subparallel to the basal contact of the SIC, and a good example of the structural complexity caused by faulting is illustrated in the East Range by the orebodies at Capre. Faulting subparallel to the contact can interrupt the continuity of contact mineral zones, detach footwall zones, and produce unexpected repetitions in the prospective stratigraphy (Fig. 4.78B).

A third type of structural detachment discussed earlier in this chapter is Garson where south over north thrusting has re-activated earlier structures to produce shear zones along which contact mineralization has locally been displaced, but more commonly moved by kinesis from areas of compression to zones of extension.

In summary, the effects of deformation can be important modifiers to contact and footwall mineral systems, but the deformation events that matter most to the configuration of the footwall ore systems are syn-magmatic adjustments of the crater wall. The role of hydrothermal processes in the formation of contact and footwall mineral systems is considered to be negligible, although there is a local effect from volatiles that were intrinsic to the sulfide melt or remobilized by heating of the footwall. There is no sign of large-scale alteration systems that have incorporated isotopically different sulfur (Ripley et al., 2015). Although there is evidence that mineral zones have been modified by syn-magmatic and postmagmatic alteration produced by fluids (Ames and Farrow, 2007; Farrow and Watkinson, 1996), the principal mechanism responsible for the primary formation of the ore deposits involved magmatic processes. The process of magmatic differentiation of the sulfide melts will be treated in more detail in Chapter 5 in the context of the evolution of the melt sheet.

Mineralized inclusion-rich Quartz Diorite From the Stobie mineral system. The magmatic textures of the Sudbury ore deposits was first documented at Frood-Stobie where immiscibility between silicate magma and sulfide melt was recognized as a primary process control.

THE RELATIONSHIP BETWEEN THE IMPACT MELT SHEET AND THE Ni-Cu-PGE SULFIDE MINERAL SYSTEMS AT SUDBURY

INTRODUCTION AND OBJECTIVES

The most exciting and provocative geoscience at Sudbury attempts to reconcile the formation of the ore deposits to the impact event that created the Sudbury Structure. On the one hand, observations provide geoscientists with an understanding of the processes of formation of the ore deposits by the saturation of the silicate melt in sulfide, and the segregation of the dense immiscible sulfides toward the base of the melt sheet. On the other hand, the science supports a pragmatic application of empirical observations and technologies to the discovery of magmatic Ni-Cu-PGE sulfide ore bodies (Chapter 6 covers this aspect in more detail). This chapter concerns the relationships between meteorite impact, melt sheet differentiation, and the formation of the Sudbury ore deposits.

The morphology of impact craters is most easily examined in well-preserved structures on our planet (eg, various papers in a volume edited by Osinski and Pierazzo, 2010). It is also possible to use the remote sensing record of impacts on other planets and moons where plate tectonic processes, sedimentation, and fall-out from subsequent impact events have not destroyed the record. These extraterrestrial impact structures help to establish models for impact processes that are not easily unpicked from the record provided by ancient structures like Sudbury. The record on Earth provides evidence of the tremendous energy released during an impact event, and the unusual conditions that give rise to crustal melt sheets, dykes, breccias, and the complex topography of the crater floor beneath a melt sheet. Efforts to understand Sudbury must be founded on at least some level of understanding of impact structures on other planets and moons as well as the record of large impacts on Earth (eg, http://www.unb.ca/fredericton/science/research/passc/).

An understanding of the mechanism by which the melt sheet evolved within a crater structure that underwent a significant degree of postimpact modification is supported by the record of the igneous rocks in the Offset Dykes and the Main Mass of the SIC. The embayments and troughs at the base of the SIC are part of a crater-floor topography that controlled the localization of mineralization. The pre-, syn-, and postimpact crater re-adjustment phases are important as they allow the system to be modified during the cooling, crystallization, and solidification of the silicate and sulfide melts.

The ore deposits at Sudbury exhibit a wealth of classic evidence for the formation and accumulation of immiscible magmatic sulfide ores as illustrated by the blebby sulfides that comprise part of the Frood-Stobie mineral system. Compelling evidence exists for the development of immiscible blebs of magmatic sulfide in an inclusion-bearing QD matrix (front image of this chapter), and conversely, immiscible segregations of QD can occur within semimassive sulfide (eg, Fig. 5 in Hawley, 1965). The dominant control of magmatic processes in ore formation is the nexus of the discussion presented in this chapter, and the relationships between the silicate rocks of the melt sheet and the ore deposits are investigated.

The composition of the melt sheet has been overprinted by assimilation of residual mafic clasts, melting of the underlying country rocks, and crystal differentiation after the magmatic processes that formed the ore deposits. The record of metal depletion in the Main Mass is consistent with sulfide saturation and segregation coupled with crystallization (Keays and Lightfoot, 2004). The data presented in Keays and Lightfoot (2004) are used to establish the principal process controls on the PGE distribution in the Main Mass, and the data are used to refine models for the sulfide saturation history of the SIC.

A major focus of this chapter is to examine why some parts of the SIC are endowed with economic ore deposits yet other parts contain no economically significant mineralization. To explain these features, the linkage between melt sheet thickness and the development of mineral systems is examined. It is shown that the mineral systems developed below the thicker parts of the melt sheet have greater economic significance than those developed beneath thin domains of the melt sheet.

Mechanisms of sulfide differentiation are discussed in the context of both the elemental assay record of geochemistry of the ores as well as the studies of the geochemistry of representative samples of mineralization undertaken by Naldrett et al. (1994a, 1999). The extent of fractionation of the sulfide melt, and the geochemical and mineralogical record of fractionation processes are examined and discussed using examples of the mineral systems discussed in Chapter 4. The footwall mineralization of both the North and South Ranges of the SIC contains an important contribution of sulfide derived by segregation of MSS to form the contact mineralization and the expulsion of residual fractionated sulfides into the footwall.

This chapter briefly examines the importance of hydrothermal processes and kinesis of sulfide during deformation in the formation of the ore deposits at Sudbury. Chapter 3 provided evidence that much of the metal value in Sudbury ores is controlled by magmatic processes, so this section is included in order to place recent studies of smaller deposits and occurrences in the footwall into the perspective of the larger mineral systems at Sudbury. Even though some hydrothermal modification is evident at the fringes of footwall mineral systems, there is no compelling requirement for ore-forming hydrothermal events related to the alteration recorded in the upper part of the Main Mass and the overlying Onaping Formation. Although some mineral zones are modified by sulfide kinesis linked to deformation, the vast majority of the ore deposits were localized by primary magmatic processes.

There are four principal objectives at the core of this chapter:

1. To relate the morphology of the Sudbury Structure to an understanding of the geometry of planetary impact structures, and to address an important question. Is Sudbury unique among terrestrial impact structures in terms of the world-class endowment in ore deposits?
2. To provide a unified description of the process controls that created the relationships between the melt sheet, the breccias, and the ore deposits.
3. To establish where the metals and S in the ore deposits actually originate, and why the ore deposits have an irregular distribution at the lower contact of the SIC.
4. To provide an explanation of the sulfide differentiation process controls that created a wide range in sulfide composition within individual mineral systems.

THE IMPACT RECORD OF THE SOLAR SYSTEM

Chapter 1 provides much of the evidence that supports the hypothesis of an impact origin of the Sudbury Structure and the SIC. An understanding of Sudbury can draw on information derived from both well-exposed and less-deformed impact structures on Earth such as Vredefort and Morokweng in South Africa, and Manicouagan in Canada, as well as the record from other planets and moons in the Solar System.

IMPACT CRATER RECORD BEYOND EARTH

A wealth of new information derived from remote sensing studies of planets and moons in the Solar System provides a path to help understand terrestrial impact structures. In this short synthesis, attention is drawn to a number of features described in craters from the Moon, Mercury, and Venus that help to inform models depicting the primary configuration of the Sudbury Structure at the time of impact.

Many of the extraterrestrial craters are young, undeformed, and most importantly not covered in young sediments or lavas, so their detailed morphology can be imaged by the remote sensing equipment on spacecraft missions.

The comet Shoemaker–Levy 9 which consisted of many pieces (~2 km in diameter, or less in size) collided with Jupiter in July 1994, providing the remarkable first direct observation of an extraterrestrial collision of a comet with a planet; the impact events were captured in a series of spectacular images from the Galileo spacecraft. The geological record of multiple impact events is also found in linear chains of impact craters (Wichman and Wood, 1995), as exemplified by the lunar Davy crater chain and the Enki Catena crater chain on the Jovian moon, Ganeymede (Weitz et al., 1997).

Much of the recent high-resolution and low-angle imaging of the Moon and Mars has added a wealth of detail that allows the recognition of features that can be used to unpick process controls on crater development.

Record of crater embayment structures

Morrison (1984) first described features of the lunar impact craters in relation to the morphology of the Sudbury Structure. In particular he drew attention to the well-developed geometric irregularities in the crater walls of the large lunar impact craters such as Copernicus. It is now clear that both peak-uplift structures and collapse features exist adjacent to the melt sheet in the crater wall (Fig. 5.1A,B). These features are also found in images of the lunar crater Vavilov (98 km diameter; Fig. 5.1C) where well-developed irregularities describing polygonal patterns are present in the crater walls.

Morrison (1984) suggested that the geometry of the crater walls was produced by excavation of the crater as ejected material was removed during primary impact. The removal of material resulted in the development of distal ejecta horizons that form rays around the impact structure such as those exhibited by the Kuiper impact structure on Mercury (Fig. 5.1D). A combination of erosion of material along the contacts between different country rocks and slumping of the crater walls controled the geometry of the inner crater and the basal contact of the melt sheet.

When viewed in the context of the Sudbury Structure, the geometric distribution of the primary troughs, embayments and Offset Dykes is illustrated in a reconstruction of the crater immediately after the formation of the Sudbury melt sheet (Fig. 5.1E). The distribution of the principal troughs is highlighted, and their location broadly matches the distribution of landforms at the margins of the Copernicus and Vavilov craters (Fig. 5.1A–C).

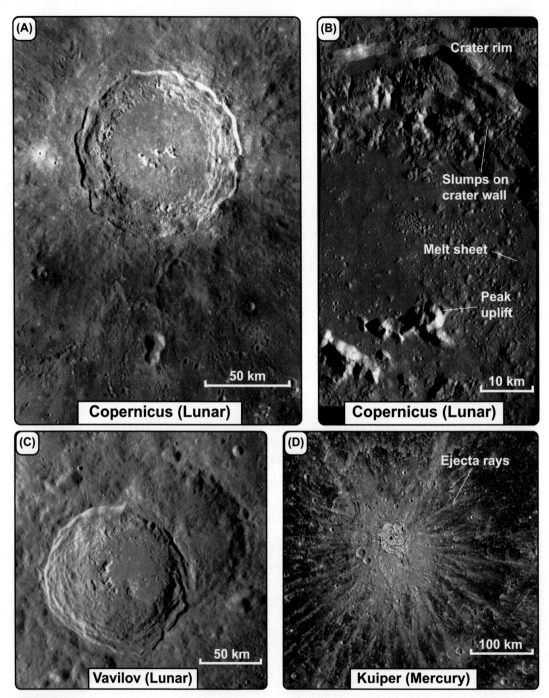

FIGURE 5.1

(A) Photograph of the morphology of the Copernicus impact structure on Earth's moon. The interior, slumped terraced walls, rim and some of the ejecta blanket of Copernicus as seen in this mosaic image (source: NASA Lunar Reconnaissance Orbiter). (B) Detailed view of part of the Copernicus crater showing the slump features on the crater wall, the peak-uplift features and the distribution of the melt sheet on the crater floor as seen in this mosaic image. (C) View of the lunar crater Vavilov (0.8°S, 138.8°W) which has a depth of 7 km from the crater rim. This 98 km diameter crater exhibiting chaotic terraces reflecting a high degree of slumping and massive landslides after the crater formed. (D) View of the Kuiper impact crater on Mercury showing the development of ejecta rays that result from scouring of the crater floor, and the development of a slumped crater wall.

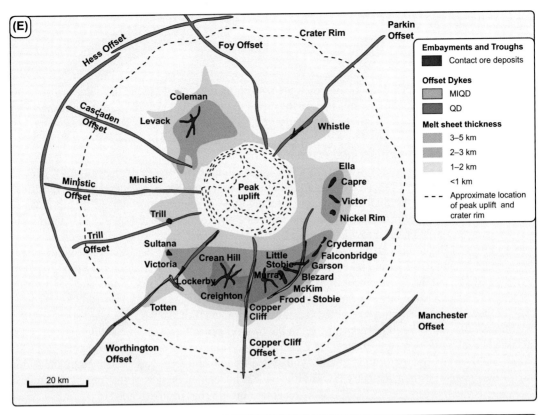

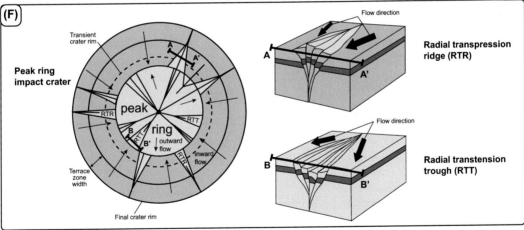

FIGURE 5.1 *(cont.)*

(E) Schematic diagram showing the primary configuration of the melt sheet, Offset Dykes, and contact ore deposits at Sudbury. A polygonal shape for the un-deformed Sudbury Structure was proposed by Morrison (1984), but the new model shown here focuses attention on the distribution of mineralization, the Offset Dykes, and the variation in thickness of the melt sheet. (F) Diagram illustrating the development of radial depressions around the crater rim and also around the central peak uplift of a large crater. In this model, transpressional ridges would be developed on the walls inside the crater rim, but transtensional features would be developed around the central peak-uplift.

(A,B) NASA Lunar Reconnaissance Orbiter. (C) http://lunarnetworks.blogspot.ca/2012/01/view-from-vavilov.html

(D) NASA MESSENGER satellite (F) Modified from Kenkmann and Dalwigk (2000) with the permission of John Wiley and Sons.

As discussed in Chapter 3, the ore deposits are irregularly developed around the outer margin of the SIC. One possible explanation for this rests in the distribution and thickness of the overlying melt sheet as shown in Fig. 5.1E; more about this topic later in this chapter.

Morphology of the crater floor and walls

Large impact craters often have walls with a low-angle slope (typically 25 degrees or less), which is considerably less than the range of slopes found at the base of the SIC (~35 degrees to overturned). There is no doubt that the Sudbury Structure was modified by deformation during the terminal Penokean and Mazatzal deformation events, and this accounts for some of these features (the relative timelines of deformation are provided in Fig. 2.40). The original crater shape and heterogeneity in the thickness of the competent SIC likely helped to control the regional deformation. Evidence from impact craters on the moon and other planets points to the development of a complex topography on the crater floor, and the development of extensive slump features on the crater walls and possibly also around the central peak uplift of large impact craters (Fig. 5.1A,B).

The development of pseudotachylite breccias in the footwall of the SIC provided surfaces along which the crater walls could fail and subside, in the process generating a chaotic assemblage of blocks of country rock at the base of the melt sheet which solidified to form the megabreccias. Morrison (1984) suggested that the primary location of embayment and trough structures was controlled by the slumping of the crater walls in a geometry that is now evident in detailed images of the impact structures at Copernicus and Vavilov (Fig. 5.1A–C). More recently, this idea has been extended to postimpact crater stabilization where collapse of the crater walls was accompanied by the injection of fractionated sulfide melts into the footwall beneath the contact type ore bodies developed in troughs and embayments at the base of the melt sheet (Lightfoot, 2015).

The possible role of slumping and the development of embayed crater walls is discussed in Kenkman and Dalwigk (2000), and this model for the crater rim wall at Sudbury has been discussed in Chapter 4. The process responsible for the development of slump features around the crater rim and the peak-uplift is shown in Fig. 5.1F. Transpression tends to be associated with the crater rim, but the flanks of the peak uplift structure would be subject to transtension (Fig. 5.1F; Kenkman and Dalwigk, 2000). The exact position of the Sudbury melt sheet in the context of this model remains unresolved at this time as the depth of erosion is not fully understood, but it is possible that radial transtensional troughs surrounding the peak uplift contributed to the development of the troughs and embayments together with the development of fragments present in the breccias developed at the base of the melt sheet.

Extraterrestrial impact melts and breccias

Impact melts are recognized in association with craters on other planets and moons in the Solar System. For example, Copernicus is a 96 km diameter complex crater with terraced walls and a cluster of central peaks (Fig. 5.1A), located on the Earth-facing side of the Moon. The crater has an extensive ray system and it represents one of the youngest formations on the Moon. The heterogeneous target rocks modified by the Copernicus impact event consist of a feldspathic northern domain and more mafic southern domain. Impact melts at Copernicus cover extensive parts of the crater floor and on the walls in association with slump blocks, as well as beyond the crater rim. There is a range of melt and megabreccia features, and spectral imaging points to a heterogeneity in mineralogy that is possibly related to the diversity in lithology of the country rocks (Dhingra et al., 2013; Dhingra and Pieters, 2013). High-resolution imaging by NASA using the Lunar Reconnaissance Orbiter Camera points to the development of breccias that are associated with impact melts in the Copernicus crater (http://lroc.sese.asu.edu/posts/473).

The oblique impact at the lunar Tyco crater created heterogeneity in the distribution of impact melt (Krüger et al., 2013). The geological map presented in Krüger et al. (2013) shows pooling of melt and indicates that pre-existing topography is an important control on the spatial distribution of the melt. A study by Darling et al. (2010b) indicated that the mass balance requirements for impact at Sudbury may require oblique impact, but the extent to which the melt sheet thickness is a function of impact angle versus the diversity in target rocks is not fully established at this time.

Vaughan and Head (2014) show evidence that a massive impact melt sheet was developed in the Aitken Basin at the Lunar South Pole. The rocks comprise a sequence of differentiated cumulates consisting of a 12.5 km thick layer of norite above ultramafic pyroxenite and dunite layers. Vaughan and Head (2014) suggest that this stratigraphy could be produced by differentiation and crystal settling in a 50 km-thick impact melt sheet formed by an oblique impact.

The record of impact fractures on other planets and moons provides a good framework of crater structure and impact melt distribution, but there is presently no convincing evidence for the development of melt infill features on other planets that may be analogous to the Sudbury Offset Dykes that comprise parts of the Sudbury melt sheet. A more careful examination of modern high resolution images offers an opportunity to identify similar features in association with extraterrestrial impact structures and their associated melt sheets.

There is much to be learnt from Sudbury and applied in the investigation of extraterrestrial impact events. Clearly the Sudbury event produced enormous mineral wealth; if other planets are to be colonized in the far future, a search for mineral wealth might best focus on the impact record in the absence of other plate tectonic and hydrothermal processes that drive the development of ore deposits.

THE GLOBAL TERRESTRIAL IMPACT CRATER RECORD

The global impact record contains a growing number of exposed and buried impact craters varying in age from Archean to Anthropocene, and transient craters up to 300 km in diameter. There are various terrestrial impact datasets with compilations of names, locations, ages, and scales, among which the PASSC database provides a useful summary of the craters that are related to impact events (http://www.passc.net/). Other databases provide an assessment of whether the crater features are due to impact processes or endogenic events (http://impacts.rajmon.cz/index.html). The PASSC dataset is used to generate Fig. 5.2A where the global distribution of impact craters is shown with the symbols sized to crater diameter, and those craters with a diameter greater than 50 km are named in Fig. 5.2A. In Fig. 5.2B the planetary record of impact structures through time is shown. The record of impact events in Archean and some Proterozoic belts is modified by crustal re-working. Younger impact events include a large number of small craters, but there are also large impact events such as Chicxulub. The distribution of craters through time is shown in the context of the ages of major impact craters, the large igneous province (LIP) record, and major extinction events in Fig. 5.2C–E (Kelly and Sherlock, 2013; Rhode and Muller, 2005). The extent to which discrete impact events can be linked to mass extinction is not clear-cut; there actually appears to be a better relationship between extinction events and major LIPS rather than between extinction events and impacts.

The PASSC dataset summarizes crater characteristics and published information for each of these impact sites, and the reader can also consult Osinski and Pierazzo (2010) for papers that describe terrestrial impact processes and products.

The energy involved in the impact process that produced the transient crater (shown in Fig. 5.3A) is enormous, and the Sudbury, Vredefort, and other large impact events are many orders of magnitude larger than natural short-lived geological events such as eruptions and earthquakes, natural heat flow, and anthropogenic effects of human ingenuity (Fig. 5.3B). Not all terrestrial impact structures develop preserved contiguous layers of impact melt, although small craters from atomic tests contain layers of impact glass (eg, the 1–2 cm-thick layer of green glass, termed Trinitite, generated by the Trinity atomic test in Nevada).

Scale of terrestrial impact melt sheets

In a review of impact models for the SIC, Grieve (1994) and Cintala and Grieve (1992a,b) reported the relationship between the diameter of the initial transient crater cavity (shown in Fig. 5.3A) and the volume of impact melt produced assuming the impact of a chondritic bolide with a granitic basement at 25 km/s. Grieve superimposed the estimated melt volumes and transient crater diameters of a number of terrestrial impact events on this diagram. Sudbury is the largest impact-generated melt sheet with a modeled transient crater diameter of over 100 km and a melt volume of 35,000 km^3 (Fig. 5.3C).

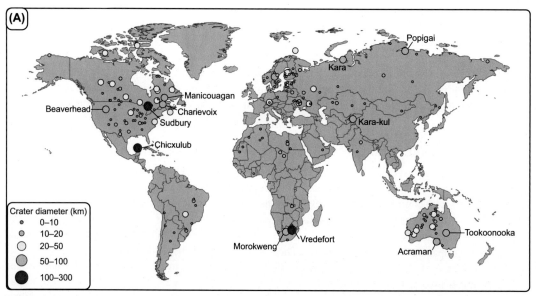

FIGURE 5.2

(A) Distribution impact craters on Earth with symbols sized to crater diameter. (B) Diagram showing the absence of any simple relationship between the age of major impact events and the diameter of the transient crater. (C) Relationship between large impact events and major mass extinction events; the age difference between the extinction event and the impact event is shown in millions of years. (D) Relationship between the timing of eruption and emplacement of large igneous province in relation to mass extinction events; the age difference between the eruption of the major volcanic center and the extinction event is stronger than the relationship between impact events and extinctions shown in subpart C. (E) Diversity of genera through time showing major mass extinction events.

(A) Modified from data available at http://www.passc.net/. (B) modified from data available at http://www.passc.net/; (C) modified after Kelly and Sherlock (2013); (D) modified after Kelly and Sherlock, 2013; (E) modified after Rhode and Muller (2005).

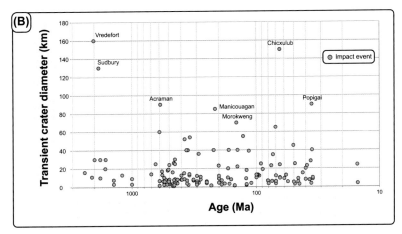

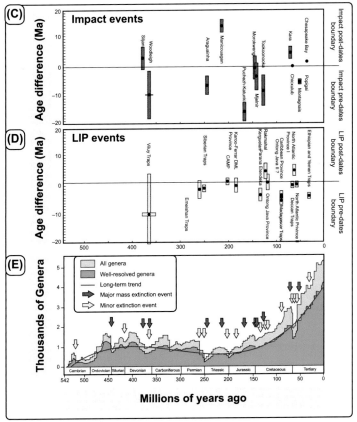

FIGURE 5.2 (*cont.*)

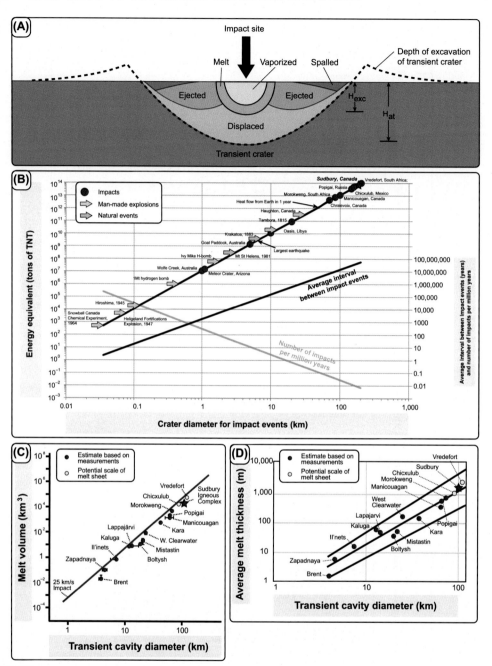

FIGURE 5.3

(A) Nomenclature used to describe crater shape and scale; H_{exc} is the excavated depth of the crater and H_{at} is the depth of the transient crater. (B) Relationship between impact crater size and amount of energy imparted by the bolide; the plot compares impact events with natural terrestrial processes (where the crater diameter is clearly hypothetical) and man-made explosions that created craters. The plot also shows the interval between impact events at the scale shown and the number of impact events each million years. This plot is largely based on Table 2.1 of French (1998). (C) Relationship between the size of the transient crater and the volume of impact melt. (D) Relationship between the diameter of the transient impact crater and the average melt sheet thickness.

(A-C) Modified from publications of the Ontario Geological Survey, and used under license to Elsevier; (D) modified after Golightly (1994).

Golightly (1994) placed the total maximum volume of the igneous rocks of the SIC at <13,900 km^3. A more reasonable estimate of the melt sheet volume can be anchored with average thickness data established from surface mapping of the units and 3D wireframe models of contacts derived from diamond drill core logs (Gibson, 2003). Based on a three-dimensional model of the North Range, the distance between the top of the Sublayer and the base of the Onaping Formation indicates that the Main Mass has an average thickness of 2,225 m, which is consistent with a minimum melt volume of 25,000 km^3. Increasing this to allow for a peak-ring melt is indicative of a total volume in excess of 30,000 km^3 which indicates a transient crater diameter similar to that estimated by Grieve (1994).

Golightly (1994) presented the relationship between transient crater diameter (Fig. 5.3A) and average melt sheet thickness using data from Cintala and Grieve (1992a,b), and showed a reasonable linear relationship anchored by Sudbury with a transient crater diameter of ~100 km and a melt sheet with an average thickness of 2 km (Fig. 5.3D). The actual variation in thickness of the Sudbury melt sheet vary from an estimate of 5 km in the South Range at Creighton to less than 1 km in parts of the North Range. The variation in thickness of the melt sheet approaches a factor of 5, and the eroded portions originally would have recorded the thinning of the melt at the edge of the crater and through radial connections with the outer ring structure (Fig. 5.1E).

Topography of the base of terrestrial melt sheets

A question of particular relevance to Sudbury is the degree to which the existing basal topography of the SIC is indicative of primary morphology of the crater floor. An uneven and sloped floor is indicated by the development of embayments, troughs, and funnels at the base of the SIC, but it is less certain whether the variation in the initial thickness of the melt sheet at Sudbury (≤5 km) is related to basal topography, postimpact deformation, or both.

Data for the undeformed and young (214 Ma) Manicouagan impact structure in Quebec, Canada is especially informative (Fig. 5.4A; Spray and Thompson, 2008). This 90 km diameter impact crater has provided new insight into the internal structure of this complex crater's central peak uplift region. Floran et al. (1976) showed that the impact event generated a ~55 km diameter melt sheet with a thickness of ~400 m, but recent drill data has revealed that the melt sheet is locally at least 1500 m thick, with kilometer-scale lateral and substantial vertical variations in the geometry of the crater floor beneath the melt sheet (Spray and Thompson, 2008). The thickest melt section occurs in a 1500 m deep central trough encircled by a horseshoe-shaped uplift of Precambrian basement shown schematically in Fig. 5.4B (O'Connell-Cooper and Spray, 2011). The uplift constitutes a modified central peak structure, at least part of which breached the roof of the melt sheet as shown in the section in Fig. 5.4A. The thicker part of the melt sheet is differentiated, but the thinner flanks are relatively undifferentiated (Fig. 5.4B). These features are interpreted by Spray and Thompson (2008) to indicate that the footwall geometry and associated trough structure were in place prior to the solidification of the melt sheet. Moreover, the marked lateral changes in basement relief beneath the melt sheet indicate the existence of a high-angle displacement in the crater floor with an offset of 100s to 1000s of meters. This observation indicated to Spray and Thompson (2008) that deformation during the modification stage of the cratering process was primarily facilitated by faults around the central peak-uplift.

The recognition of this primary crater-floor topography, and its relationship to crater modification prior to melt sheet differentiation has important implications for the understanding of the Sudbury melt sheet.

The primary variation in melt sheet thickness at Sudbury was highlighted in Chapters 2 and 4 where it is suggested that not only embayments, troughs, and funnels might localize magmatic sulfide

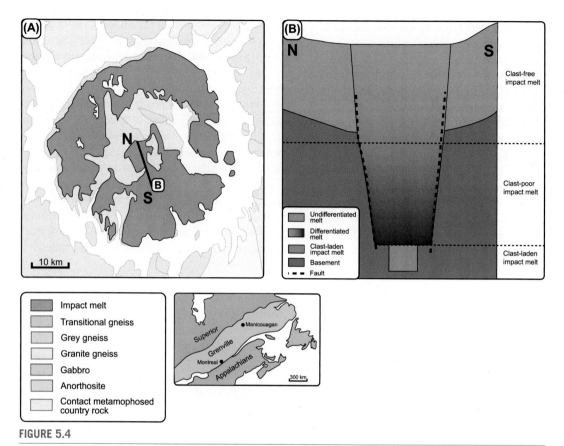

FIGURE 5.4

(A) Location and geology of the Manicouagan impact structure. (B) Simplified cross section based on drill data showing the structural controls on melt sheet thickness adjacent to the central uplift of the Manicouagan impact structure.

(A) Modified after Spray and Thompson (2008); (B) modified after O'Connell-Cooper and Spray (2011).

mineralization at the base of the melt sheet, but that the wide variation in thickness of the melt sheet is a function of the slope and topography of the crater floor. In this model, the sulfide melts accumulated from a large volume of magma; the sulfide melts may have been able to flow across the crater floor into depressions at the base of the melt sheet.

Terrestrial Impact Melt Dykes

The Vredefort impact structure in South Africa is widely considered to be the largest impact event in Earth's geological record. The 2.023 Ga impact event produced a crater with a diameter of ~300 km. The level of erosion of the impact structure is very deep, and it provides an excellent view into the basement beneath a large impact event, with excellent exposures of the peak uplift structure, crater floor pseudotachylite, and crater rim structures (Reimold and Gibson, 2006). The geology of the impact structure is shown in Fig. 5.5A, and the regional geological setting in relation to the Witwatersrand Basin is shown in Fig. 5.5B.

The eroded floor of the Vredefort impact structure is cut by granophyric dykes that are widely believed to be generated by crustal melting (Lieger and Riller, 2012). These dykes are narrower than those developed

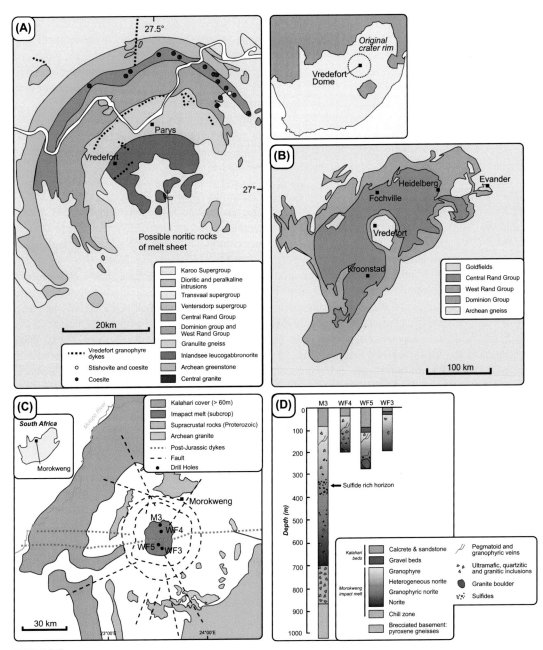

FIGURE 5.5

(A) Geological map of the Vredefort Impact Structure showing the location of granophyre dykes and the norites that Cupelli et al. (2014) consider to be foliated impact melts. (B) Geological relationship between the Vredefort impact structure and the Witwatersrand Basin which hosts the gold deposits. (C) Geological map of the Morokweng melt sheet. (D) Comparison of geological information from drill core sections through the Morokweng impact melt sheet and associated country rocks.

(A) Modified after Gibson and Reimold (2001); (B) modified after Reimold and Gibson (2006); (C) modified after Andreoli et al. (1999); (D) modified after Andreoli et al. (1999).

at Sudbury, but they contain magmatic breccias, and extend for quite large distances in the footwall of the Vredefort impact structure. These granophyres have a composition that is close to that of felsic upper crust (Fig. 5.6A). In the case of the Vredefort granophyre, the abundance of Ni or Cu is also close to that of average crust (Fig. 5.6B), and there is no evidence for sulfide saturation of the original melt sheet prior to the injection of the dykes. One unusual feature of the dykes is the high Cr concentration (185–200 ppm Cr at 3 wt.% MgO) that is indicative of an unusually Cr-rich contribution in the genesis of these melts.

The Morokweng impact melt sheet in Botswana is another example of a melt sheet, but the basal topography of this body is not understood due to poor exposure and limited drilling (Fig. 5.5C–D), and there are no known examples of dykes injected from the melt sheet.

Terrestrial Melt Sheet Compositions

Fig. 5.6A shows the average compositions of the Manicouagan and Morokweng melt sheets normalized to the composition of average upper crust. Manicouagan shows a slight Ni- and Cu-depletion, but these are very sensitive to estimates of target rock composition, and there is no evidence at this point for the development of magmatic sulfide mineralization at Manicouagan (Fig. 5.6B).

In contrast to the Manicouagan melt sheet, the 145 Ma Morokweng melt sheet shows very strong Ni enrichment, but no commensurate Cu-enrichment. The gabbronoritic rocks contain trace dissemination of magmatic sulfide and occasional narrow veins of millerite (Koeberl and Reimold, 2003). The identification of clasts derived from the impact projectile (Maier et al., 2006), coupled with the variations in PGE geochemistry are indicative of a L-type chondrite meteorite source for the metals (McDonald et al., 2001).

Cupelli et al. (2014) proposed that noritic rocks associated with the Vredefort Structure formed at the time of the Vredefort impact. Based on U-Pb geochronology studies of zircon, Cupelli et al. (2014) suggested that the deformation features in these rocks were created by postimpact relaxation of the crust.

The compositional average of samples of norite from the discovery site of these proposed Vredefort-aged melt sheet rocks is shown in Fig. 5.6E where the composition is normalized to the the average of granophyre dykes from the Vredefort Structure. The norites are clearly quite different in comparison to the granophyres, and they exhibit a sloped pattern that is an uncommon feature in crustal melts (Fig. 5.6E). Further work is required to reconcile the geochemistry of the noritic rocks with other rocks of the igneous rocks of the Vredefort Structure.

Pseudotachylite

The pseudotachylites of the Vredefort dome provide classic examples of impact breccias that have many similarities to SUBX (Fig. 2.5A–C in comparison to Fig. 5.7A,B). The geology of these impact breccias is described in French (1998), Gibson and Reimold (2001), and Mohr-Westheide et al. (2009), and references therein. Fig. 5.7A show a subhorizontal zone of pseudotachylite breccia cutting

FIGURE 5.6   ▶

(A) Compositional variations normalized to average crust (Taylor and McLennan, 1985) for samples of granophyre from the 2.0 Ga Vredefort impact structure, the average composition of the melt sheet in the Morokweng impact structure (Koeberl and Reimold, 2003) and the average composition of the melt sheet in the Manicouagan impact structure (Spray and Thompson, 2008). (B) Variations in the degree of Ni enrichment/depletion in terrestrial melt sheets expressed as Ni/Ni* (see Chapter 3 for an explanation of the importance of Ni/Ni* as a measure of nickel depletion and enrichment; data sources include those referenced in this book; Andreoli et al., 1999; Koeberl and Reimold, 2003; Spray and Thompson, 2008).

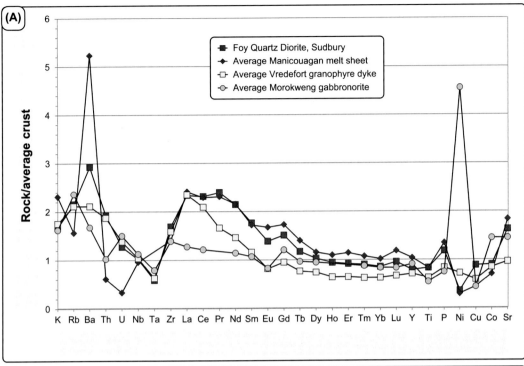

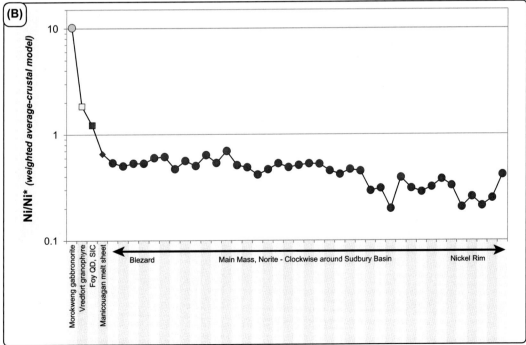

(Continued)

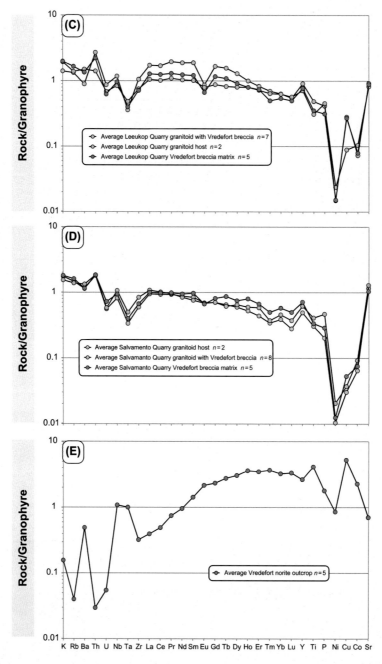

FIGURE 5.6 (*cont.*)

(C) Composition of the matrix of pseudotachylite breccias from the Vredefort impact structure at Leeukop Quarry where the pseudotachylite cuts granitoid country rocks that comprise the basement of the Vredefort Dome. (D) Composition of pseudotachylite breccias from the Vredefort impact structure at Salvamento Quarry where the psudotachylite cuts granitoid country rocks in the basement of the Vredefort Dome. Note the close correspondence of breccia matrix, country rocks, and mixed breccia-country rock in both subparts C and D. (E) Composition of average norite from the Vredefort impact structure at the site described in Cupelli et al. (2014) normalized to the average composition of the granophyre dykes of the Vredefort impact structure.

FIGURE 5.7

(A) Photograph of a subhorizontal vein of pseudotachylite breccia from Leeukop Quarry in the Vredefort Impact Structure showing the localization of large blocks of granite toward the base of the pseudotachylite. (B) Photograph of a vertical vein of pseudotachylite cutting granitoid rocks at Salvamento Quarry located within the Vredefort Structure. See Gibson and Reimold (2001) for detailed descriptions of these locations.

granitoid rocks of the Vredefort crater floor; the pseudotachylite is laden with clasts, and the clasts are strongly imbricated toward the base of the breccia, and they tend to be smaller toward the top of the breccia zone. The distribution of clasts is consistent with the settling of the fragments within a molten matrix.

Fig. 5.7B shows an example of a vertical pseudotachylite zone cutting granitoid rocks where there is a more even distribution of fragments within the matrix of a subvertical zone of breccia. The effect of settling of fragments in the matrix may explain the difference between the geometry of the flat and vertical breccia zones in Fig. 5.7A,B. This geometric relationship indicates that the pseudotachylite matrix was unable to support the clasts, perhaps because of the degree of melting.

The geochemistry of the breccias and host rocks has been discussed in Reimold et al. (2015). Fig. 5.6C,D shows data collected from samples of breccia matrix, mixed breccia matrix with fragments, and country rocks. These data indicate that the pseudotachylite breccia is a product of melting of the adjacent crustal rocks as suggested in Reimold et al. (2015). There is no requirement for a melt sheet contribution to the pseudotachylite matrix as suggested by Lieger et al. (2011).

The evidence pointing to a local source for the matrix component is aligned with the geochemical information presented in Chapter 3 for the SUBX. O'Callaghan et al. (2015, 2016) suggest that there is no geochemical evidence for a melt sheet contribution to the matrix of the SUBX.

DO OTHER TERRESTRIAL IMPACT CRATERS HOST ORE DEPOSITS?

Some of the larger impact structures on Earth contain mineral resources that are either currently mined or have the potential to pass economic hurdles in the future (Grieve and Masaitis, 1984; Naumov, 2002). Grieve (2003) estimated that 25% of all known terrestrial impact structures have associated mineral resources and they are currently exploited in about half of the cases.

There are three types of mineral systems spatially related to terrestrial impact structures (Reimold and Gibson, 2006).

1. Preimpact mineralization modified and preserved because of an impact event: the best example for this class of deposits is provided by the Archean gold-and-uranium ores of the Witwatersrand Basin, adjacent to the Vredefort impact structure.
2. Mineralization produced by the impact event: these are syngenetic with regard to an impact event, and the magmatic ore deposits at Sudbury are the best known examples.
3. Epigenetic mineralization developed after impact: this is essentially mineralization that was localized as a result of the geological environment produced by the impact structure. Many hydrocarbon deposits are related to impact structures. The volcanogenic deposits in sedimentary rocks located above the SIC were likely generated in response to heat flux from the melt sheet and formed during cooling of the melt sheet.

Modification of pre-existing mineralization

The Vredefort impact structure is adjacent to the Witwatersrand Basin where the world's largest known gold deposits are located (Fig. 5.5A,B). A synclinal structure localizes the distribution of Witwatersrand and the Transvaal Supergroups around part of the Vredefort Dome (McCarthy et al. 1986).

It is clear that the gold has a detrital origin (Robb and Robb 1998; Foya 2002; Minter 1999; Minter et al. 1993; Frimmel and Minter 2002), and there can be no doubt that much of the gold mineralization was in place prior to the impact event. Moreover, there is no evidence that the distribution of the gold mineralization was modified or upgraded by the impact process. However, the Vredefort event may have contributed to the preservation of the detrital Archean gold deposit in two ways: (1) The ring syncline of the impact structure protected the prospective stratigraphy from erosion, and (2) ejecta covered and protected the gold-bearing strata from erosion (Reimold and Gibson, 2006).

Several other examples of ore deposits that were modified or preserved as a result of impact process are described in Reimold and Gibson (2006). These include the following:

1. *The Krivoi Rog impact crater in the Ukraine*: The crater floor of this 15–18 km diameter impact structure is now deeply eroded and preserves iron and uranium mineralization associated with preimpact hydrothermal activity (Reimold and Gibson, 2006).
2. *The Carswell impact crater in Canada*: This 50–55 km diameter impact cratering event modified the contact between basement rocks and the Athabasca Group sedimentary rocks. The contact is a principal control on the unconformity type uranium mineralization (Grieve, 2003).

Syngenetic impact-related mineralization

The best known example is documented in this book; these are the Ni-Cu-PGE sulfide ore deposits at Sudbury. Other examples of syngenetic mineralization include the following:

1. *Impact diamonds*: Impact diamonds are the result of shock transformation of graphite or coal at pressures in excess of 30 GPa (Masaitis 1993; Koeberl et al. 1997; Gilmour 1998). Occurrences of impact diamond are known from several impact structures (review by Gilmour, 1998). In the 1970s diamonds were discovered in impactite of the Popigai Structure (Masaitis and Selivanovskaya 1972; Masaitis 1993). Impact diamonds have since been discovered at the Kara, Puchezh-Katunki, Ries, Ternovka, Zapadnaya, Sudbury, and Chicxulub impact sites (Masaitis 1993; Masaitis et al. 1999; Montanari and Koeberl 2000).
2. *Ni-Cu-PGE sulfides*: The 70–80 km diameter (Reimold et al. 2002) Morokweng impact structure in South Africa is known to host traces and small showings of Ni-PGE sulfide mineralization. The impact melt sheet is indeed enriched in Ni and PGE, due to a significant admixture of up to 5% of the meteoritic projectile to the melt rock (Koeberl et al. 1997; Hart et al. 2002; Maier et al. 2003). Evidence for an equivalent cosmic contribution of nickel Ni and precious metals to the SIC has not been identified, and the composition of the ores would require an unusually Cu-rich projectile.

Despite a global search by mining companies for impact melt sheets with potential to contain Sudbury type mineralization, no equivalent example has yet been found. The possibility that a large impact site exists under cover is not completely evaluated at this time, and it is also possible that additional Archean impact sites will be recognized in Archean granite–greenstone belts.

Epigenetic impact-related mineralization

The infiltrate of fluids into hot country rocks of the central uplift and the impactites of the crater fill sequence is believed to result in hydrothermal activity (Kirsimäe and Osinski, 2013), and alteration that may generate mineralization. The strong deformation by fracturing and brecciation and the near-instantaneous increase in temperature in large volumes of the target rocks provide ideal conditions for the initiation of hydrothermal fluid convection from surface and groundwater sources. Examples of relatively weak hydrothermal mineralization that may be related to impact events include those associated with the 24 km diameter Haughton impact structure in Canada (Osinski et al., 2001). The Pb-Zn sulfide deposits located in the 377 Ma old Siljan impact structure in Central Sweden represent an economically significant example of impact-related hydrothermal mineralization (Reimold and Gibson, 2006 and references therein).

Ames and Gibson (1995) and Ames et al. (1997, 1998) described extensive, regional hydrothermal alteration that pervasively affected the breccia accumulations in the Sudbury Basin of the SIC, in particular the Onaping and lower Onwatin Formations. They reported evidence for a regional subseafloor hydrothermal system that included vertically stacked, basin-wide, semiconformable alteration zones, in which the rocks underwent silicification, albitization, chloritization, calcitification, and complex feldspathification. Extensive hydrothermal alteration is also recognized in the upper part of the Main Mass stratigraphy (Ripley et al., 2015).

Massive Zn-Cu-Pb sulfide deposits within the Vermilion Member of the Onwatin Formation have been mined at the Errington and Vermilion Deposits. Ames et al. (1998) obtained isotopic evidence that times this alteration to within 4 Ma of the 1.85 Ga Sudbury impact event.

The hydrothermal system generated by the SIC impact melt may also have been responsible for remobilization and redeposition of metals in fracture zones and breccia occurrences to produce local zones of mineralization, and some authors propose that the precious metal enriched mineralization in the footwall to the SIC was produced by extensive hydrothermal modification of primary magmatic sulfide mineralization (Farrow and Watkinson 1997; Molnár et al. 1999).

Large volumes of hydrocarbons have been identified and in some cases exploited from impact structures. Commercial hydrocarbon production from North American impact structures is estimated to have generated from 2 to 30 million barrels per day, plus more than 1.4 billion cubic feet of gas per day (Donofrio, 1997). Hydrocarbon reservoirs exist in all parts of an impact structure, including central uplifts, rim structures, slump terraces, and ejecta. In the case of very large impact structures, such as Chicxulub in Mexico, disrupted and fractured rocks surrounding the impact structure represent favorable hydrocarbon exploration targets (Reimold and Gibson, 2006).

REGIONAL CONTROLS ON ORE DEPOSIT ENDOWMENT AT SUDBURY

The next section of this chapter looks at the differentiation history of the Main Mass of the SIC, and the composition of the melt sheet prior to differentiation. This section provides a foundation on which the principal controls on ore deposit endowment at Sudbury can be understood. It also attempts to reconcile the distribution and quality of the mineralization to features of the Main Mass that was created by differentiation of the impact melt.

MELT SHEET DIFFERENTIATION

The Main Mass of the SIC comprises a sequence of noritic, gabbroic, and granophyric rocks in both the North and South Ranges. Although there are some important differences in the detailed chemostratigraphy of the noritic rocks that indicate differences in the details of the crystallization history (Chapter 3), the similarity in the overall rock sequence, the constant ratios of highly incompatible trace elements, and the development of metal depleted noritic rocks in the Main Mass stratigraphy require a regional scale crystal differentiation and sulfide saturation process (Chapter 3).

Five process models have been proposed to explain the compositional variation through the Main Mass:

1. Simple Rayleigh Fractionation of a single homogenous melt derived by melting of the upper crust (Thode et al. 1962; Lightfoot et al. 2001; Therriault et al. 2002).
2. Crystallization of two or more melts. The lower mafic portion of the melt sheet composed of norite and Quartz Gabbro (QGAB), crystallized from mantle-derived basaltic magma, whereas the upper granophyric portion of the melt sheet was possibly the product of a granitic magma produced by crustal melting (Phemister 1937; Thomson 1969; Ariskin et al. 1999; Chai and Eckstrand 1993).
3. Separation of immiscible melts of different density from an impact-generated melt to form noritic rocks in the lower part and granophyric rocks in the upper part (Zieg and Marsh 2005).
4. Progressive modification of the chemical composition of the melt sheet by assimilation of the underlying crust and the fallback breccias. This process may have involved incorporation of partial melts of less dense footwall that were initially generated from country rocks beneath the melt sheet and then ascended upward through the melt sheet to accumulate toward the top of the Main Mass (Golightly, 1994). Fragments of dense mafic and ultramafic rock in the lower part of the melt sheet may have been progressively assimilated to create a more melanocratic composition. In this model, a density-stratified melt sheet is required, but it must also have very effectively mixed the melts in order to generate the uniform isotope and incompatible trace element ratio signatures (Keays and Lightfoot, 2004).
5. Generation of the melt sheet from similar target rocks, but crystallization of the magmas within domains that show differences in degree and/or pathway of differentiation. Melts created in different subbasins (Fig. 5.1E) may interact during postimpact relaxation of the crater (Lesher et al., 2014).

Chemostratigraphic modeling of the Main Mass

Attempts to simulate a simple model for the differentiation of the Main Mass of the SIC have been made by Ariskin et al. (1999) based on data for samples collected through the Main Mass of the North Range. Using parental magma compositions corresponding to the bulk SIC (granophyre + quartz gabbro:norite = 2:1), they attempted to reproduce the differentiation trends of the magma using COMAGMAT-3.5. The calculated liquid line of descent indicates that the norites of the SIC can be produced with a volumetrically dominant (70%) residual liquid containing an average of 67.5% SiO_2. At the time of this modeling, there were few systematic chemostratigraphic traverses through different segments of the Main Mass of the SIC.

Lavrenchuk et al. (2010) modeled the chemostratigraphic variations through the North Range Main Mass assuming closed system crystallization and convective crystal accumulation (Frenkel et al. 1989).

The models used the combined major element data from seven different traverses that were compiled together and fitted using a software program named PLUTON.

Lavrenchuk et al. (2010) used the bulk composition of the Main Mass which corresponds to a granodioritic parental magma. The liquidus temperature of orthopyroxene-saturated magma was estimated to be 1095°C. The settling velocity of minerals was accepted to be 100 times higher than the rate predicted by Stokes Law to ensure that the content of crystals suspended in the magma is no higher than 0.05 vol.% up to the appearance of significant amounts of quartz in the Granophyre Unit (GRAN). This model effectively simulated in situ crystallization. Allowing for the shallow setting of the melt sheet, the pressure was assumed to be 0.5 kbar. The activity of oxygen was assumed to be 100 times higher than the quartz–fayalite–iron (QFM) buffer. The residual porosity of the cumulate rocks is taken to be 40% in the model (Lavrenchuk et al., 2010).

To illustrate the models with reference to a single East Range traverse from Chapter 3, the thickness of various units in the East Range traverse from Chapter 3 has been normalized to the ratios of Norite:QGAB: GRAN in Lavrenchuk et al. (2010) which are 25:10:65. The actual proportions of the Norite, QGAB and GRAN have been estimated in the North Range based on three dimensional models of the Main Mass that respects surface geology and drill data. A ratio of 18:7.5:74.5 (based on 8,514, 6,028, and 5,855 measurements from wireframe nodes, respectively) is a better indicator of the proportions of these rocks above ~2 km depth, but the estimates used by Lavrenchuk et al. (2010) are the basis for the models shown in Fig. 5.8A–G.

Crystallization commenced with the formation of a basal orthopyroxene-rich cumulate, and this cumulate together with the interstitial liquid then crystallized to form the Mafic Norite (MNR) (Fig. 5.8A–G). The appearance of cumulus plagioclase (An_{55}) above the MNR is marked by an abrupt decrease in MgO and increase in Al_2O_3 and CaO (Fig. 5.8C,E,F,H) which corresponds to the transition through the Lower Felsic Norite (FNR) into the FNR. Through this transition, plagioclase changes from intercumulus to cumulus in texture. The composition of the FNR evolved in response to accumulation of orthopyroxene and plagioclase and the removal of trapped liquid. The base of the QGAB is marked by the increase in CaO content (Fig. 5.8F). The appearance of cumulus magnetite is expressed in peaks of FeO_{tot} and TiO_2 in this unit (Fig. 5.8B,D,H). The subsequent appearance of quartz and K-feldspar is marked by an increase in SiO_2 and K_2O contents in the GRAN Unit (Fig. 5.8A,G).

In contrast to the modeling by Ariskin et al. (1999), Lavrenchuk et al. (2010) succeeded in reproducing relative proportions of GRAN and mafic rocks, thereby circumventing one of the previous obstacles to modeling the formation of the SIC from a single pulse of magma. The success of this model is largely related to the introduction of quartz and K-feldspar in the model as well as to taking into account the progressive change in liquid viscosity during crystallization of magma in the melt sheet (Lavrenchuk et al., 2010).

The model adequately reproduces the phase and chemical compositions of the profile from the East Range, and it also fits with other North and East Range chemostratigraphy profiles. However, the model still requires further refinement as it does not explain the variations evident in different traverses, nor does it adequately explain the trend toward more mafic GRAN at the top of the melt sheet, the development of an anomalously thick QGAB Unit at Windy Lake, and it is based entirely on traverses through the North Range. In Chapter 4 the chemostratigraphy of the Main Mass in the North Range is shown to be different when compared to that of the South Range. In the South Range, the norite stratigraphy becomes more primitive up to a lens of melanorite developed two-thirds of the way through the stratigraphy of the Main Mass in the Creighton Traverse (Chapter 3; Lightfoot and Zotov, 2006). These variations have yet to be reconciled with a simple in-situ crystal differentiation model.

Composition of the parental magma to the SIC

The integrated composition of the sections through the Main Mass of the SIC provides a range in estimates of bulk composition depending largely on the relative proportion of the MNR relative to the other units of the Main Mass. The irregular shape of the topography of the basal contact means that the different segments of the SIC may have different proportions of the principal rock units.

A different approach that uses the composition of the QD unit of the Offset Dykes was discussed in Chapters 2 and 3, where it is shown that this unit is quenched and represents a melt composition that may provide the best estimate of the composition of the initial melt sheet (Lightfoot et al., 1997a;

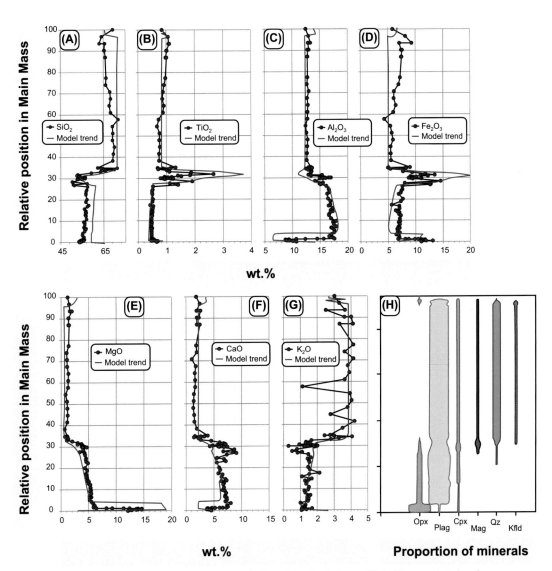

FIGURE 5.8 Geochemical Modeling of the Differentiation of the North Range Main Mass undertaken by Lavrenchuk et al. (2010) Superimposed on the Compositional Data for the East Range Traverse From Fig. 3.4A–H Where the Relative Stratigraphic Thickness in the Profile is Normalized to 25% a Stratigraphy Comprising FNR, 10% QGAB, and 65% GRAN

(A) SiO_2. (B) TiO_2 (C) Al_2O_3, (D) Fe_2O_3, (E) MgO, (F) CaO, (G) K_2O. (H) Inset shows the mineralogy of the rocks produced in the model. The model describes the broader aspects of the chemostratigraphy, but it does not produce a good fit the absolute concentrations of SiO_2 in the FNR, and it fails to produce the reverse differentiation trend in the MPEG Unit. Based on Lavrenchuk et al. (2010).

Lightfoot et al., 2001). There are small but significant systematic differences in the composition of the QD from different Offset Dykes as shown in Chapter 3, but these variations are likely indicative of the heterogeneity in melt composition between the North and South Ranges at the time of initial melt sheet formation. These estimates are more robust than estimates based on the calculation of the weighted average composition of the Main Mass from individual traverses.

Numerical modeling of the crystallization of QD as a parental magma has been undertaken by Lavrenchuk et al. (2010). Although there is some variation in the composition of the QD unit between the different Offset Dykes, Lavrenchuk et al. (2010) illustrated that the composition of this unit can be used to model the observed rock sequence in the Main Mass. However, the models in Lavrenchuk et al. (2010) also predict that the GRAN unit should be thinner than observed (as suggested by data presented in Fig. 3.13). The results indicate that the melt sheet has evolved to a more felsic composition after the emplacement of the QD unit of the Offset Dykes.

Summarizing the implications of these observations to process models for the melt sheet:

1. *Fractional crystallization of a single silicate melt*: The results of the modeling indicate that crystallization sequence, mineral compositional trends, and geochemical variations of the SIC can be adequately reproduced by fractional crystallization of a single body of granodioritic magma. This is consistent with the constant ratios of incompatible elements discussed in Chapter 3.
2. *Anatexis of the floor of the melt sheet*: Assuming an average QD parental magma, the volume of the GRAN Unit is underestimated; this can be explained if contributions of felsic magma are added to the melt sheet by anatexis of the footwall.
3. *Models requiring two or more different parental melts can be discounted*: There is no requirement in the modeling or the variations in trace element ratio shown in Chapter 3 for multiple contributions of magma from different sources (Chai and Eckstrand, 1993). This does not discount the possibility of progressive assimilation of inclusions and felsic melts from the immediate footwall (Golightly, 1994), but the process model requires that the final melt sheet has a very small range in incompatible element ratios.
4. *No requirement for density segregation*: There is no evidence to suggest that the melt sheet underwent an early phase of density segregation as proposed by Zieg and Marsh (2005). The possible effect of progressive assimilation of residual mafic-ultramafic inclusions in the lower part of the melt sheet and felsic melts in the upper part may have created some primary heterogeneity in melt sheet composition (Golightly, 1994).
5. *Differing process models in the Main Mass of the North and South Ranges*: The models do not account for the changes in chemistry in the South Range Norite stratigraphy or the development of a lens of melanorite approximately two-thirds of the way through the norite stratigraphy in the Creighton traverse (Chapter 3).

The models support the idea that a single parental magma at Sudbury gave rise to the Main Mass. However the models did not factor in the effect of sulfide saturation and segregation of immiscible sulfide from the melt sheet. The next section of this chapter investigates this process in more detail.

PROCESS MODELS FOR MELT SHEET EVOLUTION AND ORE FORMATION

Metal-depletion signature of the Main Mass

The ubiquitous development of a stratigraphy of noritic rocks in the lower part of the Main Mass with low concentrations of Ni, Cu, and PGE has been established in Chapter 3. It has also been shown that the degree of depletion of the norite stratigraphy in Ni (as indicated by the Ni/Ni* ratio) in the South Range norite stratigraphy is not as strong as that developed in the North Range (stratigraphically- weighted average values for different profiles around the Sudbury Basin are shown in Figs. 3.4A–H and 3.8A–J).

The Ni/Cu ratio of the norites is ~1 (Fig. 3.30B); this is a somewhat unexpected finding given the higher published partition coefficient for Cu into sulfide magma relative to Ni (Keays and Lightfoot, 2004).

The depletion of the Main Mass in base and precious metals was established in Lightfoot et al. (2001) and Keays and Lightfoot (2004). Examples of the stratigraphic control on Pt and Pd abundance in the Main Mass are shown in Fig. 5.9A–D for the North Range profile at Nickel Rim (see also Fig. 3.4A–H) and for a drill core from Levack (drill core 93656).

At Nickel Rim, the Pt concentration falls from 7.3 to <0.20 ppb, and Pd falls from 7.7 to <0.20 ppb with increasing distance upward through the norite stratrigraphy (Fig. 5.9A,B). At Levack, the Pt concentration falls from 10.7 to <0.06 ppb and Pd falls from 15.5 to <0.03 ppb (Fig. 5.9C,D). Palladium

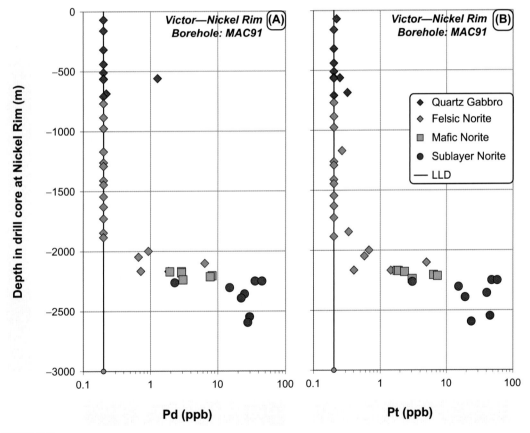

FIGURE 5.9

(A) Variation in Pd abundance through Victor-Nickel Rim Core MAC91. (B) Variation in Pt abundance through drill core MAC91.

(*Continued*)

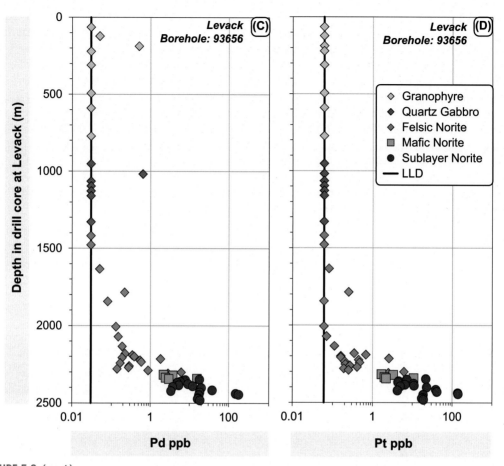

FIGURE 5.9 (*cont.*)

(C) Variation in Pd abundance through Levack drill core 93656. (D) Variation in Pt abundance through drill core 93656. Data source: Keays and Lightfoot (2004). The lower limit of detection for Pt and Pd is shown for each data series. Based on Keays and Lightfoot (2004).

shows a decline by a factor of >700 through the Main Mass norites from Nickel Rim and >1050 in the Main Mass norites from Levack (Keays and Lightfoot, 2004). Likewise Pt falls by factors of >770 and >1550 at Nickel Rim and Levack, respectively.

The variations in PGE concentrations contrast with those of Ni and Cu which decline by factors of ~80 and ~170 in Main Mass norites at Nickel Rim, and ~40 and ~50 in Main Mass norites from Levack (Keays and Lightfoot, 2004).

There are two quandaries raised by these variations that will be discussed later in this chapter:

1. The constant Ni/Cu ratio of the Main Mass rocks indicates that the sulfide stripped similar proportions of Ni and Cu (Chapter 3, Fig. 3.3), yet the partition coefficients for these elements into sulfide are quite different (D_{Cu}~700; D_{Ni}~250; Keays and Lightfoot, 2004).

2. Although the upward decline in Pt and Pd concentrations through the norite stratigraphy is very dramatic, all of the PGE should have been stripped from the bulk of the FNR magma because of their very high partition coefficients into the sulfide melt. Not only are detectable concentrations of PGE present through much of the norite stratigraphy, but there is evidence that the concentrations increase slightly in the upper part of the norite stratigraphy and the QGAB Unit (Keays and Lightfoot, 2004).

Siderophile and chalcophile metal tenor variations in the Main Mass

The concentrations of Ni, Cu, and PGE in the weakly mineralized MNR and FNR are calculated based on the assumption that 100% of the metal is contained in the sulfide component of the rock, and they are provided with the knowledge that the calculation can introduce artifacts in rocks with low sulfide abundance or where there is higher analytical uncertainty associated with the determination of analytes at levels approaching detection limits (Kerr, 2003). The trace sulfide assemblage changes from Po-Cpy-Pn in the MNR and basal FNR to Py through much of the FNR stratigraphy, so the calculation is predicated on the likelihood that the primary sulfide assemblage was a magmatic assemblage, but that some Fe-loss has taken place during low grade metamorphism. Despite this, the variations through the norite stratigraphy at Victor-Nickel Rim (Fig. 5.10A,B) and Levack (Fig. 5.10C; see Keays and Lightfoot, 2004) indicate a remarkably systematic change in metal tenor with stratigraphic position in the Main Mass. As Keays and Lightfoot (2004) point out, this is unlikely to be an artifact of the approach used in the calculation.

In 100% sulfide, the Pt and Pd concentrations systematically decline upward through the stratigraphy. For example, in Levack drill core 93656, both $[Pd]_{100}$ and $[Pt]_{100}$ decline from ~ 1100 to <30 ppb through the Main Mass noritic rocks (Fig. 5.10C). In contrast, $[Cu]_{100}$ decreases from 5.34 to 0.77 wt.% before it exhibits a sharp increase at a depth of 1450 m, whereas $[Ni]_{100}$ decreases from 6.7 to 1.2 wt.%. Hence, while the Ni and Cu tenor of the sulfides decrease by factors of 5.6 and 6.9, respectively, the Pd and Pt tenors of the sulfides decrease by more than a factor of 37.

Fig. 5.11A shows the variation in Cu versus Pd for the samples from Nickel Rim and Levack. Because Pd has a much larger partition coefficient than Cu, Pd/Cu ratios can be used to track the sequence of segregation of sulfides from the magma; the first sulfides to segregate will have a high Pd/Cu ratio whereas the last to segregate will have a very low Pd/Cu ratio. The noritic rocks of the Main Mass show a broad decline in Pd/Cu ratio as Pd and Cu concentrations decrease. The Pd/Cu ratios of the Sublayer Norite (SLNR) samples overlap with, or are higher than those of the MNR. The MIQD from Offset Dykes samples have Pd/Cu ratios that overlap with the SLNR samples having some of the highest Pd/Cu ratios. If all of the rocks (and their sulfides) had formed from the same magma, their Pd/Cu ratios indicate the following order of crystallization: Offsets > Sublayer > MNR > FNR. The Onaping Formation samples and the un-mineralized QD samples have Pd/Cu ratios that are significantly higher than the Pd-depleted FNR samples but somewhat lower than those of the HMIQD and MIQD from the Offset Dykes.

To further highlight the changing composition of the melt sheet during it's evolution, the average Pd/Ni of samples from units that crystallize at different times (early to late) are shown in Fig. 5.11B as averages of different rock units and mineral system styles. The QD has an intermediate Pd/Ni ratio, and this ratio likely reflects the initial composition of the melt sheet prior to sulfide saturation. Mineralized samples from Offsets have higher Pd/Ni relative to the QD and mineralization from contact deposits has a lower Pd/Ni ratio. There is then a progression in North Range rocks toward progressively lower Pd/Ni in the sequence SLNR matrix, MNR, and then samples from the base and top of the FNR. This

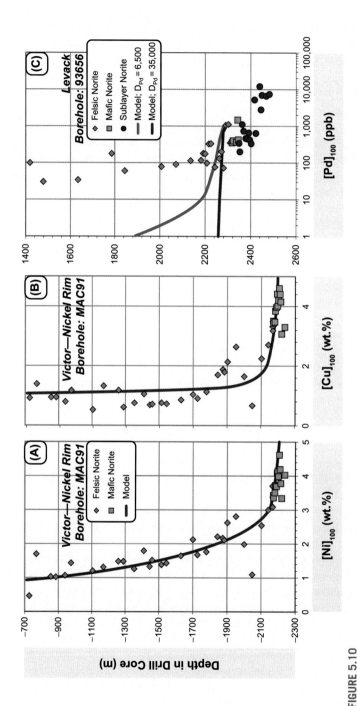

FIGURE 5.10

(A) Variations in $[Ni]_{100}$ modeled as a function of depth in the Mafic and FNR at Victor–Nickel Rim (drill core MAC91). (B) Variations in $[Cu]_{100}$ modeled as a function of depth in in the Mafic and Felsic Norites at Victor–Nickel Rim (drill core MAC91). (C) Variations in $[Pd]_{100}$ modeled as a function of depth in the Mafic and FNR at Levack (drill core 93656). In all three of the plots, the composition of the samples has been plotted against their true depth in drill core following the approach documented in Keays and Lightfoot (2004).

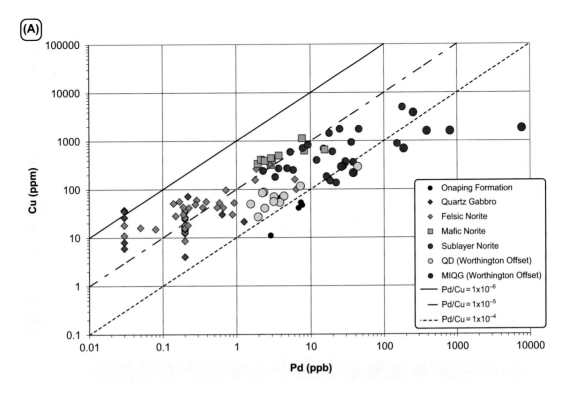

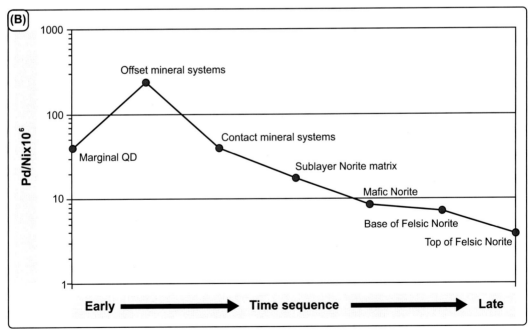

FIGURE 5.11

(A) Variation in Cu versus Pd for Main Mass samples from Victor-Nickel Rim and Levack. Also shown are compositional data for samples of QD and MIQD from Offset Dykes and samples representing melt compositions from the Onaping Formation. The plot shows thee variation in ratio between Offsets, Sublayer, and the Main Mass. (B) Variation in Pd/Ni with sequence of crystallization of the rocks that host mineralization in the SIC and form as result of differentiation of the melt sheet.

Modified from Lightfoot and Evans-Lamswood (2015) with permission of Elsevier.

plot illustrates a progressive change in Pd/Ni that is consistent with the early removal of Pd into sulfide to generate the Offset Dyke mineral systems, later crystallization of the Contact mineral systems from melts that are depleted in Pd, and finally crystallization of the sulfides entrained in the basal part of the Main Mass that have very low ratios of Pd/Ni.

Ni and Cu tenors of sulfide in Mafic Norite above barren versus mineralized contacts

The empirical relationships between the formation of the ore deposits and the sulfide saturation history of the Main Mass led Lightfoot (2009) to investigate the metal tenor variations in the MNR above barren versus mineralized embayments to establish whether these common Main Mass rocks have a geochemical signature that reflects the mineral potential of the Sublayer. This pilot project sampled MNR from four different trough structures of which two were very well-mineralized and the others contain subeconomic mineralization. The samples were chosen to maximize the sulfide content and the analyses were undertaken to establish the variations in Ni and Cu tenor in these samples.

Fig. 5.12 shows $[Cu]_{100}$ versus $[Ni]_{100}$ where the samples are assigned different symbols based on the location. The plot illustrates that the MNR has a wide range of metal tenor that define a broad 1:1 array that has the typical Ni/Cu ratio of 1 that characterizes the Main Mass (see Chapter 3). The samples from above weakly mineralized contacts exhibit a narrow range in metal tenor (Trillabelle and Cascaden),

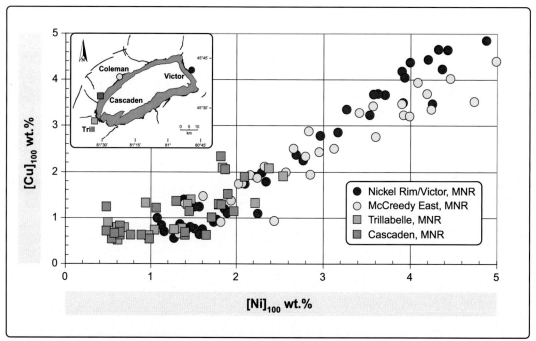

FIGURE 5.12 Variation in $[Ni]_{100}$ Versus $[Cu]_{100}$ for Samples of MNR Developed Above Four Different Embayment Structures (After Lightfoot, 2009)

The calculation of metal tenor is applied for mineralized samples with >0.5%S. The difference in the range of metal tenor developed in MNR above mineralized versus barren troughs or embayments is highlighted in this plot. It appears that the Main Mass of the SIC provides an indication of the mineral potential of the immediate contact.

Modified from Lightfoot and Evans-Lamswood (2015) with permission of Elsevier.

whereas those collected above ore bodies (Nickel Rim, Victor, and McCreedy East) define a wide range in Ni and Cu tenor. A newly completed study by the author of 1200 samples of MNR has confirmed these relationships, and provides a basis to use the geochemistry of the Main Mass to evaluate the mineral potential of adjacent Sublayer.

Modelling of the siderophile and chalcophile metal tenor variations in the Main Mass

Following Keays and Lightfoot (2004), the variations in the $[\text{Ni}]_{100}$ and $[\text{Cu}]_{100}$ in samples of FNR and the MNR have been modeled in order to estimate the initial Cu and Ni contents of the magma that produced these rocks. The approach is also designed to establish whether the sulfides in these rocks are comagmatic with the sulfides in the Sublayer. The modeling uses the Rayleigh Law:

$$C_\text{L} = C_\text{o} * F^{(D-1)}$$

where C_L is the concentration of element in the fractionated silicate melt, C_o is the concentration of the element in the initial silicate melt, F is the fraction of silicate melt remaining after fractionation, and D is the distribution coefficient of the element.

It is also necessary to use the R factor equation that describes the effect of magma/sulfide ratio on the metal content of the sulfide melt:

$$C_\text{s} = C_\text{o} * D(R+1)/(R+D)$$

Where C_s is the concentration of the element in the sulfide melt, C_o is the concentration of the element in the initial silicate melt, D is the distribution coefficient, and R is the mass ratio of silicate melt to sulfide melt (Campbell and Naldrett, 1979).

From an investigation of Sudbury ore deposits, Naldrett (1997) showed that they had R factors that varied from 140 to 1600. For the purposes of their study, Keays and Lightfoot (2004) selected an initial R factor of 1000, which decreased as the volume of silicate melt declined during the solidification of the Main Mass. It was also assumed that the MNR and FNR crystallized as a closed system, except for the loss of 5% of the total volume of silicate melt. The bulk distribution coefficient for the sulfide melt and silicate minerals must also be assigned. In the case of Cu and Ni, this was established based on the modal percentages of sulfide (calculated from the S contents of each sample), the orthopyroxene content (orthopyroxene also extracted a small amount of Ni from silicate melt), and the distribution coefficients of the elements into sulfide melt and orthopyroxene. For the sulfide melt, a partition coefficient of 250 was used for Ni whereas a value of 700 gave the best fit for Cu.

The result of the modeling by Keays and Lightfoot (2004) is summarized in Fig. 5.10 A,B, where the model curves are superimposed on the chemostratigraphy profile of drill core MAC-91 from the Victor-Nickel Rim area, and Fig. 5.10C for the Levack drill core. The variations through the stratigraphy in $[\text{Ni}]_{100}$ and $[\text{Cu}]_{100}$ are reproduced by the models, but there is a weakly developed enrichment in Cu at the top of the FNR in some profiles, and the overlying QGAB commonly shows higher Cu concentrations than the most Cu- depleted part of the FNR (Keays and Lightfoot, 2004). However, most QGAB and GRAN samples are altered and so the Cu in these rocks was likely introduced by the fluids involved in the alteration process.

Reconciling the composition of the melt sheet and the mineralization

The modeling undertaken by Keays and Lightfoot (2004) requires that the initial silicate magma contained 250 ppm Ni and 120 ppm Cu. These estimates do not incorporate the Ni and the Cu present in the sulfide ore deposits nor the non-economic accumulations of sulfides in the Sublayer and Footwall environments, so they must be treated as minimum values. These estimates are also different when compared to the composition of unmineralized QD which has a minimum concentration of 69 ppm Ni and 59 ppm Cu.

Using realistic partition coefficients of $D_{Pd} = 35,000$ and $D_{Pt} = 10,000$, it is not possible to model the distribution of Pd and Pt in the noritic part of the Main Mass at Victor (Fig. 5.10C). Although the FNR is very strongly depleted in the PGE, the use of these partition coefficients indicates that it should become depleted far more rapidly than would have been possible if the norites had formed in a closed system. It is possible to model the observed variations half way through the FNR using $D_{Pd} = 2500$, but above this height the model predicts that Pd should decrease more rapidly than it does (Fig. 5.10C). However, a partition coefficient of 2500 for Pd into sulfide is not supported by experimental studies (Keays and Lightfoot, 2004). Although the distribution of Pt and Pd through the noritic stratigraphy at Victor could not be modeled, the approach does provide an estimate of the Pd and Pt contents of the magma at the start of crystallization of the FNR (ie, 1.1 ppb Pd and 1.2 ppb Pt). A considerable amount of the PGE in the initial SIC magma would have been removed in the sulfide melts that segregated from the magma to produce the Ni-Cu-PGE sulfide ore bodies as well as the significant amount of noneconomic Ni-Cu-PGE sulfides in the Sublayer and Offset Dykes.

Using an R factor of 1000, the Ni, Cu and PGE of the silicate magma from which the sulfides formed can be calculated. This calculation assumes the composition of the Ni-Cu-PGE sulfide ore bodies and the MNR as shown in Table 5.1 from Keays and Lightfoot (2004).

The Totten mineralization has a composition consistent with a parental silicate magma containing 310 ppm Ni, 310 ppm Cu, 18 ppb Pd and 19 ppb Pt (Table 5.1). The Creighton mineralization, which is a member of the contact type South Range ore bodies, required a silicate melt with 265 ppm Ni and 170 ppm Cu, 2.7 ppb Pd and 2.7 ppb Pt. The compositions of the sulfide in the MNR indicates that the SIC magma contained 210 ppm Ni, 110 ppm Cu, 1.1 ppb Pd and 1.3 ppb Pt at the commencement of crystallization of the Main Mass norites. The model calculations also indicate that the unmineralized inclusion-free QD of the Worthington Offset, that contains a minimum of 69 ppm Ni and 59 ppm Cu,

Table 5.1 Observed Ni, Cu, Pd, and Pt Contents of SIC Ores and Calculated Composition of Silicate Melts From Which They Formed

	Observed Averages						Calculated Parental Magma			
	$[Ni]_{100}$ (wt.%)	$[Cu]_{100}$ (wt.%)	$[Pd]_{100}$ (g/t)	$[Pt]_{100}$ (g/t)	$[Ir]_{100}$ (g/t)	10000* Pd/Ni	C_o (Ni) (ppm)	C_o (Cu) (ppm)	C_o (Pd) (ppb)	C_o (Pt) (ppb)
Offset Deposits	6.2	12.8	18	19	0.28	2.90	310	310	18	21
Contact Deposits	5.3	7.1	2.2	2	0.01	0.42	265	170	2.3	2.2
Mafic/FNR	4.2	4.5	0.6	0.45	0.02	0.14	210	110	1.1	1.2
SIC Magma							61*	59*	3.9*	4*

Composition of the silicate liquid believed to be in equilibrium with the sulfide mineralization contained in the Offset Dykes, Sublayer, and Main Mass sulfides at Sudbury (from Keays and Lightfoot, 2004). Co is the initial concentration in the parental magma.

*Observed minimimum metal concentrations in QD and Onaping Formation melts (Keays and Lightfoot, 2004).

could not have been in equilibrium with the massive sulfide of the Totten Deposit (Table 5.1). This is consistent with the field evidence that indicates that the unmineralized QD is a separate intrusive phase of the Offset Dyke (see Chapter 2).

Lightfoot et al. (1997b, 2001) demonstrated that the QD which comprises the quenched marginal phase of the Offset Dykes provides an indication of the bulk initial composition of the Main Mass of the SIC. Although the compositions of the QDs do vary between Offset Dykes (Chapter 3), the samples contain a minimum of 69 ppm Ni, 59 ppm Cu, 3.9 ppb Pd and 4.0 ppb Pt (Table 5.1).

An independent estimate of the bulk composition of the SIC is provided by the suevite breccias, glassy fragments, and intrusions in the Onaping Formation (Ames et al., 2002). The composition of these rocks represents melts, and they should represent the average composition of the country rocks impacted by the meteorite. The match will not be perfect because the source rocks for the Onaping Formation would have been at somewhat higher crustal level than the crustal rocks that formed the melt sheet. The whole rock samples of suevite from the Onaping Formation have averages of 55 ppm Ni, 48 ppm Cu, and 4.9 ppb Pd (Keays and Lightfoot, 2004). These values are close to the QD of the Offset Dykes. Ames et al. (2002) state that the best estimate of the parental composition of SIC magma is provided by the least altered vitric material in the Onaping Formation; for these they obtained average values of 62 ppm Ni and 51 ppm Cu. The starting parental composition of the melt sheet as estimated from the whole rock samples and the least altered vitric material in the Onaping Formation and the QD of the Worthington Offset is estimated to be 61 ppm Ni, 59 ppm Cu, 3.9 ppb Pd and 4.0 ppb Pt (Keays and Lightfoot, 2004).

Because of their very high partition coefficients, it is expected that the PGE would have been removed from the SIC magma once it became sulfide-saturated at a much faster rate than the base metals, which have high but nevertheless much smaller partition coefficients than the PGE. The relative timing of sulfide segregation from the SIC magma should be provided by ratios such as Pd/Ni, with the earliest formed sulfides having the highest Pd/Ni ratios and the latest-formed sulfides having the lowest Pd/Ni ratios (Fig. 5.11B). As indicated by their Pd/Ni ratios (Table 5.1 and Fig. 5.11B), the sulfides in the SIC system appeared to have formed in the following sequence: Offset Ores > Contact Ores > SLNR > MNR > FNR (base) > FNR (top). Although Pd/Ni ratios drop off rapidly through this sequence, they do not drop off nearly as quickly as predicted from the partition coefficients of Pd ($\sim$35,000) and Ni ($\sim$250). In particular, the segregation of the Offset and Contact sulfides from the melt sheet should have reduced the Pd/Ni ratio of the melt to almost zero as well as removing the bulk of the Cu and Ni, and all of the PGE, from the SIC magma. The sulfide contents of the Sublayer in drill cores MAC-91 and 93656 are 2.9 and 2.1%, respectively (Keays and Lightfoot, 2004). As the Sublayer can be up to 500 m thick, it contains a very significant amount of Ni-Cu-PGE-rich sulfides. The fact that the MNR contains 1.2 ppb Pt and 1.1 ppb Pd after the SIC magma commenced crystallization means either that the PGE were not scavenged and removed efficiently in sulfide from the magma (that is, the partition coefficients are not applicable or the sulfides were not removed from the melt). Alternatively, it is possible that Pd and Pt were added to the magma after the formation of mineralization hosted in the Offset Dykes and at the lower contact of the SIC.

The observed sulfide ore bodies as well as modeling of the initial chalcophile metal content of the Main Mass magma require parental magmas with higher Ni, Cu and PGE concentrations, as well as different Cu/Ni ratios when compared to the quenched margins of the Offset Dykes. Keays and Lightfoot (2004) noted several possible factors that may help to reconcile the results of the mass balance calculations with possible observed melt compositions; these are discussed below.

Sulfur Capacity of the Melt Sheet

The high S contents of the Sublayer and the FNR, and the strong chalcophile metal depletion of the norite stratigraphy indicate that the magma from which these rocks formed was sulfide-saturated. The geological relationships and trends in geochemistry between the rocks of the Main Mass and the mineral systems provide strong evidence that the ores segregated from the melt sheet.

The high PGE content of the sulfide ores requires the SIC parental magma to have been S-undersaturated when it was formed (Keays, 1995). The SIC magma therefore contained a higher concentration of sulfur than most other silicate magmas at a similar Mg-number. Although Latypov et al. (2010) suggest otherwise, it is likely that the initial melt sheet was superheated, possibly having a temperature of 1700°C (Ivanov and Deutsch, 1997). The solubility of sulfur and sulfides in silicate melts increase significantly with increasing temperature (Haughton et al., 1974). Naldrett (1989b) estimated that the sulfur capacity of a silicate melt increases by a factor of 2.5 over the temperature range 1200–1450°C. Extrapolating Naldrett's estimate of the temperature effect on the S capacity of the SIC melt, it is probable that the S capacity at 1700°C was a factor of five times greater than at 1200°C, the approximate liquidus temperature of the Sudbury melt sheet (Keays and Lightfoot, 2004).

Effect of Superheat on Sulfide Solubility in the Melt Sheet

It was proposed by Keays and Lightfoot (2004) that the SIC melt was S-oversaturated at its liquidus temperature and so sulfide melts formed well in advance of the crystallization of any silicate or oxide minerals. The S capacity of the melt that formed the Main Mass norites would have been ~400 ppm S at the liquidus temperature of the melt. However, the bulk integrated S content of the FNR and MNR at Nickel Rim is 1070 ppm, assuming FNR and MNR in the proportions 95:5. If the norites crystallized as a closed system (as indicated by the modeling in Keays and Lightfoot, 2004), then their parental magma contained 2.7 times more sulfur than it should have at 1200°C, which was nominally the liquidus temperature of the melt. However, all of the sulfur present in the norites would have been dissolved in the superheated melt sheet. This melt would have become sulfide-saturated at ~1450°C. Hence, sulfides commenced segregation from the magma at ~250°C above the temperature at which silicates appeared on the liquidus. These sulfides would have segregated to form the sulfide mineral systems in the Sublayer and the Offset Dykes. In terms of the timing of segregation of the sulfides from the magmas, the SIC magma was therefore quite unique when compared to any other known terrestrial magma.

Additional Assimilation of Mafic Fragments and Footwall Rocks

Assimilation of mafic inclusions and footwall rocks into the melt sheet may have continually changed the composition of the melt (Golightly, 1994). Because of their much higher melting temperatures, mafic rocks in the target area would have melted more slowly than the felsic rocks. Moreover, because of their high density, they would have accumulated along the base of the superheated melt sheet as it thermally eroded its way downward into the un-melted, but very hot recrystallized country rocks below. It is envisaged that at least some of the mafic rocks remained as un-melted refractory blocks (eg, the diabase hornfels inclusions in the Sublayer at Whistle Mine described in Chapters 2 and 4) whereas others melted to form a significant proportion of the Sublayer matrix (Lightfoot et al., 1997c).

Possible Contribution of Magma from the Mantle

There is no evidence that the Sudbury impact event cracked the crust, and triggered decompression melting in the mantle, although this has been suggested as a mechanism for the formation of large igneous provinces (Jones et al., 2002). The base of the transient crater at the time of a Sudbury-scale impact was likely at ∼10–15 km below the surface, and in an active cratonic margin setting the crustal thickness was likely much larger.

Ivanov and Melsoh (2003) undertook numerical modeling of the impact of an asteroid with a diameter of 20 km striking at 15 km/s into a target with a near-surface temperature gradient of 13 K/km (cold crust) or 30 K/km (hot crust). The impact creates a 250–300-km-diameter crater with an impact melt sheet. However, the crater collapses rapidly to a flat structure, and the pressure field rapidly approaches the initial lithostatic condition. Even an impact this large cannot raise mantle material above the peridotite solidus by decompression.

If mantle-derived magmas were added to the lower part of the melt sheet prior to formation of the first sulfides, then the metal content and ratios of the parental magma would be different when compared to estimates established based on the composition of the QD.

Lightfoot et al. (1997b) argue that the addition of up to 20 % of a mantle-derived magma of picritic composition to the melt sheet would not be detectable in the chemistry of the SIC because the geochemical signature would be swamped by the strong crustal signature of the melt sheet. High-MgO tholeiitic magma or a picrite would be denser than the melt sheet and so would be emplaced into its lower portion. The addition of this type of magma would not only increase the Ni/Cu ratio of the resulting hybrid magma but it might also help to account for the presence of fresh primary mafic to ultramafic inclusions and the high proportion of orthopyroxene and chrome spinel in the Sublayer and Norites (Lightfoot et al., 1997b, Zhou et al., 1997). However, as discussed in Chapter 3, the geochemistry of these rocks has an overwhelming contribution from the crust.

There is no geological or geochemical evidence to indicate that the cratering process provided the energy required to trigger adiabatic melting in the mantle, created the pathways required for efficient migration of the magma from the mantle and the surface, or explains the localization of these melts within the Offset Dykes and the Sublayer that represent an intrinsic part of the impact-derived melt sheet. The Offset Dykes which might be the closest analogue to feeders from the mantle contain quartz diorite that is geochemically similar to the Main Mass in terms of incompatible trace element and isotope ratios, so there is no evidence for a different chemical contribution to a hypothetical series of feeder dykes. Moreover, there is no geochemical evidence to prove that a contribution from the mantle was incorporated into the Main Mass or Offset Dykes. Despite the location of Sudbury at the southern margin of the Superior Province (as noted by Begg et al., 2011), there is no evidence of other magmatic events at 1.85 Ga that might be due to explosive emplacement of highly contaminated mantle-derived melts along transcurrent faults inboard of a cratonic margin (Lightfoot and Evans-Lamswood, 2015). There is every reason to believe that the coincidence of the impact site with the cratonic margin is serendipitous.

Proto-ores Tapped by the Sudbury Event?

Although the mafic intrusive rocks in the immediate footwall of the SIC tend to have a relatively low sulfide content, both the Nipissing Gabbro and the East Bull Lake type intrusions of the Southern Province of the Canadian Precambrian Shield contain appreciable levels Cu-Pd-Pt-Ni sulfide mineralization, and some of these intrusions in the Sudbury Region contain small sulfide occurrences (see Chapter 2 for a detailed review of these rock types and their distribution). It is possible that there were significant amounts of this type of mineralization in the SIC target rocks and that this provided proto-ores to the melt sheet (Keays, 1995). However, Cu-PGE-Ni sulfide mineralization in East Bull Lake has Cu/Ni ratios of ∼3 (Peck et al., 2001) whereas the Cu-PGE-Ni sulfide mineralization in the Nipissing Gabbro has Cu/Ni ratios of ∼2 (Lightfoot et al., 1997b). On this basis, the Cu/Ni ratios of possible proto-ores are too high to be major contributors of Cu and Ni to the SIC melt sheet, unless their contribution was counter-balanced by contributions from either dissolution of mafic rocks in the target area or high-Mg magmas.

Any contribution that proto-ores made to the melt sheet must have been made before it became sulfide-saturated and therefore while it was still capable of dissolving sulfides. These proto-ores would have had to be dissolved before the SIC magma became sulfide-saturated and would have shifted the composition of the melt sheet in the wrong direction relative to the observed Cu/Ni ratios of average Sudbury mineralization.

Onset of Sulfide Saturation in the Melt Sheet

The Sudbury melt sheet varied in thickness up to 5000 m, and the base of the melt sheet had a primary topography that may have approached the scale of the melt sheet thickness. Unlike normal sills, the melt sheet probably had a large thermal gradient from base to top. Much of the Onaping is composed of hyaloclastites that are probably products of explosive magmatic processes; this meant that the top of the melt sheet cooled rapidly by interaction with water. The melt sheet is a product of decompression melting; and the temperature at the base of the melt sheet was clearly sufficient to generate partial melts as recorded in the textures of the rocks in the wide metamorphic halo beneath the SIC (Chapters 2 and 3).

Keays and Lightfoot (2004) suggested the possibility that sulfides first started forming at the top of the melt sheet, where it was cooler, and then, due to their density, sank through the sulfide-undersaturated melt sheet toward the base. The sulfides would probably acquire additional chalcophile metals as they settled through the sulfide-undersaturated melt, but this process would produce less efficient depletion of the Ni, Cu, and PGE concentrations in the magma. In this model the sulfide-saturation interface would move downward through the melt sheet as it cooled from the top toward the center (as indicated by the reverse fractionation trend in the GRAN described in Chapter 3). If they sank quickly, they would not interact with the magma, and all of the chalcophile metals would not be stripped from the melt.

SULFIDE SATURATION HISTORY OF THE SUDBURY MELT SHEET

The origin of the Ni-Cu-PGE sulfides at the lower contact of the Sudbury melt sheet is now explained by sulfide saturation, and segregation of magmatic sulfides from a melt sheet generated by meteorite impact of crustal rocks. This model is shown in Chapter 3 (Fig. 3.33A–F).

The variations in PGE abundance have several implications with respect to the sulfide saturation history of the SIC:

1. The degree of depletion of the PGE through the norites is much more rapid and extreme than the depletion in Ni and Cu. This is partly because the partition coefficients for Pd and Pt into sulfide liquid are ~1.5 orders of magnitude higher than those of Ni and Cu.
2. The high PGE contents and Pd/Cu ratios of the SIC ores indicates that the melt sheet was initially sulfide-undersaturated and PGE-undepleted.
3. The high Pd/Cu ratios of the unmineralized QD phase of the Offset Dykes indicate that the silicate melts which formed these rocks were sulfide-undersaturated.
4. The evidence points to the fact that the PGE were not entirely stripped from the Main Mass magma when it first became S-saturated and formed the Offset ores. The process of segregation and accumulation appears to have been inefficient because the initial saturation and segregation of the sulfides that formed the Offset and Sublayer ores should have entirely depleted the overlying

melt sheet. The fact that the melt sheet was not totally depleted in PGE implies either that metals were continually available to the melt sheet, perhaps by ongoing assimilation of mafic fragments or by incorporation of metals from a S-undersaturated portion of the melt sheet.

5. The marked increase in Pd, Pt, Ni, and Cu concentrations mid-way through the FNR of the Levack traverse indicates that a new supply of metals was available to the Main Mass magma at this level in the chamber.

The metal concentration data for traverses (surface and drill core) collected around the Sudbury Basin indicates that metal depletion is a widespread feature that is found in 40 different traverses (Fig. 3.29A–E). The metal depletion signature is developed in both thick and thin FNR sequences above embayments on the North, East, and West Ranges, as well as above thick and thin domains of barren contact. The noritic rocks of the South Range are also Ni- and Cu-depleted (Fig. 3.29A–E).

The PGE chemistry of the HMIQD and Sublayer mineralization indicates a wide range in sulfide metal tenors. Keays and Lightfoot (2004) showed that data from the Worthington and Copper Cliff Off-set ores have very high Pt and Pd tenors in sulfides that have Cu/Ni of ~1.5-2 (Table 5.1). In fact there is a considerable variation in metal tenor in different mineral systems within the Copper Cliff Offset as discussed in Chapter 4. The same is true of other mineral systems such as Frood-Stobie. It appears unlikely that the different mineral zones formed from a single parental sulfide melt composition.

Unfractionated sulfides in the contact Sublayer have Cu/Ni ~1, and they have much lower $[Pt]_{100}$ and $[Pd]_{100}$ than the average composition of the mineralization in the deposits hosted by the Offset Dykes (Table 5.1). The sulfides hosted in the Offset Dykes were injected with a subvertical component along broadly plunging shoots from the overlying melt sheet (Chapter 4), and so it is possible that they represent a range of early formed sulfide melts with abundance levels reflecting primary partitioning of the metals into different sulfide melts (Chapter 4). The possibility that the high Cu/Ni ratios and the high PGE contents of sulfides in the Offset Dykes may be a product of fractionation of a sulfide melt has been discussed in Chapter 4, and it appears likely that vertical variations in ore shoots from Frood-Stobie, Copper Cliff, and Totten mineral systems are due to increased degrees of sulfide fractionation of the sulfide melt with depth in the Offset Dyke.

The high Cu, Pd and Pt content of mineralization in the footwall environments described in Chapter 4 are believed to be the products of extreme fractionation of a sulfide melt, the Fe-Ni- and Ir-rich but Pd-poor monosulfide solid solution "restite" remaining in the Sublayer while the Cu-Pd-Pt-rich residual melt was injected into the footwall (Keays and Crocket, 1970; Li et al., 1992). For example, contact mineralization at Strathcona contains 71 ppb Pd and 46 ppb Ir whereas chalcopyrite stringers in the deep footwall at Strathcona have 926 ppb Pd and 1.0 ppb Ir (Keays and Crocket, 1970). The Offset mineralization has an average Pd/Ir ratio of ~60, which is a factor of 15 less than that of the Cu-rich footwall mineralization (Table 5.1). On these empirical grounds, the sulfide mineralization in Offset Dykes is not as strongly fractionated as the mineralization developed in contact-footwall mineral systems such as Levack, Coleman, and Victor (see Chapter 4).

Sequence of events in the formation of the SIC

The following constraints are added to the model shown in Fig. 3.33A-F that accounts for the observed variations in the silicate host rocks and mineral systems at Sudbury:

1. Meteorite impact generates a superheated (~1700°C) melt sheet that completely dissolves all sulfide from the target rocks; the melt sheet initially contained ~61 ppm Ni, ~59 ppm Cu, ~3.9 ppb Pd

and ~4 ppb Pt as reflected in the composition of QD from Offset Dykes and melts from the Onaping Formation.

2. Fragments of refractory mafic country rocks accumulate along the floor of the melt sheet, which continued to thermally erode the underlying country rocks to extend primary depressions in the country rocks and form the trough and embayment structures that would eventually contain the Sublayer.

3. As it cooled, the upper part of the melt sheet achieved sulfide-saturation; the resultant Ni-Cu-PGE-rich sulfides settled through the melt sheet but re-dissolved on reaching the lower part of the melt sheet, thereby increasing the concentration of Cu, Ni and PGE in the silicate melt. Additional Ni (and other metals) was introduced, either by dissolution of the mafic country rock fragments or digestion of sulfide.

4. The lower portion of the melt sheet became sulfide-saturated and Cu-PGE-rich Ni-Fe sulfides accumulate along with the refractory mafic country rock fragments. The fragment-laden sulfide was redistributed down the sloping melt sheet floor into depressions (troughs, embayments and funnels) at the base of the melt sheet.

5. The floor of the chamber was breached and the sulfide melts, with both exotic and locally derived country rock fragments were swept into the Offset Dykes by the still unfractionated SIC magma. These magmatic breccias would form the MIQD, HMIQD and ore deposits which now occupy the Offset Dykes.

6. Additional Cu, Ni and PGE are added to the lower part of the melt sheet from metal un-depleted magma and/or assimilation of mafic country rocks.

7. Additional sulfide melts segregate from the melt sheet, accumulating in depressions in the chamber floor, along with additional mafic rock fragments, to form the contact ore bodies that occupy embayments and troughs at the base of the SIC; these sulfides have lower Cu, and PGE contents than the first sulfides to segregate from the melt sheet.

8. Silicate minerals finally started to crystallize and sulfides and silicates continued to crystallize together as the melt sheet differentiated.

9. During this process, additional amounts of the PGE are contributed from sulfide-undersaturated segments of the melt to the underlying system where silicates and sulfides were undergoing cotectic crystallization.

MELT SHEET THICKNESS: PROSPECTIVITY OF THE CONTACT AND OFFSET DYKES

The established fact that the norite stratigraphy of the Main Mass is depleted in Ni and Cu by a factor of about 50% and the PGE by a factor of >90% supports the idea that the formation of the magmatic sulfide ore deposits occurred by stripping the siderophile and chalcophile metals out of the silicate magma before it began to crystallize. The lateral extent of depletion in the Main Mass documented in Chapter 3 (Fig. 3.29E) is indicative of a sulfide saturation event on the scale of the whole lower noritic part of the melt sheet.

Distribution of mineralization

Much of the SIC is explored and understood to a depth of 1 km, but as depth increases, the amount of drilling becomes more focused around known mineral systems or it is more sparsely distributed. Notwithstanding, the information is good enough to build a three dimensional rendition of the SIC to this depth. Fig. 5.13A is a three dimensional rendition of the base of the SIC and Offset Dykes where the extent of the contacts to a depth of 3 km is shown based on the three dimensional model of the Sudbury

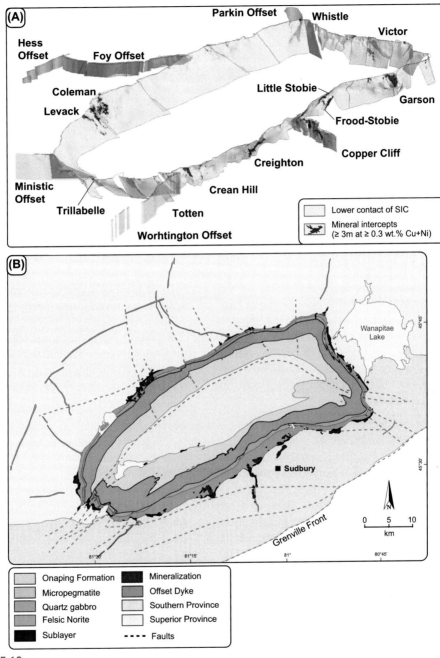

FIGURE 5.13

(A) Three-dimensional model of the SIC showing a wireframe of the basal contact of the Sublayer and the Offset Dykes to 3 km depth. The composited Cu + Ni data are from Lightfoot et al. (2002). (B) The surface distribution of ore bodies around the SIC.

Modified from Lightfoot and Evans-Lamswood (2015) with permission of Elsevier.

Basin created by Gibson (2003). Superimposed on this diagram are the composited Cu + Ni assays over 3 m where the samples from drill core have a Cu + Ni grade exceeding 0.3% (Lightfoot et al., 2002 and Lightfoot, 2009).

Fig. 5.13A shows the distribution of mineral intercepts around the Sudbury Basin and along the length of the Offset Dykes. The heterogeneous distribution of composited Cu + Ni grade values is a reflection of two factors:

1. The availability of data for the whole Sudbury Basin: because much of the data is not in the public domain, the distribution of assays has used the historic information from Lightfoot et al. (2002) which included the dominant land positions held by Inco and Falconbridge in 2003.
2. The extent of drilling which tends to be 2–3 orders of magnitude more dense in areas with known mineral systems.

Looking at the distribution of assays associated with the base of the SIC and the Offset Dykes, two main points can be made:

1. The distribution of abundant Cu + Ni assays at the base of the SIC broadly correlate with the location of the embayment and trough structures which contain Sublayer at the base of the SIC, but not all embayment and trough structures appear to be equally endowed with metal. For example the record of assays at Creighton is indicative of a large mineral system in a series of troughs, but the record at Trillabelle shows only two smaller clusters of assays within an equally large trough structure at the base of the SIC.
2. The assays from the Offset Dykes illustrate a tendency for some Offset Dykes to have heavy mineralization (eg, Copper Cliff and Frood-Stobie), other Offset Dykes to have fewer mineral zones that are more episodically distributed along the dykes (eg, Worthington), and yet other Offset Dykes have small deposits (eg, Parkin and Foy) or mineral occurrences (eg, Manchester, Hess, and Ministic).

The surface projections of most of the ore bodies were provided by the Sudbury mining companies to Ames et al. (2005) in support of an effort to update the geology of the Sudbury Region. Their distribution projected to surface is shown on a simplified map of the Sudbury Basin in Fig. 5.13B. This map, when used in conjunction with the three dimensional spatial distribution of assays, shows that the majority of the clusters in Fig. 5.13A correspond to the known ore deposits, and Fig. 5.13B also shows that the distribution of the ore deposits is heterogeneous. In both the cases of Fig. 5.13A,B, the information is dependent on the level of exploration, but both diagrams show an extreme heterogeneity in the distribution of the known mineral systems.

There are two possible explanations for this distribution. In the first, the variations would be attributed to the extent of exploration and/or availability of data. This is exceptionally unlikely as the Sudbury Basin has received extensive exploration to a depth of ~1.5 km, and most major mineral zones and significant mineral intercepts are shown on Fig. 5.13B. A second explanation would have a heterogeneous primary distribution of mineralization at the base of the SIC. The second alternative is more plausible because it is well-established that major mineral zones are linked to the troughs and embayments at the base of the SIC and the Offset Dykes. But the distribution of mineralization associated with these features is still heterogeneous. If there are primary controls on metal distribution, then this is of paramount importance to the companies that explore for mineralization in the Sudbury Basin.

Thickness of the metal-depleted Norite stratigraphy

The observation that the norite stratigraphy of the Main Mass is depleted in Ni, Cu, and PGE lends credence to a model for the formation of the ore deposits by sulfide saturation and settling within the melt sheet as discussed in Chapter 3 (Fig. 3.33A–F). In this hypothesis, the largest ore deposits would be present in areas where there is either a greater volume of overlying magma from which the metals can be extracted to form immiscible sulfide, or the distribution would reflect the effectiveness of transportation of dense sulfide melt into and along trough structures at the base of the SIC, and more locally through funnels into the Offset Dykes.

Looking first at the thickness of the norite stratigraphy, Fig. 5.14 shows a simplified map of the SIC where the outline of the Main Mass is colored based on the estimated thickness of the Main Mass noritic units (Lightfoot et al., 2002) based on surface outcrop, drill core data, and compiled sections. The norites included in this analysis are those of the Main Mass that are developed above the Sublayer and below the QGAB.

In some segments of the SIC, it is not possible to estimate the thickness of the norite stratigraphy because the rocks are pervasively faulted and sheared; this is the explanation for the unconstrained thickness in the SW and SE corners of the SIC where the South Range Shear Zone, and major fault structures

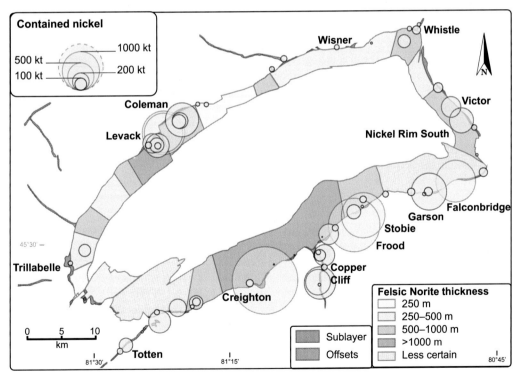

FIGURE 5.14 Variations in Thickness of the Norites of the Main Mass Around the Sudbury Basin and the Distribution of the Ore Deposits Where the Symbols are Scaled to Total Contained Metal in Production Plus Unmined Reserves and Resources

Modified after Lightfoot et al. (2002).

enter the Sudbury Basin. The extent of deformation in the East Range is also significant along faults and shears, so the estimates of thickness of the norite stratigraphy should be treated with some caution insofar as they are interpreted based on sections that contain splays of the curved fault system that dissects the base of the SIC in the East Range. The estimates of thickness of the Main Mass norites in the North Range are robust as there is no significant fabric or faulting in the norite stratigraphy. The thickness in the South Range are reasonably robust estimates based on drilling and surface data, but again, there is some fault and shear-related deformation that results in removal or repetition of stratigraphy. This is especially true of the southernmost major structure of the South Range Shear Zone, the Cliff Lake Fault which penetrates the norite stratigraphy at the western and eastern ends of the South Range.

The break-out of norite thickness into major groupings (0–500, 500–1000, 1000–1500 m true thickness) is also designed to side-step the issue of small deviations in thickness due to structural effects; at this scale, the variations are primary and not as influenced by deformation.

The variations in Main Mass norite thickness in the primary melt sheet (Fig. 5.14) shows quite a significant range around the Sudbury Basin. Areas with very thick norite stratigraphy are located in the South Range reaching a maximum for the SIC in the areas of Creighton, Copper Cliff, and Frood-Stobie. In the North Range, thick norite stratigraphy is recognized in the area of Levack and Coleman. Thin norite stratigraphy is developed in the North Range at Trillabelle and Wisner. Intermediate thickness of norite is developed in the Whistle and East Range, but here there is some influence from deformation, so the original thickness is less certain.

One of the striking features of Fig. 5.14 when compared to Fig. 5.13A,B is the fact that the clusters of known ore bodies located both at the base of the SIC and within the Offset Dykes tend to be localized in areas where the primary melt sheet was thickest. This point is also illustrated in Fig. 5.14 where the scale of the Ni resources is shown. Those segments of the basal contact of the SIC with no significant mineral systems and the Offset Dykes developed proximal to the SIC with minor sulfide occurrences tend to be located in association with segments of the SIC that had a thin primary norite thickness.

Relationship between norite thickness and mineralization

The relationship between the thickness of the SIC and the extent of known mineral systems can be evaluated in areas with least deformation and most geological information from outcrop, drilling, and underground mapping. The most reasonable way to establish the relationship is to develop an algorithm that relates the thickness of the norite stratigraphy to the historic contained nickel from the adjacent contact and Offset Dyke deposits. As explained above, this model is dependent on the extent of drilling, but it serves a purpose in illustrating the relationship between mineral potential and thickness of the Main Mass norite.

The variation in historic Ni production and un-mined metal in reserves and resources was presented in Chapter 1. The mineral systems that contain the ore deposits comprise a large amount of mineralization that is not included in the production, or reserves and resources. There is an overall tendency for large ore deposits to have a large associated footprint of subeconomic mineralization, whereas smaller ore bodies tend to occur within the footprint of smaller domains of subeconomic mineralization. As a first approximation, the use of contained Ni is a basis for evaluating the relationship between mineral potential and norite thickness.

Scale of sulfide localization in the melt sheet

One of the principal controls on the distribution of mineralization in the SIC is the effect of gravitational accumulation of dense sulfides into troughs and embayments at the base of the sloping sides of the primary crater. Mineral systems developed under a very thin melt sheet will be smaller by virtue of not only the thickness of the overlying norite stratigraphy, but also because of the thickness of the adjacent norite stratigraphy. The slope of the crater floor and the controlling effects of embayments and troughs collectively govern how efficiently the sulfide melt is collected into one location (Fig. 3.33F). The importance of this control can be illustrated by comparing the mineral potential in areas of thick and thin melt where the radius of the collection area of sulfide is scaled to the thickness of the norite stratigraphy.

Fig. 5.15A shows a simple model where the thickness of the melt sheet is assumed to be related to the collection area at the base of the SIC. In this model, a simple cylindrical model is applied to calculate the quantity of nickel in sulfide that would accumulate within a small system versus a large system. The model assumes 50% efficiency in nickel depletion of the Main Mass norite with an initial melt represented by average QD (60 ppm Ni). The model allows the sulfide to be distributed in one or more physical depressions at the base of the SIC, but it is indicative of the scale of mineral system that might be expected in areas of thin versus thick norite stratigraphy (Fig. 5.15B,C).

Models of sulfide localization are sensitive to the volume of melt from which sulfide was collected and the thickness of the melt sheet. Fig. 5.15B,C illustrates the effect of varying the scale of the melt body that gave rise to the sulfides for large, medium, and small volumes of silicate magma. This effect can also be expressed as a plot of thickness of the Main Mass norite (above the MNR) and estimated contained metal in the associated mineral system. The curve in Fig. 5.15D show the contained metal in a cylindrical footprint stripped from the melt at a constant ratio of thickness:diameter (Fig. 3.15A) where the system is 50% efficient. If localization of the sulfide is only 50% efficient when the sulfide reaches the base of the SIC, the lower curve in Fig. 5.15D gives an indication of the available metal to form ore deposits.

Superimposed on Fig. 5.15D are examples of mineral systems developed in the Offset Dykes and at the basal contact of the SIC. The estimated size is a combination of historic production plus un-mined reserves and resource (Chapter 1). There is a general tendency for the ore deposits beneath thin segments of norite to be small, whereas the large and very large ore deposits reside in areas of thick norite stratigraphy. Clearly this diagram is designed to show empirical relationships, and it does not take into account the fact that some of the mineral systems are partially removed by erosion and others may extend to depths that have not yet been discovered, and may be very challenging to mine.

Also shown in Fig. 5.15D is the average $[Ni]_{100}$ in samples from different mineral systems with $1 \geq S$ (wt.%) ≤ 5 based on rocks forming the Offset Dykes (dominantly MIQD), and igneous-textured noritic rocks from contact environments (see later in this chapter and Table 5.4). There is a broad tendency for higher metal tenor disseminated sulfides to develop in areas of thick norite and lower metal tenor mineral systems to develop beneath domains of thin norite. The same general relationships are evident from the descriptions of metal tenor provided in Chapter 4.

Very simple mass balance calculations (Table 5.2) based on estimates of the composition of the melt sheet indicate that the magma contained more than enough metals to form the known ore deposits. Mineralization at Sudbury probably extends to depth below the deepest exploration that has been undertaken up to this time.

Based on the relationships shown in Fig. 5.15D it is evident that thick melt sheet domains have a greater potential to form sulfide mineralization with elevated $[Ni]_{100}$ when compared to thin domains of the melt sheet.

In summary:

1. *There is a linkage between melt sheet thickness and quantity of mineralization at the base of, or beneath, the SIC*: The variations in norite thickness are broadly consistent with the amount of

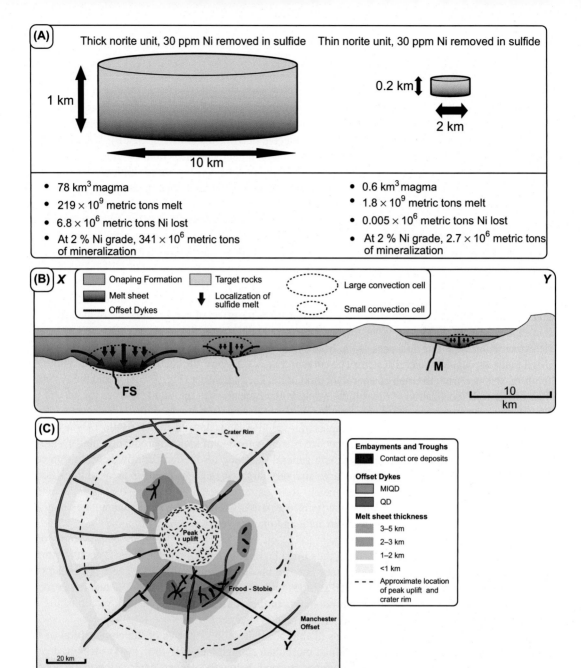

FIGURE 5.15

(A) Model showing the amount of metal lost to sulfide in cyclindrical portions of the norite stratigraphy. The significance of the cyclinder size is emphasized with respect to the simplified cross section and cartoon of relationships in a sheet of melt. The calculation assumes a starting melt with 60 ppm Ni (QD), with 50% of this metal removed in sulfide. This is a minimum estimate of parental magma composition if the interpretations of melt composition from sulfide chemistry is correct (Keays and Lightfoot, 2004) (B) Schematic section through the Sudbury melt sheet showing volumetrically large versus small collection cells in relation to the primary distribution of the melt sheet. M- Manchester Offset; FS - Frood-Stobie Deposit (C) Model illustrating one possible scenario for the primary distribution of the convection cells in the primary impact melt sheet. (D) Plot showing the relationship between norite thickness and expected scale of mineralizing system; the ratio of melt thickness to diameter of

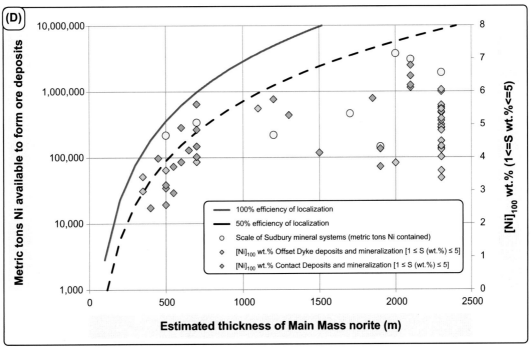

FIGURE 5.15 (*cont.*)

collection cell is set to 10; the density is set to 2.9 g/cc; the starting melt is a QD with two cases illustrated (100% and 50% of metal collected in sulfide). Examples of Sudbury mineral systems based on historic production and known mineral potential are shown where the ore deposits fall in relation to the model. The Ni tenor of disseminated sulfide mineralization is shown for mineral systems in Offset and contact settings.

metal available from the melt, and this calculation supports the hypothesis that the ore deposits were formed by the segregation of immiscible sulfide from the overlying melt sheet. The variations in thickness of the norite stratigraphy are therefore of practical value as they highlight domains of the SIC where large mineral systems may have formed.

2. *There is a broad tendency for mineralization developed under thicker parts of the melt sheet to have a higher sulfide metal tenor than mineralization developed under thinner parts of the melt sheet:* There are large local variations in the tenor of mineralization (eg, within different mineral systems in the Copper Cliff Offset), but there is a tendency for the disseminated sulfides accumulated at the base of a thick melt sheet to have higher Ni tenor than those accumulated from a thin melt sheet. This is consistent with the scavenging of metals from larger volumes of melt (high R factor).

THE DIFFERENTIATION OF SULFIDE MELT

The crystallization of sulfide magma by differentiation was first discussed by Hawley (1965). The importance of fractionation in the concentration of precious metals through the initial crystallization of monosulfide solid solution (MSS) and the production of a liquid enriched in Cu, Pt, and Pd was recognized by Keays and Crocket (1970) and Naldrett et al. (1982).

Table 5.2 Mass Balance Calculation Showing That the Noritic Units of the Melt Sheet Contained More Than Enough Metal to Form the Known Ore Deposits

Parameter	Low Estimate	Middle Estimate	High Estimate	References and Assumptions
Volume of SIC melt sheet (km³)	15,000	25,000	35,000	Golightly (1994); Grieve (1994)
Fraction of SIC comprising noritic rocks (%)	23	23	23	This book; 23%
Volume of noritic rocks (km³)	3,450	5,750	8,050	Calculated
Parental magma Ni content based on QD (ppm)	60	60	60	This book; average QD
Density of noritic rocks (assumed constant) (g/cm³)	2.9	2.9	2.9	This book; average
Nickel content of noritic magma before S saturation (million metric tons)	600	1,001	1,401	Calculated
Proportion of nickel missing (assumed to be constant) (%)	50	50	50	This book; extent of metal depletion
Nickel available to make ores (million metric tons)	300	500	700	Calculated
Amount of Ni produced + future mining (million tons)	20	20	20	This book; Chapter 1
Ratio subeconomic:economic mineralization (contained Ni)	5	5	5	Difficult to quantify; uses overestimate
Amount of subeconomic and uneconomic Ni in SIC (million metric tons)	100	100	100	Calculated
Amount of Ni not yet discovered or eroded away (million metric tons)	180	380	580	Calculated
Ni grade of sulfide (wt.%)	2	2	2	Typical grade of semi-massive sulfide mineralization
Quantity of sulfide mineralization yet to be discovered (+) or eroded away (million metric tons)	9,008	19,013	29,018	Calculated
Percentage of nickel inventory not yet found (%)	60	76	83	Calculated
Percentage of nickel that may be in economic ores (%)	12	15	17	Calculated
Percentage of nickel that has been mined or is subeconomic (%)	40	24	17	Calculated

(+) There is no certainty that this material is located at depths that can be mined at a profit with existing technologies.

Naldrett (2004) provides a detailed review of the partitioning of metals between sulfide and silicate magmas, the phase equilibrium conditions of crystallization of sulfide melts, and the effects of magma/sulfide ratio (R-factor) in controlling the composition of silicate and sulfide melts. Chapter 1 provides an overview of these concepts.

A FOUNDATION IN UNMETAMORPHOSED NATURAL SYSTEMS

As Naldrett (2004) points out, the best guide to the path that a sulfide liquid takes through the history of solidification is the succession of sulfide mineral assemblages fractionated from the melt, and represented by the sulfides created at different stages of fractionation. Some of the best documented examples of the paragenetic sequence of fresh unmetamorphosed sulfide minerals produced by fractionation come from the ores at Noril'sk and Talnakh (Genkin et al., 1981; Duzhikov et al. 1992; Distler, 1994; Stekhin, 1994). These studies recognize high, intermediate and low sulfur mineral assemblages. The high sulfur mineralization ranges from Po (monoclinic [m] > hexagonal [h]) + Cpy + Pn(Fe < Ni) to Cpy + Po(m + h) + Pn(Fe < Ni). The intermediate sulfur assemblage ranges from Po(h) + Cpy + Pn(Fe > Ni) to Cub ± Cpy + Pn(Fe > Ni) through Po(h > troilite) + Cpy + Pn(Fe > Ni) and Po(h > troilite) + Cpy + Cub + Pn(Fe > Ni) to Cub + Cpy + Pn(Fe > Ni) ± troilite. The low sulfur assemblage is an extreme development of the intermediate sulfur assemblage, and contains talnakhite +(moihoekite, putoranite) + Cub + Pn, but rarely any bornite or millerite. Through this sequence of mineral assemblages, the sulfides show systematic changes in metal ratios and metal tenors, and the Po changes in composition from monoclinic, through hexagonal, to troilite in stoichiometry (Stekhin, 1994, Torgashin, 1994, Distler, 1994).

Naldrett et al. (1994a, 1996a) report the details of the base and precious metal concentrations in representative samples of many of the ore types from Talnakh. Through the entire sequence of mineralization styles, the proportion of Pn in the sulfides remains very similar (Stekhin, 1994), and this indicates that nickel is not significant enriched in the high sulfur mineral assemblage (ie, MSS), nor is it significantly depleted in the intermediate sulfur assemblage. Partition coefficients for Ni into monosulfide solid solution (MSS) are consistent with these variations (Mungall and Brenan, 2003).

BASE AND PRECIOUS METAL PARTITIONING BETWEEN MONOSULFIDE SOLID SOLUTION AND SULFIDE LIQUID

Experimental studies indicate that D_{Ni} increases with the Cu concentration of the sulfide from values of ~0.6 at <2 wt.% Cu to 1.1 at 25 wt.% Cu (Ebel and Naldrett, 1996, 1997). The values of D_{Cu} are 0.28 at low Cu and 0.19 at 15wt.%Cu (Ebel and Naldrett, op cit).

Partition coefficients for the PGE between MSS and sulfide liquid have been the subject of studies by Li et al. (1992), Mackovicky et al. (1986), Fleet et al. (1993), Li et al. (1992), Barnes et al. (1997), and Mungall and Brenan, 2003). Partition coefficient values summarized in Naldrett (2004) indicate that Au is very incompatible in MSS (D_{Au}~0.001), Pt and Pd are not as incompatible in MSS as Au ($D_{Pt} = 0.05$, $D_{Pd} = 0.12$), but they partition very weakly into MSS, but values of $D_{Rh} = 4$, $D_{Os} = 4.2$, and $D_{Ir} = 4.4$ are within the range of experimental runs and are consistent with the composition of samples of MSS from Sudbury.

MODELS OF SULFIDE DIFFERENTIATION

Following Li et al. (1992) and Naldrett et al. (1994a,b, 1996a), the fractionation of MSS from the sulfide melt can best be resolved by plotting an element that is compatible in MSS (Rh) against one that is incompatible (eg, Cu). Models depicting the variation in Rh versus Cu in the MSS, liquid, and a mixture of MSS + liquid are shown in Fig. 5.16A for different degrees of Rayleigh crystal fractionation assuming $D_{Cu}^{(MSS/Sul.Liq)} = 0.2$ and $D_{Rh}^{(MSS/Sul.Liq)} = 4$ and a 50:50 mixture of MSS and sulfide liquid. As the degree of fractionation increases from MSS to residual sulfide melt, the Rh/Cu ratio increases as shown in Fig. 5.16B.

Naldrett et al. (1994b) used this approach to examine the compositional diversity in representative samples of Sudbury ores from contact and footwall deposits (the average compositions are shown in Table 5.3 based on Naldrett et al. (1999)). An example of samples from the Victor and Nickel Rim South mineral systems is shown in Fig. 5.17A based on data from Naldrett et al. (1999), where the vast majority

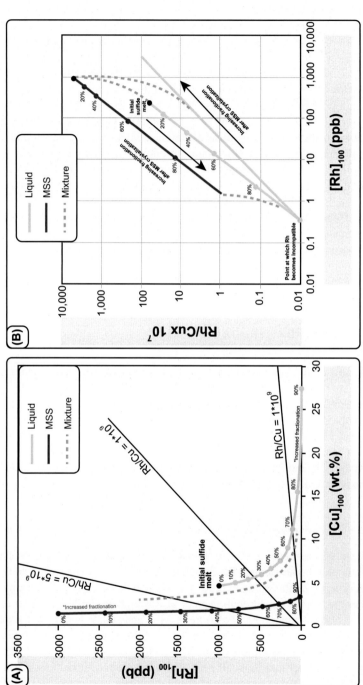

FIGURE 5.16 Comparison of Monosulfide Solid Solution and Liquid Lines Modeled by Rayleigh Fractionation

(A) Model curves showing the covariation $[Cu]_{100}$ and $[Rh]_{100}$. The plot shows a model of the effect of Rayleigh Fractionation of a starting sulfide assemblage where $D_{Cu}^{(MSS/SulLiq)} = 0.2$ and $D_{Rh}^{(MSS/SulLiq)} = 4$. A 50:50 mixture of MSS and sulfide liquid is shown. (B) Similar models for $[Rh]_{100}$ versus Rh/Cu. The modeled trends of the sulfide liquid and the monosulfide solid solution resulting from the fractional removal of the solid phase are shown, and a dotted curve illustrates the effect of mixing different proportions of unfractionated MSS and fractionated sulfide liquid.

(A) Modified from Naldrett (2004); (B) after Naldrett et al. (1996a).

Table 5.3 Average Composition of Sudbury Sulfides From Different Mineral Systems

Mineral System	Type	Range	S (wt.%)	$[Ni]_{100}$ (wt.%)	$[Cu]_{100}$ (wt.%)	$[Rh]_{100}$ (ppb)	$[Pd]_{100}$ (ppb)	$[Au]_{100}$ (ppb)	$[Pt]_{100}$ (ppb)	$[Ir]_{100}$ (ppb)	$[Os]_{100}$ (ppb)	$[Ru]_{100}$ (ppb)	Cu/(Cu + Ni)	(Cu + Ni)/S
Nickel Rim South	Contact	East	13.79	4.49	3.70	405	692	146	913	187	72	247	0.32	0.22
Victor	Contact	East	14.09	4.76	2.03	108	411	132	740	45	18	107	0.23	0.18
Coleman-Strathcona	Contact	North	22.75	3.69	1.27	27	350	52	418	11	7	17	0.22	0.13
Craig	Contact	North	12.20	5.88	4.66	104	2,046	6,835	1,348	36	4,710	66	0.38	0.28
Fraser	Contact	North	10.10	4.98	3.66	137	1,058	304	1,754	54	25	106	0.36	0.23
McCreedy East	Contact	North	25.72	3.16	26.76	4	9,693	2,256	8,342	6	8	14	0.89	0.84
McCreedy West	Contact	North	20.50	4.68	7.74	91	5,634	1,936	6,006	23	14	37	0.40	0.34
Onaping	Contact	North	15.60	5.07	1.01	188	146	15	148	47	17	101	0.15	0.16
Trilabelle	Contact	North	2.77	2.79	3.58	32	469	193	694	13	33	186	0.52	0.17
Whistle	Contact	North	27.30	3.49	0.38	205	247	26	264	60	27	53	0.10	0.10
Crean Hill	Contact	South	15.60	4.48	7.45	571	5,613	1,360	6,132	138	64	303	0.50	0.32
Creighton	Contact	South	13.05	6.58	8.31	606	3,608	1,071	3,381	193	68	340	0.47	0.41
Gertrude	Contact	South	10.55	4.75	2.70	269	220	95	397	139	57	265	0.29	0.20
Lindsley	Contact	South	14.71	3.59	4.86	357	4,680	898	2,943	94	42	151	0.51	0.22
Nickel Rim South	Footwall	East	10.40	7.52	46.46	7	10,078	6,554	22,112	4	30	53	0.85	1.71
Victor	Footwall	East	10.04	8.96	38.12	6	15,413	12,976	35,217	1	176	42	0.79	1.74
McCreedy East	Footwall	North	18.04	10.01	28.81	8	19,341	11,731	25,444	2	17	62	0.76	1.13
McCreedy West	Footwall	North	16.40	4.84	30.21	4	15,730	11,066	13,796	7	23	37	0.87	1.02
Fraser (Epidote Zone)	Footwall	North	4.42	25.16	0.44	316	1,973	98	2,393	94	42	202	0.02	0.74
Creighton	Footwall	South	6.20	10.17	23.69	84	2,579	1,428	3,430	11	4	11	0.70	0.98

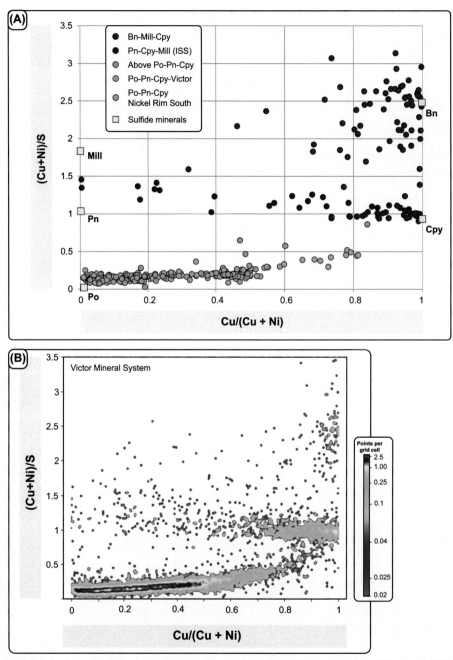

FIGURE 5.17 Comparison of a Subset of Representative Analyses From Victor and Nickel Rim South With the Compositional Field of the Victor Mineral System

(A) Plot of Cu/(Cu + Ni) versus (Cu + Ni)/S from the database of Naldrett et al. (1999), where the analyses are broken out into mineral assemblages based on the position of the assay in the diagram in relation to the dominant sulfide mineral types (see Chapter 4 for a detailed description of this plot). (B) Data density plot of Cu/(Cu + Ni) versus (Cu + Ni)/S for assay data from the Victor mineral system. These two plots illustrate that fact that the data in Naldrett et al. (1999) are representative of the diversity of mineralization types at Victor and Nickel Rim South.

of sulfides comprise mixtures of Po + Pn + Cpy, Pn-Cpy, Mill-Cpy, and Mill-Cpy-Bn-Pn. The compositions of the samples available to Naldrett et al. (1999) shown in Fig. 5.17A are compared to a data density plot of assays from the Victor mineral system in Fig. 5.17B. It can be seen that the samples carefully collected and provided by the mining companies to Naldrett et al. (1999) from Victor and Nickel Rim South cover the same compositional fields as the larger database of assays from the mineral system at Victor. The similarity of the mineralization at Victor and Nickel Rim South was highlighted by Lightfoot (2015), with the exception that the contact ores at Victor tend to plot at a slightly lower Ni tenor than those at Nickel Rim South even though the average Ni tenor from Naldrett et al. (1999) appears quite similar (Table 5.3).

Naldrett et al. (1994b, 1999) drew several important conclusions from his data, namely:

1. There is a strong compositional trend between contact and footwall mineralization in the Levack-Coleman mineral system. Contact mineralization is rich in Rh, Ir, Ru, and Os whereas footwall mineralization is rich in Cu, Pt, Pd, and Au. These variations are consistent with Rayleigh Fractionation.
2. Plots of Rh versus Cu can be used to distinguish samples that have a strong MSS affinity from those with the signature of the residual liquid.
3. The extent to which contact mineralization develops an MSS signature is an indication of how efficient the segregation of residual liquid from the system has been, and provides an indicator of footwall mineral potential.
4. The contact ore bodies from Sudbury show different degrees of MSS enrichment, and these variations are indicative of the potential for discovery of footwall style mineralization.

Naldrett (1997) discussed the compositional variation encountered in mineral systems that have undergone extensive fractionation at Sudbury and Noril'sk. They showed that the Cu content of a typical Sudbury sulfide melt can reach 32% Cu, but after this point, MSS fractionation is succeeded by intermediate solid solution (ISS) as the cumulus phase. At this point, the behavior of Cu and Rh are reversed, with Cu becoming compatible and Rh becoming incompatible as shown in Fig. 5.16B.

Naldrett et al. (1999) presented the final results of modeling based on 2500 samples collected from 12 Sudbury mineral systems. The authors showed that a plot of Rh/Cu versus $[Rh]_{100}$ helps to understand the different stages of sulfide melt fractionation. In this diagram, a progressive decrease in $[Rh]_{100}$ marks the progress of MSS fractionation, whereas the variations in Rh/Cu ratio at constant $[Rh]_{100}$ record the ratio of cumulus MSS to sulfide melt.

Though Naldrett et al. (1994b, 1999) do not discuss the Offset Dyke mineral systems, the fractionation of sulfides along the mineral shoots in the Frood, Copper Cliff, and Totten mineral systems have been illustrated in Chapter 4. These trends are also consistent with control by sulfide fractionation, where the residual sulfide was better able to penetrate deeper into the Offset Dyke than the MSS.

Variations in Cu/(Cu + Ni) versus (Cu + Ni)/S and $[Ni]_{100}$ versus $[Cu]_{100}$ for samples from Victor and Nickel Rim South are modeled in Fig. 5.18A,B using a Rayleigh Fractionation to depict the effects of melt and cumulate evolution using data from Naldrett et al. (1999). To investigate the implications of these plots with respect to the standard diagrams used in Chapter 4 to express fractionation and metal tenor, the data from Naldrett et al. (1999) have been plotted into the diagram of Cu/(Cu + Ni) versus (Cu + Ni)/S. The samples have been assigned symbols to identify them by group interval of Cu/(Cu + Ni) which reflect materials that have abundant Po through to those with abundant Pn (Fig. 5.18A). Samples along the Cpy-Pn-Mill tie lines are broken out as classified as members of one or more ISS series. Samples between the MSS-Cpy array and the ISS array are grouped together (these

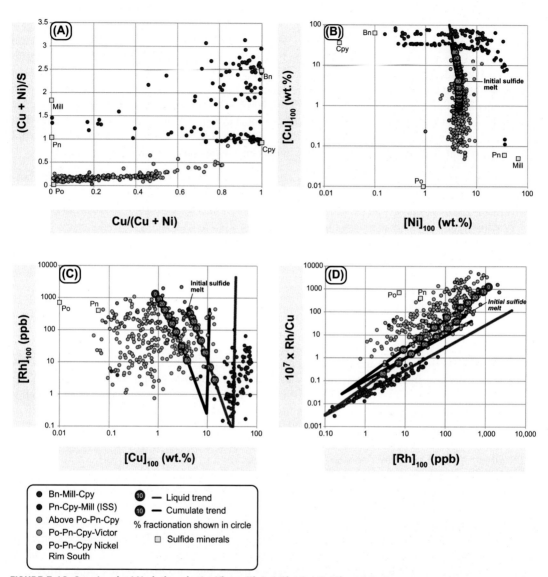

FIGURE 5.18 Geochemical Variations in the Victor-Nickel Rim South Mineral Systems

(A) Variations in Cu/(Cu + Ni) versus (Cu + Ni)/S. (B) Variation in $[Ni]_{100}$ versus $[Cu]_{100}$ for samples from the Victor and Nickel Rim South mineral systems showing the modeled effect of fractionation of an initial sulfide melt (C) Variations in $[Cu]_{100}$ versus $[Rh]_{100}$ in samples from the Victor-Nickel Rim mineral systems showing a model of fractionation of an initial sulfide melt. (D) Comparison of MSS and liquid trend lines on $[Rh]_{100}$ versus Rh/Cu with models showing the effects of Rayleigh Fractionation of an initial sulfide melt. The data are from Naldrett et al. (1999). The model trends are based on a simple Rayleigh Fractionation model with a parental sulfide melt with 4.5 wt.% Ni, 4.5 wt.% Cu, and 300 ppm Rh. $D_{Cu}^{MSS} = 0.2$, $D_{Ni}^{MSS} = 1$, and $D_{Rh}^{MSS} = 4$ (see text for discussion).

are mostly samples that comprise mixtures of contact and footwall type sulfide). Samples above the Mill-Cpy trend of the ISS cumulate array contain Bn, and are mixtures of Bn + Cpy, Bn + Cpy + Mill, Bn + Cpy + Mill + Pn, and they are grouped together (Fig. 5.18A). Using the same symbols the variations in $[Ni]_{100}$ versus $[Cu]_{100}$ and $[Cu]_{100}$ versus $[Rh]_{100}$ are shown in Fig. 5.18B,C, and the variations in $[Rh]_{100}$ versus Rh/Cu are shown in Fig. 5.18D.

Fig. 5.18C shows a model for the fractional crystallization of a Sudbury sulfide liquid containing 250 ppb Rh and 4 wt.% Cu. The first MSS to crystallize will contain 1000 ppb Rh and 0.2 wt.% Cu (assuming D's of 4 and 0.05 for Rh and Cu, respectively). With continued fractionation of MSS, the liquid and MSS evolve along the paths shown in Fig. 5.18C. Mixtures of MSS and coexisting liquid lie between the two lines shown in Fig. 5.18C.

Intermediate solid solution starts to crystallize when the Cu content of the liquid reaches ~32 wt.%, and assuming Rh is almost completely incompatible in ISS, the trend of the resulting liquid is shown.

These trends support the concepts developed in Chapter 4, where the variation in Cu/(Cu + Ni) versus (Cu + Ni)/S was introduced to illustrate the mineralogical controls on sulfide composition (Fig. 5.18A). This plot has the added practical advantage that it makes use of more extensive historic assay data to identify systems that are strongly enriched in MSS, and it helps to identify samples that have geochemical signatures produced by sulfide fractionation.

The models shown assume perfect cases of fractional crystallization which is rarely achieved in samples that come from ore bodies that have cross-cutting relationships between sulfide melts that have achieved different degrees of cumulus enrichment or degrees of fractionation. Indeed, the mineralogical controls on variations are important in very coarse-grained rocks which can comprise single Po or Cpy grains that are greater than several centimeters in grain size. Such mixtures of minerals will fall on lines between the principal sulfide minerals, and it is quite possible that a coarse-grained mixture of Po that has separated from Pn will extend the array of MSS to very low Cu/(Cu + Ni) in at least some samples. Despite these mineralogical controls, there have also been efforts to model the variations with algorithms that allow for re-melting of the MSS (Lesher et al., 1999) or equilibrium crystallization (Mungall, 2007).

The sulfides that formed from the ISS comprise Pn + Cpy and Mill + Cpy, and these minerals form very coarse-grained vein systems (e.g. Farrow and Lightfoot, 2002). These veins have an inherent heterogeneity in the distribution of the minerals that indicate that the sulfides are combinations of ISS and fractionated sulfide melt. More evolved melts crystallize to coarse-grained assemblages on Bn + Mill + Cpy, and these occupy the field above the Mill-Cpy tie line in Fig. 5.18A.

Using this approach and the data from Naldrett et al. (1999), the mineralization at Levack-Coleman can be broken out into groups based on the mineralogy of the sulfides. The diversity in Cu/(Cu + Ni) versus (Cu + Ni)/S at Levack-Coleman is very similar to the case of Victor and Nickel Rim South (Figs. 5.18A and 5.19A). Variations in $[Rh]_{100}$ versus Cu/Rh at Levack-Coleman are similar to those from Victor Nickel Rim (Figs. 5.18D and 5.19B). Modelling of these variations reproduces much of the variation in the mineral system, but once again sulfides with low Cu/(Cu + Ni) likely contain large domains of Po + Pn out of which Cpy has been locally remobilized during late magmatic and post-magmatic processes (Fig. 5.19B). Likewise the variations in Pn:Cpy, Mill:Cpy and Bn:Cpy:Mill are features of the coarse-grained textures of the sulfide mineral assemblage generated from liquids that have evolved to progressively higher Cu/(Cu + Ni) (Fig. 5.19A).

Similar models for the Creighton mineral system are shown in Fig. 5.19C where samples from Gertrude are broken out from those from the larger contact ore bodies along the Main and the Gertrude-402 Troughs (Chapter 4). These ores have quite different metal tenor, and so the models shown in

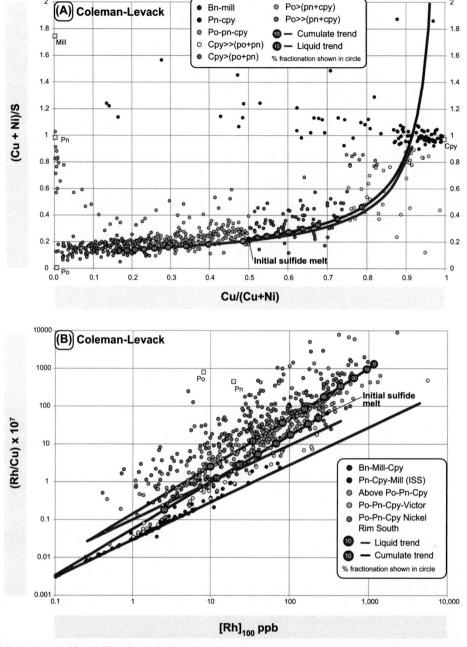

FIGURE 5.19 Compositional Diversity in Sulfides From the Coleman-Levack and Creighton Mineral Systems

(A) Variation in Cu/(Cu + Ni) versus (Cu + Ni)/S for samples from the Levack-Coleman area from Naldrett et al. (1999); the colored symbols are used to break out mineralization controlled by different sulfides as explained in Chapter 4. The Rayleigh Fractionation model is anchored by a parental sulfide melt with 4.5 wt.% Ni and 4.5 wt.% Cu. Partition coefficients for Ni into mss used in this model are 1 for Ni and 0.2 for Cu (see text). (B) Plot of $[Rh]_{100}$ versus Rh/Cu for samples from the Levack and Coleman areas showing the behavior of the samples on the MSS-Cpy, Cpy-Pn-Mill, and Bn-Mill-Cpy-Pn fields. The parental sulfide melt is chosen to have 300 ppb Rh and a partition coefficient for Rh into MSS of 4 (see text).

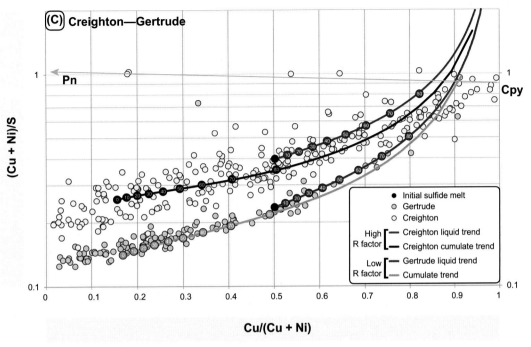

FIGURE 5.19 (*cont.*)

(C) Variation in Cu/(Cu + Ni) versus (Cu + Ni)/S for samples from the Gertrude and Creighton Deposits from Naldrett et al. (1999); the colored symbols are used to break out samples from the different deposits. The Rayleigh Fractionation model for the Gertrude samples is anchored by a sulfide melt with 4.5 wt.% Ni, 4.5 wt.% Cu whereas that for Creighton is anchored using a sulfide liquid with 8 wt.% Ni and 8 wt.% Cu.

Fig. 5.18C reflect the evolution of the sulfide liquid and cumulate formed from different initial sulfide melt compositions. As discussed in Chapter 4, the Gertrude-402 Trough exhibits a systematic increase in metal tenor with depth, and this is consistent with a progressive increase in the magma/sulfide ratio in the overlying silicate melt sheet.

EXPLORATION IMPLICATIONS OF MODELS OF SULFIDE DIFFERENTIATION

Mining companies at Sudbury have long understood that embayment and trough structures containing sulfides with low Cu/(Cu + Ni) are more likely to have associated footwall mineral zones than those with Cu/(Cu + Ni) ~0.5. Naldrett et al. (1999) introduced the concept of "wet" versus "dry" sulfide ores, where the term "wet" is applied to the sulfide melt and the term "dry" is applied to the MSS (the terms can be misleading, but there was no intention on the part of the author to imply a linkage to hydrous fluid content). Naldrett et al. (1999) suggested that very "wet" contact ore styles may relate to the points of transition between contact and footwall systems, and they use an arcane index calculated as shown in their 1999 paper called "phi" to express a measure of this property.

The ores comprising the contact deposit at Levack #4, the transitional style of mineralization at 9 Scoop, and the footwall mineralization in the 148, and 153 Ore Bodies appear to be different parts of one sulfide melt that underwent MSS fractionation at the lower contact of the SIC. The expulsion of fractionated sulfide melt from the deep part of the embayment structure along open space created in a slump plane guided by SUBX gave rise to the footwall mineral zones (Fig. 4.66). In this case, the variations in Cu/(Cu + Ni) and (Cu + Ni)/S are consistent with progressive fractionation, and the contact ores in the deepest part of the Levack #4 and 9 Scoop embayment have what Naldrett et al. (1999) refer to as a "wet" signature.

In other cases such as Victor and Nickel Rim South, the development of a transitional style of mineralization in the trough structure does not provide an obvious vector toward the adjacent footwall mineralization. The fractionated sulfides at Victor are all located in the footwall and they do not appear to have a clear transition into a massive sulfide hosted in the Sublayer as described in Chapter 4 (Fig. 4.66). The likely explanation for this decoupling of the contact and footwall systems is the reworking of the Sublayer along steep troughs, so that the spatial association between contact and footwall ores is not an immediate 1:1, but is rather a feature of the broader trough structure.

Modeling of the sulfide ores by simple Rayleigh Fractionation is an oversimplification of the processes responsible for compositional diversity but it does illustrate the fact that fractionation type process models can generate the range in ore compositions in the context of the inherent variation in grain size of the sulfide ores. More complex models that take into consideration the effect of possible re-melting of MSS cumulates were presented by Lesher et al. (1999), and there is clear evidence that mineral zones may be reworked during the evolution of the mineral systems in Offset Dykes and at the lower contact of the SIC. For example, inclusions of sulfide within MIQD at Stobie have a different metal tenor when compared to the disseminated sulfide in the host (Fig. 4.8); the sulfides in this rock were produced from two different phases of sulfide emplacement into the Stobie ore body (Hattie, 2010). Likewise, variations in the chemistry of the mineral shoots in the Copper Cliff Offset Dyke are consistent with the injection of sulfides that have evolved from melts that have different metal concentrations (Chapter 4). The presence of clear examples of sulfide (Po-Pn-rich) inclusions in GRBX at Levack and Victor (Fig. 5.20A,B) are consistent with the crystallization of MSS prior to that of the silicate melt, and the redistribution of the cumulate sulfides provides evidence for re-melting processes.

Mining companies have long understood that the contact ores in the South Range have higher Ni tenor than most contact ores in the North Range. They have also been acutely aware of the diversity in contact ore tenors within and between individual embayment structures. For example, a very low Ni tenor is found in contact mineralization at Trillabelle and Whistle relative to Levack, Coleman, and Victor. Moreover, the recognition that the metal tenor increases with depth along individual troughs was highlighted in Lightfoot (2009) for mineralization in the Gertrude-402 Trough at Creighton.

Naldrett et al. (1999) were able to use the Rh/Cu approach to identify differences in the R factors responsible for North versus South Range ore bodies (Fig. 5.21). This result is in agreement with the fact that different mineral systems around the Sudbury Basin have evolved from different starting melts. In the model presented in Naldrett et al. (1999), these differences are quantified in terms of the R factor, and the variations in R factor relate to the volumes of sulfide and silicate melt that equilibrated in the overlying Main Mass as discussed in the previous section.

Naldrett et al. (1999) also showed that mineralization from the Trillabelle embayment comprising of disseminated sulfides has a systematically different metal tenor and Rh/Cu ratio when compared to the more massive sulfides ($>10\%$S). It appears that the more massive sulfides record a signal of MSS (minus the fractionated sulfide liquid) and this signature is not recorded in the more disseminated sulfides. This observation was highlighted in Chapter 4, where the differences in $[Ni]_{100}$ and Cu/(Cu + Ni) between massive and disseminated sulfides is evident. Unfortunately, the volume of higher

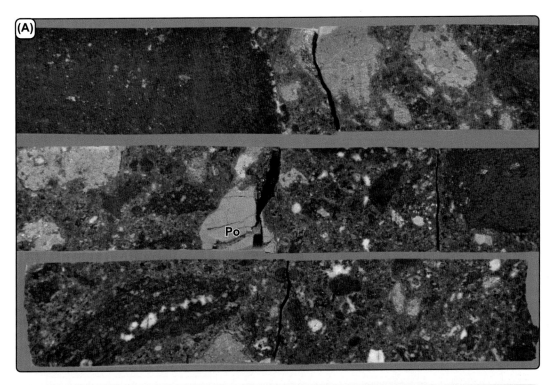

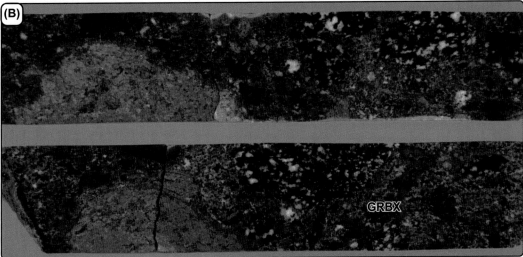

FIGURE 5.20 Example of Sulfide Inclusions That Have a Composition Close to that of Po-rich MSS

The samples come from GRBX developed in the Sublayer of the Victor Deposit. (A) Sample MG201541
from the contact mineral zone hosted in GRBX at Victor. Clasts of Po-rich sulfide are hosted in GRBX.
(B) Samples MG201540 from the contact mineral zone hosted in GRBX at Victor. A large clast of Po-rich sulfide is
hosted in GRBX.

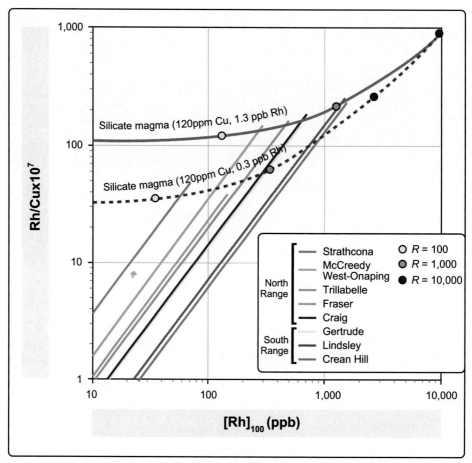

FIGURE 5.21 Comparison of the Liquid Trends of Deposits Examined in Naldrett et al. (1999)

The arrays are shown relative to two different silicate liquid trends (120 ppm Cu, 1.3 ppb Rh, and 120 ppm Cu and 0.35 ppb Rh) where the R factor is shown. North Range mineralization has evolved with a lower R factor than South Range mineralization (Naldrett et al., 1999), although Chapter 3 shows an enormous diversity within individual North and South Range mineral systems that goes a long way to undermining this conclusion.

grade MSS-rich contact mineralization in the two Trillabelle Deposits is quite small, and any footwall target associated with the differentiation of the primary sulfide melt represents quite a small exploration target.

COMPOSITIONAL VARIATION IN DISSEMINATED SULFIDES

The composition of Mafic Norite (MNR) provides a very useful indication of the metal potential of the underlying Sublayer (Fig. 5.12), but the bulk composition of mineralization hosted in MIQD, SLNR, and mineralized basal Main Mass noritic rocks also provide information about the mineral potential of the lower contact of the SIC and the Offset Dykes. Table 5.4 shows average compositional data for these rock

Table 5.4 Composition of Disseminated Sulfide Mineralization Enveloping Large Economic Mineral Zones and Small Uneconomic Mineral Zones

Mineral System (Assays ≥1 and ≤5 wt.%S)	Location	System Scale	Samples	Cu/(Cu + Ni)	(Cu + Ni)/S	$[Ni]_{100}$ (wt.%)	$[Cu]_{100}$ (wt.%)
Copper Cliff Offset	Copper Cliff Offset	Large	All samples	0.67	0.42	4.85	10.76
Kelly Lake 710	Copper Cliff Offset	Large	All samples	0.74	0.64	5.52	17.48
Kelly Lake 720	Copper Cliff Offset	Large	All samples	0.67	0.49	5.58	12.38
Kelly Lake 725	Copper Cliff Offset	Large	All samples	0.62	0.45	5.99	10.58
Kelly Lake 740	Copper Cliff Offset	Large	All samples	0.75	0.44	3.60	12.53
790–810	Copper Cliff Offset	Large	All samples	0.72	0.52	4.99	13.98
830	Copper Cliff Offset	Large	All samples	0.62	0.40	5.47	9.36
830 Flat	Copper Cliff Offset	Large	All samples	0.67	0.36	4.28	9.13
850 Flat lower	Copper Cliff Offset	Large	All samples	0.67	0.43	4.82	10.98
850 Flat upper	Copper Cliff Offset	Large	All samples	0.62	0.38	5.11	9.09
860	Copper Cliff Offset	Large	All samples	0.65	0.44	5.35	10.82
865	Copper Cliff Offset	Large	All samples	0.71	0.50	4.89	13.17
880	Copper Cliff Offset	Large	All samples	0.69	0.42	4.49	11.14
890	Copper Cliff Offset	Large	All samples	0.64	0.27	3.38	6.84
900	Copper Cliff Offset	Large	All samples	0.68	0.39	4.23	10.38
100	Copper Cliff Offset	Large	All samples	0.72	0.41	4.01	11.08
114	Copper Cliff Offset	Large	All samples	0.59	0.35	5.18	7.75
120–138	Copper Cliff Offset	Large	All samples	0.71	0.44	4.33	11.81
175	Copper Cliff Offset	Large	All samples	0.56	0.31	4.91	6.81
175–178 domain	Copper Cliff Offset	Large	All samples	0.56	0.29	4.82	6.12
178	Copper Cliff Offset	Large	All samples	0.66	0.45	5.38	11.32
191	Copper Cliff Offset	Large	All samples	0.58	0.40	6.05	8.85
Clarabelle	Copper Cliff Offset	Large	All samples	0.52	0.31	5.46	6.22
Totten	Worthington Offset	Medium	All samples	0.72	0.58	5.49	15.63
Frood	Frood-Stobie	Large	All samples	0.63	0.29	3.83	6.95
Frood-Stobie Saddle	Frood-Stobie	Large	All samples	0.52	0.22	3.87	4.41
Stobie	Frood-Stobie	Large	All samples	0.56	0.23	3.62	5.11
Foy	Foy Offset	Small	All samples	0.51	0.15	2.98	2.95
Ministic	Ministic Offset	Occurrence	All samples	0.58	0.19	3.42	3.74

(Continued)

Table 5.4 Composition of Disseminated Sulfide Mineralization Enveloping Large Economic Mineral Zones and Small Uneconomic Mineral Zones (*cont.*)

Mineral System (Assays ≥1 and ≤5 wt.%S)	Location	System Scale	Samples	Cu/ (Cu + Ni)	(Cu + Ni)/S	[Ni]$_{100}$ (wt.%)	[Cu]$_{100}$ (wt.%)
402 (deep)	Contact	Large	All noritic rocks	0.43	0.30	6.19	5.11
402 (intermediate)	Contact	Large	All noritic rocks	0.43	0.28	5.79	4.74
402 (shallow)	Contact	Large	All noritic rocks	0.43	0.20	4.14	3.54
Capre	Contact	Medium	All noritic rocks	0.36	0.18	4.02	2.88
Cascaden	Contact	Occurrence	All noritic rocks	0.42	0.13	2.47	2.59
Creighton Main (deep)	Contact	Large	All noritic rocks	0.56	0.42	6.47	8.94
Creighton Main (intermediate)	Contact	Large	All noritic rocks	0.54	0.42	6.79	8.75
Creighton Main (shallow)	Contact	Large	All noritic rocks	0.47	0.32	6.11	5.81
Ella-Selwyn	Contact	Small	All noritic rocks	0.37	0.18	4.22	2.59
Garson	Contact	Large	All noritic rocks	0.54	0.31	5.76	7.22
Gertrude west trend	Contact	Medium	All noritic rocks	0.46	0.26	5.29	4.63
Levack-Coleman	Contact	Large	All noritic rocks	0.36	0.21	4.90	3.11
Little Stobie	Contact	Large	All noritic rocks	0.56	0.24	3.72	5.26
Murray	Contact	Large	All noritic rocks	0.57	0.26	4.26	6.54
Nickel Rim South	Contact	Medium	All noritic rocks	0.47	0.29	5.62	5.13
Norman	Contact	Medium	All noritic rocks	0.43	0.15	3.07	2.67
Norman West	Contact	Small	All noritic rocks	0.42	0.15	3.17	2.40
Onaping Depth	Contact	Medium	All noritic rocks	0.40	0.22	4.84	3.47
Trill depth	Contact	Small	All noritic rocks	0.53	0.16	2.93	3.21
Trill shallow	Contact	Small	All noritic rocks	0.40	0.19	3.98	3.25
Victor	Contact	Large	All noritic rocks	0.36	0.18	4.34	2.72
Whistle Far West	Contact	Occurrence	All noritic rocks	0.52	0.12	2.57	2.23
Whistle West	Contact	Small	All noritic rocks	0.39	0.17	3.73	2.60
Whistle-Podolsky	Contact	Small	All noritic rocks	0.45	0.21	3.86	3.97

types from different mineralized environments based on assay data. Fig. 5.22A shows them plotted into the diagram of Cu/(Cu + Ni) versus (Cu + Ni)/S and Fig. 5.22B shows similar data on a plot of [Ni]$_{100}$ versus [Cu]$_{100}$. Superimposed on Fig. 5.22A are Rayleigh Fractionation models that show the evolution of the liquids and cumulate portions as a function of differing starting melt compositions that might be responsible for Sublayer mineralization. The plot illustrates the wide range in the composition of the

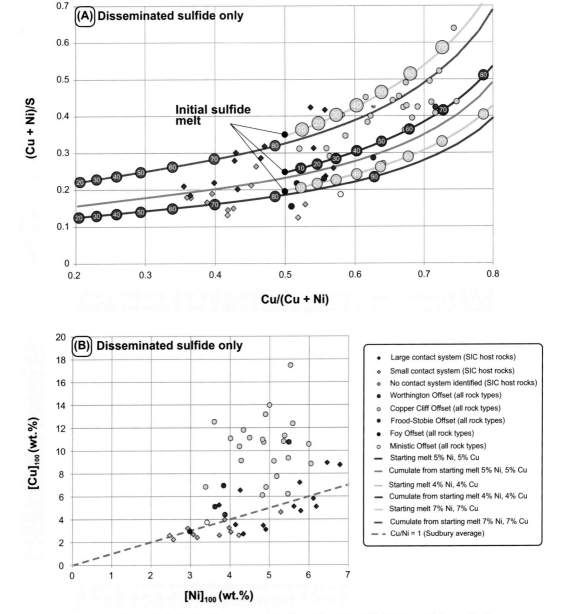

FIGURE 5.22 Compositional Variation in Disseminated Sulfides [1 > S (wt.%) ≤ 5] from Different Sudbury Mineral Systems

(A) Variation in Cu/(Cu + Ni) versus (Cu + Ni)/S in average compositional data from host rocks of different mineral systems from the Offset Dyke and Contact environments. Averages are based on sulfides from host rocks that have dioritic or noritic compositions, and are restricted to samples with 1 > S (wt.%) ≤5. Models are based on differentiation of a parental sulfide melt that has Cu/(Cu + Ni) = 0.5 and metal tenors of 3,4, and 5 wt.% Ni. (B) Variations in Ni and Cu tenor in the disseminated sulfide from mineral systems hosted in contact and Offset Dyke environments; the 1:1 line represents average undifferentiated Sudbury mineralization.

starting sulfide melts in Offset Dykes. Typically, mineral zones in Offset Dykes have higher Cu/Ni than Sublayer-hosted mineral zones, and there is a wide range in metal tenor of the starting liquid within and between different Offset Dykes. This provides strong evidence that the individual ore bodies in mineral systems such as Copper Cliff and Frood-Stobie were formed by the emplacement of several different pulses of sulfide liquid from the melt sheet. In cases of contact mineral systems with very low tenor sulfides, the ore deposits tend to be small or subeconomic. These variations are especially useful in the identification of more prospective environments in which to explore and locate additional mineralization.

Although the models using Rh data are useful in constraining petrogenesis, the break-out of mineralization styles is evident on very simple plots that use Ni, Cu, and S abundance data. Given the cost and turnaround on Rh determinations, it is suggested that the approach used in Chapter 4 is applicable in processing large datasets acquired during exploration and development of the mineral systems at Sudbury.

SULFIDE DIFFERENTIATION: DIFFERENCES BETWEEN NORTH AND SOUTH RANGE MINERAL SYSTEMS

The vast majority of North Range mineral systems exhibit pronounced sulfide differentiation (see Chapter 4). Footwall deposits like Strathcona, Podolsky, and Capre exhibit less pronounced, but significant fractionation; in these deposits the sulfide mineral assemblage consists of Cpy ± Cub ± Pn ± Mill. The apogee of fractionation is recorded in deposits such as McCreedy East 153, Nickel Rim South, and Victor 14N and 28N zones that all develop Cpy-Pn assemblages at the center, surrounded by fringe zones comprising veins of Pn + Mill + Cpy, and then Mill + Cpy + Bn, and Bn + native silver at the outer edge.

In the South Range, the diversity in mineralization types produced by differentiation of the sulfide melt is less pronounced. The best candidates for magmatic differentiation processes to form the footwall deposits of the South Range are at Little Stobie, Lindsley, the Creighton 126 ore body, Creighton Deep, Crean Hill, McKim, and perhaps parts of the Garson mineral system (details are provided in Chapter 4). The sulfides comprise (Cpy + Pn) > Po, but Po is rarely entirely absent from the footwall sulfide mineral assemblage. The relative roles of magmatic versus deformation processes in the production of the footwall mineralization in the South Range is not fully understood, but some of the mineral zones are related to breccias and structural space created under magmatic conditions. The South Range footwall mineralization does not contain any significant development of millerite or bornite.

The explanation for why South Range footwall mineralization is relatively unfractionated, but North Range mineralization is fractionated to different degrees in different deposits may rest in the relative timing of fractionation processes relative to the development of structural space in the footwall. In the case of the South Range, readjustment of the crater took place early on during initial crystallization of MSS, and some of the MSS was transported with the sulfide liquid into the footwall. This sulfide melt was able to crystallize along the MSS-liquid array, but it was not able to reach the apogee of iss crystallization. In the case of the North Range, the crater re-adjustment process may have happened later, and so the crystallization and localization of MSS was possibly complete before the residual sulfide melt was expelled into the footwall into open space crated along structural weakened domains that may link to slump features on the crater wall. These sulfide melts evolved by crystallization of a vein stock-work within a core of Pn-Cpy, surrounded by Mill-Cpy, and then admixtures of these cumulates with Bn. The crystallization process generated ore bodies that exhibit broad zonation from core domains of Pn-Cpy through veins of Mill-Cpy to peripheral development of Mill-Bn and Bn-native silver.

EFFECTS OF REGIONAL DEFORMATION ON MINERAL SYSTEMS

As explained in detail in Chapters 2 and 4, the most extreme examples of deformation that have influenced the geometry and distribution of sulfides are restricted to parts of the Garson, Falconbridge, Crean Hill, and Frood-Stobie Mineral Systems. These ore deposits rest within the broad envelope of the South Range shear zone that was likely active long after stabilization of the Sudbury Structure (see Chapter 2). There are undoubtedly other local examples of sulfide kinesis and translocation of sulfides within the ore deposits that result from deformation (see Chapters 4 and 6), but this modification is superimposed on a style of mineralization that was often present in the footwall, so the deformation effects have been ones of modification rather than creation of the ore bodies.

EFFECTS OF HYDROTHERMAL PROCESSES ON MINERAL SYSTEMS

The magmatic origin of orebodies of the Sudbury Structure was first suggested by Bell (1891a,b) and Coleman (1905a), and this mechanism of formation has become widely accepted (Hawley, 1962; Naldrett et al., 1999). The origin of the footwall ore bodies associated with contact mineralization was recognized much later (Morrison, 1984; Morrison et al., 1994). The remobilization of some of the metals from the magmatic environment into the footwall by hydrothermal processes is not as widely accepted as a possible mode of formation of Cu-Ni-PGE ore-bodies in the footwall.

Farrow and Watkinson (1997) reported that the distribution of PGM is most common in chalcopyrite near, or at, the contact between footwall sulfide veins with alteration minerals such as actinolite, quartz and epidote. Their study areas included the Barnet Property, the Deep Copper and Copper zones of the Strathcona Deposit, the McCreedy West Deposit and the epidote zone of the Fraser Deposit. The most common PGM are Pd- and Ag-rich tellurides and bismuthides. Each deposit has a characteristic assemblage of PGM, with distinct compositions of these minerals.

Among Cu-Ni-PGE mineralized zones, distinct Pt, Pd Bi and Sb variations are most common in sobolevskite, michenerite, merenskyite, moncheite and melonite. Comparison of whole-rock concentrations of Cu, Ni, Ir, Rh, Pt, Pd and Au for the Barnet Property with that of other Cu-rich footwall deposits further illustrates the individuality of each of the deposits. Based on the close spatial association of PGM with hydrous, locally Cl-bearing alteration and other rare minerals (including halides), Farrow and Watkinson (1997) suggested that these variations are is compatible with the influence of a hydrothermal fluid containing H-Cl-S-Fe-Cu-Ni-Te-Se-Bi-As. The assemblages of the PGM and the geochemical diversity in each Cu-rich footwall deposit is suggested to be a function of the interaction of contact magmatic sulfide assemblages with fluids whose compositions were buffered by rock compositions in the footwall. The metals scavenged from the primary base-metal sulfide were transported into the footwall and subsequently deposited by hydrothermal cells (Farrow and Watkinson, 1997).

Li and Naldrett (1993a) reported compositional data for fluid inclusions from Cl-rich brines at Strathcona Deep Copper Zone. They reported hydrothermal alteration selvages along the margins of massive sulfide veins containing Cl-rich hornblende. Fluid inclusions from the quartz in branching veins associated with the hornblende are highly saline and homogenized to a liquid at 260–420°C. In the view of Li and Naldrett (1993a), these features could be produced by melt-fluid separation during sulfide crystallization or convection of ambient fluids across the contact of the sulfide veins.

Molnar et al. (2001) and Tuba et al. (2014) report evidence for multiple hydrothermal processes in footwall rocks of the North and East Range. In their publications, they suggested that the primary magmatic contact ores were modified and remobilized by fluids, and then influenced by a further two stages of fluid mobilization. Tuba et al. (2014) suggested that this process was responsible for the generation of LSHPM mineralization in a part of the East Range. The details of the distribution of LSHPM mineralization was reported by Stewart and Lightfoot (2010) and Gibson et al. (2010), where it was confirmed that the localization of this mineralization occurs in association with a range of process controls that modify the distribution of metals in contact and footwall ore deposits that have a primary magmatic origin.

The effect of H_2O on the melting temperature of sulfides has been investigated in the system FeS-PbS-ZnS by Wykes and Mavrogenes (2005). The solubility of H_2O in this system is confirmed by the presence of vesicles in quenched sulfide melt. This data aligns with ideas developed in Mungall and Brenan (2003) where a magmatic origin for halogens associated with halogen-rich alteration is linked to primary sulfide melts.

Data for the McCreedy East 153 Ore Body and the Victor Deposit also indicate the presence of abnormal concentrations of halogens (Jago et al., 1994; Hanley et al., 2004; McCormick et al., 2002b). The presence of carbonate veins at the termination of Bn + native silver veins and the development of LSHPM styles of mineralization in partially melted footwall rocks is consistent with translocation of metals by volatiles derived from the sulfide melt.

The discussion on the role of hydrothermal processes in the generation of ore deposits versus the modification of magmatic systems will very likely continue, but the author's view is that the weight of evidence supports a primary magmatic origin of footwall mineralization at Sudbury. This is now generally accepted by geologists who work on the deposits, and see the extent to which metal value from primary magmatic sulfides dominate in economic value by several orders of magnitude over the hydrothermally modified or LSHPM styles of mineralization.

A GUIDE TO FUTURE APPLIED RESEARCH ON SUDBURY AND IMPACT CRATERS

Astronauts trained in NASA's Apollo program came to Sudbury in 1971 to better understand the geological formations and rock types that might be encountered on the moon during the Apollo missions (Table 1.1). The next generation of explorers will surely send more remote-controlled spacecraft beyond Earth to explore the Solar System or travel to other planets. Both space and ore deposit exploration will surely benefit from a more complete understanding of the Sudbury Structure.

The last section of this chapter brings into focus specific aspects of Sudbury geology that remain tantalizingly beyond the envelope of reasonable understanding. These topics are suggested as the basis for next steps in understanding the enigmatic Sudbury Structure and it's associated ore deposits.

BASAL TOPOGRAPHY OF IMPACT MELT SHEETS

The comparative anatomy of planetary impact structures revealed by high resolution imaging acquired during space missions brings a wealth of new observations that can be tested against the geological relationship in the SIC. In turn, the detailed three dimensional morphology of the Sudbury Structure also provides a wealth of information that underpin the modeling and interpretation of the Sudbury Structure.

There are many areas of potential future investigation, but perhaps some of the most meaningful will come from efforts to understand the geometry of pristine planetary structures that provide insights on the following features:

1. The geometry of the basal contact of the impact melt sheet.
2. Impact melt dykes and the structural controls on their distribution and relationship to the melt sheet.
3. The scale and geometry of crater slump features surrounding both the central uplift and the crater rim.
4. The details of the sequence of events and timing of processes of impact, melt generation, crystallization, and readjustment.

The recognition that the impact melt at Sudbury is well-endowed with mineral deposits at the lower contact is now well-established. There remains much uncertainty regarding exactly where the different parts of the Sudbury melt sheet resided in relation to crater wall and central uplift, and the 3D models of the Sudbury Basin tend to respect shallow information and the imaging of the deep structure remains in its infancy.

MORPHOLOGY OF THE CRATER WALL

Crater slump features, embayments, and Offsets are the main "containers" of ore deposits. There is still remarkably little information available from other impact craters or planetary structures that help understand and predict the distribution of these features, and the controls on their development.

SULFIDE SATURATION HISTORY AND MELT SHEET EVOLUTION

Two facets of Sudbury have attracted enormous attention. On the one hand there have been great advances in modeling how the SIC will differentiate, but these models tend to be restricted to a few elements in some poorly constrained sections. The 3D constraints on the geometry and deformation of the SIC at depths up to 3 km coupled with many new surface and drill core traverses offer the possibility of major advances, and these models should not be constructed independently of the controls on magmatic sulfide.

Likewise, modeling the sulfide saturation history of the SIC presently calls on a sequence of progressive changes in magma chemistry, yet there is no indication in the trace element or isotope ratio data that the melt changes composition through time. A closer reconciliation of sulfide saturation models with evidence for process controls on changing magma chemistry would be most welcome. More importantly, resolving the effect of these processes in areas of thin versus thick melt sheet and domains with large deposits versus no economic mineralization remains an objective of true economy importance to mining and exploration companies.

SULFIDE FRACTIONATION

The importance of sulfide differentiation has long been recognized at Sudbury. Trends of increasing $Cu/(Cu + Ni)$ and metal tenor with depth in the Frood Deposit was an early observation, and then the recognition of footwall mineral systems in the North Range in the 1980s has driven a massive effort to explore beneath the SIC for fractionated Cu-PGE-Ni-rich sulfide ore deposits. The recognition that contact deposits record evidence of this fractionation process has underpinned much exploration based on simple variations in base metal abundances in different rock types with different sulfur concentrations.

Much of the exploration data remains in company files, but efforts to address the problem with careful studies of representative samples have certainly helped underpin the search for the best possible models that constrain the relative roles of magmatic and post-magmatic process controls. The outcomes of new models may have impact on the details of the scientific theories that explain sulfide fractionation, but they may also provide new pathways to discovery of ore deposits at Sudbury. Both are satisfying reasons to continue the search for better models.

ANCIENT IMPACT CRATERS

Whether other terrestrial impact structures will be recognized as exploration targets remains open to debate. It is certainly interesting that small intrusions with unusual quantities of Ni-Cu sulfide mineralization are associated with norites in what may be a deeply exhumed 3.0 Ga ~100 km diameter impact structure centered on Maniitsoq in West Greenland (Garde et al., 2013; Schersten and Garde, 2013). Older impact craters often record complex geology with multiple possible explanations, so it will clearly be important to establish whether these noritic intrusions have any linkage to an impact event.

Drill Rig in the Process of Testing a Target From Surface in the Deep Part of the Copper Cliff Offset in September 2015. This image shows all of the drill rods in the vertical rod holder, and it shows the automatic system that handles the motion and connection of the drill rods.

SUDBURY NICKEL IN A GLOBAL CONTEXT

INTRODUCTION

Much of what we now know about the Sudbury Structure and its ore deposits has come from an overwhelming contribution made by geoscientists who have explored and mined the Sudbury Basin for over 100 years. This information is very important to discovery of both near-mine mineral zones and new mineral systems as well as the fundamental understanding of its origin.

The geological relationships and technologies that underpin success have changed over the years, and the first theme of this chapter examines the features of the ore deposits that are used in exploration. This section highlights how geoscientists have embraced technological developments in geophysics and geochemistry to target drilling. Drilling to depth requires technical skills in directing the drill bit toward the target (the front image in this chapter illustrates the type of drill rig required to test targets at depths > 2000 m). Once samples of mineralization are available it is important to fully understand the grade distribution, host rocks, and the response of the materials to processing (geometallurgy).

A second theme developed in this Chapter compares the magmatic sulfide ore deposits at Sudbury to other global examples. A short summary of the four very large mineral systems at Noril'sk, Voisey's Bay, Jinchuan, and Thompson is provided. Most Ni-Cu-PGE sulfide ore deposits are associated with mafic to ultramafic rocks associated with a mantle plume or continental rift, but Sudbury has a much more felsic composition and is related to an impact melt. Many facets of the Sudbury mineral system are similar to the ore deposits associated with mafic-ultramafic intrusions. The viewpoint that Sudbury is unique among the world's magmatic sulfide deposits is possibly correct from the viewpoint of classification (Naldrett, 2004), but the Sudbury mineral systems exhibits many similar geological features and process controls to other magmatic sulfide ore deposits which make them members of a common group. Using other large magmatic sulfide deposits, it is shown that many of the same processes that control ore genesis on the way from the "mantle to the bank" (Naldrett, 2010a) also control the formation of the Sudbury deposits from an impact-derived crustal melt.

The third and final theme examines Sudbury's position in the global supply chain of nickel. The enormous wealth of nickel in near-surface laterite ores has produced a major shift in the global nickel market in the past 15 years from sulfide toward laterite production. This has happened in some unexpected ways as illustrated by the growth of the nickel-pig iron market (Mitchell, 2014) while new hydrometallurgy and smelter projects are challenged to meet expectations. However, the magmatic nickel sulfide mineral systems offer opportunity to sustain production from existing process infrastructure. During periods of high metal prices, production from high-grade nickel deposits such as Thompson, and high-grade deposits that contain additional metals such as Cu, PGE, Co, and Ag (eg, Sudbury and Noril'sk) offer an economic advantage over laterite ores, as well as a pure cathode nickel product that is in high global demand (Dalvi et al., 2004).

Despite the increasing depth of many discoveries at Sudbury and the high cost of exploring the prospective geology, discoveries are still being made, and some of them are shallow. The opportunity

Nickel Sulfide Ores and Impact Melts. http://dx.doi.org/10.1016/B978-0-12-804050-8.00006-7

to sustain production from the facilities at Sudbury is strong, and given the right levels of investment in exploration and development there is a good future for the Sudbury Camp.

THE ROLE OF EXPLORATION

The Sudbury Ni-Cu-PGE-sulfide ore deposits clearly have an overwhelmingly magmatic origin, and they are related to the crystallization of a melt sheet created by impact melting of average upper crust in the Sudbury Region at 1.85 Ga (Chapter 2). Despite the felsic composition of this melt sheet (Chapter 3), the formation of the ore deposits was a very efficient process that "smelted" the metals as magmatic sulfides out of vast quantities of silicate magma to leave a residue of metal-depleted melt that crystallized to form norites in the Main Mass. The metals contained in dense sulfides were concentrated toward the lower contact of the SIC and into the Offset Dykes (Chapter 4). The largest ore deposits were developed below the thickest parts of the melt sheet (Chapter 5).

Strategies for discovery of mineralization at Sudbury are implemented by mining companies at a range of scales from near-mine to regional in objectives. Exploration is a business that has intrinsic risk and reward. The risks are normally less when exploring near to an active part of a mine, they tend to be higher when exploring in a favorable geological environment outside of the mine environment, and they are typically very high when exploring a new belt where there is little or no known mineralization.

A near-mine discovery of high-grade mineralization can alter the life of mine and/or displace material of lower value from the mine plan. An entirely new discovery of a large deposit can change the entire future production profile from an existing infrastructure. There is a long track record of discovery through prospecting, geoscience compilation followed by field programs and drilling (e.g., Woodall, 1994), and/or the most welcome influence of serendipity (Horn and Brisbois, 1998). The trade-off between risk and potential reward ensures that global mining companies explore at the scale of near-mine, camps, and new belts. Without mine- and camp-scale exploration as well as regional exploration in new geological environments, the future of magmatic sulfide nickel production is less secure.

As described below, five broad strategies exist for exploration work to define new mineralization. Exploration is a critical first step that leads to further work required to definite and delineate the ore to underpin reserve statements. Exploration contributes to future metal production through discovery at the brownfield and greenfield stages of exploration.

EXTENSIONS TO KNOWN ORE DEPOSITS

In locations close to known ore deposits, brownfield exploration can help to identify extensions of mineral zones that have the potential to extend the life of the mine. At Sudbury, good examples include the mineral zones discovered near to the ore deposits in the Copper Cliff and Worthington Offsets. The historic effort to explore the Creighton mineral system is described in Boldt (1967), and this work has continued and will secure production from this mine well into the 21st century (Chapters 1 and 4).

EXPLORATION NEAR TO KNOWN ORE DEPOSITS

The development of concepts and ideas from known mineral zones can be used to explore in geological environments that are proximal to, but beyond the immediate footprint of known ore deposits.

Examples of brownfield exploration in these environments include the identification of pathways from contact ore bodies to footwall mineral zones like those at Victor (Morrison et al., 1994; Lightfoot, 2015) and Nickel Rim South (McLean et al., 2005) as well as low sulfide mineral zones like the Denison Zone adjacent to the Crean Hill Deposit (Gibson et al., 2010). Fault-displaced segments of ore bodies like those recognized at Levack-Coleman are also good examples of exploration success (Chapter 4).

EXPLORATION BEYOND KNOWN MINERAL ZONES IN MINING CAMPS

Brownfield exploration within a mining camp also focuses on the discovery of mineralization that may support an entirely new mine complex by following-up a geological concept. This type of exploration activity has a higher risk of failure than exploring near known mineral zones, but it can also generate enormous future value, and the reward is at a level that can sustain the life of an existing integrated processing facility. Discoveries like this are less common, and require ongoing funding and work to secure success in multi-year campaigns of drilling. Discoveries at Sudbury such as Nickel Rim South (McLean et al., 2005), Capre footwall deposits (Stewart and Lightfoot, 2010), and the newly discovered Victoria deposit (Farrow et al., 2011), are examples that fall in this group.

CONVERTING A PAST DISCOVERY INTO A FUTURE MINE

The principal activity here is not so much to make a new discovery, but rather to take a past discovery that does not pass economic hurdles and turns it into a future mine. This approach often requires a fundamental rethink of the mining methods and process technology, and the success largely depends on the level of detailed geoscience information that is available to support a decision process. It also requires patience during periods of recession or modest economic growth. The largest risks are often associated with failure to fund the opportunity, and this arises because the market is often slow to respond to new ideas or places a risk premium on their implementation. The high cost associated with a project can also trigger near-term failure and result in the project being shelved until the market conditions improve and the next generation of entrepreneurs turns the discovery into a mine.

GREENFIELD EXPLORATION

Greenfield exploration focusses attention in geological environments that have theoretical potential for a discovery, but where there has been no previous discovery. This activity is the basis for the discovery of entirely new ore deposits outside of known camps, and is a technologically driven next-step to the fine art of prospecting. The approach to exploration has a higher risk of failure, but the rewards can be enormous. Greenfield exploration requires all of the basic tenets: careful compilation of available information, application of new technologies, a budget for the program of field work and drilling, and a willingness of the geologist as well as the company management to embrace new ideas, new technologies, and concepts.

It is normally the geoscientist who must live-and-breath the idea until success is achieved. Frolick (1999) describes a great example of dedication and perseverance by Charles Fipke who discovered the Ekati diamond mine in Canada's Northwest Territories using a very basic concept of tracking diamond-indicator minerals in glacial till. A willingness to embrace the scientific knowledge and understanding has also been at the heart of the discovery of important nickel deposits like the Thompson Deposit (Thompson and Beasley, 1960). Entirely new styles of nickel mineralization that do not conform to traditional ore deposit paradigms can be found like the iron-oxide nickel style of mineralization at GT-34 (Siepierski, 2008), and the hydrothermal style of nickel mineralization at the Enterprise Deposit in the Zambia Copper Belt (Wood, 2015; Capitrant et al., 2015). A technical approach to exploration for more traditional types of magmatic nickel sulfide ore deposits has resulted in many new discoveries in the last decade including examples such as Babel-Nebo (Western Australia; described in Seat et al., 2007, 2009), Sakati (Finland; Coppard, 2015), Nova-Bollinger, (Australia; Bennett, 2015), Tamarack (USA, Rossel, 2015), and Eagle's Nest in the Ring of Fire (Ontario, Canada; Mungall et al., 2010). The disseminated Ni-PGE sulfide mineralization in the South Manasan Intrusion of the Thompson Nickel Belt is an example where an entirely unexpected style of Ni-PGE-rich mineralization was found in association with an ultramafic intrusion (Franchuk et al., 2015). In the search for new deposits, practical geological knowledge, new concepts and ideas, and a focus on the objective of discovery weigh heavily in relation to an academic understanding of ore body models.

THE EXPLORATION CHALLENGE AT SUDBURY

Mining companies are faced with a very special challenge at Sudbury. It is one of risk versus reward in exploration. On the one hand, it is a relatively low-risk strategy to add mineralization along strike from a known ore body, and although the cost of finding and delineating this mineralization (so that compliant mineral reserves can be stated) is quite expensive, it is also more likely that the endeavor will meet with success than in a greenfield setting, so the risk of failure is less. The reward can be quite incremental, but it often matters to the life of the mine and the facility that it supports, so this type of exploration is often a priority for mature mining operations. On the other hand, a parochial focus on small mineral zones that have limited future impact on mine development is unwise as the expenditure may only make an incremental difference to the life of the mine and facility.

Exploration opportunities with a short-term and small reward have to be balanced against the possibility of a significant new mineral discovery in an area that is under-explored yet highly prospective. A major new discovery might support the future life of mining and facilities at Sudbury as well as potentially displace lower value mineralization that is in the mine plan with new material that has a high value. Appropriate balance is required between the short-term gain of low risk exploration at the edge of an ore body with the potentially enormous value of new discoveries that can support the long term viability of the camp. For this reason, it is critical for mining companies to foster excellence in project inception in order to benefit from the value realized over the long term from discovery success (Fig. 6.1 shows the pathway from the inception of an idea that results in the discovery of a future mine).

The application of the knowledge gained from over a 100 years of geoscience investigations at Sudbury provides a critical framework that is the basis for exploration decisions. In effect this information ensures effective and efficient exploration by providing a framework in which the geologist and manager can rank exploration opportunities in the context of the budgets. The next section of the book is therefore dedicated to the applied value of understanding the geological and geochemical controls on mineralization at Sudbury.

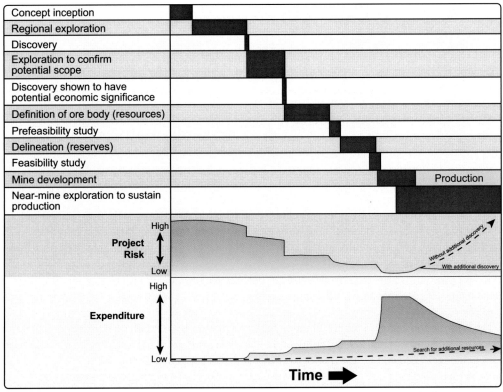

FIGURE 6.1 The Progression of Exploration From the Generation of a Concept Through Compilation, Exploration, Discovery, Delineation, Definition, and Mine Construction

Drilling is undertaken to make the discovery, bring the project to decision points that support expenditure on mine development, and commercial production. The relative risk and cost of each phase of work is illustrated, and the need to sustain exploration throughout the life of the mine is highlighted as a way to maximize value from the asset.

THE DECISION PROCESS AND THE TECHNOLOGY USED TO EXPLORE

Successful exploration demands a very complete understanding of the exploration opportunities that exist in the list of targets and prospects that are available to the company. This list is typically built by the exploration geoscientist who uses all of the compiled geological, geophysical, and geochemical data for the area of interest to build a complete understanding of the target, and the basis on which management can rank the opportunities and make decisions on exploration expenditure.

The opportunities can range from extensions of known mineral zones in mine environments, through newly recognized geological environments in existing mines or past producing mines, to regional concepts that drive exploration in the search for new discoveries that have the potential for a discovery that can support the development of a new mine. A sound exploration strategy has projects at different stages in this spectrum, and enough of these projects to provide options and choices depending on current market conditions.

The ideas are often ranked against one another, and fed into a list of opportunities where a large number of valid ideas are present at inception, and out of this approach will flow new ore deposit discoveries. This is highlighted in Fig. 6.1 where the number of ideas at the outset of the generation process will be large in regional, near-mine, and in-mine target areas. The number of projects that are implemented at any one time is often small unless exploration budgets are robust, but the projects can sit on the shelf if the ground has secure tenure, or new data can be gathered to better test the ideas before they are drill tested at a future date.

As exploration proceeds, there is normally a gradual increase in the costs associated with drilling as a discovery is made or as the economic potential of a mineral zone grows (Fig. 6.1). As the exploration reaches maturity, and a resource is identified, the project is often placed in the hands of experts who consider the mining and engineering implications, the viability of processing the material, the impact on existing mine and processing plans, and the economic viability. These studies sometimes result in a request for additional exploration, they can trigger a decision not to proceed with the project, or they can trigger mine planning through prefeasibility and feasibility studies (Fig. 6.1). At the point of a decision to develop an ore deposit, there may be requirement for expenditure of hundreds of millions of dollars through to several billion dollars to secure a viable operating mine that delivers value. The mine development phase is the most capital intensive and expensive phase of development as shown in Fig. 6.1, and it also offers a platform for additional exploration in order to extend the life of the mine which creates tremendous added value that was not originally factored into the cost of developing the mine.

The information gathered during all phases of exploration must be sufficient to support the decision process to build a new mine, and it must comply with the requirements to report resources and reserves under guidelines accepted by the mining and investment community (CIM, 2010; JORC, 2012). Exploration must also continue after a resource or reserve is declared to help sustain the future life of the mine (Fig. 6.1).

REPORTING STANDARDS FOR RESERVES AND RESOURCES

A series of regulations govern how reserve and resource information is released to the public in disclosure statements (CIM, 2010; JORC, 2012). The management and implementation of this process requires attention to detail and high ethical standards. Failure to properly report information can be a cause for investigation by the Securities Commission, so the reporting of reserve and resource data together with the underlying geoscience background is handled by geoscientists who have professional qualifications that are vetted by a Government Organization. The individual who signs-off on the information is a Professional Geoscientist (P.Geo.) who has been identified by the Company as a Qualified Person (QP) and who meets strict requirements under guidelines accepted by the mining and investment community.

The details that comprise a report that supports the declaration of a reserve or resource often involves careful analysis of all the information that provides a basis for the estimate, evaluation of the grade distribution and extent of drilling, geometallurgy of the material, and engineering considerations that relate to ground conditions or mining. The scope of this work is often large, but it is normally underpinned by the very detailed geological information that has been gathered during exploration as described in the preceding sections. For details of the modeling of ore bodies, the reader should consult appropriate specialized texts (Glacken and Snowden, 2001; CIM, 2010; JORC, 2012).

EXPLORING BEYOND THE FOOTPRINT A KNOWN ORE DEPOSIT

The identification of entirely new ore deposits at Sudbury is critical to the long-term survival of the camp as a nickel producer.

On a regional basis, information from past exploration is generally compiled using two-dimensional and/or three-dimensional software platforms, and the resulting database is used by the exploration geoscientist in support of a practical effort to discover mineralization. Empirical data often provides a first-order guide in exploration for contact, footwall, and Offset type mineralization. Layered on this are datasets of geochemical and geophysical data as well as historic drilling, and models derived by inversion of geophysical datasets constrained by geological contacts.

There are facets of mineral system and process models beneath this approach, but exploration success is largely derived from excellence in building a geological understanding of the environment, and establishing the potential trend of a target zone that has a large enough footprint for a discovery that will impact on mining.

In this process, a number of key pieces of information are examined in order to support an exploration proposal.

Favorable geology at or below the lower contact of the SIC

Virtually all of the mineralization associated with the Sudbury Structure is located at, or close to the base of the SIC, or it is associated with the continuous and discontinuous parts of radial and concentric Offset Dykes (Chapter 4). There are no economically significant examples of nickel sulfide mineralization that do not relate to these geological environments. Small mineral zones do occur outside of the footprint of the SIC, but they are likely expressions of footwall mineral zones (Broken Hammer; Hall et al., 2015) that are distal from the erosional base of the SIC or they are the tectonically detached from the SIC and possibly a product of hydrothermal modification (Amy Lake PGM occurrence; Tuba et al., 2014) of primary footwall mineralization.

The base of the SIC, the footwall beneath embayment and trough structures, the Offset Dykes, and structurally detached segments of these rocks represent the main focus of exploration.

The basal contact of the SIC

The extent of exploration of the lower contact of the SIC is quite good to a depth of about 1.0-1.5 km, but below this level, exploration is sparse and often located in areas where mineral zones extend from surface to depth and where the plunge of the mineral zone has been followed at depth. Historically, regional exploration of the contact to a depth of about 2 km has been undertaken, but the space for a new discovery increases along the basal contact of the SIC with depth. Just because a trough or embayment does not extend to surface does not mean that the base of the SIC is not prospective; the Nickel Rim South Deposit is a very good example of a deposit that is related to a small embayment at the base of the SIC that does not extend to surface. Examples of fault-repetition of the basal contact of the SIC in the East Range (eg, Victor and Capre) as well as displacement of the contact by the faults in the South Range Shear Zone (in particular, the Cliff Lake Fault) offer environments in which future discoveries can be made.

The footwall of the SIC

The identification of footwall mineralization in the North Range did not come until drilling extended beneath the Sublayer intersected Cu-rich vein style sulfide mineralization. Mid-20th century geological wisdom would have discounted the possibility of mineralization in this geological environment, yet the recognition by exploration geologists of unusually Cu-rich sulfides entirely hosted in the country rocks beneath the SIC has triggered an emphasis on exploration in the footwall over the last 30 years. Many of the most recent discoveries at Sudbury, such as Victor and Nickel Rim South, comprise important footwall mineralization, without which the associated contact ore bodies might not be economically viable. Footwall style mineral zones also occur in the South Range, and although they comprise very rich styles of mineralization relative to the contact deposits, they typically have lower base and precious metal grades than the North and East Range footwall deposits.

The recognition of footwall mineralization anywhere from a few tens of meters to many hundred meters away from the lower contact of the SIC has been a major focus for recent exploration, where drill holes and infrastructure in the contact environment is used as a platform to test the underlying footwall. Footwall mineral systems at Levack, Coleman, Podolsky, Capre, Victor, Nickel Rim South, and Creighton are described in Chapter 4, and they all point to how effective exploration of the footwall can be in areas with known contact mineralization. If there is no contact mineral system, or no sign that it has been faulted off or eroded away, then the mineral potential of the footwall is typically considered to be lower than in areas of the SIC with known contact mineralization. But the reader should keep in mind that it will only take one significant discovery to disprove the hypothesis because a large contact ore deposit may have been completely eroded away.

Expectations for the discovery of giant ore deposits in the footwall of the SIC must be tempered by the realization that the known deposits have a ratio of tons of ore in the contact relative to the footwall of about 5:1 to 10:1, and the figure 10:1 is close to that predicted by fractional crystallization modeling of sulfide liquid in Chapter 5. However, not all of the shallow footwall environments at Sudbury have been fully explored because the historic viewpoint held that the mineral zones followed the contact.

The discovery of footwall mineralization has highlighted the fact that both science and serendipity can change the exploration paradigm, and areas that previously were considered to have low mineral potential, now need to be re-evaluated.

The Offset Dykes

The heavily mineralized segments of Offset Dykes such as Copper Cliff and Frood-Stobie where the mineralization comes to surface are explored to depths of ~1.5–2.5 km depth, but not always to this extent along strike. Exploration of the southern part of the Copper Cliff Offset resulted in the discovery of the Kelly Lake Deposit (Polzer, 2000). Offset Dykes such as Worthington have enormous opportunity for new discovery, and the newest mine in the Sudbury Basin at Totten is an example of how a discovery can be made within the Offset environment by following the surface manifestation of mineralization to depth (Lightfoot and Farrow, 2002).

The distribution of the Offsets is well known from mapping and drilling, but regional mapping (Wood and Spray, 1998) and exploration (Smith et al., 2013) has identified a considerable length of previously unrecognized Quartz Diorite (QD) belonging to Offset Dykes that are now generally accepted to be part of the SIC. An opportunity exists for discovery in continuous and segmented Offset Dykes within belts of breccia that surround or underlie the SIC.

THE MORPHOLOGY OF THE BASAL CONTACT OF THE SIC AND THE OFFSET DYKES

The recognition that physical depressions at the base of the SIC control the location of contact mineral systems has long been recognized (Boldt, 1967), and the geology of important ore bodies related to this morphology has been described in Chapters 2 and 4 and in technical papers by Souch et al. (1969) and Morrison (1984).

Troughs and embayments are physical depressions at the base of the SIC (Chapter 2), and they appear to have acted as collection centers for the contact mineralization. There is no simple empirical relationship between the size of the trough or embayment and the scale of the mineral zone. The Creighton and Trillabelle troughs are broadly similar in depth, width, and length, but Creighton hosts two orders of magnitude more mineralization than Trillabelle (Chapter 1). Moreover, many important deposits such as Nickel Rim South and the Levack-Coleman Deposits occupy especially small embayment structures where the quantity of mineralization is large in relation to the size of the depression. Embayments and troughs are important, but their size is not a good guide in exploration for large deposits.

Likewise, there is no clear association between the continuity and extent of Offset Dykes and the development of mineral zones. Small discontinuous lenses of MIQD can be associated with very large deposits such as Frood-Stobie and significant recent discoveries such as the new discovery at Victoria, but continuous Offset Dykes such as Hess, Ministic, Manchester, and Foy have only small mineral zones or sub-economic occurrences of mineralization. There are clearly differences in the mineral potential of different Offset Dykes, and there is presently no reason to believe that the continuity or thickness of the dykes relates to mineral potential. Local variations in dyke morphology produced by different country rocks, Sudbury Breccia (SUBX), or anomalously wide domains of dyke within extensive zones of metabreccia are a far better guide to mineral potential than the lateral or vertical extent of the Offset Dyke.

PETROLOGY OF THE HOST ROCKS

The mineralization associated with contact ore bodies tends to be hosted in a magmatic breccia that is termed the Sublayer. These rocks are documented in detail in Chapters 2–4. The development of norite and Granite Breccia (GRBX) in the North and East Ranges are all key indicators of mineralization. Troughs and embayment structures contain Sublayer rocks, and these are the rock types that guide exploration in these depressions at the base of the SIC.

In the footwall beneath embayments and troughs, the guides to mineralization are much more difficult to identify. Empirical observations indicate that the mineral potential of the footwall is higher where there is extensive metamorphism, recrystallization, and partial melting, so a search for metabreccias and partially melted SUBX provides a good indication of mineral potential. The mineral potential of the footwall also tends to be higher in areas of extensive SUBX. As described in Chapter 4, these zones of breccia are possibly slump planes along which the crater wall was able to move during the fractionation of the sulfide melt.

In the Offset Dyke environment, the key requirement is for the development of inclusion-bearing varieties of QD. The quenched and marginal unit of QD and QD with small inclusions do not contain ore deposits, but the Offset Dyke unit comprising the MIQD and HMIQD often contains semi-massive sulfide as described in Chapters 2 and 4.

THICKNESS OF THE METAL-DEPLETED NORITE STRATIGRAPHY

As described in Chapter 3, two features of the Main Mass stratigraphy of noritic rocks are closely related to the mineral potential of the lower contact and Offset Dykes:

1. Norite thickness: a thick norite unit is developed above heavily mineralized contact and Offset Dyke domains, and a thin norite stratigraphy is developed above more weakly mineralized lower contact and Offset Dykes.

2. Metal depletion of the unmineralized norite: all around the Sudbury Basin, the Main Mass norites that do not have elevated Ni, Cu, and PGE due to sulfide control are depleted in these same elements. The degree of depletion is generally larger in the North Range Felsic Norite (FNR) stratigraphy relative to the South Range Norite.

As described in Chapter 5, these features are related to the melt volume that was processed by the dense immiscible sulfide melt before it segregated and concentrated to form the contact and Offset mineral zones.

On this empirical basis, the potential for a very large mineral discovery beneath thin norite stratigraphy is not as high as the potential beneath thick norite. Although a decision to explore in an area of opportunity is sometimes influenced by this empirical observation, there is still opportunity to make a discovery at the contact or in Offset Dykes beneath a domain of norite stratigraphy that is intermediate in thickness or thin. The decision often comes down to economics. A small near-surface ore body that can be mined with an open pit can provide more immediate return on investment than the development of a much larger deep ore body. A careful analysis must be undertaken, as the amount of metal lost from the melt sheet in areas of thin stratigraphy may still provide the economic basis to declare an ore reserve in a shallow deposit that can be mined at a profit.

APPLICATION OF SULFIDE GEOCHEMISTRY

In Chapter 5 it was illustrated that the Ni and Cu tenor of the sulfide in mineralized Mafic Norite (MNR) provides an indication of the mineral potential of the underlying Sublayer. The same approach to establishing the metal tenor of weakly mineralized rocks can also be applied to the lower FNR and Quartzose Norite, as well as the MIQD of the Offset Dykes (Lightfoot, 2009), and the derived relationships between diversity in metal tenor and known mineral potential follow the relationship established for the MNR (Fig. 5.12).

A similar approach can be applied for rocks from the footwall. Anomalous metal concentrations (Cu, and PGE) and their sulfide metal tenor in country rock, metabreccia, and SUBX can help vector toward footwall mineral zones. As discussed in Chapter 5, it is widely believed that the footprint of hydrothermal alteration associated with the footwall mineral zones is quite limited in extent, and so the metal-enrichment halo around these zones is restricted in spatial extent to a few tens or hundreds of meters. Although the halogen contents of country rocks (Jago et al., 1994; Hanley et al., 2005), and amphiboles and micas (Hanley and Bray, 2009) record evidence indicative of proximity to footwall mineralization, the signature of Ni, Cu, and PGE in the country rocks provides the most robust indication of anomalous footwall. No economic discovery has been made using halogen geochemistry, and although studies of the chemistry of alteration minerals offer potential to vector along structures and in zones of partial melting, an understanding of the paragenesis of the silicate minerals in the footwall is critical to successful application of this approach in the future.

STRUCTURAL FEATURES OF THE CONTAINER ROCKS OF THE SUDBURY ORE DEPOSITS

Chapters 2 and 4 of this book examined the deformation history of the Sudbury Structure, and highlighted the following structural controls:

1. Initial morphology of the base of the melt sheet and the Offset Dykes that was created in response to both the primary impact process and the crater modifications that happened immediately after impact. These processes largely predated sulfide saturation and crystallization.

2. Longer term modification of the crater walls by processes of slumping and reactivation of the Offset Dykes. These processes accompanied the formation of the ore bodies.
3. Postimpact deformation that modified and redistributed some of the primary magmatic sulfide ores along shear zones or either side of fault structures.

Belts of SUBX that were formed in the crater floor provided surfaces along which the crater wall may have failed. Steep walls may have been present at the inner margin of the crater rim, but they may also have developed at the outer margin of the central uplift structure. Movement along these structures possibly controlled the morphology of the embayments and troughs, and may also have modified and redistributed the sulfides before they were complete crystallized resulting in the development of inclusions of sulfide in the GRBX (Chapter 4). Domains of SUBX developed beneath mineralized embayment structure remain an attractive target for future exploration.

Primary discontinuities in the Offset Dyke are also important controls on the morphology and distribution of ore bodies in the Copper Cliff Offset Dyke as discussed in Chapters 2 and 4. The cross-linking structures between sub-vertical ore shoots controlled the lateral as well as vertical distribution of mineral zones in the Offset Dyke.

Structures that postdate the crystallization of the SIC and the stabilization of the crater are generally related to far-field tectonics that control orogenic events that deformed the Sudbury Structure. Exploration models benefit from knowing the kinematics and magnitude of displacement along structures that cut through known mineral systems or abut and appear to terminate known mineral zones as described in Chapter 4.

EXPLORING THE EXTENTS OF KNOWN ORE DEPOSITS

An extensive database of drilling and assaying provides a basis to build detailed three-dimensional models of the mineral zone, the container rocks, and the structures. This model can then be used to develop exploration plans as well as improve the strategy for delineation and definition of ore bodies.

There are still exciting opportunities to make discoveries in the shadow of the head-frames at Sudbury, and at shallow depth. A good example is provided in Chapter 4, where the geoscience approach to documenting the displacement on faults in the Coleman mineral system provided the basis for targeting a possible extension of mineralization across a fault based on a 3D kinematic restoration of postimpact displacement of the SIC (see Fig. 4.65; Gibson et al., 2010). The discovery of the Lower 170 mineral zone as an extension of the previously recognized and currently mined 170 footwall ore body is proof of this concept, and the value of careful structural studies of the mineral systems at Sudbury. Likewise, efforts to explore the footwall of the Capre contact deposit resulted in the discovery of massive footwall sulfides in the 3000, 3001, and 3002 mineral zones (Stewart and Lightfoot 2010), and efforts to explore the footwall of the Crean Hill Deposit resulted in the discovery of LSHPM styles of mineralization (Gibson et al., 2010).

EXPLORATION TECHNOLOGY

The generation of new ideas and a pipeline of exploration projects (Lightfoot, 2007) ultimately requires the technical implementation of an exploration program. The geological idea is no more than a concept unless a target is defined and tested with geophysical methodologies, diamond drilling, and further geological and geochemical evaluation. The conductive signature of a potential mineral zone is often, but not always, the trigger to drilling in order to test a target. Survey work can be undertaken at surface, after diamond drilling is complete, or down previously drilled bore holes that are open to the surface (Polzer, 2000; King, 2007).

If a geophysical survey confirms the validity of the geological concept by identifying a conductive target, then the only viable path forward to test the target is by using diamond drill technology. Both of these phases of work are often forgotten in the academic literature on ore deposits, but they underpin the assessment of the geological concepts and the discovery of new mineral zones.

Geophysical signature of the Sudbury ore deposits

Borehole electromagnetic surveying is the workhorse method employed in exploration at Sudbury (Dyck, 1991; King, 1996, 2007; West et al., 1984). These methods are the subject matter of specialized text books (Dentith and Mudge, 2014 and references therein), but many of the details of how the surveys are interpreted remain the proprietary information of exploration companies. In one of the rare exceptions, Polzer (2000) provides an account of the role played by borehole electromagnetic methods in the discovery and extension of the Kelly Lake Ore Body.

Borehole electromagnetic systems use a magnetic field established by a large transmitter loop antenna lying on the surface of the earth (Fig. 6.2A after Polzer, 2000). The primary magnetic field generated by the transmitter loop flows into the ground in a predictable pattern that can be computed at any point based on the known locations of the transmitter and down-hole receiver. Due to the extreme conductivity of magmatic sulfide, the magnetic field is largely prevented from existing inside zones of highly conductive Ni–Cu sulfide mineralization. Currents are established on the surface of the conductor in order to generate a secondary magnetic field that opposes the primary field. A 3-axis down-hole measurement system senses the direction and strength of the total magnetic field. Differences between these observations and those calculated from the primary field are attributed to the deflection caused by mineralized zones. A model of the conductive body is created using the established location, orientation, and size of the conductive plates needed to cause the observed magnetic field deflections observed in the bore-hole survey. The most important parameters available to the geologist from the interpretation are the distance and direction to the nearest conductive edge as well as the overall size and attitude of the conductive domain in three-dimensional space that is modeled as a best fit to the survey data.

In the above mentioned example of the Kelly Lake exploration program, borehole electromagnetic surveys provided information on the geometry of the mineral zone and its continuity between holes (Fig. 6.2B). As the drilling expanded the size of the mineralized zone, the geophysical information allowed subsequent holes to be more effectively positioned to identify more mineralization. The location of the nearest edge of the conductor relative to the hole can be used to establish the margin of the mineral system and ensure that each hole intersected mineralization thereby maximizing the amount of mineralization added with each drill hole. Knowledge of the overall size of the deposit in advance of drilling also provides a basis to assess the potential of the mineralization, so that decisions on expensive drilling can be properly weighed in the context of potential value.

The approach to surveying both old holes and newly drilled holes with bore-hole electromagnetic methods is part of the core activity of exploration in Sudbury. The geophysical anomalies generated from these surveys can be integrated with three-dimensional models of geology and other geophysics in order to provide a holistic basis for the exploration precision process.

A range of geophysical techniques are used in near-mine exploration. Among them, bore-hole electromagnetic methods are especially useful. Other methods such as radioimaging (RIM), bore-hole conductivity/resistivity, magnetic susceptibility, gravity, induced polarization (IP), and geophysical tools are

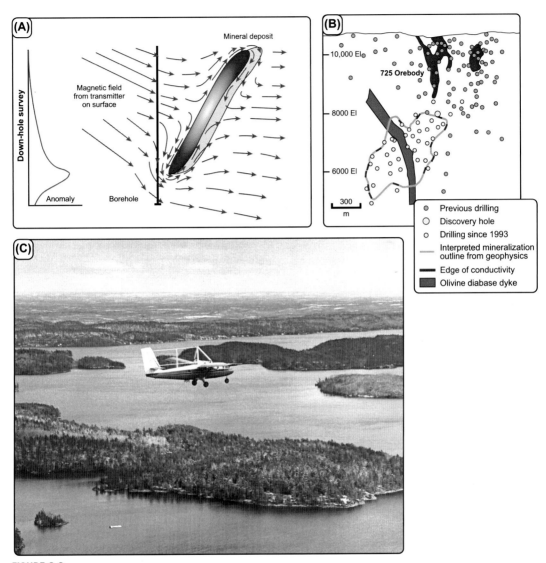

FIGURE 6.2

(A) Schematic diagram showing the primary field and its interaction with a conductive subsurface ore body as detected in a sensor lower down a nearby borehole. (B) Model showing the effectiveness of borehole electromagnetic surveys in the delineation of the Kelly Lake ore body. Drilling completed after 1983 was undertaken using a combination of geoscience information including borehole survey methods, and the diagram illustrates how this information was used to establish the edge of the conductive system and focus drilling within the economically significant part of the ore body. (C) The airborne electro magnetic system shown in this image was developed for regional exploration by Inco; this system was extensively used throughout North America in exploration and provided the basis for follow-up on the ground that resulted in discoveries in the Thompson Nickel Belt (King, 2007).

(A, B) After Polzer (2000).

used to delineate mineral zones of a specific grade (conductivity) and to map out the best approach to blasting in order to recover the ore (McDowell et al., 2004, 2007; King, 2007). Other technologies that are used in the near-mine environment to help understand the geology of the rocks that contain the mineral zones include the optical and acoustic televiewer to map ground conditions, structures, and lithologic contacts and cross-hole seismic tomography to map changes in the p-wave velocity properties of the rocks present between bore holes (at Sudbury, rocks that contain sulfides typically have anomalous p-wave velocity profiles; McDowell et al., 2007). The use of oriented core allows the geoscientist to extract structural information that can be used in interpretation and modeling (e.g., Monteiro, 2015, 2016).

Regional Geophysical Signature of Ore Deposits at Sudbury

The Sudbury Structure has been surveyed with a wide range of regional and detailed airborne and surface geophysical tools. Indeed, much of the development of the airborne method of magmatic and electromagnetic surveying was pioneered by geologists who worked on a range of nickel sulfide ore deposits, and identified the importance of associated magnetic and electromagnetic signatures that could be detected with airborne sensors. The technology developed by Inco for electromagnetic surveying is described in King (2007) and an example of the system used by Inco in the 1950s is shown in Fig. 6.2C.

Magnetic, gravity, and seismic surveys of the Sudbury Region were undertaken by government organization with the objective of increasing the understanding of the three-dimensional geometry of the rocks (Milkereit et al., 1994a,b,c; McGrath and Broome, 1994a,b; Hearst et al., 1994). Important contributions to understanding the magnetic properties of the SIC and the country rocks, and the associated paleomagnetic record of formation appear in the work of Morris (1980b, 1981a,b, 1982, 1984a,b, 1994).

Most of the high-resolution magnetic, gravity, and electromagnetic survey datasets remain outside of the public domain and they are rarely provided as a basis for academic studies (e.g., Olaniyan et al., 2014, 2015). From an exploration perspective, these datasets provide some very clear evidence of geophysical signatures that are associated with the Sudbury mineral systems. The next section looks at some of these features.

The Geophysical Signature of Offset Dykes and Associated Mineral Systems

The Offset Dykes are composed of very weakly magnetic, low-density QD that contains mineral systems that comprise Po-Pn-Cpy mineral assemblages with mafic-ultramafic inclusions that are often magnetic and dense. For this reason, mineral systems along Offset Dykes tend to have a weak to moderate positive magnetic expression but they rarely exhibit a large contrast in density with the nearby country rocks. For example, the Totten mineral system is located at the southern margin of meta-melagabbro sill, and there is no clear contrast in the gravity signature of the deposit relative to the country rocks. In other locations along the Worthington Offset Dyke, weak magnetic highs are associated with mineral zones, but there is no significant gravity contrast.

Offset Dykes such as Copper Cliff and Frood-Stobie do not have strong magnetic or gravity signatures; this is partly because of the anthropogenic effects of mining, but also a reflection of the complex magnetic stratigraphy of the host rocks. Beyond the immediate footprint of the Frood-Stobie Deposit, the mineral system is bracketed between weakly magnetic granitoid rocks of the Murray Pluton and the more dense mafic volcanic rocks and parts of the Frood Intrusion (Chapter 2). In this context, both high resolution magnetic and gravity surveys help to map out the footprint of rock types that control the locus of the mineral system.

The Geophysical Signature of Contact and Footwall Mineral Systems

The geophysical signatures of contact ore bodies developed around the margin of the SIC is very evident in high-resolution magnetic surveys. The contact mineralization tends to consist of Po-Pn-Cpy that collectively yields a strong magnetic response. Moreover, the mineralization tends to contain many fragments of mafic-ultramafic rock that contain magnetite, and occur in association with the Sublayer which contains disseminated sulfide mineralization. This combination of rock types produces a strong positive magnetic response, and the response is often amplified by the development of unusually large bodies of magnetic mafic country rock in the footwall. Collectively, these rocks provide a strong magnetic signal that is easily recognized over known ore bodies by examining the analytic signal of the magnetic data.

Although the analytic signal of the magnetic response resolves these features very clearly, it generally only maps out what is known to a depth of a few hundred meters based on drilling, but it is still a useful guide in areas with less complete exploration. The gravity response can also be important, but there is a range of density in the mineral systems, the SIC, and the country rocks that makes unconstrained interpretations quite difficult. For these reasons, the application of magnetic and gravity methods is often best achieved in conjunction with constraints from 3D geological models in geophysical inversion software that support possible models of the 3D distribution of rock types and mineral zones (King, 2007).

The footwall ore bodies tend to be poor in Po and they typically have bulk mineral compositions comprising Cpy>>Pn>Mill>Bn; they tend to have an order of magnitude weaker electromagnetic response as well a weaker magnetic response relative to contact ore bodies because of both the mineralogy and the conductivity of the sulfide assemblage.

Other technologies have been tested in areas such as Trillabelle where the footprint of the deposit has been evaluated using three-dimensional seismic tomography (Eaton et al., 2010). Likewise, a seismic traverse over the 402 Ore Body has identified a response that may be due to the mineral system or to the anthropogenic effects of mining (Eaton et al., 2010).

The application of radio-imaging (RIM) between boreholes (Stevens et al., 2000) has produced anomalies that are not always due to the presence of sulfide mineralization, but sometimes they are drill-tested on the basis of the known association of anomalies with sulfide. As the anomalies are often found in the footprint of mineral systems, it is not always clear whether the successful discovery of mineralization is due to the survey or to the willingness to take a risk and drill a hole in an area of very prospective geology.

A technology that is sometimes used for regional exploration is an application of natural field electromagnetic methods, termed the magnetotelluric survey method (Livelybrooks et al., 1996). The interpretation of the natural field response can reflect highly conductive sulfide but it can also be produced by other confounding effects such as the channeling of the signal by conductive stratigraphy, faults, and contacts.

Testing targets with diamond drill technology

Testing an exploration target is critical to the proof of concept. Without drilling there is no solid evidence of mineralization or the possibility of economically significant intervals of mineralization that can be reported to the senior management of the mining company or ultimately to the market in order to raise funds for further drilling and possible development of a discovery. The technology of drilling is illustrated in the front image of this chapter; the drill rig shown in this image is testing a geological and geophysical target at a depth of 2200 m in the southern part of the Copper Cliff Offset Dyke.

Achieving the goal of drilling the target requires attention to many factors before drilling commences. These include consultation with stakeholders, attention to the environmental impact of drilling, implementation of a contractual agreement with a drilling company to undertake the work based on a schedule of agreed costs and timelines as well as maintenance of the standards of technical excellence in drilling, utilization of modern technologies, and attention to the health and safety of the workers (Table 6.1). Near-mine drilling also requires clearance from the mine engineers so that there is no chance that the drill hole will encounter underground infrastructure.

The location of the drill site is optimized so that the target can best be tested with a minimum core length, and the location of the collar of the hole together with its dip and azimuth are conveyed to the drill company at the start of the work.

Drilling of shallow holes (<1500 m) into near surface targets can generally be undertaken with quite simple drill technology. The effectiveness of the contractor is judged based on the safety and environmental performance, cost and speed of drilling, and the recovery of complete drill core that has not been crushed, worn or ground-up during aggressive efforts to rush the drill hole to completion. Drilling of deeper holes is much more complex than shallow holes as highlighted in Table 6.1. For example, the deepest surface drill hole in the Sudbury Basin is 3444 m and was targeted at a large open area of the basal contact of the SIC with little requirement to hit an exact target in space—it was enough to simply establish whether there is a domain of Sublayer at this location in the Sudbury Basin. Other holes drilled to depths of 2000–3000 m are routinely drilled in order to target and test a domain with a precision of 25 m or less, and then surveyed to a precision of a few meters at the end of the hole. This sort of effort requires that the drill penetrate with a specific dip and azimuth through time in order to reach the target and therefore requires constant and precise monitoring of the drill-hole trajectory. Reaction time is crucial to guide a hole to target before deviation, which is often amplified by rock fabrics or faults, becomes uncontrollable. There are various methods to guide a drill hole in a specific direction. The most commonly used methods use a retrievable (Clappison) wedge that is placed at a location in the hole where a deviation from the natural path is required. This wedge can be positioned to guide the tip of the drill string in a specific direction. There are other methods of directional drilling that utilize technologies developed in the oil and gas industry to produce rapid changes in direction of the drill hole. Efforts to hit a point in space with reasonable confidence at depth often utilize these types of technology at Sudbury (Table 6.1).

The drill core collected from the drill program is recovered at surface, marked with the depth interval, and then described by the geologist in what is termed a "geological log" that records the rock type, textural relationships, quantity of mineralization, fabric, and a multitude of other properties. The core may contain mineral intervals that can be assayed to establish whether there are significant concentrations of base metals (Ni, Cu, and Co), precious metals (Pt, Pd, Au, Ag), elements that may be present in proportions that interfere with processing (eg, As, Pb, Zn), and elements that allow the calculation of element ratios [eg, (Cu + Ni)/S]. The sampling of the drill core and analytical work are controlled by strictly regulated protocols and procedures that are demanded by codes of standards (CIM, 2010; JORC, 2012). The laboratory contract must be carefully negotiated to achieve value, turnaround, precision, accuracy, and determination limits that are appropriate to the materials of interest (Potts, 1987; Riddle, 1993).

Table 6.1 Technological Approach to Drill-Testing in Exploration at Sudbury				
Stage	**Description**	**Activities**	**Description and Sources of Additional Information**	**Measures of Success**
1	Lead-up to a possible drill program	Project generation	2D and/or 3D geoscience compilation defines a target which normally requires evaluation by acquisition of drill core.	1. An inventory of targets with a sense of ranking provides a strategic pathway for a company exploration department. 2. A specific target passes preliminary hurdles and is tabled for follow-up.
		Preliminary economic evaluation	Estimated potential tonnage, grade, and value confirm that a new discovery would meet economic hurdles required to support future production.	1. A target that passes a preliminary economic evaluation is a considered an exploration opportunity. 2. Targets that fail to pass economic hurdles on the basis of present market conditions may be shelved for future consideration. 3. Expenditure is optimized to support projects that have the greatest potential economic impact.
		Survey work to improve understanding of target	Surface and/or bore hole geophysical survey work (Polzer, 2000), geological mapping, geochemical surveying, investigation of historic drill core and many other approaches are used to establish a best pathway to test the target.	1. Identify a strategy to test the target with diamond drilling. 2. Understand the opportunity to use the borehole as a geophysical platform to test the target domain. 3. Optimize time, effort, and expenditure to achieve maximum value.
		Target assessment for drilling	Optimal approach to drilling is assessed in the context of topography, drainage, roads, private property and mineral rights owner, forest cover, and a hole is designed (collar location, azimuth, dip) to best test the target at lowest cost.	1. Optimal pathway established to allow for target to be tested. 2. Potential problems are resolved before they interfere with the implementation of the plan.
		Budget allocation and project approval	Persuade management to part with company's resources by presenting an argued case that the target has a high priority, meets short- or long-term company's objectives and may alter the life of mine or facility.	Exploration budget is allocated to test the target.

(*Continued*)

Table 6.1 Technological Approach to Drill-Testing in Exploration at Sudbury (*cont.*)

Stage	Description	Activities	Description and Sources of Additional Information	Measures of Success
2	Preparation for drilling	Selection of a qualified drilling company	Drilling service provider selected to operate the program according to client requirements such as cost, safe work practices, rapid mobilization, productivity, core recovery, and deep drilling experience.	Contractor passes all hurdles and a contract is implemented to undertake the program of work.
		Consultation on surface rights or mine property	Requires legal access and must be cleared in a near-mine setting; evaluate whether the target can be tested from existing underground infrastructure or requires testing from surface.	Establish access for surface work and avoid any present or future impact to mine infrastructure.
		Site-specific hazard assessment	Potential hazards are assessed in the context of topography, drainage, roads, private property and mineral rights owner, forest cover, and presence of powerline, natural gas and rail lines, etc.	Assess hazards and implement a strategy to mitigate risk to those working on the project and to the project.
		Evaluate potential in-hole problems	Avoid impacting water table, open fault structures, underground infrastructures, natural gas, or hydrocarbon reservoirs.	Success in avoiding technical problems that delay the project or result in the project being abandoned due to problems in reaching the target.
		Environmental consultation	Drill rigs require a water supply and access trails; the rig must be operated in an environmentally sound way. Must avoid adversely affect previously reclaimed areas, monitoring wells, etc.	Minimize impact of the drill program to the natural environment.
		Consultation process with aboriginal communities, public and surface rights holder, local residents, communities	Effective engagement and support for activities that bring value and future wealth to stakeholders. Contact with Aboriginal (First Nation and Métis) communities should be made and maintained in order to ensure that Aboriginal rights and/or treaty rights are not adversely affected. Contact with the public should be made and maintained as changes to the land may influence recreational activities, raise environmental concerns or cause health or safety issues, contact with surface rights holders is maintained leading up to and throughout the program of work. See http://www.mndm.gov.on.ca/en/mines-and-minerals/mining-sequence/consultation	Good public relations established and maintained.
		Plans and permits	Normally acquired from Provincial Government following a carefully crafted application that passes approval hurdle. See https://www.ontario.ca/laws/regulation/120308	Permits issued.

Table 6.1 Technological Approach to Drill-Testing in Exploration at Sudbury (*cont.*)

Stage	Description	Activities	Description and Sources of Additional Information	Measures of Success
3	Mobilization and setup of drill rig	Secure safe access and servicing of the rig	Find safe access for crews and equipment. Locate water source and prepared drill pad with settling sump for drill water. Clear adjacent vegetation.	Viable route for access in the event of an emergency. Effective basis for rehabilitation of the drill site after work is completed.
		Position drill rig	Drill hole collar location is accurately marked using surveying equipment (Lyons, 2010).	Accurate and precise drill collar location established.
		Align drill rig	Drill alignment (azimuth and dip) is adjusted according to planned borehole using surveying equipment (Lyons, 2010).	Declination and azimuth of drill rig is optimized to test target.
4	Effective core recover	Accurate interval records	Complete record of drill intervals aligned with depth in hole.	Maintain quality assurance.
		Complete core recovery	Core recovery complete and free of grining of soft intervals of rock (eg, faults, sulfides) or mechanically grind core intervals.	Drill core recovery is optimized for logging.
5	Getting the drill hole to the target	Hole trajectory monitoring	Monitor hole trajectory (azimuth and dip) against plan using magnetic or gyroscopic survey instruments (Lyons, 2010).	Ensure that drill hole is progressing toward target
		Control and correct deviation of hole	Use of stabilization techniques (eg, round vs hexagonal core barrel), wedge or directional drilling technology to control the trajectory of the drill string affected by structures, contacts, or rock horizons (Lyons, 2010).	
		Loss of drill pressure due to bad ground	Cement hole to fill cavity and avoid loss of water pressure (Lyons, 2010).	Avoid loss of drill hole.
		Stuck drill rods	Avoid losing the drill hole by using retrieval and in-hole blasting techniques; avoid leaving rods in the hole as they may impact on future geophysical surveys (Lyons, 2010).	Avoid loss of drill hole.

(*Continued*)

Table 6.1 Technological Approach to Drill-Testing in Exploration at Sudbury (*cont.*)

Stage	Description	Activities	Description and Sources of Additional Information	Measures of Success
6	At the completion of drill hole	Survey of hole	Confirm 3D orientation of hole trajectory using in-hole surveying method such as magnetic or north-seeking gyroscopic equipment (Lyons, 2010).	Exact survey allows hole position to be accurately modeled in 3D. Exact knowledge of the location of the drill string is required to model any mineralization that is intersected.
		BH-EM survey	Open hole required for successful test for conductivity plate surrounding hole or away from hole (Polzer, 2000; King, 1996, 2007).	Hole can be surveyed with geophysical tools or entered again to deepen the drill hole or establish a branch in the hole.
		RIM, cross-hole seismic, physical property, and televiewer surveys	A variety of survey methods can be used to test between bore holes (McDowell et al., 2007). Open hole free of grease and drill cuttings required for successful test.	Optimal conditions for survey work achieved.
		Cement hole	Safeguarding of future mine openings by cementing drill hole (Lyons, 2010).	Avoid potential impact to existing or future mine infrastructure from open holes (eg, flooding).
		Demobilization of drill rig	All drilling equipment safely removed from the project.	Success in completing program to company standards of safety, health and environment.
		Environmental decommissioning of drill site	Drill site clean of debris, sump filled, and area rehabilitated.	Site returned to original state.
7	Processing of the drill core	Log lithologies	Detailed log of rock types encountered as basis for three dimensional model.	Complete and comprehensive record of observations made on drill core.
		Structural information	Detailed structural features recorded to help in the construction of 3D model (Mertie, 1943; Blenkinsop et al., 2015).	Location of faults, shears, folds, fractures, veins, etc recorded.
		Log physical properties	Support engineering studies and mine construction.	Data on density available to support resource modeling.
		Geotechnical information	Support engineering studies and mine construction project through the collection of geotechnical information (de la Vergne, 2008).	Geotechnical information available to support future mine plan.
		Assay mineral intervals	Determine metal abundance levels in intervals of core.	Representative accurate and precise drill core assays will underpin models that describe a mineral zone.

Table 6.1 Technological Approach to Drill-Testing in Exploration at Sudbury (*cont.*)

Stage	Description	Activities	Description and Sources of Additional Information	Measures of Success
	Processing of the drill core	Retention of material	Sample representative materials or retain complete or half core in core library.	Material available to support future studies without the need to drill additional holes.
		Petrology	Thin section and polished sample examination of unusual rock types in order to name rock correctly.	Understanding of the correct names that are allocated to the rock types.
		Lithogeochemistry	Specific application to establish geochemical correlations, identify metal enrichment or depletion, and characterize igneous rocks belonging to different magma series.	Establish correlations between rock units, and underpin evaluations of possible minerals and elements that may impact future processing of the ores.
		Metallurgy	Ore characteristics and metal recoveries are determined using known metallurgical processes.	Detailed investigation of mineralization in context of process technology.
8	Follow-up to a discovery or mineral intersection	Branch hole	Use existing hole to branch the hole using a wedge or directional drilling to reduce cost, time and increase effectiveness of additional targeting (Lyons, 2010).	Re-entry of historic drill hole for further branching to minimize exploration costs.
		Preservation of hole for future survey work	Ensure that hole is open for future testing as new technologies are available.	Avoid drilling expensive holes when past holes can be used for optimal target evaluation.
		Core storage	Retain representative samples, sawn core or complete core for future evaluation.	Material is available in a safe facility where it can be located through an inventory system linked to the storage location.
		Research and development	Engage research and academic community to maximize value from drill core.	Optimize value and gather input that may yield a new understanding of the vectors toward mineralization.
		Delineation and definition of the mineral zone	Once exploration work has identified a mineral zone that has economic potential, diamond drilling will continue to establish the continuity of the mineralization, and the basis for prefeasibility and feasibility studies required to mine the zone.	1. New mineral zones added in an active mine. 2. New ore deposits are brought from discovery to production in a new mine complex.

Summary of the significant steps required to undertake the testing of a target by diamond drilling. This table was constructed with assistance from Enrick Tremblay.

The more important intervals of drill core from the project is normally retained, and it may become a critical resource for future discovery or for the implementation of sampling programs to establish features of the rocks that contain the mineralization or signatures within the barren rocks that may vector toward mineralization (much of Chapter 3 covers this topic). Historic drill core is of enormous legacy value, and given the cost of drilling holes, the cost of storage and further analysis may be <<1 % of the cost of drilling a new hole.

Ore body characterization and geometallurgy

Geometallurgy studies are based on carefully selected samples that represent a mineral system or part of it. This information supports the development of a mine plan, it impacts how the sulfide ores will be processed, and most importantly it ensures that a mineral discovery can be translated into economic value.

Effective geometallurgy starts with an effort to produce a representative map of the domains that comprise the ore body as they relate to mineral processing technology. This is a requirement for prefeasibility and bankable feasibility studies that help finance a project (JORC, 2012). The process of characterization to generate this 3D understanding of the ore body occurs at different stages in the evaluation of a project:

1. On an existing ore deposit to ensure that it can be mined at a profit: This involves an evaluation of the ore using mineralogical, metallurgical, and grinding studies as they relate to the proposed concentration plant (mill).
2. In a less advanced project, the evaluation may look at the most critical risks (eg, deleterious mineralogy or geochemistry, poor recovery, or other risks to the economic success of the project), or undertake simple characterization of representative samples to establish any issues before additional drilling is undertaken.

Process technology tests include a range of geochemical, mineralogical and physical procedures that help establish the characteristics of the material. The evaluation procedure involves preparation of the material to reflect the composition of the material that will be mined; the sample must be representative and collected to reproduce reasonable conditions of oxidation. The performance of grinding methods available in a mill or mills of choice can be tested at a bench scale to see how efficient the material can be comminuted. Detailed mineralogical studies can be undertaken on the materials before and after grinding to establish whether there are any minerals that might influence the efficiency of the concentration process (eg, layer silicate minerals such as mica and talc or fibrous minerals such as asbestos) or behave unusually during routine floatation (eg, hexagonal pyrrhotite, millerite, and troilite).

A representative sample of the ground ore can be tested using bench-scale floatation cells to estimate recoveries in open and closed systems, and to assess the viability of reprocessing the residue (scavenger floatation) that might otherwise exit as waste (tailings). If the performance on a material is promising, the laboratory methods can be scaled-up to test the material with a mini-plant or a pilot plant (http://www.vbnc.com/Newsletters/December2003.pdf).

Incomplete or ineffective testing can have just as dire a consequence as failure to properly drill off the ore body and report resources and reserves to the standards of National Instrument 43–101 (CIM, 2010). The outcome of poor work or criminal activity can generate fraud that has a very negative impact on the public perception of the mining industry (eg, the Bre-X fraud; Francis, 1997).

The evaluation of ores at Sudbury tends to follow a systematic approach that is tailored to the performance of the concentration plants that are located in Sudbury. The approach is typically a comprehensive evaluation of the material following a flowsheet that is based on technical studies of mineralogy, grinding and floatation tests, and mineral and chemical impurities. The objective of this study is to develop a report that documents an assessment of the recovery potential for Ni, Cu, Co and precious metals that can support a financial analysis of the project.

The sorts of questions addressed in this study include:

1. Characterization of the composition and physical properties of pyrrhotite in terms of the relative proportions of monoclinic and hexagonal pyrrhotite.
2. Studies of sulfide mineralogy to understand whether pyrite, millerite, nicoline, gersdorffite, or other minerals are present that may influence the recovery of metals. The grain size of pentlandite and development of micron-scale lamellae of pentlandite as exsolved crystals in pyrrhotite can influence the recovery of nickel.
3. Investigation of platy or fibrous minerals such as talc, chlorite, mica, asbestos, and other silicate minerals that influence metal recovery during processing.
4. Determination of proportions, grain size, and ease of recovery of sulfide minerals in the gangue. The technologies used to investigate materials include quantitative mineral liberation analysis (MLA) and quantitative methods of scanning electron microscopy (QEMSCAN) and laser ablation mass spectroscopic analysis (LA-ICP-MS; Sylvester, 2001).
5. Understanding of the distribution and recovery of precious metals.
6. Investigation of the geochemistry of the ores. An ore with unusually high Pb, Zn, As, Sb, Bi, Sb, Cd, or other deleterious elements may create a concentrate that is not easily processed by the smelter or do not meet the quality standards for refining or marketing.

The final products of this type of study normally comprise the following:

1. Mineralogy studies of the entire ore body demonstrating the scale of variability.
2. Estimates of metallurgical performance and variability through the deposit. Normally expressed in terms of grade of concentrate and recovery in the context of a mill simulation.
3. Concentrate quality (Ni, Cu, PGE, Co, and Ag content) and studies of impurities such as As, Pb, Zn, Cr, Bi, Sb, Se, Te.
4. Ease of recovery of PGM (Au, Ag, Pt, Pd, Rh, Ir, Ru, Os) as a concentrate.
5. Viability of disposal of tailings and ease of dewatering for application in pastes in order to back-fill mine openings.
6. Sulfide self-heating studies of concentrates and tailings.
7. Tailings disposal options and process waste management

The overall outcome will be a report that helps minimize process risk and eliminate fatal flaws. This report is often a key part of not only the exploration decision process. It is supplemented with more data and test work for successive study phases as required to acceptably mitigate risk of failure in the development of the project.

SUDBURY IN THE CONTEXT OF GLOBAL NICKEL SULFIDE DEPOSITS

A number of key factors are recognized as important in the localization of economic concentration of magmatic sulfide from a magma as it travels from its point of generation (in the mantle, or in the case of Sudbury, in the crust) through to its final resting place (Naldrett, 2010a).

The pathway is quite complex, and in the case of Sudbury there are some differences when compared to other types of nickel sulfide deposit, but this section is designed to show that Sudbury shares many features in common with other nickel sulfide ore deposits.

A few of these features are chosen from the list given in Table 6.2 (see also Chapter 1), and these are used to highlight the important aspects of other mineral systems that help to inform process models of ore formation at Sudbury.

CASE STUDIES OF OTHER MAJOR MAGMATIC SULFIDE MINERAL SYSTEMS

The geology of the magmatic sulfide ore deposits at Noril'sk, Voisey's Bay, Jinchuan, and Thompson provide a very useful framework to better understand the ore deposits of the Sudbury Structure. The deposits at Noril'sk are very young (~250 Ma), and the rocks are undeformed, so the geological relationships provide some very clear insights to the geology of Sudbury. At Voisey's Bay, the distribution of mineralization is controlled by both the geometry of the intrusion that contains the ore deposits as well as the structures that created and modified this geometry. The deposit at Jinchuan resides within a very small ultramafic intrusion that was created through emplacement of sulfide-laden magma into the Longshu Fault Zone much like the ore deposits at Voisey's Bay were emplaced into an east–west trending structural corridor. Thompson is an example of a heavily deformed and metamorphosed primary magmatic ore deposit, possibly formed in a chonolith at a continental margin. Thompson now has a geology that owes far more to deformation and metamorphism than any primary magmatic processes. Collectively, much can be learnt from these world-class ore deposits that inform the ore deposit models for Sudbury.

NORIL'SK

The Noril'sk-Talnakh Ni-Cu-PGE sulfide deposits in Northwestern Siberia, Russia are the largest known accumulations of magmatic Ni-Cu-PGE sulfide ores (Chapter 1). The ore deposit at Noril'sk are associated with differentiated mafic intrusions that have been extensively studied by Russian geoscientists (e.g., Kunilov, 1994, and references therein). The Noril'sk Ni-Cu-PGE sulfide deposits are located at the northwestern margin of the Siberian Shield, and inboard of the margin of the Siberian Craton (Fig. 6.3A–B). They are juxtaposed close to the Noril'sk-Kharaelakh wrench fault beneath or within the lower part of the 250 Ma Siberian Trap basalts (Figs. 6.4A–B and 6.5A). Although there is speculation in the literature that impact events are primary drivers of magmatism in large igneous provinces (e.g., Jones et al., 2002), there is absolutely no evidence to suggest that the Siberian Trap, and by association the Noril'sk ore deposits, are the product of anything other than a mantle plume (Lightfoot and Hawkesworth, 1997; Ernst, 2014).

The basement rocks of the Siberian craton in the Noril'sk Region are unconformably overlain by Proterozoic-aged metasedimentary rocks, followed by a sequence of, Devonian and Permian-aged sedimentary rocks formed in shallow marine and sabkha environments. The Devonian and Permian-aged rocks contain intermittent deeper water marls and shales (eg, the Razvedochninsky Formation and the Tungusskaya Series, respectively), evaporates throughout the sequence, and coal measures and shales of Carboniferous age (Fig. 6.4A). These rocks are overlain by the Siberian Trap flood basalts and tuffs which comprise a stratigraphy of nine different formations in the Noril'sk Region (Fig. 6.4B; Lightfoot et al., 1990, 1993b; Wooden et al., 1993; Fedorenko et al., 1996).

The entire package of sedimentary and volcanic rocks is cut by a series of fault families, and the sequence is folded to form basins and arches. The NNE-SSW faults include the Noril'sk-Kharaelakh and Imangda Faults; these have both a strike-slip and a reverse sense of motion with the local development

Table 6.2 Summary of the Main Processes by Which a Magma Forms an Economic Concentration of Magmatic Sulfide Mineralization

Location	Process Model	Compositional Effect (Silicate)	Specific Effect on Sulfide	Best Examples to Illustrate Process Model	References	Importance in Nickel Sulfide Ore Genesis and Exploration (Author's Opinion)	Importance to Sudbury Ore Genesis (Author's Opinion)
Mantle source and degree of melting	Composition of source domain	Diversity in compositions of mantle (pyroxenite, picrite, lherzolite, etc), and degree of metasomatism with resultant depletion or enrichment	Availability of sulfide in the mantle to control metal partitioning	Global information on mantle composition and process from basaltic magmas and xenoliths in kimerlite magmas	Sobolev et al. (2009); Hawkesworth et al. (1983)	Unimportant	Important (unusually mafic crust, but no mantle contribution at 1.85 Ga)
	Degree of melting	Degree of partial melting of silicate minerals	Degree of consumption of sulfide	Global information on mantle composition and process from basaltic magmas	Keays (1995)	Important	Melting of the crustal target rocks at Sudbury gave rise to the melt sheet from which metals were stripped by magmatic sulfide
	Previous depletion or enrichment of mantle source (melting and metasomatism)	Depleted lherzolite or enriched K-richterite and phlogopite-rich sources	Previous melting has large effect on PGE abundance, then Cu, followed by Ni	Siberian Trap	Lightfoot et al. (2012); Hawkesworth et al. (1983); Zhang et al. (2008)	Important	Unimportant (unless future work shows evidence of unusual mantle contributions that are presently hard to discriminate from crustal contributions)
	Heat source—mantle plume	Trigger to extensive melting of mantle and crust	Metal inventory in melts often available	Siberian Trap	Pirajno (2004, 2013)	Important	Impact melting process resulted in the formation of SIC
	Heat source—Subduction	Metasomatism of mantle above mantle wedge, but the same source can be trapped by late transcurrent faults	Possible addition of crustal sulfide by subduction mechanism	No examples of economically significant ore deposits	Li et al. (2012); Zhang et al. (2008)	Unimportant	Unimportant at Sudbury, but some very unusual phlogopite-apatite rich inclusions and bodies are associated with the base of the SIC

(Continued)

Table 6.2 Summary of the Main Processes by Which a Magma Forms an Economic Concentration of Magmatic Sulfide Mineralization *(cont.)*

Location	Process Model	Compositional Effect (Silicate)	Specific Effect on Sulfide	Best Examples to Illustrate Process Model	References	Importance in Nickel Sulfide Ore Genesis and Exploration (Author's Opinion)	Importance to Sudbury Ore Genesis (Author's Opinion)
Transport of magma from source to near-surface	Silicate magma fractionation	Diversity in igneous rock types related by crystallization process	Influence of temperature on S solubility	No examples of economically significant ore deposits	Keays and Lightfoot, 2009	Unimportant	No evidence that fractionation process triggered sulfide segregation at Sudbury
	Role of silica-rich wall-rock	Silica-rich melts produced that contaminate mantle-derived magmas	Unless country rocks contain a source of sulfur, the main effect is to influence the capacity of the silicate melt to contain sulfur	Noril'sk— contamination and metal depletion are related	Lightfoot and Hawkesworth (1997)	Important (silica-rich composition of melts reduces sulfide solubility)	Important (silica-rich composition of melts reduces sulfide solubility)
	Crustal structures/ cratonic margins	Rapid melt transfer	Local controls on sulfide segregation at different crustal levels	Small intrusions in cross-linking structures related to strike-slip fault zones; Chinese nickel deposits	Lightfoot and Evans-Lamswood (2015)	Critical	Important in controlling mineral systems in dykes and distribution of mafic-ultramafic intrusions

Magma holding chambers	Addition of crustal sulfur	Influence on S-capacity of silicate melt	Triggers sulfide segregation	Duluth, Noril'sk, Pechenga, Kambalda, etc.	Keays and Lightfoot (2007)	Critical	Important, but country rocks are incrementally S-enriched relative to average upper crust
	Achieve sulfide saturation	Immiscible sulfide forms	Conditions met for segregation of dense immiscible sulfide melt	Sudbury, Noril'sk	Li and Ripley (2005)	Critical	Formation and segregationof dense immiscible sulfide critical port ore formation
	Contamination and brecciation	Produce a change in the composition of the silicate magma	Trigger saturation event	Voisey's Bay	Lightfoot et al. (2012)	Unimportant, but an important associated feature (geochemistry and petrology of host rocks)	Important association with brecciation and melting of country rocks
	Multiple stages of magma emplacement	Pegmatoidal reefs and compositionally unusual horizons (chromitite)	Trigger to sulfide segregation and "reef" formation	Bushveld and Stillwater Complexes	Naldrett et al. (2009); Keays and Tegner (2016)	Critical in layered intrusions	Two or more phases of magma emplacement recognized in Offset Dykes, and evidence exists for redistribution of magma between sectors of the melt sheet

(Continued)

Table 6.2 Summary of the Main Processes by Which a Magma Forms an Economic Concentration of Magmatic Sulfide Mineralization (cont.)

Location	Process Model	Compositional Effect (Silicate)	Specific Effect on Sulfide	Best Examples to Illustrate Process Model	References	Importance in Nickel Sulfide Ore Genesis and Exploration (Author's Opinion)	Importance to Sudbury Ore Genesis (Author's Opinion)
Eruption or shallow-level emplacement	Change in P, T, and/or fO_2	Crystallization history	Exact conditions of sulfide saturation	No examples of economically significant ore deposits	Wendlandt, 1982	Controls conditions of magma, but appears not to trigger major sulfide saturation events	No evidence for P, T, or fO_2 controls at Sudbury
	Morphology of dyke or chamber	Localization of cumulate silicate rocks	Channelized flows and chonoliths	Voisey's Bay, Sudbury, Noril'sk, Babel-Nebo, Karatunkg, Huangshan, Huangshandong	Lightfoot and Evans-Lamswood (2015); Ripley and Li (2011)	Critical	Concentration of segregated sulfide by lateral flow in troughs, and localization in Offset Dykes
	Gravity segregation	Density controls on silicate minerals	Sulfide melt accumulation	Sudbury	Naldrett (2003)	Critical	Principal control on distribution of sulfide mineralization
	Localization by topography	Changes in intrusion morphology or melt sheet structure	Localization of sulfide	Sudbury	Lightfoot and Evans-Lamswood (2015)	Critical	Depressions and discontinuities in intrusions control distribution of mineralization
	Partitioning and R factor	Depleted silicate rocks in metals	Upgrades sulfides to achieve economic concentrations	Virtually all magmatic sulfide ore deposits	Campbell and Naldrett (1979)	Critical	Effective concentration of metals from a large volume of magma
Fractionation of sulfide melt	Formation of monosulfide solid solution and expulsion of residual sulfide liquid	None	Creates high value sulfide zones proximal to Ni-rich sulfides	Sudbury, Noril'sk, Karatungk, Eagle	Keays and Crocket (1970)	Critical in some mineral systems	Critical in most North Range mineral systems and many in the South Range

Process							
Hydrothermal modification	Remobilization of sulfides from mafic rocks or proto-ores	Hydrothermal remobilization of metals from silicate rocks	Redistribution of metals in sulfides	GT34, Enterprise	Siepierski (2008)	Unimportant except in iron-oxide Ni-(Cu) deposits such as GT34	Important at Carajas
Deformation	Modification by shear zones	Shearing and reformation of silicates	Local variations in sulfide chemistry	Garson Deposit, Sudbury	Mukakwami (2012)	Modification of primary ore deposits	Local importance of shears in modification of ores and controlling their mineralogy
	Modification at high P-T	Container rocks surrounding sulfides	Kinesis and redistribution of sulfide; massive sulfide in Pn	Thompson, Flying Fox	Lightfoot et al. (2012); Lightfoot (2015)	Critical in high P-T systems such as Thompson	Critical at early stages, but increasingly unimportant, and minimal effect after crystallization of SIC
Weathering	Weathering of ultramafic host ± sulfide	Destruction of silicate minerals to form clay minerals	Destruction and removal of metals from profile (except PGE)	Laterite deposits in Indonesia, New Caledonia, Cuba, Philippines	Golightly (2010)	Laterite formation—a constellation of processes linked to climate and topography	None

A perfect set of conditions are required to form a large high-grade nickel sulfide ore deposit where a combination of conditions results in the collection of metal from a large volume of magma and the localization of the sulfide mineralization in an ore deposit. See Chapter 1 and Naldrett (2004) for additional information.

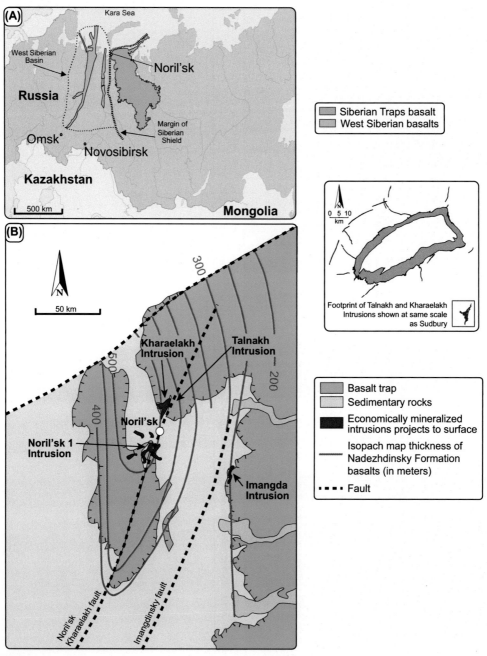

FIGURE 6.3

(A) Location of the Noril'sk Region at the northwestern margin of the Siberian Trap and the Siberian Craton. The West Siberian Basin is associated rift-related flood basalts. (B) Geological map of the Noril'sk region showing the basalt formations, the major faults and their kinematics (where known) and the location of differentiated Talnakh Type Intrusions. The red contours show the thickness of the Nadezhdinsky volcanic edifice which is a <500 m thick package of highly contaminated and Ni-Cu-PGE depleted volcanic rocks centered over Noril'sk and Talnakh.

(A) After Saunders et al. (2005); (B) after Fedorenko (1994); Fedorenko et al. (1996); Lightfoot et al. (1994a).

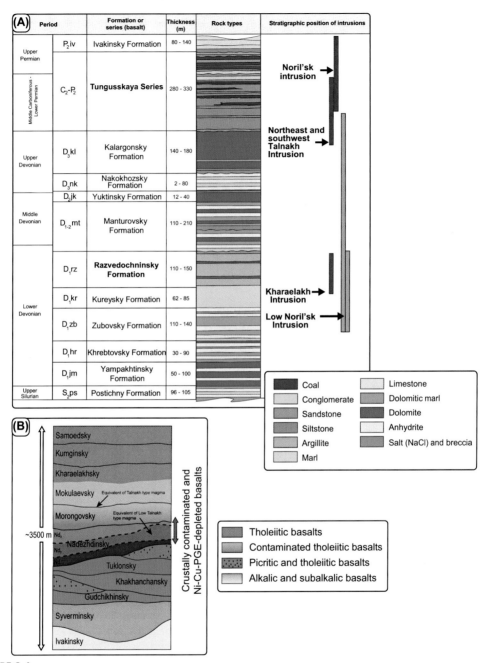

FIGURE 6.4

(A) Generalized stratigraphic column of the Silurian to Permian formations based on typical stratigraphic thickness in the area of the Talnakh Intrusion. The approximate stratigraphic positions of the Noril'sk 1, Talnakh, Kharaelakh and Low Talnakh Intrusions are shown. (B) Stratigraphy of the flood basalts of the Noril'sk Region showing the stratigraphic position of the metal-depleted and contaminated Nadezhdinsky Formation tholeiites.

(A) Modified after Zenko and Czamanske (1994); (B) modified after Lightfoot and Hawkesworth (1997).

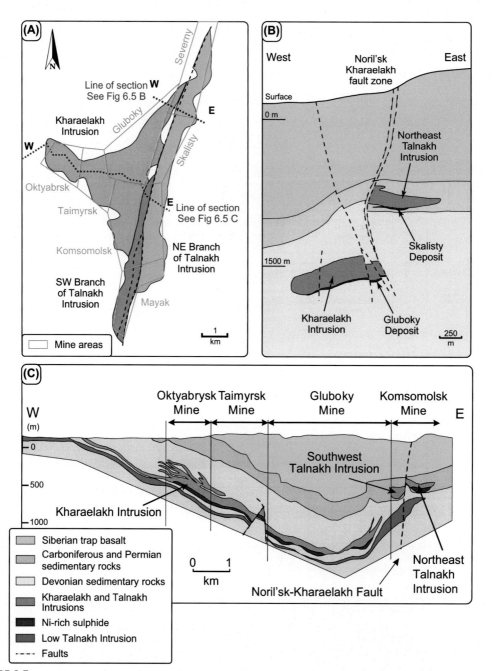

FIGURE 6.5

(A) Map showing the projected surface location of the Northwest Talnakh, Northeast Talnakh and Kharaelakh Intrusions and the configuration of the boundaries of the main mine footprints. (B) North-facing geological section through the Talnakh and Kharaelakh Intrusions in the area of the Gluboky and Severny Deposits showing the differing stratigraphic positions of the chonolith on either side of the fault. (C) North-facing geological section through the Kharaelakh, Northwest Talnakh, and Northeast Talnakh Intrusions in the area of the Oktyabrysk, Taimyrsk, and Skalisty Deposits.

(A) After Naldrett et al. (1992); (C) after Naldrett et al. (1992) and Lightfoot and Evans-Lamswood (2015).

of periclinal fold structures and fault blocks within the fault zone (Mezhvilk, 1984, 1995; Yakubchuk and Nikishin, 2004; Lightfoot and Evans-Lamswood, 2015). The direction and extent of displacement varies along the length of the Noril'sk-Kharaelakh fault (Mezhvilk, 1984, 1995).

Most geologists now agree that the disseminated and massive sulfides at Noril'sk were produced by equilibration of immiscible magmatic sulfide with silicate magma (Genkin et al., 1977; Distler et al., 1975; Naldrett et al., 1996a; Keays, 1995). The high ratio of sulfide/silicate demands an open system process where the magma equilibrated with the sulfide (Naldrett et al., 1991, 1995, 1996a, 1998) or where the sulfide melts were present within magma conduits feeding the Siberian Trap (Korzhinskii et al., 1984; Naldrett et al., 1991, 1995; Lightfoot and Zotov, 2007). These conduits are open-system magma "highways" called "chronoliths" that expedite the transfer of magma from depth toward the surface at cratonic margins (Lightfoot and Evans-Lamswood, 2015), and the intrusions now contain ore deposits that were produced from multiple pulses of silicate a and sulfide-laden magma.

The source of the sulfur in the ore deposits may be derived from country rocks; the leading contenders in present models are evaporates, shales, and possibly sour gas. Equilibration of the magma with crustal-level sulfur reservoirs can trigger sulfide saturation (e.g., Grinenko, 1985a,b; Godlevskii and Grinenko, 1963; Gorbachev and Grinenko, 1973; Keays and Lightfoot, 2010). The heavy sulfur isotope ratio of the ores of the Noril'sk Region is consistent with an evaporite source for the sulfur, but the high melting point of evaporite is not easily reconciled with this source (Lightfoot and Hawkesworth, 1997). The rocks of the Noril'sk Region have no shortage of available crustal sulfur, which contrasts with the Sudbury Region where the development of preimpact sulfide mineralization is large associated with occurrences in mafic intrusions and the local development of sulfide-bearing metasedimentary rocks within the Huronian Supergroup.

The observation that "complexity" is developed in the rocks that contain magmatic sulfide at both Sudbury and Noril'sk is a very telling feature, and it has importance when evaluating mineral potential. Complex assemblages of igneous rocks and inclusions comprise magmatic breccias like the Sublayer at Sudbury (Chapter 3) and the Taxite at Noril'sk (Fig. 6.6A–B). This group of rocks shows variability in composition, texture, and grain size from a scale of millimeters to meters, and they commonly contain xenoliths or xenocrysts derived from the country rocks (Lightfoot, 2007). It is commonly difficult to obtain a representative estimate of the mineralogy and composition of the melt component of the rock because it is packed full of xenoliths and xenocrysts.

Geology of the Talnakh and Kharaelakh Intrusions

The geology of the Kharaelakh and Talnakh Intrusions is shown in Fig. 6.5A–C, where the location of the principal mine areas are shown and the geological relationships along cross sections through the intrusions are illustrated.

The morphology of the Kharaelakh Intrusion has a sheet-like form in the central and southern segments (Fig. 6.5B), and it changes laterally into a pipe-shaped conduit extending toward the north (Fig. 6.5C). The Northeastern and Northwestern Talnakh Intrusions have a pipe-shaped cross section (Fig. 6.5B–C). The margin of the Kharaelakh Intrusion is flanked by rocks that record evidence of the abortive lateral propagation of the intrusion away from the fault, namely:

1. Sheet-like apophyses of weakly differentiated gabbrodolerite (termed "the sills of the Kharaelakh Intrusion"; Fig. 6.5C);
2. Apophyses of gabbrodolerite described in Zotov (1989) as 1–10 m thick sills within a wide domain of metamorphosed metasedimentary rock ("the horns of the Kharaelakh Intrusion"; Zotov, 1989);

FIGURE 6.6

(A) The taxite of the Talnakh Intrusion showing the development of disseminated magmatic sulfide, vesicles, inclusions of metasedimentary rock rich in hercynite spinel, and variable-textured olivine gabbro (Skalisty Mine, Northeast Talnakh Intrusion). I, possible inclusion containing black hercynitic spinel; v, vesicle. (B) Photomicrograph in polarized light of a taxitic olivine gabbrodolerite from the Talnakh Intrusion (sample RX187911) showing two different textural domains with different size and morphology of olivine and plagioclase crystals. CGOl, coarse-grained olivine; FGOl, fine-grained olivine.

3. Local lobes of fine-grained gabbrodolerite surrounded by hornfelsed sedimentary rocks; these fine-grained to quenched gabbrodolerites form lobes on the scale of 10–50 cm (termed "the ears of the Kharaelakh Intrusion"; Zotov, 1989; Lightfoot and Zotov, 2007).

The economic Ni-Cu-PGE sulfide ores are spatially related to differentiated intrusions (Zolotukhin et al., 1975) located along the flanking domains of the Noril'sk-Kharaelakh fault zone (Fig. 6.3B).

The development of mantle-penetrating structures inboard of the craton margin is well established (e.g., Rempel, 1994; Lightfoot and Evans-Lamswood, 2015). The intrusions are characterized by several key features:

1. They occur within rift blocks between faults and in the fold axes of periclinal structures (Mezhvilk, 1984, 1995; Smirnov et al., 1994; Lightfoot and Evans-Lamswood, 2015);
2. The intrusions largely replace the stratigraphy of the sedimentary package and they are volumetrically small (<2.5 km^3) and thin (<150 m) (Naldrett et al., 1991);
3. The intrusions comprise differentiated layers of picritic gabbrodolerite, tolivine gabbrodolerite, and olivine-free gabbrodolerite (Zen'ko and Czamanske, 1994);
4. The lower and sometimes the upper margins of intrusions are composed by coarse-grained to pegmatoidal olivine gabbrodolerites with inclusions (the "taxite"; Naldrett et al., 1991; Zen'ko and Czamanske, 1994);
5. The intrusions contain primary magmatic Ni-rich and disseminated Ni-Cu-PGE sulfide mineralization produced by sulfide differentiation (Naldrett et al., 1991);
6. The flanks of the intrusions comprise sills and apophyses that extend into the country rocks (Zotov, 1989);
7. Skarns are developed at the upper and lower exocontacts in association with the cuprous ores (Zotov, 1989).

There are five principal types of mineralization developed in the Talnakh and Kharaelakh Intrusions:

1. Interstitial disseminated sulfides within the taxite and blebby sulfide segregations in the olivine- and picritic-gabbrodolerite (Torgashin, 1994);
2. Massive differentiated lenses and veins of sulfide mineralization at the lower contact of the intrusion and in the underlying sedimentary rocks. The sulfide minerals comprise pyrrhotite, chalcopyrite, pentlandite, chalcopyrite, cubanite, and pentlandite, and mooihoekite-talnakhite (Torgashin, 1994);
3. Breccia-like ores rich in chalcopyrite developed in the upper exocontact (Kharaelakh Intrusion) and lower exocontact (Southeastern and Northwestern Talnakh Intrusion) (Lightfoot and Zotov, 2014);
4. Streaky disseminated sulfides in metamorphic and metasomatic rocks developed in the exocontact of mineralized intrusions (Zotov and Pertsev, 1978);
5. Disseminated sulfides enriched in PGE (termed reef horizons) associated with taxites in the upper part of the intrusion (Distler, 1994; Sluzhenikin et al., 1994).

The cuprous ore in the country rock breccias (Fig. 6.7A–C) are an economically important part of the ore deposits associated with the Kharaelakh and Northeast Talnakh Intrusion. The Cuprous ore deposits have reserve and resources of 151.7 million metric tons with grades of 0.96 wt.% Ni, 4.11 wt.% Cu, and 12.3 g/t 3E (Noril'sk Nickel, 2013; www.nornik.ru/en/our-products/MineralReservesResourcesStatement/). The differentiated nickel-rich sulfide mineralization at the lower contact of the intrusions exhibits a wider range in sulfide mineralogy than the Cuprous ores (Figs. 6.7D–E).

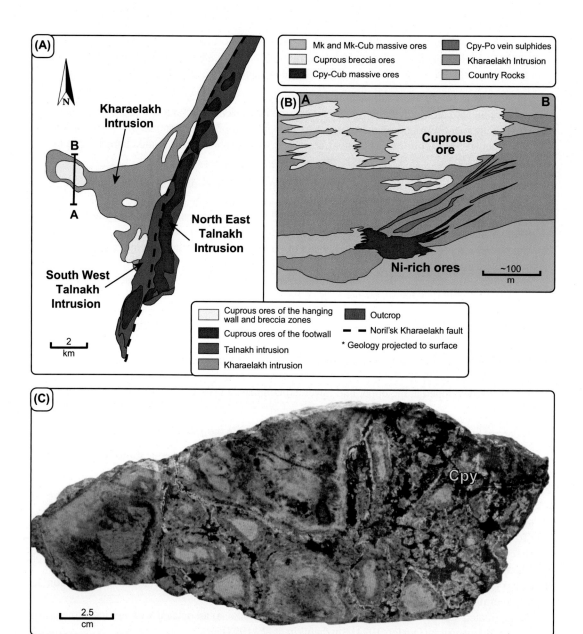

FIGURE 6.7

(A) Map showing the distribution of fractionated exocontact mineralization (Cuprous Ores) in the Talnakh and Kharaelakh Intrusions. (B) Cross-section through the Kharaelakh Intrusion showing the location of the massive Ni-rich Ores and the Cuprous Ores. (C) Sample of exocontact cuprous mineralization (chalcopyrite, Cpy) in brecciated skarns developed beneath the lower contact of the Talnakh Intrusion in the Skalisty Deposit.

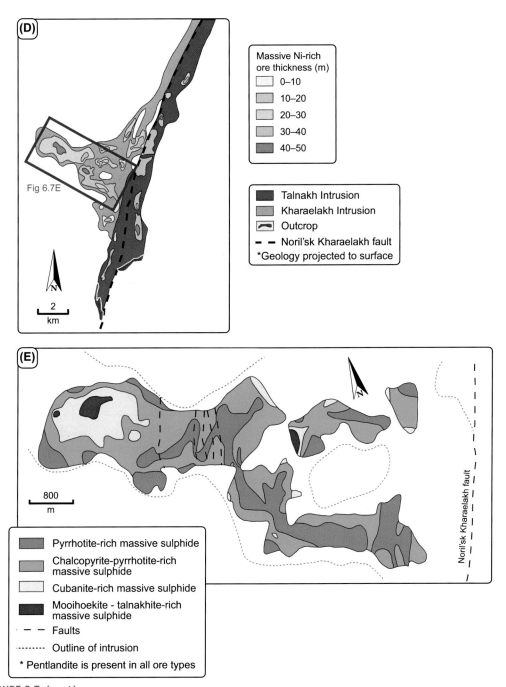

FIGURE 6.7 (*cont.*)

(D) Map of the Kharaelakh and Talnakh Intrusions showing the thickness of massive Ni-rich sulfide at the lower contact. (E) Map showing the distribution of sulfide mineral assemblages in the Oktyabrysk and Taimyrsk Deposits.

(A) Modified after Lightfoot and Zotov (2014); (B) after Lightfoot and Zotov (2014);

(D) after Lightfoot and Zotov (2014); (E) after Stekhin (1994).

The cuprous ores

A feature of both the Talnakh and Kharaelakh Intrusions is the association of Cu–PGE sulfide mineralization with brecciated country rocks in the footwall exocontact domain of the Talnakh Intrusion, and the hanging wall exocontact domain of the Kharaelakh Intrusion (Fig. 6.7A,B). The mineralization typically has elevated concentrations of Cu (Cu/Ni∼5) and PGE (10–15 ppm), but low in Ni (<1 wt.%) which contrasts with the differentiated Cu-rich mineralization developed at the lower contact of the intrusion in association with the the Ni-rich ores (Fig. 6.7D–E; Cu/Ni∼15) as described in Naldrett et al. (1992, 1995) and Stekhin (1994).

In hand sample, the cuprous ores occur in breccias and metasomatized country rocks that have the geological characteristics of skarns (Fig. 6.7C; Zotov, 1989; Lightfoot and Zotov, 2014). This mineralization style exhibits some similarity to the footwall type of mineralization at Sudbury, and the cuprous ores are likely the product of sulfide fractionation and transport of the Cu-PGE-Ni sulfide mineralization into space created adjacent to the intrusion by deformation linked to motion on the Noril'sk–Kharaelakh Fault (Lightfoot and Zotov, 2014). In this respect, late structural readjustment was important in the localization of sulfide melts that were produced by differentiation processes within the intrusion.

The Disseminated Sulfides

The interstitial and blebby disseminated sulfides at Talnakh and Kharaelakh are hosted by olivine gabbro, taxitic gabbrodolerite, and picritic gabbrodolerite. The sulfide blebs commonly exhibit a geopetal structure. Pyrrhotite and pentlandite occupy the lower part of the bleb and chalcopyrite occupies the upper part of the bleb (Fig. 6.8A–B). Surrounding the chalcopyrite there is commonly chalcedony which may have formed within a vesicle. This led Lightfoot and Zotov (2014) to suggest that these blebs occupied vesicles in a degassing shallow-level magma chamber.

Lightfoot and Zotov (2014) showed that both the massive and disseminated sulfides exhibit a zonation in composition where the disseminated sulfides have compositions that broadly match the composition of the adjacent massive Ni-rich sulfides. This relationship can be explained by interaction of the sulfide-laden silicate magma with an adjacent body of massive sulfide melt (Lightfoot and Zotov, 2014).

The Sudbury and Noril'sk mineral systems develop disseminated sulfides in association with massive sulfide ore deposits. The sulfide mineralization from both camps shows evidence of primary magmatic textures in the sulfides. However, there are complex geological relationships between the disseminated and massive sulfides in both locations that indicate that the process of formation was more complex than simple in situ segregation and accumulation of magmatic sulfides, where the samples of disseminated and massive sulfides record end-members of this process.

The Ni-rich Massive Sulfides

The massive Ni-rich sulfides at Noril'sk comprise coarse-grained crystals of pyrrhotite enclosed in loops of pentlandite+chalcopyrite (termed "loop-textured sulfide"). As the proportion of chalcopyrite increases, the mineralization changes from a pyrrhotite-rich mineralization with exsolved and granular pentlandite (Fig. 6.9A) through loop-textures sulfide (Fig. 6.9B–C), into mineralization that consists of pentlandite eyes in a matrix of chalcopyrite and cubanite (Fig. 6.9D). The massive sulfide textures in deposits at Talnakh and Kharaelakh are similar to parts of the contact–footwall mineral systems at Sudbury, but the mineralogy of the footwall mineralization at Sudbury has a much higher Cu/S and Ni/S than the cuprous ores at Noril'sk.

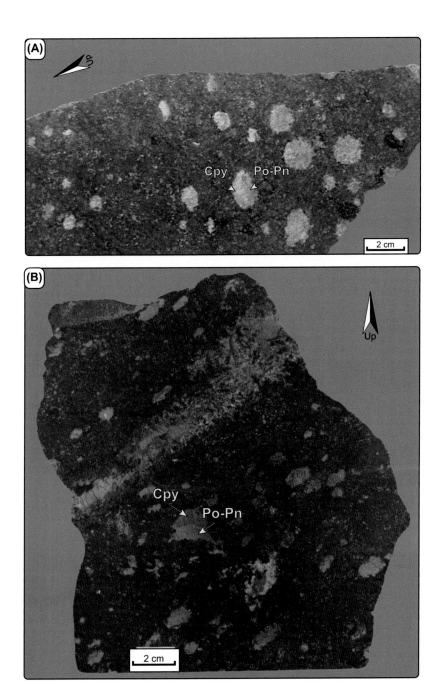

FIGURES 6.8

(A) Blebby sulfide from the picritic gabbrodolerite at Nori'sk 1. Note the development of chalcopyrite (Cpy) in the upper parts of the blebs and Po + Pn in the lower parts. (B) Blebby sulfide in taxitic gabbrodolerite from the Kharaelakh Intrusion. The sulfides exhibit geopetal features such as those described in subpart A.

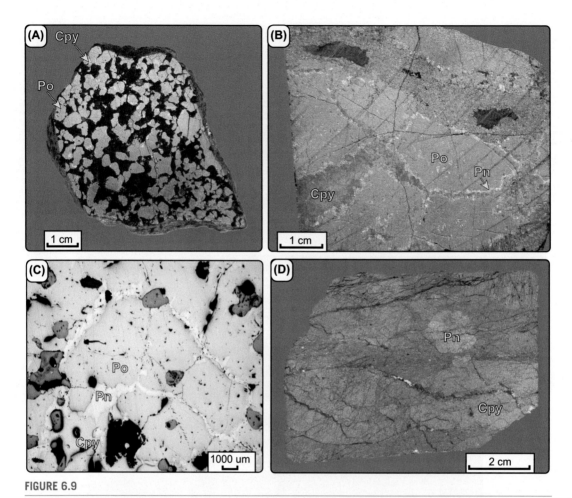

FIGURE 6.9

(A) Massive Po-rich sulfide in the exocontact sediments of the Kharaelakh Intrusion (Po, pyrrhotite; Cpy, chalcopyrite). (B) Loop-textured pyrrhotite (Po) surrounded by a net of chalcopyrite (Cpy) and pentlandite (Pn), Kharaelakh Intrusion. (C) Polished sample of loop-textured sulfide from the Kharaelakh Intrusion showing the grains of pyrrhotite (Po) separated by chalcopyrite (Cpy) with eyes of pentlandite (Pn) (D) Massive chalcopyrite (Cpy) and pentlandite (Pn) mineralization from the more fractionated parts of the Oktyabrysk Deposit, illustrating the development of pentlandite "eyes."

Compositional Diversity in Sulfides

Fig. 6.10 shows the variation in Cu/(Cu + Ni) versus (Cu + Ni)/S in samples from the Talnakh, Kharaelakh, and Noril'sk 1 Intrusions. The database of samples includes materials available in the study of Naldrett et al. (1994b, 1996a). The plot illustrates a number of features that are quite different when compared to the more fractionated mineral systems at Sudbury. Firstly, all of the samples contain at least some Pn as pointed out by Stekhin (1994), so the actual range in Cu/(Cu + Ni) is not as wide as that found in the Sudbury system. A second point is that the available samples from Kharaealakh,

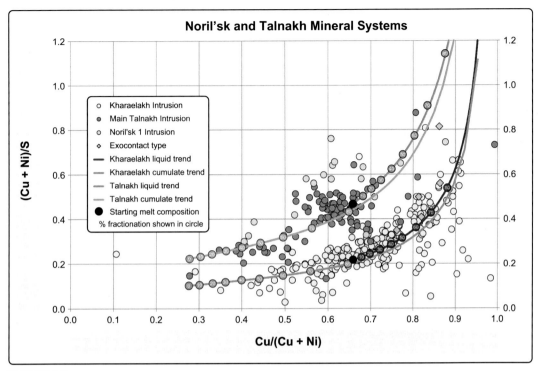

FIGURE 6.10

Compositional diversity in samples from the Talnakh, Kharaelakh, and Noril'sk 1 Intrusions. Variations in
Cu/(Cu + Ni) versus (Cu + Ni)/S for samples with >1 wt.% S and >0.3 wt.% (Cu + Ni). The modeling shows
the effect of Rayleigh Fractionation assuming that the initial sulfide melt at Kharaelakh contains 3 wt.% Ni
with Cu/Ni = 1.9, and partition coefficients as per Chapter 5. For Talnakh, the starting sulfide melt had a higher
Ni content of 6.5 wt.% with Cu/Ni = 1.9; this is consistent with the concepts developed by Naldrett et al. (1996a)
where the Talnakh and Noril'sk 1 sulfides evolved in an open system where the ratio of magma to sulfide changed as
a function of a modified form of the R factor, that is called the N factor (Naldrett et al., 1998).

Data from Naldrett et al. (1994a, 1998).

Talnakh and Noril'sk do not have compositions that pass above the Cpy-Pn tie line into the field of
bornite and millerite. However, the available samples did not include Oktyabrysk ores that are enriched
in Cu and/or Ni (Stekhin, 1994); these ores contain minerals like mooihoekite ($Cu_9Fe_9S_{16}$), talnakhite
[$Cu_9(Fe, Ni)_8S_{16}$], and putoranite ($CuFeS_2$).

The Ni-rich sulfides from Noril'sk do not fractionate to achieve the apogee of melt compositions
found in the footwall ore deposits at Sudbury.

The majority of the samples from Kharaelakh have lower metal tenor styles of mineralization than
those from the Talnakh and Noril'sk 1 Deposits (Fig. 6.10; Naldrett et al., 1998). This difference is
explained if the different ore deposits were generated from initial sulfide melts that differed in composition as shown in Fig. 6.10.

JINCHUAN

Jinchuan is the largest deposit in China with over 5,000 kt of contained Ni in association with ultra-mafic rocks is a small dyke-like intrusion. The intrusion is located in a Proterozoic belt termed the Longshushan in Gansu Province, and is contained within a deformation zone which parallels to the long axis of the belt (Fig. 6.11A).

The geology of the intrusion and distribution of mineralization has been described in Tang (1993), and a synthesis of the geology in plan and three representative cross-sections is shown in Fig. 6.11B–E. The intrusion has a surface area of less than 1.5 km², and extends to a depth of at least 2 km as a dyke-like stem or narrow cone-shaped keel in cross-section. The geometry of the intrusion comprises four segments of different thickness that are cut by late NE-SW faults. The mineralization in Number 1 Mine Area in the west cropped out at surface where the original discovery in 1959 was made after the similarity of the mineralization to the Sudbury style was recognized (Tang, 1992). The mineralization

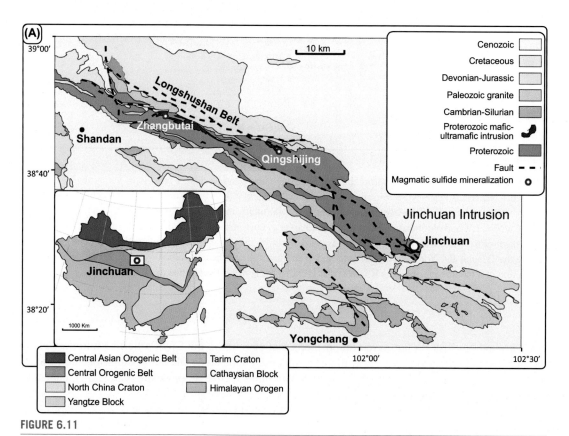

FIGURE 6.11

(A,B) Location of the Jinchuan Intrusion in the Proterozoic Longshushan Belt of Western Gansu Province, China. (C) Geological plan of the Jinchuan Intrusion. (D–F) Northwest-facing geological sections through different segments of the Jinchuan Intrusion.

(C) After Tang (1992) and Lightfoot and Evans-Lamswood (2015); (D) after Tang (1992) and Lightfoot and Evans-Lamswood (2015).

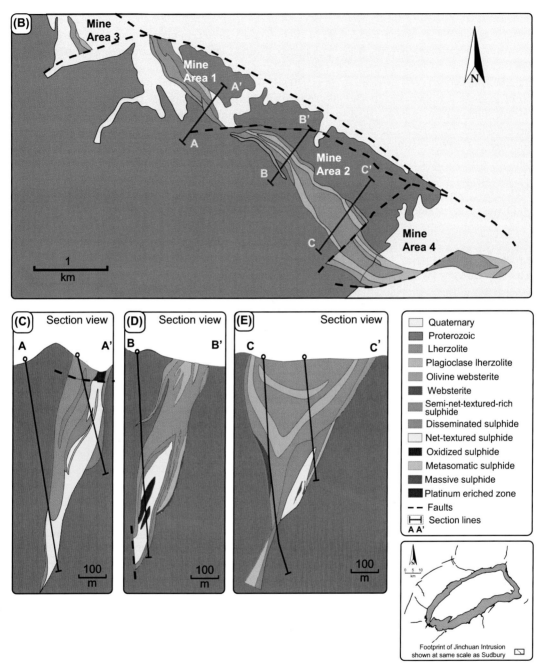

FIGURE 6.11 (*cont.*)

in Number 2 Mine Area is entirely at depth within the intrusion, and the lherzolite, plagioclase lherzolite, and dunite at surface are devoid of disseminated sulfide mineralization. The margins of the intrusion are variously described as serpentinized dunite and lherzolite through to pyroxenite, but the vast majority of the contacts are splays of regional structures that controlled the emplacement of the Jinchuan magma at 827 ± 8 Ma (Li et al., 2004). These structures were possibly related to the break-up of Rodinia (Pirajno, 2012). The geometry of the intrusion comprises four segments of different thickness that are cut by late NE-SW faults. No data are published on the sense of displacement of the structures, but the form of the intrusion and internal distribution of rocks types shown in Fig. 6.11D–F and reported by Tang (1993) are consistent with emplacement of the magma and the entrained sulfides into a vertical structure. The complex deformation history of the Longshushan Belt does not provide clear kinematic indicators for Jinchuan, but it appears possible that the intrusion morphology was controlled by cross-linking structures within a strike-slip segment of the Longshu Fault Zone (Lightfoot and Evans-Lamswood, 2015).

VOISEY'S BAY

The 1.34 Ga Voisey's Bay Intrusion is located in Eastern Labrador, Canada. The deposit was found by prospectors engaged in regional exploration for diamonds (Naldrett et al., 1996b). An active open-pit mine on the Ovoid Deposit with original reserves of 31.7 Mmt @ 2.83% Ni, 1.68% Cu, and 0.12% Co, is located within a larger mineral system that comprises five significant mineral zones (Lightfoot et al., 2011a, 2012). The structures that bound the Voisey's Bay Intrusion have a sinistral strike-slip displacement, and the same family of W-E-trending faults extends to Greenland. Their latest motion is controlled by extension in the Labrador Sea (e.g., Myers et al., 2008).

The Voisey's Bay Intrusion cross-cuts through a series of N–S trending basement gneisses comprising, from west to east, paragneisses of the Churchill Province, enderbitic orthogneisses of the Churchill Province, and quartzofeldspathic gneisses of the Nain Province. The intrusion crops out at surface in the east (the Eastern Deeps Intrusion) and is connected via a dyke-like conduit to a sub-surface chamber in the west (termed the Western Deeps Intrusion). This is shown in Fig. 6.12A,B in plan (where the mineral zones are shown projected to surface) and a long-section view (projected onto the dyke and the north walls of the intrusions).

The Eastern Deeps Intrusion is a tabular body of olivine gabbro, troctolite, and ferrogabbro containing variably assimilated blocks of orthogneiss as well as magmatic breccias laden with paragneiss fragments and magmatic Ni-Cu-Co sulfide mineralization (Lightfoot et al., 2011a, 2012; Naldrett and Li, 2007). The Eastern Deeps Deposit is located at the base of the northern sub-vertical wall of the intrusion and plunges ∼15 degrees ESE along the locus of the entry point of the dyke into the intrusion; this is shown in Fig. 6.12B in a west-facing cross-section through the Eastern Deeps Intrusion and Deposit. The Eastern Deeps Deposit zone is surrounded by a halo of variable-textured troctolite. The variable-textured troctolite has a range in plagioclase and olivine grain size, pegmatoidal patches rich in orthopyroxene and ilmenite, 0–20% disseminated magmatic-textured sulfide, inclusions of paragneiss (1–10 cm in size), and large pendants of orthogneiss (up to 200 m across), with local hybridized margins.

The Red Dog Fault marks the southern margin of the Eastern Deeps Intrusion; the fault has a sinistral offset of ∼1700 m (Fig. 6.13E). Both the Eastern Deeps and Red Dog Intrusions appear to be part of one original chamber, and the mineral occurrence at Ryan's Pond is consistent with an ESE continuation of the trend of the Eastern Deeps Deposit to the south of the Red Dog Fault, but displaced

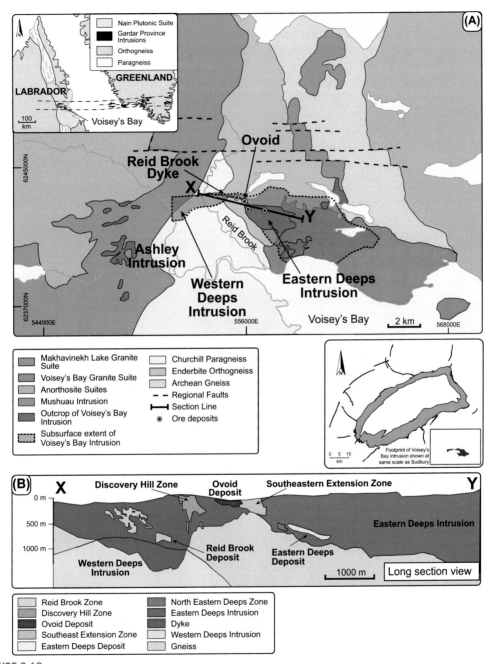

FIGURE 6.12

(A) Geology of the Voisey's Bay Intrusion. (B) Localization of magmatic sulfide mineralization shown in a north-facing long-section projected onto the dyke and walls of the Voisey's Bay Intrusion (see Lightfoot and Evans-Lamswood, 2015 for details).

~1700 m the East (Lightfoot and Evans-Lamswood, 2015). The Eastern Deeps Intrusion trends west–east, and the western end of the intrusion subcrops along a contact that trends south and then south–east relative to the eastern end of the Ovoid (Fig. 6.12A). Toward the east, the geological relationships along the wall are summarized in a cross-section in Fig. 6.13D.

West of the Eastern Deeps chamber, the dyke bifurcates, and the upper part of the dyke is connected into an east–west trending elongated bowl-shaped intrusion termed the Ovoid (Fig. 6.12A). The massive sulfide of the Ovoid is entirely devoid of gneiss fragments, but the margins have similar rock types to those developed around the Eastern Deeps, including both leopard-textured troctolite and breccia sulfides. The shape of the space occupied by the Ovoid Deposit is broadly an elongate west–east trough with a dyke connected to the base as a keel which dips to the north. The north wall of the

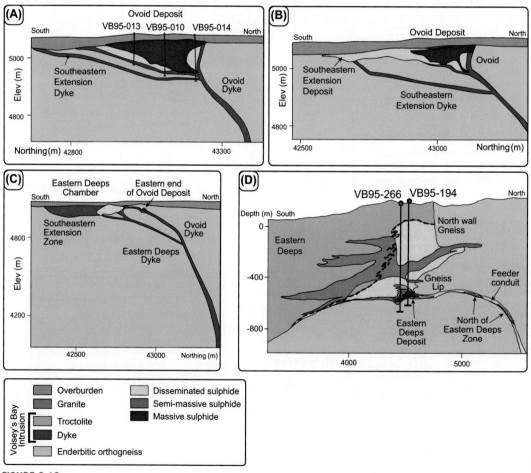

FIGURE 6.13

West-facing sections showing geological relationships at Voisey's Bay. (A) Ovoid Deposit. (B) Eastern end of void Deposit. (C) South-eastern Extension Deposit. (D) Eastern Deeps Deposit.

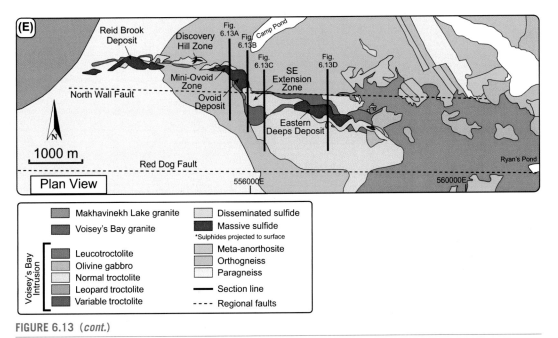

FIGURE 6.13 (*cont.*)

(E) Location map showing position of west-facing subparts A–D. (F) Photograph of a hand sample that illustrates the textures of loop-textured massive sulfide from the Ovoid Deposit. Crystals of pyrrhotite with heavily resorbed grains of magnetite are separated by chalcopyrite and granular pentlandite.

Ovoid curves toward the south at the east end of the trough, and the south wall curves toward the north at the west end, so the geometry is broadly consistent with the space created by both rifting (Cruden et al., 2008) and cross-structures between faults undergoing local dextral transtension (Lightfoot and Evans-Lamswood, 2015). The massive sulfides that comprise the economically significant heart of the Ovoid Deposit are classic examples of magnetite-rich sulfides that comprise pyrrhotite crystals with local exsolution of pentlandite that are separated by loops of chalcopyrite and granular pentlandite (Fig. 6.13F). Samples from Talnakh exhibit loops on the scale of centimeters (Fig. 6.9B), but those from the Ovoid occur on the scale of meters.

Further to the west, the Ovoid narrows into the mini-Ovoid deposit where the chamber walls are steeper, and the deposit is contained in an intrusion that has a funnel shape with a narrow keel (Evans-Lamswood, 2000; Fig. 6.13A–C). Fig. 6.13G shows the surface expression of the Ovoid, Mini-Ovoid, and Discovery Hill Dyke during mining activity in 2012.

West of the Ovoid, the intrusion has the form of a dyke (Fig. 6.13G) which dips toward the north, steepening with depth before bending over and dipping toward the south. The axis of the inflection of the dyke plunges east at ~27°, and the inflexion reaches the surface in the Reid Brook Zone to the west of the Mini-Ovoid (Fig. 6.14B). The wider zones of the dyke (≤100-m width) plunge at 45° to the east, and contains heavy Ni-Cu sulfide mineralization in association with magmatic breccias (Fig. 6.14C–D).

The dyke follows the south side of the Western Deeps Intrusion (Fig. 6.14B). Although the dyke is adjacent to the north wall of this chamber, it is not entirely clear whether the dyke joins with the

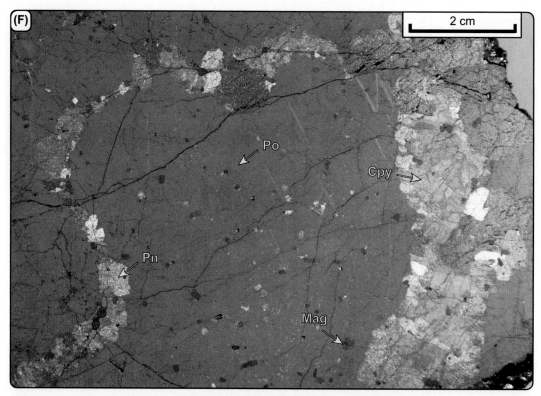

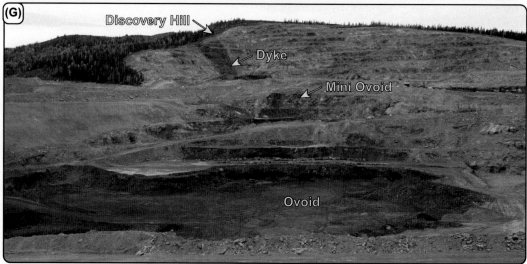

FIGURE 6.13 (*cont.*)

(G) Photograph of mine development of the Ovoid Deposit in the summer of 2012 showing the massive sulfide at the base of the open pit (exhibiting textures similar to those shown in Fig. 6.13F), the Mini-Ovoid in the wall of the open pit mine, and the Discovery Hill Dyke enclosed by enderbitic orthogneiss that is now exposed below the initial discovery site on Discovery Hill (photography by the author).

After Lightfoot et al., (2011a, 2012); Lightfoot and Evans-Lamswood (2015).

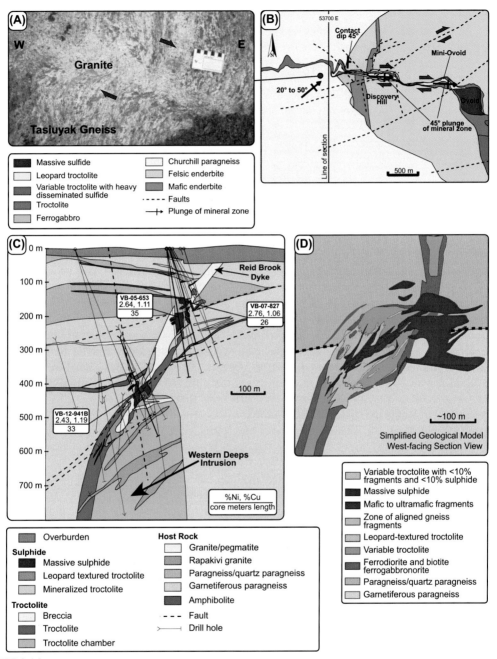

FIGURE 6.14

(A) Photograph showing displacement of granitoid rocks in a dextral zone of transpression oriented parallel to the Reid Brook Dyke, formed shortly after the Voisey's Bay Intrusion (Lightfoot and Evans-Lamswood, 2015). (B) Detailed geological map of the Reid Brook Dyke on Discovery Hill to the west of the Ovoid Deposit at Voisey's Bay. (C) West-facing geological section through the Reid Brook Dyke (prepared by Sheldon Pitman and Danny Mulrooney; Lightfoot and Evans-Lamswood, 2015). (D) Schematic diagram showing geological relationships at a discontinuity in the Reid Brook Dyke (Lightfoot and Evans-Lamswood, 2015).

chamber or just skirts the south-dipping southern margin of the chamber because the contacts are normally cut by late granitoid intrusive rocks (Lightfoot et al., 2011a, 2012).

The dyke hosts the Reid Brook Deposit, and this mineral zone plunges from west to east at about 27 degrees within the dyke as shown on the long-section in Fig. 6.15A. The dyke plunges to the north near to the surface, bends over, and then plunges toward the south at about 70 degrees at depth (Fig. 6.15A). The dyke intersects a chamber of olivine gabbro, troctolite, ferrogabbro, and leucogabbro at depth that is termed the Western Deeps Intrusion. The similarity of the plunging zones of mineralization in the Reid Brook Dyke is compared to the Copper Cliff Offset in Fig. 6.15B.

The breccia sequence rocks are commonly associated with mineralization throughout the Voisey's Bay mineral system, and there is a 1:1 association between the development of disseminated to leopard-textured sulfide and the presence of inclusions of paragneiss (Fig. 6.16A,B) and, ferrogabbro, olivine ferrogabbro, troctolite and ultramafic inclusions (Lightfoot et al., 2011a, 2012). The ferrogabbro, olivine gabbro and Troctolite inclusions have a similar mineralogy, texture and whole rock composition to the marginal rocks of the conduit, and they likely originate by erosion of previously crystallized rock from the conduit walls by the sulfide-laden magma. The paragneiss inclusions strongly resemble aluminous portions of the Tasiuyak Gneiss, and examples have been encountered in drill intervals from the Western Subchamber where incipient development of inclusions from country rock paragneiss occurs in association with mineralized variable-textured troctolite toward the base of the chamber.

Geological relationships in the mineralized part of the dyke are summarized in Fig. 6.14D. The wide zones of stronger mineralization associated with magmatic breccias tend to plunge down the dyke toward the east with a rake of about 27 degrees. Above the wider channel in the dyke, the dyke often narrows or it is absent; in many cases the only manifestation of the dyke is massive sulfide (Fig. 6.14D). The sulfide lenses follow structures that dip to the east at 20–30 degrees. It is these structures that appear to represent primary discontinuities in the dyke along which sulfide magmas (devoid of silicate inclusions of melt) have been injected into the country rock paragneiss (Fig. 6.14D). The country rocks adjacent to the dyke develop shear zones with evidence of dextral motion (Fig. 6.14A,B).

The geological relationships summarized above point to a protracted evolution of the crust in the area of the Voisey's Bay Deposit. The host intrusion appears to have been emplaced into a segment of crust that is undergoing extension. The geological evidence points to the presence of at least two main chambers in the structural corridor, the Eastern Deeps and the Western Deeps connected by a dyke (Lightfoot and Naldrett, 1999; Evans-Lamswood, 2000). The evolution of this chamber has been considered the product of caldera collapse within a west-east rift (Cruden et al. 2008) based on the structural setting, and this model has been widely used to explain the emplacement of dense magmatic sulfides and fragment laden melts from deeper chambers to shallower chambers in the crust (Lightfoot and Evans-Lamswood, 2015).

Although the data from Voisey's Bay indicate that the morphology of the intrusion has been affected by collapse within a rift structure, there is also a growing weight of evidence to point to a local development of cross-structures within the dyke that produce widening; these cross-structures are consistent with local dextral transtension (Fig. 6.14A,B). The plunging pipe of mineralization within the dyke, and the associated openings in the gneisses where the dyke is thin are small-scale manifestations resulting from trans-tensional slip that is recorded by $< \sim 100$ m displacements in lithologic units across the dyke. The introduction of dense magmatic sulfides through this plumbing system would be a response

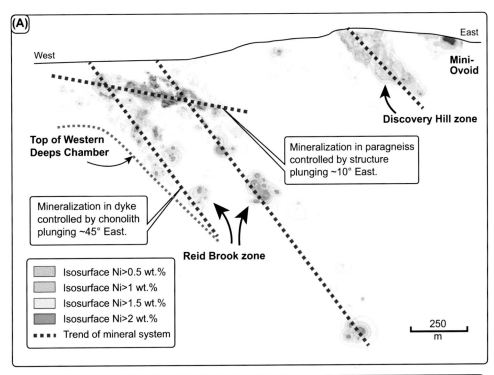

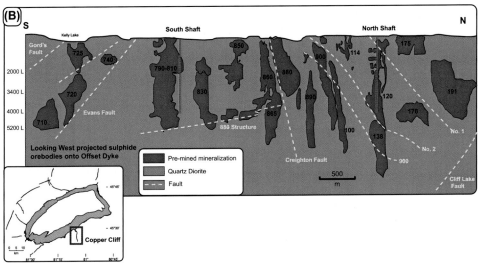

FIGURE 6.15

Comparison of the geology of the Copper Cliff Offset Dyke to the Reid Brook Dyke projected onto the dyke in longitudinal section. (A) Simplified long section of the Reid Brook Dyke showing the distribution of mineralization as grade shells generated in the Leapfrog software. (B) Long section of the Copper Cliff Offset showing the location of the ore bodies, and the intrinsic similarity in the morphology of the mineral zones to those developed at Voisey's Bay.

(A) Image generated by Lisa Gibson; modified after Lightfoot and Evans-Lamswood (2015).

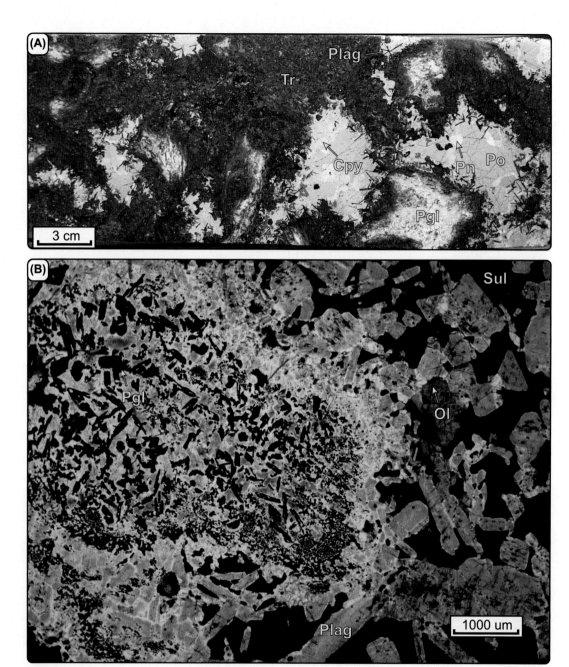

FIGURE 6.16

(A) Basal breccia sequence sample from the Ovoid Deposit, Voisey's Bay Intrusion (see Naldrett et al., 1996b and Lightfoot et al., 2011a, 2012 for details). Tr, Troctolite; PgI, paragneiss inclusion; Plag, plagioclase; Cpy, chalcopyrite; Po, pyrrhotite; Pn, pentlandite. (B) Polished thin section in plane polarized light of a paragneiss inclusion that has reacted with an olivine ferrogabbro matrix within the mineralized breccia sequence from Voisey's Bay (VTDL0010). PgI, paragneiss inclusion; Plag, plagioclase; Ol, olivine (altered); Sul, magmatic interstitial sulfide.

to lateral tectonic pumping of sulfides rather than vertical transport of entrained sulfides in a flowing magma (de Bremond d'Ars, 2001).

The geological controls on the localization of mineralization in the Reid Brook Dyke are similar to the controls on the ore deposits hosted by the Sudbury Offset Dykes. Fig. 6.15A illustrates the fact that mineralization is hosted within the dyke, and occupy two zones that plunge ~45° to the east within the dyke. Primary magmatic discontinuities in the dyke produce localization of high-grade sulfide mineralization along structures that plunge more gently at ~10° toward the east. The mineral zones in the Copper Cliff Offset Dyke (Fig. 6.15B) and the Worthington Offset Dyke (Chapter 4) share two important controls:

1. Both the Offset Dykes and the Reid Brook Dyke appear to have formed by two or more phases of emplacement of magma. At Voisey's Bay, a unit of sulfide-undersaturated ferrodiorite and ferrogabbro was formed at the wall of the dyke, whereas at Sudbury a unit of sulfide-undersaturated quartz diorite crystallized at the wall of the dyke. In the case of Sudbury, inclusions of quartz diorite derived from this unit are entrained in later pulses of MIQD (Chapter 4). In the case of Voisey's Bay, the early formed ferrogabbro and ferrodiote is often found as inclusions in the mineral zones at Voisey's Bay.
2. The mineral zones are hosted within the dyke or near to the dyke in association with discontinuities at both the Reid Brook dyke at Voisey's Bay and the Offset Dykes at Sudbury. At Sudbury, the mineralization appears to have been emplaced from the overlying melt sheet whereas at Voisey's Bay it appears more likely that the mineralization was emplaced from below along wide domains within the dyke.

THOMPSON

The Ni-Co-(PGE) sulfide deposits of the Thompson Nickel Belt (TNB) in Northern Manitoba, Canada are part of the fifth largest nickel camp in the world based on contained nickel. Past production from the TNB deposits is 2,500 kt Ni (Lightfoot et al., 2011b), which is almost double Kambalda production.

The Thompson Deposit is located on the eastern and southern flanks of the Thompson Dome structure, which is a refolded nappe structure formed during collision of the Trans-Hudson Orogen with the Canadian Shield at 1.9–1.7 Ga. The Thompson Deposit is almost entirely hosted by P2 member sulfidic metasedimentary rocks of the Paleoproterozoic Ospwagan Group. Variably serpentinized and altered dunite, peridotite and pyroxenite contain disseminated sulfides and they have a spatial association with sediment-hosted Ni sulfides that comprise the bulk of the ores at Thompson Mine. These rocks formed from rift-related komatiitic magmas that were emplaced at 1.85 Ga, and subsequently deformed by boudinage, thinning, folding, and stacking. The Thompson Deposit occurs on the flanks of an overturned dome structure, and the mineral zones are localized along the southern and eastern flanks of this dome as shown in Fig. 6.17A–C.

Disseminated sulfide mineralization in the large serpentinized peridotite and dunite intrusions (the Birchtree and Pipe Ni–Co sulfide deposits) typically has $[Ni]_{100} = 4$–6 wt.%. The disseminated sulfides in the small heavily deformed serpentinized peridotite and dunite bodies associated with the Thompson Deposit have $[Ni]_{100} = 7$–10 wt.%.

The majority of sulfides at Thompson Mine are hosted in sulfidic schist of the Pipe Formation which was developed from a shale protolith; the mineralization varies from barren sulfide (<200 ppm Ni) through sulfide with low Ni tenor (<1 wt.% Ni) to Ni-enriched sulfide with $[Ni]_{100} = 1$–18 wt.% (Fig. 6.18A,B). The semi-massive and massive sulfide schist-hosted mineralization has a multi-modal distribution of $[Ni]_{100}$ with peaks at ~<0.25, 8, 12 and 14 wt.% Ni (Lightfoot et al., 2011b).

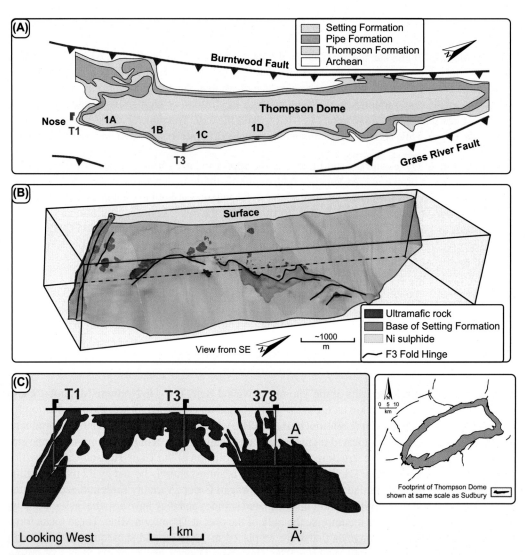

FIGURE 6.17

(A) Geological map showing the position of the Thompson Deposit in the overturned Pipe Formation stratigraphy of the Thompson Dome. (B) Three dimensional model of the base of the Setting Formation based on exploration drilling and mapping of underground openings; the location of secondary structures on the flanks of the Dome controls the location of thick zones of Po–Pn mineralization. (C) Long section showing the distribution of the mineral zones in the Thompson Deposit.

(A–C) After Lightfoot et al. (2011b).

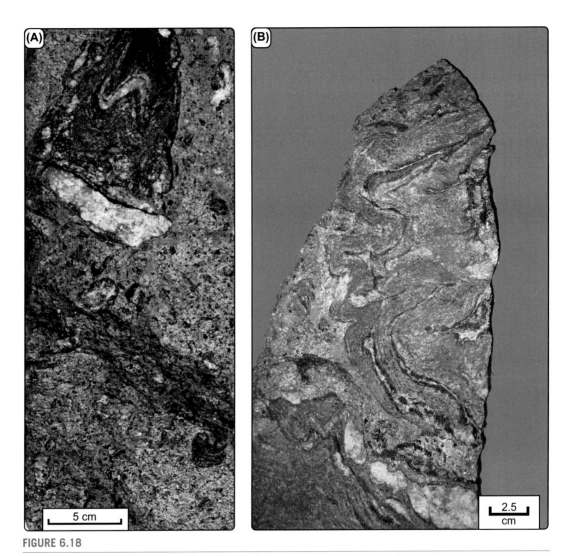

FIGURE 6.18

(A) Semi-massive sulfide with inclusions of deformed paragneiss within inclusion-bearing massive sulfide from the Thompson 1D Deposit. (B) Nickeliferous sulfide occupying the space in a strongly folded sample of the Pipe Formation, Thompson Deposit.

The Thompson Deposit forms an anastomosing domain on the south and east flanks of a first order D3 structure which is the Thompson Dome (Fig. 6.19A; Lightfoot et al., 2011b). In detail, a series of second order doubly plunging folds on the eastern and southern flank of the Thompson Dome control the geometry of the mineral zones (Fig. 6.17A–C; Lightfoot et al., 2011b). The position of these folds on the flank of the Thompson Dome is a response to the anisotropy of the host rocks during deformation; ultramafic boudins and layers of massive quartzite in ductile metasedimentary rocks control the geometry of the doubly plunging D3 structures.

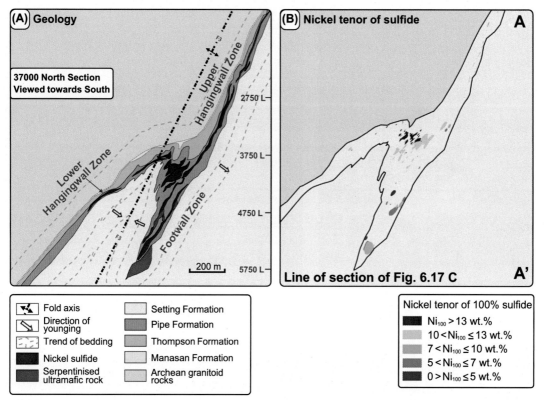

FIGURE 6.19

(A) South-facing geological section through the 1D Deposit on the eastern flank of the Thompson Dome (Lightfoot et al., 2011b). (B) Model showing the distribution of Ni tenor of sulfide based on iso-shells constructed in Leapfrog to illustrate the distribution of the different styles of mineralization.

(B) Model prepared by Lisa Gibson; Lightfoot et al., 2011b and Lightfoot, 2015.

The envelope of mineralization is almost entirely contained within the P2 member of the Pipe Formation, and the deposit is folded into doubly plunging fold hinges on the flank of the Dome. The sulfides with highest Ni tenor (>13 wt.% Ni in sulfide) define a systematic trend that matches the geometry of the doubly plunging D3 structures on the flanks of the Dome (Figs. 6.19B and 6.20). Although moderate to high Ni tenor mineralization is sometimes localized in fold hinges, more typically the highest Ni tenor mineralization is located on the flanks of the fold.

There is no indication of the mineralogical and geochemical signatures of sedimentary exhalative or hydrothermal processes in the genesis of the Thompson ores. The primary origin of the mineralization is undoubtedly magmatic.

Variations in metal tenor in disseminated sulfides hosted by ultramafic rock indicate a higher magma/sulfide ratio in the Thompson parental magma relative to Birchtree and Pipe. The variation in Ni tenor of the semi-massive and massive sulfide broadly can be interpreted to support this conclusion, but the variations in metal tenor in the Thompson ores was likely created during

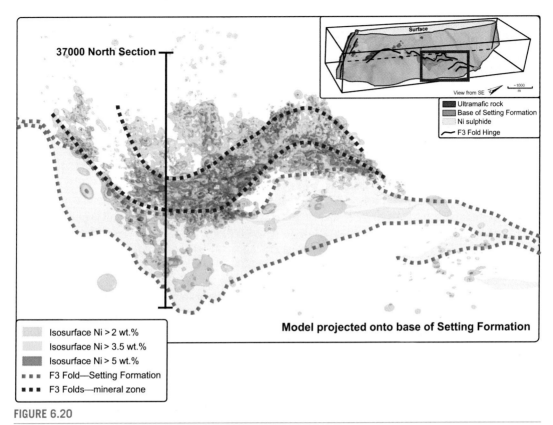

FIGURE 6.20

Distribution of mineralization on the eastern flank of the Thompson Dome shown in a model of grade shell isosurface distribution projected onto the base of the Setting Formation.

Model prepared by Lisa Gibson; Lightfoot (2015).

deformation. The sequence of rocks was modified by burial and loading of the crust (D2 events) to a peak temperature of 750°C and pressure of 7.5 kbar (Lightfoot et al., 2011b). The third phase of deformation (D3) was a sinistral transpression that generated the dome and basin configuration of the Thompson Nickel Belt.

These conditions allowed for progressive reformation of pyrrhotite and pentlandite into MSS as pressure and temperature increased; kinesis of the sulfide took place into domains of space created by deformation. Separation of MSS from residual granular Pn would produce a range in Ni content of sulfide, and the MSS would likely be spread out in the stratigraphy to form a broad halo around the main deposit to form the low Ni tenor sulfide. Reformation of Pn and Po after the peak D2 event would explain the broad mineralogical diversity of sulfide in the footprint of the mineral system. The effect of the D3 event at lower pressure and temperature was to locally redistribute the mineralization into structural space and imbricate the stratigraphy containing the lenses of sulfide (Lightfoot et al., 2011b).

Similarities between the mineralization style at Sudbury and Thompson are not overwhelming. Some of the mineral zones associated with shear zones at Sudbury share the development of contorted fragments of schist within massive sulfide veins that are entirely hosted by country rocks. Relative to Thompson, the styles of mineralization contained in these zones at Sudbury were developed at much lower pressure and temperature, and it appears less likely that primary differences in metal tenors of Sudbury sulfides are a product of physical fractionation of sulfide minerals.

STRUCTURAL PATHWAYS THAT CONTROL MAGMATIC SULFIDE ORE DEPOSITS

Deposits of Ni-Cu-Co-(PGE) sulfide commonly occur in association with small differentiated intrusions that reside within local transtensional spaces in strike-slip fault zones (Fig. 6.21A–C). These faults develop by incipient rifting of the crust associated with the formation of large igneous provinces, and in response to far-field stresses generated by continental drift.

The development of pathways along which sulfide-bearing magmas can migrate is controlled by structures that also control the shape of the magma chambers and intrusions that localize mineralization.

Many deposits of magmatic sulfide reside in dykes (eg, the Reid Brook Zone of mineralization in the Voisey's Bay mineral system), are associated with funnel-shaped domains at the base of small intrusions (eg, Jinchuan), or occupy open-system chonoliths that act as pathways for the migration of silicate and sulfide magma (eg, Noril'sk and Talnakh).

The volume of silicate magmas in small mafic-ultramafic intrusions is often insufficient to provide the metals and sulfur required to form the large magmatic sulfide ore deposits that they host. This has led to the suggestion that the intrusions represent conduits through which not only a large amount of magma could pass, but into which the magmatic sulfides were injected (Lightfoot and Evans-Lamswood, 2015). These small intrusions are important hosts to magmatic sulfide deposits.

Rhomboid-shaped intrusions with funnel cross-section

One group of small intrusions has the plan shape of an asymmetric rhomboid with a long axis parallel to the fault zone, and contacts which have often been structurally modified during and and/or after emplacement of the magma (Lightfoot and Evans-Lamswood, 2015). The typical cross-section is a downward-closing cone shape with curved walls and often a dyke-like keel at the base. The cross-section of the intrusion typically narrows away from the widest part of the intrusion. An example of this style or mineral system from China is the Jinchuan Deposit; the geological relationships are illustrated in Fig. 6.11A–F (Lightfoot and Evans-Lamswood, 2015). A second example is the Ovoid Deposit which comprises part of the Voisey's Bay mineral system in Labrador, Canada; the similarities are striking and have been discussed in detail in Lightfoot et al. (Lightfoot et al., 2011a, 2012; Lightfoot and Evans-Lamswood, 2015). This morphology is also found in the Huangshan, Huangshandong, Hong Qi Ling, Limahe, Qingkuangshan, and Jingbulake Intrusions in China, and the Eagle Deposit in the United States and the Eagle's Nest Deposit in Canada.

Discontinuous dykes and sheets

A second group of deposits is associated with magmatic conduits within dyke and sheet-like intrusions, and also where there are discontinuities in the dyke which were created in response to the emplacement mechanism. Examples include the Discovery Hill Deposit and the Reid Brook Zone of the Voisey's Bay Intrusion (Fig. 6.15A), where there are plunging domains where the dyke is wider that cross-cut gently dipping structures along which the mineralization is emplaced into the country rock gneisses (Fig. 6.14C,D).

In the Noril'sk Region, Russia, the Oktyabrysk, Taimyrsk, Komsomolsk, and Gluboky Deposits are localized at the base of the thickest part of the Kharaelakh Intrusion (Fig. 6.5B,C). The antiformal

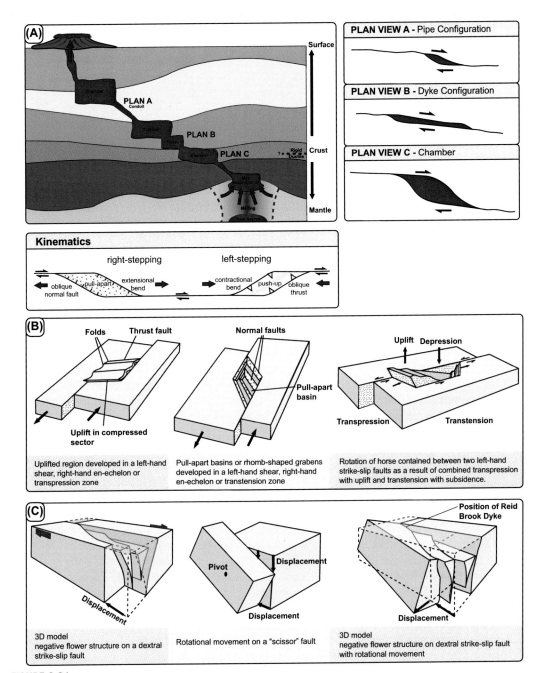

FIGURE 6.21

(A) Schematic diagram showing the pathway of magma from the mantle to the surface through a strike-slip structure with the development of diamond-shaped intrusion morphologies and dykes (Lightfoot and Evans-Lamswood, 2015). The importance of structural controls on a wide range of ore deposits has long been recognized (e.g., O'Driscoll, 1990 and papers in Bourne and Twidale, 2007). The strike-slip structures form "magma highways" from the mantle to the surface, but they offer a tortuous route that is heavily controlled by the available space in the network of faults. A series of plans (A through C) show the likely geometries of the conduits at different points in the crust. The kinematics and terminology used to describe these structures is highlighted. (B) Block diagram illustrating both pop-up and pop-down structures. The creation of space in a transtensional cross-structure within a strike-slip fault zone is critical in establishing pathways for magmas. (C) Example of the effects of scissor faults in zones of transtension, and the development of the magma chambers and dykes at Voisey's Bay.

(C) After Lightfoot and Evans-Lamswood (2015).

structures adjacent to the splays of the Noril'sk-Kharaelakh Fault have localized the distribution of the magma channel-ways in the adjacent stratigraphy. Space for these chronoliths is created by warping of the stratigraphy and block in response extension and compression on cross-linking structures in a strike-slip fault zone (Cunningham and Mann, 2007).

The unusual exocontact Cu-PGE sulfide ores are likely the product of sulfide melt differentiation, but their emplacement is probably controlled by space created by movement on faults. Some of this space was available at the base of the intrusion to the east of the Noril'sk-Kharaelakh Fault, but to the west it was generated in response to the collapse of roof rocks over the partially crystallized rocks of the chonolith along minor structures parallel to the Noril'sk-Kharaealakh Fault (Lightfoot and Zotov, 2007). In this sense, the distribution of Ni-rich and Cu-rich ores is controlled by sulfide differentiation, but their redistribution is a function of the tectonic environment in which the magma was emplaced and then crystallized.

Tubiform chonoliths
A third group of mineralized intrusions located within structural corridors have the geometry of long oblate tubes; examples include Karatungk in China, Talnakh and Noril'sk 1 (Fig. 6.5A), Babel-Nebo in Australia, and Nkomati in South Africa (Lightfoot and Evans-Lamswood, 2015). Tubiform chonoliths can occupy bridging structures between the more significant strike-slip structures. Examples include the Tamarack Intrusion in Minnesota, USA, and the Current Lake Complex in Ontario, Canada, all of which contain magmatic Ni–Cu sulfide mineralization (Lightfoot and Evans-Lamswood, 2015).

In all three types of intrusion described above, the original magma occupied an open system pathway, and so the term "chonolith" can be applied to describe them as a group. The intrusions are characterized by a high ratio of sulfide/silicate; there are 1–3 orders of magnitude more sulfide in the intrusion than the magma contained in the intrusion is capable of dissolving, and the formation of these deposits is considered to take place in open system magma conduits, where the sulfides were upgraded by equilibration of successive batches of silicate magma passing through the conduit, and equilibrating with a stationary pool of magmatic sulfide. At Voisey's Bay there appears little doubt that the sulfides were injected through a conduit dyke into higher level magma chambers, and a similar model has been proposed for the formation of the deposits at Jinchuan and Noril'sk-Kharaelakh. Economically significant nickel sulfide deposits tend to be high in Ni tenor, they are often related to the late injection of magma that form distinct parts of the intrusion; the localization of mineralization tends to be related to changes in the geometry of the magma chamber. Strongly deformed and metamorphosed komatiite-associated deposits (eg, Pechenga, Thompson, and the Yilgarn komatiite associations) appear to be the remains of small open system magma conduits now represented by segmented and boudinaged ultramafic bodies in terrains that have undergone more than four phases of postemplacement deformation.

Large igneous province magmatism at craton margins is recognized as a primary control on the genesis of magmatic sulfide deposits (Fig. 6.22A); the principal regional controls of strike-slip tectonics underpin the exact local geometry of the intrusions, and provide an explanation for why so many of the global nickel sulfide ore deposits are associated with intrusions that share common morphologies and characteristics. This model provides a framework for more detailed structural investigations of nickel sulfide deposits, and it also provides a practical framework to assist in exploration.

In the context of Sudbury, the role of transtension in controlling the morphology of the Offset Dykes has been discussed in Chapter 4, and the possible role of structural space as a control on footwall mineralization has also been highlighted in the case of the Coleman mineral system. The possibility that the target rocks at Sudbury were host to protoores associated with west–east fault structures is a possibility, but there are presently no go examples of mafic-ultramafic inclusions that rest within these shear zones.

RELATIVE ROLES OF SETTLING AND TRANSPORTATION IN SULFIDE LOCALIZATION

Two end-member models exist to explain the concentration of immiscible magmatic sulfides, namely, (1) In situ segregation and localization into depressions at the base of a magma column by gravitational processes alone; (2) physical transportation of sulfide melt from their site of segregation to the site of final deposition. The traditional view of segregation is certainly important as it allows for efficient localization of sulfide melts, and this is undoubtedly an important process at Sudbury. However, the discovery of magmatic sulfides in the Ovoid and Eastern Deeps at Voisey's Bay provided evidence that mineralization is localized where a dyke widens at or close to its entry point into a magma chamber (Fig. 6.22B). The mineral zones in the Reid Brook Dyke extend along sub-horizontal structures that represent primary discontinuities in the dyke (Fig. 6.22C). The localization was recognized by Naldrett et al. (1996b) and became the basis for a model in which magmatic sulfides are transported from depth and emplaced at locations where there is a decline in the velocity of emplacement of the magma (Fig. 6.22B; Lightfoot et al., 2011a, 2012).

Models of sulfide transportation have also been proposed by Tang (1992) to explain the different style of disseminated sulfide mineralization associated with the Jinchuan Intrusion (Fig. 6.22D).

The same question is pertinent to Sudbury. Much of the sulfide mineralization associated with the Offset Dykes has been transported as a sulfide melt into a reactivated pathway that followed the trajectory of emplacement of the initial melt into the dyke. Although the direction is evidently downward at Sudbury, the geological relationship between sulfide mineralization and dyke geometry has many similarities to the relationships found at Voisey's Bay.

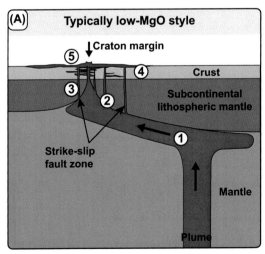

FIGURE 6.22

(A) Simplified cartoon showing the development of structural zones at the margins of cratons (Begg et al., 2011; Lightfoot and Evans-Lamswood, 2015) where mafic-ultramafic magmatism linked to mantle plumes may be localized.

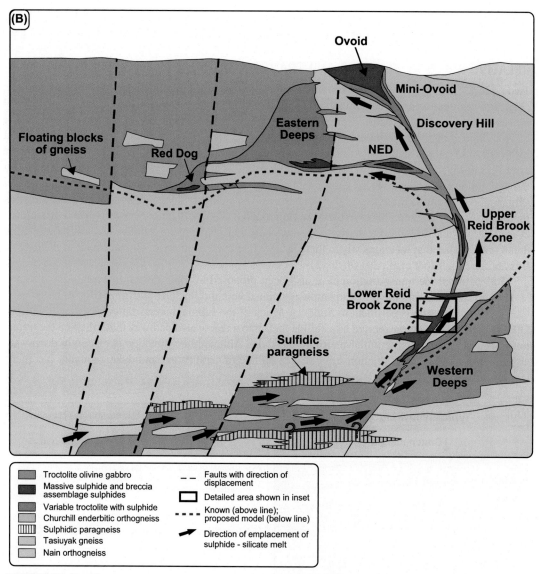

FIGURE 6.22 *(cont.)*

(B) Simplified geological model for the Voisey's Bay Intrusion showing the location of the mineralization in the chamber and dyke system. The diagram is constructed to show the broad relationships on a single west-facing section; it is not a geological section. Geological relationships around the Ovoid indicate that the deposit occupied a funnel-like opening of the dyke into a now-eroded western extension of the Eastern Deeps. Geological relationships in the deep chamber have not yet been established through drilling, but geological observations point to a deeper level of equilibration of the VBI magma with Tasiuyak paragneiss that contain both crustal sulfide and a compositionally unusual assemblage of mineralogy that give rise to the paragneiss fragments in the breccia sequence (Lightfoot, 2009).

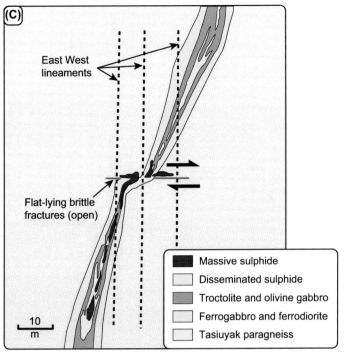

FIGURE 6.22 (*cont.*)

(C) Geological relationships in the Reid Brook Zone and conceptual model for the development of footwall and hanging-wall mineralization close to the dyke along structures that dip gently east (inset shows the detailed geology of the Reid Brook Dyke where the discontinuities resemble those found in the Copper Cliff Offset).

THE IGNEOUS ROCKS ASSOCIATED WITH MAGMATIC SULFIDES

The formation of sulfide ore deposits through sulfide saturation, segregation, and accumulation of a dense sulfide melt from a mafic or ultramafic magma has a matching influence on the composition of the rocks from which the sulfide melt was extracted. As discussed in Chapters 3 to 5, the saturation of the melt sheet in sulfide at Sudbury resulted in not only the localization of sulfide melt toward the base of the melt sheet, but also the depletion of the noritic portion of the melt sheet in Ni, Cu, and PGE. These elements that were effectively stripped out of the magma column by virtue of the elevated partition coefficients of these elements between silicate and sulfide melt. If the argument holds for the Sudbury mineral systems, then it should also be a trait found in other magmatic sulfide mineral systems.

Noril'sk is where the relationship between ore formation and metal depletion was first formally established using chemostratigraphic studies of the associated basalts (Lightfoot et al., 1990, 1993b, 1994a). The metal depletion signature of the silicate rocks is spatially and temporally associated with the mineralized intrusions that host the Noril'sk and Tanakh ore deposits (Naldrett et al., 1991).

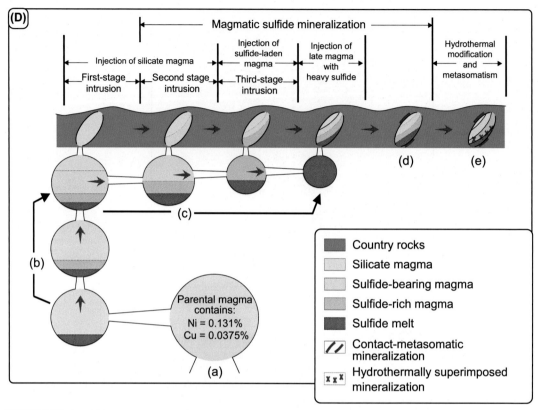

FIGURE 6.22 (*cont.*)

(D) Model for the formation of the Jinchuan Intrusion by multiple magmatic events and emplacement of sulfide-laden magma. Silicate magma from a mantle source is emplaced into the crust and undergoes progressive accumulation of sulfide melt in a deeper holding chamber (b); the emplacement of this melt as the deep chamber emptied of silicate and then progressively more sulfide created the different pulses of magma recognized by Tang (1992) in the stages labeled "c." The final stages involved contact metasomatism (d) and hydrothermal modification (e).

(A) Modified after Begg et al. (2011); (B) and (C) Modified from Lightfoot and Evans-Lamswood (2015) with permission of Elsevier; (D) modified after Tang (1992).

The evidence for sulfide segregation from comagmatic basaltic magmas at Noril'sk

The sequence of basaltic rocks in the Noril'sk Region (Fig. 6.4B) comprises a stratigraphy of subalkalic and alkalic basalts, picritic basalts and tholeiites that are described in detail in Lightfoot et al. (1990, 1993b, 1994a) and Fedorenko et al. (1996). It is the detailed chemostratigraphy of the basalts in Lightfoot et al. (1990) show that one of the tholeiitic basalt formations has a strong contamination signatures arising from the assimilation of continental crust (Fig. 6.4B); these basalts belong to the Nadezhdinsky Formation (Fedorenko, 1981), and their position in the chemostratigraphy is shown in Fig. 6.23A–F. These Nadezhdinsky Formation basalts have very low concentrations of Ni and Cu over a narrow range of MgO (Fig. 6.23A–C) (Lightfoot et al., 1990; Naldrett et al., 1991), they are exceptionally contaminated as illustrated by the Sr-, Nd-isotope and ratios of incompatible trace elements (Fig. 6.23D;

Lightfoot et al., 1990, 1993b) and they were subsequently shown to have very low abundances of the PGE and high Cu/Pd ratios (Fig. 6.23E,F; Keays and Lightfoot, 2009). The Nadezhdinsky basalts exhibit almost an order of magnitude lower Ni and Cu, and two orders of magnitude lower PGE than the other tholeiitic basalts that have a similar MgO content (Fig. 6.23A–C, D–F).

The dominant center of accumulation (the depocenter) of basaltic rocks of the Nadezhdinsky Formation of the Siberian Trap is localized along the axis of the Noril'sk-Kharaelakh Fault, and this volcanic edifice has a volume of ~5,000–10,000 km³ (Fig. 6.3B).

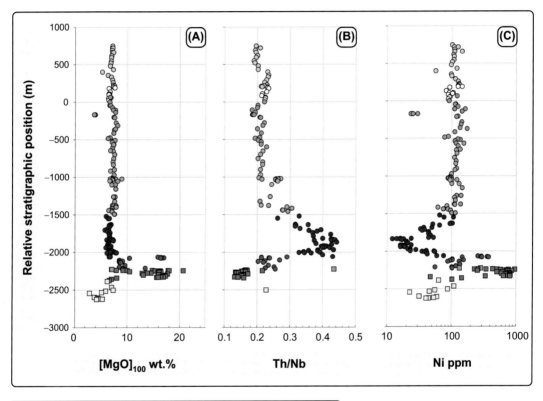

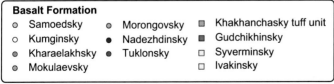

Basalt Formation

⊙ Samoedsky	◐ Morongovsky	▨ Khakhanchasky tuff unit
○ Kumginsky	● Nadezhdinsky	■ Gudchikhinsky
◑ Kharaelakhsky	◑ Tuklonsky	□ Syverminsky
◐ Mokulaevsky		□ Ivakinsky

FIGURE 6.23 Compilation of Data From Surface and Drill Core Into One Traverse Through the Stratigraphy of the Siberian Trap at Noril'sk

(A) MgO (wt.% normalized free of LOI); (B) Th/Nb ratio; (C) Ni (ppm); (D) Cu (ppm); (E) Pd (ppb); and (F) Cu/Pd ratio.

After Lightfoot et al. (1990, 1993b, 1994a); Hawkesworth et al. (1995), and Keays and Lightfoot (2007).

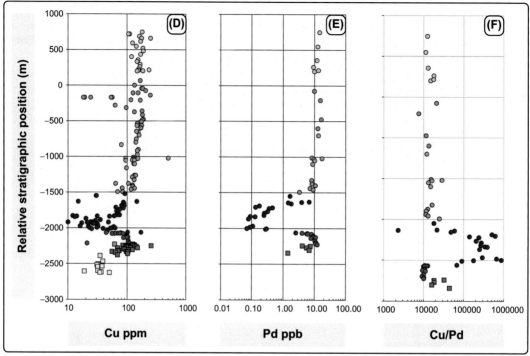

FIGURE 6.23 (*cont.*)

This volcanic edifice is underlain by sedimentary rocks that contain intrusions that are more primitive than the Nadezhdinsky Formation basaltic rocks, but have similar ages and trace element ratios which indicate that they belong to a magma series that is temporally and spatially related (Naldrett et al., 1995; Hawkesworth et al., 1995). These intrusions comprise the strongly differentiated and weakly mineralized Ni-Cu-PGE-depleted Low Noril'sk and Low Talnakh types that underlie the footprint of the economically mineralized intrusions at Kharaelakh, Talnakh and Noril'sk 1 (Fig. 6.3B; Lightfoot and Zotov, 2006). The Talnakh, Kharaelakh, and Noril'sk 1 Intrusions are heavily mineralized and are positioned stratigraphically above the Low Talnakh and Low Noril'sk Intrusions and they cross-cut through the stratigraphy of the sedimentary rocks beneath the Siberian Trap (Fig. 6.5A), and in the case of the Noril'sk 1 intrusion, the most basal Ivakinsky and Syverminsky Formation basalts of the Siberian Trap (Fig. 6.5A). The Talnakh and Kharaelakh Intrusions are located adjacent to the Noril'sk-Kharaelakh Fault, but are not fed along the main fault (Fig. 6.5B,C). The Talnakh intrusion is at a higher stratigraphic level in Permian sedimentary rocks than the Kharaelakh Intrusion which is hosted in Devonian-aged sedimentary rocks (Fig. 6.5A). The main part of each intrusion rests largely within argillites of the Razvedochninsky Formation and the Tungusskaya Series, respectively (Fig. 6.4A). These sedimentary rocks are typically missing or partially missing from the sequence along the axis of the widest part of the intrusion (Zotov, 1989).

The strong spatial and temporal relationship between the ore deposits and the metal depleted volcanic center at Noril'sk is well-established (Naldrett et al., 1991, 1996a), but there is debate on the detailed timing and the role of the chonoliths as feeders to the vast volumes of melt into the Traps (Arndt et al., 2005; Lightfoot and Zotov, 2006; Krivolutskaya, 2010, 2016; Lightfoot and Evans-Lamswood, 2015).

Examples of metal depleted mafic rocks associated with economic or subeconomic concentrations of Ni-Cu-PGE sulfide mineralization are recognized in several other large igneous provinces:

1. The Keweenawan mid-continent rift contains basaltic rocks that have anomalously low PGE concentrations in the Upper Formation of the Osler Volcanic Group (Keays and Lightfoot, 2015). Anomalous low Ni abundances in the Central Formation of the Osler Group volcanic rocks were previously linked to sulfide segregation (Naldrett and Lightfoot, 1993), but these rocks are not depleted in PGE. The linkage between the early primitive basaltic flows and magmatic sulfide ore deposits at Eagle, Tamarack, and Current Lake may be important indicators of process linkages, but it is the development of these intrusions fault zones inboard of the cratonic margin of the rift that controls the localization of the small strongly differentiated intrusions that contain the nickel sulfide mineralization (Lightfoot and Evans-Lamswood, 2015).

2. The Yangliuping Ni-Cu PGE sulfide Deposit is associated with the Permian Emeishan flood basalt in SW China that show evidence of metal depletion (Song et al., 2009). There are enormous challenges in exploring this part of the Emeishan, not least because of the active tectonics in the region (eg, the 2008 Wenchuan earthquake), the extreme topography, and the associated challenges of mining near to the footprint of the Wolong Panda Reserve in Sichuan Province.

3. The West Greenland flood basalt province crops out on Qeqertarssuaq Island and the Nuussuaq Peninsula in association with sub-economic concentrations of Ni-Cu-PGE sulfide mineralization such as the Illukunnguaq nickel showing (Keays and Lightfoot, 2007). The scale of the metal depletion signature in these volcanic rocks is consistent with the development of small high-grade mineral zones, but it pales in comparison to the volumetric extent of metal depletion found in the Nasdezhdinsky volcanic center above the Noril'sk and Talnakh ore deposits. The possibility of a future discovery of mineralization associated with the West Greenland flood basalt province was highlighted in Lightfoot and Hawkesworth (1997).

4. Voisey's Bay: The mineralized troctolite of the Eastern Deeps Intrusion at Voisey's Bay (Lightfoot et al., 2011a, 2012) contains trace sulfide with 0.1–3 ppb Pt and 0.1–3 ppb Pd, whereas weakly to heavily mineralized variable troctolites in the same unit have 1–2 orders of magnitude higher abundances of Pt and Pd. Troctolites and olivine gabbros from other parts of the Voisey's Bay Intrusion and other Nain Plutonic Suite Intrusions, including the Kiglapait, Newark Bay, Barth Island, Mushua, and Nain Bay South Intrusion, also have low platinum group element abundances. Although it is possible that this is a signature of a widespread sulfide saturation event that predated ore formation at Voisey's Bay, it is more likely that PGE depletion is a product of the mantle melting process, where low degrees of melting resulted in the retention of PGE in the mantle source (Lightfoot et al., 2011a, 2012). If so, this indicates that PGE-depletion should be used with caution as an exploration tool in the Nain Plutonic Suite.

The implications of these observations are important with respect to Sudbury. It is difficult to dismiss the coincidence of world-class Ni-Cu-PGE sulfide deposits with igneous rocks that are depleted in these same elements, but the difficulty is even more extreme when the same coincidence between ore deposits and metal depleted volcanic rocks and intrusions is found in the world's second major magmatic sulfide camp at Noril'sk-Talnakh.

MgO CONTENT OF THE PARENTAL MAGMA

The traditional view of magmatic sulfide ore deposits is that they tend to be associated with mafic or ultramafic magmas that are derived from the mantle, and make their way through the crust, fractionating, assimilating crust, and becoming sulfide saturated (Naldrett, 2010a). Viewed in the context of variations in the MgO content of the parental magma as estimated from whole-rock and olivine geochemistry (Fig. 6.24), the spectrum of nickel sulfide ore deposits extends all the way from very primitive melts that gave rise to the Yilgarn and Thompson deposits through magmas with picritic compositions that gave rise to Raglan, Pechenga, Jinchuan, Kabanga, and Kambalda, to high-MgO basaltic magmas that gave rise to the Noril'sk-Talnakh and Voisey's Bay Deposits. Sudbury is unusual in this context as the parental magma is believed to be even more evolved in composition (Fig. 6.24). When examined in the context of $Cu/(Cu + Ni)$, the most primitive mineral systems typically have the lowest ratios, whereas systems such as Noril'sk-Talnakh, Sudbury and Voisey's Bay have higher ratios (Fig. 6.24) even though they are not equally endowed with precious metals. It appears clear that high-Murray granite magmas are unlikely to be endowed with Cu, although this is far from certain for PGE with examples of high-Murray granite melts in the Thompson Nickel Belt having very high precious metal concentrations in sulfide (Franchuk et al., 2015).

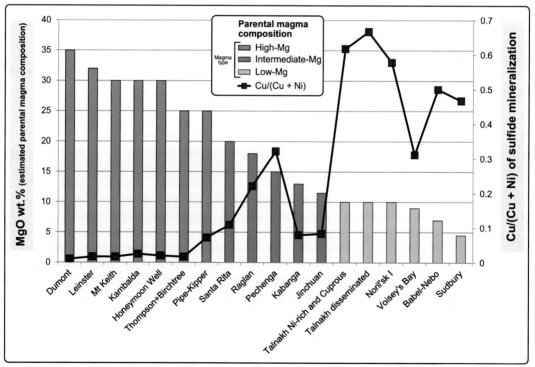

FIGURE 6.24 Variation in MgO Content of Parental Magma (Estimated Composition) and Cu/(Cu + Ni) in Sulfide Mineralization From Global Examples of Magmatic Sulfide Mineral Systems

Parental magmas generated in sources modified by subduction processes have figured more recently in models for some magmatic sulfide ore deposits (eg, Aguablanca in Spain; Tornos et al., 2001 and Kalatongke in China; Li et al., 2011). These relationships in northern China are not surprising given the extensive modification of the mantle during several phases of orogeny that predated the generation of magma. It is clear that magmas with a range in MgO content were generated from previously enriched mantle sources with no requirement for contributions of calc-alkaline magma from an active subduction zone.

SULFIDE DIFFERENTIATION MODELS

The variation in mineralogy of magmatic sulfide deposits is generally attributed to the effect of differentiation of the sulfide melt (Naldrett, 2004). Two world-class examples of this process are Sudbury (Chapters 4 and 5), and Noril'sk (Naldrett et al., 1994b, 1998). In each case a range of sulfide mineralogy is developed in response to a process of crystallization of MSS, expulsion of the trapped liquid, and further differentiation of ISS and crystallization of the residual liquid. The record of this process in the sulfides from Noril'sk is spectacular and supports many of the models developed for sulfide fractionation at Sudbury.

A second theme of importance is the geometric relationships between the sulfides and the associated igneous rocks. At Sudbury, it has been illustrated that sulfide differentiation was a top-down process where MSS remained at the contact of the SIC, whereas fractionated melts and ISS occupy positions beneath the Main Mass of the SIC. In contrast, the mss associated with the Talnakh and Kharaelakh Intrusions occur at the base of the intrusion (Naldrett et al., 1991, 1995) or in the immediately underlying sedimentary rocks (Lightfoot and Zotov, 2014). The differentiated sulfide occupies the lateral and upper parts of thick sheets of massive sulfide (Likhachev, 1994; Stekhin, 1994; Torgashin, 1994; Naldrett et al., 1991, 1995), or they occur in breccias associated with the hanging wall of the Kharaelakh Intrusion and the footwall of the Talnakh Intrusion (ie, the cuprous ores; Lightfoot and Zotov, 2014).

The comparative geology of the Noril'sk and Sudbury mineral systems highlights the following observations:

1. The differentiation of Sudbury ores involves partial isolation of the MSS at the lower contact of the SIC and injection of differentiated Cu-rich melts downward into structural space created by tectonic readjustment. In contrast, the contact Ni-rich ores at Kharaelakh show a range in compositions within the Ni-rich ore body at the lower contact. In contrast to these Ni-rich ores, a group of Cu-rich ores are developed in brecciated footwall rocks of the Talnakh Intrusion, and brecciated hangingwall rocks of the Kharaelakh Intrusion. The cuprous ores comprise sulfides that have formed by differentiation and emplacement into breccia zones created by tectonic readjustment along the margins of the Noril'sk-Kharaelakh Fault after the emplacement of magma into the chonolith.

2. The degree of fractionation of the Talnakh and Kharaelakh massive sulfides rarely reaches the apogee of millerite and bornite formation that is found at Sudbury. Examples of local enrichment in talnakhite and mooihoekite were economically important, but bornite is more rarely found in the Kharaelakh ore body. The majority of Noril'sk ores fractionate to slightly more extreme compositions than most of the South Range footwall mineral systems, but not to such extreme compositions as the North Range mineral systems.

3. The disseminated sulfides at Noril'sk vary in composition as a function of the adjacent massive Ni-rich sulfides, so they also record evidence of differentiation. The compositions of disseminated sulfides at Sudbury sometimes vary as a function of the composition of the adjacent massive sulfide mineralization (eg, Copper Cliff ore bodies), but this is not always the case (eg, the Murray Deposit).

In contrast to Noril'sk where well developed differentiated sulfide blebs are developed (Fig. 5.25A), sulfide blebs are not commonly found at Sudbury. Spectacular examples do occur in the Frood-Stobie mineral system (Figs. 4.8 and 4.9A,B), and good examples are present in the Copper Cliff and

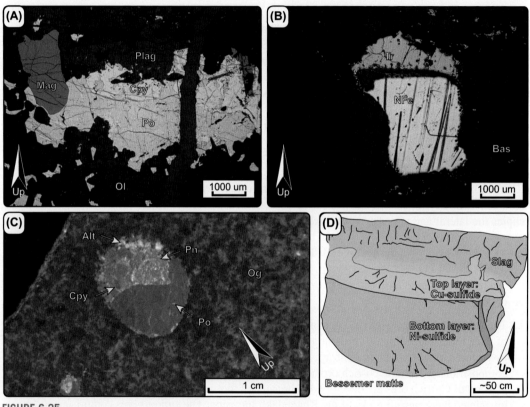

FIGURE 6.25

(A) Sulfide bleb from picritic gabbrodolerite, Talnakh Intrusion. Ol, olivine; Plag, plagioclase; Mag, magnetite; Cpy, chalcopyrite; Po, pyrrhotite. (B) Bleb of native iron and troilite from the contaminated basalts on Nuussuaq Peninsula, West Greenland (see Lightfoot and Keays, 2005 for details). Tr, trolite; NFe, native iron. (C) Blebby sulfide in olivine gabbro from Waterfall Gorge, Insizwa Complex, South Africa (Lightfoot et al., 1984). Alt – alteration zone; Cpy – chalcopyrite; Pn – Pentlandite; Po – pyrrhotite; Og- olivine gabbro. (D) Section through the charge in a Bessemer Furnace; the diagram illustrates the development of nickel-rich and Cu-rich sulfides with the Cu-rich sulfide at the top of the furnace. This type of furnace was used to process sulfide concentrates by Inco between 1891 and 1948.

(D) modified after Howard-White (1963).

Worthington Offset Dykes. The Sublayer Norite occasionally hosts sulfide blebs with diffuse margins. There is some evidence to indicate that disseminated mineralization at Sudbury varies in metal content as a function of the composition of adjacent massive or inclusion-rich sulfide mineralization.

Blebby sulfides are also developed in other magmatic systems. The example in Fig. 6.25C shows a lower Po + Pn domain and an upper Cpy domain overlain by a meniscus of an unidentified white mineral; this sample comes from the Waterfall Gorge Deposit in at the base of the Insizwa Complex, South Africa (Lightfoot et al., 1984). In more extreme cases, blebs of Ni- and PGE-rich native iron and troilite with a geopetal structure occur within crustally contaminated basaltic flows on the Nuussuaq Peninsula in West Greenland (Fig. 6.25B; Lightfoot and Keays, 2005).

Sulfide differentiation in the Bessemer Furnace

The process of separation of Ni-rich from Cu-rich sulfides has been reproduced in the products of the Bessemer Furnace (Howard-White, 1963). The Bessemer process was the first industrial process for the mass-production of steel from molten pig iron prior to the development of the open hearth furnace. The key principle is removal of impurities from the iron by oxidation using a stream of air blown through the molten iron. The oxidation raises the temperature and keeps the furnace-charge molten.

Vale utilizes the Copper Cliff Smelter (front image, Chapter 1) to process dried nickel-copper bulk concentrate from the mill. The concentrate is processed through a Flash Furnace and converters. This produces matte, sulfur dioxide, and slag. The sulfur dioxide is fixed and sold as sulfuric acid and liquid sulfur dioxide, whereas the gas stream is vented to the superstack after being cleaned of particulate matter in electrostatic precipitators. The slag from the furnaces is a waste product that is sent to the slag dump. The converter matte is cooled slowly to produce a coarse crystalline Bessemer matte (Fig. 6.25D). The matte is crushed, ground and separated into metals, nickel sulfide and copper sulfide by magnetic separation and floatation. The separation of Ni-rich from Cu-rich sulfide is exploited in the Bessemer Furnace (Fig. 6.25D), and gives rise to a similar separation to natural systems (Figs. 6.25A–C).

DEFORMATION AND SULFIDE KINESIS

The extent to which sulfides can be moved by kinesis during deformation is an important question that is best examined in nickel sulfide mineral systems that have a primary magmatic origin, yet have been extensively modified by deformation. Thompson provides an excellent example, but the reader is also pointed to descriptions of the geology and structural controls on the Pechenga mineral system (Green and Melezhik, 1999; Laverov, 1999) and studies of deformed komatiite-associated deposits in the Forrestania Belt of the Yilgarn, Western Australia (Collins et al., 2012a,b)

The relevance of Thompson to Sudbury is closest in some of the highly deformed shear-hosted sulfides in the Garson Deposit. Despite extensive deformation along these shear zones, there is no evidence of the extreme range of sulfide metal tenor found in the Thompson ore deposits, so the process of sulfide kinesis in the creation of these deposits from the contact mineralization style has less effect on the composition of the sulfide. Although it is possible that some of the Garson ores existed in the footwall before regional deformation, two pieces of evidence indicate that at least some of the shear-zone hosted mineralization was emplaced during regional deformation, namely, (1) sulfide mineralization associated with metabreccias cross-cut the Sublayer and Quartz-rich Norite of the Main Mass; (2) inclusions of Sublayer are present in the inclusion-rich sulfide mineralization hosted in the shear zones.

SUDBURY INFORMED BY GLOBAL MAGMATIC SULFIDES

Sudbury is not a unique example of magmatic sulfide ore deposits formed in association with igneous rocks, but it is a special example. It is worth summarizing the similarities and differences in the context of how information from other magmatic sulfide ore deposits has helped to inform our understanding of Sudbury.

Similarities between Sudbury and other magmatic sulfide deposits

The formation of magmatic sulfide deposits occurs by the equilibration of a sulfide melt with a silicate magma, thereby stripping the metals from the silicate, and concentrating them by gravitational forces. The same processes happen in all igneous melts, irrespective of their source (mantle- or crustal-derivation). The Sudbury magma was certainly at the felsic end of the spectrum, but the source of the magma had sufficient Ni, Cu, and PGE to form rich sulfide ores by the same processes.

Once formed, magmatic sulfides are localized by physical and structural "containers" in both intrusions and melt sheets. Physical depressions on the floor of the chamber or melt sheet are important, but sulfides are also injected into structural pathways occupied by earlier batches of magma. In the intrusive setting, these bridging structures in strike-slip fault zones are common controls on smaller mafic-ultramafic intrusions that contain magmatic sulfide, but syn-magmatic structural control is also important in controlling the development of the Offset Dykes and the Sublayer. In this respect, the similarity of the mineral system at Copper Cliff and the Reid Brook Zone point to similar controls (Fig. 6.15A,B). In both cases, the most important controls are changes in the thickness of the dyke and the development of primary discontinuities. These changes occur where the country rock types are different and where they are modified by structures or development of breccias.

There is a remarkable tendency for magmatic sulfide systems to develop in small intrusions within structural zones onboard of the cratonic margins of rift systems or within collisional settings where strike-slip faults tend to localize small intrusions. The Sudbury event occurred slightly inboard of the edge of the Superior Province, and likely tapped a crust that was richer in mafic and ultramafic rocks than typical Archean granitoid rocks and also prone to reactivation of fault structures. Although there is no evidence for a mantle contribution to the SIC, the setting of the complex resembles that of Noril'sk, Voisey's Bay, and Thompson.

Finally, the diversity and sulfide textures, and presence of inclusions are features shared by many magmatic sulfide mineral systems. These similarities and associated changes in composition accompanying the genesis (R- or N-factor) and differentiation of the sulfide share common process controls. Not all deposits show extreme differentiation of sulfide, and there is no evidence to indicate that the extent of differentiation relates to the metal tenor of the primary sulfide melt, but the development of fractionation is achieved where a spatial decoupling is achieved between mss and residual liquid, and this is often in response to the presence of space created by faulting.

Differences between Sudbury and other magmatic sulfide deposits

Perhaps the most significant difference between Sudbury and other magmatic sulfide ore deposits is the fact that the melt was generated very rapidly, and likely achieved a superheated condition. The efficiency of segregation of sulfide was likely enhanced by the superheated temperature of the melt which provided conditions equivalent to an "ideal smelter." Few other intrusions have evolved in a sulfide saturated state for long periods of time, free of the effects of crystallization of silicate minerals on the efficient accumulation of sulfide melt. Sudbury was a very efficient system from the perspective of the formation of mineral systems.

The parental magma at Sudbury is intermediate in composition, but that of the Voisey's Bay intrusion was more mafic. Despite this, the Sudbury ores have a bulk Cu/(Cu + Ni) ratio of ~1 yet the sulfide mineralization at Voisey's Bay is more copper-rich with a ratio of ~0.63. In the case of Sudbury, many of the ore deposits have high $[3E]_{100}$, yet at Voisey's Bay the mineralization has low $[3E]_{100}$. The difference may well be explained by a combination by magma sources that have different PGE concentrations (Lightfoot et al., 2011a).

At Sudbury, the direction of transport of the sulfide melt and sulfide-laden silicate magma is from the top down, so this is clearly supported by the settling process of dense sulfide. In contrast, the formation of Voisey's Bay and Jinchuan require emplacement of sulfide from below (Fig. 6.22B,C; Tang, 1992; Lightfoot et al., 2011a, 2012).

THE FUTURE OF NICKEL PRODUCTION

The final section of the book places Sudbury in the context of the past and present global supply of nickel. One of the principal changes in the past 10 years has been a ramp-up in the production of nickel from laterite sources. This production has contributed some high quality nickel to the market largely from smelters and hydrometallurgy plants, but the vast majority of the produced nickel is contained in pig iron produced in blast furnaces and electric arc furnaces (Mitchell, 2014). For this reason, the context of laterites is important in understanding the future of sulfide nickel supply.

NICKEL LATERITE DEPOSITS

Ultramafic rocks tend to have elevated primary nickel concentrations with large amount of nickel hosted in olivine and lesser amounts in orthopyroxene. These minerals are the primary source of the nickel contained in rocks created by alteration and weathering. Nickel laterites are typically developed in areas that have experienced prolonged tropical weathering of ultramafic rock containing ferro-magnesian minerals (olivine, pyroxene, and amphibole). Dunite, peridotite, pyroxenite, and serpentinite (essentially serpentine; $2H_4Mg_3Si_2O_9$) are the main rock types involved in their formation. Serpentine is a common product of hydrothermal alteration of olivine in the presence of water at 200–500°C. Serpentine has a profound influence on the chemostratigraphy of the weathering profile. If serpentine is absent or less well developed, the laterite may be upgradable in nickel content by rejecting low-grade fresh ultramafic boulders during processing (eg, the West Block type of mineralization at Sorowako in Indonesia; Dalvi et al., 2004).

Serpentinized ultramafic rocks that once formed part of the oceanic crust occur within belts in relatively young settings; these rocks are part of ophiolite sequences. Most important global nickel laterites form in hot humid tropical and subtropical areas, and they are developed by weathering of ophiolite complexes.

Large high-grade nickel laterite deposits are typically generated areas of gentle crests, spurs, and plateaus in humid environments. A typical profile showing the variability in bedrock topography, development of blocks in the profile, and thickness variations (Fig. 6.26A,B) over serpentinized versus fresh bedrock is often an important control the development of the two principal economically important types of nickel laterite, namely, (1) limonite (1.0–1.5 wt.% Ni); and (2) saprolite (1.0–3.0% Ni).

The genesis of a high-grade nickel laterite deposit depends on a range of factors. The rate and efficiency of chemical weathering is a key factor in the development of laterites, and it is controlled by the stability of the minerals, the pH of the water, the position of the water table in the profile, the redox potential, rates of fluid flow, climate conditions (temperature and rainfall), topography, tectonics, erosion history, fracturing and faulting, and time. A review of this is beyond the scope of this book, but the reader can consult a number of important contributions on laterite geology (Golightly, 2010, and references therein).

The major resources of laterites are located in Indonesia, New Caledonia, Cuba and the Phillipines. The highest grade of laterites are saprolites with >2 wt.% Ni from New Caledonia (Fig. 6.27A). However, when compared on the basis of a threshold of 1.5% Ni, Indonesia has the largest proportion of laterite resources followed by New Caledonia (Fig. 6.27B).

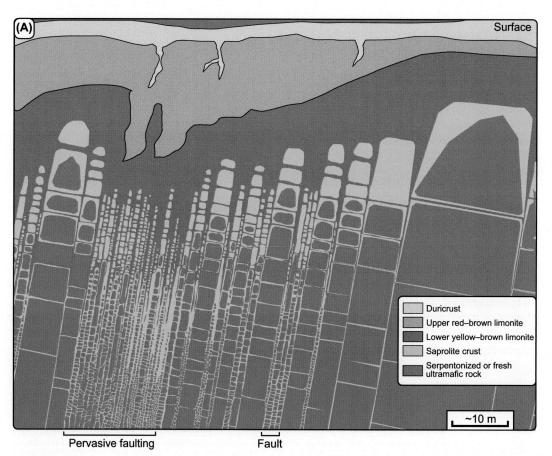

FIGURE 6.26

(A) Effect of bedrock topography and bedrock alteration on the thickness and continuity of the limonite and saprolite profiles.

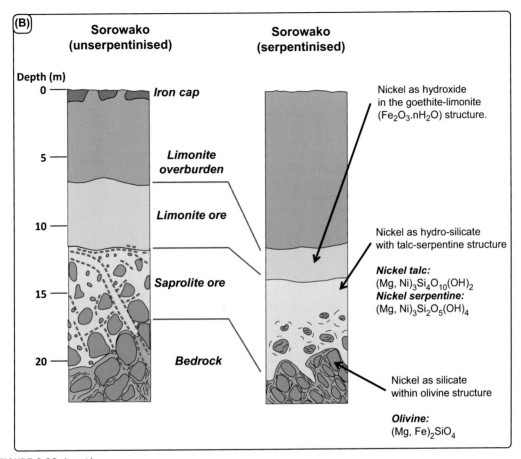

FIGURE 6.26 (*cont.*)

(B) Typical examples of laterite profiles from Sorowako (Sulawesi Island, Indonesia) showing the development of laterite on fresh bedrock (West Block Type) and serpentinized bedrock (East Block Type).

(A) Modified after Golightly (2010) (B) Modified after Dalvi et al. (2004).

In terms of nickel grade interval, Fig. 6.27C,D shows the proportions of deposits with grade intervals of ≤ 1 wt.% Ni, $1-\leq 1.5$ wt.% Ni, $1.5-\leq 2$ wt.% Ni, and >2 wt.% Ni. This diagram illustrates that at least two-thirds of global laterite reserves and resources have grades of $<1.5\%$ Ni, and the fraction with >2 wt.% Ni is a very small proportion of the global inventory. Fig. 6.27D shows a similar plot for nickel sulfide ore deposits, and illustrates a larger proportion of deposits with >2 wt.% Ni when compared to laterites, but a smaller proportion of deposits with ≤ 1 wt.% Ni when compared to the laterites.

A comparison of laterite and sulfide reserves and resources for material with $>1.5\%$ Ni indicates that the proportion of metal in laterite is slightly larger than the proportion in sulfide (Fig. 6.27E), but there is a much larger proportion of Ni in laterite when compared to sulfide if no minimum cut-off grade is used in the calculation (Fig. 6.27F).

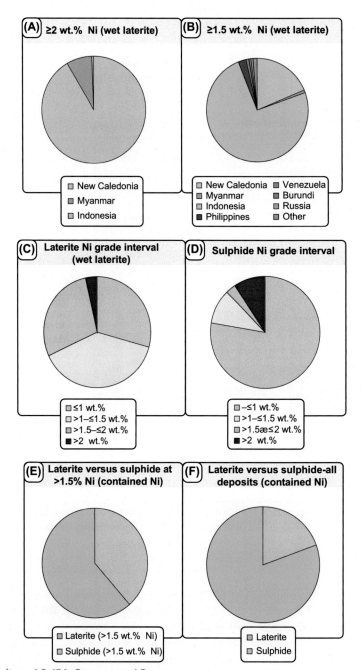

FIGURE 6.27 Laterite and Sulfide Reserves and Resources

(A) Comparison by country of contained Ni in reserves and resources for nickel laterite deposits with average grades of ≥2 wt.% Ni (in wet ore). (B) Comparison by country of contained Ni in reserves and resources of nickel laterite ores for deposits with average grades of ≥1.5 wt.% Ni (in wet ore). (C) Distribution of nickel grade in a global database of nickel laterite deposits. (D) Distribution of nickel grade in a global database of nickel sulfide deposits. (E) Comparison of global inventory of nickel in laterite versus nickel in sulfide for high-grade ores with >1.5 wt.% Ni. (F) Comparison of global inventory of nickel in laterite versus nickel in sulfide for all deposits.

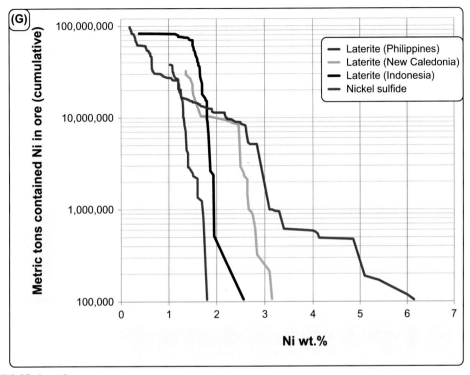

FIGURE 6.27 (*cont.*)

(G) Plot of Ni grade versus cumulate contained Ni (metric kilotons of nickel metal in ore) showing the position of global sulfide supply when compared to New Caledonia, Indonesia, and the Philippines.

Data was compiled from public domain sources by the author in 2016 for global sulfide deposits and 2015 for global laterite deposits.

    Fig. 6.27G shows a plot of grade versus cumulate tonnage for laterites from New Caledonia, Indonesia, and the Philippines, and it also shows a the trends in global sulfide deposits. This plot emphasizes the importance of higher grade Ni sulfide deposits that largely feed existing conventional smelters, but it also shows that the lions-share of higher grade nickel laterites are located in New Caledonia, and they are dominantly saprolites that feed conventional processing plants (Dalvi et al., 2004).

    There are only 10 countries that produce >50 kt Ni/annum, and this production is almost equally split between sulfides and laterites (Fig. 6.28A,B). Relative to all historic production (Fig. 6.28B), the production of Ni from sulfide sources has declined, and this decline has been greatly influenced by the impact of NPI production rather than the ramp-up of new hydrometallurgy and smelter plants.

    The growth in demand for nickel for use in stainless steel, industrial, and commercial products ramped up rapidly in China in the first decade of the 21st century, and this demand triggered an effort by Chinese companies to develop foreign nickel laterite deposits, directly ship the ore to China, and then process the ores in blast furnaces to make a low quality Ni-bearing iron, or in electric arc furnaces to produce a higher quality steel product. At the same time, many western companies were investing in multi-billion dollar projects to produce nickel using either conventional smelter technologies or newly

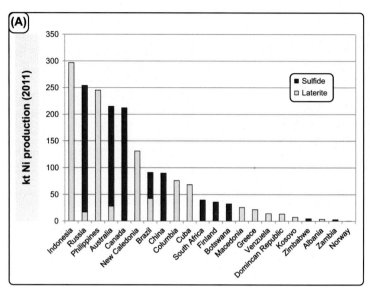

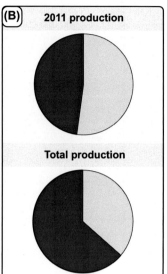

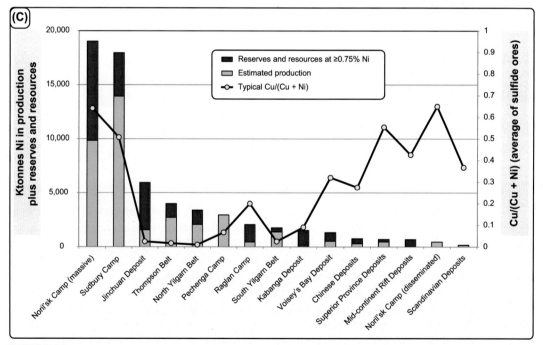

FIGURE 6.28

(A) Production of nickel from laterite and sulfide sources in 2011 by country region. (B) Relative amounts of Ni produced from laterite versus sulfide for production in 2011 and all production to end of 2011. (C) Comparison of the contained Ni and Cu in some of the largest nickel sulfide ore deposits (ranked by contained Ni). This plot also shows the Cu/(Cu + Ni) ratio of the sulfide ores.

(A) Data from Mudd and Jowitt (2014). (B) Data from Mudd and Jowitt (2014)

(C) Data was compiled from public domain sources by the author in 2016 for global sulfide deposits.

patented high pressure acid leach (HPAL) technologies. This quantum shift in strategy by China had a major impact on nickel prices, and it was only after the implementation of an export ban on untreated ores on January 12, 2014 by the Indonesian Government that the supply from this source into China started to dwindle. China rapidly shifted focus to development of projects in other locations of lower grade nickel laterites such as the Philippines.

The effect of ongoing nickel laterite production coupled with the short-term demise of China's demand for raw materials has created a major downturn in commodity prices that underpins the present depressed state of exploration for nickel as well as the development of new discoveries.

The future trends are difficult to predict. There are several competing forces that may control the future of the nickel industry:

1. *Availability of nickel in laterite versus sulfide:* A wealth of undeveloped high-grade nickel laterite deposits reside in Indonesia and New Caledonia that have not been developed. They offer enormous potential, but development of these deposits by companies that invest within the country will likely continue to be encouraged rather than direct shipment of ores to countries with space in blast furnaces and electric arc furnaces. Fig. 6.27C,D shows the global inventory of laterites versus sulfides, and highlights the contained nickel.
2. *Diversity in metal production:* Many magmatic sulfide ore deposits contain not only nickel, but a wealth of other elements that can be commercially extracted at a profit to add value to the mining operation (eg, Cu, Co, Pt, Pd, Au, Ag; Fig. 6.28C). The disadvantage of laterites is the fact that they contain only recoverable Ni and Co, with a few rare cases that have some precious metal content. A switch to nickel laterite at the expense of sulfide might satisfy part of the Ni and Co market, but it would reduce supplies of the other elements.
3. *The capital costs of new laterite projects:* The recent advantage in China has been the low cost of blast furnace and electric arc furnace infrastructure together with low production costs. This is changing as China becomes increasingly aware of the negative impact of inefficient processing on environmental and human health. The future development of nickel laterites may not involve expensive smelter and HPAL projects, but there is still a major cost implication to developing the projects and supplying them with power which is not economic at low nickel prices. A recovery in nickel prices may alter this outlook.
4. *The sunk costs associated with existing infrastructure:* One of the forces that keep large nickel camps such as Sudbury and Noril'sk alive is the enormous sunk cost associated with mines and processing plants. Many of the mines can produce at a profit with low nickel prices, so waiting out an economic downturn, and continuing to explore to supply the life of mine and facility is an important strategic focus.

SUDBURY AS A FUTURE GLOBAL NICKEL PRODUCER

The supply of Ni from new projects based on HPAL and other new technologies has not lived up to the past success of simpler proven technologies such as smelting. The impact of China-driven demand for nickel in the first decade of the 21st century highlighted the ability of companies to produce metal from laterite by smelting in blast furnaces and electric arc furnaces. The environmental and resource implications in the countries where mining has taken place (largely Indonesia), coupled with the impact on air quality in China has had a major effect on production from NPI. Moreover, many capital-expensive laterite projects have failed to ramp-up or meet market expectations (Mitchell, 2014).

The future for laterites is underpinned by large untapped resources, but the ability of companies to resolve technology issues and bring the metal to market in an environmentally sustainable way remains unproven, and the willingness of countries to export the metal value as ore for short-term gain is now diminished.

Laterites will continue to compete, but they will likely feed into production of low quality "stainless" steel, and those plants in production or moving toward boiler-plate capacity are not likely to be superseded by large amounts of new production of high quality nickel from these resources. This is good news for the sulfide producers with sunk costs in infrastructure who will continue to occupy an important position in global production as long as they maintain the infrastructure and support the future through capital investment in new mines and exploration for new deposits. This is good news for Sudbury.

Global nickel laterite resources with material grading >2% Ni are not common, and those available are largely dedicated to existing processing plants or plants in development. The sulfide inventory of materials with >2% Ni is strong with important deposits located in Sudbury and Noril'sk, undeveloped deposits such as Kabanga in central Africa, and important new discoveries of high-grade mineralization such as Nova-Bollinger, Eagle, and Tamarack. Komatiite-hosted nickel deposits in the Yilgarn such as Mt Keith contribute low value production, and companies that own deposits with similar grades and tonnages have been reluctant to develop them because the production is dominated by Ni and Co with little value-added Cu or precious metal production (Fig. 6.28C).

The komatiite targets that carry high-grade offer enormous value, and the deposits in the Forrestania Belt in Western Australia together with ongoing production from the Kambalda Dome and new discoveries in the Agnew-Wiluna Belt will continue to contribute metal, but these deposits are increasingly difficult to locate under conductive cover, and they are not the primary targets of interest as the individual deposits are sometimes small in comparison to the deposits at Noril'sk and Sudbury, but not always as illustrated by the heavily metamorphosed Thompson Deposit that have 2.5 times the historic nickel production of the Kambalda Camp (Lightfoot et al., 2011). Thompson style deposits provide important discovery opportunities into the future, and there is a substantial opportunity that deposit types with elevated PGE that belong to this group will be targeted based on the development of especially high Ni and PGE tenors in ultramafic intrusions in the Thompson Nickel Belt (Franchuk et al., 2015).

On the value scale, the deposits at Noril'sk have no equal (Fig. 6.28C). The abundances of PGE, copper and nickel will likely sustain deeper production from the Talnakh and Kharaelakh Intrusions, and underpin the large cost of developing the new ore bodies to sustain production from the existing processing plants. Into the future, there is also a substantial opportunity for low-grade disseminated sulfide mineralization in deposits such as Maslovskoe to contribute to precious metal production.

Like Noril'sk, Sudbury ores offer great value in not just contained nickel, but also copper and precious metals. This is a feature of the Offset and footwall ore deposit types that form an important part of the current production. The sunk costs of infrastructure coupled with the excellent exploration potential of the Sudbury Basin are good reasons to continue to invest in the development of new mines and the discovery through exploration of extensions to mineral zones and new ore deposits.

The future development of the Sudbury ore deposits will likely be guided by the following economic factors:

1. *Depth:* The ability to mine or deposits at great depth by remote mining methods or non-conventional solution mining processes requires the development and testing of new technologies (https://www.miningexcellence.ca/?page_id=1239).

2. *Technology for metal recovery:* Improved recovery from traditional styles of mineralization as well as improved recovery from disseminated low-grade deposits offers the potential to recover metals at lower cut-off grades and from ore bodies that do not pass present economic hurdles. These ores are attractive because of the value-added Cu and precious metal concentrations. Another challenge and opportunity is to reprocess and recover metals from historic tailings, and in the process improve the legacy of mining.

3. *New discovery:* From the perspective of this book, the greatest opportunity at Sudbury remains the opportunity to sustain the future life of mines and the existing processing plants through new discovery.

Chapter 5 provided evidence to indicate that discovery at Sudbury will continue if there is a sustained investment in exploration. Perhaps as little as 20% of the original inventory of metal lost from the melt sheet has been discovered, but much of the remainder may be located well below the depth that existing mine technologies can access. Many of the ore bodies at Sudbury show an increase in grade with depth, so the missing metal in the calculations will surely intrigue geologists for many decades to come.

The mineral systems at Sudbury are among the richest in the world, and they occur at depths where existing technologies provide a path to development. Perhaps the future will see mining at much greater depths with the development of new methods to safely extract resources in a way that is environmentally acceptable and at a profit to the investor, but it is not a stretch to get to this point. As existing mines are depleted and new mines are developed, the risk-reward paradigm will surely trigger new approaches. Some of these will involve remote mining, solution mining, or developments that reduce the power and ventilation requirements (https://www.miningexcellence.ca/?page_id=1247). Other approaches will include new ideas and concepts that help guide and target exploration at greater depth. The frontiers of the minerals industry will surely look at the wealth left in many deposits that have been mined out and they will focus on advancing methods of solution mining of the material left behind during underground mining, or the recovery of ever increasing quantities of metals from the rejected tailings or the tailings generated by historic processing.

It takes a quantum shift in confidence and faith that the technology of discovery and the ideas of the geoscientist will underpin success. This is surely not too much to expect because the history of the mining industry is replete with examples of changing technology, ideas, and confidence in geologists who can make a discovery. Those in mining companies who are risk adverse and who try to quantify every last detail are exactly those who will destroy future value.

There are few better examples than the ore deposits at Sudbury, where so many geologists have built their future in a mining camp with geological data. They have extended the known mineral zones and located entirely new ore bodies that have provided the legacy of mining, and support the future of nickel production at Sudbury. This book is a credit to their achievements and I hope that it will provide part of the foundation for future success.

References and Sudbury Bibliography

Abel, M.K., 1981. The structure of the Strathcona Mine copper zone. Can. Inst. Mining Metal. Bull. 74, 89–97.

Abel, M.K., Buchan, R., Coats, C.J.A., Penstone, M.E., 1979. Copper mineralization in the footwall complex, Strathcona Mine, Sudbury, Ontario. In: Naldrett, A.J. (Ed.), Nickel-Sulfide and Platinum-Group-Element Deposits, vol. 17, pt. 2. Can. Mineral., USA, pp. 275–285.

Abramowski, T., Stoyanova, V., 2012. Deep-sea polymetallic nodules: renewed interest as resources for environmentally sustainable development. Proceedings of Twelfth International Multidisciplinary Scientific GeoConference SGEM, Albena, Bulgaria, pp. 515–522.

Addison, W.D., Brumpton, G.R., Vallini, D.A., McNaughton, N.J., Davis, D.W., Kissin, S.A., Fralick, P.W., Hammond, A.L., 2005. Discovery of distal ejecta from the 1850 Ma Sudbury impact event. Geology 33, 193–196.

Addison, W.D., Brumpton, G.R., Davis, D.W., Fralick, Philip, W., Kissin, S.A., 2010. Debrisites from the Sudbury impact event in Ontario, north of Lake Superior, and a new age constraint: are they base-surge deposits or tsunami deposits? Geol. Soc. Am., Special Paper 465.

Ames, D.E., 1999. Geology and regional hydrothermal alteration of the crater-fill, onaping formation: association with Zn-Pb-Cu mineralization, Sudbury Structure, Canada. Unpublished PhD Thesis. Carleton University, Ottawa, Ontario, 460 pp.

Ames, D.E., Farrow, C.E.G., 2007. Metallogeny of the Sudbury mining camp. Mineral Deposits of Canada, Geological Association of Canada, Mineral Deposits Division, Special Publication, vol. 5, pp. 329–350.

Ames, D.E., Gibson, H.L., 1995. Controls on, and geological setting of, regional hydrothermal alteration within the Onaping Formation footwall to the Errington and Vermilion base metal deposits, Sudbury Structure, Ontario. Current Research 1995-E, Geological Survey of Canada, pp. 161–173.

Ames, D.E., Gibson, H.L., Watkinson, D.H., Jonasson, I.R., 1995. Large semi-conformable alteration zones within the Onaping Formation (abstract). GCS Forum '95.

Ames, D.E., Gibson, H.L., Jonasson, I.R., 1997. Impact-induced hydrothermal base metal mineralization, Whitewater Group, Sudbury structure (abstract). Large Meteorite Impacts and Planetary Evolution.

Ames, D.E., Watkinson, D.H., Parrish, R.R., 1998. Dating of a regional hydrothermal system induced by the 1850 Ma Sudbury impact event. Geology 26, 447–450.

Ames, D.E., Golightly, J.P., Lightfoot, P.C., Gibson, H.L., 2002. Vitric compositions in the Onaping Formation and their relationship to the Sudbury Structure. Econ. Geol. 97, 1541–1562.

Ames, D.E., Kajarsgaard, I.M., Douma, S.L., 2003. Sudbury Ni-Cu-PGE ore mineralogy compilation. GSC Open File Report 1787.

Ames, D.E., Davidson, A., Buckle, J.L., Card, K.D., 2005. Geology, Sudbury bedrock compilation, Ontario. Geological Survey of Canada, Open File 4570, 2 sheets.

Ames, D.E., Davidson, A., Wodicka, N., 2008. Geology of the Giant Sudbury Polymetallic Mining Camp, Ontario. Can. Econ. Geol. 103, 1067–1077.

Ames, D.E., Stoness, J.A., Roussel, D.H., 2009. Whitewater Group. In: Roussel, D.H., Brown, G.H. (Eds.), A Field Guide to the Geology of Sudbury, Ontario. Ontario Geological Survey Open File Report 6243, pp. 37–43.

Ames, D.E., Golightly, J.P., Zirenberg, R.A., 2010a. Trace element and sulfur isotope compositions of Sudbury Ni-Cu-PGE ores in diverse settings. Society of Economic Geologists 2010 Conference, Keystone Resort, Colorado. September 30 to October 9, 2010.

Ames, D.E., Golightly, J.P., Kjargaard, I.M., Farrow, C.E.G., 2010b. Minor element composition of Sudbury ores: implications for source contributions, genesis and exploration of Ni-Cu-PGE in diverse settings. Ontario Geological Survey, Miscellaneous Release of Data, vol. 269, pp. 1–4.

Ames, D.E., Farrow, C.E.G., Johansson, I.R., Pattison, E.F., Golightly, J.P., 2014. Geochemistry of 44 Ni-Cu-platinum group element deposits in the contact. footwall, offset, and breccia belt environments, Sudbury mining district, Canada. Geological Survey of Canada, Open File Report 6578, 6 pp.

Anders, D., Osinski, G.R., Grieve, R.A.F., 2013. The Onaping Intrusion, Sudbury, Canada—an impact melt origin and relationship to the Sudbury Igneous Complex. Lunar Planet. Sci. Conf. 44, 1637. Available from: http://www.lpi.usra.edu/meetings/lpsc2013/pdf/1637.pdf.

Anders, D., Osinski, G.R., Grieve, R.A., Brillinger, D.T.M., 2015. The Basal Onaping Intrusion in the North Range: roof rocks of the Sudbury Igneous Complex. Meteor. Planet. Sci. 50, 1577–1594.

Andreoli, M.A.G., Ellis, S., Webb, S.J., Pettit, W., Haddon, J., Ashwal, L.D., Gabrielli, F., Raubenheimer, E., Ainslie, L., 1999. The 145 Ma Morokweng impact, South Africa: an unusual ~90 km crater with associated multi-ring structures and Early Cretaceous mafic dykes. Meteor. Planet. Sci. 34 (4), 9.

Aniol, H.D., 1970. Geology of Hutton and Parkin townships. Ontario Department of Mines, Geological Report 80, 78 p.

Aniol, G.Z., Brown, D.L., 1979. Garson mine ore body characteristics. Unpublished report. Inco, 69 pp.

Arengi, J.T., 1977. Sedimentary evolution of the Sudbury Basin. Unpublished MSc Thesis. University of Toronto, Toronto, Ontario, 141 p.

Ariskin, A.A., 1997. Simulating phase equilibria and in situ differentiations for the proposed parental Sudbury magmas (abstract). Large Meteorite Impacts and Planetary Evolution.

Ariskin, A.A., Deutsch, A., Ostermann, M., 1999. Sudbury Igneous Complex: simulating phase equilibria and in situ differentiation for two proposed parental magmas. In: Dressler, B.O., Sharpton, V.L. (Eds.), Large Meteorite Impacts and Planetary Evolution II. Geological Society of America Special Paper 339, pp. 373–397.

Arndt, N., Lesher, C.M., Czamanske, G.K., 2005. Mantle-derived magmas and magmatic Ni-Cu-(PGE) deposits. Econ. Geol. 100th Aniversary volume, 5–24.

Artan, M.A., 1995. Lead systematics of the Sudbury nickel ores, Sudbury, Ontario, Canada. MSc Thesis. McMaster University, Hamilton, ON, Canada. 134 pp.

Avermann, M., 1992. Sudbury Project (University of Muenster-Ontario Geological Survey), 6, Origin of the polymict, allochthonous breccias of the Onaping Formation. In: Dressler, B.O., Sharpton, V.L. (Eds.), Papers presented to the International Conference on Large Meteorite Impacts and Planetary Evolution, 790, pp. 4–5.

Avermann, M., 1994. Origin of the polymict, allochthonous breccias of the Onaping Formation, Sudbury Structure, Ontario, Canada. In: Dressler, B.O., Grieve, R.A.F., Sharpton, V.L. (Eds.), Papers presented to the International Conference on Large Meteorite Impacts and Planetary Evolution. LPI Contribution. Boulder, Colorado, Geological Society of America, Special Paper 293, pp. 265–274.

Avermann, M.E., 1997. Investigations on the Green Member of the Onaping Formation, Sudbury Structure, Ontario, Canada (abstract). Large Meteorite Impacts and Planetary Evolution.

Avermann, M.E., 1999. The Green Member of the Onaping Formation, the collapsed fireball layer of the Sudbury impact structure, Ontario Canada. In: Dressler, B.O., Sharpton, V.L. (Eds.), Large Meteorite Impacts, Planetary Evolution, II. Geological Society of America Special Paper 339, pp. 323–330.

Avermann, M.E., Brockmeyer, P., 1990. The Onaping Formation of the Sudbury Structure (Canada): an example for allochthonous impact breccias (abstract). In: Pesonen, L.J., Niemisara, H. (Eds.), Symposium Fennoscandian Impact Structures, p. 42.

Avermann, M., Brockmeyer, P., 1992. The Onaping Formation of the Sudbury Structure (Canada), an example of allochthonous impact breccias. In: Pesonen, L.J., Henkel, H. (Eds.), Terrestrial Impact Craters and Craterform Structures With a Special Focus on Fennoscandia. Tectonophysics, 216 (1–2), pp. 227–234.

Bailey, J., 2011. Technical report on the North Range Properties, Sudbury, Ontario. Unpublished NI 43-101 Report. 46 pages. Available from: http://www.wallbridgemining.com/i/maps/northrange/2011-nrange-ni43-101-technical-report-w-appendix-a.pdf.

Bailey, J., 2013. Technical report on the Parkin Offset Joint Venture (POJV) Project Located near Sudbury, Ontario Including the Milnet, Parkin, CBA Parkin, and Parkin East Properties. Wallbridge Mining Company Limited. Available from: http://www.wallbridgemining.com/i/pdf/2013-Parkin-Technical-Report.pdf.

Bailey, J., Lafrance, B., McDonald, A.M., Fedorowich, J.S., Kamo, S., Archibald, D.A., 2004. Mazatzal–Labradorian-age (1.7-1.6 Ga) ductile deformation of the South Range Sudbury impact structure at the Thayer Lindsley mine, Ontario. Can. J. Earth Sci. 41, 1491–1505.

Bailey, J., McDonald, A.M., Lafrance, B., 2006. Pentlandite alteration and Ni mobility at the Thayer Lindsley mine, South Range Sudbury Igneous Complex, Ontario. Can. Mineral. 44, 1063–1077.

Bain, G.W., 1925. Amount of assimilation by the Sudbury norite sheet. J. Geol. 33, 509–525.

Baird, S., 2007. Genesis of the black norite phase of the Sudbury Igneous Complex. BSc Thesis. Laurentian University, Canada.

Bajc, A.F., 1992. Geochemical response of surficial media, north and east ranges, Sudbury Basin, Sudbury, Ontario. In: Dressler, B.O., Baker, C.L., Blackwell, B. (Eds.), Summary of Field Work and Other Activities, Ontario Geological Survey Miscellaneous Paper, 160, 147–150.

Bajc, A.F., 1993. Project Unit 92-20, Geochemical response of surficial media to nickel-copper-platinum group element mineralization, North and East Ranges, Sudbury Basin. In: Baker, C.L., Dressler, B.O., De Souza, H.A.F., Fenwick, K.G.,

Newsome, J.W., Owsiacki, L. (Eds.), Summary of Field Work and Other Activities, 1993, vol. 162. Ontario Geological Survey Miscellaneous Paper, pp. 183–187.

Ballhaus, C., Tredoux, M., Spath, A., 2001. Phase relations in the Fe-Ni-Cu-PGE-S system at magmatic temperature and application to massive sulphide ores of the Sudbury Igneous Complex. J. Petrol. 42, 1911–1926.

Barlow, A.E., 1891. On the Nickel and Copper Deposits of Sudbury, Ontario. Ottawa Naturalist, Ottawa, pp. 51–71.

Barlow, A.E., 1902. The Sudbury District (Ontario). Summary Report of the Geological Survey of Canada, Canada, pp. 143–147.

Barlow, A.E., 1903. The Sudbury Mining District (Ontario). Summary Report of the Geological Survey of Canada, Canada, pp. 254–269.

Barlow, A.E., 1904. Report on the Origin, Geological Relations and Composition of the Nickel and Copper Deposits in the Sudbury Mining District, Ontario, Canada. Geological Survey of Canada Annual Report, Canada, p. 873.

Barlow, A.E., 1906. On the origin and relations of the nickel and copper deposits of Sudbury, Ontario. Econ. Geol. 1, 454–466.

Barnes, S.-J., Lightfoot, P.C., 2005. Formation of magmatic nickel sulfide deposits and processes affecting their copper and platinum group element concentrations. Econ. Geol. 100, 179–213.

Barnes, S.-J., Makovicky, E., Makovicky, M., Rose-Hansen, J., Kraup-Moller, S., 1997. Partition coefficients for Ni, Cu, Pd, Pt, Rh, and Ir between MSS and sulfide liquid and the formation of compositionally zoned Ni-Cu sulfide bodies by fractional crystallization of sulfide liquid. Can. J. Earth Sci. 34, 366–374.

Barnes, S.J., Fiorentini, M., Duuring, P., Perring, C., Grguric, B., 2011. The perseverance and Mount Keith Ni deposits of the Agnew-Wiluna Belt, Yilgarn Craton, Western Australia. Rev. Econ. Geol. 17, 51–88.

Barnes, S.J., Cruden, S., Arndt, N., Saumur, B., 2015. The mineral system approach applied to magmatic Ni-Cu-PGE sulphide deposits. Ore Geol. Rev. 71, 673–702.

Barovich, K.M., Patchett, P.J., Peterman, Z.E., Sims, P.K., 1989. Nd isotopes and the origin of 1.9-1.7 Ga Penokean continental crust of the Lake Superior region. Geol. Soc. Am. Bull. 101, 333–338.

Basu, A.R., 1988. Impact-triggered mafic magmatism in Earth history (abstract). Canadian Continental Drilling Program Report 88-2, Scientific Drilling: The Sudbury Structure, p. 21.

Bateman, A.M., 1917. Magmatic ore deposits, Sudbury, Ontario. Econ. Geol. 12, 391–426.

Bateman, A.M., 1918. Genesis of the Sudbury nickel copper ores (discussion). Bulletin of the American Institute of Mining and Metallurgical Engineers, pp. 854–855.

Beales, F.W., Lozej, G.P., 1975. Sudbury basin sediments and the meteorite impact theory of origin for the Sudbury Structure. Can. J. Earth Sci. 12, 629–635.

Beales, F.W., Lozej, G.P., 1976. Additional note on the origin of the Sudbury structure. Can. J. Earth Sci. 13, 179–181.

Beamish, R.J., Harvey, H.H., 1972. Acidification of the La Cloche Mountain Lakes, Ontario, and resulting fish mortalities. J. Fish Res. Ed. Canada. 29, 1131–1143.

Becker, L., Bada, J.L., 1995. Fullerenes in the K/T boundary: are they a result of global wildfires (abstract). Lunar Planet. Sci. XXVI, 85–86.

Becker, L., Bada, J.L., Winans, R.E., Hunt, J.E., Bunch, T.E., French, B.M., 1994a. Fullerenes in the 1.85-billion-year-old Sudbury impact structure. Science 265, 642–645.

Becker, L., Bada, J.L., Winans, R.E., Hunt, J.E., Bunch, T.E., French, B.M., 1994b. Fullerenes in the 1.85 billion-year-old Sudbury impact structure (abstract). Sudbury Geology Workshop and Fieldtrip.

Becker, L., Poreda, R.J., Bada, J.L., 1996a. Extraterrestrial helium trapped in fullerenes in the Sudbury impact structure. Science 272, 249–252.

Becker, L., Bada, J.L., Poreda, R.J., 1996b. Extraterrestrial He trapped in fullerenes in the Sudbury impact structure. In: Sears, D.W.G. (Ed.), 59th Annual Meteoritical Society Meeting (abstracts). Meteoritics and Planetary Science, vol. 31, Suppl., pp. 12–13.

Becker, L., Bada, J.L., Poreda, R.J., Bunch, T.E., 1997. Extraterrestrial helium (He@C60) trapped in fullerenes in the Sudbury impact structure (abstract). Large Meteorite Impacts and Planetary Evolution.

Begg, G.C., Hronsky, J.A.M., Arndt, N.T., Griffin, W.L., O'Reilly, S.Y., Hayward, N., 2011. Lithospheric, cratonic, and geodynamic setting of Ni-Cu-PGE sulfide deposits. Econ. Geol. 105, 1057–1070.

Bell, R., 1890. Report on the Sudbury region. Summary Report of the Geological Survey of Canada, Ontario, 1888–1889.

Bell, R., 1891a. Report on the Sudbury District (Ontario). Annual Report—Geological Survey of Canada, Canada, pp. 1–54.

Bell, R., 1891b. Report on the Sudbury Mining District. Geological Survey of Canada, Report 6, part F, pp. 1890–1891.

Bell, R., 1891c. Summary report of work in the Sudbury region. Summary Report of the Geological Survey of Canada, Ontario, pp. 41–43.

Bell, R., 1891d. The nickel and copper deposits of Sudbury District, Can. Geol. Soc. Am. Bull. 2, 125–137.

Bell, R., 1893. On the Sudbury mining district, Geological Survey of Canada, Annual Report, 1890–1891, vol. 5, part 1, Report F, 54 pp.

Bell, J.M., 1920. The nickel-copper mines of Sudbury. Mining Mag. 23 (2), 87–94.

Bell, K., Blenkinsop, J., 1980. Ages and initial $^{87}Sr/^{86}Sr$ ratios from alkalic complexes of Ontario. Ontario Geological Survey, vol. 93. Miscellaneous Paper, pp. 16–23.

Bennett, M., 2015. Nova: The Discovery and Geology of Australia's First Globally Significant Sulfide Deposit. Prospectors and Developers Association of Canada, Toronto. Available from: http://www.pdac.ca/docs/default-source/Convention—Program—Technical-Sessions/2015/nickel—bennett.pdf?sfvrsn=2.

Bennett, G., Dressler, B.O., Robertson, J.A., 1991. The Huronian Supergroup and associated intrusive rocks. Geology of Ontario, Ontario Geological Survey, Special Volume 4, pp. 549–591.

Bennett, M., Gollan, M., Staubmann, M., Bartlett, J., 2014. motive, means, and opportunity: key factors in the discovery of the nova-bollinger magmatic nickel-copper sulfide deposits in Western Australia. SEG Spedial Publication 18EG Spedial Publication, vol. 18, pp. 301–320.

Berkey, C.P., 1918. Genesis of the Sudbury nickel-copper ores (discussion). Bull. Am. Inst. Mining Metal. Eng. 136, 855–857.

Beswick, A.E., 2002. An analysis of the compositional variations and spatial relationships within Fe-Ni-Cu sulfide deposits on the North range of the Sudbury Igneous Complex. Econ. Geol. 97, 1487–1508.

Bethune, K.M., 1993. Evolution of the Grenville Front in the Tyson Lake area, southwest of Sudbury, Ontario, with emphasis on the tectonic significance of the Sudbury diabase dykes. PhD Thesis. Queen's University, Kingston, ON, Canada. 263 pp.

Bethune, K.M., 1997. The Sudbury dyke swarm and its bearing on the tectonic development of the Grenville Front, Ontario, Canada. Precambrian Res. 85, 117–146.

Bethune, K.M., Davidson, A., 1997. Grenvillian metamorphism of the Sudbury diabase dyke-swarm: from protolith to two-pyroxene—garnet coronite. Can. Mineral. 35, 1191–1220.

Binney, W.P., Poulin, R.Y., Sweeny, J.M., Halladay, S.H., 1994. The Lindsley Ni-Cu-PGE deposit and its geological setting. In: Lightfoot, P.C., Naldrett, A.J. (Eds.), Proceedings of the Sudbury-Noril'sk Symposium, Ontario Geological Survey Special Volume 5, pp. 91–103.

Bischoff, L., Dressler, B.O., Avermann, M.E., Brockmeyer, P., Lakomy, R., Mueller-Mohr, V., 1992. Sudbury Project (University of Muenster-Ontario Geological Survey), 2, Field studies 1984–1989, summary of results. In: Dressler, B.O., Sharpton, V.L. (Eds.), Papers presented to the International Conference on Large Meteorite Impacts and Planetary Evolution. LPI Contribution, vol. 790, pp. 7–8.

Blais, E., 1994. Application de la magnetotellurique a l'exploration miniere profonde, anomalie de Trillabelle, structure de Sudbury MSc Thesis. Universite de Montreal. Montreal, PQ, Canada, 108 pp.

Blais, E., Mareschal, M., Zhang, P., 1992. An audio-magnetotelluric survey in western Sudbury: application to exploration. LITHOPROBE—Abitibi-Grenville Project, Abitibi-Grenville Transect. Report No. 33, pp. 83–90.

Blais, E., Livelybrooks, D., Mareschal,M., 1994. Analysis and 3D modelling of magnetotelluric data over a deep massive sulphide deposit, Sudbury Structure, Ontario (abstract). GCU, Program with Abstracts, Banff, AB.

Bleeker, W., 2004. Towards a 'natural' time scale for the Precambrian—a proposal. Lethia 37, 219–222.

Bleeker, W., Ernst, R.E., 2006. Short-lived mantle generated magmatic events and their dyke swarms: the key unlocking earth's paleogeographic record back to 2.6 Ga. In: Hanski, E., Mertanen, S., Rämö, T., Vuollo, J. (Eds.), Dyke Swarms—Time Markers of Crustal Evolution. Taylor & Francis Group, London, pp. 3–26.

Bleeker, W., Kamo, S., Ames, D.E., 2013. New field observations and U-Pb age data for footwall (target) rocks at Sudbury: towards a detailed cross-section through the Sudbury Structure (extended abstracts). Large Meteorite Impacts and Planetary Evolution V Meeting, August 5–8, Sudbury, Ontario, Lunar Planetary Institute contribution No. 1737, p. 13.

Bleeker, W., Kamo, S.L., Ames, D.E., Davis, D., 2015. New field observations and U-Pb ages in the Sudbury area: toward a detailed cross-section through the deformed Sudbury Structure. In: Ames, D.E., Houlé, M.G. (Eds.), Targetted Geoscience Initiative 4 Canadian Nickel-Copper-Platinum Group Elements-Chromium Ore Systems—Fertility, Pathfinders, New, Revised Models. Geological Survey of Canada, Open File 7856, pp. 151–166.

Blenkinsop, T., Doyle, M., Nugus, M., 2015. A unified approach to measuring structures in oriented drill core. Geol Soc London Special Publication 421. Available from: http://dx.doi.org/10.1144/SP421.1.

Boast, M., Spray, J.G., 2002. Hidden ore: using thermal metamorphism to determine shortening on thrust faults cutting the Sudbury Igneous Complex. GSA Abstr. Programs 34 (6).

Boast, M., Spray, J.G., 2003. A comparison of four transects through the North Range metamorphic aureole of the Sudbury Igneous Complex, Sudbury, Ontario, 1st Joint Meeting, Northeastern Section, GSA and Atlantic Geoscience Society (abstracts).

Boast, M., Spray, J.G., 2006. Superimposition of a thrust-transfer fault system on a large impact structure: implications for Ni-Cu-PGE exploration at Sudbury. Econ. Geol. 101, 1583–1594.

Boerner, D.E., Milkereit, B., 1999. Structural evolution of the Sudbury impact structure in the light of seismic reflection data. Geol. Soc. Am., Special Paper 339, pp. 419–429.

Boerner, D.E., Kellett, R., Mareschal, M., 1992. Inductive source EM sounding of the Sudbury Structure. Society of Exploration Geophysicists, 62nd Annual International Meeting (expanded abstracts). SEG Annual Meeting Expanded Technical Program Abstracts with Biographies, vol. 62, pp. 393–396.

Boerner, D.E., Kellett, R., Mareschal, M., 1993. Lithoprobe EM studies of the Sudbury structure (abstract). Canadian Geophysical Union, Banff, Alberta.

Boerner, D.E., Kellett, R., Mareschal, M., 1994a. Inductive source EM sounding of the Sudbury Structure. In: Boerner, D.E., Milkereit, B., Nadrett, A.J. (Eds.), Lithoprobe Sudbury Project. Geophys. Res. Lett. 21 (10), 943–946.

Boerner, D.E., Milkereit, B., Naldrett, A.J., 1994b. Introduction to the special section on the Lithoprobe Sudbury Project. In: Boerner, D.E., Milkereit, B., Nadrett, A.J. (Eds.), Lithoprobe Sudbury Project. Geophys. Res. Lett. 21, 919–922.

Boerner, D.E., Milkereit, B., Wu, J., Salisbury, M., 2000a. Seismic images and three-dimensional architecture of a Proterozoic shear zone in the Sudbury Structure (Superior Province, Canada). Tectonics 19, 397–405.

Boerner, D.E., Milkereit, B., Davidson, A., 2000b. Geoscience impact: a synthesis of studies of the Sudbury Structure. Can. J. Earth Sci. 37, 477–501.

Bohor, B.F., Betterton, W.J., 1992. Shocked zircons in the Onaping Formation: futher proof of impact origin. In: Dressler, B.O., Sharpton, V.L. (Eds.), Papers presented to the International Conference on Large Meteorite Impacts and Planetary Evolution. LPI Contribution 790, pp. 8–9.

Boldt, J.R., 1967. The Winning of Nickel. Longmans Canada Ltd, Canada, 487 pp.

Bonney, T.G., 1888. Notes on a part of the Huronian series in the neighbourhood of Sudbury, Canada. Quart. J. Geol. Soc. Lond. 44, 32–45.

Bouma, A.H., 1962. Sedimentology of some Flysch Deposits. Elsevier, Amsterdam, London, New York, 168 p.

Bourne, J.A., Twidale, C.R. (Eds.), 2007. Crustal Structures and Mineral Deposits: E.S.T. O'Driscoll's contribution to mineral exploration. Rosenberg Publishing Pty Ltd.

Bowen, N.L., 1925. The amount of assimilation by the Sudbury norite sheet. J. Geol. 33, 825–829.

Bowen, N.L., 1928. Evolution of the Igneous Rocks. Princeton University Press, Princeton.

Bray, J.G., Geological Staff, 1966. Shatter cones at Sudbury. J. Geol. 74, 243–245.

Brockmeyer, P., 1990. Petrogaphie, Geochemie und Isotopenuntersuchungen an der Onaping-Formation im Nordteil der Sudbury-Struktur und ein Modell zur Genese der Struktur. PhD Thesis. Westfelischen Wilhems University, Muenster, Germany, 228 pp.

Brockmeyer, P., Deutsch, A., 1989. The origin of the breccias in the lower Onaping Formation, Sudbury Structure (Canada), evidence from petrographic observations and Sr-Nd isotope data. Abstracts of Papers Submitted to the Lunar and Planetary Science Conference, 20, Part 1, pp. 113–114.

Brockmeyer, P., Lakomy, R., 1986. Footwall Breccia and breccias of the Onaping Formation in Dowling, Levack, and Morgan townships, District of Sudbury. In: Thurston, P.C., White, O.L., Barlow, R.B., Cherry, M.E., Colvine, A.C. (Eds.), Summary of field work and other activities, 1986. Ontario Geological Survey Miscellaneous Paper 132, pp. 123–124.

Brockmeyer, P., Buhl, D., Deutsch, A., 1989. Die Herkunft der Brekzien in der Onaping-Formation (Sudbury-Struktur, Kanada), Sr- und Nd-Isotopenuntersuchungen (The origin of breccias in the Onaping Formation Sudbury Structure, Canada, Sr and Nd isotope studies). In: Referate der Vortraege und Poster, 67. Jahrestagung der Deutschen Mineralogischen Gesellschaft (Abstracts of lectures and posters, the 67th annual meeting of the German Mineralogical Society) Berichte der Deutschen Mineralogischen Gesellschaft, 1989, vol. 1, p. 18.

Brockmeyer, P., Deutsch, A., Buhl, D., 1990. Sudbury impact structure (Ontario, Canada) isotope systematics. In: Pesonen, L.J., Niemisara, H. (Eds.), Symposium, Fennoscandian impact structures, programme and abstracts, p. 43.

Brocoum, S.J., Dalziel, I.W.D., 1974. The Sudbury Basin, the Southern Province, the Grenville Front, and the Penokean Orogeny. Geol. Soc. Am. Bull. 85, 1571–1580.

Brocoum, S.J., Dalziel, I.W.D., 1976. The Sudbury Basin, the Southern Province, the Grenville Front, and the Penokean Orogeny: discussion. Geol. Soc. Am. Bull. 87, 958.

Brooks, E.R., 1976. The Sudbury Basin, the Southern Province, the Grenville Front, and the Penokean Orogeny: discussion and reply. Geol. Soc. Am. Bull. 87, 954–958.

Browne, D.H., 1906. Notes on the origin of the Sudbury ores. Econ. Geol., 467–475.

Buchan, K.L., Ernst, R.E., 1994. Onaping fault system: age constraints on deformation of the Kapuskasing structural zone and units underlying the Sudbury Structure. Can. J. Earth Sci. 31, 1197–1205.

Buchan, K.L., Card, K.D., Chandler, F.W., 1989. Paleomagnetism of Nipissing diabase and associated rocks in the Englehart area, Ontario. Can. J. Earth Sci. 26, 427–445.

Buchan, K.L., Mortensen, J.K., Card, K.D., 1993. Northeast-trending early Proterozoic dykes of southern Superior Province: multiple episodes of emplacement recognized from integrated paleomagnetism and U-Pb geochronology. Can. J. Earth Sci. 30, 1286–1296.

Buddington, A.F., 1935. Review of "Life history of the Sudbury nickel irruptive, Part 1, Petrogenesis" by W.H. Collins, 1934. Econ. Geol. Bull. Soc. Econ. Geol. 30, 578–579.

Buhl, D., Deutsch, A., Lakomy, R., 1988. Sr- and Nd-isotope homogenization in a heterogeneous breccia, an example from Sudbury, Canada. In: Bottinga, Y. (Ed.), International Congress of Geochemistry and Cosmochemistry. Chem. Geol 70 (1–2), 66.

Buhl, D., Deutsch, A., Lakomy, R., Brockmeyer, P., Dressler, B., 1992. Sudbury Project (University of Muenster-Ontario Geological Survey), 7, Sr-Nd in heterolithic breccias and gabbroic dikes. In: Dressler, B.O., Sharpton, V.L. (Eds.), Papers presented to the International Conference on Large Meteorite Impacts and Planetary Evolution. LPI Contribution, vol. 790, pp. 11–12.

Buhl, D., Deutsch, A., Ostermann, M., 1994. Isotope systematics support the impact origin of the Sudbury Structure (Canada).

Bullen, C.S., 1946a. Geology (of the Sudbury, Ontario, District). The operations and plants of international Nickel Company of Canada, Limited. Can. Mining J. 67 (5), pp. 322–331.

Bullen, C.S., 1946b. Seeing Sudbury (Ontario), a vacation tour. Mineralogist 14 (5), 234–235.

Bunch, T.E., Becker, L., Schultz, P.H., Wolbach, W.S., 1997. New potential sources for Black Onaping carbon (abstract). Large Meteorite Impacts and Planetary Evolution.

Burrows, A.G., Rickaby, H.C., 1930. Sudbury basin area. Ontario Department of Mines, Annual Report for 1929, vol. 38, Part 3, p. 55.

Burrows, A.G., Rickaby, H.C., 1935. Sudbury nickel field restudied. Annual Report. Ontario Department of Mines, Ontario, vol. 43, Part 2, p. 49.

Burwasser, G.J., 1979. Quaternary geology of the Sudbury Basin area, District of Sudbury, Ontario. Ontario Geological Survey Report, vol. 181, 103 pp., including Map 2397, Scale 150,000, 2 charts.

Bush, E.R., 1894. The Sudbury nickel region. Eng. Mining J., 245–246.

Butler, H.R., 1988. Sudbury as a multi-ringed impact structure. In: Drury, M.J. (Ed.), Scientific Drilling, the Sudbury Structure, Proceedings of a Workshop. Canadian Continental Drilling Program Report, vol. 88-2, p. 14.

Butler, H.R., 1994. Lineament analysis of the Sudbury multi-ring impact structure. In: Dressler, B.O., Grieve, R.A.F., Sharpton, V.L. (Eds.), Papers presented to the International Conference On Large Meteorite Impacts and Planetary Evolution. LPI Contribution. Boulder, Colorado, Geological Society of America, Special Paper 293, pp. 319–329.

Butler, H.R., Spray, J.G., 1999. Tectonics of impact basin formation: the Sudbury example. Geological Association of Canada/Mineralogical Association of Canada, Joint Annual Meeting, Sudbury, ON, Field Trip A4 Guidebook, 28 pp.

Bygnes, L.C., 2011. Emplacement of metabreccia and Cu-PGE-rich sulfide veins along the Whistle offset of the Sudbury impact structure. MSc thesis. Laurentian University, Sudbury, Canada. 131 pp.

Cabri, L.J., 1981. Mineralogy and distribution of the platinum group in mill samples from the Cu-Ni deposits of the Sudbury, Ontario, area. In: McGachie, R.O., Bradley, A.G. (Eds.), Precious Metals. International Precious Metals Institute Conference, vol. 4, pp 23–34.

Cabri, L.J., Laflamme, J.H.G., 1974. Sudburyite: a new palladium-antimony mineral from Sudbury, Ontario. Can. Min. 12, 275–279.

Cabri, L.J., Laflamme, J.H.G., 1976. The mineralogy of the platinum-group elements from some copper–nickel deposits of the Sudbury area, Ontario. Econ. Geol. 71, 1159–1195.

Cabri, L.J., Laflamme, J.H.G., 1984. Mineralogy and distribution of platinum-group elements in mill products from Sudbury. In: Park, W.C., Hausen, D.M., Hagni, R.D. (Eds.), Proceedings of Second International Congress on Applied Mineralogy in the Minerals Industry. American Institute Mining Metallurgy, USA, pp. 911–922.

Cabri, L.J., Harris, D.C., Gait, R.I., 1973. Michenerite (PdBiTe) and Froodite (PdBi2) confirmed from the Sudbury area. Can. Min. 11, 903–912.

Cabri, L.J., Blank, H., El Goresky, A., Laflamme, J.H.G., Nobiling, R., Sizgoric, M.B., Traxel, K., 1984. Quantitative trace element analyses of sulfides from Sudbury and Stillwater by proton microprobe. Can. Mineral. 22, 521–542.

Campbell, I.H., Naldrett, A.J., 1979. The influence of silicate:sulfide ratios on the geochemistry of magmatic sulfides. Econ. Geol. 74, 1503–1506.

Cannon, W.F., Schulz, K.J., Horton, J.W., Kring, D.A., 2010. The Sudbury impact layer in the Paleoproterozoic iron ranges of northern Michigan, USA. Geol. Soc. Am. Bull. 122, 50–75.

Cantin, R., Walker, R.G., 1972. Was the Sudbury Basin circular during deposition of the Chelmsford Formation? In: Guy-Bray, J.V. (Ed.), New Developments in Sudbury Geology. Geological Association of Canada, Special Paper 10, pp. 93–101.

Capes, P.C., 2001. A petrological investigation of the Copper Cliff embayment structure, Sudbury, Canada. Unpublished MSc Thesis. University of Toronto, Toronto, Ontario. pp. 162.

Capitrant, P.L., Hitzman, M.W., Wood, D., Kelly, N.M., Williams, G., Zimba, M., Kuiper, Y., Jack, D., Stein, H., 2015. Geology of the Enterprise hydrothermal Nickel Deposit, North-Western Province, Zambia. Econ. Geol. 110, 9–38.

Card, K.D., 1965. Geology of Hyman and Drury Townships. Ontario Department of Mines, Geological Report, vol. 34, 38 pp., with map 2055.

Card, K.D., 1968. Geology of Denison-Waters Area. Ontario Geological Survey Report, vol. 60, 61 pp., with map 2119.

Card, K.D., 1978. Metamorphism of Middle Precambrian (Aphebian) rocks of the eastern Southern Province. Metamorphism in the Canadian Shield. Geological Survey of Canada, Paper 78–10, 269–282.

Card, K.D., 1979. Regional geological synthesis. central Superior Province. Current Research, Geological Survey of Canada, Paper 79-1A, pp. 87–90.

Card, K.D., 1989. Geology of the north shore of Lake Huron, Penokean fold belt and Sudbury Structure. In: Hanshaw, P.M. (Ed.), Mineral deposits of North America, vol. 2. Early Proterozoic rocks of the Great Lakes region, pp. 3–19.

Card, K.D., 1990. A review of the Superior Province of the Canadian Shield, a product of Archean accretion. Precambrian Res. 48, 99–156.

Card, K.D., 1994. Geology of the Levack Gneiss Complex, the northern footwall of the Sudbury Structure. Current Research 1994-C. Geological Survey of Canada, pp. 269–278.

Card, K.D., 2009. Superior Province. In: Roussel, D.H., Brown, G.H. (Eds.), A Field Guide to the Geology of Sudbury, Ontario. Ontario Geological Survey Open File Report, vol. 6243, pp. 7–10.

Card, K.D., Hutchinson, R.W., 1972. The Sudbury Structure: its regional geological significance. In: Guy-Bray, J.V. (Ed.), New Developments in Sudbury Geology. Geological Association of Canada, Special Paper 10, pp. 67–78.

Card, K.D., Jackson, S.L., 1995. Tectonics and metallogeny of the Early Proterozoic Huronian fold belt and the Sudbury Structure of the Canadian Shield. Geological Survey of Canada, Open File 3139, Field Trip Guidebook, 55 p.

Card, K.D., Pattison, E.F., 1973. Nipissing Diabase of the Southern Province, Ontario, Geol. Assoc. Can. Spec. Paper 12, pp. 7–30.

Card, K.D., Wodicka, N., 2009. Geology, geochronology, and tectonic history of the Levack Gneiss Complex, the northern footwall of the Sudbury Igneous Complex, Ontario; Geological Survey of Canada, Open File Report with Map 4266.

Card, K.D., Church, W.R., Franklin, J.M., Frarey, M.J., Robertson, J.A., 1972. The Southern Province. Variations in Tectonic Styles in Canada, Geological Association of Canada, Special Paper Number 11, pp. 335–380.

Card, K.D., Innes, D.G., Debicki, R.L., 1977. Stratigraphy, sedimentology and petrology of the Huronian Supergroup in the Sudbury–Espanola area; Ontario Division of Mines, Geoscience Study 16, 99 p.

Card, K.D., Gupta, V.K., McGrath, P.H., Grant, F.S., 1984. The Sudbury Structure, its regional geological and geophysical setting. In: Pye, E.G., Naldrett, A.J., Giblin, P.E. (Eds.), The Geology and Ore Deposits of the Sudbury Structure. Ontario Geological Survey Special Volume 1, pp. 25–43.

Card, K.D., Sanford, B.V., Card, G.M., 1997. Controls on the emplacement of Kimberlites and Alkalic rock-carbonatite complexes in the Canadian Shield and surrounding regions. Explor. Min. Geol. 6, 285–296.

Carignan, R., Nriagu, J.O., 1985. Trace metal deposition and mobility in the sediments of two lakes near Sudbury, Ontario. Geochim. Cosmochim. Acta 49, 1753–1764.

Carter, W.M., Watkinson, D.H., Ames, D.E., Jones, P.C., 2005. Quartz dioritic magmas and Cu-(Ni)-PGE mineralization, Podolsky Deposit, Whistle Offset Structure, Sudbury, Ontario. Geological Survey of Canada, Open File 6134, 2009, 58 pages; 1 CD-ROM.

Chai, G., Eckstrand, R., 1993. Origin of the Sudbury Igneous Complex, Ontario—differentiate of two separate magmas. Current Research, Part E. Geological Survey of Canada, Paper 93-1E, pp. 219–230.

Chai, G., Eckstrand, R., 1994. Rare-earth characteristics and origin of the Sudbury Igneous Complex, Ontario, Canada. Chem. Geol. 113, 221–244.

Chai, G., Eckstrand, R., 1996. Rare Earth Element characteristics of the Sudbury Igneous Complex and its country rocks: new constraints on genesis. Chem. Geol. 120, 303–325.

Chai, G., Eckstrand, R., Grégoire, C., 1993. Platinum Group Element concentrations in the Sudbury rocks, Ontario—an indicator of petrogenesis. Current Research, Part C, Geological Survey of Canada, Paper 93-1C, pp. 287–293.

Chen, Y., 1993. Precious-metal mineralization and sulfide-silicate relationships in some Canadian Ni-Cu sulfide deposits: Thompson Mine, Manitoba; Sudbury, Ontario; Dundonald Beach, Ontario. PhD Thesis. University of Western Ontario, London, 292 pp.

Chen, L.C., Ran, Y.K., Wang, H., Li, Y.B., Ma, X.Q., 2013. The Lushan M$_s$7.0 earthquake and activity of the southern segment of the Longmenshan fault zone. Chinese Sci. Bull. 58, 3475–3482.

Chewaka, S., 1975. The petrochemistry and structure of Elsie Mountain metabasalts, possible connection with some Sudbury sulfide mineralization, Sudbury, Ontario. MSc Thesis. University of Western Ontario. London, ON, Canada. 104 pp.

Choudhry, Abdul, G., 1982. Hart, Ermatinger, and Totten townships, District of Sudbury. In: Wood, J., White, O.L., Barlow, R.B., Colvine, A.C. (Eds.), Summary of Fieldwork, 1982. Ontario Geological Survey Miscellaneous Paper 106, pp. 70–72.

Chute, N.E., 1937. The upper contact of the Sudbury nickel intrusive. PhD Thesis. Harvard University, Cambridge, MA, United States.

Chyi, L., 1972. Distribution of some noble metals in sulphide and oxide minerals in Strathcona Mine, Sudbury. PhD Thesis. McMaster University, Hamilton, ON, Canada. 183 pp.

Chyi, L.L., Crocket, J.H., 1976. Partition of platinum, palladium, iridium and gold among coexisting minerals from the Deep Ore Zone, Strathcona Mine, Sudbury, Ontario. Econ. Geol. 71, 1196–1205.

Ciborowski, T., 2013. The geochemistry and petrogenesis of the early proterozoic matachewan large igneous province. PhD Thesis. University of Cardiff, Wales. 481 pp.

CIM, 2010. CIM Definition of Standards for mineral resources and reserves. Available from: http://web.cim.org/standards/MenuPage.cfm?sections=177,181&menu=229.

Cintala, M.J., Grieve, R.A.F., 1992a. Melt production in large-scale impact events: planetary observations and implications (abstract). In: Large Meteorite Impacts and Planetary Evolution, Lunar and Planetary Institute, Contribution 790, p. 13.

Cintala, M.J., Grieve, R.A.F., 1992b. Melt production in large-scale impact events; calculations of impact-melt volumes and crater scaling (abstract). In: Large Meteorite Impacts and Planteary Evolution, Liunar and Planetary Institute, Contribution No. 790, p. 14.

Cintala, M.J., Grieve, R.A.F., 1994. The effects of differential scaling of impact melt and crater dimensions on lunar and terrestrial craters, some brief examples. In: Dressler, B.O., Grieve, R.A.F., Sharpton, V.L. (Eds.), Papers presented to the International Conference on Large Meteorite Impacts and Planetary Evolution. LPI Contribution. Geological Society of America Special Paper 293, pp. 51–59.

Clark, J.F., 1983. Magnetic survey data at meteoritic impact sites in North America. Geomagnetic Service of Canada, Earth Physics Branch Open File # 83-5, 30 p.

Clark, B.R., Kelly, W.C., 1973. Sulfide deformation studies: I. Experimental deformation of pyrrhotite and sphalerite to 2,000 Bars and 500°C. Econ. Geol. 68, 332–352.

Clark, M.D., Riller, U., Morris, W.A., 2012. Upper-crustal, basement-involved folding in the East Range of the Sudbury Basin, Ontario, inferred from paleomagnetic data and spatial analysis of mafic dykes. Can. J. Earth Sci. 49, 1005–1017.

Clarke, W.J.G., 1940. Some comparisons between the Sudbury Basin and Bushveld Igneous Complex. MSc Thesis. University of Toronto, Toronto, ON, Canada.

Clarke, A.M., Potapoff, P., 1959a. Geology of McKim Mine. Can. Mining J. 80, 82.

Clarke, A.M., Potapoff, P., 1959b. Geology of the McKim Mine. Proc. G.A.C. 11, 67–80.

Clauson, V., 1947. Geology of the Sudbury Basin area, Ontario, Canada. PhD Thesis. University of Washington, Seattle, WA, United States, 135 pp.

Clendenen, W., Kligfield, R., Hirt, A., Lowrie, W., 1988. Strain studies of cleavage development in the Chelmsford Formation, Sudbury Basin, Ontario. Tectonophysics 145, 191–211.

Clifford, P.M., 1990. Mid-Proterozoic deformational and intrusive events along the Grenville Front in the Sudbury-Killarney area, Ontario, and their implications. In: Gower, C.F., Rivers, T., Ryan, B. (Eds.), Mid-Proterozoic Laurentia-Baltica. Special Paper—Geological Association of Canada, vol. 38, pp. 335–350.

Clow, G.G., Routledge, R.E., Reddick, J.R., Cox, J.J., 2005. NI 43-101 Technical report on Lockerby Mine, Sudbury Ontario, Prepared for First Nickel Inc. 122 pp. Available form: http://www.infomine.com/index/pr/Pa281942.PDF.

Coats, C.J.A., Snajdr, P., 1984. Ore deposits of the North Range, Onaping-Levack area, Sudbury. In: Pye, E.G., Naldrett, A.J., Giblin, P.E. (Eds.), The Geology and Ore Deposits of the Sudbury Structure. Ontario Geological Survey Special Volume 1, pp. 327–346.

Cochrane, L.B., 1982. Geochemical relations in the South Range of the Sudbury Nickel Irruptive. Unpublished MSc Thesis. Laurentian University, Sudbury, Ontario, 90 pp.

Cochrane, L.B., 1983. The Creighton Fault. Report for Science North, Sudbury, 51 pp.

Cochrane, L.B., 1984. Ore deposits of the Copper Cliff offset. In: Pye, E.G., Naldrett, A.J., Giblin, P.E. (Eds.), The Geology and Ore Deposits of the Sudbury Structure. Ontario Geological Survey Special Volume 1, pp. 347–359.

Cochrane, L.B., 1991. Analysis of the structural and tectonic environments associated with rock-mass failures in the mines of the Sudbury District. PhD Thesis. Queen's University, Kingston, ON, Canada. 284 pp.

Cohen, A.S., Burnham, O.M., Hawkesworth, C.J., Lightfoot, P.C., 2000. Pre-emplacement Re-Os ages from ultramafic inclusions in the Sublayer of the Sudbury Igneous Complex, Ontario. Chem. Geol. 165, 37–46.

Coker, W.B., Dunn, Colin, E., Hall, G.E.M., Rencz, A.N., DiLabio, R.N.W., Spirito, Wendy, A., Campbell, J.E., 1991. The behaviour of platinum group elements in the surficial environment at Ferguson Lake, N.W.T., Rottenstone Lake, Sask. and Sudbury, Ontario, Canada. In: Rose, A.W., Taufen, P.M. (Eds.), Geochemical exploration 1989, Part 1, Selected Papers From the 13th International Geochemical Exploration Symposium. Journal of Geochemical Exploration, vol. 40, pp. 1–3, pp. 165–192.

Coleman, A.P., 1893. The rocks of Clear Lake near Sudbury (Ontario). Can. Rec. Sci. 5, 343–346.

Coleman, A.P., 1903. The Sudbury nickel deposits. Ontario Geological Survey Report, pp. 235–299.

Coleman, A.P., 1904. Sudbury nickel-bearing eruptive. Geol. Soc. Am. 15, 551.

Coleman, A.P., 1905a. Geology of the Sudbury District. Eng. Min. J. 79, 189–190.

Coleman, A.P., 1905b. The Sudbury Nickel Field. Ontario Bureau of Mines, Annual Report, 5, part 3.

Coleman, A.P., 1905c. The Sudbury Nickel Region. Ontario Bureau of Mines, Annual Report, 14, part 3, 183 pp.

Coleman, A.P., 1907a. Die Sudbury-Ni ckelerze. Zeitschrift fuer Praktische Geologie 15, 221.

Coleman, A.P., 1907b. The Sudbury laccolithic sheet. J. Geol. 15, 759–782.

Coleman, A.P., 1908. The Sudbury nickel ores. Geol. Mag. 5, 18–19.

Coleman, A.P., 1912. Summary report on the Sudbury nickel field. Can, Mines Br, Sum Rp 1911, 87–89.

Coleman, A.P., 1913a. Classification of the Sudbury ore deposits. Trans. Can. Min. Inst., 283–288.

Coleman, A.P., 1913b. Sudbury to Cartier, annotated guide. Rep. Int. Geol. Cong., 13–14.

Coleman, A.P., 1913c. The Nickel Industry, with special reference to the Sudbury region, Ontario. Canada Department of Mines, Mines Branch, vol. 170, p. 206.

Coleman, A.P., 1913d. The Sudbury area (Ontario). Rep. Int. Geol. Cong. 7, 11–48.

Coleman, A.P., 1914a. The pre-Cambrian rocks north of Lake Huron with special reference to the Sudbury series. Annual Report—Ontario Department of Mines, pp. 204–236.

Coleman, A.P., 1914b. The Sudbury series and its bearing on pre-Cambrian classification. Rep. Int. Geol. Cong. XII, 387–398.

Coleman, A.P., 1915. The origin of the Sudbury nickel deposits. Econ. Geol. 10, 390–393.

Coleman, A.P., 1916a. Chief minerals of the Sudbury nickel ores. Can. Min. J. 37, 386–389.

Coleman, A.P., 1916b. Geological relations of the Sudbury nickel ores. Eng. Min. J. 102, 104–105.

Coleman, A.P., 1917. Magmas and sulphide ores (Sudbury, Ontario, deposits). Econ. Geol. 12, 427–434.

Coleman, A.P., 1924. Geology of the Sudbury nickel deposits. Econ. Geol. Bull. Soc. Econ. Geol. 20 (6), 565–576.

Coleman, A.P., 1926a. The magmatic origin of the Sudbury nickel ores. Geol. Mag. 63, 108–112.

Coleman, A.P., 1926b. The Sudbury ore deposits (Ontario). Can. Min. J. 47 (46), 1080–1081.

Coleman, A.P., 1928. The Anthraxolite of Sudbury (Ontario). Am. J. Sci. 15, 25–27.

Coleman, A.P., Moore, E.S., Walker, T.L., 1929. The Sudbury nickel intrusive University of Toronto Studies in Geology Series, vol. 28, pp. 1–54.

Collins, J.H., 1888. On the Sudbury copper deposits. Quart. J. Geol. Soc. Lond. 4, 834–838.

Collins, W.H., 1914. Geology of a portion of Sudbury map area, south of Wanapitel Lake. Ontario Summary Report of the Geological Survey of Canada, pp. 189–195.

Collins, W.H., 1917. Onaping map-area. Department of Mines, Geological Survey of Canada, Memoir 95, No.77, Geological Series, 157 pp.

Collins, W.H., 1930. Southwestern part of Sudbury nickel irruptive. Summary Report of the Geological Survey of Canada, pp. 12–16.

Collins, W.H., 1935. Life history of the Sudbury nickel irruptive. II Intrusion and deformation, vol. 29, pp. 27–47.

Collins, W.H., 1936a. Life history of the Sudbury nickel irruptive. III Environment. Royal Society of Canada Transactions, 3rd series, vol. 30, pp. 29–53.

Collins, W.H., 1936b. Sudbury series. Geol. Soc. Am. Bull. 47 (11), 1675–1690.

Collins, W.H., 1937a. Life history of the Sudbury nickel irruptive. IV Mineralization. Royal Society of Canada Transactions, 3rd series, vol. 31, pp. 15–43.

Collins, W.H., 1937b. Life history of the Sudbury nickel irruptive. II Intrusion and deformation. Royal Society of Canada Transactions, 3rd series, vol. 29, pp. 27–47.

Collins, W.H., Camsell, C., 1913. Sudbury, Ontario, to Dunmore, Alberta. Rep. Int. Geol. Cong., 11–15.

Collins, Z.J.E., Barnes, S.J., Hagemann, S.G., Campbell, T., McCuag, T.C., Frost, K.M., 2012a. Postmagmatic variability in ore composition and mineralogy in the T4 and T5 ore shoots at the high-grade Flying Fox Ni-Cu-PGE Deposit, Yilgarn Craton, Western Australia. Econ. Geol. 107, 859–879.

Collins, J.E., Hagemann, S.G., McQuaig, T.C., Frost, K.M., 2012b. Structural controls on sulfide mobilization at the high-grade Flying Fox Ni-Cu-PGE sulfide deposit, Forrestania Greenstone Belt, Western Australia. Econ. Geol. 107, 859–880.

Colony, R.J., 1920. A norite of the Sudbury type in Manitoba, a reconnaissance. Bull. Can. Inst. Min. Metallurgy 13 (24), 862–872.

Conard, B.R., Sridhar, R., Warner, J.S., 2007. High-temperature thermodynamic properties of chalcopyrite. J. Chem. Thermodyn. 12, 817–833.

Connors, S.-T., 1967. Sudbury Saturday Night. The Northlands' own Tom Connors.

Conrod, D.M., 1988. Petrology, geochemistry, and PGE potential of the Nipissing intrusions. Unpublished MSc Thesis. University of Toronto, Toronto, Ontario, 494 pages.

Conrod, D.M., 1989. The petrology and geochemistry of the Ducan Lake, Beaton Bay, Milner Lake and Miller Lake Nipissing intrusions within the Gowganda area, District of Timiskiming; Ontario Geological Survey, Open File Report 5701, 210 p.

Conroy, N., Kramer, J.R., 1995. History of geology, mineral exploration and environmental damage. In: Gunn, J.M. (Ed.), Resortation and Recovery of an Industrial Region. Springer-Verlag, New York, 10013.

Cooke, H.C., 1939. New interpretation of the geology of Sudbury District, Ontario. R. Soc. Can. Proc. 33, 198.

Cooke, H.C., 1941. New Pre-cambrian correlations indicated from recent work at Sudbury, Ontario. Trans. R. Soc. Can. 35 (35), 1–15.

Cooke, H.C., 1943. The older rocks of Sudbury District, Ontario. Am. J. Sci. 241 (9), 553–578.

Cooke, H.C., 1946. Problems of Sudbury geology. Geol. Surv. Can. Bull. 3, 77.

Cooke, H.C., 1948. Regional structure of the Lake Huron-Sudbury area (Ontario). Canadian Inst. Mining and Metallurgy, Geol. Di, Structural geology of Canadian ore deposits, pp. 580–589.

Cook, N.J., Ciobanu, C.L., Wagner, T., Stanley, C.J., 2007. Minerals of the system Ni-Te-Se-S related to the tetradymite archetype: review of classification and compositional variations. Can. Min. 45, 665–708.

Cooper, R.P.R., 1976. Ni-As mineralization in ores of the Garson Mine. BSc Thesis. University of Western Ontario, London, Ontario, 32 pages.

Cooper, M., 2000. The Sudbury Igneous Complex: Insights into Melt Sheet Evolution and Ore Genesis. PhD thesis, Open University, UK. 260 pages.

Coppard, J., 2014. Discovery history of the Sakatti Cu-Ni-PGE deposit, Finland. Available from: https://www.segweb.org/SEG/_Events/Conference_Website_Archives/2014/Abstracts/data/papers/abstracts/0393-000039.pdf.

Coppard, J., 2015. The Sakatti Magmatic Cu-Ni-PGE Deposit. Prospectors and Developers Association of Canada, Toronto, Lapland, Finland. Available from: http://www.pdac.ca/docs/default-source/Convention—Program—Technical-Sessions/2015/nickel—coppard.pdf?sfvrsn=2.

Corfu, F., Andrews, A.J., 1986. A U-Pb age for mineralized Nipissing diabase. Gowganda, Ontario. Can. J. Earth Sci. 27, 107–109.

Corfu, F., Easton, R.M., 2001. U-Pb evidence for polymetamorphic history of Huronian rocks underlying the Grenville Front tectonic zone east of Sudbury, Ontario. Chem. Geol. 172, 149–171.

Corfu, F., Lightfoot, P.C., 1997. U-Pb geochronology of the Sublayer environment, Sudbury Igneous Complex, Ontario. Econ. Geol. 91, 1263–1269.

Corless, C.V., 1916. Origin of Sudbury nickel-copper deposits. Eng. Min. J., 517–518.

Corless, C.V., 1917. On the origin of Sudbury nickel deposits. Can. Min. J., 268–269.

Corless, C.V., 1929. The Frood ore deposit (Sudbury, Ontario), a suggestion as to its origin. Canadian Min. and Met. Bull. 32, 140–150.

Corlett, M., 1971. Minor element variation in pyrrhotite from Falconbridge Mine; an aid in geological interpretation. Geol Ass. Can-Min Ass Can Joint Annual Meeting, Abstracts, pp. 15–16.

Courtillot, V., 1994. Mass extinctions in the last 300 million years: one impact and seven flood basalts? Israel J. Earth Sci. 43, 255–266.

Coveney, R., Paiva, J., 2005. Origins of Au-Pt-Pd-bearing Ni-Mo-As-(Zn) deposits hosted by Chinese black shales. Mineral Deposit Research, Meeting the Global Challenge, Springer, pp. 101–102. Available from: http://rd.springer.com/chapter/10.1007%2F3-540-27946-6_26.

Cowan, J.C., 1968. The Geology of the Strathcona ore deposit. Can. Inst. Min. Metal. Bull. 61, 38–54.

Cowan, J., 1996. Deformation of the eastern Sudbury Basin. PhD Thesis. University of Toronto, Toronto, ON, Canada.

Cowan, E.J., Schwerdtner, W.M., 1992. Intrusive origin of the Sudbury Igneous Complex, structural and sedimentological evidence. In: Dressler, B.O., Sharpton, V.L. (Eds.), Papers presented to the international conference on large meteorite impacts and planetary evolution, LPI Contribution 790, pp. 17–18.

Cowan, E.J., Schwerdtner, W.M., 1994. Fold origin of the Sudbury Basin. In: Lightfoot, P.C., Naldrett, A.J. (Eds.), Proceedings of the Sudbury-Noril'sk Symposium. Ontario Geological Survey Special Volume 5, pp. 45–55.

Cowan, E.J., Shanks, W.S., Schwerdtner, W.M., 1992. Geometrical significance of the geological map pattern of the Sudbury structure (abstract). The Canadian Mineralogist. J. Meteor. Assoc. Can. 30, 486–487.

Cowen, E.J., Riller, U., Schwerdtner., W.M., 1999. Emplacement Geometry of the Sudbury Igneous Complex, in Large meteorite impacts and planetary evolution II. In: Dressler, B.O., Sharpton, V.L. (Eds.), Geological Society of America Special Paper 339, pp. 399–418.

Cox, K.G., 1980. A model for flood basalt volcanism. J. Petrol. 21, 629–650.

Cox, K.G., Bell, J.D., Pankhurst, R.J., 1979. The Interpretation of Igneous Rocks. Allen and Unwin, London, 450 pp.

Craig, J.R., 1983. Metamorphic features of in Appalachian massive sulphides. Mineral. Mag. 47, 515–525.

Craig, J.R., Solberg, T.N., 1999. Compositional zoning in ore minerals at the Craig mine, Sudbury, Ontario, Canada. Can. Mineral. 37, 1163–1176.

Craig, J.R., Vaughan, D.J., 1981. Ore Microscopy and Ore Petrography. John Wiley and Sons, New York, 406 pp.

Crocket, J.H., Kabir, A., 1981. Geochemical pathway studies of heavy metals. Lake sediments from the Sudbury-Temagami area, Ontario. J. Great Lakes Res. 7 (4), 455–466.

Crocket, J.H., Dickin, A.P., McNutt, R.H., Richardson, J.M., Benetau, S.B., Frape, S.K., 1988. Isotopic studies at Sudbury. In: Drury, M.J. (Ed.), Scientific Drilling, the Sudbury Structure, proceedings of a workshop. Canadian Continental Drilling Program Report, pp. 88–2 22.

Crone, D., Watts, T., 1992. Case history of borehole pulse EM surveys at the Falconbridge Lindsley discovery. In: Sudbury. CIM Bulletin (1974), vol. 85, pt. 957, pp. 45–49.

Cruden, A.R., Evans-Lambswood, D.E., Burrows, D., 2008. Structural, tectonic and fluid mechanical controls on emplacement of the Voisey's Bay troctolite and its Ni–Cu–Co mineralization. GAC Program Abstracts 33, 59.

Cundiff, W.E., 1979. Will Sudbury survive the impact of ocean mining? Can. Min. J. 100 (5), 55–67.

Cunningham, D.P.K., 1954. Structural geology of Ontario pyrites deposits, Sudbury, Ontario. MSc Thesis. University of Toronto, Toronto, ON, Canada.

Cunningham, W.D., Mann, P., 2007. Tectonics of strike-slip restraining and releasing bends. Geol. Soc. Lond. 290, 1–12, Special Publication.

Cupelli, C.L., Moeser, D.E., Barker, I.R., Darling, J.R., Bowman, J.R., Dhuime, B., 2014. Discovery of mafic impact melt in the center of the Vredefort dome: archetype for continental residua of early Earth cratering? Geology 42, 403–406.

Cutifani, M., 2005. Two proud histories, one great future: a new and stronger Inco Presentation to Sudbury City. Available from: http://www.greatersudbury.ca/content/div_councilagendas/documents/PresidentsBreakfastslides-Oct-05.pdf.

Cuvier, G.B., 1813. Essay on the Theory of the Earth (R. Kerr, Trans.). With Mineralogical Notes, and an Account of Cuvier's Geological Discoveries, by Professor Jameson. William Blackwood Publishing, Edinburgh. 13 pp.

Czamanske, G.K., 2002. Petrographic and geochemical characterization of ore-bearing intrusions of the Noril'sk type, Siberia: with discussion of their origin, including additional datasets and core logs. USGS Open File Report 02-74.

Dallmeyer, R.D., Taylor, W.E.G., 1973. Photogrammetric survey of the structural geology of the Sudbury-McGregor Bay district, Ontario, Canada. Geologische Rundschau 62 (2), 350–356.

Dalvi, A.D., Bacon, W.G., Osborne, R.C., 2004. The past and the future of nickel laterites, in PDAC 2004 International Conference Trade Show and Investors Exchange, Toronto, Canada, March 7–10, 2004. Proceedings: Toronto, Canada, Prospectors and Developers Association of Canada, 27 p.

Dare, S.A.S., Barnes, S.-J., Prichard, H.M., Fisher, P.C., 2010a. The timing and formation of platinum-group minerals from the Creighton Ni-Cu-platinum-group element sulfide deposit, Sudbury, Canada: early crystallization of PGE-rich sulfarsenides. Econ. Geol. 105, 1071–1096.

Dare, S.A.S., Barnes, S.-J., Prichard, H.M., 2010b. The distribution of platinum group elements (PGE) and other chalcophile elements among sulfides from Creighton Ni-Cu-PGE sulfide deposit, Sudbury, Canada, and the origin of palladium in pentlandite. Min. Dep. 45, 765–793.

Dare, S.A.S., Barnes, S.-J., Prichard, H.M., Fisher, P.C., 2011. Chalcophile and platinum-group element (PGE) concentrations in sulfide minerals from the McCreedy East deposit, Sudbury, Canada, and the origin of PGE in pyrite. Min. Dep. 46, 381–407.

Dare, S.A.S., Barnes, S.-J., Beaudoin, G., 2012. Variation in trace element content of magnetite crystallized from a fractionating sulfide liquid, Sudbury, Canada: implications for provenance discrimination. Geochim. Cosmochim. Acta 88, 27–50.

Dare, S.A.S., Barnes, S.-J., Prichard, H.M., Fisher, P.C., 2014. Mineralogy and geochemistry of Cu-rich ores from the McCreedy East Ni-Cu-PGE deposit (Sudbury, Canada): implications for the behavior of platinum group and chalcophile elements at the end of crystallization of a sulfide liquid. Econ. Geol. 109, 343–366.

Darling, J.R., Hawkesworth, C.J., Lightfoot, P.C., Storey, C.D., Tremblay, E., 2010a. Isotopic heterogeneity in the Sudbury impact melt sheet. Earth Planet. Sci. Lett. 289 (3–4), 347–356.

Darling, J.R., Hawkesworth, C.J., Lightfoot, P.C., Storey, C.D., 2010b. Shallow Impact: isotropic insights into crustal contributions to the Sudbury impact melt sheet. Geochim. Cosmochim. Acta 74, 5680–5696.

Darling, J.R., Storey, C.D., Hawkesworth, C.J., Lightfoot, P.C., 2012. In-situ Pb isotope analysis of Fe-Ni-Cu-Zn sulphides by laser ablation multi-collector ICPMS: new insights into ore formation in the Sudbury impact melt sheet. Geochim. Cosmochim. Acta 99, 1–17.

Dasti, I.R., 2014. The geochemistry and petrogenesis of the Ni-Cu-PGE Shakespeare Deposit, Ontario, Canada. MSc Thesis. Lakehead University, Thunder Bay, Canada.

Davidson, A., 1984. Relationship between faults in the Southern province and the Grenville front south-east of Sudbury, Ontario., vol. 92-1C, pp. 121–127.

Davidson, A., Ketchum, J.W.F., 1993. Grenville Front studies in the Sudbury region, Ontario. Current Research, Part C. Geological Survey of Canada, Paper 93-1 C, pp. 271–278.

Davidson, A., van Breemen, O., 1994. U-Pb ages of granites near the Grenville Front, Ontario; in Radiogenic Age and Isotopic Studies, Report 8, Geological Survey of Canada, Current Research 1994-F, pp. 107–114.

Davies, J.F., Leroux, M.V., Whitehead, R.E., Goodfellow, W.D., 1990. Oxygen isotope composition and temperature of fluids involved in deposition of Proterozoic sedex deposits, Sudbury Basin, Canada. Can. J. Earth Sci. 27, 1299–1303.

Davis, G.C., 1984. Little Stobie Mine: a South Range contact deposit. In: Pye, E.G., Naldrett, A.J., Giblin, P.E. (Eds.), The Geology and Ore Deposits of the Sudbury Structure. Ontario Geological Survey Special Volume 13P, pp. 361–369.

Davis, D.W., 2008. Sub-million-year age resolution of Precambrian igneous events by thermal extractionthermal ionization mass spectrometer Pb dating of zircon: application to crystallization of the Sudbury impact melt sheet. Geology 36, 383–396.

de Bremond d'Ars, J., Arndt, N., Hallot, E., 2001. Analog experimental insights into the formation of magmatic sulfide deposits. Earth Planet. Sci. Lett. 186, 371–381.

de la Vergne, J., 2008. Hard Rock Miner's Handbook, fifth ed. Stantec Consulting Ltd, North Bay, Canada, 314 p.

Dence, M.R., 1964. A comparative structural and petrographic study of probable Canadian meteoritic craters. Meteoritics 2, 249–270.

Dence, M.R., 1968. Shock zoning at Canadian craters: petrography and structural implications. In: French, B.M., Short, N.M. (Eds.), Shock Metamorphism of Natural Materials. Mono Book Corporation, Baltimore, pp. 169–184.

Dence, M.R., 1972. Meteorite impact craters and the structure of the Sudbury Basin. In: Guy-Bray, J.V. (Ed.), New Developments in Sudbury Geology. Geological Association of Canada, Special Paper 10, pp. 7–18.

Dence, M.R., Popelar, J., 1972. Evidence for an impact origin for Lake Wanapitei, Ontario. In: Guy-Bray, J.V. (Ed.), New Developments in Sudbury Geology. Geological Association of Canada, Special Paper 10, pp. 117–124.

Dentith, M., Mudge, S.T., 2014. Geophysics for the Mineral Exploration Geoscientist. Cambridge University Press, Cambridge, 454 pp.

Desborough, G.A., Larson, R.R., 1970. Nickel-bearing iron sulfides in the Onaping Formation, Sudbury Basin, Ontario. Econ. Geol. 65, 728–730.

Deutsch, A., 1994. Isotope systematics support the impact origin of the Sudbury Structure (Ontario, Canada). In: Dressler, B.O., Grieve, R.A.F., Sharpton, V.L. (Eds.), Papers Presented to the International Conference on Large Meteorite Impacts and Planetary Evolution. LPI Contribution. Geological Society of America, Special Paper 293, pp. 289–302.

Deutsch, A., Grieve, R.A.F., 1994. The Sudbury Structure, constraints on its genesis from Lithoprobe results. In: Boerner, D.E., Milkereit, B., Nadrett, A.J. (Eds.), Lithoprobe Sudbury Project. Geophysical Research Letters, vol. 21, part 10, pp. 963–966.

Deutsch, A., Buhl, D., Lakomy, R., 1988. A small scale Sr-Nd study of the footwall breccia (Sudbury, Canada), a case study for isotope systematics of polymict granulitic breccias. Abstracts of papers submitted to the Nineteenth Lunar and Planetary Science Conference. Abstracts of Papers Submitted to the Lunar and Planetary Science Conference, vol. 19, Part 1, pp. 275–276.

Deutsch, A., Lakomy, R., Buhl, D., 1989. Strontium- and neodymium-isotopic characteristics of a heterolithic breccia in the basement of the Sudbury impact structure, Canada. Earth Planet. Sci. Lett. 93, 359–370.

Deutsch, A., Brockmeyer, P., Buhl, D., 1990. Sudbury again, new and old isotope data. Abstracts of Papers Submitted to the Twenty-First Lunar and Planetary Science Conference, vol. 21, pp. 282–283.

Deutsch, A., Buhl, D., Brockmeyer, P., Lakomy, R., Flucks, M., 1992. Sudbury Project (University of Muenster-Ontario Geological Survey), 4, Isotope systematics support the impact origin. In: Dressler, B.O., Sharpton, V.L. (Eds.), Papers Presented to the International Conference on Large Meteorite Impacts and Planetary Evolution. LPI Contribution 790, pp. 21–22.

Deutsch, A., Ostermann, M., Schärer, U., Agrinier, P., 1995a. On the formation of impact melt rocks: a geochemical case study at Foy offset dike, Sudbury impact structure (Canada) (abstract). IV International Conference on Advanced Material, Symposium 19, Cancún, Mexico.

Deutsch, A., Grieve, R.A.F., Avermann, M., Bischoff, L., Brockmeyer, P., Buhl, D., Lakomy, R., Müller-Mohr, V., Ostermann, M., Stöffler, D., 1995b. The Sudbury Structure (Ontario, Canada): a tectonically deformed multi-ring basin. Geologische Rundschau 84, 697–709.

Dhingra, D., Pieters, C.M., 2013. Mineralogy of impact melt at Copernicus Crater: insights into melt evolution and diversity. Large Meteorite Impacts and Planetary Evolution, vol. 5, p. 3036. Available from: http://www.hou.usra.edu/meetings/sudbury2013/pdf/3036.pdf.

Dhingra, D., Pieters, C.M., Head, J.W., Isaacson, P.J., 2013. Large mineralogically distinct impact melt feature at Copernicus crater—evidence for retention of compositional heterogeneity. Geophys. Res. Lett. 10, 1–6.

Dickin, A.P., 1991. The role of the crust as a source of PGE in the Sudbury nickel deposit, Re/Os isotope evidence. In: Milne, V.G. (Ed.), Geoscience Research Grant Program, summary of research 1990–1991. Ontario Geological Survey Miscellaneous Paper 156, pp. 99–104.

Dickin, A.P., Crocket, J.H., 1997. Sudbury: geochemical evidence for a single-impact melting event (abstract). Large Meteorite Impacts and Planetary Evolution.

Dickin, A.P., Nguyen, T., Crocket, J.H., 1999. Isotopic evidence for a single impact melting origin of the Sudbury Igneous Complex, in Large meteorite impacts and planetary evolution II. Dressler, B.O., Sharpton, V.L. (Eds.), Geological Society of America Special Paper 339, p. 361–371.

Dickin, A.P., Richardson, J.M., Crocket, J.H., McNutt, R.H., Peredery, W.V., 1992. Osmium isotope evidence for a crustal origin of platinum group elements in the Sudbury nickel ore, Ontario, Canada. Geochim. Cosmochim. Acta 56, 3531–3537.

Dickin, A.P., Artan, M.A., Crocket, J.H., 1996. Isotopic evidence for distinct crustal sources of North and South Range ores, Sudbury Igneous Complex. Geochim. Cosmochim. Acta 60 (9), 1605–1613.

Dickman, M., Agbeti, M.D., Fortescue, J., 1987. Sudbury Lake deacidification, fact or fiction? In: international Association for Great Lakes Research, 30th Conference on Great Lakes Research, Program and Abstracts. Evans, M.S., Eadie, B.J. IAGLR Program, 30, A-15.

Dickson, C.W., 1902. Note on the condition of nickel in nickeliferous pyrrhotite from Sudbury (Ontario). Eng. Min. J. 73, 660.

Dickson, C.W., 1903. Note of the condition of platinum in the nickel-copper ores from Sudbury. Am. J. Sci. 15, 137–139.

Dickson, C.W., 1904. The ore deposits of Sudbury, Ontario Transactions of the Society of Mining Engineers of American Institute of Mining, Metallurgical and Petroleum Engineers, Incorporated (AIME), pp. 3–67.

Dickson, C.W., 1913. The ore deposits of Sudbury. In: Emmons, S.F. (Ed.), Ore Deposits, pp. 455–516.

Dietz, R.S., 1959. Shatter cones in cryptoexplosion structures (meteorite impact?). J. Geol. 67, 496–505.

Dietz, R.S., 1962. Sudbury Structure as an astrobleme. Trans. Am. Geophys. Union 43, 445–446.

Dietz, R.S., 1964. Sudbury Structure as an astrobleme. J. Geol. 72, 412–434.

Dietz, R.S., 1971. Sudbury astrobleme: a review (abstract). Meteoritics 6, 259–260.

Dietz, R.S., 1972. Sudbury astrobleme, splash emplaced sub-layer and possible cosmogenic ores. In: Guy-Bray, J.V. (Ed.), New Developments in Sudbury Geology. Geological Association of Canada, Special Paper 10, pp. 29–40.

Dietz, R.S., 1988. Sudbury Structure as an astrobleme, updated. In: Drury, M.J. (Ed.), Scientific drilling, the Sudbury Structure, proceedings of a workshop. Canadian Continental Drilling Program Report, pp. 88–2 13.

Dietz, R.S., Butler, L., 1964. Orientation of shatter cones at Sudbury, Canada. Nature 204, 280–281.

Dietz, R.S., McHone, J.F., 1992. Shatter coning in astroblemes defines event sequence, Sudbury example. Geological Society of America, 1992 annual meeting. Abstracts with Programs—Geological Society of America, 24 (7), 196.

Ding, T.P., Schwarcz, H.P., 1984. Oxygen isotopic and chemical compositions of rocks of the Sudbury Basin, Ontario. Can. J. Earth Sci. 21, 305–318.

Distler, V.V., 1994. Platinum mineralization of the Noril'sk Deposits. In: Lightfoot, P.C., Naldrett, A.J. (Eds.), Proceedings of the Sudbury-Noril'sk Symposium. Ontario Geological Survey, vol. 5, pp. 243–262.

Distler, V.V., Genkin, A.D., Filimonova, A.A., and others, 1975. Zoning of copper-nickel ores of Talnakh and Oktyabrskoe deposits. Geologiya rud. Mestorzhdenii 2, 16–27.

Donofrio, R.R., 1997. Survey of hydrocarbon-producing impact structures in North America: exploration results to date and potential for discovery in Precambrian basement rock. In: Johnson, K.S., Campbell, J.A. (Eds.), Ames Structure in

Northwest Oklahoma and Similar Features: Origin and Petroleum Production (1995 Symposium). Oklahoma Geological Survey Circular 100, pp. 17–29.

Drake, A., 1992. The geology alteration and mineralization of the Simmons Lake Pb-Zn Showing, Sudbury Basin, Ontario. Unpublished BSc Thesis. Carleton University, Ottawa, Ontario, 63 pp.

Dresser, M.A., 1917. Some quantitative measurements of minerals of the nickel eruptive at Sudbury (Ontario). Econ. Geol. 12, 563–580.

Dressler, B., 1979. Footwall of the Sudbury Irruptive. In: Milne, V.G., White, O.L., Barlow, R.B., Kustra, C.R. (Eds.), Summary of field work, 1979. by the Ontario Geological Survey. Ontario Geological Survey Miscellaneous Paper, 90, pp. 109–111.

Dressler, B.O., 1981. Footwall of the Sudbury Irruptive, District of Sudbury. In: Wood, J., White, O.L., Barlow, R.B., Colvine, A.C. (Eds.), Summary of field work, 1981. Ontario Geological Survey Miscellaneous Paper 100, pp. 84–87.

Dressler, B.O., 1982a. Geology of the Wanapitei Lake area, District of Sudbury. Ontario Geol. Surv. Rep. 213, 131.

Dressler, B.O., 1982b. Footwall of the Sudbury Igneous Complex, District of Sudbury. In: Wood, J., White, O.L., Barlow, R.B., Colvine, A.C. (Eds.), Summary of field work, 1982. Ontario Geological Survey Miscellaneous Paper 106, pp. 73–75.

Dressler, B.O., 1983. Breccias in the footwall of the Sudbury impact structure, terrestrial equivalents of lunar breccias? Abstracts of Papers Presented to the Fourteenth Lunar and Planetary Science Conference, 14, pp. 167–168.

Dressler, B.O., 1984a. Sudbury Geological Compilation; Ontario Geological Survey Map 2491, Precambrian Geology Series, scale 150,000 geological compilation 1982–1983.

Dressler, B.O., 1984b. General geology of the Sudbury area. In: Pye, E.G., Naldrett, A.J., Giblin, P.E. (Eds.), The Geology and Ore Deposits of the Sudbury Structure. Ontario Geological Survey Special Volume 1, pp. 57–82.

Dressler, B.O., 1984c. The effects of the Sudbury Event and the intrusion of the Sudbury Igneous Complex on the footwall rocks of the Sudbury Structure. In: Pye, E.G., Naldrett, A.J., Giblin, P.E. (Eds.), The Geology and Ore Deposits of the Sudbury Structure. Ontario Geological Survey Special Volume 1, pp. 97–136.

Dressler, B.O., 1986. Falconbridge Township, District of Sudbury. In: Thurston, P.C., White, O.L., Barlow, R.B., Cherry, M.E., Colvine, A.C. (Eds.), Summary of field work and other activities, 1986. Ontario Geological Survey Miscellaneous Paper 132, pp. 127–130.

Dressler, B., 1988. The Sudbury Structure, a review. In: Drury, M.J. (Ed.), Scientific Drilling, the Sudbury Structure. Proceedings of a workshop. Canadian Continental Drilling Program Report, vol. 88-2, p. 8.

Dressler, B.O., 1988. The Sudbury structure—a review (abstract). Canadian Continental Drilling Program Report 88-2, Scientific Drilling: The Sudbury Structure, p. 8.

Dressler, B., 1990. Gabbroic rocks of the Sudbury Basin and their possible economic significance. Ontario Geol. Surv. Open File Rep. 5732, 13.

Dressler, B.O., Peredery, W.V., 1984. The Sudbury Breccia, its distribution, nature and origin. GAC/AGC-MAC/AMC Joint Annual Meeting (abstract), vol. 9, p. 58.

Dressler, B.O., Reimold, W.U., 1988. The Sudbury structure (Ontario, Canada) and Vredefort structure (South Africa), a comparison. Global catastrophes in Earth history, an interdisciplinary conference on impacts, volcanism, and mass mortality. LPI Contribution 673, pp. 42–43.

Dressler, B.O., Sharpton, V.L., 1997. The Sudbury structure, Ontario, Canada: a persistent enigma (abstract). Large Meteorite Impacts and Planetary Evolution.

Dressler, B.O., Sharpton, V.L., 1998. Comment on "Isotopic evidence for distinct crustal sources of North and South Range ores, Sudbury Igneous Complex" by A.P. Dickin, M.A. Arten, and J.H. Crocket. Geochim. Cosmochim. Acta 62, 315–317.

Dressler, B.O., Morrison, G.G., Peredery, W.V., Rao, B.V., 1987. The Sudbury Structure, Ontario, Canada—a review. In: Pohl, J. (Ed.), Research in Terrestrial Impact Structures. Earth Evolution Sciences, Braunschweig, Friedrich Vieweg, pp. 39–68.

Dressler, B.O., Gupta, V.K., Muir, T.L., 1991. The Sudbury structure. In: Thurston, P.C., Williams, H.R., Sutcliffe, R.H., Stott, G.M. (Eds.), Geology of Ontario. Ontario Geological Survey Special Volume 4, pt. 1, pp. 593–625.

Dressler, B.O., Peredery, W.V., Muir, T.L., 1992. Geology and Mineral Deposits of the Sudbury Structure: Ontario Geological Survey Guidebook 8, 33 pp. 19 Figs., including a Generalized Precambrian Geology map that is modified from OGS Map 2491 of OGS Special Volume 1, 1984.

Dressler, B.O., Grieve, R.A.F., Sharpton, V.L. (Eds.), 1994. Papers presented to the international conference on large meteorite impacts and planetary evolution. LPI Contribution Boulder, Colorado, Geological Society of America, Special Paper 293, 348 pp.

Dressler, B.O., Weiser, T., Brockmeyer, P., 1996a. Recrystallized impact glasses of the Onaping Formation and the Sudbury Igneous Complex, Sudbury Structure, Ontario, Canada. Geochim. Cosmochim. Acta 60 (11), 2019–2036.

Dressler, B.O., Weiser, T., Brockmeyer, P., 1996b. Recrystallized glasses and other melts of the Onaping Formation and the origin of the Sudbury Igneous Complex, Ontario, Canada. In: Sears, D.W.G. (Ed.), 59th Annual Meteoritical Society Meeting, abstracts. Meteoritics & Planetary Science, vol. 31, Suppl., p. 40.

Dreuse, R., Doman, D., Santimano, T., Riller, U., 2010. Crater-floor topography and impact melt sheet geometry of the Sudbury impact structure, Canada. Terra Nova 22, 463–469.

Drury, M.J., 1989. The Sudbury Structure, a CCDP workshop report. Geosci. Can. 16 (1), 29–31.

Dudas, F.O., Davidson, A., Bethune, K.M., 1994. Age of the Sudbury diabase dikes and their metamorphism in the Grenville Province, Ontario. Radiogenic age and isotopic studies, Report 8, Geological Survey of Canada Current Research 1994-F., pp. 97–106.

Duke, J.M., 1985. An overview of the Sudbury-Timmins-Algoma Mineral Program (STAMP), Ontario. Current research, Part A. Paper—Geological Survey of Canada, 85-1A, pp. 723–725.

Dupuis, L., 1972. Metallogenic study of the Sudbury Mining Division, Unpublished BSc Thesis. Laurentian University, Sudbury, Ontario.

Dupuis, L., 1979. The nature and origin of Sudbury Breccia near Lake Laurentian. Unpublished MSc Thesis. Laurentian University, Sudbury, Ontario, 149 pp.

Dupuis, L., Whitehead, R.E.S., Davies, J.F., 1982. Evidence for a genetic link between Sudbury breccias and fenite breccias. Can. J. Earth Sci. 19 (6), 1174–1184.

Dupuis, L., Whitehead, R.E.S., Davies, J.F., 1990. Alkali gabbro fragments in Sudbury breccia. Can. J. Earth Sci. 27 (6), 784–786.

Durazzo, A., Taylor, L.A., 1982. Exsolution in the MSS-pentlandite system: textural and genetic implications for the Ni-sulfide system. Min. Dep. 17, 313–332.

Dutch, S.I., 1979. The Creighton Pluton, Ontario, and its significance to the geologic history of the Sudbury region. PhD Thesis. Columbia University, Teachers College, New York, NY, United States. 141 pp.

Dutrizac, J.E., 1976. Reactions in cubanite and chalcopyrite. Can. Mineral. 14, 172–181.

Duuring, P., 2003. Remobilization of Ni-Cu-PGE sulphides. AMIRA P710, p. 22.

Duzhikov, V.V., Distler, V.V., Strunin, B.M., Mkrtychyan, A.K., Sherman, S.S., Sluzhenikin, S.S., Lurye, A.M., 1992. Geology and Metallogeny of Sulfide Deposits, Noril'sk Region, USSR. Econ Geol Special Publication 1.

Dyck, A.V., 1991. Drill-hole electromagnetic methods. In: Nabighian, M.N. (Ed.), Electromagnetic Methods in Applied Geophysics. Soc. Expl. Geophys., vol. 2, pp. 881–930.

Easton, R.M., 1992. The Grenville Province and the Proterozoic history of central and southern Ontario. In: Thurston, P.C., Williams, H.R., Sutcliffe, R.H., Stott, G.M. (Eds.), Geology of Ontario. Ontario Geological Survey Special Volume 4, part 2, pp. 715–904.

Easton, R.M., Jobin-Bevins, L.S., James, R.S., 2004. Geological guidebook to the paleoproterozoic East Bull Lake intrusive suite plutons at East Bull Lake, Agnew Lake and River Valley; Ontario Geological Survey, Open File Report 6135, 84 p.

Easton, R.M., James, R.S., Jobin-Bevans, L.S., 2010. Geological guidebook to the Paleoproterozoic East Bull Lake intrusive suite plutons at East Bull Lake, Agnew Lake and River Valley: a field trip for the 11th International Platinum Symposium; Ontario Geological Survey, Open File Report 6253, 108 p.

Eaton, D., Milkereit, B., Salisbury, M., 2003. Hardrock seismic exploration; Mature technologies adapted to new exploration targets, in Eaton, D., Milkereit, B., Salisbury, M. (Eds.), Hardrock Seismic Exploration: Society of Exploration Geophysicists, Geophysical Development Series, vol. 10, pp. 1–6.

Eaton, D.W., Adam, E., Milkereit, B., Salisbury, M., Roberts, B., White, D., Wright, J., 2010. Enhancing base-metal exploration with seismic imaging. Can. J. Earth Sci. 47, 1–20.

Ebel, D.S., Naldrett, A.J., 1996. Experimental fractional crystallization of Cu- and Ni-bearing Fe sulfide liquids. Econ. Geol. 91, 607–621.

Ebel, D.S., Naldrett, A.J., 1997. Crystallization of sulfide liquids and the interpretation of ore composition. Can. J. Earth Sci. 34, 352–365.

Eneva, M., Young, R.P., 1993. Evaluation of spatial patterns in the distribution of seismic activity in mines: a case study of Creighton Mine, northern Ontario (Canada). In: Young, R.P. (Ed.), Rockbursts and Seismicity in Mines. Balkema, Rotterdam, pp. 175–180.

Ernst, R.E., 1994. Mapping the magma flow pattern in the Sudbury dyke swarm in Ontario using magnetic fabric analysis. Current Research—Geological Survey of Canada, 1994-E, pp. 183–192.

Ernst, R.E., 2014. Large Igneous Provinces. Cambridge University Press, Cambridge, 666 pp.

Ernst, R.E., Bleeker, W., 2010. Large igneous provinces (LIPs), giant dyke swarms, and mantle plumes: significance for breakup events within Canada and adjacent regions from 2.5 Ga to present. Can. J. Earth Sci. 47, 695–739.

Ernst, R.E., Grosfils, E.B., Bethune, K.M., Davidson, A., 1994. Lateral magma flow in the Sudbury diabase dyke swarm. GAC/AGC-MAC/AMC Joint Annual Meeting (abstract), vol. 19, p. 34.

Evans-Lamswood, D.M., 2000. Physical and geometric controls on the distribution of magmatic and sulphide bearing phases within the Voisey's Bay nickel–copper–cobalt deposit, Voisey's Bay, Labrador. Unpublished MSc thesis, Memorial University of Newfoundland, 212 pp.

Eve, A.S., 1931. Geophysical investigations at the Mammoth Caves, Kentucky, and in Sudbury Basin. Memoir—Geological Survey of Canada, Ontario, pp. 78–160.

Everest, J.O., 1999. Magmatic ore at the McCreedy West Mine, Sudbury. MSc Thesis. Carleton University, Ottawa, 143 pages.

Everitt, R., 1979. Jointing in the Sudbury Basin, Sudbury, Ontario. MSc Thesis. Laurentian University, Sudbury, Sudbury, ON, Canada. 57 pp.

Faggart, B.E., 1984. Sm-Nd study of the Sudbury Complex, Ontario, Canada. MSc Thesis. University of Rochester, Rochester, NY, United States. 61 pp.

Faggart, B.E., Basu, A.R., Tatsumoto, M., 1984. Regional cross section of the Southern Province adjacent to Lake Huron, Ontario: implications for tectonic significance of the Murray fault zone. Can. J. Earth Sci. 21, 447–456.

Faggart, B.E., Basu, A.R., Tatsumoto, M., 1985a. Origin of the Sudbury Complex by meteoritic impact, Nd-isotopic evidence. Trans. Am. Geophys. Union 66 (10), 116.

Faggart, B.E., Basu, A.R., Tatsumoto, M., 1985b. Origin of the Sudbury Structure by meteorite impact: neodymium isotopic evidence. Science 230, 436–439.

Faggart, B.E., Basu, A.R., Tatsumoto, M., 1985c. Sm-Nd study of the Sudbury Complex, Ontario. The Geological Society of America, 98th annual meeting. Abstracts with Programs. Geol. Soc. Am. 17 (7), 577.

Fahrig, W.F., 1987. The tectonic settings of continental mafic dyke swarms, failed arm and early passive margin. Geological Association of Canada, Special Paper 34, pp. 331–348.

Fahrig, W.F., Wanless, R.K., 1963. Age and significance of diabase dyke swarms of the Canadian Shield. Nature 200, 934–937.

Fahrig, W.F., West, T.D., 1986. Diabase dike swarms of the Canadian Shield. Geological Survey of Canada, Map 1627A, scale 14,873,900.

Fairbairn, H.W., Robson, G.M., 1942. Breccia at Sudbury, Ontario. J. Geol. 50 (1), 1–33.

Fairbairn, H.W., Pinson, W.H., Hurley, P.M., 1958. Sudbury, Ontario, age program. US Atomic Energy Commission Report, pp. 51–57.

Fairbairn, H.W., Pinson, W.H., Hurley, P.M., 1959. Rb-Sr feldspar ages in granitic rocks of Sudbury-Blind River, Ontario, Canada. Geol. Soc. Am. Bull., 70(12), Part 2, pp. 1599–1600.

Fairbairn, H.W., Faure, G., Pinson, W.H., Hurley, P.M., 1968. Rb-Sr whole-rock age of the Sudbury lopolith and basin sediments. Can. J. Earth Sci. 5, 707–714.

Fairbairn, H.W., Hurley, P.M., Card, K.D., Knight, C.J., 1969. Correlation of radiometric ages of Nipissing diabase and Huronian metasediments with Proterozoic orogenic events in Ontario. Can. J. Earth Sci. 6, 489–497.

Fairbanks, E.E., 1980. Sudbury Basin, a geological enigma. Earth Sci. 33 (2), 73.

Falconbridge staff, 1959. The Falconbridge Story. Can. Min. J. 80, 103–230.

Fan, XM., Clifford, P.M., 1993. Tectonic implication of the structural studies in the Killarney igneous complex, Ontario, Canada. Geological Society of America, 1993 annual meeting. Abstracts with Programs. Geological Society of America, vol. 25, pt. 6, pp. 168.

Farrell, K.P., 1997. Mafic and ultramafic inclusions in the Sublayer of the Sudbury Igneous Complex at Whistle Mine, Sudbury, Ontario, Canada. MSc Thesis. Laurentian University, Canada, 179 pp.

Farrell, K.P., Lightfoot, P.C., Keays, R.R., 1993. Project Unit 93-08, Ultramafic intrusions in the Sublayer of the Sudbury Igneous Complex, Whistle Mine, Sudbury, Ontario. In: Baker, C.L., Dressler, B.O., De Souza, H.A.F. (Eds.), Summary of Field Work and Other Activities 1993.

Farrell, K.P., Lightfoot, P.C., Keays, R.R., 1995. Mafic-ultramafic inclusions in the Sublayer of the Sudbury Igneous Complex, Whistle Mine, Sudbury, Ontario. Summary of Field Work and Other Activities, Ontario Geological Survey, Miscellaneous Paper 164, pp. 126–128.

Farrow, C.E.G., 1995. Geology, alteration, and the role of fluids in Cu-Ni-PGE mineralization of the footwall rocks to the Sudbury Igneous Complex, Levack and Morgan townships, Sudbury District, Ontario. PhD Thesis. Carleton University, Ottawa, ON, Canada. 407 pp.

Farrow, C.E.G., 1997. Diversity of precious-metal mineralization in footwall Cu-Ni-PGE deposits, Sudbury, Ontario: implications for hydrothermal models of formation. Can. Mineral. 35, 817–839.

Farrow, C.E.G., Lightfoot, P.C., 2002. Sudbury PGE revisited: towards an integrated model. In: Cabri, L.J. (Ed.), The Geology, Geochemistry, Mineralogy and Mineral Beneficiation of Platinum-Group Elements; Canadian Institute of Mining, Metallurgy and Petroleum, Special Volume 54, pp. 273–297.

Farrow, C.E.G., Morrison, G., 2012. Rapid growth of a mid-tier miner: lessons for future exploration. Toronto Geology Discussion Group. Available from: http://www.tgdg.net/Resources/Documents/TGDG%20Farrow%20Nov8.pdf.

Farrow, C.E.G., Watkinson, D.H., 1992. Alteration and the role of fluids in Ni, Cu and platinum-group element deposition, Sudbury Igneous Complex contact, Onaping-Levack area, Ontario. Mineralogy and Petrology, vol. 46, pt. 1, pp. 67–83.

Farrow, C.E.G., Watkinson, D.H., 1996. Geochemical evolution of the Epidote Zone, Fraser Mine, Sudbury, Ontario, Ni-Cu-PGE remobilization by saline fluids. Explor. Min. Geol. 5 (1), 17–31.

Farrow, C.E.G., Watkinson, D.H., 1997. Diversity of precious-metal mineralization in footwall Cu-Ni-PGE deposits, Sudbury, Ontario: implications for hydrothermal models of formation. Can. Mineral. 35, 817–839.

Farrow, C.E.G., Hattori, K., Watkinson, D.H., Fouillac, A.M., 1992a. Isotopic study of alteration minerals associated with Ni, Cu and PGE mineralization in the North Range of the Sudbury intrusive complex. AGU 1992 spring meeting. Eos Trans. Am. Geophys. Union 73 (14), 344.

Farrow, C.E.G., Watkinson, D.H., Hattori, K., Fouillac, A.M., 1992b. Mineralogical and isotopic characteristics of alteration associated with Ni, Cu and PGE deposition in the North Range of the Sudbury Structure. GAC/AGC-MAC/AMC Joint Annual Meeting (abstract), 17, pp. 32–33.

Farrow, C.E.G., Watkinson, D.H., Jones, P.C., 1994. Fluid inclusions in sulfides from the North and South Range Cu–Ni-PGE deposits, Sudbury Structure, Ontario. Econ. Geol. 89, 647–655.

Farrow, C.E.G., Everest, J.O., King, D.M., Jolette, C., 2005. Sudbury Cu(-Ni)-PGE systems: refining the classification using McCreedy West mine and Podolsky project case studies. Mineralogical Association of Canada Short Course Volume 35, pp. 163–180.

Farrow, C.E.G., Everest, J., Frayne, M., 2008a. Technical report on mineral properties in the Sudbury Basin. 163 pp. Available from: http://www.infomine.com/index/pr/Pa740688.PDF.

Farrow, C.E.G., Frayne, M., Ramnath, S.M., 2008b. Technical report on mining properties in the Sudbury Basin, Ontario. Annual Information Report of FNX Mining Company Inc., issued under National Instrument 43–101, 159 pp. Available from: www.infomine.com/index/pr/Pa610908.PDF.

Farrow, C.E.G., Everest, J., Gibbins, S. Jolette, C., 2011. Technical report on the Victoria Project Deposit, Sudbury, Ontario, Canada. Quadra FNX Mining Ltd. 43–101 Compliant Report, Ontario Securities Commision, 124 pp.

Faure, G., Fairbairn, H.W., Hurley, P.M., Pinson, Jr., W.H., 1964. Whole-rock Rb-Sr age of norite and micro-pegmatite at Sudbury. J. Geol. 72, 848–854.

Fedorenko, V.A., 1979. Paleotectonics of Late Paleozoic-early Mesozoic volcanism of the Noril'sk Region and paleoptectonic control of the nickel-bearing intrusions. In Geology and Mineralization of the Taimyr-Severnaya-Zemlya folding area. NIIGA, Leningrad, pp. 16–23 (in Russian).

Fedorenko, V.A., 1981. Petrochemical series of extrusive rocks of the Noril'sk Region. Soviet Geol. Geophys. 22 (6), 66–74.

Fedorenko, V.A., 1991. Tectonic control of magmatism and regularities of Ni-bearing localities on the northwestern Siberian platform: Soviet. Geol. Geophys. 32 (1), 41–47.

Fedorenko, V.A., 1994. Evolution of magmatism as reflected in the volcanic sequence of the Noril'sk region. In: Lightfoot, P.C., Naldrett, A.J. (Eds.), Proceedings of the Sudbury-Noril'sk Symposium: Ontario Geol. Surv. Spec. Vol. 5, pp. 171–183.

Fedorenko, V.A., Lightfoot, P.C., Naldrett, A.J., Czamanske, G.K., Hawkesworth, C.J., Wooden, J.L., Ebel, D.S., 1996. Petrogenesis of the flood-basalt sequence at Noril'sk North Central Siberia. Int. Geol. Rev. 38, 99–135.

Fedorowich, J.S., 1995. Detailed structural observations at Barnet Property, Strathcona Mine area, North Range, Sudbury. Falconbridge Ltd. Exploration Internal Report, Bulletin No. 477, 23 pp.

Fedorowich, J.S., 1996a. Structural controls and tectonic setting for Cu-Ni-PGE mineralization at Strathcona Mine, Sudbury (abstract). GAC/MAC, p. A28.

Fedorowich, J.S., 1996b. Detailed structural analysis of Sudbury breccia at Strathcona Mine, Sudbury, Canada (abstract). GAC/MAC, p. A28.

Fedorowich, J., Morrison, G.G., 1999. Sudbury Ni-Cu-PGE deposits—South Range (A1) and North Range (B1). Geological Association of Canada/Mineralogical Association of Canada, Joint Annual Meeting, Sudbury, ON, Field Trips A1 & B1 Guidebook, 48 pp.

Fedorowich, J.S., Roussel, D.H., Peredery,W.V., 1997. Classification and distribution of Sudbury breccia (abstract). Large Meteorite Impacts and Planetary Evolution.

Fedorowich, J.S., Roussel, D.H., Peredery, W.V., 1999. Sudbury breccia distribution and orientation in an embayment environment. Large Meteorite Impacts and Planetary Evolution II, Geological Society of America, Special Paper 339, pp. 305–315.

Fedorowich, J.S., Parrish, R.R., Sager-Kinsman, A., 2006. U-Pb dating of a diabase dike resolves the problem of mutually crosscutting relationships within the Fraser-Strathcona Deep Copper vein system, Sudbury basin. Econ. Geol. 101, 1595–1603.

Fenner, C.N., 1935. Life history of the Sudbury nickel irruptive. Geol. Mag. 8, 381–382.

Feuten, F., Redmond, D., 1992. Structural studies in the Southern Province, south of Sudbury, Ontario. Current Research, Part C. Geological Survey of Canada, Paper 92-1C, pp. 179–187.

Feuten, F., Seabright, R., Morris, B., 1992. A structural transect across the Levack Gneiss Cartier Batholith Complex, northwest of the Sudbury Structure. LITHOPROBE—Abitibi-Grenville Project, Abitibi-Grenville Transect, vol. 33, pp. 11–15.

Finn, G.C., 1993. Field relationships within the granophyre phase, North Range, Sudbury Igneous Complex. GAC/AGC-MAC/AMC Joint Annual Meeting (abstract), vol. 18, p. 30.

Finn, G.C., Edgar, A.D., Rowell, W.F., 1982. Petrology, geochemistry, and economic potential of the Nipissing Diabase. Ontario Geological Survey, Miscellaneous Paper 103, pp. 43–58.

Fleet, M.E., 1979. Tectonic origin for Sudbury, Ontario, shatter cones. Geol. Soc. Am. Bull. 90, 1177–1182.

Fleet, M.E., 1980. Tectonic origin for Sudbury, Ontario, shatter cones: reply. Geol. Soc. Am. Bull. 91, 755–756.

Fleet, M.E., Barnett, R.L., 1978. Aliv/Alvi partitioning in calciferous amphiboles from the Frood Mine, Sudbury, Ontario. Can. Mineral. 16, 527–532.

Fleet, M.E., Barnett, R.L., Thomson, M.L., Kerrich, R., Morris, W.A., 1985. Metabasite assemblages of the Sudbury Igneous Complex. GAC/AGC-MAC/AMC Joint Annual Meeting (program with abstracts), vol. 10, p. A18.

Fleet, M.E., Barnett, R.L., Morris, W.A., 1987. Prograde metamorphism of the Sudbury Igneous Complex. Can. Mineral., 25, Part 3, 3, pp. 499–514.

Fleet, M.E., Chryssoulis, S.L., Stone, W.E., Weisner, C.G., 1993. Partitioning of platinum-group elements and Au in the Fe-Ni-Cu-S system: experiments on the fractional crystallization of sulfide melt. Contrib. Miner. Petrol 115, 36–44.

Floran, R.J., Simonds, C.H., Grieve, R.A.F., Phinney, W.C., Warner, J.L., Rhodes, M.J., Jahn, B.M., Dence, M.R., 1976. Petrology, structure and origin of the Manicouagan Melt Sheet, Quebec, Canada: a preliminary report. Geophys. Res. Lett. 3 (2), 49–52.

Fouillac, A.M., Watkinson, D., Farrow, C., Hattori, K., 1993. Stable isotope study of alteration associated with Ni, Cu and PGE deposition in the North Range of the Sudbury Structure. In: Seventh meeting of the European Union of Geosciences, abstract supplement. Terra Abstracts, 5, Suppl. 1, 370.

Foullon, H.B., 1892. Ueber einige Nickelerzvorkommen (Riddle in Oregon, Sudbury in Ontario). Verhandlungen der Geologischen Bundesanstalt (Wien), pp. 223–310.

Foya, S.N., 2002. Mineralogical and geochemical controls on gold distribution in the Kimberley Reefs, South Africa. Unpublished PhD Thesis. University of the Witwatersrand, Johannesburg, 383 pp.

Fralick, P.W., Davis, D.W., Kissin, S.A., 2002. The age of the Gunflint Formation, Ontario, Canada: single zircon U-Pb age determinations from reworked volcanic ash: Can. J. Earth Sci. 39, 1085–1091.

Franchuk, A., Lightfoot, P.C., Kontak, D., 2015. High tenor Ni-PGE sulfide mineralization in the South Manasan Ultramafic Intrusion, Paleoproterozoic Thompson Nickel Belt, Manitoba, Canada. Ore Geol. Rev. 72, 434–458.

Francis, G.G., 1881. The Smelting of Copper in the Swansea District, from the Time of Elizabeth to the Present Day. 193 pp.

Francis, D., 1997. Bre-X: The Inside Story. Key Porter Books, Toronto.

Franklin, J.A., Pearson, D., 1985. Rock engineering for construction of Science North, Sudbury, Ontario. Can. Geotech. J. 22 (4), 443–455.

Frape, S.K., Fritz, P., 1982. The chemistry and isotopic composition of saline groundwaters from the Sudbury Basin, Ontario. Can. J. Earth Sci. 19 (4), 645–661.

Frarey, M.J., Card, K.D., Richardson, J.A., 1979. Geology of the Sudbury-Sault Ste. Marie area, Ontario. In: Sears, D.W.G. (Ed.), Field Trip Guidebook for the Archean and Proterozoic Stratigraphy of the Great Lakes area, United States and Canada. Guidebook Series—Minnesota Geological Survey, vol. 13, pp. 7–28.

Frarey, M.J., Loveridge, W.D., Sullivan, R.W., 1982. A U-Pb age for the Creighton granite, Ontario. Shock Metamorphism of Natural Materials, Monography Book Corporation, Baltimore, pp. 383–412.

Frarey, M.J., Loveridge, W.D., Sullivan, R.W., 1995. A U-Pb zircon age for the Creighton Granite, Ontario. Geological Survey of Canada, Paper 82-1C, pp. 129–132.

Freeman, B.C., 1934. The Long Lake diorite and associated rocks, Sudbury District, Ontario. J. Geol. 42 (1), 23–44.

Freeman, R.M., Muir, J.A., 1979. Orebody characteristics: Stobie Mine. Unpublished Inco Report. 154 pages.

French, B.M., 1967. Sudbury Structure, Ontario: some petrographic evidence for origin by meteorite impact. Science 156, 1094–1098.

French, B.M., 1968a. Shock metamorphism as a geologic process. In: French, B.M., Short, N.M. (Eds.), Shock Metamorphism of Natural Materials. Mono Book Corporation, Baltimore, pp. 1–17.

French, B.M., 1968b. Sudbury Structure, Ontario: some petrographic evidence for an origin by meteorite impact. In: French, B.M., Short, N.M. (Eds.), Shock Metamorphism of Natural Materials. Mono Book Corporation, Baltimore, pp. 383–412.

French, B.M., 1970. Possible relations between meteorite impact and igneous petrogenesis, as indicated by the Sudbury Structure, Ontario. Bull. Volcanologique 34, 466–517.

French, B.M., 1972a. Shock-metamorphic features in the Sudbury Structure, Ontario: a review. In: Guy-Bray, J.V. (Ed.), New Developments in Sudbury Geology. Geological Association of Canada, Special Paper 10, pp. 19–28.

French, B.M., 1972b. Production of deep melting by large meteorite impacts: the Sudbury structure, Canada. International Geological Congress, 24th, Planetology, pp. 125–132.

French, B.M., 1988. Shock-metamorphic features at Sudbury, Ontario, their use in a scientific deep-drilling project. In: Drury, M.J. (Ed.), Scientific drilling, the Sudbury Structure, proceedings of a workshop. Canadian Continental Drilling Program Report, vol. 88-2, p. 10.

French, B.M., 1990. Twenty five years of the impact-volcanic controversy: is there anything new under the Sun or inside the Earth? EOS 71 (17), 411–414.

French, B.M., 1998. Traces of Catastrophe: a handbook of shock-metamorphic effects in terrestrial meteorite impact structures. LPI Contribution Number 954. Lunar and Planetary Science Institute, Hauston, TX, 120 pages.

Frenkel, M.Y., Yaroshevsky, A.A., Ariskin, A.A., Barmina, G.S., Koptev-Dvornikov, E.V., Kireev, B.S., 1989. Convective-cumulate model simulating the fgormation process of satratified intrusions. In: Bonin, N., Didier, J., LeFort, P., Propach, G., Puga, E., Vistelius, A.B. (Eds.), Magma-crust interactions and evolution. Theophrastus, Athens, pp. 3–38.

Frimmel, H.E., Minter, W.E.L., 2002. Recent developments concerning the geological history and genesis of the Witwatersrand gold deposits, South Africa. Soc. Econ. Geol. Spec. Pub. 9, 17–45.

Frolick, V., 1999. Fire Into Ice: Charles Fike and the Great Diamond Hunt. Raincoast Books, Vancouver, 354 pp.

Fueten, F., 1990. Deformation of the Mississaga Formation south of the Sudbury Basin. Abitibi-Grenville Project, workshop report III. Lithoprobe Report, vol. 19, pp. 77–80.

Fueten, F., Redmond, D.J., 1997a. Documentation of a 1450 Ma contractional orogeny preserved between the 1850 Ma Sudbury impact structure and the 1 Ga Grenville orogenic front, Ontario. GSA Bulletin, vol. 109, pp. 268–279.

Fueten, F., Redmond, D.J., 1997b. Structural history between the Sudbury Structure and the Grenville Front, Ontario, Canada. Geol. Soc. Am. Bull. 109, 268–279.

Fueten, F., Seabright, R., and Morris, B., 1992. A structural transect across the Levack Gneiss-Cartier Batholith Complex, northwest of the Sudbury Structure. Lithoprobe, Abitibi-Grenville Project, Abitibi-Grenville Transect. Lithoprobe Report 33, pp. 11–15.

Fueten, F., Seabright, R., and Morris, B., 1994. A structural transect across the Levack Gneiss-Cartier batholith complex, northwest of the Sudbury Structure. Lithoprobe, Abitibi-Grenville Project, Abitibi-Grenville transect. Lithoprobe Report 41, pp. 135–139.

Fullagar, P.D., Botting, M.L., French, B.M., 1971. Rb-Sr study of shock-metamorphosed inclusions from the Onaping Formation, Sudbury, Ontario. Can. J. Earth Sci. 8, 435–443.

Gaal, G., 1992. Global Proterozoic tectonic cycles and early Proterozoic metallogeny. South African J. Geol. 95, 80–87.

Gait, R.I., Harris, D.C., 1972. Hauchecornite, antimonian, arsenia and tellurian varieties. Can. Min. 11, 819–825.

Galong, W., 1991. Interpretation of gravity and magnetic data, Sudbury Structure. MSc Thesis. Laurentian University, Sudbury. Sudbury, ON, Canada. 184 pp.

Garde, A.A., Pattison, J., Kokfelt, T.F., McDonald, I., Secher, K., 2013. The norite belt in the Mesoarchaean Maniitsoq structure, southern West Greenland: conduit-type Ni-Cu mineralization in impact-triggered, mantle-derived intrusions? Geol. Surv. Denmark Greenland Bull. 28, 45–48.

Garnier, J., 1891. Mines de nickel, cuivre, et platine du district de Sudbury, Canada. Soc Ing Civils France, Mem 5 (44), 239–259.

Gates, B.l., 1991. Sudbury Mineral Occurrence Study. Ontario Geological Survey, Open File Report 5771, 235 pp.

Genkin, A.D., Kovalenker, V.A., Simirnov, A.V., Muravitskaya, G.N., 1977. Peculiarities of mineral composition of Noril'sk sulfide disseminated ores and their genetic significance. Geol. Ore Dep. 19, 17–20.

Genkin, A.D., Distler, V.V., Gladyshev, G.D., Filimonova, A.A., Evstigneeva, T.L., Kovalenker, V.A., Laputina, I.P., Smirnov, A.B., Grokhovskaya, T.L., 1981. Sulfide Ni-Cu ores of the Noril'sk Deposits. Nauka, Moscow, 234 pp (in Russian).

Geraud, Y., Caron, J.M., Faure, P., 1995. Porosity network of a ductile shear zone. J. Struct. Geol. 17, 1757–1769.

Gibbins, W.A., 1974. Rubidium-strontium mineral and rock ages at Sudbury, Ontario. PhD Thesis. McMaster University, Hamilton, ON, Canada. 230 pp.

Gibbins, S.F.M., 1994. Geology, geochemistry, stratigraphy and mechanisms of emplacement of the Onaping Formation, Dowling area, Sudbury Structure, Ontario, Canada. Unpublished MSc Thesis. Laurentian University, Sudbury, Ontario, 314 pp.

Gibbins, S.F.M., 1997. The Sudbury Structure with emphasis on the Whitewater Group. Institute on Lake Superior Geology, 43rd Annual Meeting, May 6–11, 97, Sudbury, Ontario. Field Trip Guidebook, 43, pt. 4, 53 pp.

Gibbins, W.A., McNutt, R.H., 1975a. The age of the Sudbury Nickel Irruptive and the Murray Granite. Can. J. Earth Sci. 12, 1970–1989.

Gibbins, W.A., McNutt, R.H., 1975b. Rubidium-Strontium mineral ages and polymetamorphism at Sudbury, Ontario. Can. J. Earth Sci. 12, 1990–2003.

Gibbins, S.F.M., Gibson, H.L., Whitehead, R.E.S., Watkinson, D.H., Jonasson, I.R., 1994. Geology, geochemistry, stratigraphy and mechanisms of emplacement of the Onaping Formation, Dowling area, Sudbury Structure, Ontario, Canada. GAC/AGC-MAC/AMC Joint Annual Meeting (abstracts), vol. 19, p. 41.

Gibbins, S.F.M., Gibson, H.L., Jonasson, I.R., 1996. The Onaping Formation: a product of passive and explosive hydroclastic fragmentation of an impact melt? (abstract). GAC/MAC, A35.

Gibbins, S.F.M., Gibson, H.L., Ames, D.E., Jonasson, I.R., 1997. The Onaping formation: stratigraphy, fragmentation, and mechanisms of emplacement (abstract). Large Meteorite Impacts and Planetary Evolution.

Giblin, P.E., 1982. Garson Township, District of Sudbury. In: Wood, J., White, O.L., Barlow, R.B., Colvine, A.C. (Eds.), Summary of field work, 1982. Ontario Geological Survey Miscellaneous Paper 106, pp. 51–52.

Giblin, P.E., 1984a. History of Exploration and Development, of Geological Studies and Development of Geological Concepts. In: Pye, E.G., Naldrett, A.J., Giblin, P.E. (Eds.), The Geology and Ore Deposits of the Sudbury Structure. Ontario Geological Survey Special Volume 1, pp. 3–23.

Giblin, P.E., 1984b. Glossary of Sudbury Geology Terms. In: Pye, E.G., Naldrett, A.J., Giblin, P.E. (Eds.), The Geology and Ore Deposits of the Sudbury Structure. Ontario Geological Survey Special Volume 1, pp. 571–574.

Giblin, P.E., 1987. The Sudbury Structure, Sudbury, Ontario. In: Roy, D.C. (Ed.), Northeastern section of the Geological Society of America, pp. 323–326.

Giblin, P.E., Martins, J.M., 1981. 1980 report of the Sudbury Resident Geologist. In: Kustra, C.R. (Ed.), Annual Report of the Regional and Resident Geologists, 1980. Ontario Geological Survey Miscellaneous Paper 95, pp. 116–122.

Giblin, P.E., Martins, J.M., 1982. 1981 report of the Sudbury resident geologist. In: Kustra, C.R. (Ed.), Annual report of the regional and resident geologists, 1981. Ontario Geological Survey Miscellaneous Paper 101, pp. 139–147.

Giblin, P.E., Martins, J.M., 1983. 1982 report of the Sudbury Resident Geologist. In: Kustra, C.R. (Ed.), Report of Activities, Regional and Resident Geologists, 1982. Ontario Geological Survey Miscellaneous Paper 107, pp. 153–161.

Gibson, T.W., 1922. The alleged coal beds at Sudbury (Ontario). Can. Min. J. 43 (33), 554–555.

Gibson, H.M., 1994. Shatter cone morphology and orientation: implications for their formation at Sudbury and other impact structures (abstract). European Science Foundation, Third International Workshop. Shock Wave Behaviour of Solids in Nature and Experiments, Limoges, France.

Gibson, L., 2003. Sudbury Basin 3D Model. Virtual Reality Laboratory Presentation. ICDP Planning Meeting, Sudbury, Canada.

Gibson, R.L., Reimold, W.U., 2001. The Vredefort impact structure, South Africa (The scientific evidence and a two-day excursion guide). Memoir 92, Council for Geoscience, Pretoria, 110 pp.

Gibson, H.M., Spray, J.G., 1993. Orientation, morphology and classification of shatter cones from the Sudbury impact structure, Canada. AGU 1993 fall meeting. Eos Trans. Am. Geophys. Union, 74 (43), 389.

Gibson, H.M., Spray, J.G., 1994. Detailed morphology of striated, conical fracture surfaces (shatter cones) and their orientation with respect to the Sudbury Structure, a possible ground zero indicator for the Sudbury impact event? GAC/AGC-MAC/AMC Joint Annual Meeting (abstracts), vol. 19, p. 41.

Gibson, H.M., Spray, J.G., 1998. Shock-induced melting and vaporization of shatter cone surfaces: evidence from the Sudbury impact structure. Meteorit. Planet. Sci. 33, 329–336.

Gibson, H.L., Gibbins, S.F.M., Gray, M.J., Stoness, J.A., Rogers, D., Jonasson, I.R., Ames, D., 1994. Origin of the Onaping Formation and Zn-Cu-Pb massive sulphide deposits of the Proterozoic Sudbury structure. Summary Report 1993–1994, northern Ontario Development Agreement, pp. 90–94.

Gibson, H.L., Gibbins, S.F., Stoness, J.A., Gray, M.J., Paakki, J.J., Ames, D.E., Jonasson, I.R., 1995. The origin of the Palaeoproterozoic Onaping Formation and Errington and Vermilion Zn-Cu-Pb massive sulphide deposits, Sudbury, Ontario (abstract), GSC Forum '95.

Gibson, A.M., Lightfoot, P.C., Evans, T.C., 2010. Contrasting Styles of Low Sulphide High Precious Metal Mineralisation in the 148 and 109 FW Zones: North and South Ranges of the Sudbury Igneous Complex, Ontario, Canada. 11th International Platinum Symposium, Sudbury, Ontario, Canada. Available from: https://www.researchgate.net/publication/259005788_Contrasting_Styles_of_Low_Sulphide_High_Precious_Metal_Mineralisation_in_the_148_and_109_FW_Zones_North_and_South_Ranges_of_the_Sudbury_Igneous_Complex_Ontario_Canada.

Gilmour, I., 1998. Geochemistry of carbon in terrestrial impact processes. In: Grady, M.M., Hutchison, R., McCall, G.J.H., Rothery, D.A. (Eds.), Meteorites: Flux with Time and Impact Effects. Geological Society of London, Special Publication 140, pp. 205–216.

Ginn, R.M., 1958. A study of granitic rocks in the Sudbury area. MSc Thesis. Queen's University, Kingston, ON, Canada. 137 pp.

Ginn, R.M., 1965. Nairn and Lorne Townships. Ontario Department of Mines; Geological Report 35, 46 pages with map 2062.

Glacken, I.M., Snowden, D.V., 2001. Mineral resource estimation in mineral resource and ore reserve estimation. In: Edwards, A.C. (Ed.), The AusIMM Guide to Good Practice. pp. 189–197.

Glass, B.P., 1990. Tektites and microtektites: key facts and inferences. Tectonophysics 171, 393–404.

Glikson, A.Y., 1996. Mega-impacts and mantle-melting episodes: tests of possible correlations. AGSO J. Aust. Geol. Geophys. 16, 587–607.

Godlevskii, M.N., 1959. Traps and ore-bearing intrusions of the Noril'sk Region. Gosgeoltekhixdat, Moscow, 68 pp (in Russian).

Godlevskii, M., Grinenko, L.N., 1963. Some data on the isotopic composition of sulfur in the sulfides of the Noril'sk deposit. Geochemistry 1, 335–341.

Goldthwait, J.W., 1905. The sand plains of glacial Lake Sudbury (eastern Massachusetts). Bulletin of the Museum of Comparative Zoology, Harvard University, pp. 263–301.

Golightly, J.P., 1988. The Sudbury Structure as a meteorite crater, where is its centre and what should we expect there? In: Drury, M.J. (Ed.), Scientific drilling, the Sudbury Structure, proceedings of a workshop. Canadian Continental Drilling Program Report, vol. 88-2, p. 12.

Golightly, J.P., 1992. Fitting a crater to the Sudbury structure (abstract). Can. Mineral. 30, 484–485, 1992.

Golightly, J.P., 1994. The Sudbury Igneous Complex as an impact melt: evolution and ore genesis. In: Lightfoot, P.C., Naldrett, A.J. (Eds.), Proceedings of the Sudbury-Noril'sk Symposium. Ontario Geological Survey Special Volume 5, pp. 105–118.

Golightly, J.P., 2010. Progress in understanding the evolution of nickel laterites. In: Goldfarb, R.J., Marsh, E.E., Monecke, T. (Eds.), The Challenge of Finding New Mineral Resources—Global Metallogeny, Innovative Exploration, and New Discoveries. Society of Economic Geologists Special Publication No. 15, pp. 451–485.

Golightly, J.P., Pattison, E.F., Lightfoot, P.C., 2010. Ni-Cu-PGE mineralization in the South Range of the Sudbury Igneous Complex: a field trip for the 11th International Platinum Symposium; Ontario Geological Survey, Open File Report 6252, 41 p.

Goodchild, W.H., 1918. Magmatic ore deposits of Sudbury. Ontario Econ. Geol. 13, 137–143.

Goodfellow, W.D., Geldsetzer, H., Gregoire, C., Orchard, M., Cordey, F., 2010. Geochemistry and origin of geographically extensive Ni(Mo, Zn, U)-PGE sulphide deposits hosted in Devonian black shales, Yukon. TGI-3 Workshop: Public geoscience in support of base metal exploration programme and abstracts, Geological Association of Canada, Cordilleran Section, pp. 15–18.

Goodwin, W.L., 1893. A highly nickeliferous pyrite (Sudbury, Ontario). Canadian Record of Science. pp. 346–347.

Gorbachev, N.S., Grinenko, L.N., 1973. The sand plains of glacial Lake Sudbury (eastern Massachusetts-ore deposit, Noril'sk region, in the light of sulfide and sulfate sulfur isotope compositions. Geochem. Int. 10, 843–851.

Grainger, C.J., Groves, D.I., Tallarico, F.H.B., Fletcher, I.U.R., 2007. Metallogenesis of the Carajas Mineral Province, Southern Amazon Craton, Brazil: varying styles of Archean through Paleoproterozoic to Neoproterozoic base- and precious-metals mineralization. Ore Geol. Rev. 39, 451–489.

Grant, R.W., Bite, A., 1984. Sudbury quartz diorite offset dikes. In: Pye, E.G., Naldrett, A.J., Giblin, P.E. (Eds.), The Geology and Ore Deposits of the Sudbury Structure. Ontario Geological Survey Special Volume 1, pp. 275–300.

Gray, M.J., 1995. The geological setting of the Vermilion Zn-Cu-Pb-Ag-Au massive sulfide deposit, Sudbury Basin, Ontario, Canada. Unpublished MSc Thesis. Laurentian University, Sudbury, Ontario, 244 pp.

Gray, M.J., Gibson, H.L., 1993. Geological setting of the Vermilion Cu-Zn-Pb-Au-Ag massive sulphide deposit, Sudbury Basin. GAC/AGC-MAC/AMC Joint Annual Meeting (abstract), p. 37.

Green, A.H., Melezhik, V.A., 1999. Geology of the Pechenga Ore Deposits—a review with comments on ore forming process. Keays, R.R., Lesher, C.M., Lightfoot, P.C., Farrow, C.E.G. (Eds.), Dynamic Processes in Magmatic Ore Deposits and their Application to Mineral Exploration, 13, GAC Short Course, pp. 287–328.

Green, A.G., et al., 1988. Crustal structure of the Grenville front and adjacent terranes. Geology 16, 788–792.

Greenman, L., 1970. The petrology of the footwall breccias in the vicinity of the Strathcona Mine, Levack, Ontario. PhD Thesis. University of Toronto, Toronto, Ontario, 153 p.

Gregory, J.W., 1908. Origin of the Sudbury nickel ores. Geol. Mag., 139–140.

Gregory, J.W., 1925. Magmatic ores. Transactions of the Faraday Society 20, 449–458.

Gregory, S.K., 2005. Geology, mineralogy and geochemistry of transitional contact/footwall mineralization in the McCreedy East Ni-Cu-PGE deposit, Sudbury Igneous Complex. Unpublished MSc Thesis. Sudbury, Ontario, Laurentian University, 138 p.

Grieve, R.A.F., 1982. The record of impact on Earth: implications for a major Cretaceous/Tertiary impact event. Geological Society of America, Special Paper 190, pp. 25–37.

Grieve, R.A.F., 1987. Terrestrial impact structures. Ann. Rev. Earth Planet. Sci. 15, 245–270.

Grieve, R.A.F., 1992. An impact model for Sudbury (abstract). Can. Mineral. 30, 496.

Grieve, R.A.F., 1994. An impact model of the Sudbury Structure. In: Lightfoot, P.C., Naldrett, A.J. (Eds.), Proceedings of the Sudbury-Noril'sk symposium. Ontario Geological Survey Special Volume 5, pp. 119–132.

Grieve, R.A.F., 2003. Extraterrestrial triggers for resource deposits [ext abs]. Appl. Earth Sci. 112, B145–B147.

Grieve, R.A.F., Cintala, M.J., 1992. An analysis of differential impact melt-crater scaling and implications for the terrestrial impact record. Meteoritics 27, 526–538.

Grieve, R.A.F., Deutsch, A., 1994. The Sudbury Structure, additional constraints on its origin and evolution. Abstracts of Papers Submitted to the Twenty-Fourth Lunar and Planetary Science Conference, 25, Part 1, pp. 477–478.

Grieve, R.A.F., Masaitis, V.L., 1984. The economic potential of terrestrial impact craters. Int. Geol. Rev. 36, 105–151.

Grieve, R.A.F., Pesonen, L.J., 1992. The terrestrial impact cratering record. Tectonophysics 216, 1–30.

Grieve, R.A.F., Richardson, P.B., 1988. The Sudbury impact structure, a context for deep drilling. In: Drury, M.J. (Ed.), Scientific drilling, the Sudbury Structure, proceedings of a workshop. Canadian Continental Drilling Program Report, vol. 88-2, p. 9.

Grieve, R., Therriault, A., 2000. Vredefort, Sudbury, Chicxulub: three of a kind? Ann. Rev. Earth Planet. Sci. 28, 305–338, 2000.

Grieve, R.A.F., Stoeffler, D., Deutsch, A., 1991a. The Sudbury Igneous Complex, an impact melt sheet. GAC/AGC-MAC/AMC Joint Annual Meeting (abstract), 16, p. 48.

Grieve, R.A.F., Stöffler, D., Deutsch, A., 1991b. The Sudbury Structure: controversial or misunderstood? Journal of Geophysical Research, 96, E5, 22,753–22,764.

Grieve, R.A.F., Stoeffler, D., Deutsch, A., 1993. Clarification to The Sudbury Structure, controversial or misunderstood? J. Geophys. Res. E Planet. 98 (11), 20,903–20,904.

Grieve, R.A.F., Deutsch, A., Stöffler, D., 1995a. A self-consistent model of the origin and evolution of the Sudbury structure (abstract). IV International Conference on Advanced Material, Symposium 19, Cancún, Mexico.

Grieve, R.A.F., Rupert, J., Smith, J., Therriault, A., 1995b. The record of terrestrial impact cratering. GSA Today, 5, 10, p. 189, and 194–196.

Grieve, R.A.F., Deutsch, A., Therriault, A.M., Ostermann, M., 1996. The Sudbury Igneous Complex, basic arguments for an impact origin. In: Sears, D.W.G. (Ed.), 59th Annual Meteoritical Society Meeting, abstracts. Meteoritics & Planetary Science, vol. 31, Suppl., pp. 54–55.

Grieve, R.A.F., Reimold, W.U., Morgan, J., Riller, U., Pilkington, M., 2008. Observations and interpretations at Vredefort, Sudbury and Chicxulub: towards a composite kinematic model of terrestrial impact basin formation. Meteor. Planet. Sci. 43, 855–882.

Grinenko, L.N., 1985a. Sources of sulfur of the nickeliferous and barren gabbro-dolerite intrusions of the northwest Siberian platform. Int. Geol. Rev. 27, 695–708.

Grinenko, L.N., 1985b. Hydrogen sulfide-containing gas deposits as a source of sulfur for sulfurization of magma in ore-bearing intrusives of the Noril'sk area. Int. Geol. Rev. 27, 290–292.

Gunn, J.M., 2011. Restoration and recovery of an industrial region: progress in restoring the smelter-damaged landscape near Sudbury, Canada. Springer, 358 pp.

Guo, L., Vokes, F.M., 1996. Intergrowths of hexagonal and monoclinic pyrrhotites in some sulfide ores from Norway. Min. Mag. 60, 303–316.

Gupta, V.K., 1981. Bouguer gravity and generalized geological map of the Sudbury-Onaping Lake area. Ontario Geological Survey, Preliminary Map, P. 2482, scale 1100000, 1.

Gupta, V.K., 1983. On the Sudbury gravity anomaly. In: Wood, J., White, O.L., Barlow, R.B., Colvine, A.C. (Eds.), Summary of Field Work, 1983. Ontario Geological Survey Miscellaneous Paper 116, pp. 145–147.

Gupta, V.K., Grant, F.S., 1985. Mineral-exploration aspects of gravity and aeromagnetic surveys in the Sudbury-Cobalt area, Ontario. In: Hinze, W.J. (Ed.), The Utility of Regional Gravity and Magnetic Anomaly Maps, pp. 392–411.

Gupta, V.K., Grant, F.S., Boud, A., 1981. A gravity study in North central Ontario, districts of Sudbury, Nipissing, and Timiskaming. In: Wood, J., White, O.L., Barlow, R.B., Colvine, A.C. (Eds.), Summary of Field Work, 1981. Ontario Geological Survey Miscellaneous Paper 100, pp. 160–164.

Gupta, V.K., Grant, F.S., Card, K.D., 1984a. Gravity and magnetic characteristics of the Sudbury Structure. In: Pye, E.G., Naldrett, A.J., Giblin, P.E. (Eds.), The Geology and Ore Deposits of the Sudbury Structure. Ontario Geological Survey Special Volume 1, pp. 381–410.

Gupta, V.K., Grant, F.S., Card, K.D., McGrath, P.H., 1984b. Gravity and magnetic studies of the Sudbury Structure. In: GAC/AGC-MAC/AMC Joint Annual Meeting (abstract), vol. 9, p. 69.

Guy-Bray, J.G., Geological Staff, 1966. Shatter cones at Sudbury. J. Geol. 74, 243–245.

Guy-Bray, J.V. (Ed.), 1972. New Developments in Sudbury Geology. The Geological Association of Canada, Special Paper No.10, 124 pp.

Guy-Bray, J.G., Peredery, W.V., 1971. Field guide for excursion (1), Geological Association of Canada/Mineralogical Association of Canada, Joint Annual Meeting, Laurentian University, Sudbury, Ontario, 19 pp.

Hall, M., Lafrance, B., Gibson, H., 2015. Deformation and structural controls at the Broken Hammer deposits, Sudbury. Program with Abstracts, GAC/MAC/AGU Joint-Annual Meeting, Montreal.

Hanbury, P.M., 1982. A petrographic and petrochemical comparison of exotic gabbroic inclusions of the Sudbury Sublayer and Nipissing Diabase-Sudbury Gabbro. MSc Thesis. Washington State University, Pullman, WA, United States. 50 pp.

Hanley, J.J., 2002. The distribution of the halogens in Sudbury breccia matrix as pathfinder elements for footwall Cu-PGE mineralization at the Fraser Cu Zone, Barnett Main Copper Zone, and surrounding margin of the Sudbury Igneous Complex, Onaping-Levack area, Ontario, Canada. MSc Thesis. University of Toronto, Toronto, Ontario. 255 pages.

Hanley, J.J., Bray, C.J., 2009. The trace metal content of amphibole as a proximity indicator for Cu-Ni-PGE mineralization in the footwall of the Sudbury Igneous Complex, Ontario, Canada. Econ. Geol. 104, 113–125.

Hanley, J.J., Mungall, J.E., 2003. Chlorine enrichment and hydrous alteration of the Sudbury breccia hosting footwall Cu-Ni-PGE mineralization at the Fraser Mine, Sudbury, Ontario, Canada. Can. Mineral. 41, 857–881.

Hanley, J.J., Mungall, J.E., Bray, C.J., Gorton, M.P., 2004. The origin of bulk and water-soluble Cl and Br enrichments in ore-hosting Sudbury Breccia in the Fraser Copper Zone, Strathcona Embayment, Sudbury, Ontario, Canada. Can. Mineral. 42 (6), 1777–1798.

Hanley, J.J., Mungall, J.E., Pettke, T., Spooner, E.T.C., Bray, C.J., 2005. Ore metal redistribution by hydrocarbon-brine and hydrocarbon-halide melt phases, North Range footwall of the Sudbury Igneous Complex, Ontario, Canada. Miner. Deposita 40, 237–256.

Harcourt, G.A., 1933. Distribution of nickel in the Sudbury norite-micropegmatite. PhD Thesis. Queen's University, Kingston, ON, Canada, 53 pp.

Harker, A., 1926. The Sudbury laccolith. Geol. Mag. 63, 192.

Harland, W.B., Armstrong, R.L., Cox, A.V., Craig, L.E., Smith, A.G., Smith, D.G., 1989. A Geologic Time Scale—1989. Cambridge University Press, Cambridge, New York, Sydney, 263 pp.

Hart, R.J., Cloete, M., McDonald, I., Carlson, R.W., Andreoli, M.A.G., 2002. Siderophile-rich inclusions from the Morokweng impact melt sheet, South Africa: possible fragments of a chondritic meteorite. Earth Planet. Sci. Lett. 198, 49–62.

Hattie, K., 2010. Sulfide saturation history of the Frood-Stobie system, Blezard and McKim Townships, District of Sudbury, Ontario, Canada. MSc Thesis. Laurentian University, Canada, 114 pp.

Haughton, D.R., Roeder, P.L., Skinner, B.J., 1974. Solubility of sulfur in mafic magmas. Econ. Geol. 69, 451–467.

Haw, V.A., 1948. Further studies of nickel ores of the Sudbury type. MSc Thesis. Queen's University, Kingston, ON, Canada. 105 pp.

Haw, V.A., 1953. Further studies of nickel ores of the Sudbury type. Can. Min. J. 74 (6), 98.

Hawkesworth, C.J., Erlank, A.J., Marsh, J.S., Menzies, M.A., Van Calsteren, P., 1983. Evolution of the continental lithosphere: evidence from volcanics and xenoliths in Southern Africa. In: Hawkesworth, C.J., Norry, M.J. (Eds.), Continental Basalts, Mantle, Xenoliths. Nantwich, Shiva. pp. 111–138.

Hawkesworth, C.J., Lightfoot, P.C., Fedorenko, V.A., Blake, S., Naldrett, A.J., Doherty, W., Gorbachev, N.S., 1995. Magma differentiation and mineralization in the Siberian continental flood basalts. Lithos. 34, 61–88.

Hawley, J.E., 1962. The Sudbury ores, their mineralogy and origin. Can. Mineral. 7, 207.

Hawley, J.E., 1965. Upside-down zoning at Frood, Sudbury, Ontario. Econ. Geol. 60, 529–575.

Hawley, J.E., Berry, L.C., 1958. Michenerite and froodite, palladium bismuthite minerals. Can. Min. 6, 200–209.

Hawley, J.E., Stanton, R.L., 1962. The Sudbury ores: their mineralogy and origin. Part II. The facts: the ores, their minerals, metals and distribution. Can. Mineral. 7, 30–145.

Hawley, J.E., Lewis, C.L., Wark, W.J., 1951. Spectrographic study of platinum and palladium in common sulphides and arsenides of the Sudbury District, Ontario. Econ. Geol. Bull. Soc. Econ. Geol. 46 (2), 149–162.

Hearst, R.B., Morris, W.A., 1992. Gravity and magnetic modelling and interpretation across the North Range, Sudbury Structure, exploration implications. AGU 1992 Spring Meeting. Eos Trans. Am. Geophys. Union 73(14 supplement), 305.

Hearst, R.B., Morris, W.A., 1993. Interpretation of the Sudbury structure through Euler deconvolution. Society of Exploration Geophysicists, expanded abstracts, 1993 technical program, 63rd Annual Meeting and International Exhibition. SEG Annual Meeting Expanded Technical Program Abstracts with Biographies, vol. 63, pp. 421–424.

Hearst, R.B., Morris, W.A., 1994. The North Range contact of the Sudbury intrusive complex, an integrated interpretation. Society of Exploration Geophysicists, 64th annual international meeting, expanded abstracts. SEG Annual Meeting Expanded Technical Program Abstracts with Biographies, vol. 64, pp. 524–527.

Hearst, R.B., Morris, W.A., 1995. Rock magnetic properties, why bother? Society of Exploration Geophysicists, 65th annual international meeting, expanded abstracts. SEG Annual Meeting Expanded Technical Program Abstracts with Biographies, vol. 65, pp. 274–277.

Hearst, R.B., Morris, W.A., Thomas, M.D., 1992a. Comparison of detailed ground profile data and airborne magnetic and regional gravity data along the Sudbury Structure Lithoprobe transect. Lithoprobe, Abitibi-Grenville Project, Abitibi-Grenville Transect. Lithoprobe Report, vol. 33, pp. 105–107.

Hearst, R.B., McGrath, P.H., Morris, M.D., Thomas, M.D., Broome, H.J., Tanczyk, E.I., Keating, P., Halliday, D.W., 1992b. Gravity and magnetic interpretation along the Lithoprobe Transect of the Sudbury Structure. Lithoprobe Abitibi Grenville Transect Workshop proceedings volume, Lithoprobe Report no, 25, pp. 69–74.

Hearst, R., Morris, W., Thomas, M., 1994. Magnetic interpretation along the Sudbury Structure Lithoprobe transect. In: Boerner, D.E., Milkereit, B., Nadrett, A.J. (Eds.), Lithoprobe Sudbury Project. Geophys. Res. Lett., 21, 10, pp. 951–954.

Hearst, R.B., Morris, W.A., Thomas, M.D., 1994a. Magnetic interpretation along the Sudbury Structure LITHOPROBE Transect. In: Lightfoot, P.C., Naldrett, A.J. (Eds.), Proceedings of the Sudbury-Noril'sk Symposium. Ontario Geological Survey Special Volume 5, pp. 33–43.

Hearst, R.B., Morris, W.A., Fueten, F., Seabright, R., 1994b. Geophysics of the Levack Gneiss Cartier Batholith complex. Society of Exploration Geophysicists, 64th Annual International Meeting, expanded abstracts. SEG Annual Meeting Expanded Technical Program Abstracts with Biographies, vol. 64, pp. 528–530.

Heath, A.J., Karrow, P.F., 1990. Algonquin-Nipissing lake levels, Manitoulin Island to Sudbury Basin, Ontario. Canadian Quaternary Association-American Quaternary Association, First Joint Meeting. Morgan, A.V., p. 19.

Hebil, K.E., 1978. A petrographic study of the granite breccia, Levack Mine, Sudbury, Ontario. MSc Thesis. McGill University, Montreal, PQ, Canada. 152 pp.

Hecht, L., Wittek, A., Riller, U., Mohr, T., Schmitt, R.T., Grieve, R.A.F., 2008. Differentiation and emplacement of the Worthington Offset Dike of the Sudbury Impact Structure, Ontario. Meteor. Planet. Sci. 43, 1659–1679.

Hellinger, T.S., 1977. Natural microscopic deformation and annealing features of some massive sulfide ores, Sudbury, Ontario. MSc Thesis. University of Michigan, Ann Arbor, MI, United States.

Helms, W., 1987. Nickel-Kupfererzbergbau im Sudbury District von Ontario (Nickel-copper-ore mining in the Sudbury District, Ontario). Erzmetall 40 (2), 96–101.

Hemming, S.R., McDaniel, D.K., McLennan, S.M., Hanson, G.N., 1996. Pb isotope constraints on the provenance and diagenesis of detrital feldspars from the Sudbury Basin, Canada. Earth Planet. Sci. Lett. 142 (3–4), 501–512.

Hewins, R.H., 1971. The petrology of some marginal mafic rocks along the North Range of the Sudbury Irruptive. PhD Thesis. University of Toronto, Toronto, ON, Canada.

Hewins, R.H., Schuyler, T.K., Szuwalski, D.R., 1984. Origin of Sudbury Sublayer. GAC/AGC-MAC/AMC Joint Annual Meeting (abstract), vol. 9, p. 73.

Heymann, D., Dressler, B.O., 1997. Origin of the carbonaceous matter in rocks from the Whitewater Group of the Sudbury structure (abstract). Large Meteorite Impacts and Planetary Evolution.

Heymann, D., Buseck, P.R., Knell, J., Dressler, B.O., 1997a. The search for fullerenes in rocks from the Whitewater Group of the Sudbury structure, Ontario, Canada (abstract). Large Meteorite Impacts and Planetary Evolution.

Heymann, D., Dressler, B.O., Thiemens, M.H., 1997b. Origin of native sulfur in rocks from the Sudbury Structure, Ontario, Canada (abstract). Large Meteorite Impacts and Planetary Evolution.

Heymann, D., Dressler, B.O., Dunbar, R.B., Mucciarone, D.A., 1997c. Carbon isotopic composition of carbonaceous matter in rocks from the Whitewater Group of the Sudbury structure (abstract). Large Meteorite Impacts and Planetary Evolution.

Heymann, D., Dressler, B.O., Knell, J., Thiemaens, M.H., Buseck, P.R., Dunbar, R.B., Mucciarone, D., 1999. Origin of carbonaceous matter, fullerenes and elemental sulfur in rocks of the Whitewater Group, Sudbury impact structure, Ontario, Canada. In Large Meteorite Impacts and Planetary Evolution II. In: Dressler, B.O., Sharpton, V.L. (Eds.), Geological Society of America Special Paper 330, Boulder, pp. 345–360.

Hirt, A., Lowrie, W., Kligfield, R., 1984. Fabric development in the Chelmsford Formation, Sudbury, Ontario, Canada. AGU 1984 fall meeting. Eos Trans.Am. Geophys. Union 65 (45), 1099.

Hirt, A.M., Lowrie, W., Clendenen, W.S., Kligfield, R., 1988. The correlation of magnetic anisotropy with strain in the Chelmsford Formation of the Sudbury Basin, Ontario. Tectonophysics 145, 177–189, 3–4.

Hirt, A.M., Lowrie, W., Clendenen, W.S., Kligfield, R., 1993. Correlation of strain and the anisotropy of magnetic susceptibility in the Onaping Formation: evidence for a near circular origin of the Sudbury basin. Tectonophysics 225, 231–254.

Hitchcock, C.H., 1921. The great fault of the Sudbury nickel district (Ontario). Can. Min. J. 42 (11), 215.

Hixon, H.W., 1905. Geology of the Sudbury District (Ontario). Eng. Min. J., 334–335.

Hixon, H.W., 1906a. The ore deposits and geology of the Sudbury District (Ontario). Transactions of the Canadian Mining Institute 9, 223–235.

Hixon, H.W., 1906b. The Sudbury nickel region. Eng. Min. J. 1, 313–314.

Hixon, H.W., 1909. Origin of the Sudbury, Ontario ores. Inst. M. Met. Tr. 18, 196–198.

Hoffman, E.L., 1978. The platinum group element and gold content of some nickel sulphide ores. PhD Thesis. University of Toronto. Toronto, ON, Canada.

Hoffman, P.F., 1988. United Plates of America, the birth of a craton: early Proterozoic assembly and growth of Laurentia. Annu. Rev. Earth Planet. Sci. 16, 543–603.

Hoffman, P.F., 1989. Precambrian geology and tectonic history of North America, Ch.16. In: Bally, A.W., Palmer, A.R. (Eds.), The Geology of North America—An Overview, Boulder, Colorado, Geological Society of America, The Geology of North America, pp. 447–512.

Hoffman, P.F., 1990. On accretion of granite-greenstone terranes. In: Robert, F., Sheahan, P.A., Green, S.B. (Eds.), Nuna Conference on Greenstone Gold and Crustal Evolution. Proceedings of a workshop, Val d'Or, Quebec, May 24–27, 1990, Geological Association of Canada, Mineral Deposits Division, pp. 32–45.

Hoffman, E.L., Naldrett, A.J., Alcock, R.A., Hancock, R.G.V., 1979. The noble-metal content of ore in the Levack West and Little Stobie mines, Ontario. Can. Mineral. 17, 437–451.

Holliger, K., 1996. Upper-crustal seismic velocity heterogeneity as derived from a variety of P-wave sonic logs. Geophys. J. Int. 125, 813–829.

Holliger, K., Green, A.G., Juhlin, C., 1996. Stochastic analysis of sonic logs from the upper crystalline crust, methodology. In: White, D.J., Ansorge, J., Bodoky, T.J., Hajnal, Z. (Eds.), Seismic reflection probing of the continents and their margins. Tectonophysics, vol. 264, pt. 1–4, pp. 341–356.

Hood, P.J., 1961. Paleomagnetic study of the Sudbury Basin. J. Geophys. Res. 66, 1235–1241.

Hore, R.E., 1912. Origin of the Sudbury nickel and copper deposits (Ontario). Min. World 36, 1345–1349.

Hore, R.E., 1913. Magmatic origin of Sudbury nickel copper deposits. Trans. Can. Min. Inst. 5, 85–96.

Horn, R., 2002. Metals exploration in a changing industry. CIM Bull. 95, 35–48.

Horn, R., Brisbois, C., 1998. Exploration Success-luck or skill? In: Walton, G., Jambor, J. (Eds.), Pathways '98. BC and Yukon chamber of Mines, pp. 92–93.

Hornsby, J.K., Bruce, B., 1985. A preliminary analysis of Landsat MSS and TM data in the Levack area, Sudbury, Canada. In: McIntosh, R.E. (Eds.), Remote Sensing Instrumentation, Technology for Science And Applications, pp. 131–140.

Hornsby, J.K., Bruce, B., 1986. Regional geobotany with TM, a Sudbury case study. In: Thompson, M.D., Brown, R.J. (Eds.), Proceedings of the Tenth Canadian symposium on Remote sensing. Proceedings—Canadian Symposium on Remote Sensing, 10, pp. 601–609.

Howard, F.B., White, H., 1963. Nickel: an historical review. Van Nostrand, New York, 218 pp.

Howard-White, F.B., 1963. Nickel, A Historical Review. Van Nostrand, 350 pp.

Howe, E., 1914. Petrographical notes on the Sudbury nickel deposits. Econ. Geol. 9, 505–522.

Hriskevich, M.E., 1968. Petrology of the Nipissing Diabase sill of the Cobalt area, Ontario, Canada. Geol. Soc. Am. Bull. 79, 1387–1404.

Hsieh, S.S., 1967. Analysis of ruthenium and osmium abundances in sulfide minerals from the Sudbury ores, Ontario. MSc Thesis. McMaster University, Hamilton, ON, Canada, 68 pp.

Hsu, M.Y., 1968. Structural analysis along the Grenville Front, near Sudbury, Ontario. MSc Thesis. McMaster University, Hamilton, ON, Canada, 104 pp.

Huhn, F.J., 1985. Lake sediment records of industrialization in the Sudbury area of Ontario, Canada. In: The Geological Society of America, 98th Annual Meeting. Abstracts with Programs—Geological Society of America, vol. 17, pt. 7, p. 615.

Huminicki, M.A.E., Sylvester, P.J., Cabri, L.J., Lesher, C.M., Tubrett, M., 2005. Quantitative mass balance of platinum-group elements in the Kelly Lake Ni-Cu-PGE deposit, Copper Cliff offset, Sudbury. Econ. Geol. 100, 1631–1646.

Hurst, R.W., 1975. Geochronologic studies in the Precambrian Shield of Canada, Part I, The Archaean of coastal Labrador, Part II, The Sudbury Basin, Sudbury, Ontario. PhD Thesis. University of California, Los Angeles. Los Angeles, CA, United States. 144.

Hurst, R.W., Fahart, J., 1977. Geochronologic investigations of the Sudbury Nickel Irruptive and the Superior Province granites north of Sudbury. Geochemim. Cosmochim. Acta 41, 1803–1815.

Hutton, J., 1795. Theory of the Earth with Proofs and Illustrations, p. 297.

Inco and Falconbridge, 2006. Sudbury – Inco and Falconbridge. Filed by Inco Limited. Pursuant to Rule 425 under the Securities Act of 1933. Subject Company: Falconbridge Limited. Commission File No. 1-11284 Inco Limited Commission File No. 1-1143. Available from: http://www.sec.gov/Archives/edgar/data/49996/000090956706001043/o31988e425.htm.

Innes, D.G., 1978a. Proterozoic volcanism in the Southern Province of the Canadian Shield. Unpublished MSc Thesis. Laurentian University, Sudbury, Ontario, 161 p.

Innes, D.G., 1978b. McKim Township, District of Sudbury; Ontario Geological Survey, Preliminary Map P. 1; 1978, scale 115,840.

Innes, D.G., Colvine, A.C., 1984. The regional metallogenetic setting of Sudbury. In: Pye, E.G., Naldrett, A.J., Giblin, P.E. (Eds.), The Geology and Ore Deposits of the Sudbury Structure. Ontario Geological Survey Special Volume 1, pp. 45–56.

Irvine, T.N., 1975. Crystallization sequences in the Muskox intrusion and other layered intrusions—II. Origin of chromitite layers and similar deposits of other magmatic ores. Geochim. Cosmochim. Acta 39, 1009–1020.

Ivanov, B.A., 1998a. Large impact crater formation: thermal softening and acoustic fluidization. Meteor. Planet. Sci. 33 (4), A76.

Ivanov, B.A., 1998b. Large impact crater formation: thermal softeneing and acoustic fluidization, 61st Annual Meteoritical Society Meeting, Dublin, Ireland.

Ivanov, B.A., Deutsch, A., 1997. Sudbury impact event: Cratering mechanics and thermal history (abstract). Large Meteorite Impacts and Planetary Evolution.

Ivanov, B.A., Deutsch, A., 1999. Sudbury impact event: cratering mechanics and thermal history. Geological Society of America Special Paper 339, p. 389–397.

Ivanov, B.A., Melosh, H.J., 2003. Impacts do not initiate volcanic eruptions: eruptions close to the crater. Geology 31, 869–872.

Ivanov, B.A., Deutsch, A., Ostermann, M., Ariskin, A., 1997. Solidification of the Sudbury impact melt body and nature of the Offset dikes—thermal modeling (abstract). Lunar and Planetary Science XXVIII, pp. 633–634.

Jago, B.C., 2007. Assessment report on the Trill Property, Totten and Trill Townships, Sudbury, Ontario. Available from: http://www.geologyontario.mndmf.gov.on.ca/mndmfiles/afri/data/imaging/20000003373//20005107.pdf.

Jago, B.C., Morrison, G.G., Little, T.L., 1994. Metal zonation patterns and microtextural and micromineralogical evidence for Alkali- and Halogen-rich fluids in the genesis of the Victor Deep and McCreedy East footwall copper orebodies, Sudbury Igneous Complex. In: Lightfoot, P.C., Naldrett, A.J. (Eds.), Proceedings of the Sudbury–Noril'sk Symposium. Ontario Geological Survey Special Volume 5, pp. 65–75.

Jambor, J.L., 1971. The Nipissing Diabase. Can. Mineral. 42, 743–770.

James, R.S., Dressler, B.O., 1992. Nature and significance of the Levack Gneiss Complex—Footwall rocks of the north and east ranges of the Sudbury Igneous Complex. Can. Mineral. 30, 487–488.

James, R.S., Golightly, J.P., 2009. Chapter 10: Metamorphism and Metasomatism. Field Guide to the Geology of Sudbury, Ontario,Ontario Geological Survey, Open File Report 6243, p. 14–30.

James, R.S., Peredery, W., Sweeny, J.M., 1992a. Thermobarometric studies on the Levack Gneisses, footwall rocks to the Sudbury Igneous Complex. In: Dressler, B.O., Sharpton, V.L. (Eds.), Papers Presented to the International Conference on Large Meteorite Impacts and Planetary Evolution. LPI Contribution 790, 41.

James, R.S., Sweeney, J.M., Peredery, W.V., 1992b. Thermobarometry of the Levack gneisses—Footwall rocks to the Sudbury Igneous Complex. Lithoprobe Report No. 25, Abitibi-Grenville Transect, pp. 179–182.

James, R.S., Peredery, W.V., Sweeney, J.M., 1994. Thermobarometric studies of the Levack Gneisses, footwall rocks to the Sudbury Igneous Complex. GAC/AGC-MAC/AMC Joint Annual Meeting (abstract), vol. 19, p. 54.

James, R.S., Easton, R.M., Peck, D.C., Hrominchuk, J.L., 2002a. The East Bull Lake intrusive suite; remnants of a ~2.48 Ga large igneous and metallogenic province in the Sudbury area of the Canadian Shield. Econ. Geol. 97, 1577–1606.

James, R.S., Jobin-Bevans, S., Easton, R.M., Wood, P., Hrominchuk, J.L., Keays, R.R., Peck, D.C., 2002b. Platinum-group element mineralization in Paleoproterozoic basic intrusions in central and northeastern Ontario, Canada. In: Cabri, L.J. (Ed.), Geology, Geochemistry, Mineralogy and Mineral Processing of Platinum Group Elements, CIM Special Publication 54, p. 339–366.

Jamieson, R.A., Culshaw, N.G., Wodicka, N., Corrigan, D., 1992. Timing and tectonic setting of Grenvillian metamorphism—constraints from a transect along Georgian Bay, Ontario. J. Metamorph. Geol. 10, 321–332.

Jessop, A.M., 1988. Geothermal experiment in the Sudbury Basin 1967–1969. In: Drury, M.J. (Ed.), Scientific Drilling, the Sudbury Structure, proceedings of a workshop. Canadian Continental Drilling Program Report, vol. 88-2, p. 17.

Jirsa, M., Fralick, P., 2010. Field Trip 4: geology of the Gunflint Iron Formation and the Sudbury Impact Layer, Northeastern Minnesota. Available from: http://www.d.umn.edu/prc/workshops/Guidebooks/BIF%20Guidebook4.pdf.

Johns, G.W., 1996. Precambrian geology of Garson and Blezard Townships. Ontario Geological Survey, Open File Report 5950, 47 pp.

Johns, G.W., Dressler, B.O., 1995. The Sudbury Igneous Complex, an impact melt sheet? Abstracts of papers submitted to the Twenty-sixth lunar and planetary science conference. Abstracts of Papers Submitted to the Lunar and Planetary Science Conference, 26, Part 2, pp. 679–680.

Johnson, M.D., Armstrong, D.K., Sanford, B.V., Telford, P.G., Rutka, M.A., 1992. Paleozoic and Mesozoic geology of Ontario. Ontario Geological Survey, Special Volume 4, Part 2, pp. 907–1008.

Jones, W.A., 1930. A study of certain xenoliths occurring in gabbro at Sudbury, Ontario. University of Toronto Studies, Geological Series, pp. 61–73.

Jones, A.P., Price, D.G., Pric, N.J., DeCarli, P.S., Clegg, R.A., 2002. Impact induced melting and the development of large igneous provinces. Earth Planet. Sci. Lett. 202, 551–561.

JORC, 2012. Australasian code for reporting of exploration results, mineral resources and ore reserves. Available from: http://www.jorc.org/docs/JORC_code_2012.pdf.

Joreau, P., French, B.M., Doukhan, J.-C., 1996. A TEM investigation of shock metamorphism in quartz from the Sudbury impact structure (Canada). Earth Planet. Sci. Lett. 138, 137–143.

Jørgensen, T.R.C., Tinkham, D.T., Lesher, C.M., 2013. Major and trace element geochemistry of the Elsie Mountain Formation volcanic rocks, Southern Province, Canada. Implication for contact metamorphism along the southern margin of the Sudbury Igneous Complex. GAC-MAC Abstracts, Volume 36.

Jørgensen, T.R.C., Tinkham, D.T., Lesher, C.M., 2014. Zirconium mobility and constraints on partial melting and melt segregation within the contact metamorphic aureole of the 1.85Ga Sudbury Igneous Complex (abstract), Ontario, Canada. GAC-MAC Fredericton, Canada.

Kamo, S.L., Krogh, T.E., Kumarapeli, P.S., 1995. Age of the Grenville dyke swarm, Ontario–Quebec: implications for the timing of Iapetan rifting. Can. J. Earth Sci. 32, 273–280.

Karvinen, W.O., Horst, R., 1979. Report of Sudbury Resident Geologist. In: Kustra, C.R. (Ed.), Annual report of the regional and resident geologists 1978. Ontario Geological Survey Miscellaneous Paper 84, pp. 99–107.

Katsube, T.J., Salisbury, M., 1994. Implications of laboratory electrical measurements on interpretation of EM-surveys and origin of the Sudbury Structure. In: Boerner, D.E., Milkereit, B., Nadrett, A.J. (Eds.), Lithoprobe Sudbury Project. Geophys. Res. Lett., vol. 21, pt. 10, pp. 947–950.

Kay, G.M., 1951. Taconic thrust in Sudbury, Vermont. Geol. Soc. Am. Bull., vol. 62, 12, Part 2, p. 1554.

Keays, R.R., 1995. The role of komatiitic and picritic magmatism and S-saturation in the formation of ore deposits. Lithos 34, 1–18.

Keays, R.R., Crocket, J.H., 1970. A study of precious metals in the Sudbury nickel irruptive ores. Econ. Geol. 65, 438–450.

Keays, R.R., Jowitt, S.M., 2013. The Avebury Ni deposit, Tasmania: a case study of an unconventional nickel deposit, Ore Geol. Rev. [P], vol. 52. Elsevier, Amsterdam Netherlands, pp. 4–17.

Keays, R.R., Lightfoot, P.C., 1999. The role of meteorite impact, source rocks, protores and mafic magmas in the genesis of the Sudbury Ni-Cu-PGE sulfide ore deposits. In: Keays, R.R., Lesher, C.M., Lightfoot, P.C., Farrow, C.E. (Eds.), Dynamic Processes in Magmatic Ore Deposits and Their Application in Mineral Exploration. Geol. Assoc. Can., short course notes, vol. 13, pp. 329–366.

Keays, R., Lightfoot, P.C., 2004. Formation of Ni–Cu–Platinum Group Element sulfide mineralization in the Sudbury Impact Melt Sheet. Mineral. Petrol. 82, 217–258.

Keays, R.R., Lightfoot, P.C., 2007. Siderophile and chalcophile metal variations in Tertiary picrites and basalts from West Greenland with implications for the sulphide saturation history of continental flood basalt magmas. Min. Dep. 42, 319–336.

Keays, R.R., Lightfoot, P.C., 2009. Crustal sulfur is required to form magmatic Ni–Cu sulfide deposits: evidence from chalcophile element signatures of Siberian and Deccan Trap Basalts. Min. Dep. 45, 241–257.

Keays, R.R., Lightfoot, P.C., 2010. Crustal sulfur is required to form magmatic Ni-Cu sulfide deposits: evidence from chalcophile element signatures of Siberian and Deccan Trap basalts. Min. Dep. 45, 241–257.

Keays, R.R., Lightfoot, P.C., 2015. Geochemical stratigraphy of the Keweenawan Midcontinent Rift volcanic rocks with regional implications for the genesis of associated Ni, Cu, Co and platinum group element sulfides. Econ. Geol. 220, 1235–1268.

Keays, R.R., Tegner, C., 2016. Magma chamber processes in the formation of the low-sulphide magmatic Au–PGE mineralization of the Platinova Reef in the Skaergaard Intrusion, East Greenland. J. Petrology. 56 (12), 2319–2339.

Keays, R.R., Lightfoot, P.C., Hamlyn, P.R., 2010. Sulfide saturation history of the Stillwater Complex, Montana: chemostratigraphic variation in the platinum group elements. Min. Dep. 47, 151–173.

Keller, G., Abramovich, S., Berner, Z., Adatte, T., 2009. Biotic effects of the Chicxulub impact, K–T catastrophe and sea level change in Texas. Palaeogeogr. Palaeoclimatol. Palaeoecol. 271 (1–2), 52–68.

Kellet, R.L., Rivard, B., 1996. Characterization of the Benny deformation zone, Sudbury, Ontario. Can. J. Earth Sci. 33, 1256–1267.

Kellett, R.L., Rivard, B., Long, D.T., Spray, J.G., 1994a. Combining remote sensing and geophysical data to characterize faulting and dike intrusion around the Sudbury impact structure, Canada. Geological Society of America, 1994 annual meeting. Abstracts with Programs—Geological Society of America, vol. 26, pt. 7, p. 266.

Kellett, R.L., Rivard, B., Spray, J.G., 1994b. Characterisation of faulting and dyke intrusion in the North Range of Sudbury using remote sensing data. In: Lithoprobe, Abitibi-Grenville Project, Abitibi-Grenville transect. Lithoprobe Report, vol. 41, pp. 141–144.

Kelly, S.P., Sherlock, S.C., 2013. The geochronology of impact craters. In: Osinski, G.R., Pierazzo, E. (Eds.), Impact Cratering. Wiley Blackwell, Oxford.

Kenkmann, T., Dalwigk, I.V., 2000. Radial transpression ridges: a new structural feature of complex impact craters. Meteoritics and Planetary Science 35, 1189–1201.

Kerr, A., 2003. Guidelines for the calculation and use of sulphide metal contents in research and mineral exploration. Current Research, Newfoundland Department of Mines and Energy, Geological Survey. Report 03-1, pp. 223–229.

Kerr, A., 2012. The Pants Lake Intrusions, Central Labrador: geology, geochemistry and magmatic Ni-Cu-Co sulfide mineralization. Report 12-02. Geological Survey, Newfoundland and Labrador. 135 pp.

Ketchum, K.Y., Heaman, L.M., Bennett, G., Hughes, D.J., 2013. Age, petrogenesis and tectonic setting of the Thessalon volcanic rocks, Huronian Supergroup, Canada. Precambrian Res. 233, 144–172.

Khryanina, L.P., 1978. The structure of meteorite craters and their central uplifts (in Russian). Doklady Akademii Nauk SSSR 238, 195–198.

Kindle, E.D., 1933. An analysis of the structural features of the Sudbury Basin and their bearing on ore deposition (Ontario). PhD Thesis. University of Wisconsin-Madison. Madison, WI, United States.

King, A., 1996. Deep drillhole electromagnetic surveys for nickel/copper sulphides at Sudbury, Canada. Explor. Geophys. 27, 105–118.

King, A., 2007. Geophysics for Ni-Cu-PGE sulfide deposits. In: Milkereit, B. (Ed.), Proceedings of Expoloration 2007, Toronto. Ore Deposits and Exploration Technology, p. 647–665.

Kirsimäe, K., Osinski, G.R., 2013. Imact-induced hydrothermal activiaty. In: Osinski, G.R., Pierazzo, E. (Eds.), Impact Cratering. Wiley Blackwell, Oxford.

Kissin, S.A., 1988. Fluid inclusions in meteorite impact structures and their relevance to the Sudbury Structure. In: Drury, M.J. (Ed.), Scientific drilling, the Sudbury Structure, proceedings of a, workshop. Canadian Continental Drilling Program Report, vol. 88-2, p. 18.

Kissin, S.A., 1990. Fluid inclusions in meteorite impact structures and their relevance to the Sudbury Structure. In: Abitibi-Grenville Project, workshop report III. Lithoprobe Report, 19, pp. 109–110.

Klimczak, C., Wittek, A., Doman, D., Riller, U., 2007. Fold origin of the NE-lobe of the Sudbury Basin, Canada: Evidence from heterogeneous fabric development in the Onaping Formation and the Sudbury Igneous Complex. J. Struct. Geol. 29 (11), 1744–1756.

Knight, C.W., 1917. Geology of the Sudbury Area and description of Sudbury Ore bodies. Report of the Royal Ontario Nickel Commission, pp. 104–211.

Knight, C.W., 1923. The chemical composition of the norite micropegmatite, Sudbury, Ontario, Canada. Econ. Geol. Bull. Soc. Econ. Geol. 18, 592–594.

Knight, C.J., 1965. A petrographic study of the Spragge Group and description of its correlation with the Sudbury Series (Precambrian), Ontario. MSc Thesis. University of Toronto. Toronto, ON, Canada.

Koeberl, C., Milkereit, B., 2007. Continental drilling and the study of impact craters and processes—an ICDP perspective. In: Koeberl, C., Zoback, M.D., Continental Scientific Drilling. Springer, Berlin. pp. 95–162.

Koeberl, C., Reimold, W.U., 2003. Geochemistry and petrography of impact breccias and target rocks from the 145 Ma Morokweng impact structure, South Africa, Geochim. Cosmochim. Acta 67 (10), 1837–1862.

Koeberl, C., Masaitis, V.L., Shafranovsky, G.I., Gilmour, I., Langenhorst, F., Schrauder, M., 1997. Diamonds from the Popigai impact structure, Russia. Geology 25, 967–970.

Koeberl, C., Sharpton, V.L., 2015. Terrestrial Impact Craters, second ed. Available from: http://www.lpi.usra.edu/publicaCons/slidesets/craters/.

Kontak, D.J., MacInnis, L.M., Ames, D.E., Rayner, N.M., Joyce, N., 2015. A geological, petrological, and geochronological study of the Grey Gabbro unit of the Podolsky Cu-(Ni)-PGE deposit, Sudbury, Ontario, with a focus on the alteration related to the formation of sharp-walled chalcopyrite veins. In: Ames, D.E., Houlé, M.G. (Eds.), Targetted Geoscience Initiative 4 Canadian Nickel-Copper-Platinum Group Elements-Chromium Ore Systems – Fertility, Pathfinders, New, Revised Models. Geological Survey of Canada, Open File 7856, pp. 287–301.

Korzhinskii, D.S., Pertsev, N.N., Zotov, I.A., 1984. Transmagmatic fluids and magmatogenic ore formation. A problem of mantle ore sources. Proceeding of the Sixth Quadrennial IAGOD Symposium, Stuttgart, Germany.

Krivolutskaya, N.A., 2016. Siberian Traps and Pt-Cu-Ni Deposits of the Noril'sk Area. Springer. 364 pp.

Krogh, T.E., McNutt, R.H., Davis, G.L., 1982. Two high precision U-Pb Zircon ages for the Sudbury Nickel Irruptive. Can. J. Earth Sci. 19 (4), 723–728.

Krogh, T.E., Davis, D.W., Corfu, F., 1984. Precise U-Pb zircon and baddeleyite ages for the Sudbury area. In: Pye, E.G., Naldrett, A.J., Giblin, P.E. (Eds.), The Geology and Ore Deposits of the Sudbury Structure. Ontario Geological Survey Special Volume 1, pp. 431–446.

Krogh, T.E., Corfu, F., Davis, D.W., Dunning, G.R., Heaman, L.M., Kamo, S.L., Machado, N., Greenough, J.D., Nakamura, E., 1987. Precise U-Pb isotopic ages of diabase dykes and mafic to ultramafic rocks using trace amounts of baddeleyite and zircon. In: Halls, H.C., Fahrig, W.F. (Eds.), Mafic Dyke Swarms. Geological Association of Canada, Special Paper 34, pp. 147–152.

Krogh, T.E., Kamo, S., Nunn, G., 1992a. Dating the first metamorphic and frontal thrust near Killarney, Sudbury, North Bay (Ontario), Chibougamau (Quebec) and Churchill Falls (Labrador): LITHOPROBE—Abitibi-Grenville Project, Abitibi-Grenville Transect, Report No. 33, p. 165.

Krogh, T.E., Kamo, S.L., Bohor, B.F., 1992b. U-Pb isotopic results for single shocked and polycrystalline zircons record 550-65.5 Ma ages for a K-T target site and 2700-1850-Ma ages for the Sudbury impact events. In: Dressler, B.O., Sharpton, V.L. (Eds.), Papers Presented to the International Conference on Large Meteorite Impacts and Planetary Evolution. V.L. LPI Contribution 790, pp. 44–45.

Krogh, T.E., Kamo, S.L., Bohor, B.F., 1996. Shocked metamorphosed zircons with correlated U-Pb discordance and melt rocks with concordant protolith ages indicate an impact origin for the Sudbury Structure. In: Earth Processes: Reading the Isotopic, Code., American Geophysical Union Monograph 95, pp. 343–352.

Krüger, T., van der Bogert, C.H., Hessinger, H., 2013. New high-resolution melt distribution map and topographic analysis of Tycho Crater. 44th Lunar and Planetary Science Conference. Available from: http://www.lpi.usra.edu/meetings/lpsc2013/pdf/2152.pdf.

Kull, U., 1986. Entstehung der Erzlagerstaette von Sudbury (Genesis of the ore deposits of Sudbury). Naturwissenschaftliche Rundschau 39 (12), 535–536.

Kullerud, G., 1963. Thermal stability of pentlandite. Can. Mineral. 7, 353–366.

Kullurud, G., 1963. The Fe-Ni-S System. Carnegie Instituion, Annual Report of the Director. Geophys. Lab. 62, 175–189.

Kullerud, G., Bell, P.M., England, J.L., 1965. High-Pressure Differential Thermal Analysis. Carnegie Institution of Washington Year Book, Washington, 64, 197–199.

Kullerud, G., Yund, R.A., Moh, G.H., 1969. Phase relationships in the Cu-Fe-S, Cu-Ni-S and Fe-Ni-S systems. Econ. Geol. Monogr. 4, 323–343.

Kunilov, V.Ye., 1994. Geology of the Norilsk Region: The History of the Discovery, Prospecting, Exploration and Mining of the Norilsk Deposits. In: Lightfoot, P.C., Naldrett, A.J. (Eds.), Proceedings of the Sudbury—Norilsk Symposium. Ontario Geological Survey Special Publication 5, pp. 203–216.

Kunz, G.F., 1918. Genesis of the Sudbury nickel-copper ores (discussion [occurrence of palladium and platinum]). Bulletin of the American Institute of Mining and Metallurgical Engineers, 848–849.

Kuo, H.Y., Crocket, J.H., 1979. Rare earth elements in the Sudbury Nickel Irruptive: comparison with layered gabbros and implications for Nickel Irruptive petrogenesis. Econ. Geol. 79, 590–605.

Kwak, T.A.P., 1968. Metamorphic petrology and geochemistry across the Grenville Province-southern Province boundary, Dill Township, Sudbury, Ontario (Canada). PhD Thesis. McMaster University, Hamilton, ON, Canada. 200 pp.

L'Heureux, E., Ugalde, H., Milkereit, B., Boyce, J., Morris, W., Eyles, N., Artemieva, N., 2005. Using vertical dikes as a new approach to constraining the size of buried impact craters: an example from lake Wanapitei, Canada. Geol Soc America Special Paper 384, pp. 43–50.

Lafleur, J., 1981. Cascaden, Dowling, and Trill townships, District of Sudbury. In: Wood, J., White, O.L., Barlow, R.B., Colvine, A.C. (Eds.), Summary of field work, 1981. Ontario Geological Survey Miscellaneous Paper 100, pp. 80–83.

Lafleur, J., Dressler, B.O., 1985. Geology of Cascaden, Dowling, Levack, and Trill Townships, District of Sudbury. Ontario Geological Survey, Open File Report 5533, 135 pp.

Lafrance, B., Kamber, B.S., 2010. Geochemical and microstructural evidence for in situ formation of pseudotachylitic Sudbury breccia by shock-induced compression and cataclasis. Precambrian Res. 180, 237–250.

Lafrance, B., Legault, D., Ames, D.E., 2008. The formation of the Sudbury breccia in the North Range of the Sudbury Impact Structure. Precambrian Res. 165, 107–119.

Lafrance, B., Bygnes, L., McDonald, A., 2014. Emplacement of metabreccia along the Whsitle offset dike, Sudbury: implications for post-impact modification of the Sudbury impact structure. Can. J. Earth. Sci. 51, 466–484.

Lakomy, R., 1988. Petrographische und geochemische Untersuchungen an der Footwall-Breccie,Sudbury, Kanada (Petrographic and geochemical studies of footwall breccia, Sudbury, Ontario, Canada). Referate der Vortraege und Poster, 66. Jahrestagung der Deutschen Mineralogischen Gesellschaft. Fortschritte der Mineralogie, Beiheft, vol. 66, pt. 1, p. 94.

Lakomy, R., 1990a. Distribution of impact induced phenomena in complex terrestrial impact structures: implications for transient cavity dimensions. Lunar Planet. Sci. XXI, 676–677.

Lakomy, R., 1990b. Implications for cratering mechanics from breccias in the basement of the Sudbury impact crater, Canada. In: Abstracts of papers submitted to the Twenty-first lunar and planetary science conference. Abstracts of Papers Submitted to the Lunar and Planetary Science Conference, vol. 21, pp. 678–679.

Lakomy, R., 1990c. Implications for cratering mechanics from a study of the Footwall Breccia of the Sudbury impact structure. Can. Meteor. 25, 195–207.

Lakomy, R., 1990d. Sudbury (Canada): Implications for cratering mechanics from the footwall breccia (abstract). In: Pesonen, L.J., Niemisara, H. (Eds.), Symposium Fennoscandian Impact Structures, p. 60.

Lakomy, R., Deutsch, A., Buhl, D., 1988. Sr-Nd-Kleinbereichsuntersuchungen an der Footwall-Breccie (Sudbury, Kanada) (Sr-Nd adjustment studies of footwall breccia, Sudbury, Canada.) In: Referate der Vortraege und Poster, 66. Jahrestagung der Deutschen Mineralogischen Gesellschaft. Fortschritte der Mineralogie, Beiheft, vol. 66, pt. 1, p. 95.

Lamothe, D., 2007. Lexique stratigraphique de l'orogène de l'Ungava. Ministère des Ressources naturelles, de la Faune et des Parcs, DV 2007-03, Québec.

Langenhorst, F., Shafranovsky, G., Masaitis, V.L., 1998. A comparative study of impact diamonds from the Poigai, Ries, Sudbury, and Lappajarvi craters. Meteor. Planet. Sci. 33, 90–91.

Langford, F.F., 1960. Geology of Levack Township and the northern part of Dowling Township, District of Sudbury. Ontario Department of Mines Preliminary Report 1960-5, 78 pp.

Latypov, R.M., Lavrenchuk, A., Lightfoot, P.C., 2010. Was the parental magma of the Sudbury Igneous Complex really superheated. 11th International Platinum Symposium, Sudbury.

Lausen, C., 1930. Graphic intergrowth of niccolite and chalcopyrite, Worthington Mine, Sudbury (Ontario). Econ. Geol. Bull. Soc. Econ. Geol. 25 (4), 356–364.

Laverov, N.P., 1999. Copper-nickel deposits of the Pechenga Camp. GEOS Moscow, 235 pp.

Lavrenchuk, A., Latypov, R.M., Lightfoot, P.C., 2010. The Sudbury Igneous Complex, Canada: numerical modeling confirms fractionation of a single parental magma. 11th International Platinum Symposium, Sudbury.

Lee, C., Siddorn, J., 2006. Garson structural geology review. Unpublished SRK report for Vale, 12 p.

Le Maitre, R.W. (Ed.), 2002. Igneous Rocks. A classification and glossary of terms. Recommendations of the International Union of Geological Sciences Subcommission on the Systematics of Igneous Rocks, second ed.:xvi + 236 pp. Cambridge University Press, Cambridge, New York, Melbourne.

LeFort, D.T.R., 2012. A mineralogical and fluid inclusion study of modified, contact-style Ni-Cu-PGE ores in the #1 and #4 shear zones, Garson Mine, Sudbury, Ontario, Canada. MSc Thesis. Saint Mary's University, Halifax, N.S. 189 pp.

Legault, D., LaFrance, B., Ames, D.E., 2003. Structural study of Sudbury Breccia and sulfide veins, Levack embayment, North Range of the Sudbury Structure, Ontario. Geol. Surv. Canada Curr. Res. 2003-C1, 1–9.

Lenauer, I., 2012. Structural analysis and 3D kinematic restoration of the Southern Sudbury Basin. PhD Thesis. McMaster University, Hamilton, Canada. 122 pp.

Lenauer, I., Riller, U., 2012a. Geometric consequences of ductile fabric development from brittle shear faults in mafic melt sheets: Evidence from the Sudbury Igneous Complex, Canada. J. Struct. Geol. 35, 40–50.

Lenauer, I., Riller, U., 2012b. Strain fabric geometry within and near deformed igneous sheets: the Sudbury Igneous Complex, Canada. Tectonophysics 558–559, 45–57.

Lennon, J., 2007. The Chinese nickel outlook and the role of nickel pig iron. McQuire Research Commodities. International Nickel Study Group. Available from: http://www.insg.org/presents/Mr_Lennon_May07.pdf.

Lesher, C.M., Golightly, J.P., Keays, R.R., 1999. Dynamic incongruent melting in magmatic sulphide deposits. 1999 Geological Association of Canada Annual Meeting, Sudbury, ON, 24 (Abstract Volume), p. 70.

Lesher, C.M., Walker, J.A., Lightfoot, P.C., 2014. Geochemistry and petrogenesis of the Main Mass of the Sudbury Igneous Complex (abs): Geological Association of Canada—Mineralogical Association of Canada Joint Annual Meeting, p. 160.

Lewis, C.L., 1950. The minor elements of the Sudbury ore minerals. MSc Thesis. Queen's University, Kingston, ON, Canada. 92 pp.

Lewis, C.R., 1951. The age relationship of the Murray granite and "Sudbury norite" (Ontario). Can. Min. J. 72, 55–62.

Lewis, A., 1981. Sudbury, the world's largest Ni district. In: Dayton, S.H., White, R.L., Burger, J.R., Jackson, D., Lewis, A., Sassos, M.P., Edelman, B. (Eds.), Spotlight on Canada's resourceful mining industry. E & MJ, Eng. Min. J. 182, 84–85.

Li, C., 1993. A quantitative model for the formation of sulfide ores at Sudbury and a study on the distributions of platinum-group elements in the Strathcona copper-rich zones, Sudbury, Ontario. PhD Thesis. University of Toronto. Toronto, ON, Canada, 231 pp.

Li, C., Naldrett, A.J., 1989. Grant 326, PGE studies in the footwall at Sudbury. In: Geoscience Research Grant Program, summary of research, 1988–1989. In: Milne, V.G. (Ed.), Ontario Geological Survey Miscellaneous Paper 143, pp. 104–113.

Li, C., Naldrett, A.J., 1990. PGEs at Sudbury, their concentration by sulphide-liquid fractionation. In: Milne, V.G. (Ed.), Geoscience Research Grant Program, summary of research 1989–1990. Ontario Geological Survey Miscellaneous Paper 150, pp. 37–46.

Li, C., Naldrett, A.J., 1992. PGE studies in the Footwall at Sudbury. Ontario Geological Survey, Open File Report 5830, 118 pp.

Li, C., Naldrett, A.J., 1993a. High chlorine alteration minerals and calcium-rich brines in fluid inclusions from the Strathcona Deep Copper Zone, Sudbury, Ontario. Econ. Geol. 88, 1780–1796.

Li, C., Naldrett, A.J., 1993b. Sulfide capacity of magma, a quantitative model and its application to the formation of sulfide ores at Sudbury, Ontario. Econ. Geol. Bull. Soc. Econ. Geol. 88 (5), 1253–1260.

Li, C., Naldrett, A.J., 1994. A numerical model for the compositional variations of Sudbury sulfide ores and its application of exploration. Econ. Geol. Bull. Soc. Econ. Geol. 89 (7), 1599–1607.

Li, C., Naldrett, A.J., 2000. Melting reactions of gneissic inclusions with enclosing magma at Voisey's Bay, Labrador, Canada: Implications with respect to ore genesis. Econ. Geol. 95, 801–814.

Li, C., Ripley, E.M., 2005. Empirical equations to predict the sulfur content of mafic magmas at sulfide saturation and application to magmatic sulfide deposit. Min. Deposita 40, 218–230.

Li, C., Rucklidge, J.C., Naldrett, A.J., 1990. PGE at Sudbury, in discrete minerals or dissolved in base metal sulfides? GAC/AGC-MAC/AMC Joint Annual Meeting (abstract), vol. 15, p. 76.

Li, C., Naldrett, A.J., Rucklidge, J.C., Kilius, L.R., 1991. Partitioning of platinum-group elements and gold in sulfides from Strathcona Mine, Sudbury, Ontario. Geological Society of America, 1991 annual meeting. Abstracts with Programs—Geological Society of America, vol. 23, pt. 5, p. 291.

Li, C., Naldrett, A.J., Coats, C.J.A., Johannessen, P., 1992. Platinum, palladium, gold, copper-rich stringers at the Strathcona Mine, Sudbury, their enrichment by fractionation of a sulfide liquid. Econ. Geol. Bull. Soc. Econ. Geol. 87 (6), 1584–1598.

Li, C., Naldrett, A.J., Rucklidge, J.C., Kilius, L.R., 1993. Concentrations of platinum-group elements and gold in sulfides from the Strathcona Deposit, Sudbury, Ontario. Can. Mineral. 31, 523–531.

Li, C., Naldrett, A.J., Ripley, E.M., 2001. Critical factors for the formation of a Ni-Cu deposit in evolved magmatic system: Lessons from a comparison of the Pants Lake and Voisey's Bay sulfide occurrences in Labrador. Min. Dep. 36, 85–92.

Li, C., Zhang, M., Fu, P., Qian, Z., Hu, P., Ripley, E.M., 2012. The Kalatongke magmatic Ni-Cu deposit in the Central Asian Orogenic Belt, NW China: product of slab window magmatism? Mineral. Deposit. 47, 51–67.

Li, X., Su, L., Song, B., Liu, D., 2004. SHRIMP U-Pb-zircon age of the Jinchuan ultramafic intrusion and it's geological significance. Chin. Sci. Bull. 49, 420–422.

Li, X.H., Su, L., Chung, S.-L., Li, Z.X., Liu, X., Song, B., Liu, D.Y., 2005. Formation of the Jinchuan ultramafic intrusion and the world's third largest Ni-Cu sulfide deposit: associated with the ~825 Ma south China mantle plume? Geochem. Geophys. Geosys. 11, 1–16.

Lieger, D., Riller, U., 2012. Emplacement history of Granophyre dikes in the Vredefort Impact Structure, South Africa, inferred from geochemical evidence. ICARUS 219, pp. 168–180.

Lieger, D., Riller, U., Gibson, R., 2009. Generation of fragment-rich pseudotachylite bodies during central uplift formation in the Vredefort Impact Structure. South Africa. Earth Planet. Sci. Lett. 279, 53–64.

Lieger, D., Riller, U., Gibson, R.L., 2011. Petrographic and geochemical evidence for an allochthonous, possibly impact melt, origin of pseudotachylite from the Vredefort Dome, South Africa. Geochim. Cosmochim. Acta 75, 4490–4514.

Lightfoot, P.C., 1995. The West Greenland Gabbronorite Province. Unpublished Report. November 12, 1995. Cominco Exploration Limited, Toronto.

Lightfoot, P.C., 1996. The Giant Nickel Deposits at Sudbury and Noril'sk. Nickel in a Nutshell presented at 64th Annual Convention and Trade Show of the Prospectors and Developers Association of Canada. March 10–13, 1996.

Lightfoot, P.C., 2007. Advances in Ni-Cu-PGE deposit models and exploration technologies, exploration. In: Milkereit, B. (Ed.), Proceedings of Exploration 07 Fifth Decennial International Conference on Mineral Exploration, p. 629–646. Available from: http://www.dmec.ca/ex07-dvd/E07/pdfs/44.pdf.

Lightfoot, P.C., 2009. Mechanisms of Ni-Cu-PGE sulphide ore formation in the impact-generated Sudbury melt sheet. GAC-MAC-AGU Annual Meeting, Toronto. Program and Abstracts. Available from: http://www.agu.org/meetings/ja09/program/index.php.

Lightfoot, P.C., 2015. Structural controls on ore deposits at Sudbury, Thompson, and Voisey's Bay, Canada. Prospectors and Developmers Association of Canada, Nickel a Global View. Available from: http://www.pdac.ca/docs/default-source/Convention—Program—Technical-Sessions/2015/nickel—lightfoot.pdf?sfvrsn=2.

Lightfoot, P.C., Evans-Lamswood, D., 2015. Structural controls on the primary distribution of mafic-ultramafic intrusions containing Ni-Cu-Co-(PGE) mineralization in the roots of large igneous provinces. Ore Geol. Rev. 64, 354–386.

Lightfoot, P.C., Farrell, K.P., 1993. Project Unit 93-08, Geochemistry of the Sudbury Igneous Complex. In: Baker, C.L., Dressler, B.O., De Souza, H.A.F., Fenwick, K.G., Newsome, J.W., Owsiacki, L. (Eds.), Summary of Field Work and Other Activities 1993. Ontario Geological Survey Miscellaneous Paper 162, pp. 79–82.

Lightfoot, P.C., Farrow, C.E.G., 2002. Geology, geochemistry and mineralogy of the Worthington Offset Dyke: towards a genetic model for offset mineralisation in the Sudbury Igneous Complex. Econ. Geol. 97, 1419–1446.

Lightfoot, P.C., Hawkesworth, C.J., 1997. Flood basalts and magmatic Ni, Cu, and PGE sulfide mineralization: comparative geochemistry of the Noril'sk (Siberian Trap) and West Greenland sequences. Am. Geophys. Union 100, 357–380.

Lightfoot, P.C., Keays, R.R., 2005. Siderophile and Chalcophile Metal Variations in Flood Basalts from the Siberian Trap, Noril'sk Region: implications for the Origin of the Ni-Cu-PGE Sulfide Ores. Econ. Geol. 100, 439–462.

Lightfoot, P.C., Naldrett, A.J., 1989. Assimilation and crystallization in basic magma chambers: Trace-element and Nd- isotopic variations in the Kerns sill, Nipissing diabase province, Ontario. Can. J. Earth Sci. 26, 737–754.

Lightfoot, P.C., Naldrett, A.J. (Eds.), 1994. Proceedings of the Sudbury-Noril'sk Symposium. Ontario Geological Survey Special Volume 5, 423 pp.

Lightfoot, P.C., Naldrett, A.J., 1996. Petrology and geochemistry of the Nipissing Gabbro: exploration strategies for nickel, copper, and platinum group elements in a large igneous province. Ontario Geological Survey Study, vol. 58, 80 pp.

Lightfoot, P.C., Naldrett, A.J., 1999. Geological and geochemical relationships in the Voisey's Bay intrusion Nain Plutonic Suite Labrador Canada. In: Keays, R.R., Lesher, C.M., Lightfoot, P.C., Farrow, C.E.G. (Eds.), Dynamic Processes in Magmatic Ore Deposits and Their Application in Mineral Exploration. Geol Ass Canada 13, pp. 1–31.

Lightfoot, P.C., Noble, S.R., 1992. U-Pb baddeleyite ages of the Kerns and Triangle Mountain Intrusions, Nipissing Diabase, Ontario. Can. J. Earth Sci. 29, 1424–1429.

Lightfoot, P.C., Zotov, I.A., 2006. Geology and geochemistry of the Sudbury Igneous Complex, Ontario, Canada: origin of nickel sulfide mineralization associated with an impact-generated melt sheet. Geol. Ore Dep. 47 (5), 349 (in Russian and English).

Lightfoot, P.C., Zotov, I.A., 2007. Ni-Cu-PGE sulphide deposits at Noril'sk, Russia. Short Course; International Polar Year; Prospectors and Developers Association of Canada. Short Course Notes.

Lightfoot, P.C., Zotov, I.A., 2014. Geological relationships between the intrusions, country rocks, and Ni-Cu-PGE sulfides of the Kharaelakh Intrusion, Noril'sk Region: implications for the role of sulfide differentiation and metasomatism in their genesis. Northwest. Geol. 47, 1–35.

Lightfoot, P.C., Naldrett, A.J., Hawkesworth, C.J., 1984. The geology and geochemistry of the Waterfall Gorge Section of the Insizwa Complex with particular reference to the origin of the nickel sulfide deposits. Econ. Geol. 79, 1857–1879.

Lightfoot, P.C., Naldrett, A.J., Gorbachev, N.S., Doherty, W., Fedorenko, V.A., 1990. Geochemistry of the Siberian Trap of the Noril'sk Area, USSR, with implications for the relative contributions of crust and mantle to flood basalt magmatism. Contrib. Mineral. Petrol. 104, 631–644.

Lightfoot, P.C., Hawkesworth, C.J., Hergt, J., Naldrett, A.J., Fedorenko, V.A., Gorbachev, N.S., Doherty, W., 1992. Remobilisation of the continental lithoepshere by a mantle plume: major-, trace-element and Sr-, Nd- and Pb-isotope evidence from picritic and tholeiitic lavas of the Noril'sk District, Siberian Trap, Russia. Contrib. Mineral. Petrol. 114, 171–188.

Lightfoot, P.C., De Souza, H., Doherty, W., 1993a. Differentiation and source of the Nipissing Diabase Intrusions, Ontario, Canada. Can. J. Earth Sci. 30, 1123–1140.

Lightfoot, P.C., Hawkesworth, C.J., Hergt, J., Naldrett, A.J., Gorbachev, N.S., Fedorenko, V.A., Doherty, W., 1993b. Remobilisation of the continental lithosphere by mantle plumes: major-, trace element, and Sr-, Nd-, and Pb-isotope evidence from picritic and tholeiitic lavas of the Noril'sk District, Siberian Trap, Russia. Contributions to Mineralogy and Petrology, vol. 114, pp. 171–188.

Lightfoot, P.C., Naldrett, A.J., Hawkesworth, C.J., Gorbachev, N.S., Fedorenko, V.A., Hergt, J., Doherty, W., 1994a. Source and evolution of Siberian Trap Lavas, Noril'sk District, Russia: Implications for the origin of the sulfides. Sudbury-Noril'sk Symposium Volume. Special Publication No. 5. Ontario Geological Survey.

Lightfoot, P.C., Keays, R.R., Moore, M., Farrell, K., Pekeski, D., 1994b. Geochemistry of the Sudbury Igneous Complex. In: Summary of Field Work and Other Activities, Ontario Geological Survey, Miscellaneous Paper 163, pp. 81–86.

Lightfoot, P.C., Farrell, K., Moore, M., Pekeski, D., Crabtree, D., Keays, R.R., 1995. Geochemical relationships among the Main Mass, Sublayer and Offset of the Sudbury Igneous Complex, Ontario, Canada. Ontario Geol. Surv. Misc. Pub. 164, 116–121.

Lightfoot, P.C., Hawkesworth, C.J., Olshefsky, K., Green, T., Doherty, W., Keays, R.R., 1997a. Geochemistry of Tertiary tholeiites and picrites from Qeqertarssuaq (Disko island) and Nuussuaq, West Greenland with implications for the mineral potential of comagmatic intrusions. Contrib. Mineral. Petrol. 128, 139–163.

Lightfoot, P.C., Keays, R.R., Morrison, G.G., Bite, A., Farrell, K.P., 1997b. Geological and geochemical relationships between the contact sublayer, inclusions, and main mass of the Sudbury Igneous Complex: a case study of the Whistle Mine embayment. Econ. Geol. Bull. Soc. Econ. Geol. 92, 647–673.

Lightfoot, P.C., Keays, R.R., Morrison, G.G., Bite, A., Farrell, K.P., 1997c. Geochemical relationships. In: The Sudbury Igneous Complex: Origin of the Main Mass and Offset Dykes. Econ. Geol., vol. 92, pp. 289–387.

Lightfoot, P.C., Doherty, W., Farrell, K., Keays, R.R., Moore, M., Pekeski, D., 1997d. Geochemistry of the Main Mass, Sublayer, Offsets, and Inclusions from the Sudbury Igneous Complex, Ontario. Ontario Geological Survey, Open File Report 5959, 231 pp. Available from: http://www.geologyontario.mndmf.gov.on.ca/mndmfiles/pub/data/imaging/OFR5965/OFR5965.pdf.

Lightfoot, P.C., Doherty, W., Farrell, K., Keays, R.R., Moore, M., Pekeski, D., 1997e. Analytical Data for Geochemistry of the Main Mass, Sublayer, Offsets and Inclusions from the Sudbury Igneous Complex, Ontario—Tables 5, 6 and 7; Ontario Geological Survey, Miscellaneous Release-Data 30. Available from: http://www.geologyontario.mndmf.gov.on.ca/mndmfiles/pub/data/imaging/MRD030/MRD030.pdf.

Lightfoot, P.C., Keays, R.R., Doherty, W., 1997f. Can impact-generated melts have mantle contributions? Geochemical evidence from the Sudbury Igneous Complex (abstract). Large Meteorite Impacts and Planetary Evolution.

Lightfoot, P.C., Keays, R.R., Morrison, G.G., Bite, A., 1997g. Geological and geochemical relationships between the contact sublayer, offsets, and main mass of the Sudbury Igneous Complex (abstract). Large Meteorite Impacts and Planetary Evolution.

Lightfoot, P.C., Keays, R.R., Doherty, W., 2001. Chemical Evolution and Origin of Nickel Sulfide Mineralization in the Sudbury Igneous Complex, Ontario, Canada. Econ. Geol. 96, 1855–1875.

Lightfoot, P.C., Gibson, L., Golightly, J.P., Morrison, G., 2002. Geometry of the Sudbury Structure and distribution of the Ni, Cu, and PGE Sulfide Ores. ICDP Planning Meeting, Sudbury, Canada.

Lightfoot, P.C., Keays, R.R., Evans-Lamswood, D., Wheeler, R., 2011a. S-saturation history of the Nain Plutonic Suite mafic intrusions: origin of the Voisey's Bay Ni-Cu-Co sulfide deposit, Labrador. Min. Dep. 47, 23–50.

Lightfoot, P.C., Stewart, R., Gribbin, G., Mooney, S., 2011b. Relative contributions of magmatic and post-magmatic processes in the genesis of the Thompson Belt Ni-Co sulfide ore deposits, Manitoba, Canada. 12th International Ni-Cu-(PGE) Symposium, Guiyang, China.

Lightfoot, P.C., Evans-Lamswood, D., Wheeler, R., 2012. The Voisey's Bay Ni–Cu–Co sulfide deposit, Labrador, Canada: emplacement of silicate and sulfide-laden magmas into spaces created within a structural corridor. Northwest. Geol. 45, 17–28.

Likhachev, A.P., 1994. Ore-bearing intrusions of the Norilsk Region. In: Naldrett, A.J., Lightfoot, P.C., Sheahan, P. (Eds.), The Sudbury - Norilsk Symposium, Ontario Geological Survey, Special Publication No. 5, pp. 185–202.

Likhachev, A.P., Kiricov, A.D., Kirisov, A.D., LiFatu, A.V., 1994. An attempt of determination of isotope composition of Fe, Ni and Cu in minerals of PGE-Cu-Ni deposits in Noril'sk and Sudbury. International Mineralogical Association, 16th general meeting, abstracts. Abstracts of the General Meeting of the international Mineralogical Association, vol. 16, p. 244.

Lindeman, E., 1912. Magnetometric survey of a nickeliferous pyrrhotite deposit in the Sudbury District. Can Mines Br, Sum Rp, 1911, pp. 103–104.

Lindgren, W., 1918. Genesis of the Sudbury nickel-copper ores (discussion). Bull. Am. Inst. Min. Metal. Eng. 136, 857.

Livelybrooks, D.W., Mareschal, M., Blais, E., Smith, J.T., 1996. Magnetotelluric delineation of the Trillabelle massive sulfide body in Sudbury, Ontario. Geophysics 61 (4), 971–986.

Lockhead, D.R., 1955. A review of the Falconbridge ore deposit. Econ. Geol. 50, 42–50.

Lohman, P.D., 1992. Impact origin of the Sudbury Structure, evolution of a theory. In: Dressler, B.O., Sharpton, V.L. (Eds.), Papers presented to the international conference on large meteorite impacts and planetary evolution. LPI Contribution 790, p. 48.

Long, D.G.F., 1974. Glacial and paraglacial genesis of conglomeratic rocks of the Chibougamau Formation (Aphebian), Chibougamau, Quebec. Can. J. Earth Sci. 11, 1236–1252.

Long, D.G.F., 1978. Depositional environments of a thick Proterozoic sandstone, the (Huronian) Mississagi Formation of Ontario, Canada. Can. J. Earth Sci. 15, 190–206.

Long, D.G.F., 1992. Does the sedimentology of the Chelmsford formation provide evidence for a meteorite impact origin of the Sudbury structure? (abstract), International Conference on Large Meteorite Impacts and Planetary Evolution, LPI Contribution No. 790, pp. 48.

Long, D.G.F., 1995. Huronian sandstone thickness and paleocurrent trends as a clue to the tectonic evolution of the Southern Province; Can. Mineral., vol. 33, part 4, pp. 922–923.

Long, D.G.F., 2004. The tectonostratigraphic evolution of the Huronian basement and the subsequent basin fill: geological constraints on impact models of the Sudbury event. Precambrian Res. 129, 203–223.

Long, D.G.F., 2009. The Huronian Supergroup. In: Roussel, D.H., Brown, G.H. (Eds.), A Field Guide to the Geology of Sudbury, Ontario. Ontario Geological Survey Open File Report 6243, pp. 14–30.

Long, D.G.F., Lloyd, T.R., 1983. Placer gold potential of basal Huronian strata of the Elliot Lake Group in the Sudbury area, Ontario; in Summary of Field Work 1983, Ontario Geological Survey, Miscellaneous Paper 116, pp. 256–258.

Long, D.G.F., Young, G.M., Rainbird, R.H., Fedo, C.M., 1999. Actualistic and non-actualistic Precambrian sedimentary styles: examples from the Proterozoic, north shore of Lake Huron; Field Trip B5, Geological Association of Canada–Mineralogical Association of Canada, Joint Annual Meeting, Sudbury, Ontario, Guidebook, 50 p.

Lowman, P.D., 1988. The Sudbury Structure, a planetary perspective. In: Drury, M.J. (Ed.), Scientific drilling, the Sudbury Structure, proceedings of a workshop. Canadian Continental Drilling Program Report, 88-2, p. 15.

Lowman, P.D., 1992. The Sudbury Structure as a terrestrial mare basin. Rev. Geophys. 30, 227–243.

Lowman, Jr., P.D., 1992. Impact origin of the Sudbury structure: Evolution of a theory (abstract). International Conference on Large Meteorite Impacts and Planetary Evolution, LPI Contribution No. 790, pp. 48.

Lowman, P.D., 1993. Formation of the Sudbury Igneous Complex by impact-induced crustal anatexis. In: AGU 1993 spring meeting. Eos Trans. Am. Geophys. Union 74 (16 Suppl.), 301.

Lowman, P.D., 1994a. Original shape of the Sudbury Structure, Canada: a study with airborne imaging radar. Can. J. Remote Sens. 17 (2), 152–161.

Lowman, P.D., 1994b. Radar geology of the Canadian Shield: a 10-year review. Can. J. Remote Sens. 20, 198–209.

Lowman, Jr., P.D., Whiting, P.J., Short, N.M., Lohmann, A.M., Lee, G., 1992. Fracture patterns on the Canadian Shield: a lineament study with Landsat and orbital radar imagery. Int. Basement Tectonics Assoc. Publ. 7, 139–159.

Lowman, P.D., Singhroy, V.H., Slaney, V.R., 1992. Imaging radar investigations of the Sudbury Structure. In: Dressler, B.O., Sharpton, V.L. (Eds.), Papers presented to the international conference on large meteorite impacts and planetary evolution. LPI Contribution 790, p. 49.

Lowrie, W., Hirt, A.M., 1989. A rock magnetic investigation of the magnetic anisotropy of the Onaping Formation (Sudbury Basin, Ontario, Canada). AGU 1989 spring meeting. Eos, Transactions, American Geophysical Union, vol. 70, suppl 15, p. 318.

Lowrie, W., Hirt, A.M., Kligfield, R., 1994. Evidence for a near-circular origin for the Sudbury Basin (Ontario) from strain and AMS data. In: AGU 1994 spring meeting. Eos, Transactions, American Geophysical Union, vol. 75, 16, Suppl., p. 122.

Lumbers, S.B., 1975. Geology of the Burwash area, Districts of Nipissing, Parry Sound and Sudbury. Ontario Division of Mines, Geological Report 116.

Lumbers, S.B., Frarey, M.J., Card, K.D., Richardson, J.A., 1979. Day 3, Sudbury, Ontario, to Blind River, Ontario. In: Morey, G.B. (Ed.), Field trip guidebook for the Archean, Proterozoic stratigraphy of the Great Lakes area, United States, Canada. Guidebook Series—Minnesota Geological Survey, vol. 13, pp. 43–48.

Luo, Z., Mo, X.-X., Lu, X.-X., He, C.-B., Ke, S., Hou, Z.-Q., Jiang, W., 2007. Metallogeny by trans-magmatic fluids—theoretical analysis and field evidence. Earth Sci. Front. 14, 165–183.

Lyell, C., 1830. The principles of geology. Murray, London. vol. 2, pp. 20–21.

Lyons, W., 2010. Working guide to drilling equipment and operations. Elsevier, Amsterdam, 602 pp.

Mackovicky, M., Mackovicky, E. Rose-Hanen, J., 1986. Experimental studies on the solubility and distribution of platinum group elements in bnase-metal sulfides in platinum deposits. In: Gallagher, M.J., Ixer, R.A., Neary, C.R., Prichard, H.M. (Eds.), Metallogeny of Basic and Ultrabasic rocks. Inst. Min. Met. Spec. Publ. pp. 415–426.

MacLean, W.H., Shimazaki, H., 1976. The partition of Co, Ni, Cu and Zn between sulfide and silicate liquids. Econ. Geol. Bull. Soc. Econ. Geol. 71, 1049–1057.

Maddock, R.H., 1983. Melt origin of fault-generated pseudotachylites demonstrated by textures. Geology 11, 105–108.

Magyarosi, Z., 1998. Metamorphism of the Proterozoic Rocks Associated with the Sudbury Structure. MSc Thesis. Carleton University, Ottawa. 101 pages.

Magyarosi, Z., Walksinson, D., Jones, P.C., 2002. Mineralogy of the Ni-Cu-Platinum-Group Element Sulfide Ore from the 800 and 810 Orebodies, Copper Cliff South Mine: P-T-X Conditions During the Formation of Platinum Group Minerals. Econ. Geol. 97, 1471–1486.

Maier, W.D., Groves, D.I., 2011. Temporal and spatial controls on the formation of magmatic PGE and Ni-Cu deposits. Min. Dep. 46, 841–857.

Maier, W.D., Andreoli, M., Macdonald, I., Prevec, S., 2003. The Morokweng impact melt sheet, South Africa: a reconnaissance study with implications for Ni-Cu-PGE sulfide mineralization. Appl. Earth Sci. Trans. Inst. Min. Metall. 112, 150–152.

Maier, W.D., Andreoli, M.A.G., I. McDonald, I., Higgins, M.D., Boyce, A.J., Shukolyukov, A., Lugmair, G.W., Ashwal, L.D., Gräser, P., Ripley, E.M., Hart, R.J., 2006. Discovery of a 25-cm asteroid clast in the giant Morokweng impact crater, South Africa. Nature 441, 203–206.

Manning, S.E., Morris, W.A., 1994. Form and extent of the Grenville Front tectonic zone in the proximity of Coniston, Ontario as defined by aeromagnetic, paleomagnetic, and magnetic fabric studies of Sudbury olivine diabase dikes. GAC/AGC-MAC/AMC Joint Annual Meeting (abstract), vol. 19, p. 71.

Marchand, M., 1976. A geochemical and geochronologic investigation of meteorite impact melts at Mistastin Lake, Labrador and Sudbury, Ontario. PhD Thesis. McMaster University, Hamilton, ON, Canada. 142 pp.

Marshall, D., Watkinson, D., Fouillac, A.M., Molnar, F., Farrow, C., 1996. Fluid mixing during the emplacement of the Sudbury Igneous Complex, fluid inclusion, Ar, O and H isotopic evidence. In: Brown, P.E., Hagemann, S.G. (Eds.), PACROFI VI, Sixth biennial Pan-American conference on Research on fluid inclusions, program and abstracts. Program and Abstracts, Biennial Pan-American Conference on Research on Fluid Inclusions, vol. 6, pp. 83–85.

Marshall, D., Watkinson, D., Farrow, C., Molnar, F., Fouillac, A.-M., 1999. Multiple fluid generations in the Sudbury Igneous Complex: fluid inclusion, Ar, O, H, Rb and Sr evidence. Chem. Geol. 154, 1–19.

Martin, W.G., 1957. Errington and Vermilion mines, in Structural Geology of Canadian Ore Deposits Volume 2. Commonwealth Mining and Metallurgical Congress, pp. 363–376.

Martins, J.M., 1984. Geology of a suite of granitic rocks, South Range of the Sudbury Igneous Complex. GAC/AGC-MAC/AMC Joint Annual Meeting (abstract), vol. 9, p. 87.

Masaitis, V.L., 1993. Diamondiferous impactites, their distribution and petrogenesis. Reg. Geol. Metall. 1, 121–134, (in Russian).

Masaitis, V.L., Selivanovskaya, T.V., 1972. Shock-metamorphosed rocks and impactites of the Popigay meteor crater. Zapiski Vsesoyuznogo Mineralogicheskogo Obshchestva 100 (4), 385–393 (in Russian).

Masaitis, V.L., Shafranovsky, G.I., Grieve, R.A.F., Langenhorst, F., Peredery, W.V., Balmasov, E.L., Fedorova, I.G., Therriault, A., 1997. Discovery of impact diamonds at the Sudbury structure (abstract). Large Meteorite Impacts and Planetary Evolution.

Masaitis, V.L., Shafranovsky, G.I., Grieve, R.A.F., Langenhorst, F., Peredery, W.V., Therriault, A.M., Balmasov, E.L., Fedorova, I.G., 1999. Impact diamonds in the suevitic breccias of the Black Onaping Formation, Sudbury Structure, Ontario, Canada. In: Dressler, B.O., Sharpton, V.L. (Eds.), Large Meteorite Impacts and Planetary Evolution II, Geological Society of America, Special Paper 339, pp. 317–321.

Masliwec, A., York, D., 1984. 40Ar/39 Ar dating of Sudbury minerals. GAC/AGC-MAC/AMC Joint Annual Meeting (abstract), vol. 9, p. 88.

Masuoka, Penny, M., Hoppin, Andrew, D., Lowman, P.D., 1995. A study of the Sudbury Basin, Ontario, using radar combined with digital elevation data, preliminary results. In: 1995 ACSM/ASPRS annual convention & exposition, technical papers, vol. 3, ASPRS. ACSM/ASPRS Annual Convention & Exposition Technical Papers, pp. 704–713.

Mathieu, L., van Wyk de Vries, B., Holohan, E.P., Troll, V.R., 2008. Dykes, cups, saucers and sills: analogue experiments on magma intrusion into brittle rocks. Earth Planet. Sci. Letts 271, 1–13.

Mavrogenes, J.A., O'Neill, H.S.C., 1999. The relative effects of pressure, temperature and oxygen fugacity on the solubility of sulfide in mafic magmas. Geochim. Cosmochim. Acta 63, 1173–1180.

Mawdsley, J.B., 1931. Geology of part of the Falconbridge and Errington properties in the vicinity of Sudbury, Ontario. Memoir—Geological Survey of Canada 165, 82–88.

Maxwell, C.D., 1995. Acidification and metal contamination, implications for the soil biota of Sudbury. In: Gunn, J.M. (Ed.), Restoration, Recovery of an Industrial Region, Progress in Restoring the Smelter-Damaged Landscape Near Sudbury, Canada, pp. 219–231.

McCarthy, T.S., Charlesworth, E.G., Stanistreet, I.G., 1986. Post-Transvaal structural features of the northern portion of the Witwatersrand Basin, South Africa. Trans. Geol. Soc. South Africa 89, 311–324.

McCormick, K.A., McDonald, A.M., 1999. Chlorine-bearing amphiboles from the Fraser Mine, Sudbury, Ontario, Canada: description and crystal chemistry. Can. Mineral. 37, 1385–1403, 1999.

McCormick, K.A., James, R.S., Fedorowich, J.S., Roussel, D.H., McDonald, A.M., Gibson, H.L., 1997. A study of mineral and rock fragments and shock features in the late granite breccia at the Fraser Mine, North Range, Sudbury structure (abstract). Large Meteorite Impacts and Planetary Evolution.

McCormick, K.A., Fedorowich, J.S., McDonald, A.M., James, R.S., 2002a. A textural, mineralogical, and statistical study of the footwall breccia within the Strathcona Embayment of the Sudbury Structure. Econ. Geol. 97, 125–143.

McCormick, K.A., Lesher, C.A., McDonald, A.M., Fedorowich, J.S., James, R.S., 2002b. Chlorine and alkali geochemical halos in the footwall breccias and sublayer norite at the margin of the Strathcona Embayment , Sudbury Structure, Ontario. Econ. Geol. 97, 1509–1519.

McDaniel, D.K., Hemming, S.R., McLennan, S.M., Hanson, G.N., 1994a. Petrographic, geochemical, and isotopic constraints on the provenance of the early Proterozoic Chelmsford Formation, Sudbury Basin, Ontario. J. Sediment. Res. A 64, 362–372.

McDaniel, D.K., Hemming, S.R., McLennan, S.M., Hanson, G.N., 1994b. Resetting of neodymium isotopes and redistribution of REEs during sedimentary processes, the early Proterozoic Chelmsford Formation, Sudbury Basin, Ontario, Canada. Geochim. Cosmochim. Acta 58, 931–941.

McDonald, I., Andreoli, M.A.G., Hart, R.J., Tredoux, M., 2001. Platinum-group elements in the Morokweng impact structure South Africa: evidence for the impact of a large ordinary chondrite projectile at the Jurassic-Cretaceous boundary. Geochim. Cosmochim. Acta 65 (2), 299–309.

McDowell, G.M., Fenlon, K., King, A., 2004. Conductivity-based nickel grade estimation for grade control at Inco's Sudbury mines, in Proceedings of the 74th Annual Society of Exploration Geophysicists Conference, October 10–15, Denver, Colorado, USA.

McDowell, G.M., Stewart, R., Monteiro, R.N., 2007. In-mine exploration and delineation using and integrated approach. Exploration 2007. Toronto, Ontario. Available from: http://www.dmec.ca/ex07-dvd/E07/53/Presentation_Files/index.html.

McGowan, R.R., Roberts, S., Foster, R.P., Boyce, A.J., Coller, D., 2003. Origin of the copper-cobalt deposits of the zambian copperbelt: an epigenetic view from Nchanga. Geology 31, 497–500.

McGrath, P.H., Broome, H.J., 1994a, A gravity model for the Sudbury Structure. In: Lightfoot, P.C., Naldrett, A.J. (Eds.), Proceedings of the Sudbury-Noril'sk Symposium. Ontario Geological Survey Special Volume 5, pp. 21–31.

McGrath, P.H., Broome, H.J., 1994b. A gravity model for the Sudbury Structure along the Lithoprobe seismic line. In: Lithoprobe Sudbury Project. Boerner, D.E., Milkereit, B., Nadrett, A.J., Geophys. Res. Lett., 21: 955–958.

McGregor, R.G., 1994. The solid phase controls on the mobility of metals at the Copper Cliff tailings area, near Sudbury, Ontario. MSc Thesis. University of Waterloo. Waterloo, ON, Canada.

McGuinty, W.J., 1981. Petrology, geochemistry and paleomagnetism of a section from the North Range of the Sudbury Irruptive near Lavack, Ontario. University of Ottawa, Ottawa, p. 60.

McHone, J.F., Dietz, R.S., Peredery, W.V., 1992. Sudbury breccia and suevite as glacial indicators transported 800 km to Kentland astrobleme, Indiana. In: Dressler, B.O., Sharpton, V.L. (Eds.), Papers presented to the international conference on large meteorite impacts and planetary evolution. LPI Contribution 790, p. 51.

McKinley, S.D., 1992. Intrusive relationships and alteration associated with a Melt Body in the Onaping Formation, Wisner Twp., Ontario. Unpublished BSc Thesis. Queen's University, Kingston, Ontario, 63 pp.

McLean, S.A., Straub, K.H., Stevens, K.M., 2005. The discovery and characteristics of the Nickel Rim South Deposit, Sudbury Ontario. In: Exploration for Platinum-group element deposits. Mineralogical Association of Canada Short Course, vol. 35, pp. 359–368.

McLean, S.A., Straub, K.H., Stevens, K.M., 2006. The discovery and characterization of the Nickel Rim South Deposit, Sudbury, Ontario. Min. Assoc. Canada Short Course 35, Oulu, Finland, pp. 359–368.

McNamara, G., 2011. Pb isotopic variations in the Sudbury Igneous Complex Unpublished MSc Thesis. Laurentian University, Canada.

McNish, J., 1999. The Big Score: Robert Friedland, INCO, and the Voisey's Bay Hustle. Doubleday, Canada. 356 pp.

Meldrum, A., Abdel-Rahman, A.F.M., Martin, R.F., Wodicka, N., 1997. The nature, age and petrogenesis of the Cartier Batholith, northern flank of the Sudbury Structure, Ontario, Canada. Precambrian Res. 82, 265–285.

Melosh, H.J., 1989. Impact Cratering: A Geologic Process. Oxford University Press, New York, 245 pp.

Melosh, H.J., 1995. Under the ringed basins. Nature 373, 104–105.

Melosh, H.J., Ivanov, B., 1999. Impact crater collapse. Ann. Rev. Earth Planet. Sci. 27, 385–415.

Mertie, J.B., 1943. Structural determinations from diamond drilling. Econ. Geol. 38, 298–312.

Merz, B., 1976. Potassium-argon dating, paleomagnetism and geochemistry of the Sudbury diabase dykes. MSc Thesis. University of Western Ontario. London, ON, Canada.

Meyn, H.D., 1970. Geology of Hutton and Parkin townships; Ontario Department of Mines. Geological Report 80, 78 p.

Mezhvilk, A.A., 1984. The role of horizontal movements in the development of tectonic structures and deposits in the Noril'sk Region. Geotectonics 18, 70–78.

Mezhvilk, A.A., 1995. Thrust and strike-slip zones in Northern Russia. Geotectonics 28, 298–305.

Miao, X.G., Moon, W.M., Milkereit, B., 1992. Multi-offset vertical seismic profiling (VSP) experiment in Sudbury Basin. In: AGU 1992 spring meeting. Eos, Transactions, American Geophysical Union, vol. 73, 14 Suppl., p. 319.

Miao, X., Moon, W.M., Milkereit, B., Mwenifumbo, C.J., 1994. Three component vertical seismic profiling (VSP) experiment in the Sudbury Basin. In: Boerner, D.E., Milkereit, B., Nadrett, A.J. (Eds.), Lithoprobe Sudbury Project. Geophys. Res. Lett., vol. 21, pp. 939–942.

Miao, X.G., Moon, W.M., Milkereit, B., 1994. Seismic reflection imaging of the crustal structure in the northwest of the Sudbury Basin. In: Society of Exploration Geophysicists, 64th annual international meeting, expanded abstracts. SEG Annual Meeting Expanded Technical Program Abstracts with Biographies, vol. 64, pp. 735–738.

Miao, X.-G., Moon, W.-M., Milkereit, B., 1995. A multioffset, three-component VSP study in the Sudbury Basin. Geophysics 60, 341–353.

Michener, C.E., 1940. Minerals associated with large sulphide bodies of the Sudbury type. PhD Thesis. University of Toronto, Toronto, ON, Canada. 63 pp.

Michener, C.E., Peacock, M.A., 1943. Parkerite $Ni_2Bi_2S_2$ from Sudbury, Ontario, redefinition of the species. Am. Mineral. 28 (6), 343–355.

Mickle, G.R., 1897. Mineralogical notes on Sudbury anthracite. Proceedings of the Canadian Institute 1, 64–66.

Milkereit, B., 1992. Third dimension of the Sudbury structure: Results from reflection seismic profiling (abstract). Can. Mineral. 30, 481.

Milkereit, B., Boerner, D.E., 1997. Integrated geophysical study of the Sudbury structure (abstract). Large Meteorite Impacts and Planetary Evolution.

Milkereit, B., Green, A., and the Sudbury Working Group, 1992. Deep geometry of the Sudbury structure from seismic reflection profiling. Geology 20, 807–811.

Milkereit, B., Wu, J., 1994. Seismic image of an early Proterozoic rifted continental margin. In: AGU 1994 fall meeting. Eos, Transactions, American Geophysical Union, vol. 75, 44 Suppl., p. 642.

Milkereit, B., Wu, J., 1996. Seismic image of an early Proterozoic rift basin. In: White, D.J., Ansorge, J., Bodoky, T.J., Hajnal, Z. (Eds.), Seismic reflection probing of the continents and their margins. Tectonophysics, 264, 1–4, pp. 89–100.

Milkereit, B., Green, A.G., Dressler, B.O., Morrison, G.G., Naldrett, A.J., Snajdr, P., 1991. Reflection seismic profiling across the Sudbury structure. In: AGU 1991 fall meeting. Eos, Transactions, American Geophysical Union, vol. 72, pt. 44, p. 297.

Milkereit, B., Wu, J., Boerner, D.E., 1993. Seismic imaging of the Sudbury Impact Crater. In: AGU 1993 fall meeting. Eos, Transactions, American Geophysical Union, vol. 74, 43 Suppl., p. 442.

Milkereit, B., Green, A., Wu, J., White, D., Adam, E., 1994a. Integrated seismic and borehole geophysical study of the Sudbury Igneous Complex. In: Boerner, D.E., Milkereit, B., Nadrett, A.J. (Eds.), Lithoprobe Sudbury Project. Geophys. Res. Lett., vol. 21, pp. 931–934.

Milkereit, B., White, D., Adam, E., Boerner, D.E., Salisbury, M., 1994b. Implications of the Lithoprobe seismic reflection transect for Sudbury geology. In: Lightfoot, P.C., Naldrett, A.J. (Eds.), Proceedings of the Sudbury-Noril'sk symposium. Ontario Geological Survey Special Volume, vol. 5, pp. 11–20.

Milkereit, B., White, D.J., Green, A.G., 1994c. Towards an improved seismic imaging technique for crustal structures, the Lithoprobe Sudbury experiment. In: Boerner, D.E., Milkereit, B., Nadrett, A.J. (Eds.), Lithoprobe Sudbury Project. Geophys. Res. Lett., vol. 21, pt. 10, pp. 927–930.

Milkereit, B., Eaton, D., Wu, J., Perron, G., Salisbury, M., Berrer, E.K., Morrison, G., 1996. Seismic Imaging of massive sulfide deposits: Part II Reflection seismic profiling. Econ. Geol. 9, 829–834.

Milkereit, B., Berrer, E.K., King, A.R., Watts, A.H., Roberts, B., Adam, E., Eaton, D.W., Wu, J., Salisbury, M.H., 2000. Development of 3-D seismic exploration technology for deep nickel-copper deposits–A case history from the Sudbury basin, Canada. Geophysics 65 (6), 1890–1899.

Miller, W.G., 1913. The Sudbury-Cobalt-Porcupine region, preface. Rep. Int. Geol. Cong., 5–7.

Miller, W.G., 1923. The Sudbury-Cobalt-Porcupine area (Ontario). Min. Metal. 4 (202), 523–525.

Miller, A.H., Innes, M.J.S., 1955. Gravity in the Sudbury Basin and vicinity (Ontario). Publications of the Dominion Observatory, Ottawa 18 (2), 11–43.

Miller, W.G., Knight, C.W., 1913. Sudbury, Cobalt, and Porcupine geology (Ontario). Eng. Min. J. 95, 1129–1133.

Milton, D.J., 1977. Shatter cones—An outstanding problem in shock mechanics, in Roddy, D.J., Pepin, R.O., Merrill, R.B. (Eds.), Impact and Explosion Cratering. Pergamon, New York, pp. 703–714.

Minter, W.E.L., 1999. Irrefutable detrital origin of Witwatersrand gold and evidence of eolian signatures. Econ. Geol. 94, 665–670.

Minter, W.E.L., Goedhart, M., Knight, J., Frimmel, H.E., 1993. Morphology of Witwatersrand gold grains from the Basal reef: Evidence for their detrital origin. Econ. Geol. 88, 237–248.

Mitchell, A., 2014. Despicable Nickel: Challenges for a Challenging industry. Prospectors and Developers Association of Canada, Toronto. Available from: http://www.pdac.ca/docs/default-source/Convention—Program—Technical-Sessions/commodities—mitchell.pdf?sfvrsn=0.

Mitchell, G.P., Mutch, A.D., 1956a. Geology of the Hardy Mine, Sudbury District, Ontario. Canadian Mining and Metallurgical Bulletin (Canadian Institute of Mining and Metallurgy). Montreal 526, 75–81.

Mitchell, G.P., Mutch, A.D., 1956b. Geology of the Hardy Mine, Sudbury, Ontario. Can. Inst. Min. Metal. Trans. 49, 37–43.

Mohr-Westheide, T., Reimold, W.U., Riller, U., Gibson, R.L., 2009. Pseudotachylitic breccia and microfracture networks in Archean gneiss of the central uplift of the Vredefort impact structure, South Africa. S. Afr. J. Geol. 112, 1–22.

Molnár, F., Watkinson, D.H., Everest, J.O., 1996. Heavy-metal rich, highly saline fluid inclusions from the Cu-rich footwall contact, Little Stobie Ni-Cu-PGE mine, Sudbury, Canada. In: Brown, Philip, E., Hagemann, Steffen, G. (Eds.), PACROFI VI, Sixth biennial Pan-American conference on Research on fluid inclusions, program and abstracts. Program and Abstracts, Biennial Pan-American Conference on Research on Fluid Inclusions, 6, pp. 89–91.

Molnár, F., Watkinson, D.H., Jones, P.C., Gatter, I., 1997. Fluid inclusion evidence for hydrothermal enrichment of magmatic ore at the contact zone of the Ni-Cu-platinum-group element 4b deposit, Lindsley Mine, Sudbury, Canada. Econ. Geol., vol. 92, pp. 674–685.

Molnár, F., Watkinson, D.H., Everest, J.O., 1999. Fluid-inclusion characteristics of hydrothermal Cu-Ni-PGE veins in granitic and metavolcanic rocks at the contact of the Little Stobie deposit, Sudbury, Canada. Chem. Geol. 154, 279–301.

Molnár, F., Walkinson, D.H., Jones, P.C., 2001. Multiple hydrothermal processes in footwall units of the North Range, Sudbury Igneous Complex, Canada, and implications for the genesis of vein-type Cu-Ni-PGE deposits. Econ. Geol. 7, 1645–1670.

Mond, L., Langer, C., Quincke, F., 1890. Action of carbon monoxide on nickel. J. Chem. Soc. 57, 749–753.

Money, D.P., 1993. Metal zoning in the Deep Copper Zone 3700 level, Strathcona Mine, Ontario. Explor. Min. Geol. 2, 307–320.

Montanari, A., Koeberl, C., 2000. Impact Stratigraphy. Lecture Notes in Earth Sciences 93, Springer-Verlag, Berlin-Heidelberg, 364 pp.

Monteiro, R., 2015. Exploration Success… What's the Drill? http://vektore.com/2015/07/28/exploration-success-whats-the-drill/.

Monteiro, R., 2016. Structural Vectoring in Mineral Exploration: What it is and How, When and Why we Should Use it. Prospectors and Developers Association of Canada. http://www.pdac.ca/convention/programming/short-courses/sessions/short-courses/structural-vectoring-in-mineral-exploration-what-it-is-and-how-when-and-why-we-should-use-it.

Monteiro, R., Krstic, S., 2006. Solid-state deformation and translocation of massive sulfides – part 1. Struct. Econ. Geol. Newsl. 2, 12–24.

Monteiro, L.V.S., Xavier, R.P., de Cavalho, E.R., Hitman, M.W., Johnson, C.A., Souza Filho, C.R.d., Torresi, I., 2008. Spatial and temporal zoning of hydrothermal alteration and mineralization in the Sossego iron oxide–copper–gold deposit, Carajás Mineral Province, Brazil: paragenesis and stable isotope constraints. Min. Dep. 43, 129–159.

Moon, W.M., Jiao, L.X., 1998. Sudbury impact structure modeling with high resolution seismic reflection survey results. J. R. Astronom. Soc. Can. 92, 250–257.

Moon, W., Messfin, D., Kanasewich, E.R., Mereu, R., Crossley, D., Dowsett, J., Kraus, B., Berrer, E., 1985. Sudbury Seismic Project. Annual Report—Centre for Precambrian Studies, University of Manitoba, 1984, p. 28.

Moon, W.M., Sereu, M., Kublik, E.R., Krause, B.R., 1988. High resolution seismic survey in the north range of Sudbury basin, Ontario (abstract). Canadian Continental Drilling Program Report 88-2, Scientific Drilling: The Sudbury Structure, p. 19.

Moon, Wooil, M., Sereu, M., Kublick, E.R., 1988. High resolution seimic survey in the North Range of Sudbury Basin, Ontario. In: Drury, M.J. (Ed.), Scientific drilling, the Sudbury Structure, proceedings of a workshop. Canadian Continental Drilling Program Report, vol. 88-2, p. 19.

Moon, W.M., Miao, X.G., Milkereit, B., 1994. Crustal structure northwest of the Sudbury basin (LITHOPROBE Sudbury Reflection Line #42). Sudbury Geology Workshop and Fieldtrip.

Moore, E.S., 1930. Geological structure of the southwest portion of the Sudbury Basin (Ontario). Trans. Can. Inst. Min. Metal.; Min. Soc. Nova Scotia 33, 292–302.

Moore, E.S., 1941. Some comparisons of Sudbury with the Bushveld complex. Econ. Geol. Bull. Soc. Econ. Geol. 36 (1), 106.

Moore, C.M., Nikolic, S., 1994. The Craig Deposit, Sudbury, Ontario. In: Lightfoot, P.C., Naldrett, A.J. (Eds.), Proceedings of the Sudbury-Noril'sk Symposium. Ontario Geological Survey Special Volume 5, pp. 77–90.

Moore, M., Lightfoot, P.C., Keays, R.R., 1993. Project Unit 93-08, Geology and geochemistry of footwall ultramafic rocks, Sudbury Igneous Complex, Fraser Mine, Sudbury, Ontario. In: Baker, C.L., Dressler, B.O., De Souza, H.A.F., Fenwick, K.G., Newsome, J.W., Owsiacki, L. (Eds.), Summary of field work and other activities 1993. Ontario Geological Survey Miscellaneous Paper 162, pp. 85–86.

Moore, M., Lightfoot, P.C., Keays, R.R., 1994. Geology and geochemisry of footwall ultramafic rocks, Fraser mine and Levack West mine, Sudbury Igneous Complex, Sudbury, Ontario. In: Summary of Field Work and Other Activities, Ontario Geological Survey, Miscellaneous Paper 163, pp. 91–94.

Moore, M., Lightfoot, P.C., Keays, R.R., 1995. Geology and geochemisry of footwall ultramafic rocks, Sudbury Igneous Complex, Sudbury, Ontario. In: Summary of Field Work, Other Activities, Ontario Geological Survey, Miscellaneous Paper 164, pp. 122–123.

Morey, G.B., 1996. Continental margin assemblage. In: Sims, P.K., Carter, L.M.H. (Eds.), Archean and Proterozoic geology of the Lake Superior region, USA, 1993. US Geological Survey Professional Paper, vol. P 1556, pp. 30–44.

Morgan, J., Lana, C., Kersley, A., Coles, B., Belcher, C., Montanari, S., Diaz-Martinez, E., Barbosa, A., Neumann, V., 2006. Analyses of shocked quartz at the global K-P boundary indicate an origin from a single, high-angle, oblique impact at Chicxulub. Earth Planet. Sci. Lett. 251 (3–4), 264–279.

Morris, W.A., 1980a. Tectonic and metamorphic history of the Sudbury Norite, the evidence from paleomagnetism. Eos Trans. Am. Geophys. Union 61 (17), 214.

Morris, W.A., 1980b. Tectonic and metamorphic history of the Sudbury norite, the evidence from paleomagnetism. Econ. Geol. Bull. Soc. Econ. Geol. 75 (2), 260–277.

Morris, W.A., 1981a. Intrusive and tectonic history of the Sudbury micropegmatite, the evidence from paleomagnetism. Econ. Geol. 76 (4), 791–804.

Morris, W.A., 1981b. Paleomagnetism of some sulphide occurrences from the South Range of the Sudbury Basin. J. Can. Soc. Explor. Geophys. 17, 55–71.

Morris, W.A., 1982. A paleomagnetic investigation of the Sudbury Basin Offsets, Ontario, Canada. Tectonophysics 85, 291–312.

Morris, W.A., 1984a. Paleomagnetic constraints on the magmatic, tectonic, and metamorphic evolution of the Sudbury Basin region. In: Pye, E.G., Naldrett, A.J., Giblin, P.E. (Eds.), The Geology and Ore Deposits of the Sudbury Structure. Ontario Geological Survey Special Volume 1, pp. 411–427.

Morris, W.A., 1984b. Paleomagnetic constraints on the magmatic, tectonic, and metamorphic evolution of the Sudbury Basin region. GAC/AGC-MAC/AMC Joint Annual Meeting (abstract), vol. 9, p. 90.

Morris, W.A., 1992. A positive contact test from the North Range of the Sudbury intrusive complex. In: AGU 1992 spring meeting. Eos, Transactions, American Geophysical Union, 73, 14, special supplement, p. 92.

Morris, W.A., 1994. Is the Sudbury Igneous Complex a single melt sheet produced by a meteorite impact? Paleomagnetic and geochemical evidence for genesis by multiple phases. GAC/AGC-MAC/AMC Joint Annual Meeting (abstract), vol. 19, p. 79.

Morris, W.A., Hearst, R.B., 1993. Magnetic interpretation along the Sudbury structure Lithoprobe transect. In: Society of Exploration Geophysicists, expanded abstracts, 1993 technical program, 63rd annual meeting and international exhibition. SEG Annual Meeting Expanded Technical Program Abstracts with biographies, vol. 63, pp. 455–457.

Morris, W.A., Pay, R., 1981. Genesis of the Foy offset and its sulfide ores, the paleomagnetic evidence from a study in Hess Township, Sudbury, Ontario. Bull. Soc. Econ. Geol. 76 (7), 1895–1905.

Morris, W.A., Watts, A.H., 1991. Identification of a paleomagnetic halo associated with a mineral deposit located at the Onaping/Onwatin contact, Sudbury region, Ontario. GAC/AGC-MAC/AMC Joint Annual Meeting (abstract), vol. 16, p. 85.

Morris, W.A., Hearst, R.B., Thomas, M.D., 1992a. Interpretation of the deep structure of the Sudbury Basin through Euler deconvolution. In: Lithoprobe, Abitibi-Grenville Project, Abitibi-Grenville Transect. Lithoprobe Report, 33, pp. 121–127.

Morris, W.A., Hearst, R.B., Thomas, M.D., 1992b. Magnetic interpretation along the Sudbury Structure Lithoprobe transect. In: Lithoprobe, Abitibi-Grenville Project, Abitibi-Grenville Transect. Lithoprobe Report, 33, pp. 129–132.

Morris, W.A., Hearst, R.B., Thomas, M.D., 1992c. The North Range contact of the Sudbury Igneous Complex, paleomagnetic evidence for a simple intrusive contact. In: Lithoprobe, Abitibi-Grenville Project, Abitibi-Grenville Transect. Lithoprobe Report, 33, pp. 133–137.

Morris, W.A., Mueller, E.L., Parker, C.E., 1995. Borehole magnetics, navigation, vector components, and magnetostratigraphy. In: Society of Exploration Geophysicists, 65th annual international meeting, expanded abstracts. SEG Annual Meeting Expanded Technical Program Abstracts with Biographies, vol. 65, pp. 495–498.

Morrison, G.G., 1982. Impact crater morphology and its relevance to the emplacement of the Sudbury Basin ore deposits. GAC/AGC-MAC/AMC Joint Annual Meeting (abstract), vol. 7, pp. 68.

Morrison, G.G., 1984. Morphological features of the Sudbury Structure in relation to an impact origin. In: Pye, E.G., Naldrett, A.J., Giblin, P.E. (Eds.), The Geology and Ore Deposits of the Sudbury Structure, Ontario Geological Survey Special Volume 1, pp. 513–520.

Morrison, G.G., 1988. The Sudbury Basin at depth. In: Scientific drilling, the Sudbury Structure, proceedings of a, workshop., Drury, M.J. (Ed.), Canadian Continental Drilling Program Report, 88-2, p. 11.

Morrison, G.G., Jago, B.C., Little, T.L., 1992. Sudbury footwall mineralization, with particular reference to the McCreedy East and Victor deposits (abstract). Can. Mineral. 30, 483–484.

Morrison, G.G., Jago, B.C., Little, T.L., 1994. Footwall mineralization of the Sudbury Igneous Complex. In: Lightfoot, P.C., Naldrett, A.J. (Eds.), Proceedings of the Sudbury-Noril'sk Symposium, Ontario Geological Survey Special Volume 5, pp. 57–64.

Morse, S.A., 1969. The Kiglapait layered intrusion, Labrador. Geol. Soc. America Memoir 112, 204.

Mossman, D., Eigendorf, G., Tokaryk, D., GauthierLafaye, F., Guckert, K.D., Melezhik, V., Farrow, C.E.G., 2003. Testing of fullerenes in geologic materials: Oklo carbonaceous substances, Karelian shungites, Sudbury Black Tuff. Geology 31, 255–258.

Mourre, G.A., 2000. Geological relationship at discontinuities in the Copper Cliff quartz diorite offset: an investigation into offset dikes and their relationship to the Sudbury Igneous Complex, Ontario. Unpublished MSc Thesis. Laurentian University, Sudbury, Ontario. pp. 255.

Mudd, G.M., 2009. Nickel sulfide versus laterite: the hard sustainability challenge remains. In: Liu, J., Peacey, J., Barati, M., Kasani-Nejad, S., Davis, B. (Eds.), International Symposium on Pyrometallurgy of Nickel and Cobalt, 2009. 48th CIM Annual Conference of Metallurgists. Metallurgical Society, Canadian Institute Mining Metallurgy and Petroleum. Sudbury, Canada pp 23–32.

Mudd, G.M., 2010. Global trends and environmental issues in nickel mining: sulfides versus laterites. Ore Geol. Rev. 38, 9–26.

Mudd, G.M., Jowitt, S.M., 2014. A detailed assessment of global nickel resource trends and endowments. Econ. Geol. 109, 1813–1841.

Mueller, E.L., Morris, W.A., 1995. A 3-D model of the Sudbury Igneous Complex. In: Society of Exploration Geophysicists, 65th annual international meeting, expanded (abstracts). SEG., Annual Meeting Expanded Technical Program Abstracts with Biographies, vol. 65, pp. 777–780.

Mueller-Mohr, V., 1990. The Sudbury Structure (Canada), breccias in the basement of a deeply eroded impact structure. In: Pesonen, L.J., Niemisara, H. (Eds.), Symposium, Fennoscandian impact structures, programme and abstracts, p. 49.

Mueller-Mohr, V., 1992a. Breccias in the basement of a deeply eroded impact structure, Sudbury, Canada. In: Pesonen, L.J., Henkel, H. (Eds.), Terrestrial impact craters and craterform structures with a special focus on Fennoscandia. Tectonophysics 216, 219–226.

Mueller-Mohr, V., 1992b. Sudbury Project (University of Muenster-Ontario Geological Survey), 5, New investigations on Sudbury breccia. In: Dressler, B.O., Sharpton, V.L. (Eds.), Papers presented to the international conference on large meteorite impacts and planetary evolution. LPI Contribution 790, p. 53.

Muir, T.L., 1982. Geology and origin of the Onaping Formation. Ontario Geological Survey Misc. Paper 106, pp. 76–79.

Muir, T.L., 1983. Geology of the Morgan Lake–Nelson Lake area, District of Sudbury; Ontario Geological Survey, Open File Report 5426, 203 p.

Muir, T.L., 1984. The Sudbury Structure, considerations and models for an endogenic origin. In: Pye, E.G., Naldrett, A.J., Giblin, P.E. (Eds.), The Geology and Ore Deposits of the Sudbury Structure. Ontario Geological Survey Special Volume 1, pp. 449–489.

Muir, T.L., Peredery, W.V., 1984. The Onaping Formation. In: Pye, E.G., Naldrett, A.J., Giblin, P.E. (Eds.), The Geology and Ore Deposits of the Sudbury Structure. Ontario Geological Survey Special Volume 1, pp. 139–210.

Mukwakwami, J., 2012. Structural controls of Ni-Cu-PGE ores and mobilization of metals at the Garson Mine, Sudbury. PhD Thesis. Laurentian University, Sudbury, 218 p.

Mukwakwami, J., Lafrance, B., Lesher, C.M., 2012. Back-thrusting and overturning of the southern margin of the 1.85 Ga Sudbury Igneous Complex at the Garson Mine, Sudbury, Ontario. Precambrian Res. 196–197, 81–105.

Mukwakwami, J., Lesher, C.M., Lafrance, B., 2014a. Geochemistry of deformed and hydrothermally-mobilized magmatic Ni-Cu-PGE ores at the Garson Mine, Sudbury. Econ. Geol. 109, 367–386.

Mukwakwami, J., Lafrance, B., Lesher, C.M., Tinkham, D.K., Rayner, N.M., Ames, D.E., 2014b. Deformation, metamorphism, and mobilization of Ni-Cu-PGE sulfide ores at Garson Mine, Sudbury. Min. Dep. 49, 175–198.

Mungall, J.E., 2002. Late-stage sulfide liquid mobility in the Main Mass of the Sudbury Igneous Complex: examples from the Victor Deep, McCreedy East and Trillabelle Deposits. Econ. Geol. 97, 1563–1576.

Mungall, J.E., 2007. Crystallization of magmatic sulfides: an empirical model and application to Sudbury ores. Geochim. Cosmochim. Acta 71, 2809–2819.

Mungall, J.E., Brenan, J.M., 2003. Experimental evidence for the chalcophile behaviour of the halogens. Can. Mineral. 41, 207–220.

Mungall, J.E., Brenan, J.M., 2014. Partitioning of platinum-group elements and Au between sulfide liquid and basalt and the origins of mantle-crust fractionation of the chalcophile elements. Geochim. Cosmochim. Acta 125, 265–289.

Mungall, J.E., Hanley, J.J., 2004. Origins of outliers of the Huronian Supergroup within the Sudbury Structure. The J. Geol. 112, 59–70.

Mungall, J.E., Ames, D.E., Hanley, J.J., 2004. Geochemical evidence from the Sudbury structure for crustal redistribution by large bolide impacts. Nature 429, 546–548.

Mungall, J.E., Harvey, J.D., Balch, S.J., Azar, B., Atkinson, J., Hamilton, M.A., 2010. Eagle's nest: a magmatic Ni-sulfide deposit in the James Bay Lowlands, Ontario, Canada. Econ. Geol. 5, 539–557.

Murphy, A.J., Spray, J.G., 2002. Geology, mineralization, and emplacement of the Worthington Offset Dike: a genetic model for Offset Dike mineralization in the Sudbury Igneous Complex. Econ. Geol. 97, 1399–1418.

Murray, A., 1857. Report for the Year 1856. Geological Survey of Canada, Reports of Progress for the Years 1853-56, p. 180.

Mutch, A.D., 1949. Temperature-pressure conditions of deposition of base metal sulphides of the Sudbury type. Progress Report—Temperature-Pressure Research of Hydrothermal Mineral Deposits, vol. 1, pt. 2, pp. 64–66.

Mutch, A.D., 1950. The decrepitation of pyrrhotite from Falconbridge nickel mine, Sudbury, Ontario. Progress Report—Temperature-Pressure Research of Hydrothermal Mineral Deposits, vol. 2, pt. 7, pp. 124–125.

Myers, J.S., Voordouw, R.J., Tettelaar, T.A., 2008. Proterozoic anorthosite–granite Nain batholith: structure and intrusion processes in an active lithosphere-scale fault zone, Northern Labrador. Can. J. Earth Sci. 45, 909–934.

Naldrett, A.J., 1961. The geochemistry of cobalt in the ores of the Sudbury district (Ontario). MSc Thesis. Queen's University, Kingston, ON, Canada. 129 pp.

Naldrett, A.J., 1973. Nickel sulfide deposits—their classification and genesis with special emphasis on deposits of volcanic association. Trans. Can. Inst. Mining Metallurgy 76, 183–201.

Naldrett, A.J., 1981. Platinum-group element deposits. In: Cabri, L.J. (Ed.), Platinum-Group Elements: Mineralogy, Geology, Recovery. Candian Institute Mining Metallurgy Special Volume 23, pp. 197–232.

Naldrett, A.J., 1984a. Introduction to the geology of the Sudbury Igneous Complex. In: Pye, E.G., Naldrett, A.J., Giblin, P.E. (Eds.), The Geology and Ore Deposits of the Sudbury Structure. Ontario Geological Survey Special Volume 1, p. 234.

Naldrett, A.J., 1984b. Mineralogy and composition of the Sudbury ores. In: Pye, E.G., Naldrett, A.J., Giblin, P.E. (Eds.), The Geology and Ore Deposits of the Sudbury Structure. Ontario Geological Survey Special Volume 1, pp. 309–325.

Naldrett, A.J., 1984c. Ni-Cu ores of the Sudbury Igneous Complex, introduction. In: Pye, E.G., Naldrett, A.J., Giblin, P.E. (Eds.), The Geology and Ore Deposits of the Sudbury Structure. Ontario Geological Survey Special Volume 1, pp. 302–306.

Naldrett, A.J., 1984d. Summary, discussion, and synthesis. In: Pye, E.G., Naldrett, A.J., Giblin, P.E. (Eds.), The Geology and Ore Deposits of the Sudbury Structure. Ontario Geological Survey Special Volume 1, pp. 533–569.

Naldrett, A.J., 1986. Geochemistry of the Sudbury Igneous Complex, a model for the complex and its ores. In: Friedrich, G.H., Genkin, A.D., Naldrett, A.J., Ridge, J.D., Sillitoe, R.H., Vokes, F.M. (Eds.), Geology and Metallogeny of Copper Deposits, Proceedings of the Copper Symposium, 27th International Geological Congress. Special Publication of the Society for Geology Applied to Mineral Deposits, vol. 4, pp. 91–110.

Naldrett, A.J., 1989a. Contamination and the origin of the Sudbury structure and its ores. In: Whitney, James, A., Naldrett, A.J. (Eds.), Ore deposition associated withmagmas. Rev. Econ. Geol. 4, 119–134.

Naldrett, A.J., 1989b. Magmatic sulfide deposits. Oxford Monographs on Geology and Geophysics, vol. 14, Oxford University Press, Oxford, p. 178.

Naldrett, A.J., 1994. The Sudbury-Noril'sk Symposium, an overview. In: Lightfoot, P.C., Naldrett, A.J. (Eds.), Proceedings of the Sudbury-Noril'sk Symposium. Ontario Geological Survey Special Volume 5, pp. 3–8.

Naldrett, A.J., 1997. Key factors in the genesis of Ni–Cu–PGE deposits: implications for exploration. In: Naldrett, A.J. (Ed.), 17th Ore Deposits Workshop, vol. 1. University of Toronto, December 15–18, pp. 12.1–12.49.

Naldrett, A.J., 1999. World class Ni-Cu-PGE Deposits: Key factors in their genesis. Min. Dep. 34, 227–240.

Naldrett, A.J., 2003. From impact to riches: evolution of geological understanding as seen at Sudbury Canada. GSA Today 13, 4–9.

Naldrett, A.J., 2004. Nickel Sulfide Deposits. Springer, New York, 707 pp.

Naldrett, A.J., 2010a. From the mantle to the bank: the life history of a Ni-Cu-(PGE) sulfide deposit. Geological Society South Africa 113, 1–32. doi:10.2113/gssajg.113.1-1.

Naldrett, A.J., 2010b. Secular variations of magmatic sulfide deposits and their source magmas. Econ. Geol. 105, 669–688.

Naldrett, A.J., Barnes, S.J., 1986. The behavior of platinum group elements during fractional crystallization and partial melting with special reference to the composition of magmatic sulfide ores. Fortschritte der Mineralogie 64, 113–133.

Naldrett, A.J., Hewins, R.H., 1984. The Main Mass of the Sudbury Igneous Complex. In: Pye, E.G., Naldrett, A.J., Giblin, P.E. (Eds.), The Geology and Ore Deposits of the Sudbury Structure. Ontario Geological Survey Special Volume 1, pp. 235–251.

Naldrett, A.J., Kullerud, G., 1967. A study of the Strathcona Mine and its bearing on the origin of the nickel- copper ores of the Sudbury District, Ontario. J. Petrol. 8, 453–531.

Naldrett, A.J., Li, C., 2007. The Voisey's Bay deposit, Labrador, Canada. In: Goodfellow, W.D. (Ed.), Mineral Deposits of Canada: A Synthesis of Major Deposit-Types, District Metallogeny, the Evolution of Geological Provinces, and Exploration Methods. Geological Association of Canada, Mineral Deposits Division, Sp Pub 5, pp. 387–407.

Naldrett, A.J., Lightfoot, P.C., 1993. A model for giant magmatic sulphide deposits associated with flood basalts. Society of Economic Geologists, Special Publication No. 2. pp. 81–124.

Naldrett, A.J., MacDonald, A.J., 1980. Tectonic settings of some Ni-Cu sulfide ores: their importance in genesis and exploration. In: Strangway, D.W. (Ed.), The Continental Crust and its Mineral Deposits—A Volume in Honour of J. Tuzo Wilson. Geological Association of Canada Special Paper Number 20, pp. 633–657.

Naldrett, A.J., Pessaran, A., 1992. Compositional variation in the Sudbury ores and prediction of the proximity of footwall copper-PGE ore bodies. Ontario Geological Survey, Miscellaneous Paper 159, pp. 47–62.

Naldrett, A.J., Craig, J.R., Kullerud, G., 1967. The central portion of the Fe-Ni-S system and its bearing on pentlandite exsolution in iron-nickel sulfide ores. Econ. Geol. 62, 826–847.

Naldrett, A.J., Bray, J.G., Gasparrini, E.L., Podolsky, T., Rucklidge, J.C., 1970. Cryptic variation and the petrology of the Sudbury Nickel Irruptive. Econ. Geol. 65, 122–155.

Naldrett, A.J., Hewins, R.H., Greenman, L., 1972. The main Irruptive and sub-layer at Sudbury, Ontario. Proceedings of the 24th international Geological Congress, Montreal, Canada, vol. 4, pp. 206–213.

Naldrett, A.J., Hoffman, E.L., Green, A.H., Chow, C.L., Naldrett, S.R., 1979. The composition of Ni-sulfide ores with particular reference to their content of PGE and Au. Can. Min. 17, 403–416.

Naldrett, A.J., Innes, D.G., Sowa, J., Gorton, M.P., 1982. Compositional variations within and between five Sudbury ore deposits. In: von Gruenewaldt, G. (Eds.), A Further Issue Devoted to the Platinum-Group Elements. Econ. Geol. Bull. Soc. Econ. Geol., vol. 77, pp. 1519–1534.

Naldrett, A.J., Duke, J.M., Lightfoot, P.C., Thompson, J.F., 1984a. Quantitative modelling of the segregation of magmatic sulfides: an exploration guide. Can. Min. Metal. Bull. 77, 46–56.

Naldrett, A.J., Hewins, R.H., Dressler, B.O., Rao, B.V., 1984b. The contact sublayer of the Sudbury Igneous Complex. In: Pye, E.G., Naldrett, A.J., Giblin, P.E. (Eds.), The Geology and Ore Deposits of the Sudbury Structure. Ontario Geological Survey Special Volume 1, pp. 253–274.

Naldrett, A.J., Rao, B.V., Evensen, N.M., Dressler, B.O., 1985. Major and trace element and isotopic studies at Sudbury—A model for the structure and its ores. Ontario Geological Survey Miscellaneous Paper 127, pp. 30–44.

Naldrett, A.J., Rao, B.V., Evensen, N.M., 1986. Contamination at Sudbury and its role in ore formation. In: Gallagher, M.J., Ixer, R.A., Neary, C.R., Prichard, H.M. (Eds.), Metallogeny of basic and ultrabasic rocks, pp. 75–91.

Naldrett, A.J., Lightfoot, P.C., Doherty, W., Fedorenko, V., Gorbachev, N.S., 1991. The Ni-Cu ores at Noril'sk and Sudbury. In: Page, l., M., Leroy, J. (Eds.), Source, Transport and Deposition of Metals. Balkema, Rotterdam, pp. 9–10.

Naldrett, A.J., Lightfoot, P.C., Fedorenko, V.A., Doherty, W., Gorbachev, N.S., 1992. Geology and geochemistry of intrusions and flood basalts of the Noril'sk region, USSR, with implications for the origin of the Ni-Cu ores. Econ. Geol. 87, 975–1004.

Naldrett, A.J., Pessaran, R., Asif, M., Li, C., 1994a. Compositional variation in the Sudbury ores and prediction of the proximity of Footwall Copper-PGE orebodies, In: Lightfoot, P.C., Naldrett, A.J. (Eds.), Proceedings of the Sudbury-Noril'sk Symposium. Ontario Geological Survey Special Volume 5, pp. 133–143.

Naldrett, A.J., Asif, M., Gorbachev, N.S., Kunilov, V.Ye., Stekhin, A.I., Fedorenko, V.A., Lightfoot, P.C., 1994b. The composition of the Ni-Cu ores of the Oktyabry'sky Deposit, Noril'sk Region. In: Lightfoot, P.C., Naldrett, A.J. (Eds.), Proceedings of the Sudbury-Noril'sk Symposium. Ontario Geological Survey Special Volume 5, 357–371.

Naldrett, A.J., Fedorenko, V.A., Lightfoot, P.C., Kunilov, V.E., Gorbachev, N.S., Doherty, W., Johan, Z., 1995. Ni-Cu-PGE deposits of the Noril'sk region Siberia: their formation in conduits for flood basalt volcanism. Trans. Inst. Mining Metall. 104, B18–B36.

Naldrett, A.J., Fedorenko, V.A., Asif, M., Lin, S., Kunilov, V.E., Stekhin, A.I., Lightfoot, P.C., Gorbachev, N.S., 1996a. Controls on the composition of Ni-Cu sulfide deposits as illustrated by those of the Noril'sk Region, Siberia. Econ. Geol. 91, 751–773.

Naldrett, A.J., Keats, H., Sparkes, K.E., Moore, R., 1996b. Geology of the Voisey's Bay Ni-Cu-Co Deposit, Labrador, Canada. Explor. Min. Geol. 5, 169–179.

Naldrett, A.J., Fedorenko, V.A., Lightfoot, P.C., Gorbachev, N.S., Doherty, W., Asif, M., Lin. S., Johan, Z., 1998. A model for the formation of the Ni-Cu-PGE deposits of the Noril'sk region. In: Laverov, N.P., Distler, V.V. (Eds.), International Platinum. Theophrastus Publications, Athens, p. 92–106.

Naldrett, A.J., Asif, M., Schandl, E., Searcy, T., Morrison, G., Binney, P., Moore, C., 1999. PGE in the Sudbury ores: significance with respect to the origin of different ore zones and the exploration for footwall ore bodies. Econ. Geol. 94, 185–210.

Naldrett, A.J., Wilson, A., Kinnaird, J., Chunnett, G., 2006. PGE tenor and metal ratios within and below the Merensky Reef, Bushveld Complex: implications for its genesis. J Petrol 50, 625–659.

Naldrett, A.J., Wilson, A., Kinnaird, J., Chunnett, G., 2009. PGE Tenor and Metal Ratios within and below the Merensky Reef, Bushveld Complex: Implications for Genesis. J. Petrol. 50, 625–659.

Naumov, M.V., 2002. Impact-generated hydrothermal systems: data from Popigai, Kara, and Puchezh-Katunki impact structures. In: Plado, J., Pesonen, L.J. (Eds.), Impacts in Precambrian Shields, Impact Studies Series, Springer-Verlag, Berlin-Heidelberg, pp. 117–171.

Nelles, E.W., 2012. Genesis of Cu-PGE-rich footwall-type mineralization in the Morrison Deposit, Sudbury. MSc Thesis. Laurentian University, Suidbury, Ontario, 87 pp.

Nicholls, J.C., 1930. The Sudbury ore (Ontario). Eng. Min. J. 130, 433–434.

Noble, S.R., Lightfoot, P.C., 1992. U-Pb baddeleyite ages of the Kerns and Triangle Mountain intrusions, Nipissing Diabase, Ontario. Can. J. Earth Sci. 29, 1424–1429.

Norman, M.D., 1992. Sudbury Igneous Complex, impact melt or igneous rock? Implications for lunar magmatism. In: Dressler, B.O., Sharpton, V.L. (Eds.), Papers Presented to the International Conference on Large Meteorite Impacts and Planetary Evolution. LPI Contribution 790, pp. 54–55.

Norman, M.D., 1994. Sudbury Igneous Complex, evidence favoring endogenous magma rather than impact melt. In: Twenty-fifth lunar and planetary science conference, abstracts of papers. Abstracts of Papers Submitted to the Lunar and Planetary Science Conference, vol. 25, Part 2, pp. 1007–1008.

Nriagu, J.O., 1983. Arsenic enrichment in lakes near the smelters at Sudbury, Ontario. Geochim. Cosmochim. Acta 47, 1523–1526.

O'Callaghan, J.W., Weirich, J.R., Osinski, G.R., Lightfoot, P.C., 2015. Geochemical variations in Sudbury Breccia of the Sudbury Impact Structure, Canada. AGU-GAC-MAC, Monteal, Canada.

O'Callaghan, J.W., Osinski, G.R., Lightfoot, P.C., Linnen, R.L., Weirich, J.R., 2016. Reconstructing the Geochemical Signature of Sudbury breccia, Ontario, Canada: implications for its formation and trace metal content. Econ. Geol. 111 (7), (in the press).

O'Connell-Cooper, C.D., Spray, J.G., 2011. Geochemistry of the impact-generated melt sheet at Manicouagan: evidence for fractional crystallization. J. Geophys. Res. 116, B6.

O'Connor, J.P., Spray, J.G., 1997. Geological setting of the Manchester Offset Dike within the South Range of the Sudbury impact structure (abstract). Large Meteorite Impacts and Planetary Evolution.

O'Driscoll, E.S.T., 1990. Lineament tectonics of Australia ore deposits. In: Hughes, F.E. (Ed.), Geology of the Mineral Deposits of Australias and Papua New Guinea. Australian Institute of Mining and Metallurgy, Melbourne, pp. 33–41.

O'Keefe, J.D., Ahrens, T.J., 1992. Melting and its relationship to impact crater morphology. In: Dressler, B.O., Sharpton, V.L. (Eds.), Papers presented to the international conference on large meteorite impacts and planetary evolution. Lunar Planetary Institute Contribution 790, pp. 55–57.

O'Keefe, J.D., Ahrens, T.J., 1994. Impact-induced melting of planetary surfaces. In: Dressler, B.O., Grieve, R.A.F., Sharpton, V.L. (Eds.), Special Paper – Geological Society of America 293, pp. 103–109.

Olaniyan, O., Smith, R.S., Lafrance, B., 2014. A constrained potential field data interpretation of the deep geometry of the Sudbury structure. Can. J. Earth Sci. 51 (7), 715–729.

Olaniyan, O., Smith, R.S., Lafrance, B., 2015. Regional 3D geophysical investigation of the Sudbury Structure: Interpretation 3(2), SL63-SL81. Available from: http://dx.doi.org/10.1190/INT-2014-0200.1.

Oliver, T.A., 1949. A study of the effect of uralization upon the chemical composition of the Sudbury (Ontario) Norite. MSc Thesis. University of Manitoba. Winnipeg, MB, Canada.

Oliver, T.A., 1951. The effect of uralitization upon the chemical composition of the Sudbury norite (Ontario). Am. Mineral. 36, 421–429.

Osinski, G.R., Grieve, R.A., Tornabene, L.T., 2013. Excavation and impact eject emplacement. In: Osinski, G.R., Pierazzo, E. (Eds.), Impact Cratering. Wiley Blackwell, Oxford.

Osinski, G.R., Pierazzo, E., 2010. Impact Cratering: Processes and Products. Wiley Blackwell, Hoboken, NJ, pp. 32–42

Osinski, G.R., Spray, J.G., Lee, P., 2001. Impact-induced hydrothermal activity within the Haughton impact structure, arctic Canada: generation of a transient, warm, wet oasis. Meteor. Planet. Sci. 36, 731–745.

Ostermann, M., Deutsch, A., 1995. Impaktschmelzen in der Sudbury-Struktur(Kanada)—eine geochemische Fallstudie am Foy Offset Dike. Exkursionsfhrer und Verffentlichungen, GGW 195, 59–60.

Ostermann, M., Schaerer, U., Buhl, D., Deutsch, A., 1994a. U-Pb-data for baddeleyite and zircon from the Foy offset dyke (Sudbury Canada). Mineral. Mag. 58A, 678–679.

Ostermann, M., Schaerer, U., Deutsch, A., 1994b. Constraints on the origin of the offset dikes (Sudbury impact structure, Canada) from U-Pb data. Twenty-Fifth Lunar and Planetary Science Conference (abstracts), vol. 25, Part 2, pp. 1031–1032.

Ostermann, M., Schaerer, U., Deutsch, A., 1994c. Impact melting and 1850 Ma Offset Dike emplacement in the Sudbury Impact Structure: constraints from zircon and baddeleyite U-Pb ages Meteoritics, vol. 29, p. 513.

Ostermann, M., Schärer, U., Buhl, D., Deutsch, A., 1994d. U-Pb-data for baddeleyite and zircon from the Foy Offset Dike (Sudbury, Canada). Mineral. Mag, 58a, 678–679.

Ostermann, M., Schärer, U., Buhl, D., Deutsch, A., 1994e. Constraints on the origin of the Offset Dikes (Sudbury impact structure, Canada). Lunar Planet. Sci. XXV, 1031–1032.

Ostermann, M., Deutsch, A., Agrinier, P., 1995a. Geochemical variation in the Foy Offset Dike, Sudbury impact structure. Ann. Geophys. 13 (Suppl 111), C741.

Ostermann, M., Deutsch, A., Agrinier, P., 1995b. Geochemical variations (REE, d18O) in the dioritic Foy Offset Dike, Sudbury Structure (Canada). Fourth International ESF-Workshop on Impact Cratering and Evolution of Planet Earth, vol. 4, pp. 131–132.

Ostermann, M., Deutsch, A., Agrinier, P., 1995c. Geochemistry of the Sudbury drill cores 70011 and 52848 (North Range). GAC/AGC-MAC/AMC Joint Annual Meeting (abstract), vol. 20, p. 71.

Ostermann, M., Deutsch, A., Agrinier, P., 1995d. New geochemical constraints on the formation of the Foy Offset dike, Sudbury impact structure (Canada). Meteoritics 30, 559.

Ostermann, M., Deutsch, A., Therriault, A., Grieve, R.A.F., 1996a. The Sudbury impact structure: geochemistry of the drill cores 70011 and 52848. Meteor. Planet. Sci. 31A, 102.

Ostermann, M., Schaerer, U., Deutsch, A., 1996b. Impact melt dikes in the Sudbury multi-ring basin (Canada), implications from uranium-lead geochronology on the Foy offset dike. Meteoritics 31 (4), 494–501.

Owen, D.L., Coats, C.J.A., 1984. Falconbridge and East mines. In: Pye, E.G., Naldrett, A.J., Giblin, P.E. (Eds.), The Geology and Ore Deposits of the Sudbury Structure. Ontario Geological Survey Special Volume 1, pp. 371–378.

Paakki, J.J., 1990. The relationship between the upper zone of the Sudbury Igneous Complex and the Basal member of the Onaping Formation. Unpublished BSc Thesis. Laurentian University, Sudbury, Ontario, 60 pp.

Paakki, J.J., 1992a. The Errington Zn-Cu-Pb massive sulphide deposit Sudbury, Ontario: Its structural and stratigraphic setting and footwall alteration. MSc Thesis. Laurentian University, Sudbury, ON, Canada.

Paakki, J.J., 1992b. The Errington Zn-Cu-Pb massive sulfide deposit, Sudbury, Ontario: its structural and stratigraphic setting and footwall alteration. Unpublished MSc Thesis. Laurentian University, Sudbury, Ontario, 140 pp.

Paakki, J.J., Gibson, H.L., Cecchetto, J., 1992. The structural and stratigraphic setting of the Errington Zn-Cu-Pb massive sulphide deposit, #1 and #2 Shaft Area, Sudbury, Ontario. Ontario Geological Survey, Summary of Research 1991-92, pp. 80–87.

Palmer, G.R., Dixit, S.S., MacArthur, J.D., Smol, J.P., 1989. Elemental analysis of lake sediment from Sudbury, Canada, using particle-induced X-ray emission. In: Nriagu, J.O. (Ed.), Trace Metals in Lakes, The Science of the Total Environment, vol. 87–88, pp. 141–156.

Papapaplova, K., Darling, J., Storey, C., Moser, D., Lightfoot, P.C., Lasalle, S., 2015. Titanite petrochronology reveals the timing of strain localization and sulfide remobilization in the Sudbury Impact Structure, Ontario, Canada. Goldschmidt Conference, Prague, abstracts.

Park, J., 1925. Sudbury ore deposits. Econ. Geol. 20, 500–504.

Paterson, S.R., Vernon, R.H., Tobisch, O.T., 1989. A review of criteria for the identification of magmatic and tectonic foliations in granitoids. J. Struct. Geol. 11, 349–363.

Pattison, E.F., 1979. The Sudbury Sublayer, In: Naldrett, A.J. (Ed.), Nickel-Sulfide and Platinum-Group Element Deposits. Can. Mineral. 17,257–274.

Pattison, E.F., 2009. The Sudbury Igneous Complex, a field guide to the geology of Sudbury, Ontario; Ontario Geological Survey. Open File Report 6243, p. 14–30.

Pattison, E.F., Fleet, M.E., 1980. Tectonic origin for Sudbury, Ontario, shatter cones, discussion and reply. Geol. Soc. Am. Bull. 91, 1754–1756.

Peach, C.L., Mathez, E.A., Keays, R.R., 1990. Sulfide melt-silicate melt distribution coefficients for noble metals and other chalcophile elements as deduced from MORB: implications for partial melting. Geochim. Cosmochim. Acta 54, 3379–3389.

Pearson, W.N., 1983. Geology and Ni-Cu deposits of the Sudbury area. In: Pearson, W.N., Richardson, J.A., McMillan, R.H. (compilers), CIM Geology Division Field Trip Guidebook. Metallogeny of the Southern Province, Sudbury-Elliot Lake Area, Ontario, September 21–23, 1986, p. 3.

Peck, D.C., James, R.S., Chubb, P.T., 1993. Geological environments for PGE-Cu-Ni mineralization in the East Bull Lake gabbro-anorthosite intrusion, Ontario. Explor. Min. Geol. 2, 85–104.

Peck, D.C., James, R.S., Chubb, P.T., Keays, R.R., 1995. Geology, metallogeny and petrogenesis of the East Bull Lake intrusion, Ontario. Ontario Geological Survey. Open File Report 5923, 117 p.

Peck, D.C., Keays, R.R., James, R.S., Chubb, P.T., Reeves, S.J., 2001. Controls on the formation of contact-type PGE mineralization in the East Bull Lake Intrusion. Econ. Geol. 96, 559–581.

Pecock, M.A., Yatsevitch, G.M., 1936. Cubanite from Sudbury, Ontario. American Mineralogist 21 (1), 55–62.

Pekeski, D.E., Lightfoot, P.C., Keays, R.R., 1994. Geology and geochemistry of the Totten mine exposure, Worthington Offset, Sudbury Igneous Complex, Sudbury, Ontario. Summary of Field Work and Other Activities, Ontario Geological Survey, Miscellaneous Paper 163, pp. 95–96.

Pekeski, D.E., Lightfoot, P.C., Keays, R.R., 1995. Geology and geochemistry of the Totten mine section of the Worthington Offset, Sudbury Igneous Complex, Ontario. In: Summary of Field Work and Other Activities, Ontario Geological Survey, Miscellaneous Paper 164, pp. 124–125.

Penfield, S.L., 1893. On pentlandite from Sudbury, Ontario, Cananada, with remarks upon three supposed new species from the same region. Am. J. Sci. 45, 493–497.

Pentek, A., 2013. Technical Report on the Sudbury Camp Joint Venture (SCJV) Projects Located near Sudbury Ontario. Wallbridge Mining Company. Available from: http://wallbridgemining.com/i/pdf/2013-SCJV-Technical-Report.pdf.

Péntek, A., Molnar, F., Tuba, G., Watkinson, D.H., Jones, P.C., 2013. The significance of partial melting processes in hydrothermal low sulfide Cu-Ni-PGE mineralization within the footwall of the SIC, Ontario, Canada. Econ. Geol. 108, 59–78.

Percival, J.A., Card, K.D., 1983. Archean crust as revealed in the Kapuskasing uplift, Superior Province, Canada. Geology 11, 323–326.

Percival, J.A., West, G.F., 1994. The Kapuskasing uplift: a geological and geophysical synthesis. Can. J. Earth Sci. 31, 1256–1286.

Peredery, W.V., 1971. Volcanic rocks, chilled micropegmatite, or impact breccias and melt rocks in the Onaping formation? (abstract). Geological Association of Canada-Mineralogical Association of Canada, Abstracts of Papers, Sudbury, p. 55.

Peredery, W.V., 1972a. Chemistry of fluidal glasses and melt bodies in the Onaping Formation, In: Guy-Bray, J.V. (Ed.), New Developments in Sudbury Geology. Geological Association of Canada, Special Paper 10, pp. 49–59.

Peredery, W.V., 1972b. The origin of rocks at the base of the Onaping Formation, Sudbury, Ontario. PhD Thesis. University of Toronto, Toronto, ON, Canada, 366 pp.

Peredery, W.V., 1991. Geology and ore deposits of the Sudbury Structure, Ontario. Geological Survey of Canada, Open File Report 2162, 1–38.

Peredery, W.V., Morrison, G.G., 1984a. Origin of the Sudbury Structure, meteorite impact aproach. GAC/AGC-MAC/AMC joint annual meeting (abstracts), vol. 9, p. 96.

Peredery, W.V., Morrison, G.G., 1984b. Discussion of the origin of the Sudbury Structure. In: Pye, E.G., Naldrett, A.J., Giblin, P.E. (Eds.), The Geology and Ore Deposits of the Sudbury Structure. Ontario Geological Survey Special Volume 1, pp. 491–511.

Peredery, W.V., Naldrett, A.J., 1975. Petrology of the upper Irruptive rocks, Sudbury, Ontario. Econ. Geol. 70, 164–175.

Peters, E.D., 1890. The Sudbury ore deposits (Ontario). Transactions of the Society of Mining Engineers of American Institute of Mining, Metallurgical and Petroleum Engineers, Incorporated (AIME), pp. 278–289.

Peterson, D.W., 1970. Ash-flow deposits—their character, origin, and significance. J. Geol. Edu. 18, 66–76.

Petrus, J.A., Ayer, J.A., Long, D.G.F., Lightfoot, P.C., Kamber, B.S., 2012. Contributions to the Sudbury Igneous Complex and the depth of excavation: evidence from Onaping Formation Zircons (abstracts). Large Meteorite Impacts and Planetary Evolution. Available from: http://www.hou.usra.edu/meetings/sudbury2013/pdf/3056.pdf.

Phemister, T.C., 1925. A criticism of the application of the theory of assimilation to the Sudbury sheet. J. Geol. 33, 819–824.

Phemister, T.C., 1926. Igneous rocks at Sudbury and their relation to the ore deposits. Ontario Department of Mines Annual Report 34, 1–47.

Phemister, T.C., 1928. A comparison of the Keweenawan sill rocks of Sudbury and Cobalt, Ontario. Proc. Trans. R. Soc. Can. 22, 121–197.

Phemister, T.C., 1937. A review of the problems of the Sudbury Irruptive. J. Geol. 45, 1–47.

Phemister, T.C., 1939. Notes on several properties in the district of Sudbury. Annual Report—Ontario Department of Mines, vol. 48, Part 10, pp. 16–28.

Phemister, T.C., 1957. The Copper Cliff rhyolite in McKim Township, District of Sudbury, pp. 91–116.

Philpotts, A.R., 1964. Origin of pseudotachylites. Am. J. Sci. 262, 1008–1035.

Piercey, P., Schneider, D.A., Holm, D.K., 2007. Geochronology of Proterozoic metamorphism in the deformed Southern Province, northern Lake Huron region, Canada. Precambrian Res. 157, 127–143.

Pilles, E., Osinski, G.R., Smith, D., Bailey, J., 2013. Timing relationship between radial and concentric Offsey Dykes at the Sudbury Impact Structure, Ontario: a case study of the Foy and Hess Offset Dykes. Large Meteorite Impacts and Planetary Evolution V.

Pirajno, F., 2000. Ore deposits and mantle plumes. Springer, 656 pp.

Pirajno, F., 2004. Hotspots and mantle plumes: global intraplate tectonics, magmatism and ore deposits. Mineral. Petrol. 82, 183–216.

Pirajno, F., 2012. The Geology and Tectonic Setting of China's Mineral Deposits. Springer, New York, 679 pp.

Pirajno, F., 2013. Ore Deposits and Mantle Plumes. Springer, 557 pages.

Polzer, B., 2000. The role of Borehole EM in the discovery and definition of the Kelly Lake Ni-Cu Deposit, Sudbury, Canada. Soc. Econ. Geol. (expanded abstracts).

Polzero, A., Taylor, L.A., 1982. Exsolution in the MSS-pentlandite system: textural and genetic implications for Ni-sulfide ores. Min. Dep. 17, 313–332.

Popelar, J., 1972. Gravity interpretation of the Sudbury area. In: Guy-Bray, J.V. (Ed.), New Developments in Sudbury Geology. Geological Association of Canada, Special Paper 10, pp. 103–115.

Popelar, J., 1994. Gravity measurements in the Sudbury area. Gravity Map Series, vol. 138. Department of Energy, Mines and Resoures, Canada, p. 1.

Popelar, J., 1995. Discrimination of barren and mineralized Sublayer environments of the Sudbury Igneous Complex. Summary of Field Work and Other Activities, Ontario Geological Survey, Miscellaneous Paper 164, p. 129.

Potter, R.W.K., 2012. Numerical modelling of basin-scale impact crater formation. PhD Thesis. Imperial College London, 236 pages.

Potts, P.J., 1987. A Handbook of Silicate Rock Analysis. Blackwell, Oxford, 622 pp.

Prevec, S.A., 1993. An isotopic, geochemical and petrographic investigation of the genesis of early Proterozoic mafic intrusions and associated volcanism near Sudbury, Ontario. Unpublished PhD Thesis. University of Alberta, Edmonton, Alberta, 223 pp.

Prevec, S.A., Baadsgaard, H., 1993. Geochemical, isotopic and geochronological evidence regarding the origin of early Proterozoic mafic magmatism near Sudbury, Ontario. GAC/AGC-MAC/AMC Joint Annual Meeting (abstract), p. 85.

Prevec, S.A., Baadsgaard, H., 1994. Petrogenesis and enigmatic geochronology from early Proterozoic gabbros in Drury and Falconbridge Townships; shock resetting: (abstract). GAC/MAC, Program with Abstracts, Annual Meeting, Waterloo, ON, vol. 19, p. A90.

Prevec, S.A., Baadsgaard, H.B., 2005. Evolution of Paleoproterozoic mafic intrusions located within the SIC thermal aureole: isotopic, geochronological and geochemical evidence. Geochim. Cosmochim. Acta 69, 3653–3669.

Prevec, A.A., Cawthorn, R.G., 2002. Thermal evolution and interaction between impact melt sheet and footwall; A genetic model for the contact sublayer of the Sudbury Igneous Complex, Canada. J. Geophys. Res. 107, ECV5-1-ECV5-14.

Prevec, S.A., Zhou, M.-F., Lightfoot, P.C., Keays., R.R., 1995. Discrimination of Barren and mineralized sublayer environments of the Sudbury Igneous Complex. Ontario Geological Survey Miscellaneous Paper 164, p. 129.

Prevec, S.A., Corfu, F., Moore, M.L., Lightfoot, P.C., Keays, R.R., 1997a. Postmagmatic zircon growth or exotic early magmatic phases associated with the emplacement of the Sudbury Igneous Complex: Samarium-neodymium isotopic, geochemical, and petrographic evidence (abstract). Large Meteorite Impacts and Planetary Evolution.

Prevec, S.A., Keays, R.R., Lightfoot, P.C., Xie, Q., 1997b. Origin or the sublayer of the Sudbury Igneous Complex: Samarium-neodymium isotopic, geochemical, and petrographic evidence for incomplete crustal homogenization (abstract). Large Meteorite Impacts and Planetary Evolution.

Prevec, S.A., Xie, Q., Keays, R.R., Lightfoot, P.C., 1997c. Geochemical and samarium-neodymium isotopic evidence regarding the origin of offset dikes of the Sudbury Igneous Complex, Ontario (abstract). Large Meteorite Impacts and Planetary Evolution.

Prevec, S.A., Lightfoot, P.C., Keays, R.R., 2000. Evolution of the Sublayer of the Sudbury Igneous Complex: geochemical, Sm-Nd isotopic and petrological evidence. Lithos 51, 271–292.

Prevec, S.A., Cowan, D.R., Cooper., G.R.J., 2005. Geophysical evidence for a pre-impact Sudbury dome, southern Superior Province, Canada. Can. J. Earth Sci. 42, 1–11.

Primmer, G.H., 1927. The Sudbury nickel region (Ontario). Geog Society Philadelphia, Bulletin, vol. 25, pt. 1, pp. 33–42. Proceedings—Symposium on Rock Mechanics, vol. 22, pp. 103–110.

Prochaska, K.M., 1981. A geochemical study of the Sudbury Breccia. MSc Thesis. Western Michigan University, Kalamazoo, MI, United States. 70 pp.

Pye, E.G., 1984. The origin of the Sudbury Structure, introduction. In: Pye, E.G., Naldrett, A.J., Giblin, P.E. (Eds.), The Geology and Ore Deposits of the Sudbury Structure. Ontario Geological Survey Special Volume 1, p. 448.

Pye, E.G., Naldrett, A.J., Giblin, P.E. (Eds.), 1984. The Geology and Ore Deposits of the Sudbury Structure. Ontario Geological Survey Special Volume 1, 603 p.

Rae, D.R., 1975. Inclusions in the Sub-layer from Strathcona Mine, Sudbury and their significance. MSc Thesis. University of Toronto, Toronto, ON, Canada. 194 pp.

Raharimahefa, T., Lafrance, B., Tinkham, D.K., 2014. New structural, metamorphic, and U–Pb geochronological constraints on the Blezardian Orogeny and Yavapai Orogeny in the Southern Province, Sudbury, Canada. Can. J. Earth Sci. 51, 750–774.

Ramsay, J.G., 1961. Shatter cones in the rocks of the Vredefort ring—discussions. Geol. Soc. South Africa Trans. 64, 156–157.

Rao, B.V., Evensen, N.M., 1986. Trace element and Sr, Nd-isotope geochemistry of melt rocks in the Sudbury Structure, Ontario, Canada. In: AGU 1986 fall meeting and ASLO winter meeting. Eos Trans. Am. Geophys. Union, vol. 67, pp. 1267.

Rao, B.V., Naldrett, A.J., Evensen, N.M., Dressler, B.O., 1983. Grant 146, Contamination and genesis of the Sudbury ores. In: Pye, E.G. (Ed.), Geoscience Research Grant Program, Summary of Research 1982–1983. Ontario Geological Survey Miscellaneous Paper 113, pp. 139–151.

Rao, B.V., Naldrett, A.J., Evensen, N.M., 1984a. Crustal contamination in the Sublayer, Sudbury Igneous Complex: a combined trace element and strontium isotope study. Ontario Geological Survey Miscellaneous Paper 121, pp. 128–146.

Rao, B.V., Naldrett, A.J., Evensen, N.M., Dressler, B.O., 1984b. Crustal contamination in the sublayer, Sudbury Igneous Complex, a combined trace element and strontium isotopic study. GAC/AGC-MAC/AMC Joint Annual Meeting (abstract), vol. 9, p. 98.

Rao, B.V., Naldrett, A.J., Evensen, N.M., 1985. Crustal contamination of the sublayer, Sudbury Igneous Complex, and its relevance to the genesis of Ni-Cu sulfides. Magmatic Sulfide Field Conference IV (abstracts). Can. Mineral. 23, pp. 329–330.

Rayleigh, J.W.S., 1896. Theoretical considerations respecting the separatiuon of gases by diffusion and similar processes. Phil. Mag. 42, 77–107.

Redmond, D.J., 1992. Structural studies in the Southern Province, south of Sudbury, Ontario. MSc Thesis. Brock University, St. Catharines, ON, Canada, 121 p.

Reimold, W.U., 1993. Further debate on the origin of the Sudbury structure, is it relevant to the Vredefort Dome and the Bushveld Complex? South African J. Sci. 89, 546–552.

Reimold, W.U., 1995. Pseudotachylite in impact structures—generation by frictional melt and shock brecciation? A review and discussion. Earth Sci. Rev. 39, 247–265.

Reimold, W.U., Dressler, B.O., 1990. The economic significance of impact processes (abstract). Abstracts for the International Workshop on Meteorite Impact on the Early Earth, Perth, Australia, pp. 36–37.

Reimold, W.U., Gibson, R.L., 2006. The melt rocks of the Vredefort impact structure—Vredefort Granophyre and pseudotachylitic breccias: Implications for impact cratering and the evolution of the Witwatersrand Basin. Chemie der Erde—Geochem. 66 (1), 1–35.

Reimold, W.U., Armstrong, R.A., Koeberl, C., 2002. A deep drillcore from the Morokweng impact structure, South Africa: petrography, geochemistry and constraints on the crater size. Earth Planet. Sci. Lett. 201, 221–232.

Reimold, W.U., Koeberl, C., Gibson, R.L., Dressler, B.O., 2005. Economic mineral deposits in impact structures: a review. In: Koeberl, C., Henkel, H. (Eds.), Impact Tectonics. Springer, Heidelberg, pp. 479–552.

Reimold, W.U., Wannek, D., Hoffmann, M., Hansen, B.T., Hauser, N., Schulz, T., Siegert, S., Thirlwall, M., Zaag, P.T., Mohr-Westheide, T., 2015. Vredefort Pseudotachylitic Breccia and Granophyre (Impact Melt Rocks): clues to their Genesis from New Field, Chemical and Isotopic Investigations. Bridging the Gap III, Freiburg, Germany, p. 2.

Rempel, G.G., 1994. Regional Geophysics at Noril'sk. In Proceedings of the Sudbury-Noril'sk Symposium. Ontario Geological Survey Special Volume 5, pp. 147–160.

Reuther, R., Foerstner, U., Allan, R.J., 1982. Chemical forms of heavy metals in sediment cores from lakes near Sudbury, Ontario (Canada). Nriagu, J.O., Troost, R., (compilers). Int. Congress Sedimentol. 11, 33.

Reynolds, D.L., 1935. Review of "Life history of the Sudbury nickel irruptive, Part 1, Petrogenesis" by W.H. Collins, 1935. Geol. Mag. 72, 285–287.

Rhode, R.A., Muller, R.A., 2005. Cycles in fossil diversity. Nature 434, 208–210.

Ribet, I., Ptacek, C.J., Blowes, D.W., Jambor, J.L., 1995. The potential for metal release by reductive dissolution of weathered mine tailings. J. Contaminant Hydrol. 17, 239–273.

Richardson, P.B., Dence, M.R., 1979. The Meteoritical Society at Sudbury, 1978. Geosci. Canada 6, 93–96.

Rickard, J.H., Watkinson, D.H., 2001. Cu-Ni-PGE mineralization within the Copper Cliff Offset Dike, Copper Cliff North Mine, Sudbury, Ontario: evidence for multiple stages of emplacement. Explor. Min. Geol. 10 (1), 111–124.

Riddle, C., 1993. Analysis of Geological Materials. Marcel Decker Inc., New York, 463 pp.

Riller, U., 1996. Tectonometamorphic episodes affecting the southern footwall of the Sudbury Basin and their significance for the origin of the Sudbury Igneous Complex, central Ontario, Canada. PhD Thesis. University of Toronto, Toronto, ON, Canada, 135 pp.

Riller, U., 2005. Invited Review: Structural characteristics of the Sudbury Impact Structure, Canada: impact-induced and orogenic deformation—a review. Meteor. Planet. Sci. 40, 1723–1740.

Riller, U., 2009. Felsic Plutons. A Field Guide to the Geology of Sudbury, Ontario; Ontario. Geological Survey, Open File Report 6243, pp. 14–30.

Riller, U., Dressler, B.O., 2003. Structural Characteristics of the Sudbury Impact Structure, Canada, Point to a Protracted Tectonomagmatic Evolution of the Sudbury Igneous Complex. Large Meteorite Impacts.

Riller, U.P., Schwerdtner, W.M., 1994. Structural significance of deformation patterns in Huronian footwall rocks of the Sudbury basin (abstract). Sudbury Geology Workshop and Fieldtrip.

Riller, U.P., Schwerdtner, W.M., 1997. Mid-crustal deformation at the southern flank of the Sudbury Basin, central Ontario, Canada. Geol. Soc. Am. Bull. 109, 841–854.

Riller, U., Cruden, A.R., Schwerdtner, W.M., 1996. Magnetic fabric and microstructural evidence for a tectono-thermal overprint of the early Proterozoic Murray pluton, central Ontario, Canada. J. Struct. Geol. 18, 1005–1016.

Riller, U., Schwerdtner, W.M., Halls, H.C., Card, K.D., 1999. Transpressional tectonism in the Eastern Penokean Orogen, Canada: consequences for Proterozoic crustal kinematics and continental fragmentation. In: Mengel, F. (Ed.), Precambrian Orogenic Processes. Precambrian Research, vol. 93, pp. 51–70.

Riller, U.P., Doman, D., Grieve, R.A.F., 2006. The Sudbuy Igneous Complex; Canada: evidence for large-meteorite impact during paleoproterozoic orogenic activity. First International Conference on Impact Cratering in the Solar System.

Riller, U., Fletcher, S., Santimano, T., 2009. Sudbury area, Ontario: meteorite impact, ore deposits and regional geology. Field trip guide book of the Geological Association of Canada—Mineralogical Association of Canada, 43 pages.

Riller, U., Lieger, D.L., Gibson, R.L., Grieve, R.A.F., Stoffler, D., 2010. Origin of large-volume pseudotachylite in terrestrial impact structures. Geology 38, 619–622.

Ripley, E.M., Li, C., 2011. A review of conduit-related Ni-Cu-(PGE) sulfide mineralization at the Voisey's Bay Deposit, Labrador, and the Eagle Deposit, Northern Michigan. Econ. Geol. 17, 181–197.

Ripley, E.M., Li, C., 2013. Sulfide saturation in mafic magmas: Is external Sulfur required for magmatic Ni-Cu-(PGE) ore genesis? Econ. Geol. 108, 45–58.

Ripley, E.M., Lightfoot, P.C., Stifter, E.C., Underwood, B., Taranovic, V., Dunlop, III, M., Donoghue, K.A., 2015. Heterogeneity of S-isotope compositions recorded in the Sudbury Igneous Complex, Canada: Significance to formation of Ni-Cu sulfide ores and the host rocks. Econ. Geol. 110, 1125–1135.

Rison, W., Niemeyer, S., Kuroda, P.K., 1979. Superheavy elements in nature? Fissiogenic xenon in Sudbury, Ontario, norite, discussion and reply. Geochem. J. 13, 31–36.

Rivard, Benoit., Kellett, R., Saint, Jean, R., Singhroy, Vernon, H., 1994. Mapping structures and lithologic contacts in the area of Geneva Lake, Sudbury using SAR, digital topography, and airborne geophysics. GAC/AGC-MAC/AMC Joint Annual Meeting (abstract), vol. 19, p. 93.

Robb, L.J., Robb, V.M., 1998. Gold in the Witwatersrand Basin. In: Wilson, M.G.C., Anhaeusser, C.R. (Eds.), The Mineral resources of South Africa, Council for Geoscience, Pretoria, Handbook 16, pp. 294–349.

Roberts, H.M., Longyear, R.D., 1918a. Exploration of nickel-copper properties in Falconbridge township, Sudbury District. Ontario Can. Min. J. 792, 50–53.

Roberts, H.M., Longyear, R.D., 1918b. Origin of Sudbury nickel-copper deposits. Can. Min. J. 39, 135–136.

Roberts, H.M., Longyear, R.D., Grout, F.F., Bateman, A.M., Berkey, C.P., Kunz, G.F., Lindgren, W., Miller, W.G., 1918. Genesis of the Sudbury nickel-copper ores as indicated by recent explorations. Transactions of the Society of Mining Engineers of American Institute of Mining, Metallurgical and Petroleum Engineers, Incorporated, pp. 27–56.

Robertson, P.B., Grieve, R.A.F., 1975. Impact structures in Canada: Their recognition and characteristics. J.R. Astronom. Soc. Can. 69, 1–21.

Robson, G.M., 1940. Sudbury breccia. MSc Thesis. Queen's University, Kingston, ON, Canada. 47 pp.

Roest, W.R., Pilkington, M., 1994. Restoring post-impact deformation at Sudbury, a circular argument. In: Boerner, D.E., Milkereit, B., Nadrett, A.J. (Eds.), Lithoprobe Sudbury Project. Geophys. Res. Lett., vol. 21, pp. 959–962.

Roest, W.R., Pilkington, M., Grieve, R.A.F., 1993. Deformation of the Sudbury Structure, Ontario, Canada. Seventh meeting of the European Union of Geosciences. Terra Abstracts, vol. 5, Suppl. 1, pp. 208–209.

Rollinson, H.R., 1993. Using geochemical data: evaluation, presentation, interpretation. Pearson/Prentice Hall, Harlow, 352 pp.

Rondot, J., 1984. Comparaison entre les astroblemes de Charlevoix et de Sudbury (Comparison between the astroblemes of Charlevoix and Sudbury). GAC/AGC-MAC/AMC Joint Annual Meeting (abstract), vol. 9, p. 100.

Rondot, J., 1997. Charlevoix and Sudbury as simple readjusted craters (abstract). Large Meteorite Impacts and Planetary Evolution.

Rondot, J., 2000. Charlevoix and Sudbury as gravity-readjusted impact structures. Meteor. Planet. Sci. 35, 707–713.

Roscoe, S.M., Card, K.D., 1992. Early Proterozoic tectonics and metallogeny of the Lake Huron region of the Canadian Shield. Precambrian Res. 58, 99–119.

Ross, C.S., Smith, R.L., 1961. Ash-flow tufts—their origin, geologic relations, and identification. United States Geological Survey, Professional Paper 366, 81 pp.

Roussel, D., 2015. Nickel Sulfide Ore Deposits of the Keweenawan Midcontinent Rift. Prospectors and Developers Association of Canada, Toronto.

Roussel, D.H., 1972. The Chelmsford Formation of the Sudbury Basin—a Precambrian turbidite, In: Guy-Bray, J.V. (Ed.), New Developments in Sudbury Geology. Geological Association of Canada, Special Paper 10, pp. 79–91.

Roussel, D.H., 1975. The origin of foliation and lineation in the Onaping Formation and the deformation of the Sudbury Basin. Can. J. Earth Sci. 12, 1379–1395.

Roussel, D.H., 1980. Kink bands in the Onaping Formation, Sudbury Basin, Ontario, In: Schwerdtner, W.M., Hudleston, P.J., Dixon, J.M. (Eds.), Analytical Studies in Structural Geology. Tectonophysics, vol. 66, pp. 83–97.

Roussel, D.H., 1981a. Grant 96, Mineralization in the Whitewater Group, Sudbury Basin. In: Pye, E.G. (Ed.), Geoscience Research Grant Program, summary of research, 1980–1981. Ontario Geological Survey Miscellaneous Paper 98, pp. 233–242.

Roussel, D.H., 1981b. Sudbury and the meteorite theory: discussions. Geosci. Can. 8, 167–169.

Roussel, D.H., 1982a. Grant 96, Mineralization in the Whitewater Group, Sudbury Basin. In: Pye, E.G. (Ed.), Geoscience Research Grant Program, summary of research, 1981–1982. Ontario Geological Survey Miscellaneous Paper 103, pp. 171–184.

Roussel, D.H., 1982b. Mineralization of the fill of an Aphebian astrobleme, the Sudbury Basin. GAC/AGC-MAC/AMC Joint Annual Meeting (abstract), vol. 7, p. 77.

Roussel, D.H., 1983. Nature and origin of mineralization inside the Sudbury Basin. Ontario Geological Survey, Open File Report 5443, 53 pp.

Roussel, D.H., 1984a. Structural geology of the Sudbury Basin. In: Pye, E.G., Naldrett, A.J., Giblin, P.E. (Eds.), The geology and ore deposits of the Sudbury Structure. Ontario Geological Survey Special Volume 1, pp. 83–95.

Roussel, D.H., 1984b. Onwatin and Chelmsford formations, In: Pye, E.G., Naldrett, A.J., Giblin, P.E. (Eds.), The Geology and Ore Deposits of the Sudbury Structure. Ontario Geological Survey Special Volume 1, pp. 211–218.

Roussel, D.H., 1984c. Mineralization in the Whitewater Group. In: Pye, E.G., Naldrett, A.J., Giblin, P.E. (Eds.), The Geology and Ore Deposits of the Sudbury Structure. Ontario Geological Survey Special Volume 1, pp. 219–232.

Roussel, D.H., 1984d. Nature and origin of mineral occurrences inside the Sudbury Basin. Can. Inst. Min. Metal. Bull. 77, 63–75.

Roussel, D.H., Everitt, R.A., 1981. Jointing in the Sudbury Basin, Ontario. In: Leary, D.W., Earle, J.L. (Eds.), Proceedings of the International Conference on Basement Tectonics, vol. 3, pp. 381–391.

Roussel, D.H., Long, D.G.F., 1998. Are outliers of the Huronian Supergroup preserved in structures associated with the collapse of the Sudbury Impact Crater? J. Geol. 106, 407–419.

Roussel, D.H., Trevisiol, D.D., 1988. Geology of the mineralized zone of the Wanapitei complex, Grenville Front, Ontario. Min. Dep. 23, 138–149.

Roussel, D.H., Gibson, H.L., Jonasson, I.R., 1997. The tectonic, magmatic and mineralization history of the Sudbury structure. Explor. Mining Geol. 6, 1–22.

Roussel, D.H., Meyer, W., Prevec, S.A., 2002. Bedrock geology and mineral deposits. The Physical Environment of the City of Greater Sudbury, Ontario Geological Survey, Special Volume 6, pp. 21–55.

Roussel, D.H., Fedorowich, J.S., Dressler, B.O., 2003. Sudbury Breccia (Canada): a product of the 1850 Ma Sudbury Event and host to footwall Cu-Ni-PGE deposits. Earth Sci. Rev. 60, 147–174.

Roussel, D.H., Paakki, J.J., Gray, M.J., 2009. Mineralization in the Whitewater Group. In: Roussel, D.H., Brown, G.H. (Eds.), A Field Guide to the Geology of Sudbury, Ontario. Ontario Geological Survey Open File Report 6243, pp. 132–138.

Roussel, D.H., 2009. Structural Geology. In: Roussel, D.H., Brown, G.H. (Eds.), A Field Guide to the Geology of Sudbury, Ontario. Ontario Geological Survey Open File Report 6243, pp. 74–94.

Rowell, W.F., Edgar, A.D., 1986. Platinum-group element mineralization in a hydrothermal Cu-Ni sulphide occurrence, Rathbun Lake, northeastern Ontario. Econ. Geol. 81, 1272–1277.

Royal Ontario Nickel Commission., 1917. Report of the Royal Ontario Nickel Commission.

Rucklidge, J.C., Wilson, G.C., Kilius, L.R., Cabri, L.J., 1991. Trace element analysis of sulphide concentrates from Sudbury by accelerator mass spectrometry. GAC/AGC-MAC/AMC Joint Annual Meeting (abstract), vol. 16, p. 108. Rundschau, vol. 84, pt. 4, pp. 697–709.

Rucklidge, J.C., Wilson, Graham, C., Kilius, Linas, R., Cabri, Louis, J., 1992. Trace element of sulfide concentrates from Sudbury by accelerator mass spectrometry. Can. Mineral. 30, 1023–1032.

Rutstein, M.S., 1980. Nickeloan melanterite from Sudbury Basin. Am. Mineral. 65 (9–10), 968–969.

Sadler, J.F., 1958. A detailed study of the Onwatin Formation. Unpublished MSc Thesis. Queen's University, Kingston, Ontario, 184 pp.

Sage, R.P., 1987. Geology of Carbonatite—Alkalic Rock Complexes in Ontario: Spanish River Carbonatite Complex, District of Sudbury. Ontario Geological Survey, Study 30, 62 p.

Salisbury, M.H., Iuliucci, R., Long, C., 1994. Velocity and reflection structure of the Sudbury Structure from laboratory measurements. In: Boerner, D.E., Milkereit, B., Nadrett, A.J. (Eds.), Lithoprobe Sudbury Project. Geophys. Res. Lett. 21, 923–926.

Salter, A.P., 1856. Report of the Commission of the Crwon Lands of Canada 1856, 265 pp.

Sangster, D.F., 1970. Metallogenesis of some Canadian lead-zinc deposits in carbonate rocks. Proceedings of the Geological Association of Canada, vol. 22, pp. 27–36.

Santimano, T., Riller, U., 2012. Revisiting thrusting, reverse faulting and transpression in the southern Sudbury Basin, Ontario. Precambrian Res. 200–203, 74–81.

Saunders, A.D., England, R.W., Reichow, M.K., White, R.V., 2005. A mantle plume origin for the Siberian traps: uplift and extension in the West Siberian Basin. Lithos 79, 407–424.

Schandl, E.S., 1982. The feldspar mineralogy of the Sudbury Complex. MSc Thesis. McGill University, Montreal, PQ, Canada, 147 pp.

Schandl, E.S., Martin, R.F., Stevenson, J.S., 1982. Metastable feldspars in the Onaping tuffs, Sudbury, Ontario. GAC/AGC-MAC/AMC Joint Annual Meeting (abstract), vol. 7, p. 79.

Schandl, E.S., Martin, R.F., Stevenson, J.S., 1986. Feldspar mineralogy of the Sudbury Igneous Complex and the Onaping Formation, Sudbury, Ontario. Can. Mineral. 24, 747–759.

Schandl, E.S., Gorton, M.P., Davis, D.W., 1994. Albitization at 1700+ or -2 Ma in the Sudbury-Wanapitei Lake area, Ontario, implications for deep-seated alkali magmatism in the Southern Province. Can. J. Earth Sci. 31 (3), 597–607.

Scherstén, A., Garde, A.A., 2013. Complete hydrothermal re-equilibration of zircon in the Maniitsoq structure, West Greenland: a 3001 Ma minimum age of impact? Meteor. Planet. Sci. 48, 172–1498.

Schulz, R.H., Merrill, R.B. (Eds.), 1981. In: Proceedings of the Conference on Multi-ring Basins: formation and evolution. Pergamon Press, New York, 295 pp.

Schuyler, T.K., 1985. Petrology of the sub-layer along the North Range of the Sudbury Irruptive. MSc Thesis. Rutgers, The State University, New Brunswick. United States, 142 pp.

Schwarcz, H.P., 1973. Sulfur isotope analyses of some Sudbury, Ontario, ores. Can. J. Earth Sci. 10, 1444–1459.

Schwarz, E.J., Buchan, K.L., 1982. Uplift deduced from remanent magnetization, Sudbury area since 1250 Ma ago. Earth Planet. Sci. Lett. 58 (1), 65–74.

Scott, R.G., Benn, K., 2001. Peak-ring rim collapse accomodated by impact melt-filled transfer faults, Sudbury impact structure. Canada, Geology 29, 747–750.

Scott, R.G., Benn, K., 2002. Emplacement of Sulfide Deposits in the Copper Cliff Offset Dyke during Collapse of the Sudbury Crater Rim: evidence from Magnetic Fabric Studies. Econ. Geol. 97, 1447–1458.

Scott, R.G., Spray, J.G., 1999. Magnetic fabric constraints on friction melt flow regimes and ore emplacement direction within the Southern Range Breccia Belt, Sudbury Impact Structure. Tectonophysics 307, 163–189.

Scott, R.G., Spray, J.G., 2000. The South Range Breccia Belt of the Sudbury Impact Structure: a possible terrace collapse feature. Meteor. Planet. Sci. 35, 505–520.

Scribbins, B.T., 1978. Exotic inclusions from the South Range Sublayer, Sudbury. MSc Thesis. Toronto. 208 pp.

Scribbins, B.T., Rae, D.R., Naldrett, A.J., 1984. Mafic and ultramafic inclusions in the sublayer of the Sudbury Igneous Complex. In: Eckstrand, O.R., Watkinson, D.H. (Eds.), Ore Deposits and Related Petrology of Mafic-Ultramafic Suites. Can. Mineral., vol. 22, pp. 67–75.

Seat, Z., Beresford, S.W., Grguric, B.A., Waugh, R.S., Hronsky, J.M.A., Gee, M.A.M., Mathison, C.I., 2007. Architecture and emplacement of the Nebo-Babel gabbronorite-hosted magmatic Ni-Cu-PGE sulphide deposit, West Musgrave. Min. Dep., Western Australia, 42, 551–581.

Seat, Z., Beresford, S.W., Grguric, B.A., Gee, M.A.M., Grassineau, N.V., 2009. Re-evaluation of the role of external sulfur addition in the genesis of Ni-Cu-PGE deposits: evidence from the Nebo-Babel Ni-Cu-PGE deposit, West Musgrave, Western Australia. Econ. Geol. 104, 521–538.

Senzu, M.H., 1990. Application of high resolution seismic technique in Precambrian terrain (Sudbury, Ontario). MSc Thesis. University of Manitoba. Winnipeg, MB, Canada.

Shand, S.J., 1916. The pseudotachylyte of Paris (Orange Free State) and its relation to "trap-shotten gneiss" and "flinty crush rock". Quarter. J. Geol. Soc. London 72, 198–221.

Shanks, W.S., 1991. Deformation of the central and southern portions of the Sudbury Structure. PhD Thesis. University of Toronto. Toronto, ON, Canada. 239 pp.

Shanks, W.S., Schwerdtner, W.M., 1988. Grant 275, Deformation of the Sudbury Structure. In: Milne, V.G. (Ed.), Geoscience Research Grant Program, Summary of Research 1987–1988. Ontario Geological Survey Miscellaneous Paper 140, pp. 46–55.

Shanks, W.S., Schwerdtner, W.M., 1989. Grant 275, Deformation of the Sudbury Structure. In: Milne, V.G. (Ed.), Geoscience Research Grant Program, summary of research 1988–1989. Ontario Geological Survey Miscellaneous Paper 143, pp. 4–17.

Shanks, W.S., Schwerdtner, W.M., 1991a. Crude quantitative estimates of the original northwest-southeast dimension of the Sudbury Structure, south central Canadian Shield. Can. J. Earth Sci. 28, 1677–1686.

Shanks, W.S., Schwerdtner, W.M., 1991b. Structural analysis of the central and southwestern Sudbury Structure, Southern Province, Canadian Shield. Can. J. Earth Sci. 28 (3), 411–430.

Shanks, W.S., Schwerdtner, W.M., 1991c. Erratum: Crude quantitative estimates of the original northwest-southeast dimension of the Sudbury structure, south central Canadian shield. Can. J. Earth Sci. 29, 835.

Shanks, W.S., Dressler, B.O., Schwerdtner, W.M., 1990. New developments in Sudbury geology. Abstracts for the international workshop on Meteorite impact on the early Earth. LPI Contribution 746, p. 46.

Sharpton, V.L., Dressler, B.O., Schuraytz, B.C., 1994. The Chicxulub and Sudbury multi-ring impact structures, a comparison. GAC/AGC-MAC/AMC Joint Annual Meeting (abstract), vol. 19, p. 102.

Shellnutt, J.G., MacRae, N.D., 2011. Petrogenesis pf the Meosoproterozoic (1.23Ga) Sudbury dyke swarm and its questionable relationship to plate separation. Int. J. Earth Sci. 101, 3–23.

Shirt, N.N., 2006. A comparison of features characteristic of nuclear explosion craters and astroblemes. Ann. NY Acad. Sci. 123, 371–1235.

Short, N., 2006. A comparison of features characteristic of nuclear explosion craters and astroblemes. Ann. NY Acad. Sci. 123, 573–616.

Siddorn, J.P., Halls, H.C., 2002. Variation in plagioclase clouding intensity in Matachewan dykes: evidence for the exhumation history of the northern margin of the Sudbury Igneous Complex. Can. J. Earth Sci. 39, 933–942.

Siddorn, J.P., Ham, A., 2006. Review of Garson Mine phase 3 exploration program. SRK report for INCO Ltd. Unpublished, 53 pp.

Siddorn, J.P., Lee, C.B., 2005. Garson Mine structural review—summary of investigation. SRK letter report for INCO Ltd. Unpublished, 12 pp.

Siemiatkowska, K., 1972. Fenitic breccias in the Sudbury area (Ontario). MSc Thesis. McGill University, Montreal, PQ, Canada. 89 pp.

Siepierski, L., 2008. Geologia e Petrologia do prospecto GT-34: evidencia de metassomatismo de alta temperature e baixa fO$_2$, Provincia Mineral Carajas, Brasil. MSc Thesis. University of Brasilia, Brasilia. 62 pp.

Silver, L.P., 1902. The sulphide ore bodies of the Sudbury region (Ontario). Trans. Can. Min. Inst. 10, 528–551.

Sims, P.K., Card, K.D., Morey, G.B., Peterman, Z.E., 1980. The Great Lakes tectonic zone—a major crustal structure in central North America. Geol. Soc. Am. Bull. 91, 690–698.

Singhroy, V., Mussakowski, R., Dressler, B.O., Trowell, N.F., Grieve, R., 1992. SAR in support of geological investigations of the Sudbury Structure. In: Dressler, B.O., Sharpton, V.L. (Eds.), Papers presented to the international conference on large meteorite impacts and planetary evolution. LPI Contribution 790, p. 69.

Sinha, M.N., 1983. Geochemistry and petrology of mafic dikes of Mackenzie and Sudbury swarms. MSc Thesis. University of Waterloo. Waterloo, ON, Canada. 188 pp.

Slack, J.F., Cannon, W.F., 2009. Extraterrestrial demise of banded iron formations 1.85 billion years ago. Geology 37, 1011–1014.

Slaght, W.H., 1953. A petrographic study of the Copper Cliff offset, Sudbury District, Ontario. Can. Min. J. 74 (3), 110.

Slaney, V.R., Misra, K., 1988. Radar studies of the Sudbury Basin. In: Scientific drilling, the Sudbury Structure, proceedings of a, workshop. In: Drury, M.J. (Ed.), Canadian Continental Drilling Program Report, vol. 88-2, p. 16.

Slaught, W.H., 1951. A petrographic study of the Copper Cliff offset in the Sudbury District (Ontario). MSc Thesis. McGill University, Montreal, QU, Canada. 68 pp.

Sluzhenikin, S.F., Distler, V.V., Dyuzhikov, O.A., Kravtsov, V.F., Kunilov, V.E., Laputina, I.P., Turovtsev, D.M.N., 1994. Low-sulfide platinum mineralization in the Noril'sk differentiated intrusive bodies. Geol. Ore Dep. 36 (3), 171–195.

Smirnov, O.N., Lul'ko, V.A., Amosov, Yu.N., Salav, V.M., 1994. Geological Structure of the Noril'sk Region. In Proceedings of the Sudbury-Noril'sk Symposium. Ontario Geological Survey Special Volume 5, pp. 161–170.

Smith, F.G., 1951. Garnet from Sudbury region. Progress Report—Temperature-Pressure Research of Hydrothermal Mineral Deposits, vol. 3, pt. 4, p. 81.

Smith, M.D., 2002. The timing and petrogenesis of the Creighton pluton, Ontario: and example of felsic magmatism associated with Matachewan Igneous Events? MSc Thesis. Edmonton, Canada. 123 pp.

Smith, D.A., Bailey, J.M., Pattison, E.F., 2013. Discovery of new offset dykes and insights into the Sudbury Impact Structure. Large Meteorite Impacts, Planetary Evolution, Sudbury, Canada, Abstract, p. 3090.

Snelling, P.E., 2009. The influences of stress and structure on mining-induced seismicity in Creighton mine, Sudbury, Canada. MSc Thesis. Queen's University, Kindston, Canada, 209 pp.

Sobolev, A.V., Krivolutskaya, N.A., Kuzmin, D.V., 2009. Petrology of the parental melts and mantle sources of Siberian Trap magmatism. Petrology 17, 253–286.

Song, X.Y., Keays, R.R., Xiao, L., Qi, H.W., Ihlenfeld, C., 2009. Platinum-group element geochemistry of the continental flood basalts in the central Emeishan large igneous province, SW China. Chem. Geol. 262, 246–261.

Sopher, S.R., 1961. Paleomagnetic study of the Sudbury Irruptive (Ontario). MSc Thesis. Carleton University, Ottawa, ON, Canada, 93 pp.

Sopher, R.S., 1963. Paleomagnetic study of the Sudbury Irruptive. Geol. Surv. Can. Bull. 90, 34 pp.

Souch, B., Podolsky, T., Geological Staff, 1969. The sulphide ores of Sudbury: their particular relationship to a distinctive inclusion-bearing facies of the nickel irruptive. Econ. Geol. Monograph, vol. 4, pp. 252–261.

Speers, E.C., 1956. Age relations and origin of Sudbury breccias, Ontario. PhD Thesis. Queen's University, Kingston, ON, Canada. 393 pp.

Speers, E.C., 1957. The age relation and origin of common Sudbury breccia (Ontario). J. Geol. 65, 497–514.

Spray, J.G., 1995. Pseudotachylite controversy: fact or friction? Geology 23, 1119–1122.

Spray, J.G., 1997. Superfaults. Geology 25, 579–582.

Spray, J.G., 1998. Localized shock- and friction-induced melting in response to hypervelocity impact. In: Grady, M.M., Hutchinson, R., McCall, G.J.H., Rothery, D.A. (Eds.), Meteorites: Flux with Time and Impact Effects. London, England. Geological Society, Special Publication 140, pp. 195–204.

Spray, J.G., Thompson, L.M., 1993. Friction melt distribution in multi-ring impact basins, evidence from the Sudbury Structure. AGU 1993 fall meeting. Eos, Transactions, American Geophysical Union, vol. 74, pt. 43, Suppl., p. 387.

Spray, J.G., Thompson, L.M., 1995. Friction melt distribution in a multi-ring impact basin. Nature 373, 130–132.

Spray, J.G., Thompson, L.M., 2008. Constraints on central uplift structure from the Manicouagan impact crater. Meteor. Planet. Sci. 43, 2049–2057.

Spray, J.G., Butler, H.R., Thompson, L.M., 2004. Tectonic influences on the morphometry of the Sudbury impact structure: implications for terrestrial cratering and modeling. Meteor. Planet. Sci. 39, 287–301.

Springer, G., 1989. Chlorine-bearing and other uncommon minerals in the Strathcona deep copper zone, Sudbury District, Ontario. Can. Mineral. 27, 311–313.

Sproule, R., Sutcliffe, R., Tracanelli, H., Lesher, C.M., 2007. Palaeoproterozoic Ni–Cu–PGE mineralisation in the Shakespeare intrusion, Ontario, Canada: a new style of Nipissing gabbro-hosted mineralization. Appl. Earth Sci. IMM Trans. B 116, 188–200.

Spurr, J.E., 1924. Ore deposition at the Creighton nickel mine, Sudbury, Ontario. Econ. Geol. Bull. Soc. Econ. Geol. 19, 275–280.

Stanton, R.L., 1972. Ore Petrology. McGraw-Hill, New York, 714 pp.

St-Clair, S., 1914. Origin of the Sudbury ore deposits. Mining and Scientific Press 109, 243–246.

Stekhin, A.I., 1994. Mineralogical and geochemical characteristics of the Cu-Ni ores of the Oktyabr'sky and Talnakh deposits. In: Lightfoot, P.C., Naldrett, A.J., Sheahan, P. (Eds.), Proceedings of the Sudbury-Noril'sk Symposium. Ontario Geological Survey, Special Publication 5, pp. 217–230.

Stevens, K., Watts, A., Redko, G., 2000. In-Mine Applications of the Radio-Wave Method in the Sudbury Igneous Complex. Paper presented at the 2000 SEG Annual Meeting.

Stevenson, J.S., 1960. Origin of quartzite at the base of the Whitewater Series, Sudbury Basin. International Geological Congress, Report of 21st Session, Norden, part 26, pp. 32–41.

Stevenson, J.S., 1961. Recognition of the quartzite breccia in the Whitewater series, Sudbury basin, Ontario. Trans. R. Soc. Can. LV, 57–66.

Stevenson, J.S., 1963. The upper contact phase of the Sudbury micropegmatite. Can. Mineral. 7, 413–419.

Stevenson, J.S., 1972. The Onaping ashflow sheet, Sudbury, Ontario, In: Guy-Bray, J.V. (Ed.), New Developments in Sudbury Geology. Geological Association of Canada, Special Paper 10, pp. 41–48.

Stevenson, J.S., 1979. Geological concepts developed in the Precambrian of Sudbury, Ontario. In: Kupsch, W.O., Sargeant, W.A.S. (Eds.), History of Concepts in Precambrian Geology. Geological Association of Canada, Special Paper 19, pp. 224–244.

Stevenson, J.S., 1983. Trachyandesite and dacite in the Onaping Formation, Sudbury, Ontario. GAC/AGC-MAC/AMC Joint Annual Meeting (abstract), vol. 8, p. A65.

Stevenson, J.S., 1984. Sudbury problems and suggestions for further research, In: Pye, E.G., Naldrett, A.J., Giblin, P.E. (Eds.), The Geology and Ore Deposits of the Sudbury Structure. Ontario Geological Survey Special Volume 1, pp. 523–531.

Stevenson, J.S., 1990. The volcanic origin of the Onaping Formation, Sudbury, Canada. In: Nicolaysen, L.O., Reimold, W.U. (Eds.), Cryptoexplosions and catastrophes in the geological record, with a special focus on the Vredefort Structure. Tectonophysics, vol 171, pt. 1–4, pp. 249–257.

Stevenson, J.S., Colgrove, G.L., 1968. The Sudbury Irruptive: some petrogenetic concepts based on recent field work. Proc. XXII I Int. Geol. Congress 4, 27–35.

Stevenson, J.S., Stevenson, L.S., 1980. Sudbury, Ontario, and the meteorite theory. Geosci. Can. 7 (3), 103–108.

Stewart, L., 1908. The Creighton Mine of the Canadian Copper County, Sudbury District. Ontario Trans. Can. Min. Inst. 11, 567–585.

Stewart, M., 2002. Petrology and mineralogy of Cu-Ni-PGE ore, Totten area, Worthington Offset, Sudbury Igneous Complex. ntario. Carleton University, Ottawa. 192 pp.

Stewart, M.C., Lightfoot, P.C., 2010. Diversity in Platinum Group Element (PGE) mineralization at Sudbury: new discoveries and process controls. Conference: 11th International Platinum Symposium, Sudbury, Ontario. Available from: https://www.researchgate.net/publication/259005801_Diversity_in_Platinum_Group_Element_%28PGE%29_Mineralization_at_Sudbury_New_Discoveries_and_Process_Controls.

Stockwell, C.H., 1982. Proposals for time classification and correlation of Precambrian rocks and events in Canada and adjacent areas of the Canadian Shield. Geological Survey of Canada Paper 80–19, 135 p.

Stöffler, D., Avermann, M., Bischoff, L., Brockmeyer, P., Deutsch, A., Dressler, B.O., Lakomy, R., Mueller-Mohr, V., 1989. Sudbury, Canada, remnant of the only multi-ring impact basin on Earth? In: Wasson, J.T. (Ed.), Meteoritical Society, 52nd meeting (abstracts). Meteoritics, vol. 24, pt. 4, p. 328.

Stöffler, D., Deutsch, A., Avermann, M., Brockmeyer, P., Lakomy, R., Mueller-Mohr, V., 1992. Sudbury Project (University of Muenster-Ontario Geological Survey), 3, Petrology, chemistry, and origin of breccia formations. In: Dressler, B.O., Sharpton, V.L. (Eds.), Papers Presented to the International Conference on Large Meteorite Impacts and Planetary Evolution. LPI Contribution 790, pp. 71–72.

Stöffler, D., Deutsch, A., Avermann, M., Bischoff, L., Brockmeyer, P., Buhl, D., Lakomy, R., Müller-Mohr, V., 1994. The formation of the Sudbury Structure, Canada: toward a unified impact model. In: Dressler, B.O., Grieve, R.A.F., Sharpton, V.L. (Eds.), Papers Presented to the International Conference on Large Meteorite Impacts and Planetary Evolution. LPI Contribution. Boulder, Colorado, Geological Society of America, Special Paper 293, pp. 303–318.

Stokes, R., 1907. The Sudbury nickel-copper field, Ontario. Mining World 27, 507–510.

Stonehouse, H.B., 1954. An association of trace elements and mineralization at Sudbury (Ontario). Am. Min. 39 (5–6), 452–474.

Stoness, J.A., 1994. The stratigraphy, geochemistry and depositional environment of the Paleoproterozoic Vermilion and Onwatin formations and their relationship to the Zn-Cu-Pb massive sulfide deposits in the Sudbury Basin. Unpublished MSc Thesis. Laurentian University, Sudbury, Ontario, 205 pp.

Stoness, J.A., Gibson, H.L., Long, D.G.F., 1993. The stratigraphy, geochemistry and depositional environment of the Proterozoic Vermilion Member, host to the Errington and Vermilion Zn-Cu-Pb massive sulphide deposits, Sudbury, Ontario. GAC/AGC-MAC/AMC Joint Annual Meeting (abstract), p. 102.

Stout, A.E., 2009. Geology, Mineralogy, and Geochemistry of the McCreedy East 153 Cu-Ni-PGE Deposit, Sudbury, Ontario. Unpublished MSc Thesis. Utrecht, The Netherlands, Utrecht University, 39 p.

Streckeisen, A.L., 1967. Classification and nomenclature of igneous rocks. Neues Jahrbuch fur Mineralogie, Abhandlungen 107, 144–240.

Strickland, K.E., 1971. The geology of Stobie and Marconi townships (Precambrian), District of Sudbury, Ontario, Canada. MSc Thesis. Bowling Green State University, Bowling Green, OH, United States. 108 pp.

Strongman, K., 2016. Chemostratigraphy of the South Range Norite, Sudbury Igneous Complex. BSc thesis, Laurentian University, Sudbury. 53 pages.

Stutzer, O., 1908. Die Nickelerzlagerstaetten bei Sudbury in Kanada. Zeitschrift fuer Praktische Geologie, 285–287.

Sugaki, A., Kitakaze, A., 1998. High form of pentlandite and its thermal stability. Am. Mineral. 83, 133–140.

Sutcliffe, R., Tracanelli, H., Davis, D.W., 2002. Shakespeare intrusion, Abstract volume for the 2002 Ontario Prospectors Association Meeting, Toronto, Canada.

Sweeny, M., Farrow, C., 1990. Geology Report, Morgan West Project, Levack and Morgan Townships. Unpublished Report, Falconbridge Limited Exploration, 35 pp.

Sylvester, P., 2001. Laser Ablation ICP-MS in the Earth Sciences: Principles and Applications. GAC-MAC Volume 29. ISBN: 0-9212294-29-8.

Sylvester, P., 2008. Laser Ablation ICP-MS in the Earth Sciences: Current Practices and Outstanding Issues. MAC Short Course 40.

Szabo, E., Halls, H.C., 2006. Deformation of the Sudbury Structure: paleomagnetic evidence from the Sudbury Breccia. Precambrian Res. 150, 27–48.

Szentpeteri, K., Watkinson, D.H., Molnar, F., Jones, P.C., 2002. Platium-group elements-Co-Ni-Ge sulfarsenides and mineral paragenesis of some Cu-Ni-platinum group element deposits, Copper Cliff North Area, Sudbury, Canada. Econ. Geol. 97, 1459–1470.

Szuwalski, D.R., 1984. The petrology of the ultramafic rocks in the footwall Levack Complex, Fraser area, Sudbury, Canada (possible source of the Sudbury sublayer ultramafic xenoliths). MSc Thesis. Rutgers, The State University, New Brunswick. New Brunswick, NJ, United States. 116 pp.

Tang, Z., 1990. Nickel Deposits of Chinas. In Mineral Deposits of China, vol. 2. Geological Publishing House, Beijing, p. 52–99.

Tang, Z., 1993. Genetic model of the Jinchuan nickel–copper deposit. Kirkham, R.V., Sinclair, W.D., Thorpe, R.I., Duke, J.M. (Eds.), Mineral Deposit Modelling, 40, Geological Survey of Canada Special Paper, pp. 389–401.

Tang, Z.L., 1999. Jinchuan Copper-Nickel Sulfide Deposit. Geological Publishing House, Beijing, China, 84 pp.

Tanton, T.L., 1917. Reconnaissance along Canadian Northern Railway between Gogama and Oba, Sudbury and Algoma districts, Ontario Summary Report of the Geological Survey of Canada, pp. 179–182.

Taranovic, V., Ripley, E.M., Li, C., Rossel, D., 2014. Petrogenesis of the Ni-Cu-PGE sulfide-bearing Tamarack Intrusive Complex, Midcontinent Rift System, Minnesota. Lithos 16, 212–215.

Taylor, S.R., McLennan, S.M., 1985. The Continental Crust: Its Composition and Evolution. Blackwell, Oxford, 312 pp.

Therriault, A.M., Fowler, A.D., Grieve, R.A.F., 2002. The Sudbury Igneous Complex: a differentiated impact melt sheet. Econ. Geol. 97, 1521–1540.

Therriault, A.M., Grieve, R.A.F., Ostermann, M., Deutsch, A., 1996a. Sudbury Igneous Complex, how many melt systems? In: Sears, D.W.G. (Ed.), 59th Annual Meteoritical Society Meeting (abstracts). Meteoritics & Planetary Science, vol. 31, Suppl., pp. 141–142.

Therriault, A.M., Ostermann, M., Grieve, R.A.F., Deutsch, A., 1996b. Are Vredefort granophyre and Sudbury offsets birds of a feather? In: Sears, D.W.G. (Ed.), 59th annual Meteoritical Society meeting, abstracts. Meteoritics & Planetary Science, vol. 31, Suppl., p. 142.

Thode, H.G., Dunford, H.B., Shima, M., 1962. Sulfur isotope abundances in rocks of the Sudbury district and their geological significance. Econ. Geol. 57, 565–578.

Thomas, K., 1912. The Sudbury nickel district. Ontario Mining and Scientific Press 105, 433.

Thomas, K., 1914. The Sudbury nickel district of Ontario. Eng. Min. J., 152–154.

Thomas, M.D., Sharpton, V.L., Grieve, R.A.F., 1987. Gravity patterns and Precambrian structure in the North American central plains. Geology 15, 489–492.

Thompson, P., 1906. The Sudbury nickel region. Eng. Min. J. 82, 3–4.

Thompson, L.M., 1994. Ring characteristics of the Sudbury multi-ring impact basin (abstract). European Science Foundation, Third International Workshop. Shock Wave Behaviour of Solids in Nature and Experiments, Limoges, France.

Thompson, L.M., 1996. A study of pseudotachylite associated with the Sudbury Structure, Ontario, Canada. Unpublished Ph.D. Thesis. University of New Brunswick, New Brunswick, Canada, 202.

Thompson, J.F., Beasley, N., 1960. For the Years to Come: a Story of International Nickel of Canada. Longmans, Toronto, 374 pp.

Thompson, J.E., Card. K.D., 1963. Geology of Kelly and Davis Townhsips. Ontario Department of Mines Geology Report 15–20 pages with map.

Thompson, L.M., Spray, J.G., 1992. A comparison of the chemistry of pseudotachylyte breccias in the Archean Levack Gneisses of the Sudbury Structure, Ontario. In: Dressler, B.O., Sharpton, V.L. (Eds.), Papers Presented to the International Conference on Large Meteorite Impacts and Planetary Evolution. LPI Contribution 790, pp. 73–74.

Thompson, L.M., Spray, J.G., 1994. Pseudotachylytic rock distribution and genesis within the Sudbury impact structure. In: Dressler, B.O., Grieve, R.A.F., Sharpton, V., L. (Eds.), Papers Presented to the international Conference on Large Meteorite Impacts and Planetary Evolution. LPI Contribution. Special Paper – Geological Society of America, vol. 293, pp. 275–287.

Thompson, L.M., Spray, J.G., 1996. Pseudotachylyte petrogenesis: constraints from the Sudbury impact structure. Contrib. Mineral. Petrol. 125, 359–374.

Thompson, L.M., Spray, J.G., Kelley, S.P., 1998. Laser-probe 40Ar/39Ar dating of pseudotachylyte from the Sudbury structure: evidence for post-impact thermal overprinting in the North Range. Meteor. Planet. Sci. 33, 1205–1216.

Thomson, R., 1935a. A study of the nickel intrusive, Sudbury, Ontario. PhD Thesis. University of Chicago. Chicago, IL, United States, 51 pp.

Thomson, R., 1935b. Nickel eruptive of Sudbury, Ontario. Pan-Am. Geol. 63, 248–264.

Thomson, R., 1935c. Sudburite, a metamorphic rock near Sudbury, Ontario. J. Geol. 43, 427–435.

Thomson, R., 1935d. The "offset dikes" of the nickel intrusive. Sudbury, Ontario. Am. J. Sci. 30 (178), 356–367.

Thomson, J.E., 1953a. Geology and mineral deposits of the Sudbury area, Ontario. In: Geol. Soc. America, Guidebook Toronto Field Trip, vol. 7, pt. 6–7, p. 15.

Thomson, J.E., 1953b. Problems of Precambrian stratigraphy west of Sudbury, Ontario. Trans. R. Soc. Can. 47, 61–70.

Thomson, J.E., 1956. Geology of the Sudbury basin. Ontario Department of Mines Annual Report for 1956, vol. 65, pt. 3, pp. 1–56.

Thomson, J.E., 1957a. Recent geological studies in Sudbury camp (Ontario). Can. Min. J., vol. 78, pt. 4, pp. 109–112.

Thomson, J.E., 1957b. The questionable Proterozoic rocks of the Sudbury-Espanola area (Ontario). In: Gill, J.E. (Ed.), The Proterozoic in Canada. Special Publications—Royal Society of Canada, vol. 2, pp. 48–53.

Thomson, J.E., 1969. Discussion of Sudbury geology and sulfide deposits. Ontario Department of Mines. Miscellaneous Publication 30, 22.

Thomson, J.E., Allen, J.S., 1939. Nickeliferous pyrite from the Denison Mine, Sudbury District, Ontario. University of Toronto Studies. Geol. Ser. 42, 135–138.

Thomson, M.L., Barnett, R.L., 1981. High aluminum metamorphic amphibole from the South Range Norite, Sudbury Nickel Irruptive. The Geological Society of America, 94th Annual Meeting. Abstracts with Programs—Geological Society of America, vol. 13, pt. 7, p. 567.

Thomson, J.E., Williams, H., 1959. The myth of the Sudbury lopolith. Can. Min. J. 80, 57–62.

Thomson, M.L., Barnett, R.L., Fleet, M.E., Kerrich, R., 1985. Metamorphic assemblages in the South-Range norite and footwall mafic rocks near the Kirkwood Mine, Sudbury, Ontario. Can. Mineral., vol. 23, pt. 2, pp. 173–186.

Therriault, A.M., Fowler, A.D., Grieve, R.A.F., 2002. The Sudbury Igneous Complex: a differeniated impact melt sheet. Econ. Geol. 97, 1521–1540.

Torgashin, A.S., 1994. Geology of the massive and copper ores of the western part of the Oktyabrysk Deposit. In: Lightfoot, P.C., Naldrett, A.J. (Eds.), Proceeding of the Sudbury-Noril'sk, Symposium, Ontario Geological Survey Special Volume 5, pp. 231–242.

Tornos, F., Casquet, C., Galindo, C., Velasco, F., Canales, A., 2001. A new style of Ni-Cu mineralization related to magmatic breccia pipes in a transpressional magmatic arc36Mineral. Deposit., Aguablanca, Spain, 700–706.

Tuba, G., Molnár, F., Ames, D.E., Péntek, A., Watkinson, D.H., Jones, P.C., 2014. Multi-stage hydrothermal processes involved in "low-sulfide" Cu(-Ni) -PGE mineralization in the footwall of the Sudbury Igneous Complex (Canada): Amy Lake PGE zone, East Range. Min. Dep., vol. 49, pp. 7–47.

Tuchscherer, M.G., Spray, J.G., 2002. Geology, Mineralization, and Emplacement of the Whistle-Parkin Offset Dike, Sudbury. Econ. Geol. 97, 1377–1398.

Ulrych, T.J., 1962. Gas source mass spectrometry of trace leads from Sudbury, Ontario. PhD Thesis. Untersuchungen an Impaktbreccien der Sudbury-Struktur, Ontario, Kanada, und ihre Bedeutung fur die Genese der Struktur. Muenster Forsch Geol Palaeont, vol. 77, pp. 27–41.

Urbancic, T.I., 1991. Source studies of mining-induced microseismicity at Strathcona Mine, Sudbury, Canada, a spatial and temporal analysis. PhD Thesis. Queen's University, Kingston, ON, Canada, 356 pp.

USGS, 2015. Nickel: Statistics and Information. Available from: http://minerals.usgs.gov/minerals/pubs/commodity/nickel/.

Ushah, A.M.A., Moon, W., 1985. Application of 2-D Hilbert transform in 3-D potential field data (gravity and magnetic) from Sudbury. Annual Report—Centre for Precambrian Studies, University of Manitoba, pp. 12–46.

Van Breeman, O., Davidson, A., 1988. Northeast extension of Proterozoic terranes of mid-continental North America. Geol. Soc. Am. Bull. 100, 630–638.

Van Schmus, W.R., 1976. Early and middle Proterozoic history of the Great Lakes area, North America. Philos. Trans. R. Soc. Lond. Ser. A 280, 606–629.

Van Schmus, W.R., Card, K.D., Harrower, K.L., 1975. Geology and ages of buried Precambrian rocks, Manitoulin Island, Ontario. Can. J. Earth Sci. 12, 1175–1189.

Vaughan, W.M., Head, J.W., 2014. Impact melt differentiation in the South Pole-Aitken basin: some observations and speculations. Planet. Space Sci. 91, 101–106.

Vezina, J., 1992. Origin of concretions in the Proterozoic Chelmsford Formation, Sudbury Basin, Ontario. Unpublished BSc Thesis. Queen's University, Kingston, Ontario, 43 pp.

Vogel, D.C., James, R.S., Keays, R.R., 1998a. The early tectono-magmatic evolution of the Southern Province: implications from the Agnew intrusion, central Ontario, Canada. Can. J. Earth Sci. 35, 854–8790.

Vogel, D.C., Vuollo, J.I., Alapieti, T.T., James, R.S., 1998b. Tectonic, stratigraphic and geochemical comparisons between ca. 2500–2440 Ma mafic igneous events in the Canadian and Fennoscandian shields. Precambrian Res. 92, 89–116.

Vogel, D.C., Keays, R.R., James, R.S., Reeves, S.J., 1999. The geochemistry and petrogenesis of the Agnew Intrusion, Canada: a product of S-undersaturated, high-Al and low-Ti tholeiitic magmas. J. Petrol. 40, 423–450.

Vokes, F.M., 1968. Regional metamorphism of the Palaeozoic sulphide ore deposits of Norway. Inst. Mining Metallurg Trans 77, B53.

Vokes, F.M., 1969. A review of the metamorphism of sulphide deposits. Earth Sci. Rev. 5, 99–143.

Vokes, F.M., 1973. "Ball texture" in sulphide ores. Geologiska Föreningen Stockholm Förhandlingar. 95, 403–406.

Vujovic, Z., Lightfoot, P.C., Hawkesworth, C.J., 1995. Origin and geochemical evolution of mafic and ultramafic inclusions in the Sublayer of the Sudbury Igneous Complex at Whistle mine. In: Summary of Field Work and Other Activities, Ontario Geological Survey, Miscellaneous Paper 164, pp. 130–133.

Walker, L.T., 1894. Notes on nickeliferous pyrite from Murray Mine, Sudbury. Ontario Am. J. Sci. 47, 312–314.

Walker, L.T., 1897. Geological and petrographical studies of the Sudbury nickel district, Canada. Quart. J. Geol. Soc. Lond. 53, 40–66.

Walker, L.T., 1915. Certain mineral occurrences in the Worthington Mine, Sudbury, Ontario, and their significance. Econ. Geol. 10, 536–542.

Walker, L.T., 1935. Magmatic differentiation as shown in the nickel intrusive of Sudbury, Ontario. University of Toronto Studies, Geological Series, 23–30.

Walker, J., 2014. The Black Norite orthopyroxene-(sulfide)-rich layer, South Range Norite. BSc thesis, Laurentian University.

Walker, R.J., Morgan, J.W., Lambert, D.D., Naldrett, A.J., Bohlke, J.K., Rajamani, V., 1991a. The Re-Os isotope system as a tracer in the study of platinum-group-element and gold deposits. United States Geological Survey, Circular, vol. 1062, pp. 76–77.

Walker, R.J., Morgan, J.W., Naldrett, A.J., Li, C., 1991b. Re-Os isotope evidence for a major crustal component in Ni-Cu sulfide ores, Sudbury Igneous Complex, Ontario. In: AGU-MSA 1991 spring meeting. Eos, Transactions, American Geophysical Union, vol. 72, pt. 17, p. 305.

Walker, R.J., Morgan, J.W., Naldrett, A.J., Li, C., 1991c. Re-Os isotopic systematics of Ni-Cu sulfide ores, Sudbury Igneous Complex, Ontario, evidence for a major crustal component. GAC/AGC-MAC/AMC Joint Annual Meeting (abstract), vol. 16, p. 130.

Walker, R.J., Morgan, J.W., Naldrett, A.J., Li, C., Fassett, J.D., 1991d. Re-Os isotope systematics of Ni-Cu sulfide ores, Sudbury Igneous Complex, Ontario: evidence for a major crustal component. Earth Planet. Sci. Lett. 105, 416–429.

Walker, R.J., Morgan, J.W., Hanski, E., Smolkin, V.F., 1994. The role of the Re-Os isotope system in deciphering the origin of magmatic sulphide ores: a tale of three ores, In: Lightfoot, P.C., Naldrett, A.J. (Eds.), Proceedings of the Sudbury-Noril'sk Symposium. Ontario Geological Survey Special Volume 5, pp. 343–355.

Wandke, A., Hoffman, R., 1924. A study of the Sudbury (Ontario) ore deposits. Econ. Geol. Bull. Soc. Econ. Geol. 19, 169–204.

Wang, Y., Lesher, C.M., LighKoot, P.C., 2016. Shock metamorphic features of olivine, orthopyroxene, amphibole, and plagioclase in phlogopite—bearing ultramafic—mafic inclusions in Contact Sublayer, Sudbury Igneous Complex, PDAC Annual Meeting, Toronto.

Warner, S., Martin, R.F., Abdel-Rahman, A-F.M., Doig, R., 1998. Apatite as a monitor of fractionation, degassing, and metamorphism in the Sudbury Igneous Complex, Ontario. Can. Mineral. 36, 981–999.

Warren, P.H., 1992. The Sudbury-Serenitatis analogy and so-called pristine nonmare rocks. In: Ryder, G., Schmitt, H.H., Spudis, P.D. (Eds.), Workshop on Geology of the Apollo 17 landing site. LPI Technical Report, vol. 92-09, pt. 1, pp. 59–61.

Warren, P.H., Claeys, P., Cedillo, Pardo, E., 1994. Where are the Chicxulub coarse-grained, igneously layered impact melt rocks analogous to those at Sudbury? In: Papers presented at New Developments Regarding the KT Event and other Catastrophes in Earth History. LPI Contribution 825, pp. 128–130.

Warren, P.H., Claeys, P., Cedillo, P.E., 1996. Mega-impact melt petrology (Chicxulub, Sudbury, and the Moon), effects of scale and other factors on potential for fractional crystallization and development of cumulates. In: Ryder, G., Fastovsky, D., Gartner, S. (Eds.), The Cretaceous-Tertiary event and other catastrophes in Earth history. Special Paper—Geological Society of America, 307, pp. 105–124.

Watkinson, D.H., 1994. Fluid-rock interaction at contact of Lindsley 4b Ni-Cu-PGE orebody and enclosing granitic rocks, Sudbury, Canada. Institution of Mining and Metallurgy, Transactions, Section B. Appl. Earth Sci., vol. 103, pp. B121–B128.

Watkinson, D.H., Molnar, F., 1996. Remobilization of platinum-group elements by saline fluid: Evidence from fluid-inclusion studies at the Lindsley Mine, Sudbury, Ontario (abstract). GAC/MAC, p. A100.

Watts, A.B., 2010. Crust and Lithosphere Dynamics: Treatise on Geophysics, vol. 6, Elsevier, Oxford, 632 pp.

Wegener, A., 1929. Die Entstehung der Kontinente und Ozeane, fourth ed. Friedrich Vieweg & Sohn Akt. Ges., Braunschweig.

Weiser, T., Dressler, B.O., Brockmeyer, P., 1996. The Onaping Formation, Sudbury structure—A Proterozoic impact breccia (abstract). GAC/MAC, p. A101.

Weitz, J.W., Head, I.R., Pappalardo, G., Neukum, B., Giese, J., Oberst, A., Cook, B., Schreiner, R., Greeley, P., Helfenstein, C., Chapman, the Galileo SSI Team, 1997. Galileo observations of Ganymede impact crater morphology. Lunar and Planetary Sciences, vol. 28, pp. 1111.

Wendlandt, R.F., 1982. Sulfide saturation of basalt and andesite melts at high pressures and temperatures. Am. Mineral. 67, 877–885.

West, G.F., MacNae, J.C., Lamontagne, Y., 1984. A time-domain electromagnetic system measuring the step response of the ground. Geophysics 49 (07), 1010–1026.

White, C.J., 2012. Low-sulfide PGE-Cu-Ni Mineralization from Five Prospects within the Footwall of the Sudbury Igneous Complex, Ontario, Canada. PhD Thesis. University of Toronto, Canada. Available from: http://hdl.handle.net/1807/32849.

White, S., Morris, W.A., 1979. Contact test evidence of complex remanence acquisition from the Sudbury Basin. Eos, Transactions, American Geophysical Union 60 (42), 747.

White, D.J., Milkereit, B., Wu, J.J., Salisbury, M.H., Mwenifumbo, J., Berrer, E.K., Moon, W., Lodha, G., 1994. Seismic reflectivity of the Sudbury Structure North Range from borehole logs. In: Boerner, D.E., Milkereit, B., Nadrett, A.J. (Eds.), Lithoprobe Sudbury Project. Geophys. Res. Lett., vol. 21, pt. 10, pp. 935–938.

Whitehead, R.E.S., Davies, J.F., Goodfellow, W.D., 1990. Isotopic evidence for hydrothermal discharge into anoxic seawater, Sudbury Basin, Ontario. Chem. Geol. 86, 49–63.

Whitehead, R.E.S., Davies, J.F., Goodfellow, W.D., 1992. Lithogeochemical patterns related to sedex mineralization, Sudbury Basin, Canada. Chem. Geol. 98 (1–2), 87–101.

Whiteway, P., 1990. A Guide to Prospecting and Mining. Northern Miner Press, Toronto, 144 pp.

Wichman, R.W., Schultz, P.H., 1990. Implications of early crater-centered volcanism and tectonism at the Sudbury structure, Ontario. Abstracts for the international workshop on Meteorite impact on the early Earth. LPI Contribution, vol. 746, pp. 56–57.

Wichman, R.W., Schultz, P.H., 1992. Floor-fractured crater models of the Sudbury Structure, Canada. In: Dressler, B.O., Sharpton, V.L. (Eds.), Papers Presented to the International Conference on Large Meteorite Impacts and Planetary Evolution. LPI Contribution 790, pp. 79–80.

Wichman, R.W., Schultz, P.H., 1993. Floor-fractured crater models of the Sudbury Structure, Canada: implications for initial crater size and crater modification. Meteoritics 28, 222–231.

Wichman, R.W., Wood, C.A., 1995. The Davy Crater Chain: Implications for tidal disruption in the Earth-Moon System and elsewhere. Geophys. Res. Lett. 22, 583–586.

Wignall, P.B., 2001. Large igneous provinces and mass extinctions. Earth Sci. Rev. 53, 1–33.

Williams, H., 1956. Glowing avalanche deposits of the Sudbury Basin. Ontario Department of Mines, Annual Report, vol. 65, pp. 57–89.

Wilson, H.D.B., 1956a. Structure of lopoliths. Bull. Geol. Soc. Am. 67, 289–300.

Wilson, H.D.B., 1956b. Structure of lopoliths, discussion and reply. Geol. Soc. Am. 68, 1071–1075.

Wodicka, N., 1995. Late Archean history of the Levack Gneiss Complex, southern Superior Province, Sudbury, Ontario: New evidence from U-Pb geochronology international Conference on Tectonics and Metallogeny of Early/Mid Precambrian Orogenic Belts, Precambrian'95, Program with Abstracts, p. 191.

Won, J.S., Miao, X.G., An, P., Moon, W.M., Singh, V., Singhroy, V., Slaney, R.V., 1992. Preliminary investigation of Sudbury Structure using airborne multi-sensor geophysical data. 29th International Geological Congress (abstracts), vol. 29, p. 977.

Wood, D., 2015. Enterprise: A Sediment-Hosted Ni Deposit in NW Zambia. Prospectors and Developers Association of Canada, Toronto. Available from: http://www.pdac.ca/docs/default-source/Convention—Program—Technical-Sessions/2015/nickel—wood.pdf?sfvrsn=2.

Wood, C.R., Spray, J.G., 1998. Origin and emplacement of Offset Dykes in the Sudbury impact structure: constraints from Hess. Meteor. Planet. Sci. 33, 337–347.

Woodall, R., 1994. Empiricism and concept in successful mineral exploration. Aust. J. Earth Sci. 41, 1–10.

Wooden, J.L., Czamanske, G.K., Fedorenko, V.A., Arndt, N.T., Chauvel, C., Bouse, R.M., King, B.W., Knight, R.J., Siems, D.F., 1993. Isotopic and trace element constraints on mantle and crustal contributions to Siberian continental flood basalts, Noril'sk area, Siberia. Geochim. Cosmochim. Acta 57, 3677–3704.

Wren, C., 2012. Risk assessment and environmental management: a case study in Sudbury, Ontario, Canada. Maralte, Netherlands, 454 pp.

Wu, J., Milkereit, B., Boerner, D.E., 1993. The enigmatic Sudbury Structure, What does seismic say? AGU 1993 fall meeting. Eos, Transactions, American Geophysical Union, vol. 74, pt. 43, Suppl., 389.

Wu, J., Milkereit, B., Boerner, D., 1994. Timing constraints on deformation history of the Sudbury impact structure. Can. J. Earth Sci. 31, 1654–1660.

Wu, J., Milkereit, B., Boerner, D.E., 1995. Seismic imaging of the enigmatic Sudbury Structure. J. Geophys. Res. 100, 4117–4130, 3.

Wykes, J., Mavrogenes, J., 2005. Hydrous sulfide melting: experimental evidence for the solubility of H_2O in sulfide melts. Econ. Geol. 100, 157–164.

Xie, Q., Keays, R.R., Prevec, S.A., Lightfoot, P.C., 1997. Origin of the Sudbury Igneous Complex and sulfide mineralization: evidence from platinum group elements (abstract). Large Meteorite Impacts and Planetary Evolution.

Yakubchuk, A., Nikishin, A., 2004. Noril'sk–Talnakh Cu–Ni–PGE deposits: a revised tectonic model. Min. Dep. 39, 125–142.

Yates, A.B., 1938. The Sudbury Intrusive. Trans. R. Soc. Can. 32, 151–172.

Yates, A.B., 1948. Properties of international Nickel Company of Canada (Sudbury area, Ontario). Canadian Inst. Mining and Metallurgy, Geol. Structural geology of Canadian ore deposits, pp. 596–617.

York, D., 1985. The union of (40)Ar/(39)Ar with paleomagnetism. Geological Society of America, 98th Annual Meeting (abstracts), vol. 17, p. 757.

Young, J.W., 1924. The Sudbury (Ontario) ore deposits. Bull. Soc. Econ. Geol. 19, 677–681.

Young, G.M., 1970. An extensive early Proterozoic glaciation in North America? Paleogeogr. Paleoclimatol. Paleoecol. 7, 85–101.

Young, G.M., 1973a. Tillites and aluminum quartzites as possible time markers for Middle Precambrian (Aphebian) rocks of North America. Huronian Stratigraphy and Sedimentation, Geological Association of Canada, Special Paper Number 12, pp. 97–127.

Young, G.M., 1973b. Origin of carbonate-rich early Proterozoic Espanola Formation, Ontario. Geol. Soc. Am. Bull. 84, 135–160.

Young, G.M., 1981. Diamictites of the early Proterozoic Ramsay Lake and Bruce formations, north shore of Lake Huron, Ontario, Canada. Earth's Pre-Pleistocene Glacial Record. Cambridge University Press, Cambridge, UK, pp. 813–816.

Young, G.M., 1983. Tectono-sedimentary history of early Proterozoic rocks of the northern Great Lakes region. EarlyProterozoic Geology of the Great Lakes Region, Geological Society of America, Memoir 160, pp. 15–32.

Young, G.M., 1995. The Huronian Supergroup in the context of a Paleoproterozoic Wilson Cycle in the Great Lakes region. Can. Min. 33, 917–944.

Young, G.M., 2015. The Penokean orogeny of the Lake Superior Region. Precambrian Res. 157, 4–25.

Young, G.M., Nesbitt, H.W., 1985. The Gowganda Formation in the southern part of the Huronian outcrop belt, Ontario, Canada: stratigraphy, depositional environments and regional tectonic significance. Precambrian Res. 29, 265–301.

Young, R.P., Talebi, S., Hutchins, D.A., Urbancic, T.I., 1989. Analysis of mining induced microseismic events at Strathcona Mine, Sudbury, Canada. Pure Appl. Geophys. 129, 455–474, 3–4.

Young, G.M., Long, D.G.F., Fedo, C.M., Nesbitt, H.W., 2001. The Paleoproterozoic Huronian Basin: product of a Wilson Cycle accompanied by glaciation and meteorite impact. Sediment. Geol. 141, 233–254.

Zenko, T.E., Czamanske, G.K., 1994. Spatial and Petrologic Aspects of the Intrusions of the Norilsk and Talnakh Ore Junctions: in Proceedings of the Sudbury-Noril'sk Symposium. In: Lightfoot, P.C., Naldrett, A.J. (Eds.), Ontario Geological Survey Special Publication 5, pp. 263–282.

Zhang, Y.X., Luo, Y.N., Yang, C.X., 1988. The Panxi Rift and It's Geodynamics. Beijing Publishing House, Beijing, 415 pp.

Zhang, M., O'Reilly, S.Y., Wang, K.-L., Hronsky, J., Griffin, W.L., 2008. Flood basalts and metallogeny: the lithospheric mantle connection. Earth Sci. Rev. 86, 145–174.

Zhang, M.J., Kamo, S.L., Li, C.S., Hu, P.Q., Ripley, E.M., 2010. Precise U-Pb zircon-baddelysite age of the Jinchuan sulfide ore-bearing ultramafic intrusion, western China. Min. Dep. 45, 3–9.

Zhou, M.-F., Lightfoot, P.C., Keays, R.R., Moore, M.L., Morrison, G.G., 1997. Petrogenetic significance of chromian spinels from the Sudbury Igneous Complex, Ontario, Canada. Can. J. Earth Sci. 34, 1405–1419.

Zieg, M.J., Marsh, B.D., 2005. The Sudbury Igneous Complex: viscous emulsion differentiation of a superheated impact melt sheet. Geol. Soc. Am. Bull. 117 (11–12), 1427–1450.

Zietz, I., Henderson, R.G., 1955. The Sudbury (Ontario) aeromagnetic map as a test of interpretation methods. Geophysics 20, 307–317.

Zolnai, A.l., Price, R.A., Helmstaedt, H., 1984. Regional cross section of the Southern Province adjacent to Lake Huron, Ontario: implications for the tectonic significance of the Murray Fault Zone. Can. J. Earth Sci. 21, 447–456.

Zolotukhin, V.V., Rjabov, V.V., Vasiljev, J.R., Shatkov, V.A., 1975. Talnakh ore-bearing differentiated trap intrusions. Science, Novosibirsk.

Zotov, I.A., 1989. Transmagmatic fluids in magmatism and ore formation. Nauka, Moscow. 256 pp. (in Russian).

Zotov, I.A., Persev, N.N., 1978. Genesis of metasomatic Cu-Ni ores at Talnakh. Science, Moscow. 86–95 (in Russian).

Zurbrigg, H.F., 1948. The development of mining geology in the Sudbury District. Can. Min. Metal. Bull. 51, 345–349.

Zurbrigg, H.F., 1957. The Frood-Stobie mine. Struct. Geol. Can. Ore Dep. 2, 341–350.

Index

653